MAGNETIC RECORDING HANDBOOK

Valdemar Poulsen (b. 1869 Nov. 23, d. 1942 July 23) inventor of magnetic recording, with his colleague P. O. Pedersen (1874–1941) who also made important contributions to magnetic recording (Fredriksborg Museum, Knud Larsen).

MAGNETIC RECORDING HANDBOOK

Marvin Camras
Research Professor
Illinois Institute of Technology
Chicago, Illinois

Formerly Senior Scientific Advisor
IIT Research Institute
Chicago, Illinois

VNR VAN NOSTRAND REINHOLD COMPANY
New York

Library of Congress Catalog Card Number 86-24762
ISBN 0-442-26262 0

Printed in the United States of America

Van Nostrand Reinhold Company Inc.
115 Fifth Avenue
New York, New York 10003

Van Nostrand Reinhold Company Limited
Molly Millars Lane
Wokingham, Berkshire RG11 2PY, England

Van Nostrand Reinhold
480 La Trobe Street
Melbourne, Victoria 3000, Australia

Macmillan of Canada
Division of Canada Publishing Corporation
164 Commander Boulevard
Agincourt, Ontario M1S 3C7, Canada

16 15 14 13 12 11 10 9 8 7 6 5 4 3 2 1

Library of Congress Cataloging-in-Publication Data

Camras, Marvin.
Magnetic recording handbook.

Bibliography: p.
Includes index.
1. Magnetic recorders and recording—Handbooks, manuals, etc. I. Title.
TK7881.6.C28 1987 621.389'3 86-24762
ISBN 0-442-26262-0

To

George Ziegler, Harold Vagtborg, and Donald Richardson whose boundless enthusiasm spurred the renaissance of magnetic recording; and to my patient wife, Isabelle

Preface

When I started in magnetic recording nearly fifty years ago, it was easy to perceive the common sense of it. There was very little mathematics and every new finding was a source of wonder. I have tried to recapture this spirit with simple explanations, while maintaining a high density of information and covering the entire field.

This book introduces a novice to magnetic recording and its many branches. It includes reference data for designers and users. Each chapter stands by itself; no prerequisites are essential. For a quick survey, the equations and worked-out examples can be disregarded.

The magnetic recording art is changing so rapidly that new advances are announced almost every month. These are properly covered by journal articles and manufacturers' catalogs. This book will fulfil its purpose if it gives a background for easily comprehending the new advances. I have included subjects and devices not found elsewhere, and some unconventional viewpoints. I would welcome comments from readers.

To Jay McKnight I am deeply grateful for important suggestions and helpful comments. I appreciate also the help of BASF, John Boyers, Joseph Dundovic, Charles Ginsburg, Peter Hammar, Yasuo Imaoka, Hal Kaitchuk, Otto Kornei, Harold Miller, Jack Mullin, Jim Novak, Lenard Perlman, Carl Powell, Sidney Rubens, John Sherman, Shigeo Shima, Heinz Thiele, Yoshimi Watanabe and many others; and to my daughter Ruth for typing.

Instead of a book which should be on everyone's shelf, I would like this book to be off the shelf and in frequent use.

MARVIN CAMRAS

Contents

MAGNETIC RECORDING HANDBOOK

1. Magnetic Recording History and Early Recorders

The year 1898 was ending. Valdemar Poulsen, a research engineer of the Copenhagen Telephone Company in Denmark had just applied for patents on a new invention—a device that could record the human voice on a steel wire. He could listen to the wire as many times as he liked, or wipe it off at any time and put on a new record. Most remarkable of all, the wire was never used up. Poulsen did not change it physically in any way—it could be played and replayed, recorded, erased, and rerecorded millions of times.

As a demonstration for his friends, Poulsen fastened one end of a steel wire to the top corner of his laboratory. He ran it across the room to a lower part of the opposite corner, stretched it taut, and fastened it to the wall. On a trolley arrangement (Figure 1-1), he hung a little electromagnet that resembled the magnet of an ordinary electric buzzer. To the electromagnet Poulsen connected a battery and telephone transmitter in series. Then he started the trolley carriage rolling down the inclined wire, and as he followed it, he shouted into the microphone.

When he came to the end of the wire, Poulsen disconnected his battery and transmitter. He slid the carriage to the top again and now connected a telephone receiver across the electromagnet. He allowed the carriage to roll down the wire while one of his friends listened to the telephone receiver. The friend heard a faint reproduction of Poulsen's voice in the earpiece. Again and again, the electromagnet was slid down the wire as each of the party listened in turn. After everyone had listened, Poulsen ran a strong magnet across the wire and showed that it had wiped out the record. On the cleaned wire he could record again, just as if it never had a record before.

Several months of concentrated effort preceded this first demonstration of magnetic recording in 1898. During the summer of that year Poulsen was trying to conceive an apparatus that could record a telephone message when the subscriber was not at home. In the course of his experiments, using a small magnet, he made some marks on the side of a steel tuning fork. When he dipped the tuning fork in iron filings, he noticed that particles would adhere in such a way that they outlined the parts touched by the magnet. This gave Poulsen the idea that he might be able to record on a wire "spots" of magnetism that would correspond to a human voice.

In August 1898, Poulsen went on vacation in the country. He did not get much rest or fresh air, for he stayed shut in his room, and repeated "Yakob,

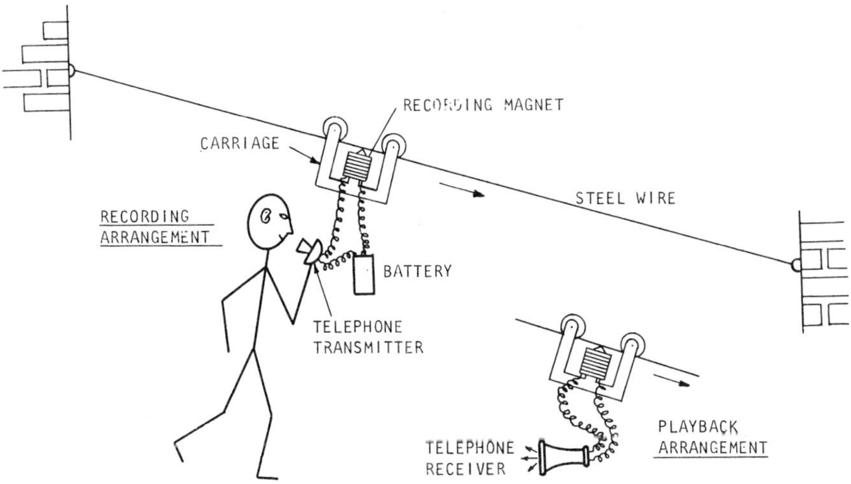

Figure 1-1. Poulsen's demonstration. Poulsen demonstrated recording to his friends with a steel wire stretched across the laboratory.

Yakob'' into a microphone all day. Poulsen's host and the guests had their doubts about the peculiar young man who was talking to himself.

In later years Poulsen explained why he had chosen the word *Yacob*, which is the Danish equivalent of Jacob or Jake in English. This word has two vowels and a very distinctive sound that Poulsen could recognize when it came back faintly through his crude apparatus.

The record that Poulsen made on his steel wire was nothing more than a multipolar magnet similar to the one shown in Figure 1-2 but greatly reduced in diameter. When he spoke into the transmitter (Figure 1-1) he varied the electric current that the battery sent through the electromagnet; thus he varied the magnetic field at the tip of the magnet where it touched the wire. Each point on the steel wire became permanently magnetized according to the strength of the electromagnet at the moment it travelled by. One might therefore consider the recording magnet as a stylus that wrote a magnetic pattern on the steel wire. The pattern corresponded to the condensations and rarefactions of the air waves that made up the original sound.

In playing back the record, the magnetic patterns of the wire set up a changing field in the electromagnet. This changing field induced a voltage in the magnet winding. The receiver connected to the magnet was operated by this voltage, and converted the fluctuating electric energy back into sound waves.

Poulsen called his invention the *telegraphone*, a combination of ``telegraph'' and ``telephone.'' He believed that one of its most important uses would be the

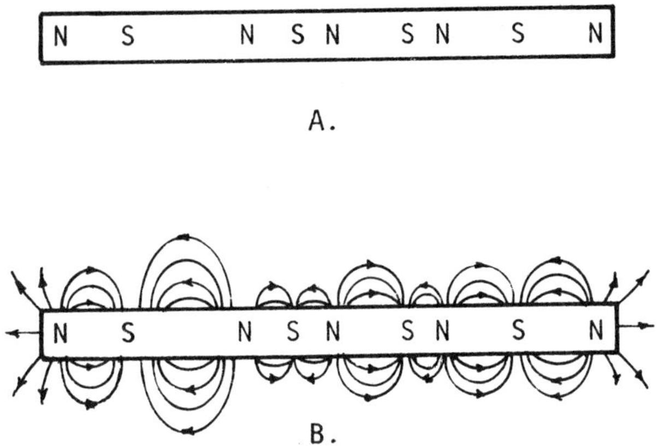

Figure 1-2. Multipole magnets. A. A bar magnet can be magnetized so as to have multiple poles. B. Field patterns surrounding a multipole bar magnet.

recording of telephone messages, and the name telegraphone signified "writing down a distant voice."

Poulsen applied for patents in all major countries of the world. He exhibited his telegraphone at the Paris Exposition of 1900 where it caused a sensation and won the coveted Grand Prix. Figure 1-3 is from his first patent, Danish Patent No. 2653.

The telegraphone seemed especially useful for dictation purposes because the wire could be erased and used over again. Hence there would be no expense for new cylinders as in Edison's dictating machine. Poulsen's first USA Patent No. 661,619 (Figure 1-4) was adaptable for this purpose. A later USA Patent 789,336 (Figure 1-5) shows an arrangement for recording a telephone message, an application that had inspired Poulsen's original experiments.

In spite of great scientific interest in the telegraphone principle, attempts to set up a European enterprise for manufacturing the telegraphone were turned down. Peter Jensen, as a young man, was hired by Poulsen to demonstrate the machine to crowned heads of European countries, but no one was inclined to risk any capital. Jensen later founded the famous Jensen Loudspeaker Company.

The United States proved to be a more fertile territory, and in 1903 the American Telegraphone Company was incorporated. The enthusiasm for this venture was expressed by its president, C. K. Fankhauser, when he spoke before the Franklin Institute in Philadelphia on 16 December 1908:

> We have here a discovery, which, in many respects is the most remarkable and unique ever made. . . . It is my belief that what type has been to the

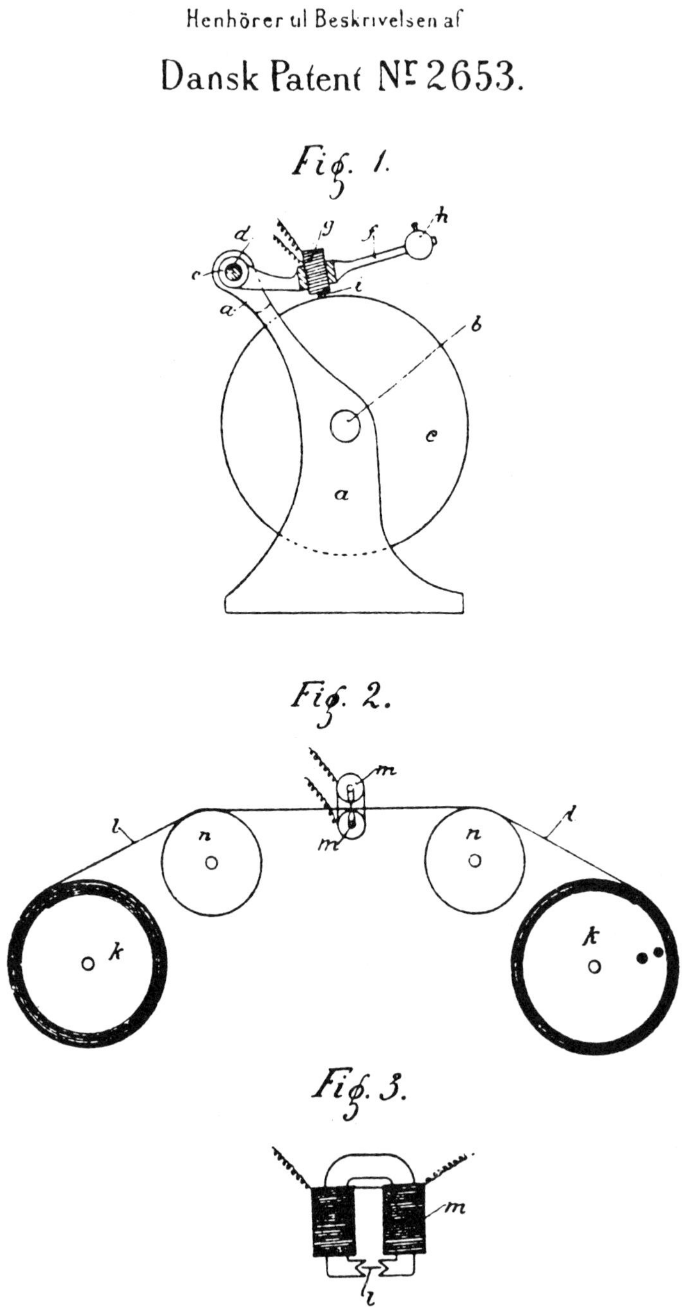

Figure 1-3. Poulsen's first patent. Drawing from Danish Patent No. 2653.

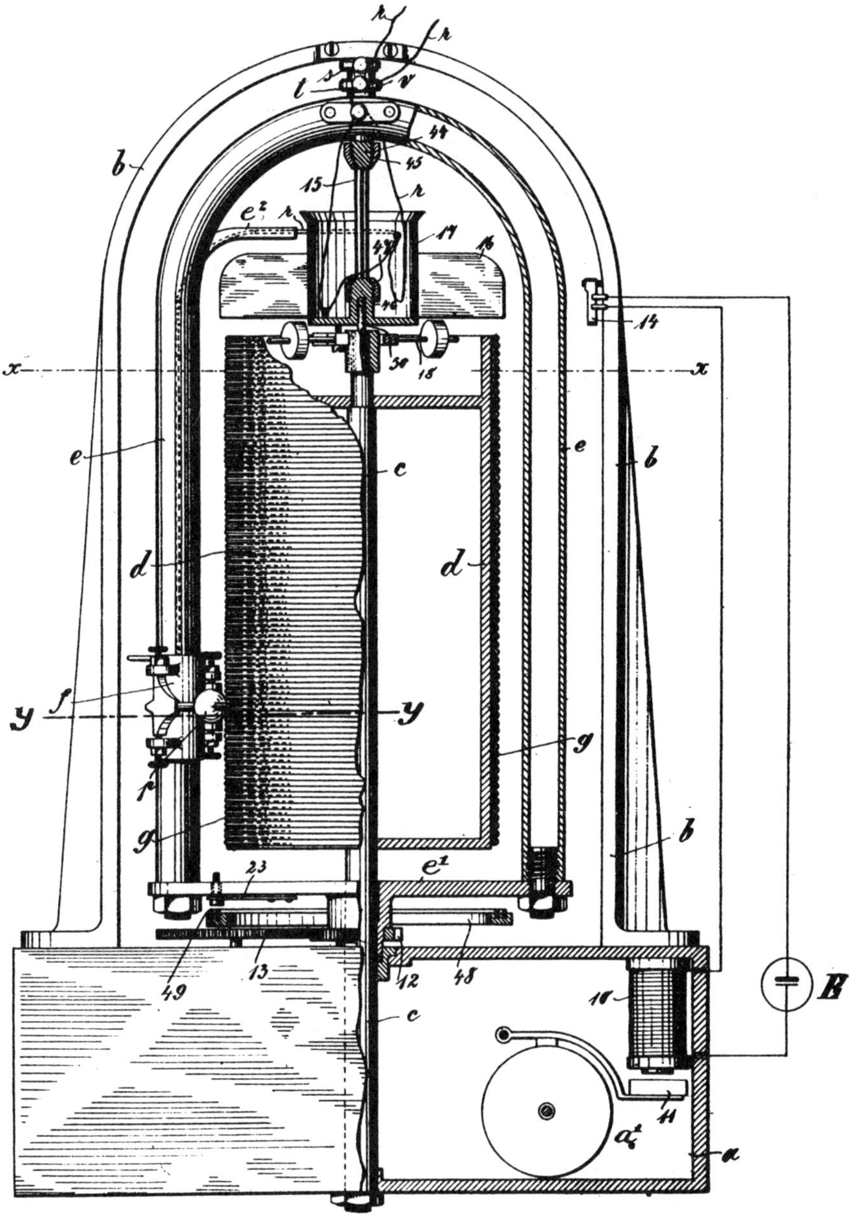

Figure 1-4. An early Poulsen machine of U.S. Patent 661,619 shows the head revolving around a stationary drum of wire.

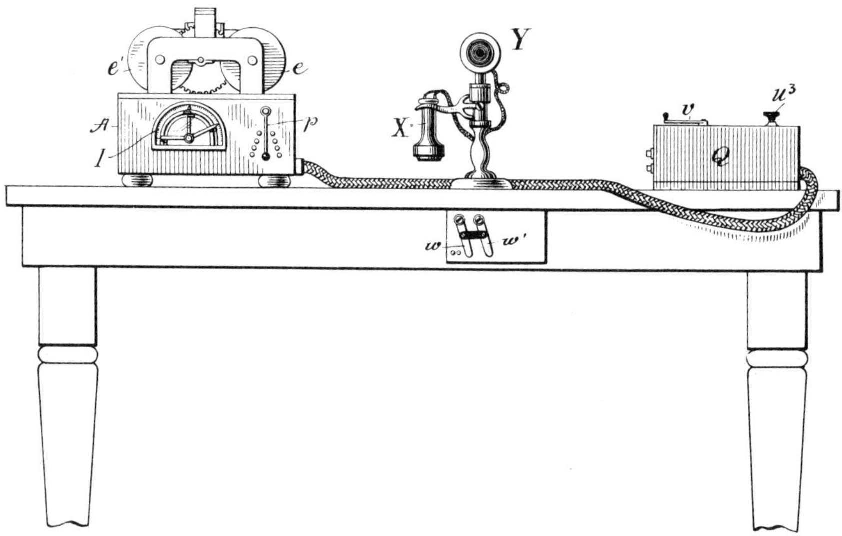

Figure 1-5. Telephone recorder. This reproduction from an early patent shows a telegraphone used for answering a telephone (from U.S. Patent No. 789,336 to Poulsen, Pedersen, and Schou, dated May 9, 1905).

> spoken word, the telegraphone will be to the electrically transmitted word. . . . The next few years will see a telegraphone installed in the office of every doctor, every lawyer, every banker, in the counting rooms of every trust company, and of every industrial or commercial establishment, large or small. [It would be used for telephone recording, for dictation, for court reporting, for train dispatching.] Everywhere in the field of human endeavor, where an accurate record of the spoken word is required or desired, the machine will be found silently, but accurately doing its work, leaving nothing to imagination, nothing to chance (Fankhauser 1909).

But Mr. Fankhauser was before his time. The machines produced by the American Telegraphone Company were heavy, expensive, and difficult to operate. Even so, the idea of sound on a wire was so intriguing that one could have overlooked these shortcomings. If only the telegraphone had given loud, clear, reliable sound, it would have met with public acceptance. But the reproduction was weak and spotty. Attempts to use De Forest's audion vacuum tube for amplification were unsuccessful. Although it made the sound louder, the audion also brought out noises and distortions heretofore inaudible. The net result was no better than before.

A few years later, Fankhauser gave up. He admitted that of all the ventures he had ever sponsored, this one showed the brightest prospects and gave the most bitter disappointments. After these efforts magnetic recording was dormant, but by no means dead, for it had so many inherent advantages over other methods that attempt after attempt was made to revive it both in the U.S.A. and in Europe.

HISTORICAL HIGHLIGHTS, DEVICES, AND PIONEERS

Appendix A documents important events in the development of magnetic recording, giving dates, companies, people who were involved, and pictures of historical devices. Early machines were used for sound recording and Morse code telegraphy. No one dreamed of their role in video and computers until the late 1940s, but today video and computer discs are household items. Progress in video and computers is documented in Appendix A and in Table 14-14. Pictures of pioneers who developed magnetic recording are collected in Appendix B.

AMERICAN DEVELOPMENTS

Very seldom is an invention completely unprecedented, and this is true also of Poulsen's Telegraphone. Ten years before Poulsen, in 1888, Oberlin Smith, an American engineer, published the concept of recording and playing back magnetically, either on iron dust particles held by a nonmagnetic carrier or on a solid steel strand. Smith complained that other business prevented him from continuing his experiments, and therefore he was publishing his ideas in the hope that others could carry them out (Smith 1888) (Figure A-1, see Appendix A).

As previously described, Valdemar Poulsen, in 1898, was the first to demonstrate that magnetic recording was indeed possible, and his early success became legendary. He experimented with tape recorders, wire recorders, magnetic discs, and magnetic cylinders, hoping that one or more of these would meet an unfilled need and become accepted commercially (Figure A-2). But his devices remained expensive toys, and the American Telegraphone Company, organized in 1903 to exploit his invention, floundered until 1908, when it came under the control of an industrialist C. Dexter Rood who bought a stake in the company for $188,000 and became its new president. Rood had just retired from the presidency of the Hamilton Watch Company but was still full of energy; he expected to repeat his successful career and to make the Telegraphone a household word.

By 1912 the company had moved from Washington, D.C., to Wheeling, West Virginia, and finally to Rood's home town in Springfield, Massachusetts. It settled down to producing the definitive version of Telegraphone pictured in Figure A-3A, based on Tiffany's U.S.A. Patent 1,142,384 filed in 1909. Advertising pamphlets featured Phoebe Snow, a famous model of that era, posing as a stenographer enthusiastically operating her office Telegraphone for dictation. But the American Telegraphone was unattractive in appearance, unreliable in operation, heavy, clumsy, and expensive. Very few were manufactured. Customers were mainly experimenters and companies who thought of unusual applications.

Mr. Rood was a colorful character who involved his company in one deal after another, some of which led to civil and criminal lawsuits. In the 1914–1918 years of World War I, Rood was accused of discouraging the sale of Telegraphones to American military agencies and of sending them defective machines. At the same time, he sold Telegraphones freely to German interests who installed them in submarines and who also used them for coding messages transmitted at high speed to receiving stations in Germany by means of a superpower wireless transmitter in Sayville, Long Island (Angus 1973).

E. I. DuPont de Nemours & Company bought a fleet of recorders intended for an ambitious central dictation system in their Wilmington, Delaware, offices. The installation was a failure, resulting in another costly lawsuit. Stockholders became disgruntled and appointed attorney George Sullivan to represent their interests against management. Management responded by promising stockholders a bright future of innovations and new wonders, as in the advertisement shown in Figure 1-6, circulated in 1917.

By the 1920s production had ceased at the American Telegraphone Company, and litigation remained its sole activity. Its most important patent, Tiffany's 1,142,384, was about to expire in 1932 and Mr. Rood worried that this would spell the end of his company. Patents are not renewable, but Mr. Rood's attorneys petitioned for an act of U.S. Congress to extend the Tiffany patent, arguing that the company did not have a fair chance to develop it during its normal life. Congress rejected the petition, but even a favorable action would not have saved this shell of a once-glorious enterprise.

EUROPEAN DEVELOPMENTS

Europe too had its share of enthusiastic promotors. In the early 1920s, Kurt Stille of Germany organized a cartel, the Telegraphie-Patent-Syndikat, that acquired rights to important magnetic recorder patents, and encouraged European production under the blanket of its patent protection. One of Stille's first customers was the Vox Company of Berlin, which in 1925 developed a central dictation machine remotely controlled by telephone (Figure A-3*B* and *C*). It was widely publicized but commercially unsuccessful.

In about 1930 Ludwig Blattner in Germany acquired Syndikat rights for entertainment purposes and especially for talking pictures. Sound movies using phonographic and photofilm methods had just taken the world by storm. Blatt-

Latest Development

Telegraphone Fast Coming Into Its Own

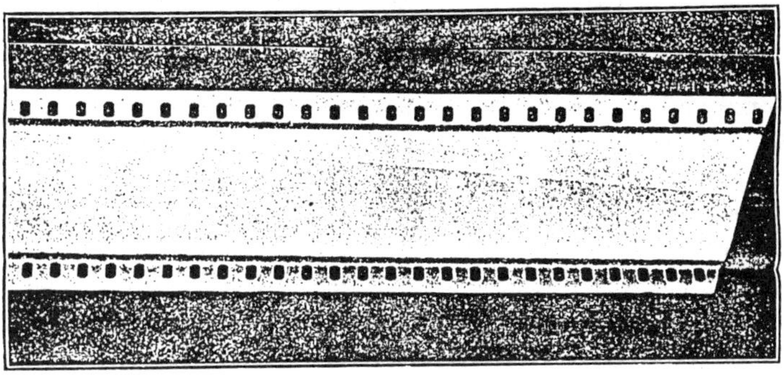

THE ABOVE ILLUSTRATION is that of the first moving picture film in the world to talk. The Telegraphone principle has solved the problem. The black stripe on either side of the film are steel filings embedded into and is part of the film itself, therefore the synchronism is perfect, and the films may be any length desired. A few minor details must be worked out—it then only awaits the audion amplifier (De Forest type) which made possible transcontinental telephony. The voice of the actor whether spoken or sung is perfectly recorded and may be brought to any part of a building regardless of size so that those seated farthest from the stage or curtain will hear every word spoken just as clearly as if seated in the orchestra row.

The moving picture industry is today the third largest in the world, we are told, and with a perfect talking picture the proceeds from this field alone will undoubtedly make a tremendous amount of money for our stockholders. The principle is the local magnetization of steel filings and are covered by the Telegraphone patent, which belong to our stockholders. A more complete description of this wonderful amusement feature will be given general publicity very soon, and those holding Telegraphone stock should not listen to brokers who might want to buy your stock or induce you to exchange for "something better."

To give to our stockholders all information regarding developments being made, Telegraphone News will be published quarterly.

For further particulars, address

H. P. O'REILLY & CO.

DISTRIBUTORS

40 Harrison Avenue — Springfield, Massachusetts

Figure 1-6. Motion picture film with sound tracks on magnetic stripes were promised in this circular of about 1917 sent by the American Telegraphone Company to stockholders who were becoming restless after years of unfulfilled promises. (Ampex Museum, Peter Hammar).

ner expected that magnetic recording would be even better, and after considerable effort introduced the Blattnerphone, a tape recorder using huge reels of heavy steel tape. England seemed more receptive to Blattner's device than Germany, so Blattner moved his experiments there, trying to synchronize his tape with movies to give talking pictures. The British Broadcasting Company (BBC) and the Marconi Company became interested. They saw the tape recorder's future in radio broadcasting rather than in motion pictures and bought out Blattner's rights and his equipment. By 1934, after considerable improvement with the help of the German Stille Laboratories, the Marconi-Stille recorders were in regular use for BBC broadcasting (Figure A-4). In a parallel development, the Lorenz Company in Germany made a similar steel-tape recorder called the Lorenz-Stahltonmaschine, which was adopted in 1935 by the German Broadcasting Company (Begun 1949).

Karl Bauer of Germany associated himself with the Syndikat in 1932 and formed the Echophone Company to build a small office-type dictation machine called the Dailygraph. A year later he sold his interests to the Lorenz Company, who improved the Dailygraph and changed its name to the Textophone (Figure A-3*D*. This voice-quality machine met with limited success.

Meanwhile an independent German inventor, Fritz Pfleumer, was trying a different concept: a lightweight paper tape coated with a thin layer of magnetic particles. He interested the I.G. Farben Company (BASF) to make experimental tapes while the German General Electric Company (AEG) worked on a suitable recorder. Their Magnetophone, announced in 1935, had mediocre sound quality. In performance, it was hardly a challenger to the refined Stille steel-tape machines, but was much smaller, lighter, and cheaper. Within the next years a better tape using iron oxide made the Magnetophone a more formidable competitor. The final touch, high-frequency bias, introduced in the 1942 model, improved its sound quality so greatly that all previous steel tape machines became obsolete.

Development of high-frequency bias is a story in itself: the early concepts in England in the 19th century, experiments and patents in America in the 1920s, and the further work in the 1930s in Japan, Germany, and America all led to commercial use in the 1940s (Appendix A).

LATER AMERICAN DEVELOPMENTS

Through the years magnetic recording activity shifted among different countries, but revived again in the United States in the 1930s. The American Telephone Company (AT&T) developed a superior metal tape alloy, vicalloy, and manufactured a weather-announcing machine and a voice-training recorder called the Mirrophone. The Brush Development Company of Cleveland pioneered similar machines under the name Soundmirror; they later introduced electroplated magnetic media and, in 1946, a home tape recorder using magnetite-coated paper tape. In 1941, Armour Research Foundation (now IITRI) manufactured wire recorders using high-frequency bias and was responsible for the popular wire recorders of World War II and the postwar years. It developed magnetic sound for motion pictures, stereophonic tape recorders, and the acicular gamma oxide still used on most tapes. It also did early research on rotating head scanners for videotape recorders (Appendix A).

CONTEMPORARY DEVELOPMENTS

By the late 1940s and throughout the 1950s many of the now-famous recorder companies got their start: Ampex, 3M, Magnecord, Webcor, IBM, Philips, Sony, Revox, and countless others, as documented in Appendix A. Japanese manufacturers became especially prominent in the 1960s because of their high quality, reasonable pricing, and attractive designs. More recently they designed and became the exclusive makers of Betamax and VHS home video recorders. European manufacturers were most successful in expensive precision professional recorders. The low-cost cassette and cartridge recorder, mass produced, is the specialty of Taiwan, Korea, Hong Kong, and other low labor-cost countries.

Part 1
THEORY, TECHNOLOGY, AND MANUFACTURE

2. Fundamentals of Magnetic Recording

In recent years magnetic recording has become a preferred medium for storing information such as printed language, mathematics, sound, and pictures. It is compact, economical, easily updated, and instantly retrievable. Magnetic data banks equivalent to large libraries can be scanned in a few seconds, and the selected information displayed or printed immediately.

The magnetic recorder itself is an assembly of many components working together. Each major component requires a different discipline for its development and manufacture, and therefore each has become a separate specialized art in itself, with its own research and production facilities. A separate chapter is therefore devoted to each of these specialized units.

The recorder as a whole, and the recording process are described in this chapter, including general theory of magnetic recording, patterns of recorded information, nonlinear magnetic effects, high frequency bias, the head-tape interface, contact printing, permanence of recordings, and unconventional methods of recording. All of these subjects require some knowledge of heads, tapes, and so forth. Therefore this chapter includes a brief introduction to essential components, although later chapters treat them in depth. A familiarity with magnetism is, of course, basic to understanding magnetic recording, so this chapter begins with a review of magnetic concepts and terminology that are used throughout this book.

MAGNETISM

Magnets are sources of magnetism, which is a mysterious force that can attract or repel matter, even through a perfect vacuum. This so-called *action at a distance*, a property of magnetism, electric charge, and gravity, has been a source of wonder to the best minds of past centuries, including Newton and Einstein, but has not been explained further and is accepted as a given property of our universe.

All substances are magnetically susceptible, that is, they are attracted to or repelled by a magnet, but the force is so weak in most cases that it can only be observed with delicate instruments. Substances that are attracted to a magnet are *paramagnetic* (positive susceptibility), and those that are repelled are *diamagnetic* (negative susceptibility). Of the paramagnetic materials there are an exceptional few that show extremely strong magnetic attraction; these are called *ferromagnetic*. In common imprecise language, when someone calls a substance *magnetic* he implies that it is *ferromagnetic* like iron, cobalt, and nickel. When he calls something *nonmagnetic* he really means not-ferromagnetic, and is ignoring the weak but universal magnetic qualities of all substances. Nowadays physicists apply the name *ferrimagnetic* to the strongly magnetic behavior of ferrites which have a different arrangement of electron spins than *ferromagnetic* substances. However, in the following discussion on applied magnetics no distinction need be made between them as practical magnetic materials.

Certain magnetic substances become magnets themselves (magnetized) after being subjected to the influence (field) of another magnet. This property is the basis for magnetic recording, where the record is magnetized by a magnetic field and thereafter retains the magnetic impressions.

Over the past century, magnetism has branched into two main studies: basic (or theoretical) magnetism and applied magnetism.

Basic magnetism is an essential part of those theories of matter in which magnetic relations must harmonize with electron orbits, electron spins, molecules, atoms, domains, exchange forces, crystal structure, and so on. Magnetic susceptibility has been one of the most important tools used by physicists and chemists for investigating atomic structure of elements and compounds. Although basic magnetic phenomena are of great interest and are responsible for most significant advances in the magnetic arts, the needs of a technologist are more directly answered by information in applied magnetism.

Applied magnetism is the province of the engineer and is concerned with strongly magnetic materials, mostly iron alloys or iron compounds, and their application to useful devices and machinery such as loudspeakers, motors and transformers. Magnetic recording is definitely in this category.

ELECTROMAGNETISM

An important aspect of applied magnetism is *electromagnetism* where ferromagnetic materials are surrounded by electrical windings; and also *permanent magnets*, where magnetism persists even though windings are absent and currents are zero.

The action of electromagnetism illustrated in Figure 2-1 is that a coil of (nonmagnetic) wire becomes a magnet when a current flows through the coil. The magnetism disappears if the current ceases. The inverse relation is that changing the magnetic flux through the axis of a coil produces a voltage across the coil terminals. The voltage persists only as long as the flux is changing

Electromagnetism as in Figure 2-1 occurs in air alone and even in a vacuum, but is greatly enhanced when a core of ferromagnetic material is introduced inside the coil of wire, or along the path of the magnetic field; and in fact the ferromagnetic material is most effective when it forms a closed path linking the coil. Some ferromagnetic materials enhance the magnetization more than others. Their effectiveness relative to air is measured by their relative permeability

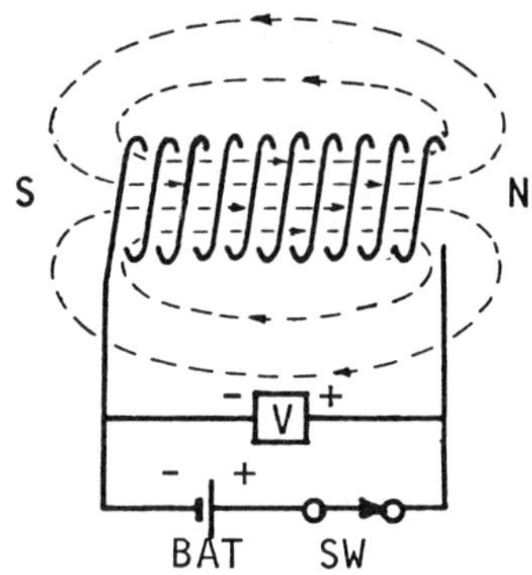

Figure 2-1. Electromagnetism. Electric current through the coil causes it to become a magnet with N and S poles as indicated. When the switch is opened the changing magnetic field generates a voltage of reverse polarity at V.

μ_r. Thus a closed core having $\mu_r = 1000$ will produce a thousand times as much flux as a core of air when the same magnetic field is applied.

The Magnetic Circuit.

The closed magnetic path of Figure 2-2 is called a magnetic circuit. In such a magnetic circuit, the magnetomotive force F is analogous to emf in an electric

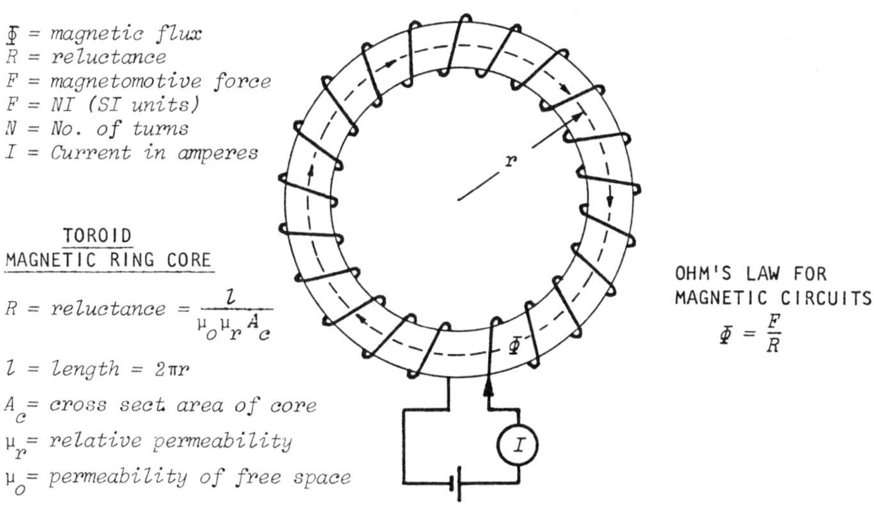

Figure 2-2. Ohm's law for magnetic circuit of a ring core. Φ is analogous to electric current, F is analogous to electromotive force, and R is analogous to resistance.

circuit, flux (Φ) is analogous to current flow, and reluctance (R) is analogous to resistance.

The simple magnetic circuit shown in Figure 2-2 is a ring (toroid) core of ferromagnetic material wound with wire that can be connected to an external circuit. Ohm's law for magnetic circuits is

$$\Phi = \frac{F}{R} \tag{2-1}$$

where F is the magnetomotive force. In the SI system the magnetomotive force is

$$F = NI \tag{2-2}$$

where N is the number of turns wound around the ring and I is the current in amperes through the winding. Φ is the magnetic flux in webers circulating around the ring. The reluctance R of the magnetic circuits is

$$R = \frac{\ell}{\mu_o \mu_r A_c} \tag{2-3}$$

where ℓ is the length (average circumference) of the ring in meters, μ_o is the permeability of space ($\mu_o = 4\pi \times 10^{-7}$), μ_r is the relative permeability of the core material, and A_c is the cross sectional area of the ring in square meters. Relative permeability μ_r is a measure of how strongly the material becomes magnetized per unit of applied field, compared to the magnetization that would exist in the same space if the core material were absent. In the cgs system $F = 0.4\pi NI$ gilberts, Φ is in maxwells (lines), and the reluctance is $R = \ell/\mu A_c$, where ℓ is in centimeters, A_c is in square centimeters, and $\mu = \mu_o \mu_r = \mu_r$, since $\mu_o = 1$ in cgs units.

Equation (2-1) is a good approximation even if the core is not a toroid and even if the coil is concentrated rather than distributed, as long as μ_r is high (about 100 or more). If the magnetic circuit has a gap, the gap reluctance is added to the core reluctance as explained in detail in Chapter 4.

Magnetic Units.

Although the official system of units is now the SI (Système Internationale d'Unités) which is essentially the same as the MKS (meter-kilogram-second) system, many workers are more comfortable and familiar with the older cgs (centimeter-gram-second) system. In this book both will be shown most of the time, with cgs units normally in parentheses. Table 2-1 lists common magnetic units and conversion factors.

Table 2-1. MAGNETIC UNITS AND CONVERSION FACTORS.

SYMBOL	QUANTITY	UNIT	SYSTEM	GIVEN NAME (AND EQUIVALENT)	MULTIPLY BY	TO OBTAIN NAME (AND EQUIVALENT	UNIT
B	Flux density	T	SI	tesla (Wb/m^2)	10^4	gauss ($maxwells/cm^2$)	G
B	Flux density	mT	SI	millitesla (mWb/m^2)	10	gauss	G
B	Flux density	G	CGS	gauss (Mx/cm^2)	10^{-1}	millitesla (mWb/m^2)	mT
B	Flux density	kG	CGS	kilogauss (kMx/cm^2)	10^{-1}	tesla (Wb/m^2)	T
H	Field strength*	A/m	SI	ampere-turns/m	0.01257 $=4\pi\times10^{-3}$	oersteds (gilberts/cm)	Oe
H	Field strength*	kA/m	SI	kiloamp-turns/m	12.57	oersteds (gilberts/cm)	Oe
H	Field strength*	Oe	CGS	oersteds (gilberts/cm)	79.58	ampere-turns/m	A/m
H	Field strength*	kOe	CGS	kilo-oersteds (kGi/cm)	79.58 ($\cong 80$)	kiloampere-turns/m	kA/m
H	Field strength*	Oe	CGS	oersteds (Gi/cm)	0.07958	kiloampere-turns/m	kA/m
F	Magnetomotive force	A	SI	ampere-turns	1.257	gilberts	Gi
U	Magnetic potential	A	SI	ampere-turns	1.257	gilberts	Gi
F	Magnetic potential	Gi	CGS	gilberts	0.7958	ampere-turns	A
Φ	Flux	Wb	SI	webers	10^8	maxwells	Mx
Φ	Flux	Mx	CGS	maxwells	10^{-8}	webers	Wb
μ_o	Permeability of free space	-	SI	henrys/meter	7.958×10^5	$\mu_o = 1$ (cgs)	μ_o
μ_o	Permeability of free space	-	CGS	$\mu_o = 1$	1.257×10^{-6}	henrys/meter	μ_o
R	Reluctance	$1/H$	SI	ampere-turns/weber	1.257×10^{-8}	reluctance units	$\ell/\mu\mathcal{A}$
R	Reluctance	$\ell/\mu\mathcal{A}$	CGS	reluctance units	7.958×10^7	ampere-turns/weber	$1/H$
M	Magnetization**		SI	webers/meter^2	795.8	emu/cm^3 ($ergs/Oe\text{-}cm^3$)	

*Ampere-turns/meter can be designated as At/m or A/m. Recommended usage omits the t, in which case A denotes ampere-turns.
**Based on $M = B - \mu_o H$ in SI units.

MAGNETIC MEASUREMENTS

Important magnetic properties of materials are determined by magnetic measurements of permeability, saturation density, coercivity, remanence, B vs H, and core loss. These are measured under various conditions of field intensity, flux density, frequency, lamination thickness, temperature, mechanical stress, heat treatment, etc. Most of these properties can be measured on a plotted hysteresis loop of the material. Instruments have also been designed which read selected properties directly on a meter (or oscilloscope) connected to equipment that receives the test specimen.

Hysteresis Loop.

When the resulting flux Φ is plotted against the applied magnetomotive force, the flux Φ follows a certain path while F is increasing. It does not retrace the same path when F decreases again, but takes a path which lags behind the previous one. This effect is called *hysteresis* (Greek for *lagging behind*). When F is cycled back and forth between specified values the resulting plot encloses an area called a *hysteresis loop.*

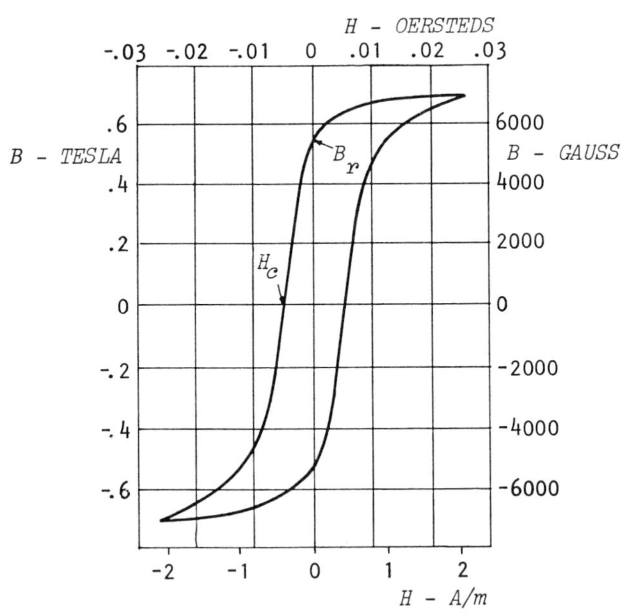

Figure 2-3. Hysteresis loop (B vs H curve) for supermalloy material. As the magnetizing field is cycled between $\pm$ 2A/m the flux density B changes between $\pm$ 0.7 tesla, tracing the loop as shown. The area of the loop represents energy loss per unit volume of material per cycle of H. Supermalloy is a magnetically soft material of low coercivity H_c and low energy loss.

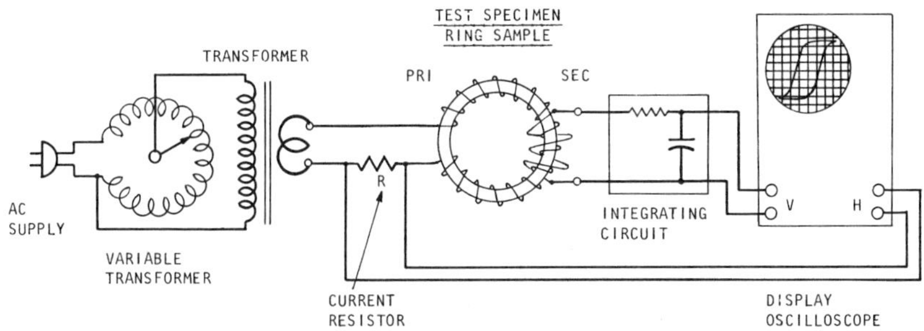

Figure 2-4. Hysteresis loop tracer (*B* vs *H*) for ring samples of magnetic materials. The horizontal deflection reads field intensity *H* which is proportional to the current through *R*. The vertical deflection reads *B* which is proportional to the integrated secondary voltage.

A hysteresis loop characteristic of the magnetic material alone, regardless of its physical dimensions, is obtained by scaling the gross readings to a unit length for F and to a unit area for Φ. We then use $H=F/\ell$ for the *field intensity* and $B=\Phi/A_c$ for the *flux density*. The characteristic hysteresis loop is a plot of B vs H as in Figure 2-3.

For measurements a ring sample is an almost ideal test specimen. It is wound with a primary and secondary connected in a circuit, as in Figure 2-4. When the primary is excited by an ac current of frequency 60 Hz, a hysteresis loop will appear on the cathode ray oscilloscope screen. The loop can be measured directly or can be photographed for permanent reference. Loops can be displayed for different applied field intensities and frequencies.

Ring samples individually wound and connected are expensive and cumbersome compared to strip samples that can be inserted into a holder and read immediately, as described in Chapter 3. Recording tape and other high coercivity materials about 100mm long and 6.3mm wide are suitable for strip samples.

MAGNETIC MATERIALS

Magnetic recording depends on "hard" or permanent ferromagnetic materials for a tape that will retain magnetism permanently once it is recorded and that can be erased or changed only by a very strong magnetic field (high coercive force). It is interesting that the active coating on most magnetic tape is a form of the oldest known magnetic material, magnetite, and is also related to rust (hydrated ferric oxide). There is another class of ferromagnets that becomes strongly magnetized very readily, and demagnetized just as readily (low coercive force). These are called "soft" magnetic materials and are needed in heads: playback, recording, and erasing.

The magnetic properties of a gamut of hard and soft materials are shown on

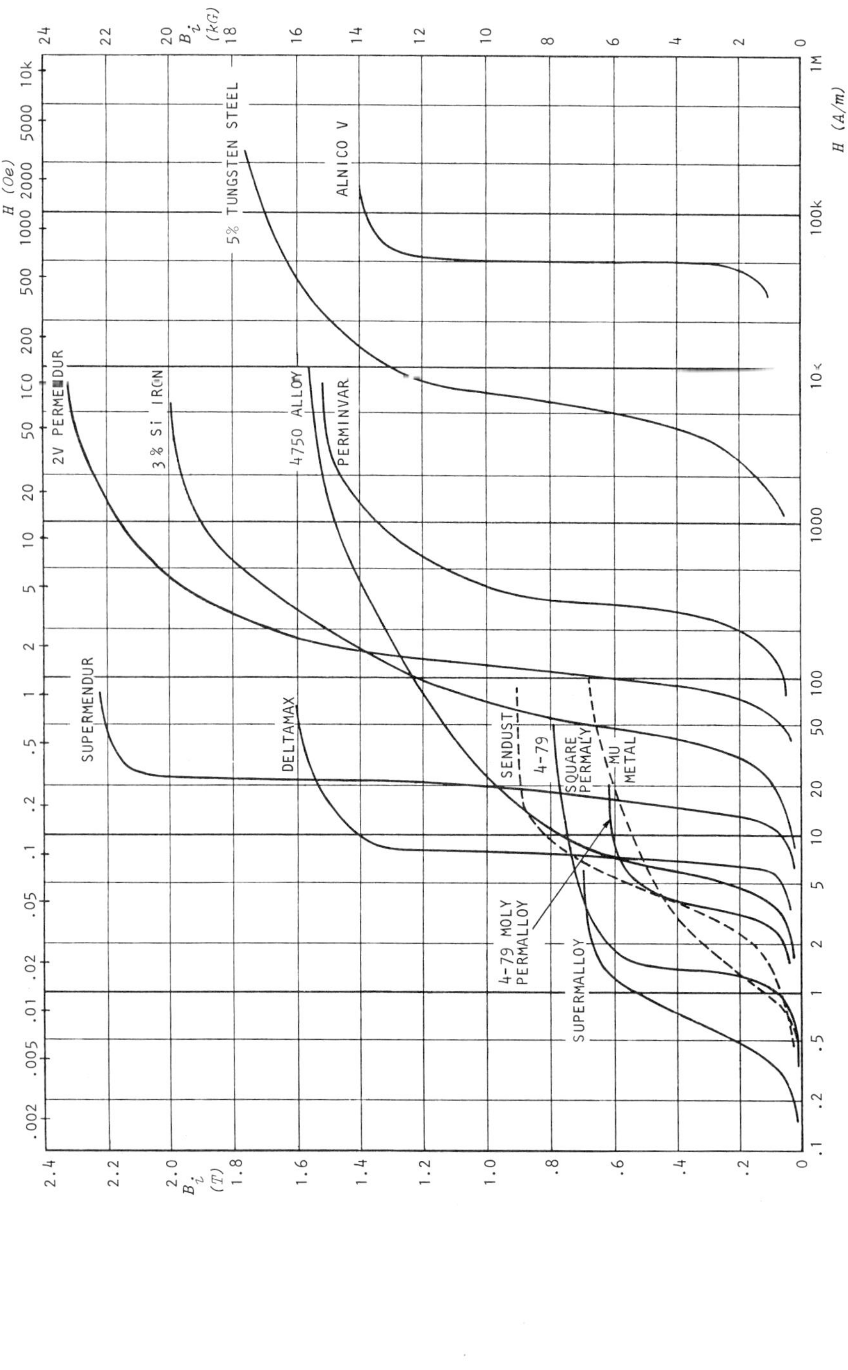

Figure 2-5. Magnetic properties of hard and soft magnetic materials.

the semilogarithmic B vs H plot in Figure 2-5. From these curves and the hysteresis loops some basic insights are gained of the behavior of ferromagnetic materials:

1. The magnetization (flux density) of ferromagnetic materials is not a linear function of the applied field, but follows an S-shaped growth curve. Below a certain threshold the material contributes less to the total B; above the threshold the B rises more and more steeply until saturation sets in and the B curve levels off.
2. All ferromagnetic materials saturate at a flux density less than 2.4 tesla (24000 gauss), some at considerably less.
3. There is a general resemblance of the magnetization curves to each other, except for differences in scale
4. The coercivity H_c of a given material is larger than the H corresponding to the steepest rise of its initial magnetization curve, being roughly 1.5 to 2 times as high.
5. The curves are not reversible. They follow a different path when H decreases than when it increases (hysteresis effect).

Permanent or Hard Magnetic Materials.

The permanent magnetic qualities of hard magnetic materials are depicted by their broad hysteresis loop as for example the B vs H plot of Figure 2-6 for a gamma ferric oxide tape coating. When an increasing field ($+H$) is applied to this material its flux density B increases along a curve a b c, slowly at a, rapidly at b, then slowly again at c as the material approaches saturation. Curve a b c, called the *initial magnetization curve*, is typical of all ferromagnetic materials that are initially in a virgin or demagnetized state.

When the magnetizing flux ($+H$) is decreased from its maximum value, the flux density B in the gamma ferric oxide does not decrease along its original path, but follows a path c d e always above the initial magnetization curve, until when the field H becomes zero a residual magnetization B_r remains, which is a measure of the permanent magnetization.

Demagnetization Curve and Energy Product. If the field is now reversed ($-H$), the B will decrease along B_r, f, g. The reverse field which reduces B_r to zero is called the coercive force H_{ci} which is another measure of permanent magnetization. The useful energy that a permanent magnet can supply is equal to the product of $B \times H$ for any point on the curve B_r, f, g, H_c. There is a point where the product is maximum $(BH)_{\max}$. In tape recording, however, the operating point varies with the wavelength, and $(BH)_{\max}$ is of secondary consideration.

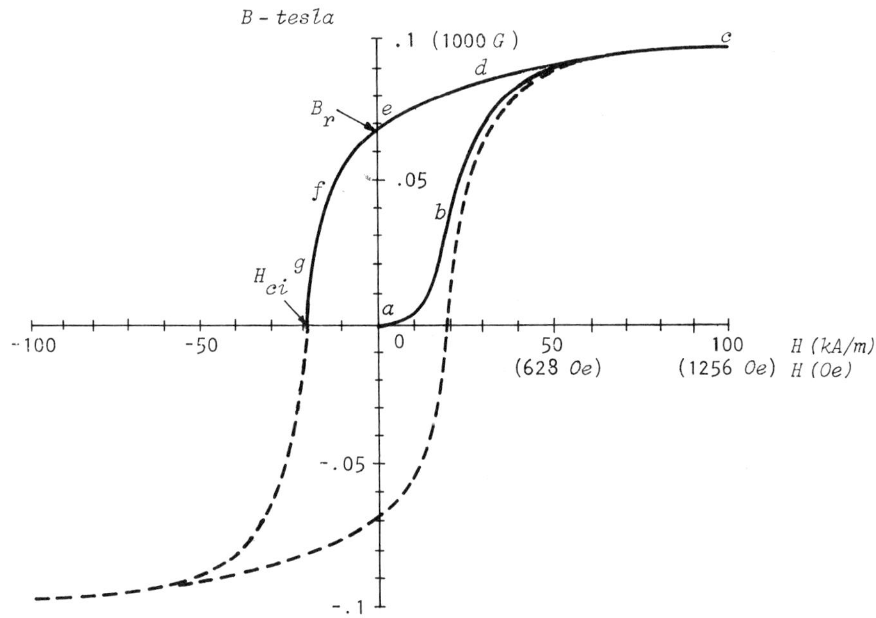

Figure 2-6. *B* versus *H* loop of gamma ferric oxide tape coating.

Permanent magnet materials are therefore characterized by the *B-H* curve in the second quadrant after saturation. From this quadrant the remanence B_r, the coercivity H_{ci}, and the maximum energy product $(BH)_{\max}$ can be determined. Figure 2-7 gives the second quadrant path and the energy product of several permanent magnet materials. The magnetic field in an air gap of length ℓ_g and area A_g manifests itself as a negative field $(-H_D)$ apportioned along the length of the magnet according to the relation:

$$\frac{\ell_m}{\ell_g} = \frac{B_g A_m}{-\mu_o H_D A_g} \tag{2-4}$$

In a simple magnet where the gap area A_g equals the magnet area and leakage and fringing are neglected:

$$\frac{\ell_m}{\ell_g} = -\frac{B_g}{\mu_o H_D} = \text{negative slope of line through origin} \tag{2-5}$$

For most efficient operation ℓ_m is chosen so that the line passes through the

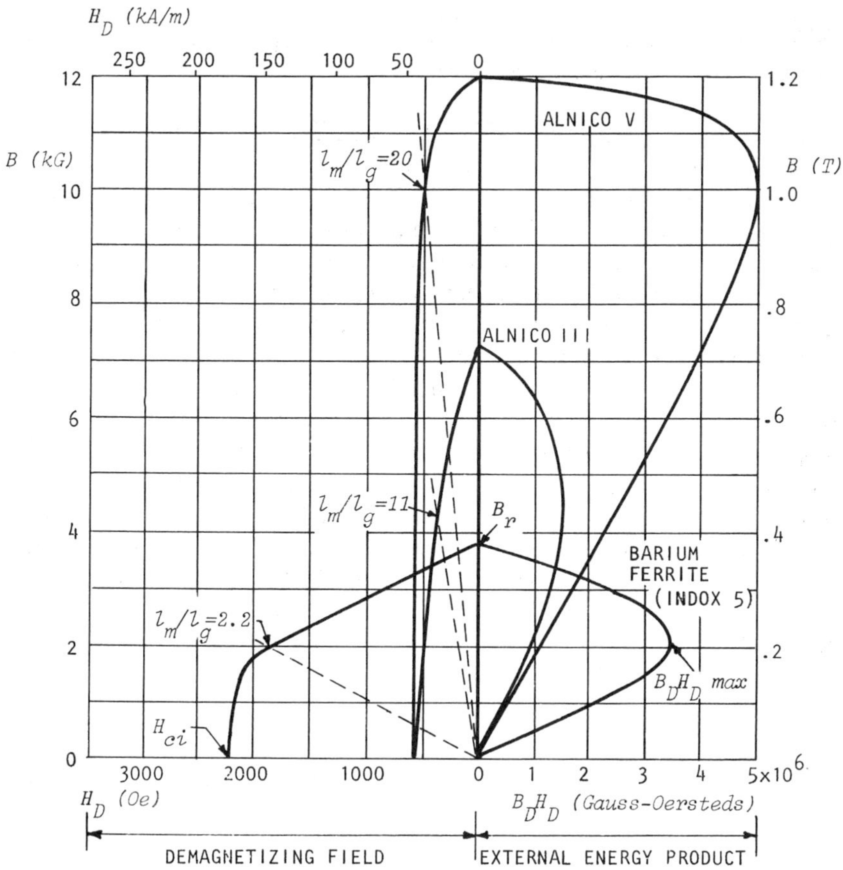

Figure 2-7. Second quadrant demagnetizing curve and energy product of some permanent magnet materials. The slope of the dashed line through the B vs H_D curve at the point corresponding to maximum energy product is equal to the ratio of magnet length (ℓ_m) to gap length (ℓ_g) for best efficiency of a simple magnet.

$(B_DH_D)_{\max}$ point. As seen in Figure 2-7 a ceramic magnet is best with a low ratio of 2.2, while an alnico V magnet requires a ratio of 20.

Major and Minor Hysteresis Loops. When the alternating field $\pm H_1$ applied to a material is enough to saturate it, as in Figure 2-8, its magnetization B will trace a hysteresis loop of a maximum area *ABCDEFA*. The area of this curve in *BH* units represents the energy lost per unit volume of material per cycle of AC. Small excursions of the field $\pm H_2$ trace a small area *abcda* while fields beyond saturation $\pm H_3$ elongate the tips of the loop but hardly increase the area *ABCDEFA*. If a small variation of $\pm H_4$ is superposed on a large steady component, a minor loop *ghijg* is traced.

Intrinsic Hysteresis Loop, Intrinsic Induction, and Intrinsic Coercivity. In any magnetic circuit that contains a magnetic core the total flux density B consists of two components: (1) the portion due to the field of the coil, which would still be present even if there were no magnetic core; and (2) the extra or enhanced portion due to the presence of the core. The B due to the coil alone is $B_{coil} = \mu_o H$. Therefore the remaining portion due to the core is (B minus $\mu_o H$) = B_i (or J), which is called the intrinsic induction. In cgs units $B_i = 4\pi M$, where M is the magnetization of the core material. The B_i vs H curve (intrinsic hysteresis loop), is shown by the dashed line in Figure 2-8. The M vs H curve is identical to it except for the scale factor. Note that the intrinsic hysteresis loop coincides with the B vs H loop on the B axis so that B_r is the same for either loop. Also the areas of both loops are the same. The height of the intrinsic hysteresis loop (B minus H value) in the first quadrant is always lower than for the ordinary loop, and its maximum value B_{mi} is less. In the second quadrant the intrinsic loop (B minus H value) is always higher than the ordinary loop, and consequently the *intrinsic coercivity* H_{ci} or ${}_iH_c$ is higher than the ordinary H_c. In SI units the intrinsic curve is a plot of (B minus $\mu_o H$) versus H, where $\mu_o = 4\pi \times 10^{-7}$.

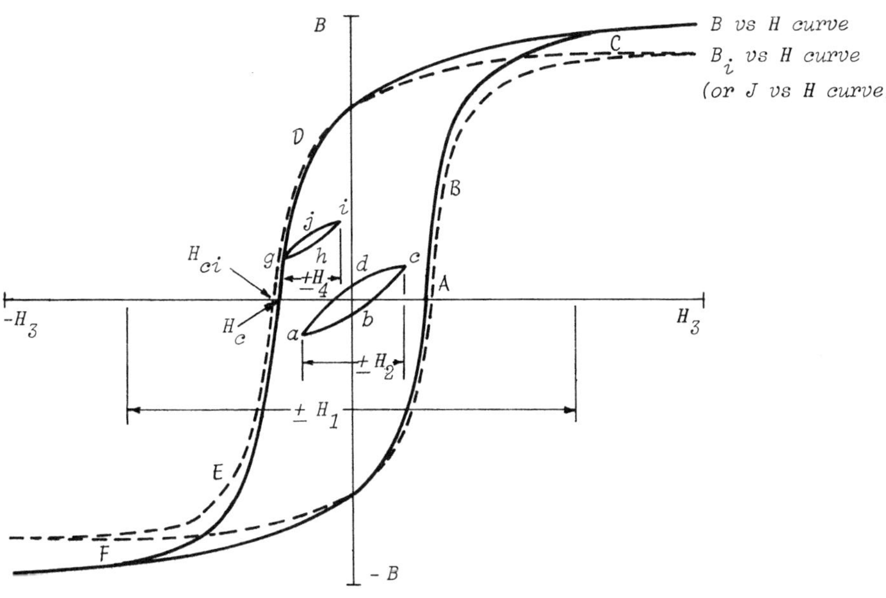

Figure 2-8. Major and minor hysteresis loops. The excursion of applied field H determines the size and position of the hysteresis loops.

Soft Magnetic Materials.

Soft magnetic materials have narrow hysteresis loops, ideally of smallest possible area for low losses, yet of maximum height representing highest induction. Figure 2-3 shows the hysteresis loop for supermalloy. Although a remanance of about 0.55 *T* (5500 Oe) is indicated, this can occur only in a closed magnetic circuit, because even a microscopic gap will reduce the residual magnetization to essentially zero when the coercivity is only 0.0004 kA/m (0.005 Oe). Supermalloy is therefore considered a soft magnetic material because for all practical purposes it loses its magnetization *B* as soon as the field *H* is removed. Other soft materials are silicon iron (used principally in transformers, motors, and relays); and permalloy, sendust, and ferrites used in magnetic heads.

Magnetic Recording Materials.

Certain ferromagnetic materials having properties especially advantageous for magnetic recording (tapes, heads, motors, transformers) are described in detail for the appropriate applications as in Chapters 3 and 4. Table 2-2 is a partial list of such materials.

SOME GENERAL CONCEPTS OF MAGNETIC RECORDING

Nowadays we know that magnetic recording can provide high quality reproduction with distortion and noise levels as low as 0.1% or less, but these results are not predictable from simple theory or previous experience. In fact magnetic nonlinearity and hysteresis should give high distortion; the inherent domain structure should give high noise level; and the instability of short magnets should

Table 2-2. MAGNETIC MATERIALS USEFUL IN MAGNETIC RECORDERS

MAGNETICALLY HARD (high coercivity)	MAGNETICALLY SOFT (low coercivity)
Gamma ferric oxide (Fe_2O_3)	Supermalloy
Ferrosoferric oxide (magnetite)(Fe_3O_4)	Molybdenum permalloy
Chromium dioxide (CrO_2)	49-Permalloy
Metal particle suspension (ferromagnetic)	
Nickel cobalt alloy electroplated	Alfenol
Cobalt chromium (CoCr) vacuum deposited	
Stainless steel, type 304, cold worked	Sendust
Stainless steel, type 420, heat treated	Manganese-zinc ferrite
Vicalloy	Nickel-zinc ferrite
Cunife	Amorphous alloys
Barium ferrite	Silicon iron
Alnico	
Samarium-cobalt	

prevent permanence. Altogether, if magnetic recording didn't yet exist, the experts would predict very negative pessimistic prospects for this unnatural method of recording.

Prior to demonstration of magnetic recording there was doubt that it would be possible to locate magnetic poles along a magnetizable bar at desired positions. At the time physicists knew that in a magnetized bar the opposite poles appeared at some distance inward from the ends, and they believed that no matter how the bar was magnetized the poles would locate themselves in certain preferred positions. It is of fundamental interest, therefore, to demonstrate that it is indeed possible to represent any complex waveform by a corresponding flux distribution in a magnetizable bar, as shown in Figure 2-9. Here the *longitudinal flux* (number of lines) inside the record is everywhere equal to the

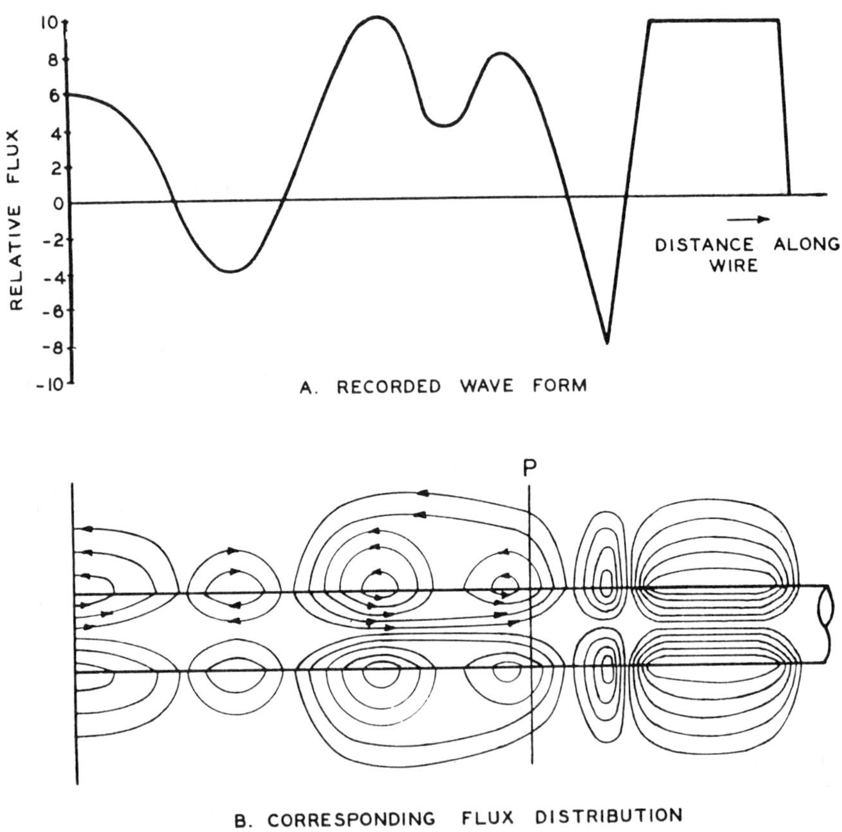

Figure 2-9. Flux distribution in a magnetic record. Corresponding to any arbitrarily chosen waveform, a flux distribution of exact accordance with the waveform can be set up in the record.

flux depicted by the arbitrary flux plot A. The total external flux through any plane P perpendicular to the axis of the record is equal and opposite to the internal flux. The pole strength, surface induction, or flux emanating from an elemental area surrounding the record is proportional to the space rate of change of internal flux through the corresponding perpendicular plane. A similar analysis can be made for other forms of records, and for different modes of magnetization.

MAGNETIC RECORDINGS MADE VISIBLE

The theory just discussed is verified dramatically by actually rendering a tape recording visible. When iron filings are sprinkled on or adjacent to a magnet, the particles align themselves to show the magnetic field direction. In the same way magnetic recordings, though much smaller, make themselves visible by attracting microscopically small particles under the proper conditions. The results are favorable when the particles are of micron size in a dilute liquid suspension. The finest grades of carbonyl-iron are suitable for the particles. Petroleum naptha is suitable for the liquid, as it dries quickly and is relatively non-toxic. The mixture is shaken vigorously and a few drops are flowed on to the surface of a magnetic record. Patterns of the recorded tracks develop out in a few seconds. Contrast improves after the liquid dries by evaporation. The pattern may be preserved by spraying with varnish or lacquer, or it may be wiped off with a cloth without harm to the recording. An alternative procedure that gives more uniformity is to dip a strip of the tape into the suspension for a short time, shake off the excess liquid, and allow to dry face up. Proprietary mixtures like the above are sold under such names as ''Magnasee'' (Reeves Soundcraft Co., New York, NY). Figures 2-10 and 2-11 show the patterns around wire recordings and tape recordings made visible by these techniques. Colloidal suspensions of much finer particles may be prepared to bring out detail according to Bitter techniques by which magnetic domains were first discovered.

The force acting on particles attracted by the tape is essentially proportional to the square of the field strength. Hence the particle density is non-linear, favoring the strongest field regions. Weak recordings may not show up at all. The patterns are strengthened by placing the tape in a dc field while the magnetic suspension is being applied (Yeh 1980). The DC field adds to the strength of the recorded north poles (for example) and suppresses the south poles, giving a clearer pattern that shows the enhanced polarity.

An ingenious gadget for making recordings visible is a pill-box shaped container about an inch in diameter, hermetically sealed with a magnetic suspension inside, a window on top, and a very thin non-magnetic bottom. When placed on a recorded tape, a pattern develops on the bottom inside surface. The pattern is erased by shaking the container. The price paid for this convenience is a

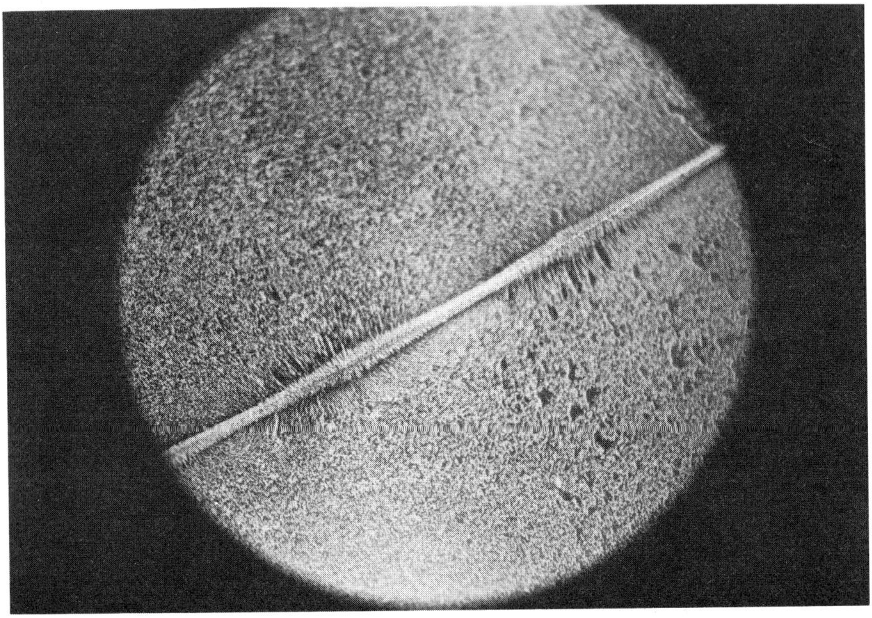

Figure 2-10. Field patterns surrounding a magnetic recording wire. Wire .004 inches diameter. Recorded frequency 100 Hz. Iron dust suspended in mineral oil was used to form these patterns. (Taken by Mr. A. V. Appel, Armour Research Foundation)

reduction in sensitivity and in detail compared to the liquid method (Recorded Pattern Viewer, Minnesota Mining and Mfg. Co., St. Paul, MN).

BASIC COMPONENTS OF A MAGNETIC RECORDER

Poulsen's simple recorder described in Chapter 1 and in Figure 1-3 contains the essentials of all magnetic recorders: a record medium; a recording and playback head; provision for erasing; and a drive for the tape

Record Medium.

Record medium is a general term for magnetic tapes, discs, drums, belts, flat sheets, wires, and any other form that the record may take. For simplicity the term *tape* will also be used to denote a record medium regardless of its form. At present, most record media consist of a thin magnetic layer deposited on a smooth strong substrate, as in Figure 2-12, since the magnetic layer is too fragile to support itself. Formerly solid wires, tapes, and discs were used which were self-supporting (Figure 2-13*B* and *C*), but were heavy, expensive, and poor in recording qualities.

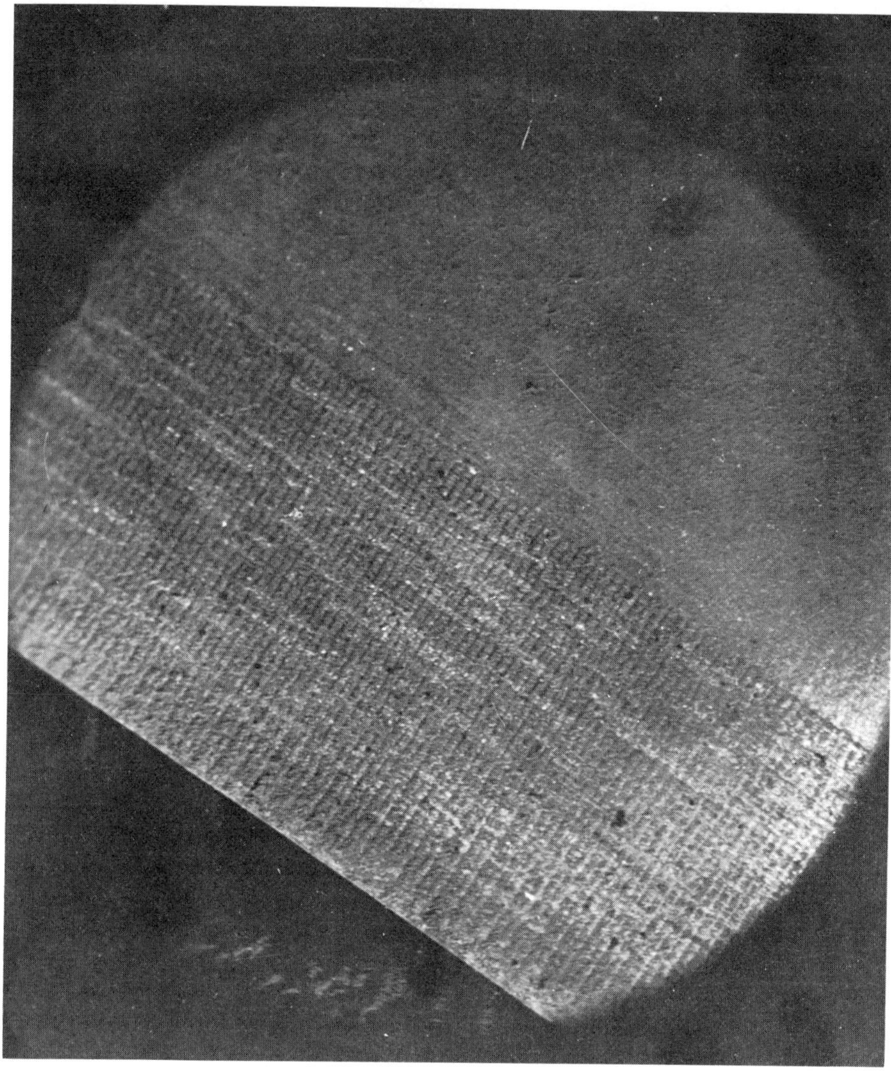

Figure 2-11. Field pattern of half track recording on $\frac{1}{4}$ inch tape.

Magnetic Properties. As expected, permanent magnetic properties are necessary in a record medium, including high coercivity (H_{ci}) and high retentivity (B_r), as in Figure 2-6. High coercivity improves the short wavelength capability and therefore the high frequency response, while high retentivity improves the output especially at long wavelengths and low frequencies, as explained in Chapter 3. It is interesting however, that the best materials for ordinary permanent magnets have proved to be useless for magnetic recording, and con-

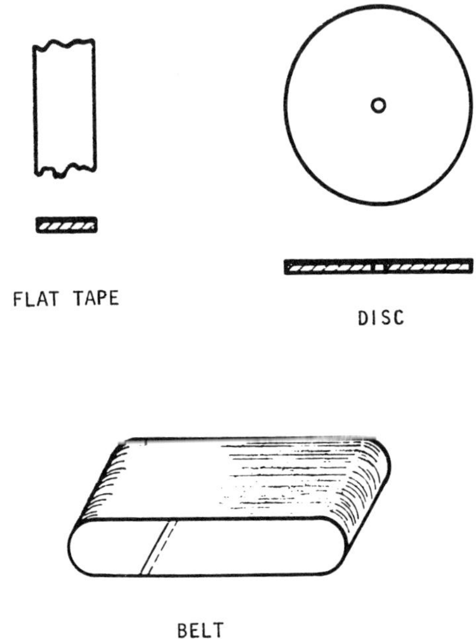

Figure 2-12. Coated magnetic records.

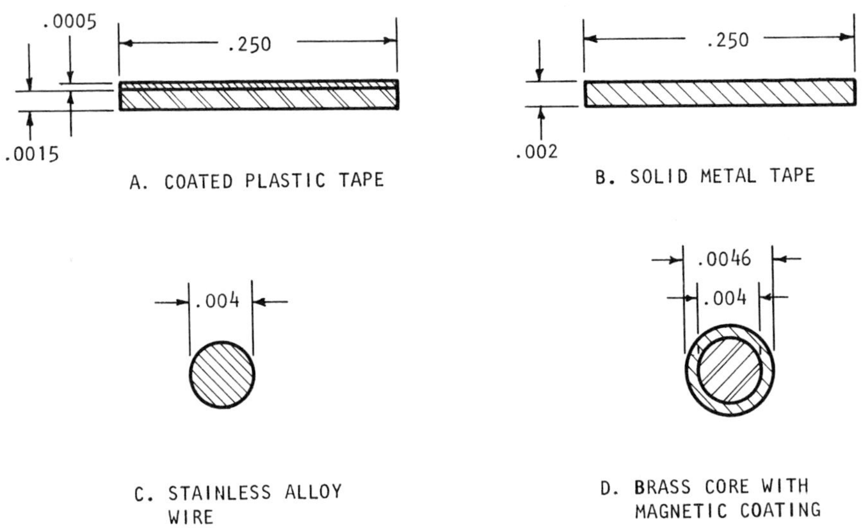

Figure 2-13. Cross section of some magnetic recording media.

Table 2-3. MAGNETIC PROPERTIES OF SOME RECORD MEDIA

RECORDING MATERIAL	YEAR INTRODUCED (APPROX)	COERCIVITY	RETENTIVITY
Carbon steel, cold worked, wire or tape	1900	10	10,800
Tungsten steel tape	1936	27	12,200
Vicalloy tape	1937	260	7,800
High carbon steel, heat treated	1940	30	10,100
420 Stainless steel	1943	60	7,000
18-8 Stainless steel	1942	270	2,250
Plated bronze wire, alloy of 80% Co, 20% Ni	1944	220	10,400
NiCo plated discs	1946	250	10,000
Carbonyl iron coated tape	1935	5	100
Low coercivity gamma Fe_2O_3 impregnated tape	1942	90	112
Low coercivity gamma Fe_2O_3 coated tape	1942	110	300
Low coercivity magnetite Fe_3O_4 coated tape	1946	120	375
Hyflux (metal particle) coated tape	1946	500	2,500
High coercivity magnetite Fe_3O_4 coated tape	1946	340	800
High coercivity gamma Fe_2O_3 coated tape	1946	250	1,200
Metal particle coated tape (Fe, Co, Ni)	1962	830	2,800
Cobalt-doped gamma Fe_2O_3 tape I	1965	290	960
Chromium dioxide coated tape	1966	550	1,400
Cobalt doped gamma Fe_2O_3 coated tape II	1978	600	1,100
Cobalt adsorbed iron oxide coated tape	1978	580	1,550
Metal particle high coercivity tape	1979	1100	3,100
Berthollide iron oxide coated tape	1979	550	1,400

versely the best recording materials are not used for bulk permanent magnets. Table 2-3 gives magnetic properties of some record media as the recording art progressed. The coercivity grew until (with some exceptions) it reached about 250-350 in 1946. It remained in this area for many years because ordinary heads could readily record and erase such material, while higher coercivity tapes gave difficulty.

Chromium dioxide tapes of 400 to 550 coercivity, announced in 1966, achieved some popularity in the 1970s, requiring recorders with special high-bias settings. Modified iron oxide compositions of similar coercivity also were marketed. In 1979, major tape manufacturers announced metal particle tapes with coercivity of about 1100, and made them available on a limited basis. Equipment manufacturers, mainly of high fidelity cassette recorders, developed recorders for such tapes. Ultimate acceptance of high potency tapes depend on their cost-effectiveness, for performance seems to rise more slowly than magnetic properties or cost. Also some troublesome side effects have occurred, such as chemical instability, and deterioration of the recordings with temperature and with mechanical stressing. Even with improvements in new materials we can expect the old proven gamma iron oxide tapes to coexist with the new for a long time.

Mechanical Properties. A good tape should have high tensile strength to withstand reeling stresses, good dimensional stability, corrosion resistance, and immunity to humidity and temperature. It should be thin to reduce bulk and weight, and should be pliable to conform with the heads. It should be uniform in thickness and have a smooth surface to minimize noise. Coated tapes should have a uniform dispersion of particles in the binder which holds them to the substrate. Cost should be reasonable.

Erase, Record, and Playback Heads.

The heart of a tape recorder is the magnetic head that makes the recording, plays it back, or erases it. The head of an early tape recorder invariably was a stylus of soft iron surrounded by an electric coil, as shown in Figure 2-14*A*, and spring pressed against the tape. Such a head could also be used with a wire, but if the wire rotated on its axis between the time of recording and playback, the output level would fluctuate. This condition was alleviated by using a pair

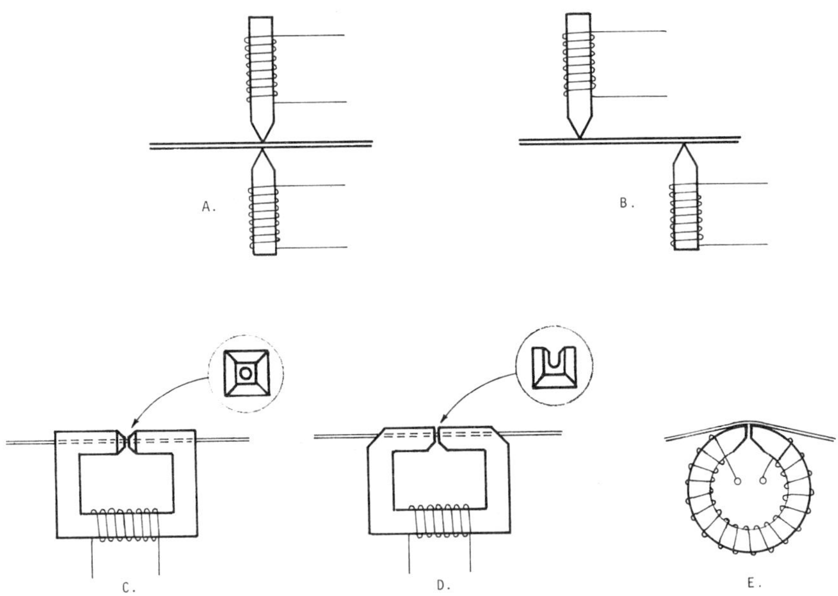

Figure 2-14. Early recorder heads. A. Opposed (stylus) polepieces. B. Staggered polepieces for quasi-longitudinal recording. C. Hole-in-gap head for symmetrical magnetization. D. Knot-bypassing wire recorder head. E. Ring head for tape recording.

of staggered polepieces on opposite sides of the wire to give a longitudinal component of magnetization, as in Figure 2-14*B*, which also increased the output but sacrificed the recording resolution, and thus made high wire speeds necessary. A more efficient head (Figure 2-14*C*), developed in the 1930s, used a gap for recording instead of thin polepieces. The wire entered through a hole bored in one pole-piece, passed through a very small gap, and exited through a hole in the second polepiece. The result was a symmetrical longitudinal magnetization of high resolution.

For tapes, especially those which were coated with a thin magnetic layer on one side only, the ring head shown in Figure 2-14*E*, was advantageous, eventually leading to the modern opposed-polepiece ring head which has practically replaced all other types.

By connecting its winding to appropriate electric circuits, the same head can serve as a playback head, recording head, or erasing head; but certain modifications will improve the efficiency for a specialized function. Thus a playback head gives best resolution with a very small gap; a recording head functions better with a larger gap, as its resolution does not depend entirely on gap size; an erase head needs a still larger gap for a strong field through the recorded layer and its core is advantageously made of high saturation material not necessarily suitable for recording or playback. The erase head is generally turned on whenever the recording head is used, to simultaneously remove an old recording as a new one is made.

Erase Heads. Modern recorders erase the tape by demagnetizing it with a strong ac field. The field alternates through positive and negative values great enough to blot out the recorded information, and then decreases gradually to zero after a sequence of many diminishing cycles, as shown in Figure 2-15*A*. The relation $\lambda = v/f$ tells us that f must be very high to give sufficient cycling of a tape element that passes over a relatively short erase gap. For example, when a tape travels 15 in./sec (38.1 cm/sec) with an erase frequency of 100 kHz, the erase wavelength is 0.15 mil. In an erase gap 5 mils long the tape element is cycled $5/0.15 = 33$ times while in the gap, and probably as many times in the decreasing fringing field after it leaves the gap.

Erasing efficiency is measured by the number of dB that a strong signal is reduced by the erase head, and by the noise levels on the tape after erasure. A good erase head should reduce the signal 60 dB or more, and leave it with a noise level within a dB or two of the "bulk erased" noise level. The best erasure, used as a standard of comparison, is achieved by a bulk-eraser where the reel of tape is immersed in the field of a large ac magnet operated at power line frequency, the reel being turned slowly as it is withdrawn from the field.

Rerecording Effect in an Erase Head. A puzzling situation often arises where an erase head will erase the signal to a certain level, and no further, even when

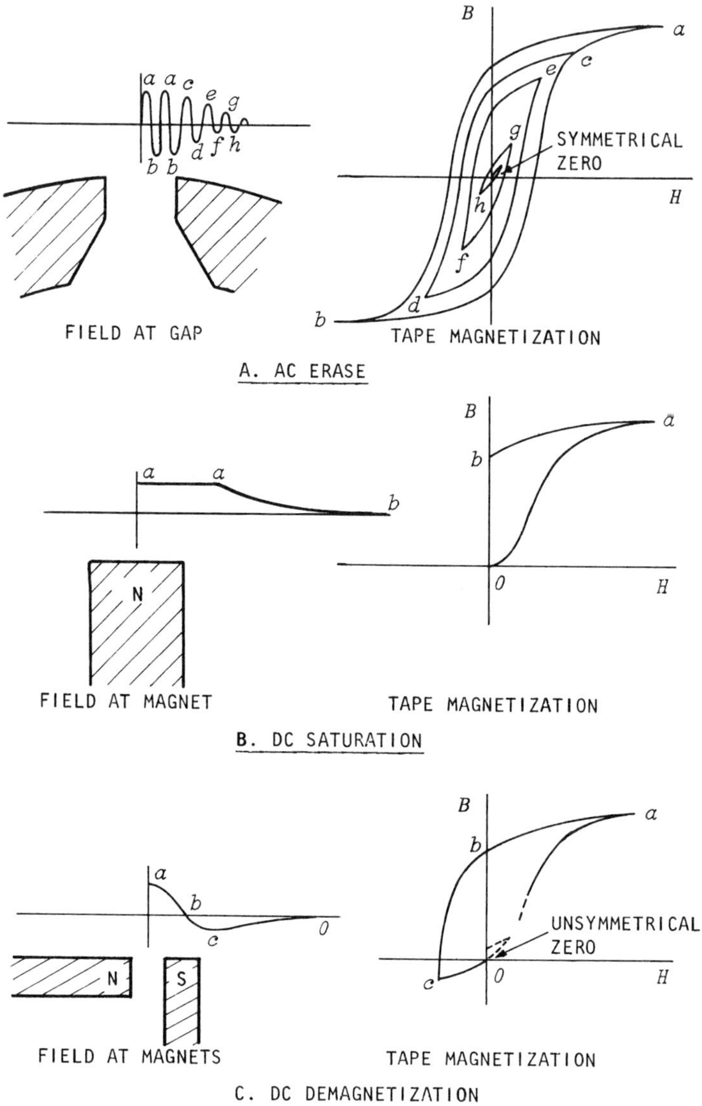

Figure 2-15. Erasing methods.

the head current is increased very considerably. The obvious answer is that the head core becomes saturated so that its field does not increase with more head current. But if the semi-erased tape is run over the head a second time, a great deal of additional erasure takes place, usually enough to remove all traces of the recording. The explanation is that the tape, just before it is erased, sets up a weak magnetization in the erase head at signal frequency. As the tape leaves the erase gap the erase frequency cycling acts as a high frequency bias that rerecords the weak signal. When one is examining for residuals that are 60 dB down (0.1%), this slight effect becomes important. The remedy is to use a succession of erase heads. In practice, heads are often built with two erase gaps, which generally give adequate results (McKnight 1963).

Special Erase. Sometimes the erasing is done by permanent magnets that leave the tape saturated or in a quasi-neutral condition (*B* and *C* in Figure 2-15). Such magnets must be moved away physically from the tape when not in use and their stray field must be prevented from harmful effects on the record and play heads. In certain machines, the erasing function is omitted; if erasing is required, it is done with a separate bulk-eraser. In computers the heads may record digitally by saturating the tape with the recording head in one direction or the other. The saturating process automatically blots out any previous recording so that separate erasing is unnecessary. Similarly a recording head will erase an old recording while a new one is made if excessive high frequency bias is used. Finally there are playback-only machines, as in eight-track stereo cartridge players, where erasing is never done.

Record Heads. As seen in Figure 2-16, during recording the tape passes over a small gap that interrupts the magnetic circuit of a high permeability core surrounded by a signal coil. A current through the coil produces a strong field, sharply defined at the record gap and concentrated on a small length of the moving tape. The signal changes in amplitude from one instant to the next, so

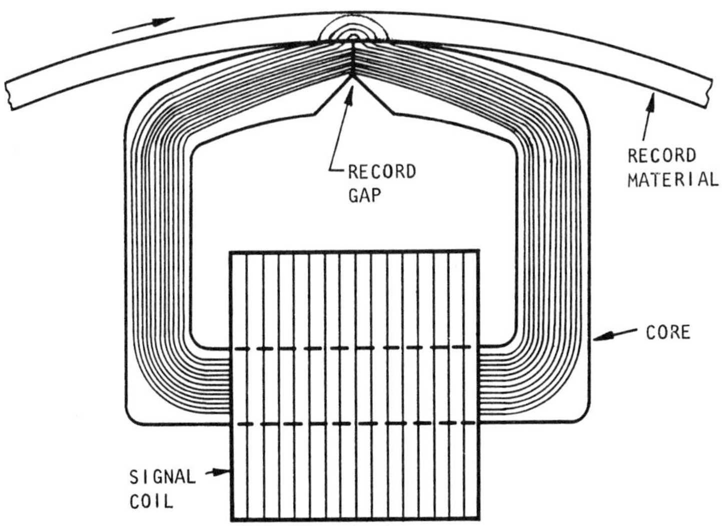

Figure 2-16. Flux paths in a magnetic head during recording.

that each element of the tape as it passes the gap "sees" and remembers a different amplitude and polarity of magnetization, becoming magnetized in a pattern of which Figure 2-9 is an example.

Playback Heads. Playback, shown in Figure 2-17, can utilize the identical head used previously for recording, but with the signal coil connected to the input of an amplifier. The magnetic flux emanating from the record favors the high permeability path through the core. The useful component *a* passes through the coil, inducing a voltage at the coil terminals proportional to $d\phi/dt$. But part of the tape flux is lost, the component *b* being short circuited across the pickup gap, while portion *c* finds its easiest path through the back of the tape.

Combination Heads. In moderately priced recorders the same head is used for both recording and for playback, its design being compromised for econ-

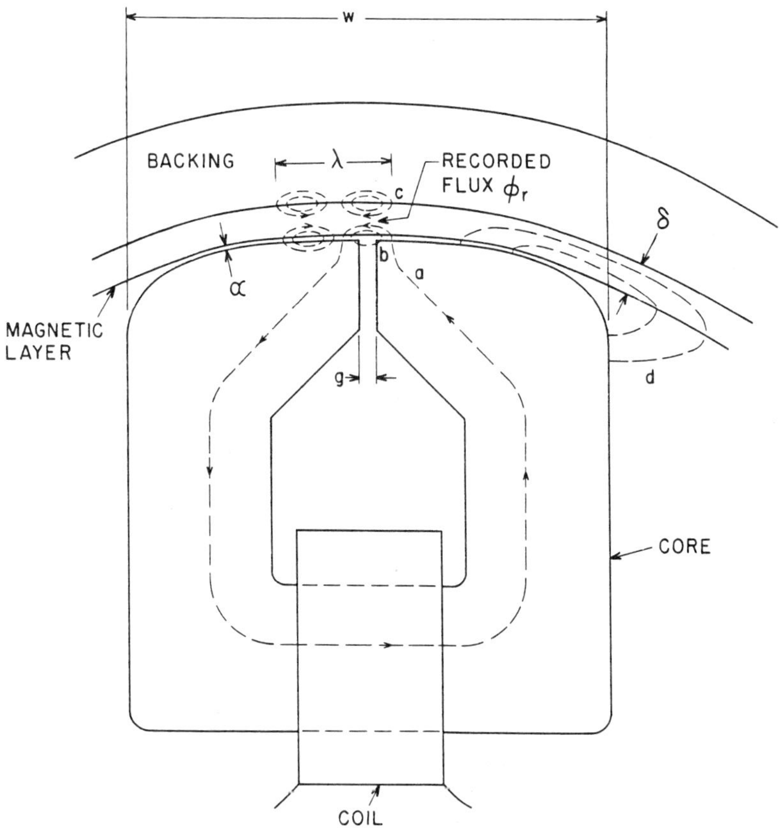

Figure 2-17. Flux paths in a magnetic head on playback.

omy. A combined record-play head does have the advantage that if its gap is slightly irregular or tilted or misaligned laterally, the recorded pattern will fit the same head on playback and thus compensate for imperfections. An erase head can also be built on the same core structure used for the record-play head, but with an independent coil and gap for erasing.

Even when separate heads are used for each function, all of the heads are often enclosed in a single shield can. The combined function head economizes on mounting and adjustment. It simplifies the tape path and reduces the frictional load on the tape. Combined heads can become quite elaborate, containing many heads if they have to perform with the tape running in either direction, and if two or four channel stereo is required.

Tape Drives.

It is often said in jest that 99 percent of a successful magnetic recorder is attributable to its mechanical design, but this is not far from the truth when in addition to the drive itself we consider such things as tape guidance, accurate winding, preventing excessive tape stress during stop and start, convenient indexing, and so forth. Also, there are mechanical aspects of heads, reels, mountings, interlocks, indexing, and ultimate "packaging."

Tape Handling and Tape Speeds. The art of handling the tape with smooth, stress-free starting, stopping, rewinding, and fast forward speeds has developed over many decades of experimentation and ingenuity. Low-cost machines use mechanical methods alone, but expensive drives take advantage of pneumatic, hydraulic and electromagnetic devices (servomechanisms) as well as electronics and microprocessors for control.

Tape speed is governed by practical recording density (bits or wavelengths per mm). A multiple or submultiple of 7.5 in/s (190.5 mm/s) is favored, with 15 in/s (381 mm/s) typical for professional audio, 1.875 in/s (47.6 mm/s) for compact cassettes, and 120 in/s (3048 mm/s) for wide band instrumentation.

Wow and Flutter. Regardless of other considerations, analog audio recorders must drive the tape at a steady speed with short term variations not more than about 0.1 percent for high quality music, or about 0.25 percent for speech. The 0.1 percent figure is attained by driving the tape past the heads with a precision roller called a capstan, with a fly-wheel mounted on the capstan shaft to iron out irregularities from rollers, belts, gears, motors, or from imperfections in the tape winding or the reels (see Figure 7-1*C*.) Less precision is tolerated in low cost recorders. The resulting unsteadiness causes sustained musical tones to have a "wow" sound superposed on them if the fluctuation is about once per second (as might occur with a bent tape reel), or a fluttery sour-note sound if the fluctuation is more rapid, as with a slipping capstan (See Chapter 7).

Endless Loops, Discs, and Drums. Specialized applications use endless loops, discs, or drums which usually need only a steady one way drive with no provision for rewinding, stopping or starting. Tape loops are often troubled with bad fluctuations in tape tension which are difficult to neutralize for high quality music. Disc and drum drives are about as steady as corresponding phonograph turntables.

Indexing. Such a vast amount of information is stored on a reel of tape that quick accurate access to a desired portion is a problem. A crude index is to mark the tape level on a supply reel. A refinement is the use of a revolution counter on the supply reel shaft. Some tapes have been printed with footage markings on the back; although accurate, these marks are inconvenient because the tape must be stopped and visually inspected. The ultimate index is a magnetically recorded code on the tape, either on a separate track, or between selections, or as part of the data itself. The indexing is then computerized automatically without effort on the part of the operator.

Discs and drums also store large amounts of data, but here the data is rapidly accessed by indexing the head radially for the disc and axially for the drum. Computer systems have developed sophisticated servo controlled electromechanical indexers for their disc memories.

Amplifiers and Equalizers.

Playback signals from magnetic heads are in the order of millivolts, requiring amplification before they are useful. Their frequency response and waveforms are deficient because of the record-playback process and therefore must be corrected by equalization or other signal processing as detailed in Chapter 6. Amplification and correction are readily done with solid state components which are so straightforward and reliable that they seldom are a problem for the designer.

THEORY OF MAGNETIC RECORDING

Theory of magnetic recording deals with:

1. The magnetic field set up by recording heads
2. The magnetization that occurs throughout the tape layer during recording
3. Biasing methods to overcome nonlinearity
4. Overall response, including playback effects

5. Special conditions such as print-through, memory, etc.
6. Miscellaneous methods of recording and playback.

These major subjects and others are discussed in the remainder of this chapter.

Magnetic Field around Recording Heads.

Magnets dipped in iron fillings show complex field patterns that might seem to defy mathematical treatment. Yet the field near the gap of a confronting pole (ring) head can be plotted by conformal mapping taught in a branch of mathematics: *Functions of a complex variable*. The result of Figure 2-18 is shown in greater detail in Figure 4-3, where Karlqvist's equation for field mapping is discussed (Karlqvist 1954). The ring head is selected here because it is the most widely used head. It is by no means the only possible form of head, and indeed an entirely different head was used in days of the old telegraphones (*A* and *B* in Figure 2-14). Many different heads are described in Chapters 4 and 5.

Magnetization in the Tape Layer.

Through the field shown in Figure 2-18 a magnetic tape passes along a line *m*, *n*, *o*, *p*, *q* and emerges magnetized if the head field and tape characteristics are appropriate. For insight into the recording process some simplifying assumptions are useful at this point and will be corrected later for more quantitative results. It is useful to assume a very thin record compared to the gap length,

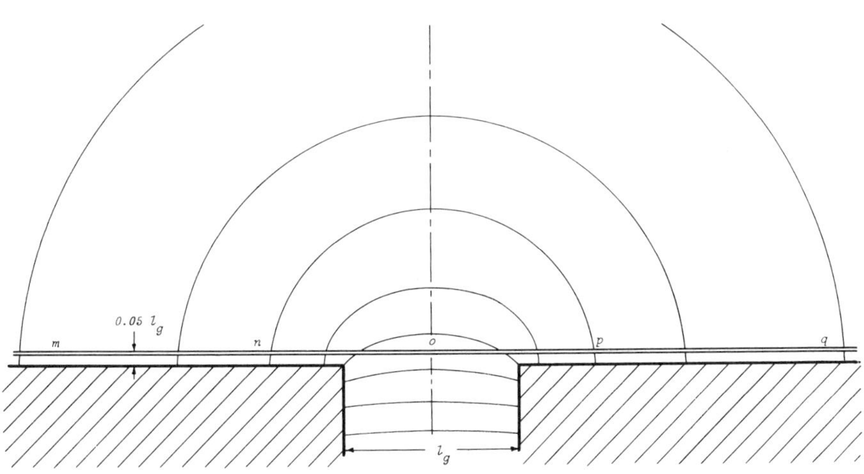

Figure 2-18. Field encountered by tape as it passes the gap of a recording head.

running for example, along a line spaced 0.05 ℓ_g above the head surface and further to consider only the longitudinal component of the head field and that the field has not been altered by presence of the tape. If the head field does not change during the time that an element of tape passes along *m,n,o,p,q*, the element will encounter a field *H* shown in Figure 2-19, which is a plot of horizontal field intensity vs. distance. The magnetization of the tape element will follow a rather devious path *m',n',o',p',q'* along a *B* vs *H* plot of Figure 2-20 as expected from hysteresis effects, reaching a maximum of B_m and ending up with a permanent residual magnetization of B_r when the applied field goes down again to zero.

Since the head field *H* does change in time, we need to know what the residual magnetization would have been for lesser and greater values of H_{max} from zero to tape saturation (assuming that all parts of Figure 2-19 change proportionally to its maximum value). Such a plot is given by the B_r vs. *H* curve of Figure 2-21, which is obtained from a family of curves like Figure 2-20. Figure 2-21 is typical of the gamma ferric oxide commonly used in magnetic tape coatings, and represents the transfer characteristic or output vs. input curve obtained by recording directly without bias.

The transfer characteristic shown in Figure 2-21 has very high distortion for an input (*H*) symmetrical about the zero axis, giving almost no output for inputs of less than ± 8 kA/m(100 Oe), and cutting off all but the peaks for higher inputs of say ± 24 kA/m(300 Oe) inputs. Such distortion is indeed obtained in audio tape recorders when the bias is disabled. The distortion might be overcome by adding DC bias to the signal input, thus operating about point *a* in Figure 2-21. Linear range is limited however, since more than half of the pos-

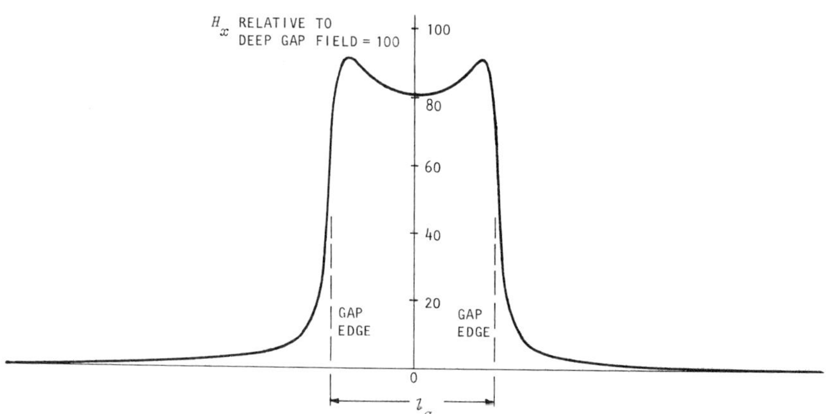

Figure 2-19. Horizontal field intensity along tape path 0.05 ℓ_g above head.

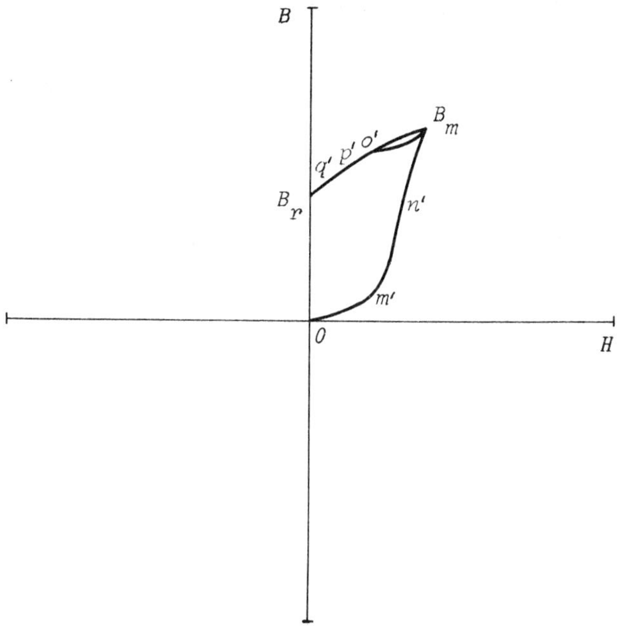

Figure 2-20. B versus H plot of element moving along tape path 0.05 ℓ_g above head

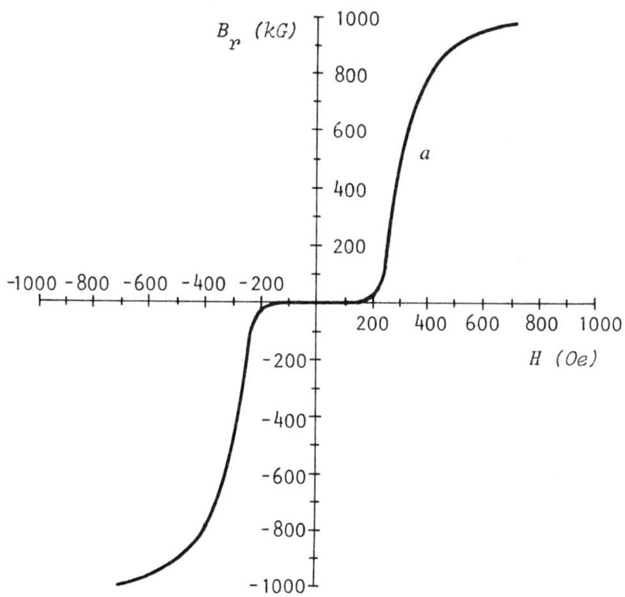

Figure 2-21. Residual magnetization vs applied field (B_r vs H) for gamma oxide coated tape of H_{ci} = 260 B_r = 1000. The plotted values of H are understood to be the maximum H to which the tape was exposed, after which the H was reduced to zero without being reversed.

sible magnetization of the record is sacrificed. The earliest telegraphones probably operated on this basis.

DC Biasing.

In about 1905, Pederson and Poulsen devised a DC biasing method for overcoming distortion which was a great improvement and became standard practice for almost forty years afterwards. In this DC biasing method shown in Figure 2-22, a DC erase head saturates the tape in one direction to a value B_m before it passes to the recording head, reaching the recording head in a state of maximum residual magnetization B_r. The recording head has an opposite steady dc

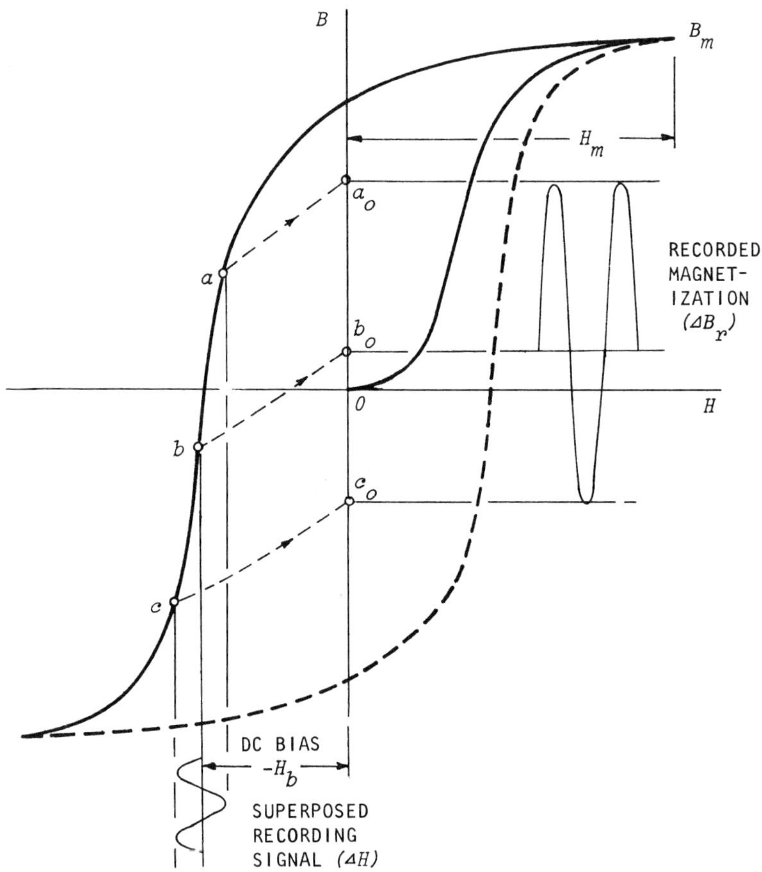

Figure 2-22. DC bias for overcoming distortion.

bias $-H_b$ that brings the tape magnetization down along the hysteresis curve along a path B_r, a, b such that the tape magnetization, after the tape leaves the recording head, falls back along b, b_o to a practically neutral condition at b_o when no signal is present. If a signal is superimposed on the recording head bias as shown, then the negative signal peaks will enhance the DC bias to the point c, leaving the tape in an ultimate condition c_o; positive signal peaks will reduce the bias to point a, leaving the tape in condition a_o.

Elements of a saturated tape moving over a head energized as in Figure 2-22 with DC bias and superposed signal will be recorded with magnetization ΔB_r as indicated, which is a reasonably faithful reproduction of the recording sine wave signal ΔH. Note that b_o is at a somewhat positive magnetization when bias is adjusted for least distortion. This condition leads to an increase in noise level compared to that of a "neutral" tape.

AC OR HIGH-FREQUENCY (HF) BIASING

One of the most important advances in analog recording is the use of high-frequency bias, which gives almost distortion-free recording in spite of the highly nonlinear, nonreversible character of magnetic record media. When high-frequency bias is used, the record is first demagnetized with an erasing head, which subjects it to an alternating field that diminishes to zero over many cycles. The record then passes through a recording head energized by a mixture of the signal to be recorded and a steady high-frequency, high-intensity current. For audio recorders, the high-frequency component may be 30 to 400 kHz, and its intensity is usually about five to ten times as high as the average recording current. Figure 2-23 shows the components of signal and bias fields and their resultant. The resultant is not an amplitude modulated signal but merely a mixture of high- and low-frequency fields.

When the record medium is passed through the composite field of Figure 2-23, it acquires a residual magnetization according to the output versus input curve of Figure 2-24, which is linear and symmetrical. The bias is not recorded since its high frequency is beyond the capability of the system.

Several different kinds of explanations have been offered for the action of high-frequency bias. An early suggestion was that the high frequency had "the effect of agitating the recording element so as to greatly increase the sensitiveness of said elements to feeble signal impressions" (Carlson and Carpenter 1927). This may explain the increase of sensitivity, but does not explain that there should be improved linearity between output and input.

A different class of explanations operate graphically or mathematically on the B_r vs. H curve of Figure 2-21. The simplest graphical explanation, shown in Figure 2-25, applies the waveform of Figure 2-23C as the input (H) of the B_r vs. H curve, and plots the recorded flux (B_r) as the output (Holmes and Clark

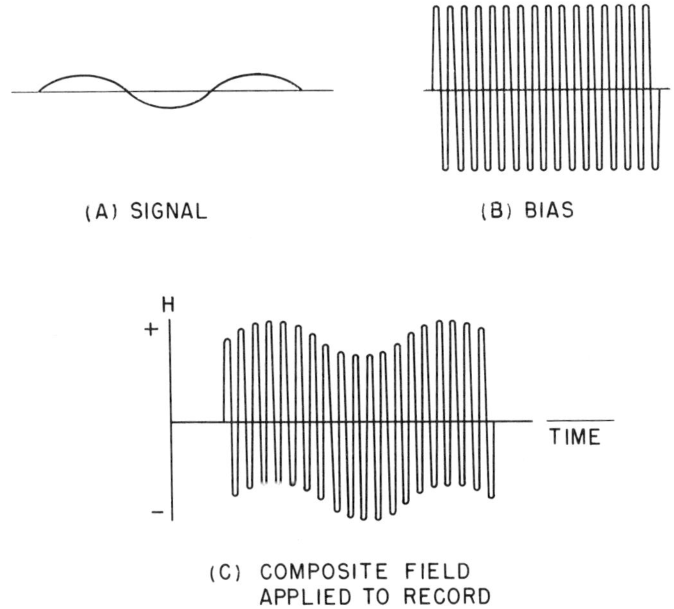

Figure 2-23. High frequency bias components and resultant.

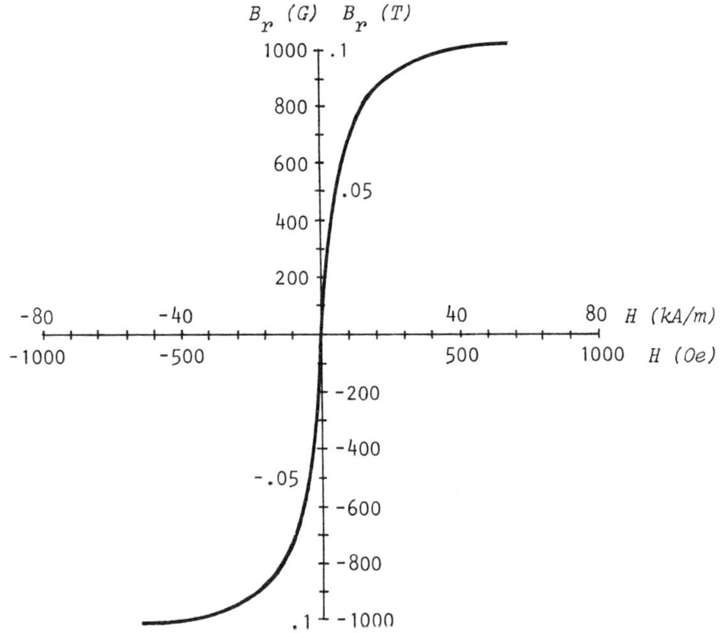

Figure 2-24. Output-input characteristic (residual magnetization vs applied field) when high frequency bias is used. (SI units are omitted in similar graphs that follow, but may be scaled by referring back to this figure).

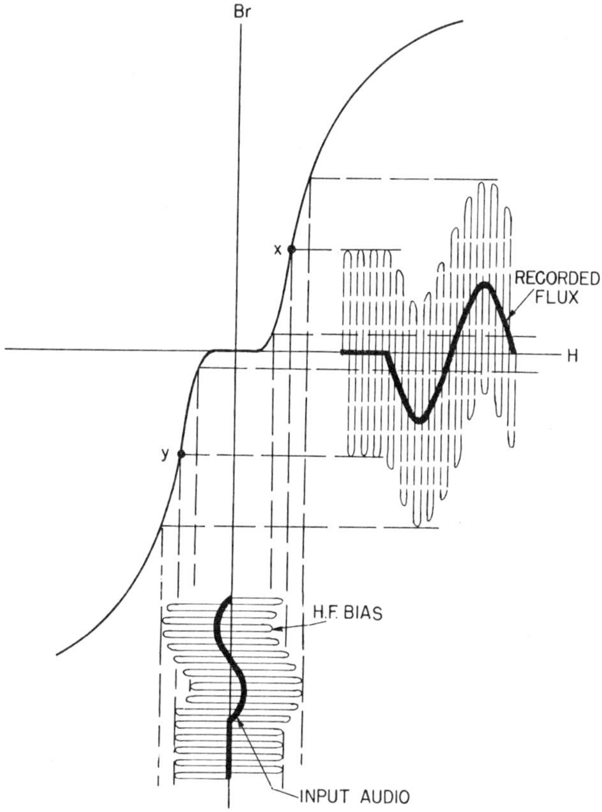

Figure 2-25. Action of high frequency bias.

1945). If the proper working point x is chosen, then the excursions of the high-frequency bias peaks are confined to linear parts of the B_r vs. H curve, and the average values of the positive and negative output peaks are a faithful reproduction of the input audio wave. Such averaging is easily done electronically by a low-pass filter that blocks out the bias frequency but passes the audio. During the record-playback process, the filtering takes place automatically since the recorder is incapable of registering the high-bias frequencies.

Output-input curves based on the averaging theory can be constructed graphically as in Figure 2-26 (Camras 1949). Here the entire B_r vs. H curve of Figure 2-22 has been displaced horizontally to the right by the positive bias peaks (curve B_r'), while an identical curve is displaced horizontally to the left by negative bias peaks (curve B_r''). The average of B_r' and B_r'' on the vertical scale gives the output-input response curve B_r when high frequency bias is present. Linearity of B_r through the origin is remarkable. There is an extra bend in the

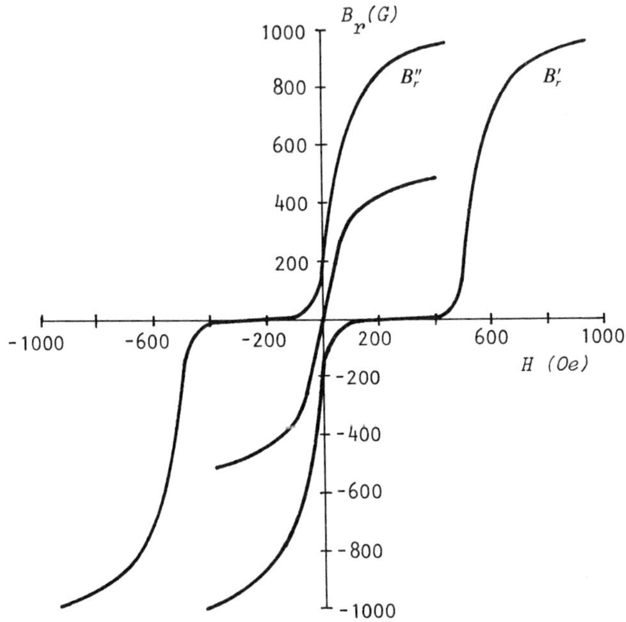

Figure 2-26. Graphical construction of output-input curve for various bias levels (H_B = 250 Oe shown).

B_r curve for applied fields above 32 kA/m (400 Oe). This can be ignored at usual recording levels which are considerably less than 6 dB below saturation.

Sensitivity and Distortion. Figure 2-27 is a family of output-input curves for different values of bias. As the bias is increased, the bend at the origin disappears and the slope of the curve becomes steeper, indicating higher sensitivity (more output for a given small value of input). Maximum sensitivity is reached with a peak bias of about 20 kA/m (250 Oe). Higher bias gives less sensitivity but good linearity. Figure 2-28 is a typical output versus bias (sensitivity) characteristic of a high-frequency biased tape recorder for a fixed small-signal input. The output rises rapidly when the bias reaches point *a* but has considerable waveform distortion until the maximum sensitivity point *c* is reached. The distortion continues to decrease as the bias is increased further although the output falls somewhat, especially at higher signal frequencies. Bias is usually set at a compromise value *d* where the output has fallen off about 10%, or one dB below maximum sensitivity.

Third Harmonic Cancellation. A peculiar effect is sometimes noticed where third harmonic distortion is minimized at low-bias setting *b*, although

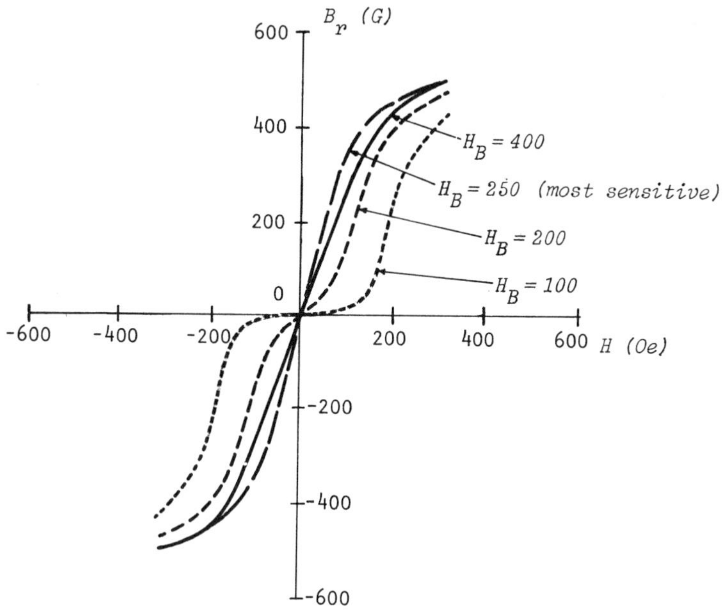

Figure 2-27. Output-input curves for various levels of HF bias.

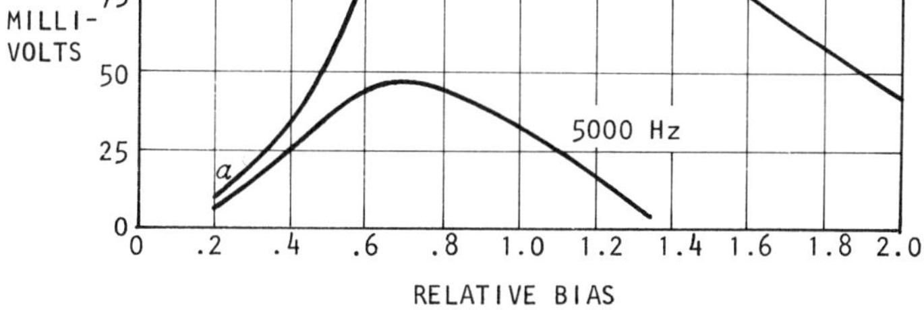

Figure 2-28. Playback output vs HF bias amplitude of a typical recorder. (A) 100 Hz fundamental output vs bias. (B) 5000 Hz fundamental output vs bias (Zenner 1951).

the waveform is still distorted otherwise. One explanation is that the recorded tape layer is not magnetized uniformly through its thickness and that the distorted magnetization pattern spaced away from the gap happens to compensate for the distorted pattern directly at the gap, with the particular bias b. Another explanation is derived from mathematical analysis which follows.

Mathematical Analysis. Most of the results obtained with high-frequency bias are nicely explained by a nonrigorous analysis that has appeared in many forms and variations (Zenner 1951). For example, the B_r versus H curve can be represented mathematically by a polynomial of the form:

$$B_r = aH + bH^3 + cH^5 + \ldots \qquad (2\text{-}6)$$

The mixture of signal and hf bias are:

$$H = A \sin\omega_s t + B \sin\omega_b t \qquad (2\text{-}7)$$

where A and B are amplitudes of the signal and bias respectively, and ω_s and ω_b are their angular frequencies. By substituting equation 2-7 into 2-6, a number of cross products are obtained. If the bias frequency ω_b is at least five times as high as the signal frequency ω_s, the cross products in the low-frequency range (ω_s or lower) are negligible at normal recording levels. When the high-frequency terms are "filtered out," the remaining B_r is a (linear) multiple of H, the same result as obtained with the graphical analysis.

The third harmonic term $3\omega_s$, which represents signal distortion, changes as the bias amplitude increases. As seen in Figure 2-29, at low bias the third harmonic is in phase with the fundamental; at high bias it is of opposite phase; and at an intermediate point it passes through zero. The zero point occurs at a lower bias than optimum, as has been noted before. The mathematical analysis shows that this zero point changes with the signal level as well as with the bias level and with the magnetic properties of the tape, and hence is not a stable operating point.

Still other related analyses that give similar results are based on pulse-width modulation that occurs when the bias and signal waves are combined (Bauer and Mee 1961).

Bias Beats. Harsh, unpleasant distortion, breakup, splatter, and rough sibilants can occur in hf-biased audio recordings at overload levels. High audio frequencies recorded at low tape speeds (as in cassette recorders) are especially vulnerable because they are often preemphasized, while the dynamic range of the tape is limited at these high frequencies. The distortion is due to spurious

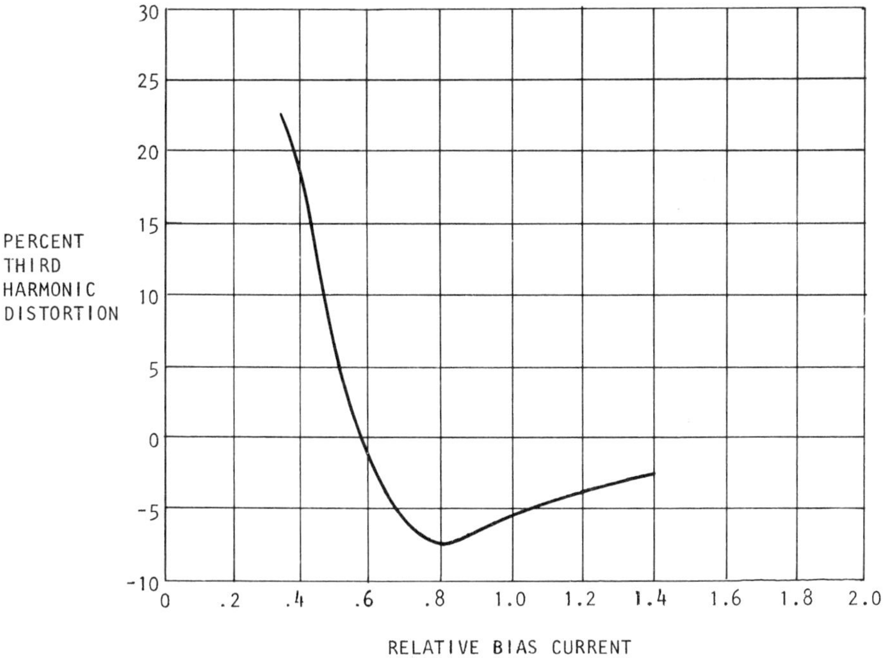

Figure 2-29. Third harmonic distortion as HF bias is increased. The third harmonic is in phase with the fundamental at relative bias below 0.6, and opposite in phase above 0.6. (Zenner 1951).

beat notes or difference frequencies generated by the hf bias beating against harmonics of the audio. Thus, an audio input of 13 kHz to the head will produce a strong third harmonic component of 39 kHz under the nonlinear overload conditions of tape saturation. High-frequency bias of 40 kHz simultaneously applied to the tape will generate a difference frequency (cross product) of 40 − 39 = 1.0 kHz. Neither the 39 kHz nor the 40 kHz will record on the slow-speed tape, but the 1.0 kHz can be so prominent that it blots out the fundamental 13 kHz signal we are trying to record. A complex audio wave of speech or music can generate many nonharmonic cross products that account for the harsh sound quality.

When a frequency response curve is run of a tape recorder, it is wise to observe the playback wave to confirm that it is indeed at the recorded frequency, rather than at a spurious beat frequency. Figure 2-30*A* is the frequency response of a recorder taken at low levels, while Figure 2-30*B* is the response taken at an overload level. Although there appears to be output at high frequencies, this output is at beat frequencies, and not at the frequency marked on the horizontal scale.

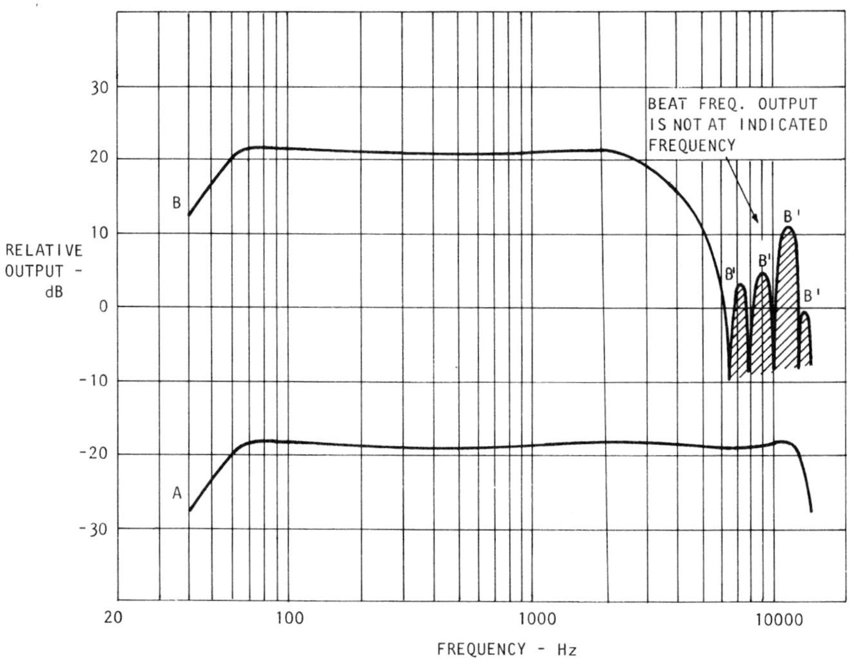

Figure 2-30. Bias beats. A. Frequency response at low signal levels. B. Response at overload levels with beats B′ generated at higher input frequencies.

Preventing Bias Beats. The best way to prevent bias beats is to choose a bias frequency at least four times higher than the highest signal frequency to be recorded, since higher audio harmonics are weak. In professional recorders a 10:1 ratio is typical, with 200 kHz bias for a 20 kHz audio limit. In domestic recorders the bias may be at 60 to 100 kHz because there are problems with erase head losses at higher frequencies. Erasing and biasing are usually done at the same frequency, but in sophisticated recorders the biasing may be at multiple of the erase frequency.

The graphical and mathematical explanations of high-frequency bias have been criticized because their B_r is derived from the B_r versus H curve shown in Figure 2-21 where H is never reversed. But in the usual hf-biased recording, H is reversed many times while an element passes through the recording head, and the resulting B_r can be less than or even opposite to that predicted by the B_r versus H curve shown in Figure 2-21. In other words, these theories do not explain why high-frequency bias linearizes the recording under conditions where there is no recording of the bias itself at any time.

Linearization when HF Bias is not Recorded. Since the simple B_r versus H curve shown in Figure 2-21 is not valid under ordinary conditions, a more accurate B_r versus H curve can be plotted by varying the H field through an actual cycling as seen by a tape element passing through a recording gap. This involves reaching a maximum H followed by several decreasing reversals of a composite bias-plus-signal field. Figure 2-31 illustrates the magnetic cycling conditions for obtaining a point on new B_r versus H curve. A surprising result, shown in Figure 2-31, is that after this complex magnetic cycling, experimentally traced, the final B_r point is determined only by the signal component of H (Camras 1944). For example, in Figure 2-31 the magnetization of a tape element as it leaves the gap may take a path *a b c d e* B_r in decreasing minor hysteresis loops; while a neighboring element somewhat out of phase with the identical excitation takes a path *g h i j k l* B_r. Yet, both end up at B_r, and the same result is obtained experimentally for other possible paths. Similar cycling for the same high-frequency bias with a smaller signal H gives a lower invariant B_r, while cycling with a larger signal H yields a larger invariant B_r. When the B_r points thus derived are plotted against the signal field H, the result is a straight line through the origin. This linearization is obtained from authentic B versus H cycling and does not assume recording and averaging of the bias.

HF BIAS AND DOMAIN THEORY

While the above explanations of high-frequency bias are in terms of the measured hysteresis properties of recording materials, a deeper analysis takes account of the causes of hysteresis and why it is modified by introduction of an AC bias component. There is evidence that on a submicroscopic level, the mag-

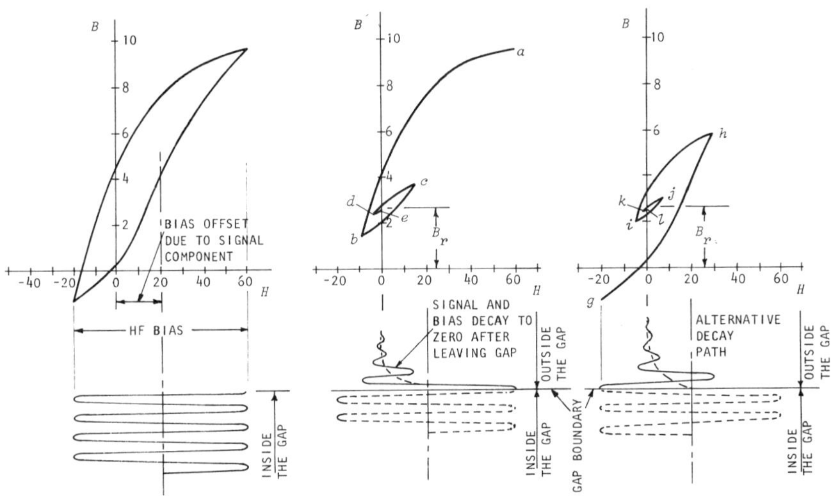

Figure 2-31. Cycling with reversible decaying H function gives a *B* that is invariant with phase of the HF bias (Camras 1944).

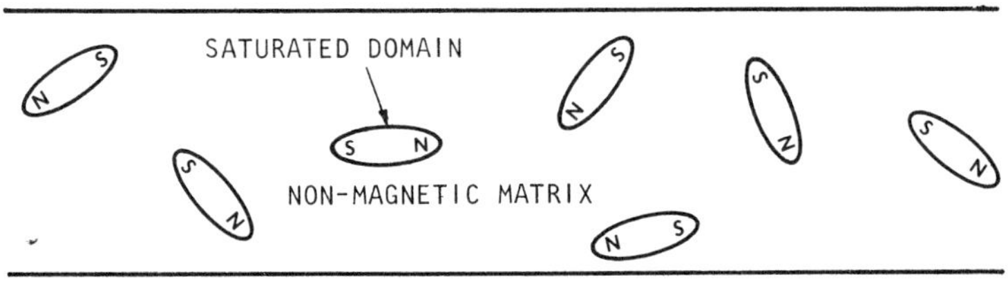

Figure 2-32. Magnetic domains in the recording layer. Magnetically saturated domains are randomly distributed in a nonmagnetic matrix. On a large scale their magnetizations cancel out to give a demagnetized medium.

netic record consists of billions of magnetic *domains* held in a nonmagnetic matrix that separates the domains from one another. Each domain is always saturated magnetically in a preferred direction or opposite to that direction, even in a demagnetized record where the positive and negative magnetizations statistically cancel out to give a neutral condition on a macroscopic scale. Figure 2-32 shows an array of such domains in which the magnets are random in strength and direction. If such an array is subjected to a reversing magnetic field, $\pm H$, a hysteresis loop will be generated. When H is rising, the magnetization of each domain will flip over as its coercive force is overcome, until all of the domains are aiding the magnetization, B. When H is falling, the domains retain their positive magnetization until H is sufficiently negative to flip them in the reverse direction, after which the cycle repeats to close the hysteresis loop.

In a recording head the domains would be subjected to a field in the gap, as in Figure 2-33 where high-frequency bias alternations $\pm H$ are offset from the

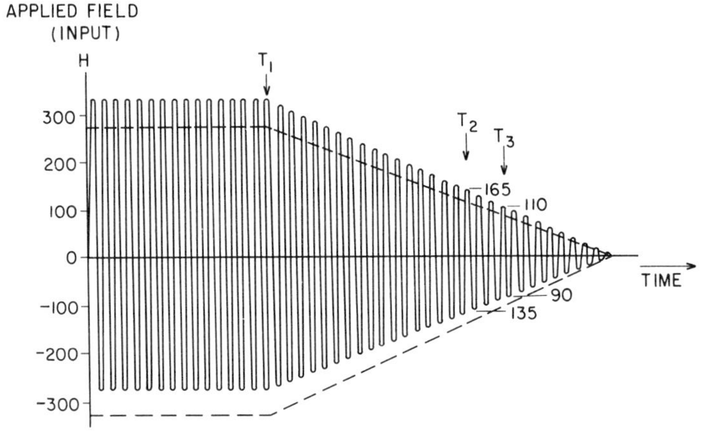

Figure 2-33. Field at the recording head gap as seen by an element of tape as it departs from the gap.

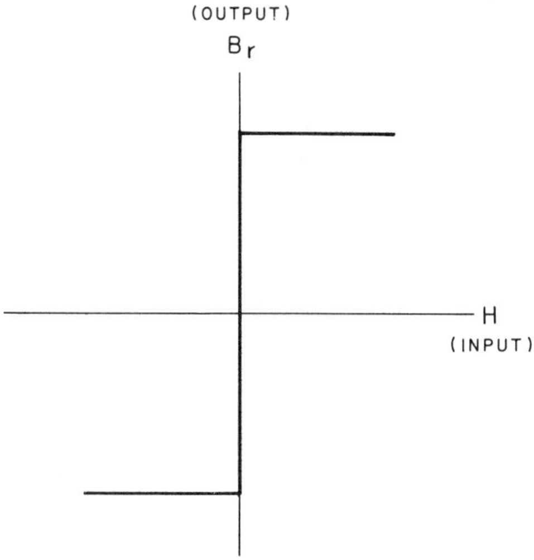

Figure 2-34. Output versus input curve in a record where there is no magnetic interaction among negative domains.

zero axis by a slowly changing signal (Camras 1962). At a time T_1 the domains pass out of the gap and the field decreases to zero over many cycles as the domains get farther and farther from the gap. What would be the final state of the domains after they came out of such a cycling? Theory predicts that if the domains are separated enough and do not influence each other, they all become saturated in the positive direction or all in the negative direction, depending on the offset (Westmijze 1953). In other words, the output-input (B_r versus H_{signal}) curve would be a vertical line at the origin, as shown in Figure 2-34. Although each particle might have a different coercivity H_c from all others, there would come a time when the field would no longer be strong enough to reverse the magnetization of that domain. For example, at T_2 in Figure 2-33 a domain with $H_c = 135$ is flipped positive, but the next half-cycle where H drops below 135 cannot reverse the domain any longer. After T_3, a domain of $H_c = 110$ remains positive and so on. Eventually, all domains remain positive.

We know that the situation shown in Figure 2-34 does not occur in practical recording; instead, we obtain an output-input curve such as the one shown in Figure 2-24. A behavior such as the one shown in Figure 2-24 is explained by taking into account the magnetic interaction between particles (Woodward and Della Torre 1959). Whenever a domain becomes permanently magnetized, it exerts a field on all its lateral neighbors, which helps them become magnetized in a direction opposite of the first domain, as indicated in Figure 2-35. The influence of magnetic interaction and of different coercivities of particles is taken into account by the Preisach diagram given in Figure 2-36, where each

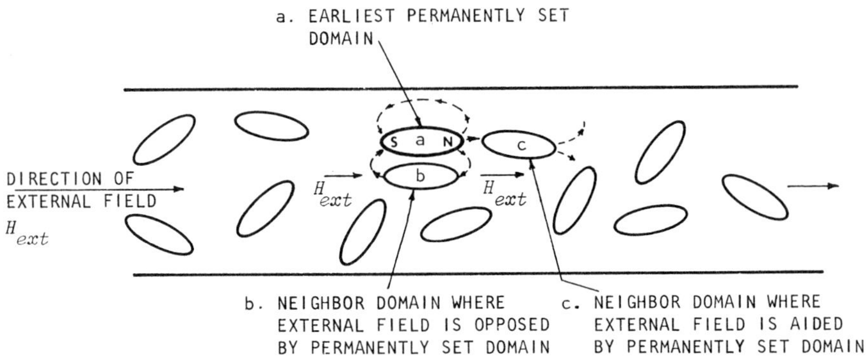

Figure 2-35. Magnetic interaction between particles. Domain *a* already magnetized exerts an opposing influence on its lateral neighbor *b*, while exerting an aiding influence on its longitudinal neighbor that sets *c* in the aiding direction.

domain is represented by a point whose abscissa shows the field required to switch it negative (H_-) and whose ordinate shows the field required to switch it positive (H_+). For example in the Preisach diagram of Figure 2-36, the domains having symmetrical hysteresis loops with equal positive and negative coercivities will lie along a 45° line, the magnetically harder domains of higher

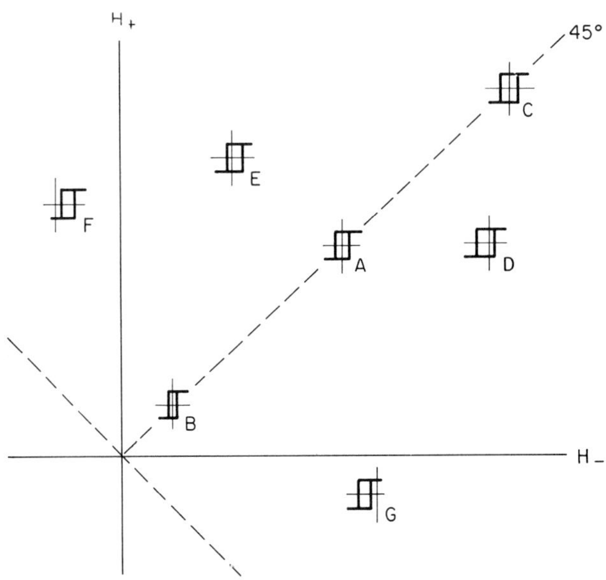

Figure 2-36. Preisach diagram. This coordinate system shows hysteresis loops of particles plotted in various areas.

coercivity being plotted farther from the origin. Domains D and E have hysteresis loops offset from the origin because their magnetized neighbor domains exert influences that modify the externally applied field. Domain F is an extreme case: The domain is switched only by positive external fields, while domain G is switched only by negative external fields. Since the magnetization of the neighboring domains is changing during the cyclic magnetizing field of Figure 2-33, a permanent point on the diagram is not assignable to each individual domain; but, statistically, each area on the Preisach diagram will be represented.

For any tape with a specific dispersion of magnetic particles, a distribution function is derivable from experimental data obtained by measuring the residual magnetization after the tape has been cycled magnetically with various proportions of AC and DC. Figure 2-37 shows the distribution function for a gamma ferric oxide magnetic tape. The two concentrations are away from the 45° line, signifying an unbalanced interaction among the magnetic domains (Woodward and Della Torre 1959). Another tape, Figure 2-38, shows a pear-shaped distribution about the 45° line, with a maximum at 20 kA/m (240 Oe). From the distribution function, we can predict the recording sensitivity, the optimum AC bias field, and the maximum output.

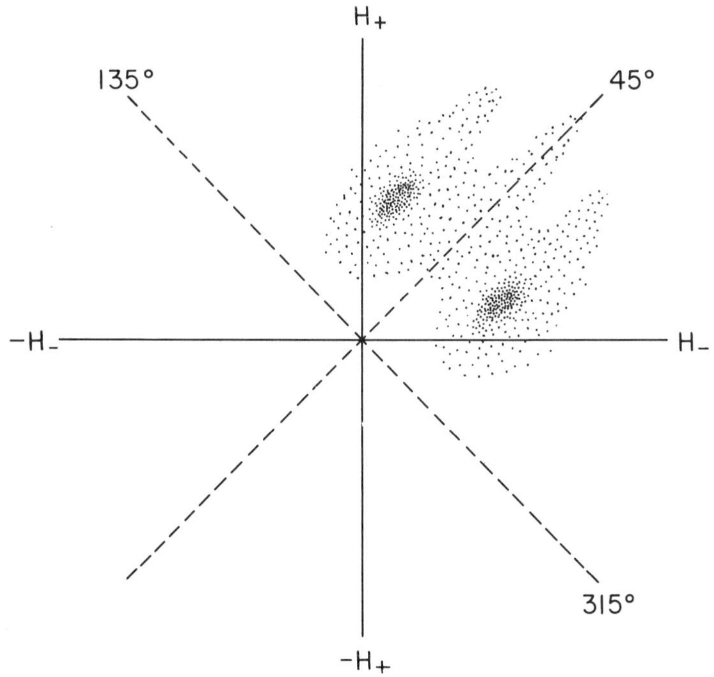

Figure 2-37. Magnetic distribution function of particles in a gamma ferric oxide tape (after Woodward and Della Torre 1959).

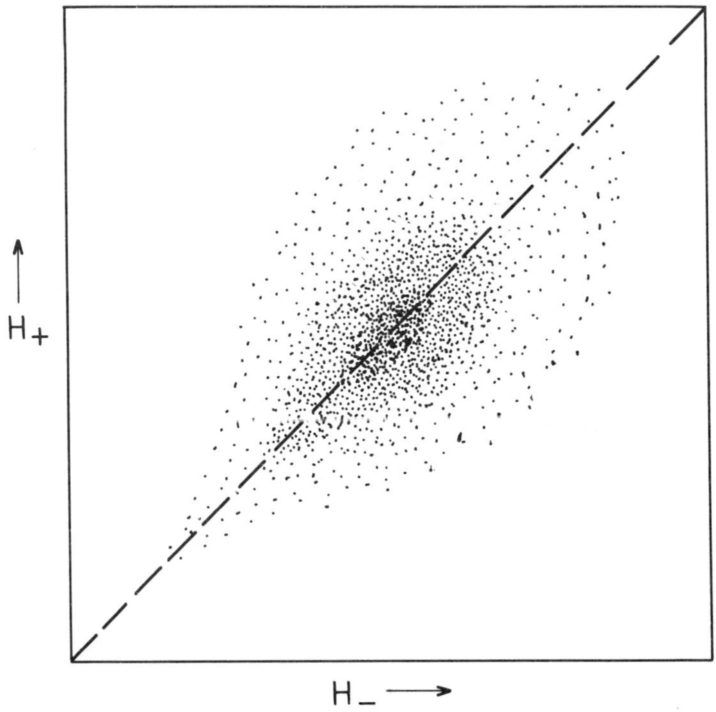

Figure 2-38 Distribution function of particles in a magnetic tape (after Daniel and Levine 1960).

Tape Layers of Finite Thickness. In the analyses above, we assumed that the tape layer had infinitesmal thickness, as in Figure 2-18. In practice the tape layer, although very thin (1 to 6 μm), is thick enough relative to the head gap so that the front side of the tape riding over the gap encounters a stronger and more concentrated magnetic field than the back side of the coating, where the field is weaker and more diffuse. One method for taking this variation into account is to subdivide the magnetic coating into thin layers and to solve for magnetization of each layer separately. The solutions are then coordinated to give an overall picture of magnetization throughout the coating thickness.

Magnetic Bubble Model. Instead of analyzing the tape coating layer by layer, we may look at the overall coating as affected by contours of equal magnetic strength at the head gap. Figure 2-39*B* shows a stationary tape over a head gap that is energized momentarily to a field strength that will permanently magnetize the entire layer, leaving a magnetized or recorded semicircular zone. The boundary between recorded and unrecorded material depends on the sharpness of rise of the B_r versus H curve of the recording material and also on the field

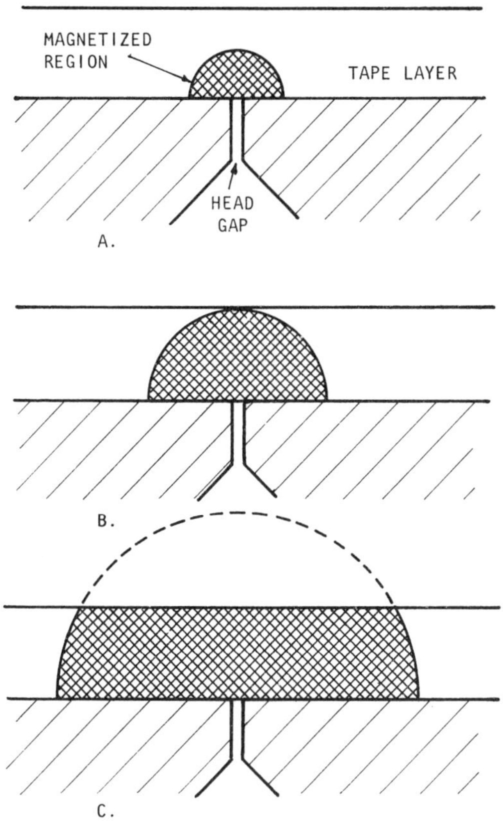

Figure 2-39. Effect of a magnetic pulse on a stationary tape. A. Gap field barely strong enough to magnetize half way through tape layer. B. Stronger field magnetizes entire layer. C. Excessive field magnetized broad area.

gradient in the tape. The field gradient is a function of head geometry, tape layer thickness, and tape permeability. If the head is energized to half the previous field strength, as in Figure 2-39*A*, the magnetized zone penetrates about half way through the tape layer; whereas if the field strength is doubled, the magnetization spreads out as shown in Figure 2-39*C*. In biasless recording, we thus have a *half-bubble* of effective magnetization at the gaps of Figure 2-39, expanding, contracting, and reversing polarity at the frequency of the signal field. On this bubble we superimpose the motion of the tape and obtain an elongated magnetized region that looks like Figure 2-40*A* when the field is increasing and like Figure 2-40*B* when the field is decreasing. A sinusoidal signal applied to a moving tape give arc-shaped magnetized regions, as shown in Figure 2-40*C*, as well as unmagnetized regions and regions of overlap where the

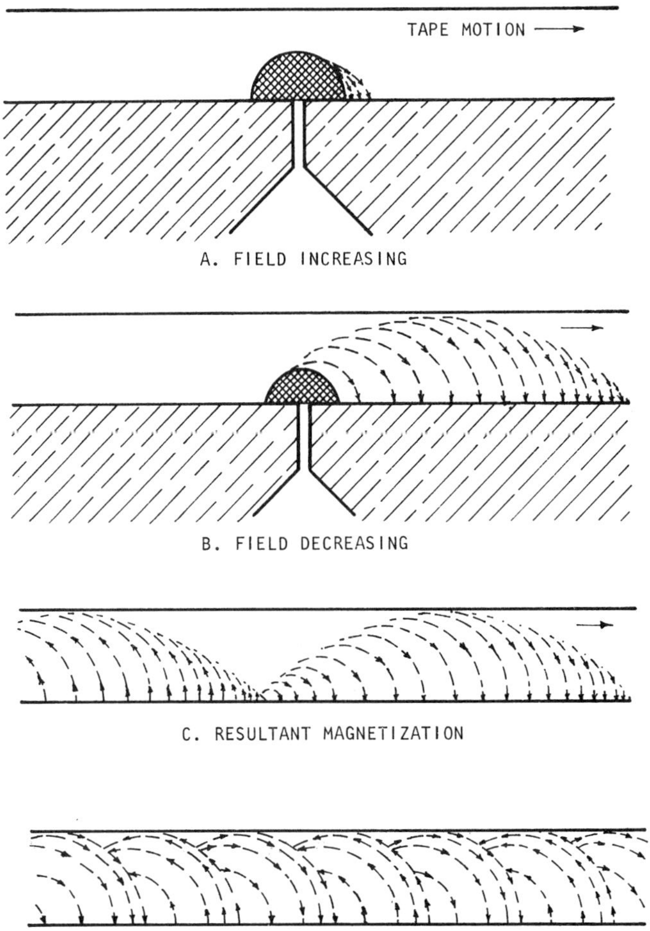

Figure 2-40. Effect of a sinusoidal recording field on a moving tape.

head end of the next wave of signal wipes out the tail of the previous wave. When the wavelength becomes shorter, the overlap is greater (Figure 2-40*D*). The trailing edge of the head gap is then of prime importance, since the tape remembers the magnetization it acquired last. There are, however, several special cases, depending on whether the application is *FM* (frequency modulation) in video or instrumentation; biasless recording in computers; or HF-biased recording in audio and analog recording.

Model of FM Recording. In FM recording, the carrier frequency is usually near the upper limit of head-tape resolution. The side of the tape layer spaced from the head gap undergoes many decreasing AC cycles before it passes out of the gap field and ends up essentially demagnetized. The effective recording then resides in a thin layer at the surface of the tape, as shown in Figure 2-41*A* and verified by actual measurements in the large head model shown in Figure 2-41*B* (Tjaden and Leyten 1963–64).

Model of NRZ Digital Recording. In NRZ digital recording, the head records a saturated signal with reversals of polarity at designated intervals. Figure

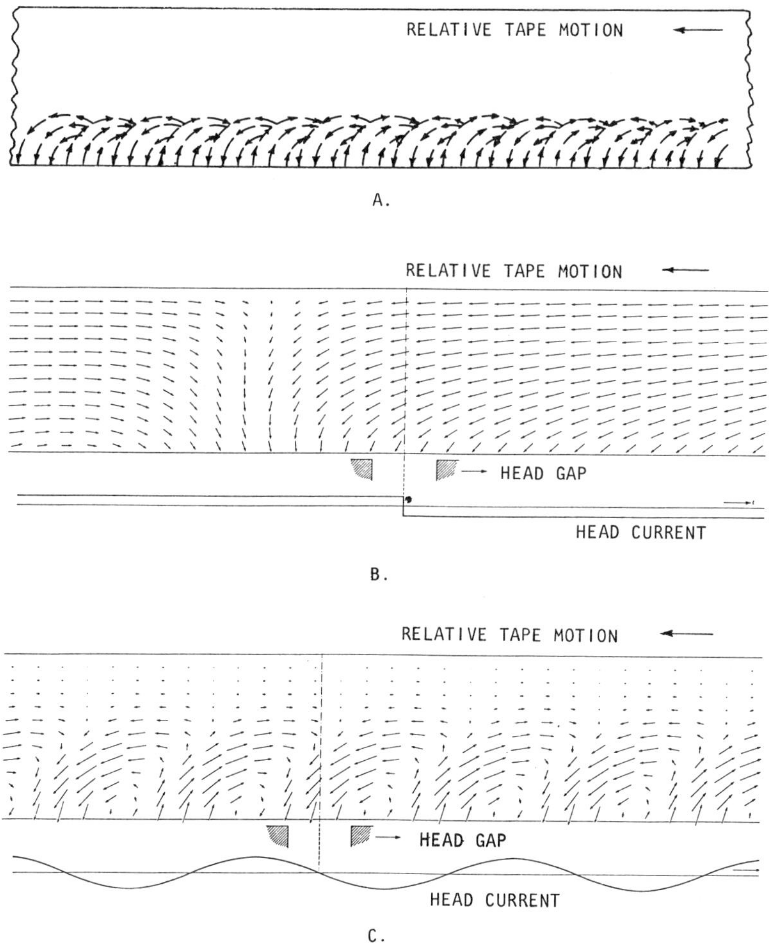

Figure 2-41. Effective magnetization of a short wavelength recording is in a thin layer of the tape surface. A. Short wave magnetization of a typical FM recording. B. Map of measured magnetization due to a step function of head current in a large scale model of head-tape interface. C. Map of short wavelength magnetization measured in a large scale model (Tjaden and Leyten 1963).

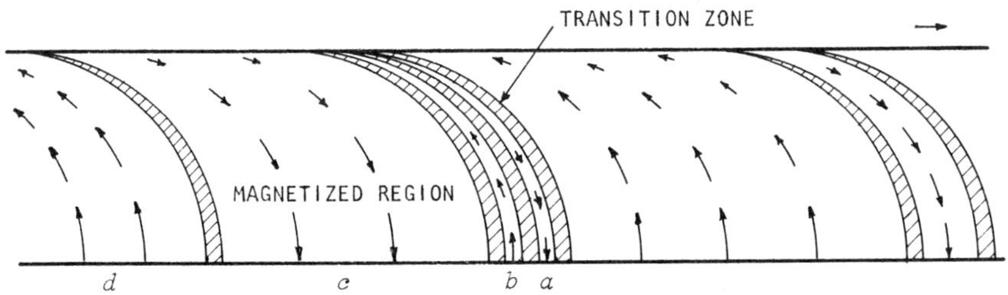

Figure 2-42. Digital NRZ recording. Short wavelengths give peak shifts. Long wavelengths give playback sag.

2-42 shows the magnetization for a succession of reversals with different spacing. When pulses are close together, as at *a* and *b*, the later pulse, *b*, crowds the earlier pulse, *a*. Their amplitude is reduced and the position of playback output peaks is changed, giving the well known *peak shift* phenomenon. When pulses are far apart (due to a series of zeros or ones), the playback head is inefficient at long wavelengths, causing the output to sag between pulses. An experimental determination of tape magnetization at a transition is shown in Figure 2-41*B*.

Model of HF-Biased Analog Recording. The following conditions are usual for high-frequency-biased analog recording: (1) The bias frequency is too high to be resolved; (2) bias is effective throughout the entire layer thickness; and (3) the signal field is small relative to the bias field. Under these conditions there is an arc-shaped critical zone beyond the trailing edge of the gap where the recording takes place, as in Figure 2-43. As an element of tape passes the head gap, it experiences a magnetic field (Figure 2-43*B*), with the final recording occurring in *C* of Figure 2-43. Unlike the recording shown in Figure 2-41, the tape layers spaced from the gap are not demagnetized at long wavelengths of the signal, but are strongly magnetized in one direction or the other, depending on the bias offset by the signal. Their magnetization therefore resembles the NRZ recording shown in Figure 2-42, except that the transitions are more gradual in accordance with analog variations of the signal. At very short wavelength, there is an erasing effect because the signal is changing while the tape passes through the critical zone. The erasing effect, together with other demagnetizing influences, causes the short-wave magnetization to abide in a shallow layer at the tape surface, as shown in Figure 2-43*C*.

Improving the Recording Resolution. The recording resolution (high-frequency response) thus depends on the size of the critical zone. With conventional heads and gamma Fe_2O_3 tapes, the critical zone length is approximately

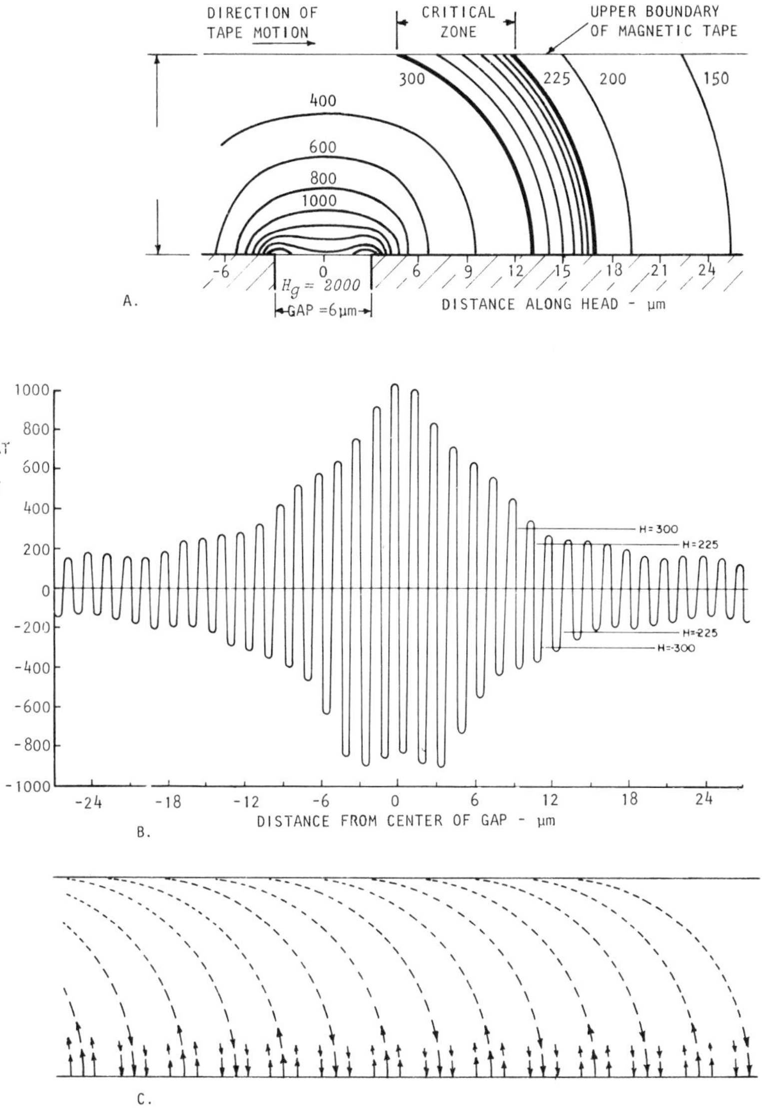

Figure 2-43. HF biased analog recording. A. Critical zone where recording occurs. B. Composite field experienced by element passing the gap when $\lambda = 79\mu m$. C. Final recorded magnetization. Longer wavelength recordings penetrate farther into tape layer along dotted lines to give higher residual intensity (Camras 1964).

one third the thickness of the tape layer. To improve the resolution by decreasing the critical zone size, one can:

1. Decrease the thickness of the tape layer. This is commonly done where output at low and medium frequencies can be sacrificed.
2. Increase the steepness of the B_r versus H characteristic. The steepness follows the squareness of the hysteresis loop, and the B_{mi}/B_r ratio. The squareness is improved by more uniform domains that switch over a narrow range of field. Oriented particles and anisotropy are also helpful.
3. Decrease the high-frequency bias. ''Starving the bias'' is a time-honored method for meeting a stringent high-frequency response specification, but the increased distortion due to underbiasing is usually not worth the better high-frequency response. When a tape is underbiased, the critical zone penetrates only part way into the tape layer, causing distortion by the expanding and contracting of the signal component.
4. Optimize the recording head gap. Decreasing the head gap below an optimum size may actually increase the critical zone length because the head core saturates. The arc-shaped zone past the trailing gap edge is rather insensitive to gap size variation, but larger gaps reduce the flux leakage across the interior of the head gap.
5. Decrease tape permeability. While this consideration is often neglected (or cannot be helped), a low tape permeability decreases the length of the critical zone.
6. Improve the head design. A cross-field head decreases the length of the critical zone.

Anisotropic Media and Recording Demagnetization. The above discussion assumes that the record medium is *anisotropic*, without a preferred direction of magnetization, and also that the demagnetizing effects are negligible in a direction perpendicular to the thin tape layer. Neither is true in practical recording media. Coated tapes are a matrix of needle-shaped (acicular) particles magnetically oriented in the longitudinal direction during the coating process while their binder is still liquid. Magnetization perpendicular to the tape surface forms poles that exert a strong self-demagnetizing field when the magnetized areas are considerable, as is true for medium- and long-wave recordings. Both of these situations tend to suppress the vertical component of magnetization at longer wavelengths.

Many experimenters have found that the horizontal component predominates in their measurements at moderate wavelengths and have concluded that the vertical component may be neglected completely (Bauer and Mee 1961, Eldridge 1960). If this is true, the models described above would have to be modified in that the semicircular bubbles become circular bubbles emanating from

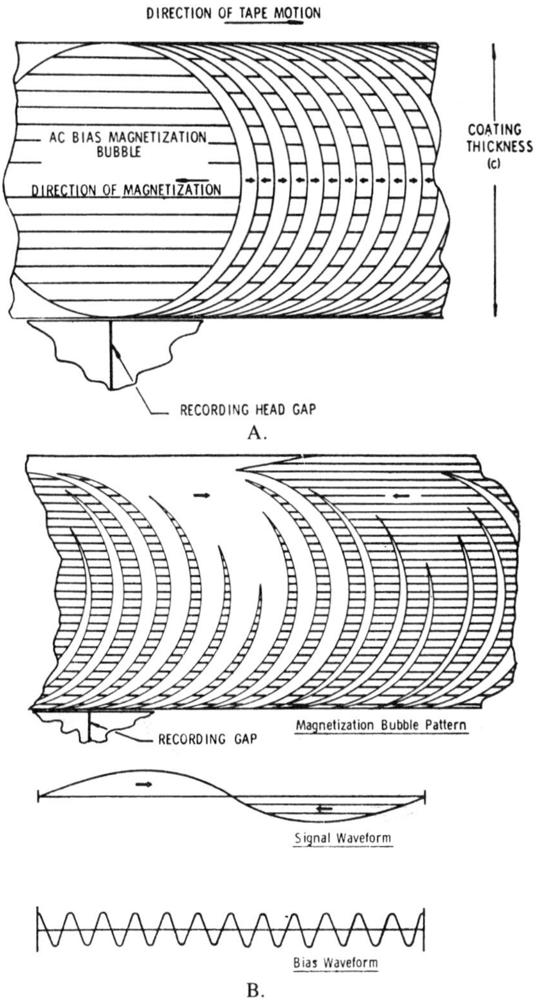

Figure 2-44. AC biased recording. A. No signal. B. Long wavelength signal (Bauer and Mee 1961).

the gap (Figure 2-44). The results in Figures 2-40 through 2-43 would then have circular shapes as in Figure 2-44, with the magnetization vectors all in the horizontal direction, while the vertical components would be suppressed (Bauer and Mee 1961).

The truth lies somewhere in between. At long wavelengths the horizontal component is all-important. At short wavelengths whose size is comparable to coating thickness the self-demagnetization is no longer unfavorable to the perpendicular direction. Also the anisotropy of tape favoring the longitudinal direction is not as much as supposed, usually being less than 2 to 1, so that the preferred magnetic directions of short-wave recordings take place in spite of mildly unfavorable orientation of the tape coating. Later analyses have recognized the importance of the perpendicular component at short waves, and there has been considerable activity in introducing discs and tapes with a vertically oriented surface layer that accepts extremely high recording densities, as described in Chapter 3.

OVERALL RESPONSE

After the recording has been made by one of the processes discussed, it may, at a later time, be converted to useful electrical waves by a playback system. The usual playback method is to move the tape over the gap in a head such as is shown in Figure 2-17, with a core that channels at least part of the recorded flux through a coil. The flux changes corresponding to the recorded signal induce a voltage across the coil terminals of:

$$E = mN \frac{d\phi_r}{dt} \tag{2-8}$$

where E is the induced voltage, m is the fraction of recorded flux that links the coil, N is the number of coil turns, and ϕ_r is the recorded tape flux in webers. Playback theory deals mainly with determining the fraction m from the shape of the recorded fields and their relation to the pickup head.

Head-Tape Interface. As previously described in Figure 2-17, much of the tape flux that does not thread the coil is lost by shunting across the gap, and this fraction is found by a straightforward calculation of gap reluctance relative to that of the magnetic circuit. But the geometry of flux paths at the head-tape interface requires a more complex theory.

The Reciprocity Principle. The *reciprocity principle* is a powerful tool for theoretical analysis. It states that the sensitivity of a playback head to a magnetic element located and oriented anywhere in space is proportional to the field that the head would produce in that space if its coil were energized by a unit of current. Since we have already mapped the head field when the coil is energized (Figure 2-18), we can use the same map for determining playback sensitivity and resolution. The overall induced voltage is found by integrating the contribution from all the magnetized tape elements within range of the head, as indicated by:

$$E = kvN \frac{d}{dt} \int_{-\infty}^{\infty} \int_{-\infty}^{\infty} \int_{-\infty}^{\infty} \overline{M} \cdot \overline{H}\, dxdydz \tag{2-9}$$

where

E = playback voltage
$\overline{M}$ = vector magnetization at a given point
$\overline{H}$ = vector field of the head at the given point when the head coil is energized with unit current. The dot product is the product of $\overline{M}$ and $\overline{H}$ vector components in the appropriate direction
N = number of turns on pickup coil
v = tape velocity
k = a constant of proportionality

Contributions to $\overline{M}$ are from the magnetized layer, of course, and the integration limits can be adjusted accordingly. The problem becomes two dimensional for the usual situation where the magnetized track is much wider than its thickness. It may be simplified further to one dimension by assuming a tape layer of negligible thickness. Tape magnetization $\overline{M}$ and factors leading to it have been considered previously (Figures 2-9, 2-31, 2-41). It remains to combine $\overline{M}$ with playback head effects according to geometry of the head field map.

Interaction of Sensitivity and Recording. An interesting theoretical problem arises from the fact that the playback sensitivity function is symmetrical about the center of the playback gap whereas the recording function is not. In other words, the leading and trailing edges of the head are both operative on playback, in contrast to the recording process, where the trailing edge is of prime importance and leading edge effects are mainly wiped out as the tape layer passes the gap. Considering that the sensitivity function follows somewhat semicircular lines, as shown in Figure 2-18, and that the recorded magnetization tends to be in quarter circle shapes, as in Figure 2-42, it appears that on playback there is a better fit between sensitivity and recorded magnetization at the trailing edge of the gap than at the leading edge. This effect improves the resolution over what would be obtained with a pure longitudinal recording.

Effect of Finite Gap Length. As the recording frequency becomes higher, the recorded wavelengths become smaller, approaching the size of the gap. The playback output drops to zero when the wavelength equals the effective gap length, since the number of positive and negative magnets in the gap is then equal and opposite. A similar condition occurs in scanning the sound track of a motion picture film. The simplest expression for gap effect is:

$$m' = \frac{\sin \pi \frac{g}{\lambda}}{\pi \frac{g}{\lambda}} \tag{2-10}$$

where m′ is the gap loss factor, g is the gap length, and λ is the recorded wavelength. A more exact gap loss function, given in Figure 2-45, predicts that the effective gap size is 1.12 times its physical length, with a playback null at $g/\lambda = 0.89$ and subsequent nulls also at less than $g = n\lambda$ of the simple formula (Westmijze 1953). These nulls occur at the specified gap length regardless of spacing between head and tape. In practice it is difficult to maintain and to measure gaps of micron size within the 12% discrepancy factor. It is also difficult to verify the above relationships because the signal is lost in the noise when wavelengths approach micron gap size. Narrow band filters must be used to detect the signal, and the exact null frequency becomes uncertain.

The existence of response peaks beyond the g/λ null has intrigued workers in magnetic recording who would like to circumvent the need for precision and the losses in heads with very short gaps. One application has been the *long-gap* method for measuring surface induction of magnetic tapes, where the theoretical response of Figure 2-46 is used as a standard, and departures in playback response from this standard are attributed to deviations from constant surface induction. In practice, it was found that the long-gap standard heads required gaps as straight and parallel as in small-gap heads of equivalent range, so this

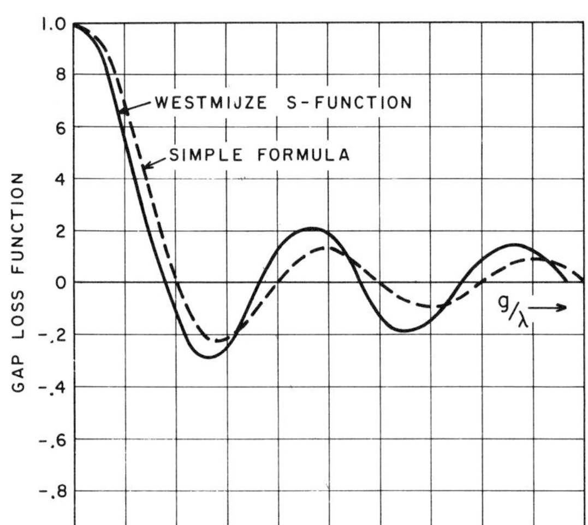

Figure 2-45. Loss in output at short wavelengths due to finite gap size (Westmijze 1953).

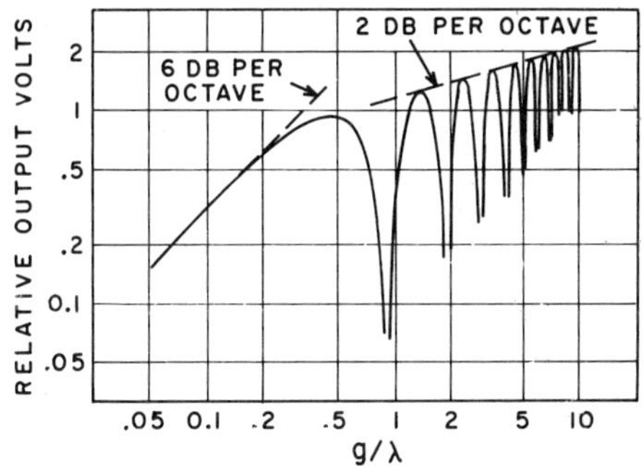

Figure 2-46. Output of a long gap head, showing peaks and valleys beyond the first g/λ null.

method did not become popular. For other uses it has also been true that the precision required in a long-gap head is better spent in making a short-gap head and avoiding the dips in frequency response. Another suggestion has been *wavelength filters*, which utilize the peaks in output at certain wavelengths and nulls at others, regardless of tape speed. The need for wave-length filters of this kind, however, is very limited.

Tapered Gap Heads. One recurring idea, a favorite of inventors, is to taper the head gap from one edge to the other. The longer gap at one edge, considered alone, gives a first series of peaks and nulls similar to the ones shown in Figure 2-46. The somewhat shorter gap adjacent to this edge gap gives a set of peaks and nulls displaced from the first series, so in principle the two sets together would mutually fill in their separate deficiencies, and, similarly, the contributions from other parts of the gap would lead to a smooth response. Unfortunately, the contribution from the first output peak beyond the null is opposite in phase from the primary response. The oppositely phased component necessarily cancels the primary component when the two are equal, thereby generating valleys and an irregular response such as that of a head with a uniform gap. A variation of the above idea is to combine the outputs of several different-sized gaps in cores or heads.

Spacing Loss. Even the slightest separation of the tape from the head gap gives a great loss in output at short wavelengths but not at long wavelengths, according to the Wallace relation (Wallace 1951):

$$\text{spacing loss, dB} = 54.6 \frac{\text{separation}}{\text{recorded wavelength}}. \qquad (2\text{-}11)$$

For example a 15 kHz recording on $1\frac{7}{8}$ in/sec cassette has a wavelength of about

3 μm (micrometer). A spacing of only 1 μm would drop the output about 18 dB. (This is playback loss only. In addition the recording loss might be as great or greater.)

Dead Layers. Since the earliest days of magnetic recording, unaccountable spacing losses have mystified those who sought higher recording densities. One explanation has been the presence of a surface layer of low permeability (dead layer) on heads made of permalloy, because this material loses permeability where its surface is stressed by tape abrasion. Another has been the presence of an air film between head and tape, or that the tape bounces enough to give slight separation. In most instances the causes are purely mechanical imperfections. As smoother tapes and better head surfaces have become available, the thickness of the absolute minimum barrier has diminished to submicron size. But evidence has also been presented that certain ferrites, particularly nickel-zinc ferrites, used for heads may have a low permeability surface that is removable by etching to give lower spacing loss (Sugaya 1968).

Thickness Loss. In a magnetic tape layer of appreciable thickness, the portions away from the gap are coupled to it less efficiently and there is a kind of spacing loss within the tape layer itself. A theoretical expression for the three kinds of losses is:

$$\phi = \phi_r \left[\frac{1 - e^{-2\pi\delta/\lambda}}{2\pi\delta/\lambda}\right] \left[e^{-2\pi a/\lambda}\right] \left[S\left(\frac{\pi g}{\lambda}\right)\right] \cos 2\pi\frac{v}{\lambda}t, \qquad (2\text{-}12)$$

where

ϕ = the flux threading the coil for a sinusoidal recording
ϕ_r = the maximum remanent flux, assumed constant regardless of wavelength
S = the gap loss function of Figure 2-45
a = the spacing between tape and head
g = the gap length
δ = the thickness of magnetic layer
λ = the recorded wavelength
v = tape velocity, and
t = time.

Thickness loss is represented by the first term in brackets, *spacing loss* by the second term, and *gap loss* by the third (Westmijze 1953). Figure 2-47 summarizes the effects of different losses as each parameter becomes comparable to the recorded wavelength. These losses are superposed on an ideal 6 db per

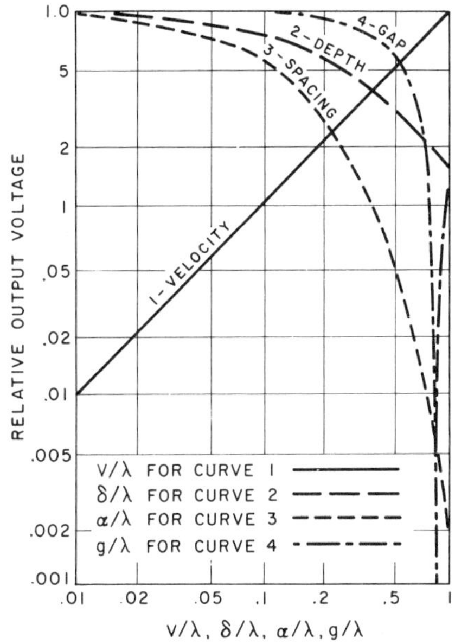

Figure 2-47., Effect of tape velocity (1), depth losses (2), spacing losses (3), and gap losses (4) on playback output voltage from a tape recording.

octave rise due to voltage output being proportional to frequency (velocity effect).

Constant Current Response.

The overall result of velocity effect combined with losses is the familiar *constant-current* playback response curve of magnetic recording (Figure 2-48) where it is assumed that the recorded flux (or surface induction) is constant at all frequencies. This response is subdivided into four regions. In region I, the wavelengths are longer than the head polepieces, which therefore pick up only a fraction of the tape flux. The fractional pickup varies inversely with the wavelength, giving an extra 6 db per octave loss for a total slope of 12 db per octave at lowest frequencies. In region II, the head-to-tape flux coupling is better than in any other region and the output voltage rises in direct proportion to the frequency with a 6 db octave slope, as predicted by theory. In region III the losses associated with gap, thickness, and spacing are evident. The output comes to a peak, then falls at an increasing rate toward the first null. Region IV represents the response, if any, beyond the first null. It is usually omitted since it seldom has any practical use. Experimental response curves are close to the ideal cal-

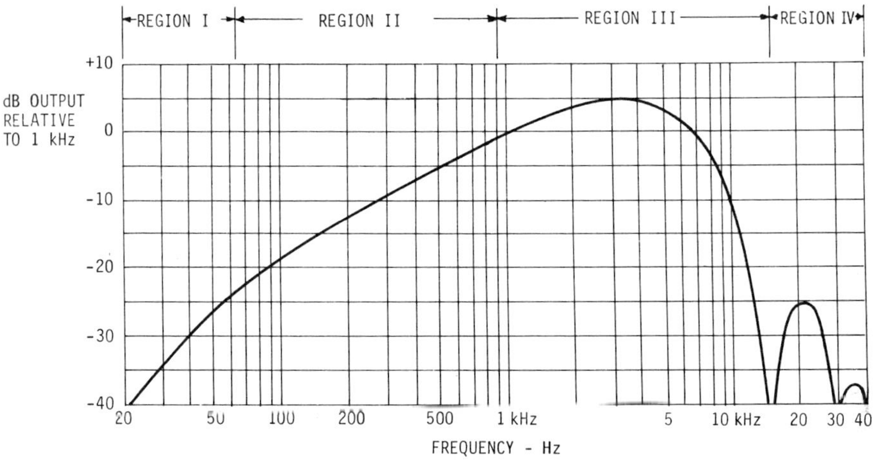

Figure 2-48. Playback response of a constant current recording (constant current response).

culated response but are lower at high frequencies because of frequency dependent core losses not previously considered, and because the assumption of constant recorded flux at short wavelengths is usually not valid.

Demagnetization

The magnetic poles on a recorded tape surface oppose the recorded magnetization within the tape. Figure 2-7 shows demagnetization of simple magnets with short gaps; a dashed line may be drawn for every ratio of magnet-length to gap-length, and its intersection with the second quadrant hysteresis curve gives the resultant magnetization.

Demagnetizing effects on a tape are much more complex, but in longitudinal recording the short wavelengths suffer more than the long because short, thick magnets have a greater demagnetization factor. A recording made with constant flux at the gap of a ring head therefore drops off in retained flux density at wavelengths comparable to the magnetic layer thickness. With vertical recording the situation is reversed and demagnetization decreases as wavelengths become shorter. In vertical recording the magnets are the same length at all wavelengths (equal to the magnetic layer thickness) but their width decreases at short wavelengths, thus increasing the ℓ/w ratio. The great interest in vertically oriented media for high density recording is based on this fact. At extremely short wavelengths, less than the magnetic layer thickness, the above models fail because the entire layer does not participate in the anticipated direction of magnetization.

Recording Demagnetization. Recording demagnetization is a name given to an effect which is different from the demagnetization just discussed. Recording demagnetization is the partial erasure of a recorded element on tape at the trailing edge of the recording gap. This happens if the field reverses before the element has moved far enough to be out of the gap influence when high-frequency bias is used, or at short signal wavelengths, as in Figure 2-41*C*. The overall effect of recording demagnetization is a loss at short wavelengths.

Equalization.

The response curve shown in Figure 2-48 covers a range of about seven to ten octaves, as for example the 20 to 20,000 Hz audio band, with useful output at the band edges at a somewhat reduced dynamic range. For *audio recordings* a flat frequency response is required within specified limits and is usually achieved by shaping the amplifier response during recording (preequalization) and also during playback (postequalization). *Computer applications,* on the other hand, pay little attention to flatness of response, but apply equalization that will give maximum distinctiveness to playback pulses. In *FM recording* the amplitude variations are removed by limiters, and the frequency spectrum is usually much narrower percentage-wise than for audio. The main concern in FM is to have accurate zero crossings of the wave, since the timing of these crossings carries the essential information that is used to recover the signal on playback.

Audio recorders with their wide spectrum require the most careful attention to equalization. During recording the audio equalization is usually designed to give a rising response in head current for frequencies above 5 kHz and sometimes for frequencies below 100 Hz. Playback equalization gives a 6 dB per octave low-frequency boost below 1 kHz to compensate for the 6 dB per octave slope shown in Figure 2-48, plus additional boost, near top and bottom ends of the spectrum. Figure 2-49 shows an equalized audio response. Equalization theory and practice are treated in more detail in Chapter 6.

Pickup from Adjacent Tracks During Playback

A playback head is sensitive to recorded flux from unshielded nearby recordings roughly according to the Wallace spacing equation:

$$\phi = \phi_m \left(-55 \frac{d}{\lambda} \right) \tag{2-13}$$

where ϕ is the stray pickup level in dB compared to the maximum level ϕ_m of in-contact pickup, d is the spacing distance from the stray flux, and λ is the recorded wave-length. Such stray pickup is of concern in audio and wideband

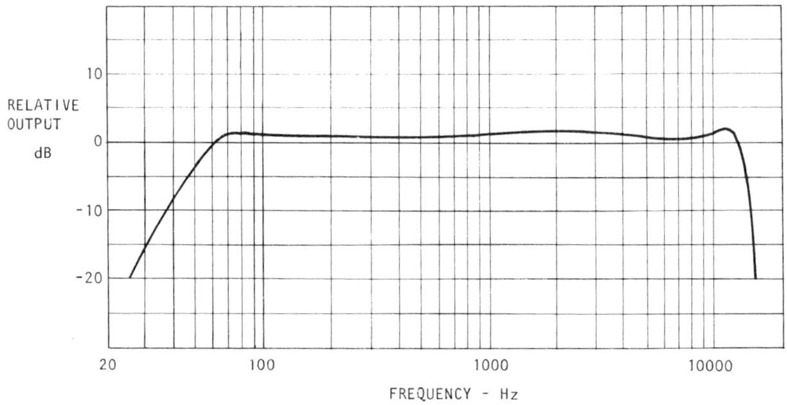

Figure 2-49. Overall response of an audio recorder (equalized).

recording where fairly long wavelengths are on adjacent tracks. Figure 2-50*A* shows how the output falls off when a typical unshielded playback head is shifted laterally until it is out of register with its recorded track. The output falls most steeply after the head no longer contacts the track ($d > w$) but there can still be considerable pickup a track-distance away. The pickup in a practical situation is reduced greatly by intertrack shields in the heads and by running the undesired track over a magnetic shunt or *keeper* (which can be its own head). These complicating factors make it advisable to determine experimentally the amount of pickup for each situation. In stereo cassettes the cross coupling at low frequencies is not of great concern because stereo separation depends mainly on medium and high frequencies.

Often a wide recording such as a calibration tape is played with a narrow head. The long wavelength output will then be too high because of fringing pickup at the sides of the head, as shown in Figure 2-50*B*. Correction by such graphs or by calibration tape instructions are approximate because commercial heads vary in their side exposure, their shielding, and their keeper effects (McKnight 1971; van Herk 1978).

SPECIAL CONDITIONS IN MAGNETIC RECORDING

A number of effects peculiar to magnetic recording are interesting in their own right as well as for the conditions they sometimes impose on practical applications of recording.

Print-Through

Whenever magnetic recordings are reeled layer upon layer, it would be expected that the external flux from one layer might magnetize the adjacent layer to give

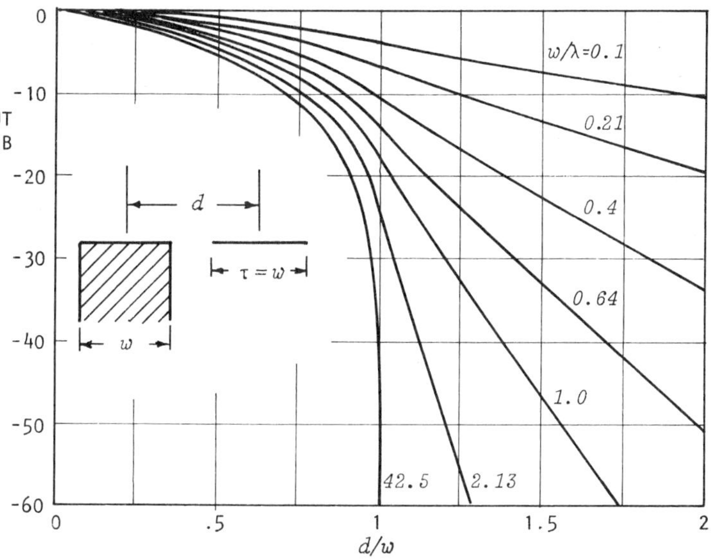

A. Registration loss (d/w <1), or playback crosstalk (d/w>1), for thin recorded layer in plane of head face (Lindholm 1978).

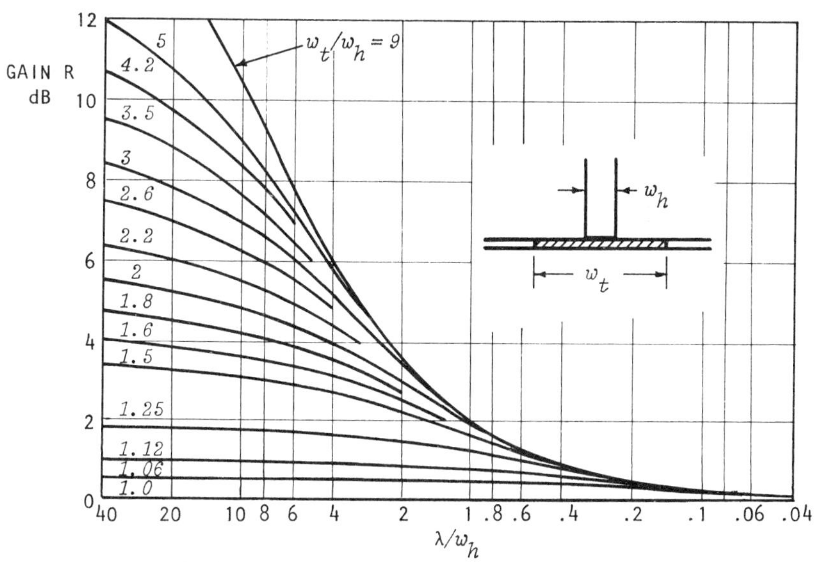

B. Gain R (dB) due to pickup of long recorded wavelengths by sides of a narrow head (w_h) centered on a wider track (w_t)(van Herk 1978).

Figure 2-50. Playback output when playback-head is laterally off-register with a recorded track, or when the recorded track is wider than the playback head.

a so-called print-through or cross-talk or interlayer transfer. Print-through is most noticeable on silent or unrecorded turns of tape adjacent to turns recorded with loud passages. On layers played back prior to the loud passage it gives a *pre-echo,* whereas on playback following the loud passage it gives a *post-echo*. The pre-echo is by far the most annoying effect in audio because it is unnatural, whereas post-echo is so common to our experience in natural sounds that it is often ignored by our hearing mechanism. In data recording both are equally important. Print-through is ordinarily 40 to 50 db or more below the signal and quite negligible. It should therefore be kept in mind in the following discussion that the effects are minor.

Studies of print-through indicate that forward printing from a recorded face is somewhat less intense than printing from the back of a strongly recorded passage, so that when tape is wound in the usual manner with the active side toward the hub, pre-echo is favored. An explanation of this effect is shown in Figure 2-51. On quiet tapes the echo can be noticed through five or more layers. The thicker the tape backing (nonmagnetic polyester, etc.), the less the print-through; super-thin double and triple playing-time tapes, on the other hand, are more susceptible. Worst of all are solid metal tapes or wires where there is no separation between layers or turns. Print-through seems to be a nonlinear effect that may be completely absent at low and moderate degrees of magnetization but becomes increasingly prominent at recording levels near tape saturation. Avoiding such high levels will therefore minimize the effects. Print-through is maximum for the midrange of recorded wavelengths, falling off rapidly and disappearing for shorter wavelengths while tapering off slowly for longer wavelengths.

Temperature, time, and possibly reeling stresses affect the degree of print-through. Storage of reeled tapes at elevated temperatures will increase the print-

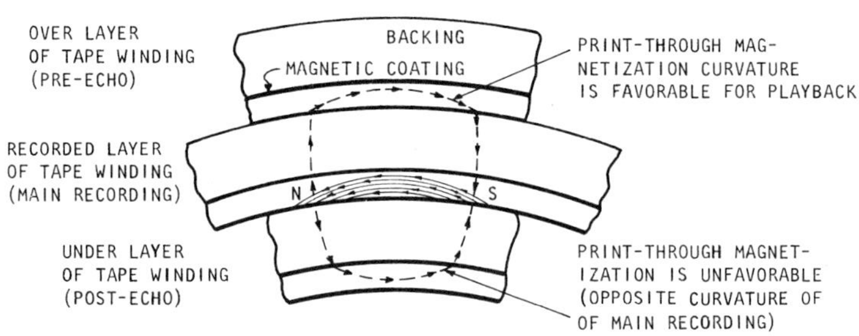

Figure 2-51. Pre-echo and post-echo print-through for inside wound tapes. The overlayer is printed with the same curvature as the magnetization of the main recording which fits the playback head. The underlayer has opposite curvature (Daniel and Axon 1951, Greinder 1953).

ing. Tapes stored for a long time show more print than tapes that are played immediately after recording. External magnetic fields, both AC and DC, enhance the printing effect and in fact are used in contact printing systems. Different kinds of tapes vary greatly in their immunity or susceptibility to contact printing. Some manufacturers have advertised a line of special low-print tapes, while other manufacturers have claimed that all of their tapes are low-print.

Reducing Print-Through. There are several ways to minimize, reduce or eliminate print-through, some obvious and some unexpected. For a tape that has been stored for some time the simplest recommendation is that before playback it should be wound and rewound one or more times. The unwound tape between reels seems to lose part of the printed signal; also the layers are matched with different neighbors when they are on the takeup reel instead of the storage reel. Some operators always store the tape on its takeup reel and therefore must rewind it before playback. Storing the tape backwards also has the advantage that the tape winding is more even than after high-speed rewind.

A very weak erasing field has been found to erase the print-through without apparent harm to the main recording (Daniel and Axon 1951; Herr and Von Behren 1952). Such a feature can be added to a playback machine. There is a danger, however, that some of the highest frequencies of the main recording may also be erased at least partially since these short wavelengths are most easily affected by weak fields. Great care must therefore be exercised because even on the same kind of tape the erasability of a recording varies with the bias that was used in making that recording, and also with different recording heads and with head-tape interfacing.

Where no print-through at all is permissible, a shielding layer of high magnetic permeability may be wound between recorded turns, the extra layer being diverted from the heads during recording or playback. Tapes with a built-in permeable backing have been suggested. Reasonable lengths of tape may be wound on drums or carried over pulleys where there is no contact between turns. For ordinary usage, adequate precautions against print-through are to choose a product having low print properties and to avoid unfavorable conditions such as saturation recording levels, extra thin base tapes, elevated storage temperatures, and strong external magnetic fields.

Causes of Print-Through. An unbiased tape has a threshold that protects it against magnetization by weak fields (see Figure 2-21); that is, fields less than about 7.9 kA/m (100 Oe) leave no apparent magnetization. It is apparent, therefore, that if the recorded level is kept low enough so that its external field strength in a spaced layer is below 100 Oe, there will be no transfer; but if the recording level is increased, the transfer rises rapidly, as confirmed by experience. In Figure 2-21, if the scale near the origin were magnified it would show some low amount of retentivity, which would increase if the applied field remained for a long time (months or years instead of seconds), and if the tests were run at elevated temperatures. The extent to which the B_r versus H curve remains horizontal at low fields depends on the uniformity of magnetic particles in the recording layer. The presence of particles that are switched by low fields is detrimental, for they give an upward slope to the curve near the origin. Hence the particles in the coating and their interaction are important, and a Preisach diagram such as the one shown in Figure 2-38, is preferred when the plotted points have a strong concentration in one area and almost none at low fields.

Temperature Effects.

A higher temperature increases the vibration of the molecular lattice in which magnetism resides. This thermal agitation reduces the coercivity of a tape and increases its susceptibility to print-through. Some tape materials have a percentage of domains that are thermally sensitive, such that high room temperatures causes their recorded level to go down, especially at short wavelengths. This effect was noticed in some early kinds of cobalt-doped tapes. A rule of thumb is that thermal effects double for every 10° C of temperature rise and are halved for every 10° C of temperature reduction, but the rule is not valid near critical points where rapid transitions take place.

Every magnetic material becomes nonmagnetic when heated to a temperature high enough for thermal agitation to destroy the ordering of its molecules. This Curie temperature is quite high for tape oxides (about 600° C) and for permalloy head cores (about 400° C). But chromium dioxide tape material has a Curie point of only 126° C, and some head cores made of ferrites have a similar low Curie point. No adverse effects have been found with chromium dioxide tapes in spite of their low Curie temperature. However, erase heads with ferrite cores may become heated at high-power inputs and the cores will then lose their effectiveness. Magnetically soft materials generally show an increase of permeability and a reduction in coercivity at elevated temperatures below their Curie point. Cores of playback and record heads may therefore improve when they become warm.

Curie Point Recording.

The usual way to magnetize something is to subject it to a strong field. An alternative method is to heat a material above its Curie point and then cool it through its Curie point, at which time even a weak field will magnetize it strongly. Under special circumstances Curie point recording may be of advantage. One application is to record bits of information with an electron beam on a magnetizable target inside a cathode ray tube. An experimental target made of manganese-bismuth alloy was readily heated in microscopic areas by a focused electron beam to above its Curie temperature of about 300° C. Upon cooling, these areas became magnetized oppositely to that of their surrounding neighbors (Mayer 1958). Similar results are obtained by scanning with a modulated laser beam. Another application of Curie point recording is for contact printing of tapes, as described below.

Magnetic Contact Printing.

A printed magnetic record can be made by pressing the magnetic recorded surface of a master record into close contact with the unmagnetized surface of a copy tape, applying an AC field above the threshold value of the B_r versus H curve of Figure 2-21, decreasing the field gradually over many cycles to zero, and finally separating the copy from the master, when the copy will be found magnetized in a mirror image replica of the master (Camras and Herr 1949).

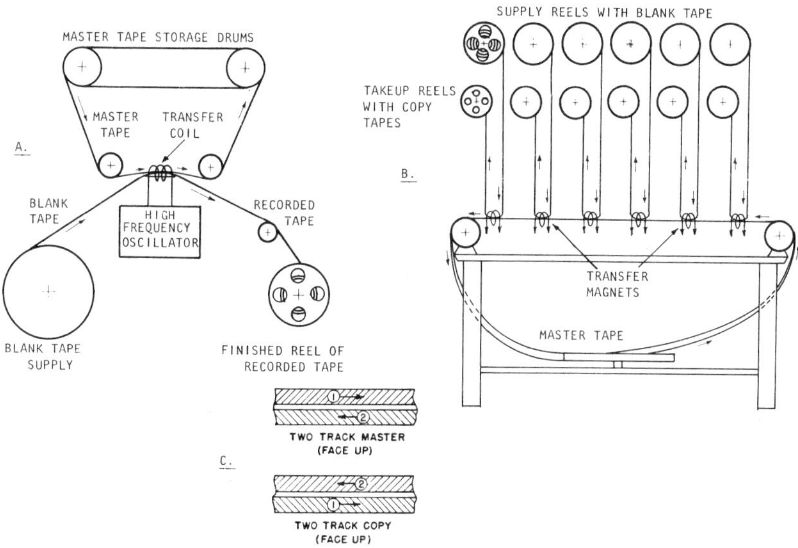

Figure 2-52. Contact printing of magnetic tapes. A. Single reel of copy tape. B. Multiple copying stations. C. Mirror image reversal of copy compared to the master.

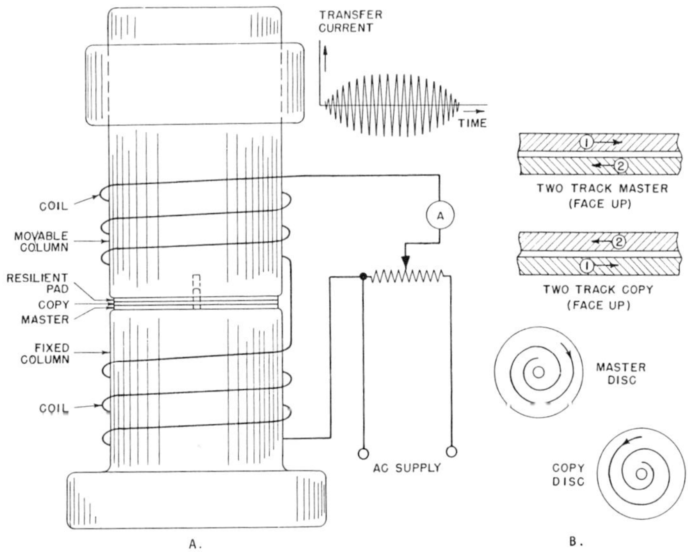

Figure 2-53. Duplicator for magnetic discs. A. Disc press with 60Hz printing field. B. Mirror image negative-to-positive copies.

Figure 2-52 shows how a continuous loop master can make a series of copies by contact printing to reels of a copy tape. Figure 2-53 shows a duplicator for magnetic discs. Because of the mirror image effect the master is usually made with a reversed recorder, processor, or scanner such that its mirror image copy will play back on ordinary machines.

The master record of Figure 2-54*A* is preferably made of higher coercivity material than the copy record, allowing an AC transfer field strong enough to cycle the copy material over nearly its full range of magnetization, while the master record will hardly be affected. Figure 2-54*B* compares the relative hysteresis loops of master and copy. Figure 2-54*C* shows the cycling (*KLMNPQ*) that takes place in the copy during the duplicating process and the final magnetization (*R*) of a copy element.

Curie Point Printing.

Chromium dioxide tape has a remarkable combination of high coercivity and low Curie point, making possible some interesting methods of contact printing:

1. Chromium dioxide (CrO_2) tape of 400 or higher coercivity is recorded in the usual manner to make a master suitable for copying on to ordinary

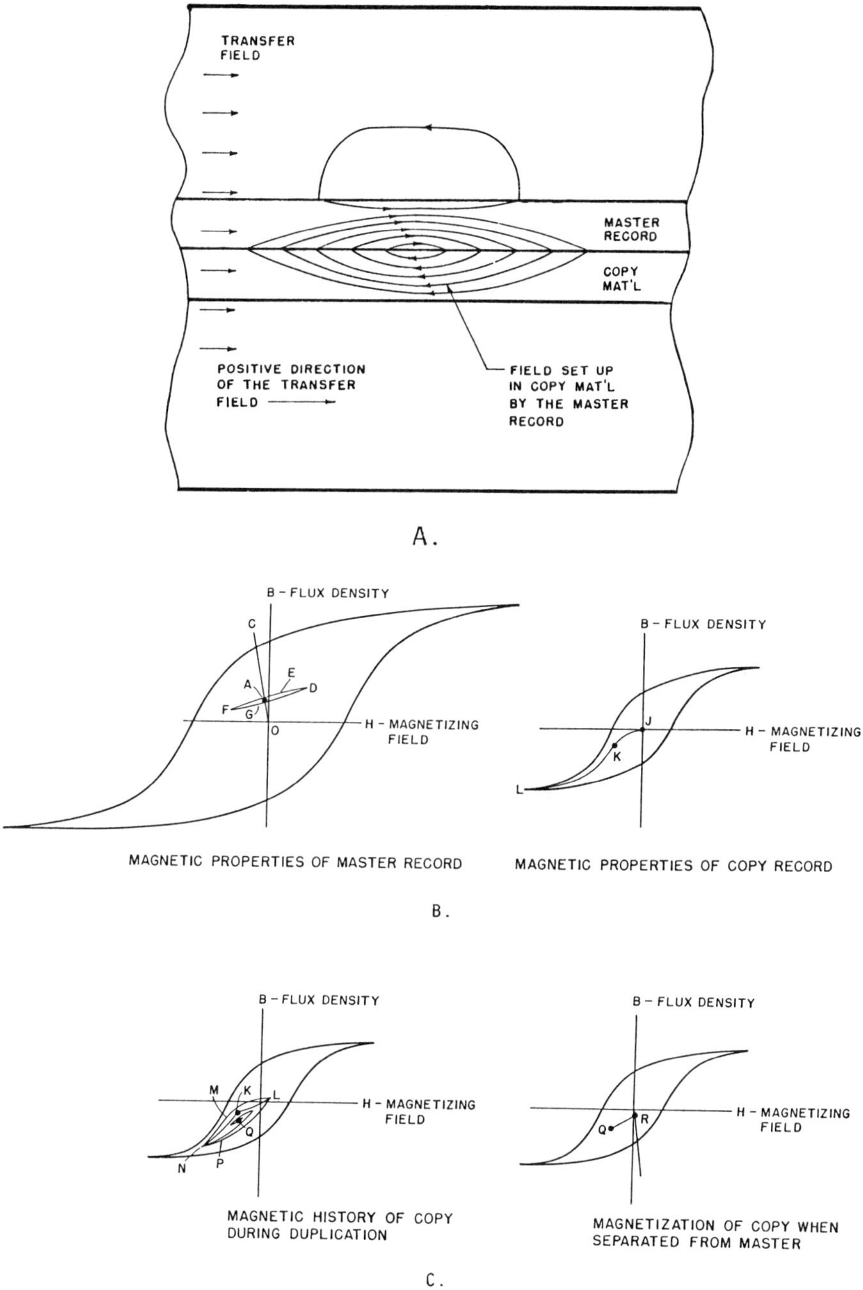

Figure 2-54. Magnetic influences in the copying process. A. Fields acting on a copy during duplication. B. Hysteresis loops of master and copy materials. C. Magnetization paths during duplication.

gamma Fe_2O_3 of 250 coercivity by a magnetic transfer field as described above.

2. The master is recorded on ordinary gamma Fe_2O_3 of 250 coercivity. The copy material is chromium dioxide tape in contact with the master. Heat that exceeds the 126° C Curie temperature of chromium dioxide is applied; then both tapes are cooled before they are separated. At its Curie point the normally high coercivity of chromium dioxide vanishes and it is readily magnetized by the Fe_2O_3. The Fe_2O_3 is unaffected by the 126° C temperature since its Curie point is 600° C. After cooling to room temperature, the chromium dioxide holds its magnetization more tenaciously than the Fe_2O_3 master because the CrO_2 has much higher coercivity. During the Curie point printing an AC magnetic field may be applied in addition to the heat-cool cycle, to enhance the transfer process.
3. The chromium dioxide copy made by process (2) just described can be used as a high coercivity master for process (1) where it is copied by an AC magnetic field alone. Since this master is reversed when it is made from the original, and its copies will be reversed again, the final copies will be the same as the original recording. The master can be considered as an intermediate "negative" analogous to negatives used in photography.

Time Effects.

How permanent are magnetic recordings? The fact that records can be erased has an implication of impermanence. Yet there are magnetic recordings made a half century ago that audibly seem as good as when they were first made. No archival quantitative experiments were set up in those early days, when no one was sure that magnetic recording was a viable method. We can make an indirect inference, however, from old instruments such as D'Arsonval voltmeters whose accuracy depends on the permanence of their magnets; and whose calibration has not changed a dB in a century. And of course the natural lodestones chemically similar to tape oxide have retained their magnetism for millions of years.

In designing experiments to measure the loss of magnetization with time, a major difficulty is to separate such loss from the effects of winding the tape and of running it over head cores and head gaps that exert magnetic influence on the recording. Controlling temperature and environment over long periods is difficult; subtle chemical and physical changes can take place in the polymer binders and bases of the tapes and even in solid metal materials. Best evidence shows that time degradation is small; that degradation is most rapid in early times with exponential approach to an asymptotic limit; that short wave-lengths are more vulnerable than long ones are; and that recordings made with stronger HF bias fields are most stable.

Memory Effects.

A mysterious phenomenon that has become a legend in magnetic recording is the memory effect, where a completely erased recording is said to appear again at a later time. It was first reported around 1947 when two new kinds of tapes were introduced: black Fe_3O_4 (magnetite) of 350 Oe coercivity, and tan γFe_2O_3 (gamma ferric oxide) of 250 Oe coercivity. The Fe_3O_4 material was superior magnetically but was said to have a memory effect. For this reason the γFe_2O_3, which did not exhibit memory effect, was chosen, and also because the γFe_2O_3 was more easily erased with heads that were available at the time.

The revival of an erased record would, at best, be very weak. Careful experimenters who have tried to duplicate the original experiments have failed. Some have concluded that in the original experiments the erasure was not quite complete and that the unerased signal was detected later under more sensitive conditions. This is plausible, because it was very difficult to demagnetize tapes completely with the heads available in 1947. Also when we are dealing with levels 50 or 60 db below normal, we are detecting energies of a few parts per million, and some very subtle higher order effects are possible. Today, memory effects are of concern to those who would like to dispose of erased tapes that at one time were recorded with classified information.

Possible Causes of Memory Effects. Assuming that the erase head can clean the track completely, what happens if it is offset laterally so that a few microns at an edge of the recorded track escapes full erasure? The playback head that monitors erasure may not cover the unerased track edge because the head is too narrow, or because of a very slight shift in tape guidance. We would thus assume that the track was completely erased. On a different machine or later on the same machine the reels may ''sit'' differently and the heads and guides can change, causing the unerased edge of the track to ride on the playback head and thus to give a memory effect. A more subtle possibility is that the binder system in the tape can flow slightly and that the forces between magnetized elements can therefore cause mechanical condensations and rarefactions in the tape coating. Time and elevated temperatures might increase such flow. After erasure the mechanical waviness should modulate a DC field by its variable reluctance.

Mysterious Erasure.

The opposite of memory effects, mysterious erasure, is reported when a recording disappears from a previously ''recorded'' reel or cassette. Often the recorded tape is shipped to some destination and upon arrival there is no recording on the tape. The suspected culprits might be storage near diesel electric motors of trains, engine rooms, dynamos of ships, superchargers in jet aircraft, the ionosphere, cosmic rays, X rays, metal-detectors, radar, force fields, and bolts of lightning, to name a few. As with UFOs, a little detective work usually produces a simple explanation, but a certain percentage of cases remain unsolved. Most of the time, however, it is found that a blank reel of tape was shipped through mislabeling or other confusion, and the recorded reel is found at home. Sometimes the recording was not made in the first place because the recorder was not operating properly, and the tape was never checked afterwards to verify that there was indeed a recording on it. Once in a while the recorded part begins at a distance from the start of the winding, and sometimes the heads of the playback machine may track to a different standard from the recorder.

Accidental erasure by strong AC fields is very uncommon because electrical machines confine their fields to small internal gaps for efficiency reasons. To erase a tape even partially requires an ambient field of about 16 kA/m (200 Oe). Even the largest electrical machines do not produce such fields at any approachable distance. Permanent magnets are a different story because their fields are unconfined and because they may inadvertently be brought close to a tape. Magnetic door latches are often found in storage cabinets and should be prohibited from rooms where tapes are handled, as should magnetic bulletin boards, toy magnets, and so on. Refrigerators and freezers use magnetized rubber gaskets around the edge of their doors. Loudspeakers sometimes have exposed magnets of considerable strength. Unknowingly one may brush a reel of tape past a door gasket or other magnet. Usually the damage is slight or undetectable. It is very unlikely that an entire record would be erased without a trace.

Protection Against Accidental Erasure. Since magnetic fields usually fall as the inverse square of distance or even faster, a reasonable distance is the simplest protection for tape recordings. Shipping them in large cartons in which

Figure 2-55. Bulk erasers. An entire reel is erased without winding the tape, by turning the reel slowly through the field of a powerful electromagnet. The magnet inside the base is energized by 60 Hz AC current.

the tapes are packed away from the sides assures that they will be spaced from any fields. For the most cautious, a container of fairly thick gauge steel gives almost complete protection. Since only high fields are of interest, the shield need not be of mumetal or similar expensive alloy.

Noise Levels in Magnetic Recording.

Noise is the ultimate limitation of recording systems, and tape noise due to domain size gives an ultimate limit to noise in analog magnetic recording. Smaller domains give less noise, but below a certain size they cannot retain magnetism. In studies of noise during tape playback, experience shows that:

1. The quietest tape is one that has been *bulk erased* on an erasing device, as shown in Figure 2-55, where the entire reel is subjected to a strong AC field that decreases gradually to zero.

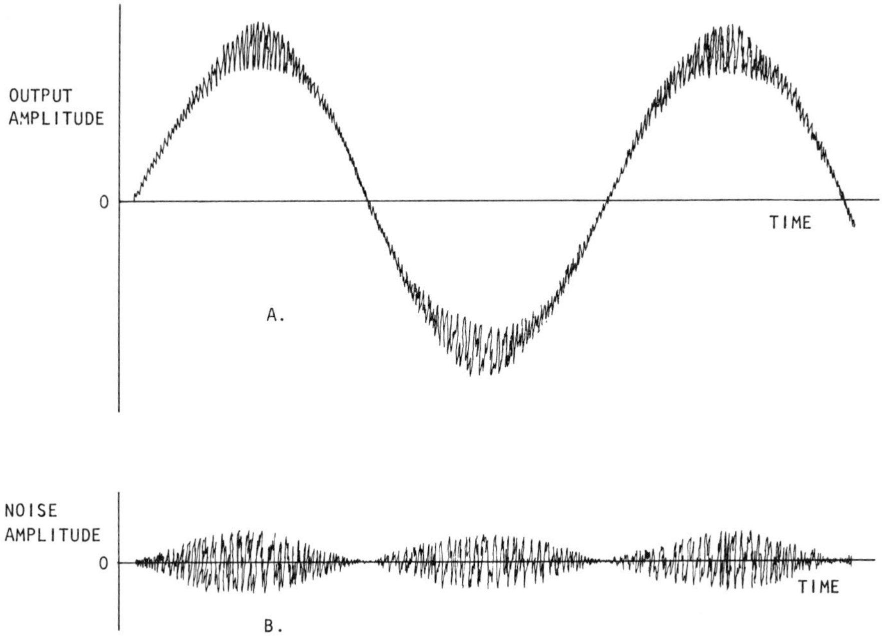

Figure 2-56. Modulation noise on a playback waveform. A. Modulation noise is associated with coating non-uniformities and with DC noise. B. removal of signal shows double frequency nature of modulation noise.

2. The erase head of a recorder gives noisier erased levels than bulk erasing does.
3. The noise level is still higher by 2 dB or more after the tape has been run over an HF-biased recording head.
4. If the tape is saturated with DC, or recorded with DC plus HF bias, the noise level is 10 to 30 dB higher than on bulk erased tape.
5. A record made on a tape having a high DC noise will have a noise component whose level is proportional to the signal amplitude shown in Figure 2-56, the noise being highest on signal peaks and practically disappearing where the signal wave goes through zero. This effect is called *modulation noise*.
6. Noise goes up 3 dB if track width is doubled, while the signal increases 6 dB, so doubling the track width improves the signal-to-noise ratio by 3 dB.
7. The noise spectrum and the high-frequency capability of the recording system both increase in direct proportion to tape speed.

Significance of AC and DC Noise Levels. The bulk-erased (AC) noise is a measure of the ultimate domain structure of the magnetic tape coating and is to some extent independent of the mechanical details of particle dispersion or irregularities in the coating. The DC-saturated noise level, on the other hand, increases greatly if the coating particles are clumped or clotted instead of being uniformly distributed. It also is increased by variations in coating thickness, as when the base material and tape surface are not smooth. The modulation noise associated with a high DC noise level is reduced greatly by good AC erasing so that the background noise during quiet passages is not excessive, but the overall sound reproduction has a fuzzy quality. Additional discussion of noise levels is found in Chapter 8.

MAGNETOOPTICAL AND MAGNETOELECTRONIC RECORDING AND PLAY-BACK

Magnetooptical and magnetoelectronic methods eliminate the usual magnetic recording and playback heads which are really a severe constraint because they scan one "gap size" element at a time, requiring small gaps, narrow tracks, and high mechanical speeds for high information densities and transfer rates. Magnetic heads must also maintain close contact with the record, giving rise to problems of wear, mechanical damage, and close tolerances of moving parts. Although very attractive, the magnetooptical and electronic methods are still experimental because they depend on unexplored techniques, exotic materials, and weak effects.

Magnetooptical Systems.

Today, magnetooptical systems use laser beams of submicron size focussed on the record surface for recording. The beam itself as well as the record can move to scan the information. Curie point recording raises the temperature of a tiny spot on the record above its Curie point, causing it to lose its premagnetization or sensitizing it to a recording magnetic field when it cools. The record surface is therefore a thin film of low Curie point. Some suitable materials are MnBi, GdCo, GdFeGarnet, MnCuBi, GdTbFe, TbDyFe, etc. (Togami 1982).

Magnetooptic playback depends on changes in polarization of a light beam when subjected to a magnetic field. The Faraday effect rotates the polarization of a transmitted light beam; the Kerr effect operates on a reflected beam requiring a highly polished record surface, or a polished auxiliary surface affected by the recording. A delicate optical system is needed to detect the small rotations caused by tiny magnetized areas. The signal-to-noise problems are formidable.

Magnetoelectronic Systems.

Magnetoelectronic recording systems use an electron beam for writing and/or reading a magnetic record. Curie point writing is feasible since a fine electron beam localizes considerable energy on a selected spot. The writing is then analogous to laser beam magnetooptical recording, except that electron beams operate in a vacuum, where it is disadvantageous to enclose the record medium (Mayer 1958).

Playback depends on deflecting an electron beam with the recorded magnetization, either when the beam is passing through the magnetized region or when it is reflected while under its influence. Although electron beam recording and playback have been demonstrated in a vacuum, attempts have been made to use electron scanning of records outside the vacuum either by a thin window or by an array of pins in the face of an electron tube (see Chapter 5) (Camras 1963).

Hybrids of magnetooptical and magnetoelectronic are possible, for example the advantages of laser beam recording with electron beam playback, and even combinations with conventional recording or playback.

MAGNETOFERROGRAPHY

Magnetic particles are attracted to a magnetized surface in a manner quite analogous to the electrostatic attraction used in xerox machines, and therefore the various printing possibilities of xerography can be carried out magnetically, especially if the data are already in video form. Where the original data are optical, they must be converted to a magnetic image, usually by scanning, as in facsimile or in video recording. Earlier records of this kind were made with boundary displacement heads and magnetic scan heads (Daniels 1952, Camras 1967). Recent equipment gives high-speed printing and even images in color (Eltgen and Magnenet 1980).

3. Magnetic Record Media

Since the 1940s many special materials have been developed which are advantageous for magnetic records. These range from the popular gamma iron oxide to sputtered cobalt-chromium for perpendicular recording. The manufacture, properties, testing and uses of important record media are discussed in detail in this chapter. Tables and graphs specify physical and magnetic characteristics of particles, depositions, solid media and substrates.

When magnetic recording first became known, very little thought was given to magnetic qualities of the record. The art of permanent magnets was primitive in the early 1900s. At that time the best magnets for magnetos, compasses, and electrical instruments were made of hardened high-carbon steels, which were sometimes alloyed with chromium or tungsten for better mechanical toughness. For magnetic recording hard-drawn music wire (piano wire) or hard-rolled steel tape was considered adequate until the 1930s, when more sophisticated alloys were tried. The most notable of these was vicalloy (Fe, Co, V alloy) developed by Bell Laboratories.

In the 1930s, I. G. Farben in Germany, revived an old idea of using a paper ribbon coated with magnetic ''paint,'' mainly because they were looking for new uses for carbonyl iron, an extremely fine iron powder that their chemists had developed. Carbonyl iron powder turned out to be one of the poorest magnetic recording materials that could possibly be chosen, but the cheapness, light weight, and spliceability of coated paper or plastic tape was so attractive that in the next few years other powders were tried. The most successful early powders were precipitated cubic magnetites (Fe_3O_4) and their anamorphic cubic gamma ferric oxides (γFe_2O_3). Both had low coercivity and were troubled with excessive print-through.

The era of modern recording tapes began in 1947 when the 3M Company in the United States marketed high-coercivity tapes coated with acicular (needle-shaped) particles developed at Armour Research Foundation. These tapes could run at half speed or even at quarter speed, compared to previous tapes, with high output and full response over the entire audio range. Further improvements in output and reduction of background noise were obtained by orienting the acicular particles to all point in the same longitudinal direction for longitudinal recording.

A better understanding of desirable qualities along with improved technology led to recording tapes of higher coercivies, smoother surfaces, smaller particles, and more uniform dispersions. Alternatives to the standard gamma oxide coatings now include chromium dioxide, cobalt modified oxides, Berthollide oxides, metal particles (fine alloy powders of Fe, Co, Ni), evaporated or sputtered metallic films, electroplated films, and electroless plated films.

DESIRABLE QUALITIES OF A MAGNETIC RECORD

Although magnetic properties have a prime influence, it is essential for the coating to be uniform both mechanically and magnetically. The tape surface should be smooth and nonabrasive, with no tendency to adhere to heads or guides or to adjacent tape layers. The record should be limp so that it will conform to the head; thin tapes have this quality and also are more compact for a given length of recording. Obviously the tape should be strong enough not to break or stretch under normal operating conditions, it should not deteriorate with time, and its coating should not flake off. Table 3-1 lists different kinds of media useful for magnetic recording and designates how each one rates in various qualities. Magnetic properties of various records are given in Table 2-5 of the previous chapter.

EFFECTS OF MAGNETIC AND PHYSICAL PROPERTIES

The first systematic investigation of how magnetic properties affect the performance of record media was carried out in the early 1940s when it became possible to make recording materials with a wide range of magnetic characteristics. As expected from theory, the long wavelength (low-frequency) response was proportional to retentivity (B_r), while the short wavelength (high-frequency) response improved as the coercivity (H_c) increased. Correlation was difficult to verify experimentally because long wave response depends also on thickness of the record layer and on concentration of the magnetic particles. Short wavelength response depends not only on coercivity but also on the head-tape interface, including tape surface smoothness, gap size, and head-to-tape spacing.

The role of coercivity in the response at short wavelengths can be inferred from Figure 2-7. The net flux available from a permanent magnet is given by the intersection of the dashed line with the second-quadrant magnetization curve. Short magnets at short wavelengths have a small ℓ_m/ℓ_g, so the dashed line makes a small angle with the 180° axis and (by similar triangles) the net flux is proportional to H_c. Similarly at long wavelength recordings ℓ_m/ℓ_g is large and the net flux is proportional to B_r. A high permeability head reduces the demagnetization field somewhat, but its effectiveness is limited by one-sided tape contact and head losses at high frequencies.

Although one might expect that the best permanent magnet materials with high-energy products such as Alnico would be superior for magnetic recording, this has not been the case. Figure 3-1 shows that magnetic recording materials usually have a relatively low-energy product compared to Alnico, and that the

Table 3-1. Record Media—Relative Characteristics and Performance.

	Tensile Strength	Dimensional Stability	Corrosion or Chemical Stability	Thinness	Pliability	Surface Smoothness	Weight	Volume (Bulk)	Uniformity in Thickness	Uniformity (Magnetic or Dispersion)	Signal/Noise Ratio	Coercivity	Remanence	Relative High-Frequency Response	Relative Output	Relative Print-Through	Relative Head-Wear	Relative Tape-Wear	Cost
Gamma ferric oxide dispersion on Mylar base	E	E	E	E	E	E	S	E	G	E	E	E	G	E	G	G	G	E	E
Chromium dioxide dispersion on Mylar base	E	E	G	E	E	E	S	E	G	E	E	S	G	S	G	G	F	E	G
Metal particle dispersion on Mylar base	E	E	F	E	E	E	S	E	G	E	E	S	S	S	S	G	G	G	F
Evaporated metal, thin layer on Mylar	E	E	F	E	E	E	S	E	E	E	G	E	S	S	G	G	G	P	F
Gamma ferric oxide dispersion on cellulose acetate	F	G	E	G	G	E	S	G	G	E	E	E	G	E	G	G	G	E	S
Gamma ferric oxide dispersion on paper	P	G	E	G	F	F	E	G	F	E	F	E	G	G	G	G	F	E	S
Gamma ferric oxide impreg. through vinyl base	F	G	E	F	E	E	E	G	F	G	E	F	P	P	G	P	G	E	G
Stainless steel (18-8) tape, $\frac{1}{4}''$ × 0.001	S	S	S	G	P	E	P	G	G	G	P	E	G	P	S	P	F	S	P
Vicalloy tape	S	S	E	G	P	E	P	G	G	G	P	E	E	G	S	P	F	S	P
Stainless steel wire, 0.004 diameter	S	S	S	F	P	E	F	E	G	G	P	E	G	P	G	P	F	S^+	G

Table 3-1. Continued

	Tensile Strength	Dimensional Stability	Corrosion or Chemical Stability	Thinness	Pliability	Surface Smoothness	Weight	Volume (Bulk)	Uniformity in Thickness	Uniformity (Magnetic or Dispersion)	Signal/Noise Ratio	Coercivity	Remanence	Relative High-Frequency Response	Relative Output	Relative Print-Through	Relative Head-Wear	Relative Tape-Wear	Cost
Carbon steel music wire 0.004″ diameter	S	S	P	F	P	E	F	E	G	G	P	P	S	P	S	P	F	S^{+}	G
Cunife wire, 0.004″ diameter	P	S	F	F	P	E	F	E	G	G	P	E	E	P	E	P	F	S^{+}	P
Nickel cobalt plated wire, .00003″ plating on 0.004″ bronze core	F	S	G	F	P	E	F	E	G	G	P	F	S	P	E	P	F	G^{+}	P
Nickel cobalt plating 0.0004″ on brass disc	S	S	G	NA	P	E	P	P	E	E	G	E	S	E	S	NA	*	*	P

S = Superior
E = Excellent
G = Good
F = Fair
P = Poor
NA = Not Applicable
*Negligible With Flying Heads
⁺Wire Subject to Accidental Breakage

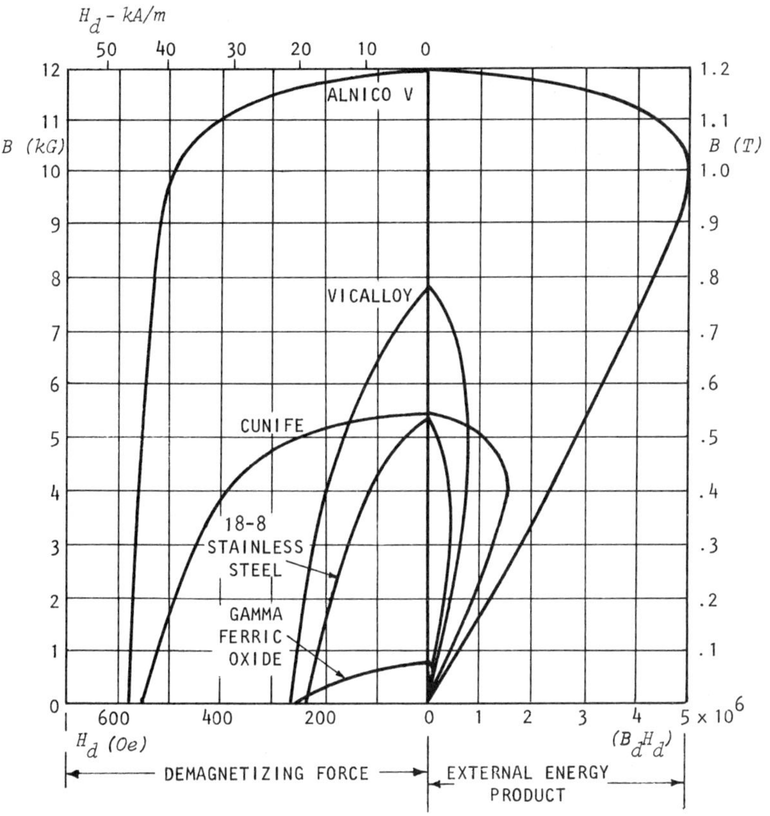

Figure 3-1. Energy product and demagnetizing curves of some magnetic record materials. Magnetic recording materials are compared to Alnico V, a strong permanent-magnet alloy. Gamma ferric oxide, an excellent recording material has a maximum energy product (B_DH_D) max of only about 70,000 (barely visible) while Alnico V reaches 5,000,000.

most popular of recording materials, gamma ferric oxide, is lowest of all. Other features, then, are more important than high-energy product.

Retentivity

Although the highest possible retentivity is desirable, there are upper limits, both theoretical and practical. The highest-known saturation magnetization of about 2.35 T (23,500 G) occurs in solid metals, and only a fraction of this, perhaps 1.0 to 1.5 T, is retained in the best metallic permanent magnets. Records made of continuous deposited metal film may approach the above values, but it is noteworthy that highest retentivity is attained only by sacrificing the coercivity and vice versa. A great reduction in retentivity occurs if the record is of discrete particles suspended in a binder, since the gaps between particles dilute the retentivity. Finally, the ferrite-type materials such as gamma ferric oxide are ferrimagnets, much weaker magnetically than metal ferromagnets, with a magnetization limit of about 400 mT (4000 gauss) and maximum practical retentivity of only about 100 mT in particulate recording tapes.

Coercivity

High coercivity records are capable of high recording densities if they also meet other requirements of smoothness, and so on, but when their coercivity exceeds 24 kA/m (300 Oe), they become difficult to erase and to record, requiring special expensive heads. This problem is alleviated somewhat if the recording layer is extremely thin, so that very high coercivity records are generally made with superthin coatings. Coercivities of about 250 have been the most common for many decades, but some machines handle chromium dioxide tapes with a coercivity of 400 and metal particle tapes of coercivity 1000. Even higher coercivities are available for special purposes.

Domains, Superparamagnetism, and Anisotropy

An interesting observation, confirmed by theory, is that most of the excellent permanent magnet materials are an array of single domain particles (Figure 3-2*C*) of optimum size suspended in a matrix that separates them somewhat from each other. If particles are too large, they become multidomain; if too small, their magnetic properties disappear. The critical size below which a particle prefers to be a single domain is where the magnetic energy of possible domain walls equals the magnetic energy of the field when completed externally without domain walls. Bulk material shown in Figure 3-2*A* is in its lowest magnetic energy state when, throughout its volume, the domains of a given magnetic direction combine with neighboring domains to form internally closed magnetic paths so that the bulk material exhibits practically no net magnetization. In Figure 3-2*B*, the small particle can still sustain a few domains, but in *C*, the even smaller particle is a single domain, which can be shown to have the lowest possible magnetic energy. For extremely small particles, as in Figure 3-2*D*, the forces aligning the electron spins that cause magnetism are overcome by thermal agitation. These particles, too small to sustain permanent magnetism, are called *superparamagnetic*.

The particle size for single domains is in the order of 0.02 to 1.0 μm. The minimum size depends on magnetic anisotropy (directional preference) that

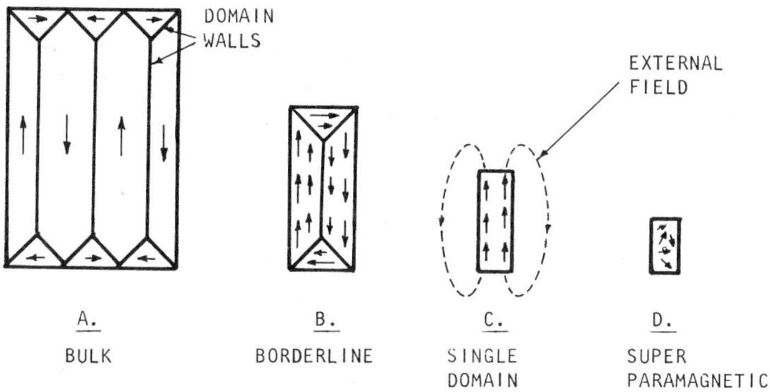

Figure 3-2. Conditions for single domain particles. A. In bulk material the magnetization divides itself into domains with closed magnetic paths for lowest energy. B. Smaller particles can sustain fewer domains. C. A single domain has lowest energy for its size. D. Extremely small particles become superparamagnetic.

maintains electron spins in a preferred direction. An anisotropy increases, the coercivity goes up, and the particles can be made smaller before they become superparamagnetic.

Anisotropy and Coercivity

In single domains, the uniform crystal lattice favors electron spin alignment (and therefore magnetization) in certain directions and discourages electron spins in other directions. The most favored magnetic direction is called the *easy axis*, whereas the least favored direction is called the *hard axis*. This directional favoritism is known as *magnetic anisotropy*.

When a domain is forced to reverse its direction of magnetization under the influence of a sufficiently strong external field, its magnetization has to pass through a hard axis. The greater the anisotropy, the more difficult the transition and the higher the coercive force. Four kinds of anisotropy have been observed: (1) shape anisotropy; (2) crystal anisotropy; (3) strain anisotropy; and (4) exchange anisotropy.

Shape anisotropy has been most important for magnetic tapes where elongated (acicular) particles of gamma ferric oxide, or chromium dioxide, or metal alloy are characterized by extremely small size for low noise level, and high coercivity for excellent high frequency response. An added benefit is the possibility of orienting the particles in a preferred direction to increase the retentivity, decrease the noise level of the tape, and increase the squareness ratio (B_r/B_{mi}).

Crystal anisotropy is increased by adding a few percent of cobalt to the lattice of gamma ferric oxide particles. The added cobalt brings up the coercivity to 500 or more in a tape that, lacking the cobalt, would measure about 250. Cobalt modified oxide (sometimes called cobalt doped oxide) can be acicular, so that advantage is taken of both anisotropies: crystal and shape. Cobalt modified cubic particles have also been tried. Since coercivity increases with the percentage of cobalt, the oxide can be tailored to a desired magnetic value. The enhancement of magnetic properties by crystal anisotropy has been found less stable than by shape anisotropy. The magnetic properties of crystal anisotropic oxides fall off at higher temperatures, and they also seem sensitive to stresses that occur when the recording tape is bent around guides of small radius. Acicular oxides, modified slightly with cobalt, have been more satisfactory for magnetic recording than strongly modified cubic oxides that rely mainly on crystal anisotropy.

Strain anisotropy has not been of advantage in magnetic recording, but exchange anisotropy is utilized for an oxide called Super Avilyn®, where the acicular gamma ferric oxide particles are coated with a thin shell of cobaltous oxide. The antiferromagnetic shell is linked to the inner single domain by exchange interaction between electron spins, resulting in higher coercivity without the disadvantages of ordinary cobalt modification. The exchange anisotropic oxides (called cobalt adsorbed oxides) have been so successful that they have largely supplanted all other oxides where high coercivities above 32 kA/m (400 Oe) are required.

Signal-to-Noise Ratio

A high level of background noise can nullify any advantages of high remanence and high coercivity. It often happens that a tape with high output also has a proportionally higher noise level. Noise level is minimized by a smooth flat substrate; a uniform magnetic layer with a smooth surface; a uniform particle dispersion free of clumps, bubbles, chips, protrusions, and nonmagnetic inclusions; and a uniform orientation of particles in the desired direction.

Assuming that all else is ideal, there is still an ultimate lower level of magnetic irregularities in a demagnetized record due to random distribution of saturated single domains. The smaller the domains, the less the magnetic granularity and the lower the noise level (in proportion to the square root of the particle volume); but when the domains become too small (less than about 0.1 μm), they become superparamagnetic and no longer can sustain permanent magnetism. Particles of lengths 0.2 to 0.5 μm are therefore commonly chosen for coated tapes.

COATED TAPES

Most tape is made by coating the surface of a plastic film base or backing with a paint or "ink" containing magnetic particles homogenized in a binder that adheres to the film and dries by evaporation. Figure 3-3 is a cross section of such a tape. The coating is usually so thin that it is only a small fraction of the overall tape thickness. The following sections give details of materials and processes used in manufacture of coated tapes.

Base

Although the base is magnetically inert, it must meet stringent requirements of thickness, strength, and stability.

Thickness. A thin base is important to give a reasonable playing time in a compact reel. Early commercial tapes were about 2 mils (50 μm) thick, filling a 7 inch (180 mm) diameter reel with 1,200 feet (365 m) of tape, which ran for a half hour at 7.5 in/s (190 mm/s). This tape was thicker, bulkier, and stronger than necessary for most applications; a thinner tape, about 1.5 mils (37 μm) in thickness, *$1\frac{1}{2}$ play*, soon became popular. The trend continued as thinner polyester films became available allowing *double play* one mil tapes, and even *triple play* and *quadruple play* tapes. Tapes less than a mil thick are rather flimsy and require a delicate drive mechanism. They are apt to warp if wound tightly on their reels and to stretch if started and stopped rapidly. They may climb the edges of guide posts, flip over, and fold in half. Thin tapes also are troubled with high print-through. Tables 3-2 and 3-3 summarize the capacity and playing time of reels and cassettes with various tape thickness.

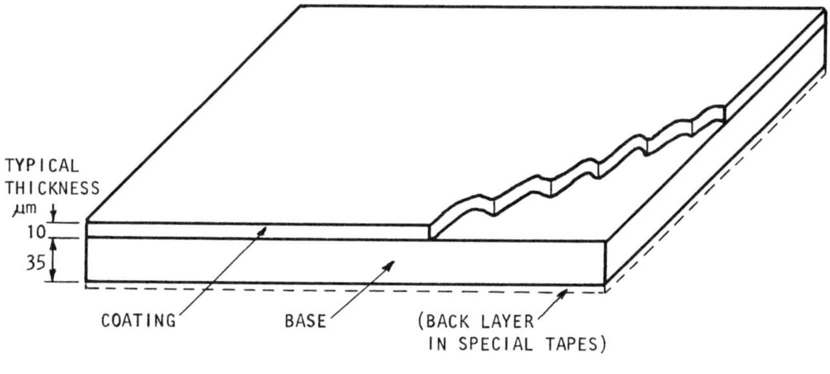

Figure 3-3. Cross section of a coated magnetic tape.

Table 3-2. Specifications of Coated Tapes Audio and Video.

Reel Type	Playing Time (minutes)	Speed (mm/s)	Base Matl*	Width (mm)	Length (m)	Breaking Strength (N)	Yield Strength (N)†	Thickness (μm) Total	Coating	Base	Back	Notes
Open reel 7″ (180 mm) Normal play	32 × 2	190	P	6.3 +.0 −.1	365			50	12	38	0	Normal play
(SAME) Extended play	48 × 2	190	P	6.3	550			35	10	25	0	Extended (1.5×) play
(SAME) Double play	64 × 2	190	P	6.3	730			25	12	13	0	Double (2×) play
(SAME) Triple play	96 × 2	190	P	6.3	1100			18	5	13	0	Triple (3×) play
Open reel 10.5 NAB (270 mm)	66 × 2	190	P	6.3[a]	760			50	12	38	0	Normal play
(SAME) Extended play	96 × 2	190	P	6.3	1100			35	12	22	1	(1.5×) play. Back treated
Cassette (C45)	23 × 2	47.6	P	3.81	67.5			18	6	12	0	
Cassette (C60)	30 × 2	47.6	TP	3.81	90	15	7.5	18	6	12	0	Gamma Fe_2O_3
Cassette (C60)	30 × 2	47.6	TP	3.81	90	15	7.5	16	4	12	0	Metal Particle
Cassette (C90)	45 × 2	47.6	TP	3.81	135	11	6	12	5	7	0	
Cassette (C120)	60 × 2	47.6	TP	3.81	180			9	3	6	0	

8-Track cartridge (45 min)	11.7 × 4	95.2	P	6.3	67			18	5	12	1	45 min. play. Lubricant layer on back
8-Track cartridge (60 min)	15 × 4	95.2	P	6.3	87			18	5	12	1	60 min. play.
								31	11	19	1	60 min. play. Alternative
8-Track cartridge (90 min)	22.5 × 4	95.2	P	6.3	87			18	5	12	1	90 min. play.
Videocassette VHS format T-60	30-60	33.3 or 16.7	P	12.7 +.0 −.1	68	>39	>24	20	5	15	0	
(SAME) T-120	60-120	″	P	12.7	128	39	24	20	5	15	0	
(SAME) T-180	90-180	″	P	12.7	192	39	24	20	5	15	0	
(SAME)	120-240	″	P	12.7	248	39	24	20	5	15	0	
Videocassette Beta format L-250	60	20	P	12.7	78							
(SAME) L-500	120	20	P	12.7	150							
U-Matic format	60	95.3	P	19	358	115		26				

*P = polyester
TP = tensilized polyester
† = 5% elongation
a = widths to 50.8 mm
‡ 190 mm/s = 7.5 in/s
47.6 mm/s = 1.875 in/s

Table 3-3. Specifications of Coated Tapes Various Applications.

Application	Thickness (μm)			
	Total	Coating	Base	Back
Instrumentation, Open Reel				
Intermediate band (normal play)	48	10	38	0
Intermediate band (extended 1.5 play)	35	10	25	0
Wide band	30	5	25	0
Wide band (back coated)	31	5	25	1
Logging tape	30	5	25	0
Logging tape (back coated)	31	5	25	1
Logging tape (extended play	18	5	12	1
Instrumentation grade, thin coat	28	3	25	0
Computer, Open Reel				
Half inch (12.6 ± 0.05 mm) wide	48	7	41	0
Sprocketed Film				
35 mm polyester	90	14	76	0
16 mm polyester	89	13	76	0
35 mm and 16 mm triacetate	137	15	122	0
Audio Tape, Special				
Thin coated	25	6	19	0
Digital audio tape	31	5	25	1
Video, Open Reel				
2 in quadruplex				
1 in (25.4 mm) helical scan	36	10	25	1.5

Material. Low cost was a major advantage of coated tapes, so a low-cost paper base was the natural choice for early tapes. These tapes were magnetically noisy because the absorbent fibrous nature of paper did not allow a uniform magnetic coating. Cellulose acetate had a much smoother surface and in the 1950s became the preferred base for tapes manufactured in the United States. Its strength was moderate, and even 1.5 mil base tape was marginal in durability. Notches and nicks at the edges of a cellulose acetate tape tended to propagate across it until the tape broke. In Germany the preferred base material was a polyvinyl chloride material (PVC). Vinyl base tapes and cellulose acetate tapes perform quite similarly, except that, when over-tensioned, the vinyl will stretch instead of breaking. Users of professional recorders would prefer to have the tape break cleanly, rather than stretch permanently, because the break can be spliced together to restore a recording, whereas a stretched part cannot be remedied.

Polyester material such as DuPont's Mylar® has been the preferred base material for tapes since the 1960s. It is far stronger, more stable, more durable, and useful in thinner gauges than any other plastic base. At first its cost was high and there were problems with surface smoothness and coating adhesion, so there was valid justification for choosing an alternative base material. By 1980 the early problems with polyester bases were largely solved and their cost was more competitive, so that polyesters have practically become the universal backing for magnetic tape.

Polyimide plastic films are usable (intermittently) at temperatures up to 480° C, which would destroy any of the previously mentioned plastic base materials. Tapes that must withstand high temperatures have been coated on this base (DuPont Kapton®) using special binders. Base materials useful for magnetic tapes are described in Tables 3-4 and 3-5.

Metals as a base for coated tapes are unusual, but are feasible where exceptional strength, durability, dimensional stability, temperature resistance, and so on, are required. But for hard magnetic discs (or drums), metals are the standard base material.

Aluminum is chosen for magnetic discs because of light weight and stability, and because its surface can be anodized and polished to give an exceptionally hard smooth substrate for the magnetic coating. Other metals and even glass have been used for magnetic coated discs.

Impregnated Bases. An interesting form of recording medium is made of plastic or rubberlike material impregnated with a well-dispersed filler of magnetic particles such as gamma ferric oxide. In this case, the base and the magnetizable material are one and the same thing. Such records have disadvantages: (1) The percentage of magnetic pigment must be much lower than in tape coatings (otherwise the record would have no tensile strength) and hence the recorded output is lower; (2) the thick layer becomes magnetized to a considerable depth at long wavelengths and is difficult to erase; (3) the milling process tends to align the acicular particles of gamma ferric oxide in such a way that disc records, for example, may present alternately favorable and unfavorable orientation that give variations in output. Nevertheless, a magnetic recording rubber of this kind was used for the surface of magnetic drums that were self-lubricating and had a very long life. For a limited time in about 1943, the I.G. Farben Company manufactured LG tape, which was a homogeneous 6.5 mm wide tape impregnated with low coercivity oxide in a polyvinyl chloride matrix. Production was discontinued because the product had low output, poor short wave response, and high print-through.

Table 3-4. Properties of Base Materials Useful for Coated Tapes.

Material	Tensile Strength		Stress for 5% Elong.		Ultimate Elongation	Tensile Modulus		Tear Strength (Elmendorf)	Density	Coef. of Thermal Expansion	Coef. of Humidity Expansion	Max. Operating Temp
	PSI	MPa	PSI	MPa	%	PSI	MPa	gm/mil	gm/cm^3	per °C	per %RH	°C
Tensilized polyester	45000	310	23000	158	40	800000	5510	20	1.377	17×10^{-6}	3×10^{-6}	93
Polyester	25000	174	15000	104	120	550000	3790	20	1.395	17×10^{-6}	6×10^{-6}	149
Cellulose acetate	16000	110	15000	104	20	520000	3580	5	1.357	120×10^{-6}	150×10^{-6}	82
Cellulose triacetate	16000	110	11200	77	30			8		54×10^{-6}	35×10^{-6}	177
Polyvinyl chloride (38 μm th)	28000	193	10000	69	40	540000	3720		1.382	59×10^{-6}		80
Polyimide (25 μm th)	25,000	172	13000	90	70	430000	3000	8	1.42	20×10^{-6}	22×10^{-6}	400
Bronze (hard drawn)	125000	860				13×10^6	89500		8.8	18×10^{-6}		1000 (melts)
Aluminum	35000	241				10×10^6	69000		2.71	24×10^{-6}		660 (melts)
Stainless steel (18-8)	300000	2064				30×10^6	207000		7.86	17×10^{-6}		1425 (melts)

Table 3-5. Base Film Thickness and Strength of Tapes.

	Base				Tensile (Breaking) Strength		Stress for 5% Elong.		Area Factor	
	Thickness		Width						in^2	m^2
Material	mils	(μm)	mils	(mm)	lb	(N)	lb	(N)	per lb	per kg
Tensilized polyester	0.46	11.7	150	3.81	3.3	15	1.6	7	43600	62
	0.46	11.7	246	6.25	5.5	24	2.7	12		
	0.28	7.1	150	3.81	2.4	11	1.3	6	71700	102
	0.28	7.1	246	6.25	4.0	18	2.2	10		
	0.23	5.85	150	3.81	2.1	9			87300	124
	0.23	5.85	246	6.25						
Polyester	1.42	36.1	246	6.25	11.5	51	6.0	27	14000	20
	0.92	23.2	246	6.25	7.1	32	3.8	17	21500	31
	0.88	22.4	246	6.25					22500	32
	0.85	21.6	246	6.25					23200	33
	0.75	19.1	246	6.25	5.9	26	3.9	17	27100	39
	0.48	12.2	246	6.25					41600	59
	0.48	12.2	246	3.81						
Polyester (for discs and cards)	3.00	76.2	—	—					6768	9.6
	5.00	127	—	—					4032	5.7
	7.50	190	—	—					2592	3.7
Cellulose acetate	1.45	36.8	246	6.25	5.8	26	4.7	21	20300	29
Cellulose triacetate	1.0	25	246	6.25	4.1	18	3.0	13		
Polyvinyl chloride	1.0	25	246	6.25	7.1	32	2.7	12	20000	29
Polyimide	1.0	25	246	6.25	6.2	28	3.2	14	19500	28

Magnetic Particles

The essence of particle-coated tape is of course the particle itself, its magnetic properties, its relation to neighboring particles, and its relation to the transducer head. A variety of useful magnetic particles as well as their chemical composition, size, shape, and magnetic properties are given in Table 3-6. It is desirable to have the highest coercivities and remanences, and the smallest particle size, but extreme choices are usually accompanied by undesirable side effects such as erasing difficulties, print-through, or need for nonstandard heads and electronics.

Particle Size and Shape. Chemical precipitation is an effective way to achieve the small size and the uniformity required of particles for magnetic recording. Attempts have so far been unsuccessful in manufacturing a satisfactory powder by crushing or grinding bulk material such as alnico or magnetite. When particles are made synthetically, their size can usually be controlled by varying the conditions of the chemical reaction such as concentration, temperature, reaction time, and choice of the reacting materials. The particle shape usually depends on the form of the original crystallite. For example, Fe_3O_4 or gamma Fe_2O_3 have cubic crystallites when made by direct precipitation; but when made indirectly from hydrated Fe_2O_3, the crystallites are acicular (needle shaped) because the hydrated Fe_2O_3 grows as acicular crystals when manufactured as described in the next section. The range of sizes in Table 3-6 are typically 0.05 to 0.5 μm, a compromise between smallness to assure low noise level and sufficient size to avoid superparamagnetism.

Acicular Gamma Ferric Oxide (γFe_2O_3). For many decades, acicular γFe_2O_3 has been the standard particle material used for magnetic tapes (Camras 1954). Typical magnetization curves are shown in Figure 3-4, and magnetic properties are listed in Table 3-6. The process for making these particles is as follows:

Three parts of caustic soda (NaOH) are dissolved in 10 parts of water. Twelve parts of copperas, $FeSO_4 \cdot 7H_2O$, are separately dissolved in 60 parts of water. The caustic soda solution and the ferrous sulphate solution are then mixed in a vessel provided with agitation and the mixture agitated slowly for a number of hours (for example 17 hours), a new surface of the mixture constantly being exposed to the atmosphere. The reaction that takes place yields a thick greenish-yellow mass and may probably be expressed as follows:

$$4FeSO_4 \cdot 7H_2O + O_2 + 8NaOH \rightarrow 2Fe_2O_3 \cdot H_2O + 4Na_2SO_4 + 30H_2O$$

Table 3-6. Magnetic Particles for Tape Coatings.

Material	Chemical Composition	Size (μm) l × d	Shape	Coercivity H_{ci} kA/m	(Oe)	Remanence Br (In a Coating) G	mT	Squareness Ratio Bm/Br	Saturation Density B_s G	T
Gamma ferric oxide (high Hc)	γFe_2O_3	0.5 × 0.1	acicular	20	250	1000	100	0.88	1140	114
Gamma ferric oxide (Co mod.)	γFe_2O_3 + Co	0.5 × 0.1	acicular	42	530	1500	150			
Gamma ferric oxide (cobalt coat.) (adsorbed)	γFe_2O_3 + Co	0.3 × 0.05	acicular	43	540	1500	150	0.83	1810	181
Gamma ferric oxide (Cubic, Co Modified)	γFe_2O_3 + Co	0.2	cubic	40	500	1000	100	0.75	1500	150
Chromium dioxide	CrO_2	0.4 × 0.04	acicular	44	550	1800	180	0.9	2000	200
Bertholomide oxide (+Co)	γFe_2O_3 $+Fe_xO_y$	0.3 × 0.05	acicular	44	550	1400	140	0.88	1590	150
Metal particles	FeCo	0.2 × 0.02	acicular	88	1100	3050	305	0.85	3600	360
Gamma ferric oxide (low Hc)	γFe_2O_3	0.2	cubic	8.8	110	300	30	0.27	1110	111
Magnetite (high Hc)	Fe_3O_4	0.5 × 0.1	acicular	28	350	1000	100			
Magnetite (low Hc)	Fe_3O_4	0.2	cubic	9.6	120	375	38	0.25	1500	150
Carbonyl iron	Fe	1.0	spherical	0.4	5	100	10	0.05	2000	200
Barium Ferrite (modified)	Be-FeO	0.08 × 0.03	hex plates (easy vert.)	100	1250	1440	144	0.92	1570	157
Cobalt chromium	Co-Cr	thin film	vertical needles	80	1000	6900	690	≈ 1.0	6900	690

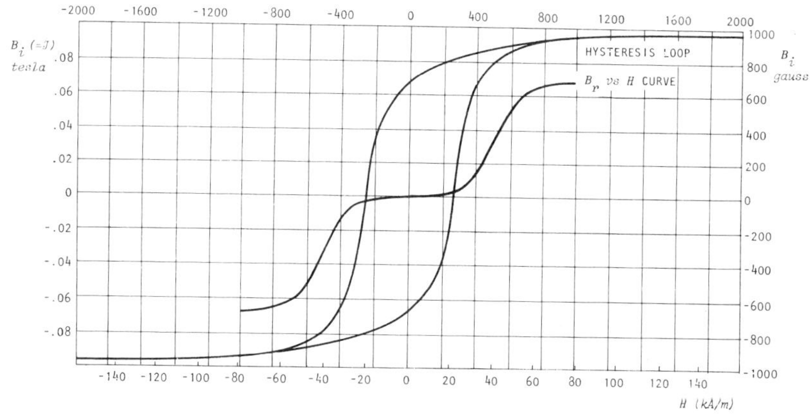

Figure 3-4. Magnetization curves of acicular γFe_2O_3 tape. Early commercial magnetic tape, Scotch brand type 111A. From these measurements H_{ci} = 21kA/m (260 Oe), B_r = 0.096T (960G), H_m = 159kA/m.

The $Fe_2O_3 \cdot H_2O$ thus produced is in colloidal form. The finely divided particles act as nuclei for growing larger crystals of $Fe_2O_3 \cdot H_2O$ in accordance with the subsequent steps to be described.

In a separate vessel, 175 parts of $FeSO_4 \cdot 7H_2O$ are mixed with 3,500 parts of water, and 1,000 parts of scrap iron are added. This mixture is then heated to 60° C after which the thick greenish-yellow mass produced above is added and air bubbled through the resulting mass for about four hours, holding the same at about 60° C.

The $Fe_2O_3 \cdot H_2O$ produced in accordance with the foregoing is obtained in the form of light yellow acicular crystals about $\frac{1}{4}$ to 1μm in length and about $\frac{1}{10}$ to $\frac{3}{10}\mu$m in width. These size determinations are obtained by electron photomicrographs at 5000 magnification. A chemical analysis of the $Fe_2O_3 \cdot H_2O$ product shows 86% Fe_2O_3, and 12% H_2O.

The suspension of $Fe_2O_3 \cdot H_2O$ obtained as described above is filtered, washed, and dried at 100° C. This gives a filter cake that is easily crushed to release the crystals in their original fine state of division. The material thus produced is placed in a closed chamber, together with sharp-edged iron lumps to facilitate mixing. Hydrogen is passed through the chamber at a temperature in the chamber of 399° C and the mixture tumbled until the powder turns almost black. At this point, the amount of water vapor, which was heretofore being released in large volume, drops off substantially, indicating that the reaction should then be stopped, since continued treatment beyond this point would result in deterioration of the material.

The material thus prepared is then cooled by quenching the chamber with cold water while the gas is still passing through the chamber. Tumbling is continued until the product is at about room temperature before the gas supply is cut off. The chamber is then sealed and tumbled for about ten minutes, with further cooling, before air is again admitted. If this precaution is not followed, the powder is apt to ignite (pyrophoric tendency).

The resulting powder is black and has an acicular crystalline structure. The particle size and shape are as described above for the light yellow crystals. The chemical analysis shows 78.7% Fe_3O_4 and 19.3% Fe_2O_3. However, the analysis may very considerably. The extent of the reduction is indicated by the color of the material, which may vary from a dark brown to a deep black at room temperature.

The black powder thus obtained is magnetite or ferrosoferric oxide, Fe_3O_4. It is converted to gamma ferric oxide (γFe_2O_3) by heating it in air at a temperature of about 232° C while mixing it to expose all parts. Temperatures of 150° C to 315° C are effective, with the reaction taking place very rapidly at the higher temperature. Too high an oxidation temperature, however, will convert the powder to nonmagnetic alpha Fe_2O_3. The gamma ferric oxide made by the above process has the familiar tan color of tape coatings. Its coercivity is about 22 kA/m (275 Oe). The electron micrograph in Figure 3-5 shows the acicular nature of this oxide.

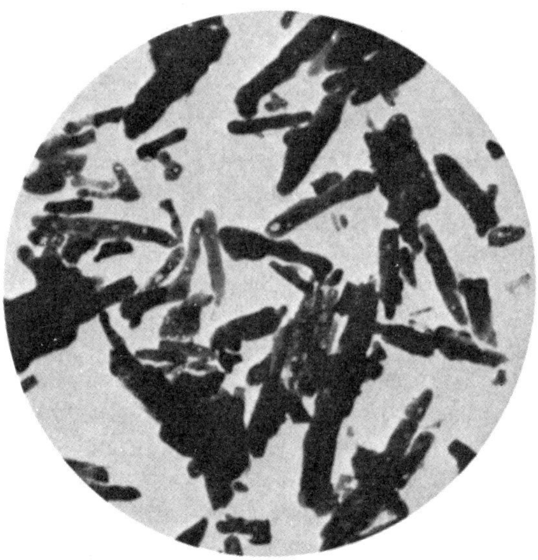

← 1 μm →

Figure 3-5. Electron micrograph of acicular gamma ferric oxide (Fe_2O_3).

Acicular Magnetite (Fe_3O_4). The process is the same as for gamma ferric oxide, except that the final step of heating in air is omitted. The black Fe_3O_4 powder has a coercivity of at least 28 kA/m (350 Oe) and its tape requires higher bias and erase fields than gamma oxide tapes.

Cobalt-Modified Ferric Oxide. When a small percentage of cobalt is substituted for some of the iron in gamma ferric oxide, changes are set up in the crystal lattice of the particles, resulting in crystalline anisotropy that increases the coercivity. The manufacturing process is similar to that of pure gamma ferric oxide except that a percentage of cobalt salt (as cobalt sulfate) is substituted for iron salt in the solutions where the crystals are formed. As a rough estimate the coercivity of the final product will rise about 133 Oe for every percent of cobalt content up to about 6%.

Cubic iron oxide that inherently has coercivity below 16kA/m (200 Oe) can be brought up to 32 kA/m (400 Oe) or more with cobalt "doping", even when the particles are smaller than 0.1 micron. The smaller particles give a quieter tape. However, acicular particles are preferred because they require less cobalt and are more stable magnetically with respect to temperature and stresses.

Cobalt-Adsorbed Ferric Oxide. Instead of forming particles of homogeneous chemical composition it is possible to make a composite particle, with a core of magnetite or gamma ferric oxide surrounded by a thin (less than 4 nm) cobalt adsorbed shell. The adsorbed cobalt produces an exchange interaction with the core material, increasing the coercivity as with cobalt modified oxides, but with greater stability of coercivity than in cobalt modified oxides. (Kubota, Morita, Tokuoka, and Imaoka 1979).

One process of preparing these particles is the same as the one used for the acicular γFe_2O_3 or acicular Fe_3O_4, with an additional step as follows: Cobalt sulfate in proportion of 5 parts of cobalt ions and ammonia in proportion of 13 parts are dissolved in 470 parts of deionized water. This is mixed with 100 parts of the prepared Fe_3O_4. The mixture is treated in an autoclave with hydrogen at a pressure of 4900 kPa (50 kg/cm^2), a temperature of 150° C, and a time of 4 hours. It is then washed and dried at a reduced pressure and 50° C. The resulting particles have a coercivity of 89 kA/m (1120 Oe).

A surface treatment with SiO_2 is recommended for the Fe_3O_4 prior to the autoclaving step. The surface treatment consists of dispersing the oxide in deionized water and adding aqueous colloidal SiO_2 of particle size 10 to 20 nm to the suspension. The final product using these treated particles has a lower coercivity of 54 kAt/m (680 Oe) but is magnetically more stable with temperature than the product using untreated particles.

Berthollide Ferric Oxide. Berthollide is an iron oxide mineral that is intermediate between Fe_3O_4 and Fe_2O_3, the former having an oxygen/iron ratio of 1.33, the latter a ratio of 1.5. As discussed previously, acicular Fe_3O_4 can be converted to acicular γFe_3O_3 by an oxidation step, during which the oxide retains its favorable magnetic properties. By halting the oxidation step before it is completed, the product will be intermediate between Fe_3O_4 and γFe_2O_3; and as with γFe_2O_3, the intermediate properties may be enhanced by a cobalt adsorption treatment to give a high coercivity oxide interchangeable with cobalt adsorbed γFe_2O_3. The tape oxide is synthetically produced, and is not made from the mineral Berthollide found in nature.

Cubic Gamma Ferric Oxide. Gamma ferric oxide is produced directly by bubbling air through a solution of tetrapyridine ferrous chloride. The precipitate of gamma ferric oxide hydrate ($\gamma Fe_2O_3 \cdot H_2O$) is purified by decantation and filtering, yielding a synthetic form of lepidocrocite. This powder is dehydrated by a current of air at 150° C to give gamma ferric oxide of tan color with cubic crystals of fractional micron size. Similar results are obtained with other iron-organo complexes as for example aniline ring compounds (Johnson and Johnson 1936).

Gamma ferric oxide may also be made indirectly from cubic magnetite, manufactured by the process described below. The magnetite is oxidized by heating it in air at temperatures of 150° C to 300° C. The product is a tan powder with cubic crystals of the same size and shape as the original magnetite. Both the direct and the indirect process give gamma oxides of coercivity about 6 to 12 kA/m (70 to 150 Oe), usually somewhat lower than for magnetite particles of corresponding size.

Although the low-coercivity oxides are of little interest at present, they were used for magnetophone tapes until about 1947. Cobalt-modified particles of these types can give high coercivities but have the disadvantages of instability associated with crystalline anisotropy.

Cubic Magnetite. A commercial process for manufacturing cubic magnetite is as follows: 26 parts of hydrated ferrous sulfate are dissolved in 114 parts of water; and 13 parts of sodium carbonate are dissolved in 43 parts of water. Both solutions are added to an autoclave containing 151 parts of water. The precipitated ferrous carbonate suspension is agitated for half an hour at a temperature of 50° C and pressure of about 3.5 atmospheres. Air is then introduced at a pressure of 7 atmospheres and the temperature increased to 150° C, the treatment being continued for 2.5 hours or until samples show an absence of carbonate. The residue is then washed, filtered, dried, and pulverized, giving a black magnetite powder with cubic particles of about 0.4 micron size. An electron micrograph of cubic magnetite is shown in Figure 3-6.

If an autoclave is not available, the magnetite can be produced in a similar manner in air by boiling the suspension at a temperature of 100° C, with the addition of potassium nitrate (KNO_3) as an oxidizing agent.

The magnetites produced as above have coercivities of about 8 to 14 kA/m (100 to 175 Oe). German magnetophone tape, type C, used such a powder. Tape for the Brush "Soundmirror" of 1946 used a paper base coated with a dispersion of this magnetite powder.

Chromium Dioxide. Since chromium is not ferromagnetic, we would not expect its oxides to have significant magnetic properties, but CrO_2 in the form of acicular single domain particles makes a recording medium that magnetically exceeds acicular gamma ferric oxide. Chromium dioxide suitable for recording particles is prepared by decomposing chromium trioxide (CrO_3) at high temperature and pressure in the presence of water and a catalyst (Swoboda et al. 1961). The CrO_3 is mixed with an equimolar proportion of water and about 2% of Sb_2O_3 or Cr_2O_3 catalyst. It is placed in a high-pressure vessel and heated to 300–400° C at 50 to 600 atmospheres for a reaction time of 5 to 10 minutes. Acicular particles 0.2 to 1.5 μm long and 0.01 to 0.03 μm wide are obtained.

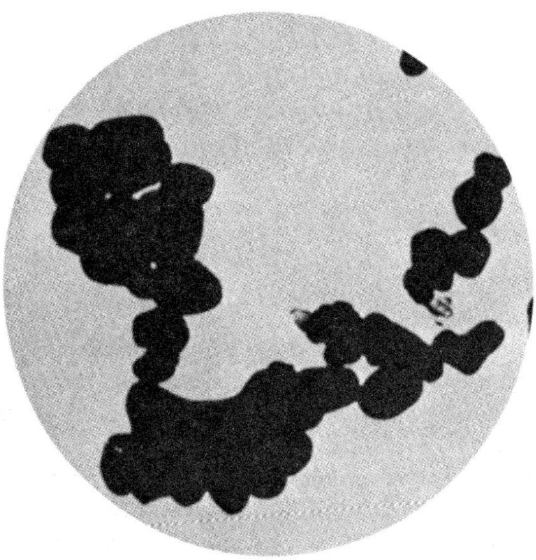

← 1 μm →

Figure 3-6. Electron micrograph of cubic magnetite particles (Fe_3O_4).

The magnetic properties are targeted at about 32 kA/m (400 Oe) with a remanence about the same as for γFe_2O_3, although these properties can be adjusted by varying the conditions of the reaction. Chromium dioxide particles are shown in Figure 3-7.

The choice of a higher coercivity than for the standard gamma ferric oxides requires special recorders with high bias (and high erase) fields and special equalization for chromium dioxide tapes. These tapes are popular for audio cassettes and for video recorders having superior high frequency response. Head wear with CrO_2 tapes is somewhat faster than with γFe_2O_3 tapes. Chemical stability should also be considered for archival applications, CrO_2 being less stable than γFe_2O_3.

Metal Particles. Advantages of metal particles for tape coatings are a remanence about three to four times as high as for oxides, a coercivity about 80 to 160 kA/m (1000 to 2000 Oe), and extremely small particles for low background noise. Though metal particle tapes were heralded as an innovation when announced in about 1979, it should be noted that metal particle coatings were

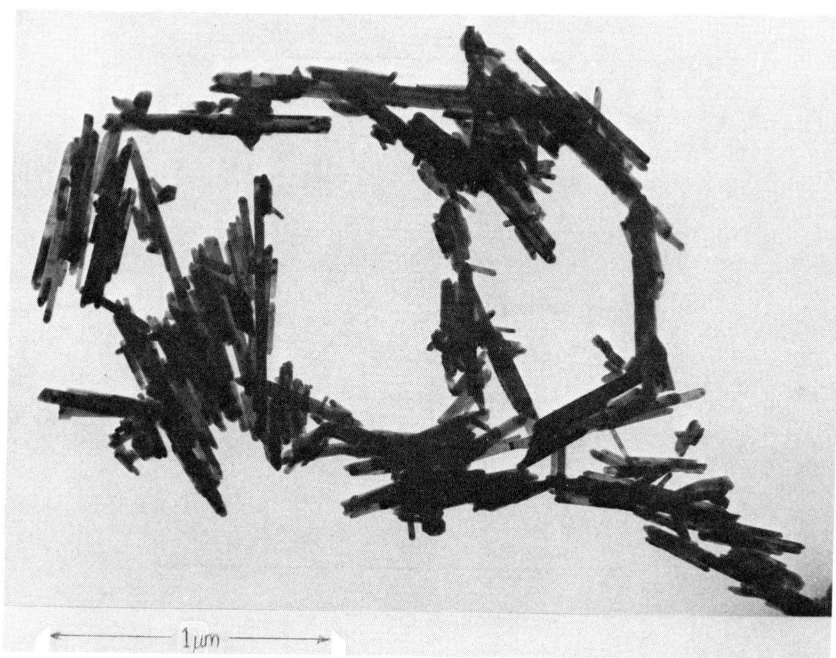

Figure 3-7. Chromium dioxide particles. A typical particle is about 0.4 μm long by 0.04 μm in diameter. (Courtesy E. I. DuPont de Nemours Co.)

contemplated in 1888 and were used in 1935, with improved versions in 1946 and in 1962 (Smith 1888, Lubeck 1937, Howell 1947, Iwasaki and Nagai 1962). The success of metal particle tapes depends on protecting the particles from chemical instability, since their small size makes them extremely reactive with air, for example. Also special heads had to be developed to advantageously record and erase these media.

The metal particles are single domains of iron, usually alloyed with cobalt or nickel. Figure 3-8 shows that the coercivity of iron increases remarkably when particle size is decreased to where single domains are magnetically the most stable form. The coercivity decreases again if the particles are made too small. Isolated particles of 0.02 μm (20 nm) have maximum coercivity of about 1000, but this ideal value decreases under practical conditions, for instance where particles are close packed. The magnetic properties of single-domain particles are enhanced by using acicular particles having shape anisotropy, and by alloying with cobalt, and so on, to give crystalline anisotropy.

Manufacture of Fine Metal Particles. Metal particles of single-domain size can be made (1) by reducing metal oxide particles, as for example acicular oxides in a hydrogen atmosphere; (2) by reducing metal salts in aqueous solu-

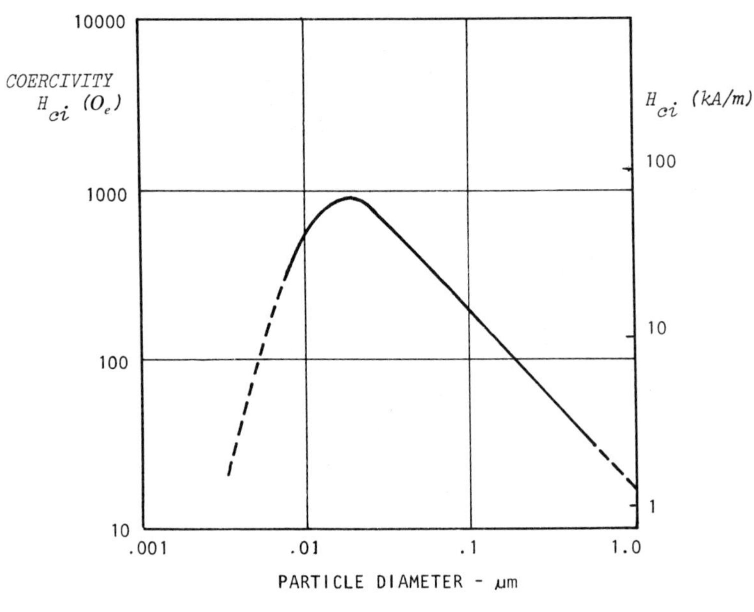

Figure 3-8. Coercivity of pure iron when subdivided into small particles. The coercivity is almost 80kA/m (1000 Oe) for equant particles 0.02 microns long. Acicular particles would reach higher coercivities. (Luborsky, *J. Chem. Phys.* 1957)

tions with a reducing agent such as sodium borohydride; (3) by electro-deposition on to a mercury cathode; and (4) by sputtering or gaseous discharge in a low-pressure atmosphere of an inert gas such as argon (Watanabe et al. 1981).

The most successful metals for high coercivity magnetic particles have been cobalt, iron, and nickel or their alloys, combined in some cases with minor ingredients such as phosphorous. The magnetic properties are especially outstanding when the particles are acicular in shape or are grouped as elongated chains-of-spheres. The particle shaping is aided in some cases by applying a magnetic field during particle formation.

Iron-Cobalt Metal Particles. The iron-cobalt particles are made by preparing a l molar aqueous solution of mixed iron and cobalt salts with cobalt in a desired proportion, 30% for example. This is reacted with a 1 molar solution of sodium borohydride in a magnetic field of 0.1T (1000G) to give elongated particles, which are washed and dried in a non-oxiding atmosphere. The particles measure about 0.03 um (300 Å) long and have a coercivity of about 95 kA/m (1200 Oe).

As shown in Figure 3-9, the coercivity can be varied by adjusting the percentage of cobalt in the starting solution. Particle size decreases as the cobalt percentage is raised, with larger and more stable particles of H_c = 1000 obtainable at about 10% cobalt content. Tapes with particles smaller than 0.025 μm show superparamagnetic behavior, with a decrease in coercivity and an increase in print-through. The best sample when made into a tape has 6 to 8 dB higher output and S/N ratio than chromium dioxide tape and has magnetic properties of H_c = 95 kA/m (1200 Oe), B_m = 0.32T (3200G), and B_r/B_m = 0.74.

Extra High Coercivity Metal Particles. A coercivity of 2200 Oe (175 kA/m) is obtained in iron-cobalt particles made as above, but by first drying the wet particles in hydrogen gas at 200° C and then applying an additional heat treatment in hydrogen at 400° C for an hour. The most favorable alloy composition responding to this treatment is 35% Co, 65% Fe (Watanabe et al. 1981).

Iron-Cobalt-Nickel Metal Particles. An aqueous solution of iron, cobalt, and nickel salts is treated with oxalic acid to deposit a solid solution of iron, cobalt, and nickel oxalate crystals of acicular form. These are dried and then reduced to metal alloy powder in an atmosphere of hydrogen (Iwasaki and Nagai 1962). The resulting particles, shown in Figure 3-10, are less than 0.1 μm in size, have a coercivity of 54 kA/m (680 Oe), a B_r/B_m of 0.76, and a B_r of 0.28T (2800G) in a tape coating 2.5 μm thick. This B_r is four times as much as is obtained in a comparable γFe_2O_3 coating.

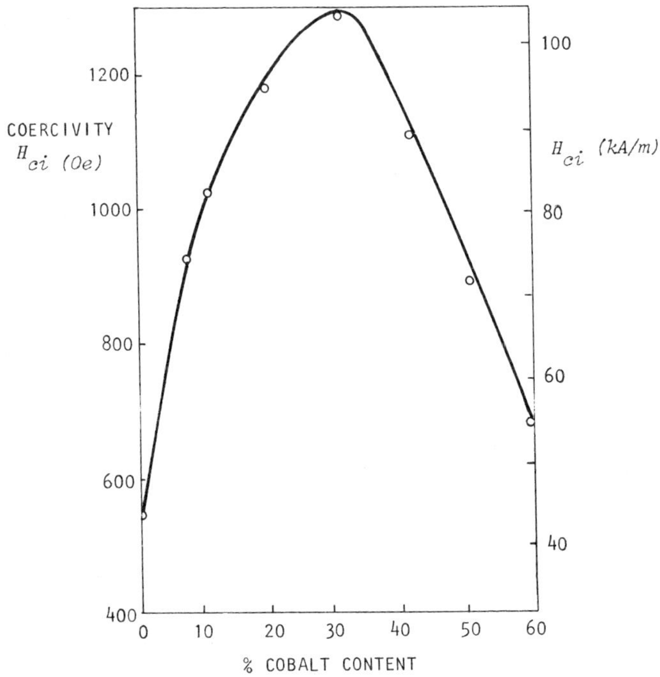

Figure 3-9. Coercivity of iron-cobalt particles vs. cobalt content. (Uehori 1978).

Comparison of Particle-coated Tapes

Tapes with different particles and magnetic properties are sold commercially under various trade names. They fall into three general groups, which are compared in Figures 3-11 and 3-12.

1. *Gamma ferric oxide tapes* require *normal bias* and *equalization* (so called "Normal bias", and 120 μs equalization in compact cassettes). Figure 3-11 shows electron micrographs of the particles and gives the *B-H* curves of tapes made with these particles.
2. *High-performance tapes* require *high bias* and *less equalization* (so-called CrO_2 (chrome) bias, and 70 μs equalization in compact cassettes). These high-performance tapes are made with special particles such as chromium dioxide, cobalt modified oxide, cobalt adsorbed oxide, Berthollide ferric oxide, and so on. Figure 3-12 (*A* and *B*) shows representative high-performance tapes. "Ferrichrome" tapes (described later in this chapter) require intermediate bias-equalization.

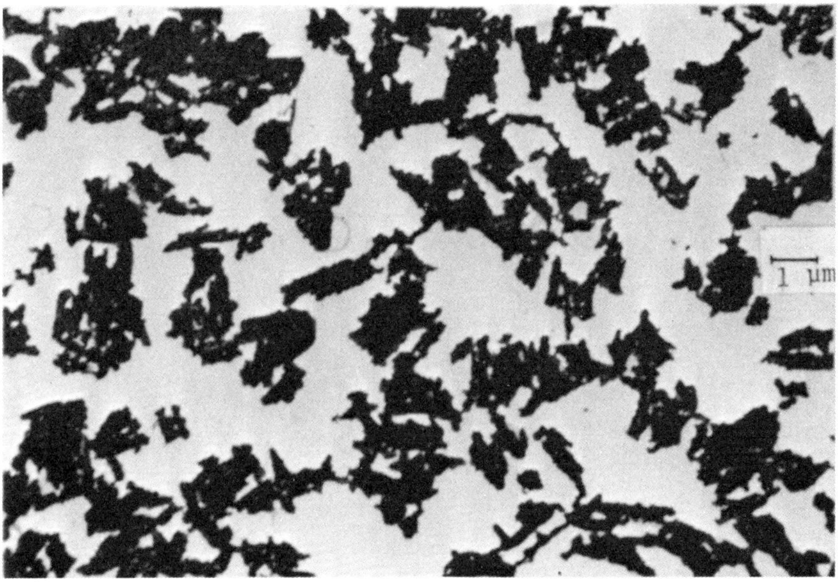

Figure 3-10. Electron micrograph of iron-cobalt-nickel alloy particles. (Iwasaki and Nagai 1962)

3. *Metal particle tapes* require *extra high bias* and *less equalization* (so-called Metal bias, and 70 μs equalization in compact cassettes). Figure 3-12*C* gives the properties of these super performance tapes, which have about twice the H_{ci} and twice the B_r of high-performance tapes.

Binder System

The binder, which holds the magnetic particles and adheres them to the base, is a synthetic varnish that must meet special requirements to make a good tape. Formulating the binder is more of an art than a science. Through trial and error, each tape manufacturer has found a more or less satisfactory mixture and procedure which he regards as proprietary. Desirable qualities in a binder include the following:

1. It must adhere firmly to the base and never come loose in spite of ordinary or accidental stressing, humidity, temperature, aging, and environmental conditions.
2. It must not shrink, warp, cup, curl, or otherwise deform the tape.

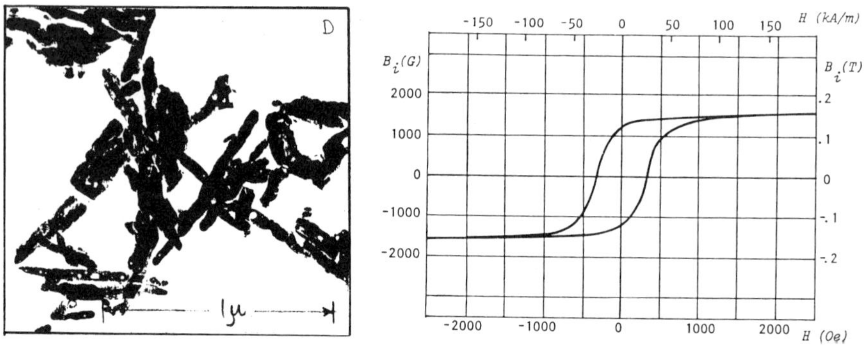

A. STANDARD GAMMA FERRIC OXIDE

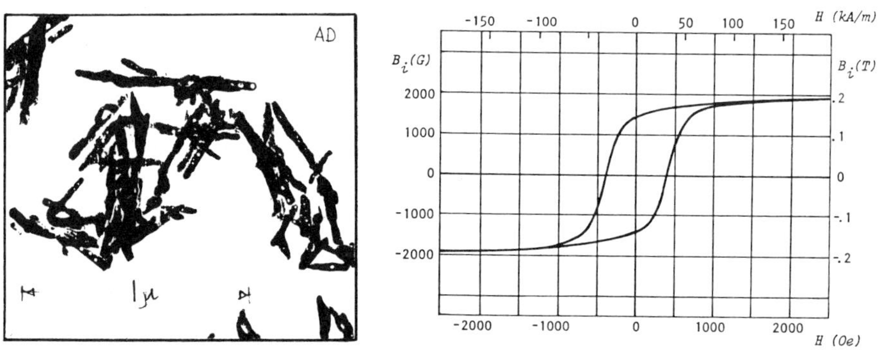

B. FINE PARTICLE GAMMA FERRIC OXIDE

Figure 3-11. Particles and magnetization curves of some commercial gamma oxide tapes. A. The particles are about 0.5 μm long by 0.05 μm wide, and have many vacuoles or pockets. The magnetic properties are better than of earlier tape of Figure 3-4. (Sample is TDK type D). B. Smaller particles give somewhat higher magnetic properties and lower noise (type AD). (Courtesy TDK Co.)

3. It must hold the magnetic particles in a homogeneous suspension, without clumps or voids that give noise and dropouts.
4. It must not react adversely with the magnetic particles.
5. It should hold the highest possible percentage of magnetic particles to give a high-output tape without loss of strength or excessive abrasiveness.

Figure 3-12. Particles and magnetization curves of high performance commercial tapes. A. The cobalt content doesn't alter the appearance but increases the coercivity to about 44 kA/m, requiring special bias and equalization (type SA tape). B. Even higher coercivity is permissible in home-use videotape (type HG tape); C. Metal particles are smallest, show evidence of chains of spheres, need special erase, bias, and equalization (type MA tape). (Courtesy TDK Co.)

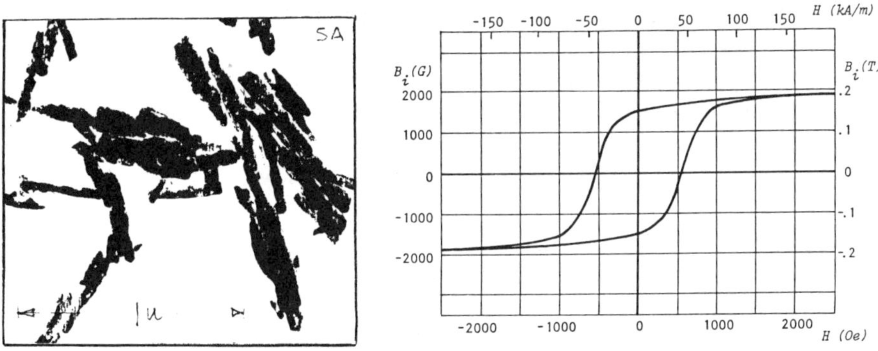

A. COBALT ADSORBED GAMMA FERRIC OXIDE (AUDIO GRADE)

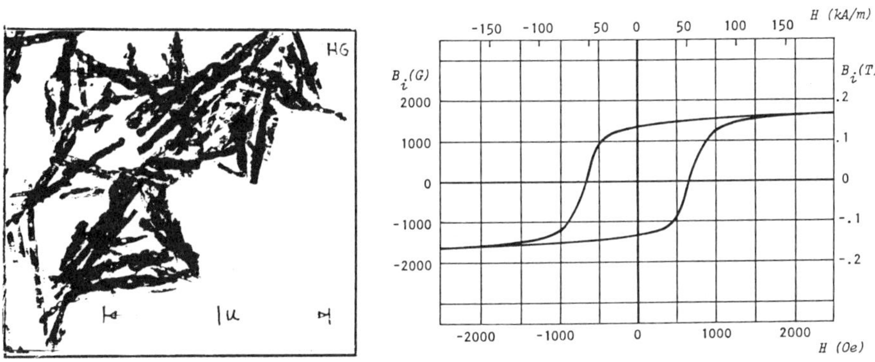

B. COBALT ADSORBED GAMMA FERRIC OXIDE (VIDEO GRADE)

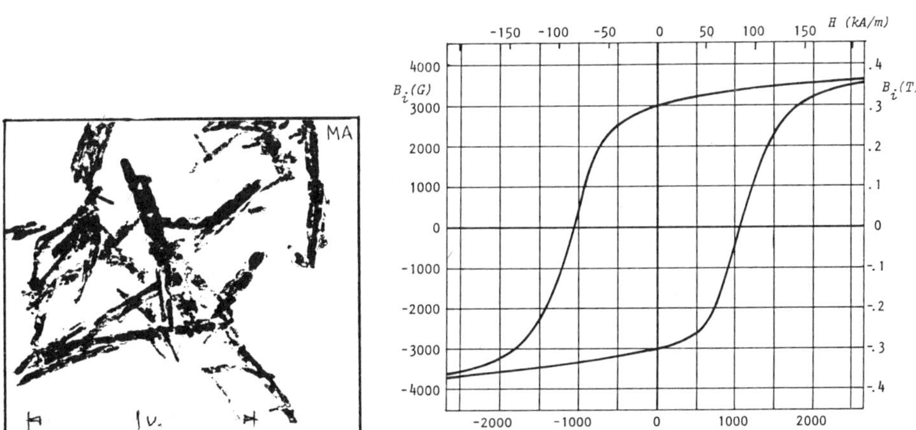

C. COBALT-IRON METAL PARTICLES FOR MAGNETIC RECORDING

6. It should have a smooth surface for good head contact.
7. It should not stiffen the tape when applied to extra thin backings.
8. It should have a low coefficient of friction.
9. It should not have an affinity for surfaces over which it slides, such as head cores, potting resins, and guides.
10. It should be nonabrasive.
11. It should not develop electrostatic charges, or otherwise tend to collect dust and debris.
12. It should not release debris, magnetic or nonmagnetic, or semiliquids such as plasticizers.
13. There should be no adherence (called blocking) between wound layers on reels.
14. It must not be flammable.
15. It must not be toxic, odorous, or stain releasing or have other harmful or unpleasant qualities.
16. It must be stable and permanent.
17. It should retain the above qualities during normal operation and storage, and preferably under adverse conditions.

Basic Binder. The choice of the basic binder is a compromise of qualities that are most important including compatibility with the base to which it will be applied, and cost of the ingredients and processing. Successful binders have been made from cellulose nitrate, polyvinyl chloride-acetate, polyurethane, acrylonitrile, epoxy, synthetic rubbers, and other adhesive and film-forming plastics. Mixtures of binders are useful. Some kinds may include a resin, with beneficial results. Stabilizers may be added to prevent deterioration due to aging or adverse conditions. Some binders used for magnetic tape coatings are listed in Table 3-7.

The binder is usually a thermoplastic material, with adhesive properties when dissolved in appropriate solvents. During drying the solvents evaporate, leaving

Table 3-7. Binders Useful for Magnetic Tape Coatings.

Cellulose nitrate	Methyl methacrylate polymer
Cellulose acetate	Polytetrafluoroethylene
Vinyl chloride polymer	Polybutadiene
Vinyl acetate	Chlorinated butyl rubber
Vinyl chloride acetate copolymer	Polyethylene terephthalate
Acrylonitrile	Polyisoprene
Polyurethane	Vinylidene chloride acrylonitrile copolymer
Polyamide	Polycarbonates
Polyethylene	Epoxy resins
Polystyrene	Butadiene styrene rubber polymers

a plastic film coating. After drying a certain degree of polymerization and cross-linking takes place in many of these binders, making them stronger and less soluble. Binders with more complete thermosetting properties have also been developed for durability and solvent resistance, but are more difficult to process consistently and give less pliable coatings.

Plasticizers. Because many binders are not of themselves flexible enough, it is necessary to mix them with plasticizers that make them less brittle and more rubbery. Suitable plasticizers have been worked out for each binder. For example, castor oil may be used with cellulose nitrates and dioctyl phthalate with vinyl binders. Plasticizers have a tendency to migrate, possibly causing sticky surfaces and embrittled coatings. Hence it is better to minimize their quantity or to avoid them entirely by choosing a binder that is inherently flexible. Unfortunately such flexible binders may not perform well in dispersing the particles or in other respects. Other plasticizers are dibutyl phthalate, linseed oil, tricresyl phosphate, and proprietary products of principal plastics producers.

Wetting Agents. Ideally each tiny particle should be surrounded and isolated from every other particle by a layer of binder, Wetting agents such as lecithin reduce surface tension at the particles, helping them become well dispersed in the binder. But after they have done their work, the agents are unnecessary fillers that tend to weaken the binder system. The same is true of other nonbinding additives such as lubricants, flow agents, antistatic agents, fungicides, and so on, which should be minimized or avoided if possible.

Flow Agents. Many binder mixtures and dispersions are thixotropic. A thixotropic mixture becomes a somewhat firm jellylike mass if allowed to stand still. Yet if the same mixture is agitated properly, it will be a thin, free-flowing liquid. When the coating is applied to the base, it is very important that it should have the correct viscosity for proper flow, and this may be improved by adding flow agents to the dispersion.

Lubricants. To reduce friction against the heads, a small percentage of lubricant may be incorporated in the binder, such as a waxy or oily material or a dry lubricant like graphite. The most famous additive has been a silicone lubricant. Although silicone is not outstanding in antifriction qualities, it is relatively unaffected by changes in temperature and it repels moisture. Graphite may serve a double purpose of lubrication and of conducting away static charges.

Antistatic Agents. Good insulators such as polyester film bases are apt to accumulate static electric charges due to friction against guides, and so on, when the relative humidity is low. The static electricity can cause tape layers to stick together while unwinding and may also cause noises due to spark discharges. The static charges can be eliminated by reducing the tape resistivity to about 10^{10} ohms per square (measured across opposite edges of a square of any size, for example a 10 mm square of tape). Carbon-black is sometimes incorporated in the coating for this purpose, and certain organic compounds are also effective. If a layer is included on the back side of the base (for endless tape or to improve winding), the antistatic agent is preferably incorporated in the back layer to avoid weakening the magnetic coating.

Abrasives. In the early years of tape manufacture, abrasiveness of tape was a problem, but with the present generation of ultrasmooth tapes it is sometimes beneficial to add a very small amount (as 1%) of extremely fine abrasive to keep the head surfaces clean and to improve tape life. Suitable additives are carborundum, alumina, or chromium oxide of submicron particle size.

Fungicides. For tropical climates, a fungicide may be added to prevent fungus from attacking the coating. Such additives are discouraged when they are not essential, because of toxicity or other side effects.

Solvents. The major proportion by volume in all thermoplastic binder coating mixtures are the solvents, which hold the other ingredients in fluid suspension. After the suspension is applied to the tape base, the solvents are allowed to evaporate. As the coating dries, it shrinks to a fraction of the wet thickness. The chosen solvents must of course be compatible with the binders and additives. Where a mixture of binders is used, each binder component may require a designated solvent which should not evaporate before the other solvents, or else a component of the binder may precipitate out. A solvent mixture is useful that contains fast-evaporating solvents (low boiling point) for fast drying initially, plus slow-evaporating solvents (high boiling point) for slower drying afterwards to allow the coating to level out and to prevent blooming, bubbling, orange peel, and other problems.

Common solvents are acetone, methylethyl ketone, methyl isobutyl ketone, ethyl acetate, butyl acetate, isopropyl acetate, amyl acetate, butyl propionate, cellosolve, cellosolve acetate, ethyl lactate, xylol, toluol, methanol, ethanol, mineral spirits (naptha), and various special solvents made by binder manufacturers for their own binders.

Diluents or Latent Solvents. Diluents are often added which by themselves would not dissolve the binder but which together with the solvents will reduce the viscosity of the dispersion. Diluents are less expensive than solvents; also they may be effective solvents for minor ingredients such as plasticizers and resinous additives. Examples of diluents are butanol, isopropanol, amyl alcohol, diethyl ether, and sometimes mineral spirits, xylol, and toluol.

Since the organic solvent mixtures are extremely flammable, enclosed drying chambers of fire and explosion preventive construction are a necessary part of the coating machinery. These chambers are used advantageously with a solvent recovery system in large-scale coating operations.

Formulations

The blending and dispersing of tape coating mixtures requires many experiments and tests before a satisfactory product is obtained. Some examples will clarify the general procedure. All parts are by weight.

Cellulose Nitrate Formulation. This is suitable for a cellulose acetate base, or soundstripe on motion picture film.

- 60 parts acicular gamma oxide (prepared as in the preceding section)
- 20 parts Duraplex A-29 phthalic alkyd resin (Resinous Products & Chemical Co.)
- 5 parts Paraplex RG-2 alkyd polyester (Resinous Products & Chemical Co.)
- 10 parts cellulose nitrate ($\frac{1}{2}$ sec RS grade)
- 50 parts solvent, consisting of 10 parts amyl acetate, 20 parts xylene, 10 parts butanol, 10 parts ethyl lactate.

The gamma oxide is mixed with the Duraplex and Paraplex and with enough solvent to make a thick paste which is blended together for a few hours. A vehicle is prepared by dissolving the cellulose nitrate in about half of the solvent mixture. The gamma oxide paste and vehicle are then combined in a ball mill (a rotating drum containing metal balls or stones) and milled for about 12 to 40 hours. Samples are taken at intervals. When the milling is considered complete the balance of the solvent mix is added and milled together for an additional few hours. The viscosity of the mixture is adjusted by adding more or less of the solvent according to requirements of the coating apparatus. After filtering through a 3 to 5 micron filter and final test sampling the dispersion is ready for coating, or it may be stored.

Ball milling is a tricky process where success depends on the ball size, the relative filling of the mill with the ball charge and with the paste, the viscosity of the paste, the speed of rotation, the length of milling time, temperature, etc. The optimum can only be determined by experiment and experience with the equipment on hand. Undermilling gives undispersed aggregates, resulting in a rougher tape surface and higher magnetic modulation noise. Overmilling can fracture the particles and harm the magnetic properties.

There are alternatives to ball mills for dispersing the particles such as multi-roller paint mills, sand mills or ultrasonic dispersers, but ball mills have been the most widely used. Figure 3-13 shows a laboratory dispersion mill.

Vinyl-chloride-acetate Copolymer Formulation. This most popular and versatile binder is often reinforced with other resins and polymers.

- 100 parts acicular gamma oxide
- 8 parts dioctyl phthalate
- 20 parts VMCH vinyl chloride-acetate copolymer (Union Carbide and Chemical Corp.)
- 12 parts VYLF vinyl chloride-acetate copolymer (Union Carbide and Chemical Corp.)
- 100 parts solvent consisting of 50 parts methyl-isobutyl ketone and 50 parts xylene

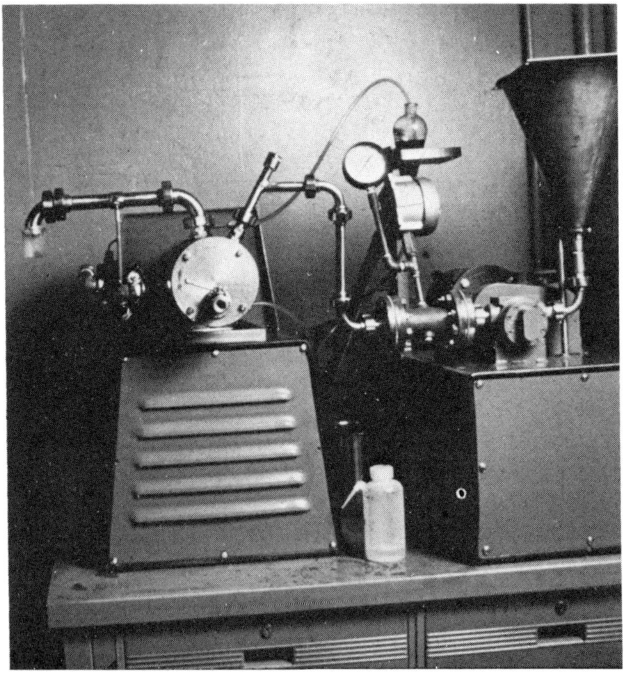

Figure 3-13. Small media mill used for laboratory dispersions of magnetic particles. (Courtesy Pfizer Inc.)

The gamma oxide is blended with the dioctyl phthalate and with enough solvent to make a paste. The VMCH and VYLF resins are dissolved in about half of the solvent mixture. Mixing and milling procedures are as for the cellulose nitrate formulation.

Polyurethane Formulation (Modified). This formulation has high resistance to wear.

- 600 parts acicular gamma oxide
- 100 parts polyurethane elastomer resin (Estane X-1)
- 100 parts vinylidene chloride/acrylonitrile copolymer (Saran F220)
- 30 parts lechitin
- 800 parts solvent consisting of 560 parts tetrahydrofuran and 240 parts toluene

The oxide and lechitin are mixed with enough solvent to make a thin slurry which is milled in a pebble mill for about 36 hours. A small amount of lubricant as silicone oil, or high melting wax may optionally be included in the slurry. The X-1 and F220 resins are dissolved in the balance of the solvent and are mixed into the milled oxide slurry. The final mixture is milled for 6 to 24 hours, after which is filtered. It is then ready for coating by any of the methods to be described.

Coating Process

In the coating process the magnetic dispersion is applied to the base. A knife coater has been the universal favorite method, although many alternatives are available including roller (gravure) coating, reverse roll coating, extrusion, spraying, and dipping. An interesting process is to coat the dispersion on a precision roll, after which the base is cast on top of the dispersion and both are stripped from the roll.

Clean room conditions must be observed at all stages of the operation, with operators wearing special clothing and with vacuum cleaning of the web at critical locations.

Knife Coating. As shown in Figure 3-14, a knife coater is essentially a trough that feeds the dispersion onto a moving tape surface. As the tape exits from the trough, it passes through a slot whose size is determined by a *doctor blade* spaced very accurately from the tape guiding surface. The blade trims the coating height to a wet thickness about six times that of the dried coating. The blade height is adjusted by experiment to give the specified coating thickness on the finished tape. For a uniform product, all coating conditions, such as viscosity,

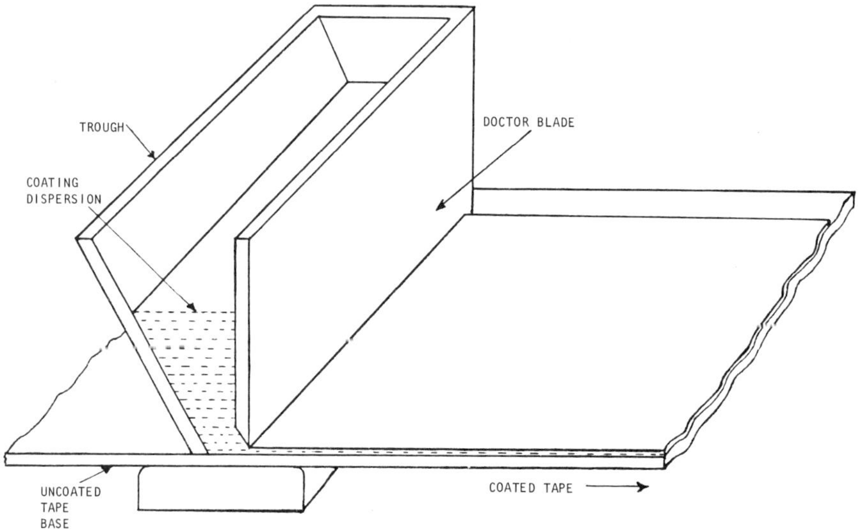

Figure 3-14. Knife coater.

concentration, temperature, mechanical tolerances, and web speed, must be controlled.

Orienting Magnet. Immediately following the coater, a large magnet is arranged to influence the liquid dispersion by orienting the suspended particles, turning the tiny elongated magnets to point in the same direction, usually in the direction that the tape will move when it is being recorded. Ordinarily this will be longitudinal orientation. The orienting treatment decreases the noise level of the tape by several dB and also increases its output level, giving a very worthwhile overall improvement in performance. In some cases, orientation may be omitted, as for example where disc records will be cut from the web.

The magnet extends across the entire width of the web, 24 inches or more, and is preferably a permanent magnet for stability, although an electromagnet can be used. Orienting magnet configurations are shown in Figure 3-15. The field that the tape sees last when it leaves the magnet should be in the desired direction of orientation. If the maximum field is too strong the liquid coating will be held back, shifted and clumped. If too weak, the orientation will be incomplete. Fields of 32 to 128 kA/m (400 to 2000 Oe) are appropriate. The field of a solenoid is considered ideal because the particles do not encounter any stray disorienting field when leaving it. The magnet shown in *C* of Figure 3-15 is designed to minimize such influences (Bate and Dunn 1980).

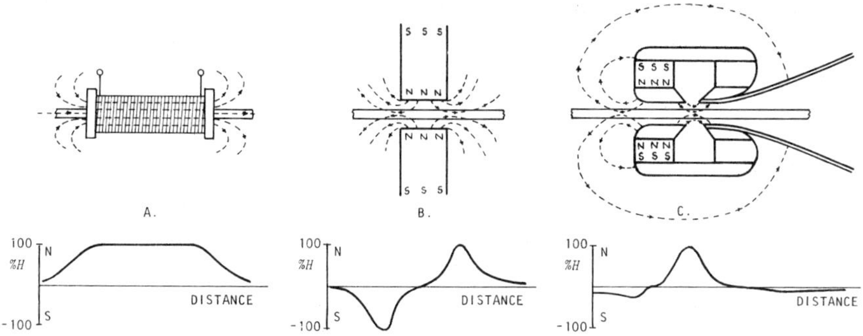

Figure 3-15. Orienting magnets and their field functions. A. Solenoid electromagnet. B. Opposed pole magnet. C. Longitudinal field magnet (Bate 1980).

Orienting magnets energized by a-c have been devised, with the idea of repeatedly agitating the particles prior to their final settling, and thus overcoming interferences that prevent maximum orientation in a one-shot orienter. Another technique is to pass the tape over a number of field disturbing bars while in the main unidirectional field (Camras 1966).

Drying Operation. A general outline of a tape-coating operation is shown in Figure 3-16. After passing the coater and orienter the liquid coating will evaporate sufficient solvents to become tacky and eventually to dry so that the web can be rolled up. To save time and space, the drying is accelerated by applying heat, infrared radiation, or rapid air convection. The processing must

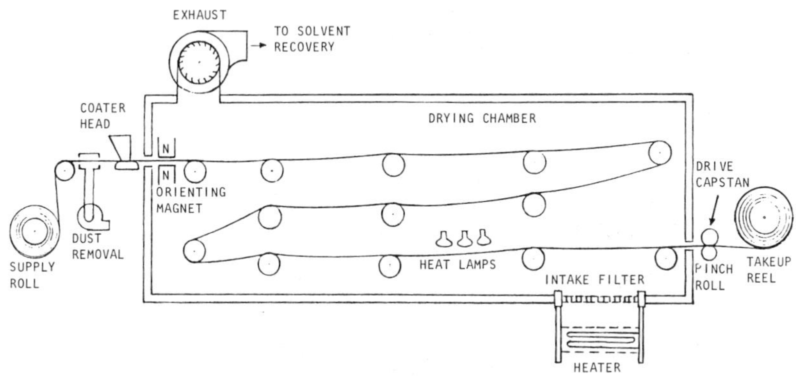

Figure 3-16. Tape-coating operation. The web is cleaned, coated, oriented, dried over a long path, heat cured, and wound on a takeup roll.

be coordinated with the solvent and binder system to give rapid drying without boiling, precipitation, irregular shrinking, or other harmful effects. Web speeds vary from 0.15 m/s to 2 m/s (30 to 400 ft/min). A small coating machine is shown in Figure 3-17.

Calendering. The optional step of calendering may now be applied to the coated film. In one type of calendering, the tape passes between rolls that have different surface speeds, to give a slipping burnishing action between the coated surface and a highly polished roll. High pressure is used between the rolls so that the tape layer is compacted at the same time by perhaps 10% to 30%, increasing the B_r value by a corresponding amount. In another form of calen-

Figure 3-17. Pilot plant tape coating machine. (Courtesy Ampex Corp.)

dering, the rollers move at the same surface speed to compact the tape without burnishing it, although tape surface defects will still be ironed out.

The main objection to calendering is the expense of adjusting and maintaining the precision rolls and the necessary cleanliness to assure that foreign matter is not ironed into the tape. Alternative methods have been used whose benefits are of lesser value, but that have cosmetic or advertising importance. Buffing gives the tape surface a high gloss but makes little or no difference in its performance. One manufacturer formerly ran the tape surface back against itself to burnish out nodules in a process called Ferrosheen®.

Back Coating. For tapes in endless loop cartridges, such as the popular stereo-eight, a lubricant coating is applied to the back of the tape, as for example a colloidal graphite dispersion or a waxy coating one to three μm thick. In these cartridges, every turn of tape slides against its neighbor and there must be no adhesion or binding. Application procedure is as for the main coating but is not as critical.

A somewhat soft rubbery layer about two μm thick is applied to the back of some professional tapes that are wound on large-size reels. Such a coating prevents telescoping or side slipping of the turns against each other and absorbs stresses that occur during winding and during storage when the reel hub expands and contracts at a different rate than the wound tape (Minnesota Mining 1970b).

Additional Front Coatings. So-called *sandwich tape* has been made with a thin layer of plastic material on top of the magnetic coating to protect it from abrasion and to reduce head wear. Such a tape has poor response at short wavelengths because of extra separation of recording layer from the head.

Multiple layer coatings have been suggested for such benefits as improved resolution, lower noise, greater output, and security (preventing forgery). These are discussed later in this chapter, but are mentioned here as a possible stage in the coating process.

A surface lubricant is more effective in preventing wear than one that is part of the coating. The surface lubricant does not weaken the binder, does not have to be internally compatible with the binder, and more lubricant can be used. On the other hand, the tape is more likely to become gummy and fouled with wear products. An advantageous surface lubricant is Krytox®, a tetrapolyfluroethylene compound that, when applied according to a "Krytoxing Process," can decrease head wear by a factor of 2 and tape wear by a factor of 10 (Wright and Tobin 1976) (Tobin and Powell 1976).

Slitting and Winding

A rotary blade slitter cuts the tape web into strips of the final designated width. Although slitting is a simple mechanical operation, all of the previous steps can be rendered worthless if the slitting is defective. The accuracy required is in the order of 0.001 inch (25 μm), but even more important is a clean straight cut that does not generate loose bits of base or coating, or irregularities that chip off afterwards. Such debris lodges between tape layers and becomes an important source of dropouts.

When the tape is wound accurately on the final reel at proper tension, it will remain in good condition for a long time. If the winding is irregular and if the tension is too high, the tape may be overstressed, becoming wavy and unsuitable for high-quality recording. For the most accurate results, the tape is wound temporarily on bulk reels as it comes from the slitter; later it is rewound onto the final reels. For economical production, some manufacturers wind the tapes directly from the slitter onto the final reels. This can be a formidable operation when a web 24 inches wide is split into 96 tapes, and all of the reels must have reliable tensioning and guiding. In this respect the 178 mm (7 inches) and smaller reels are less demanding than the larger professional sizes, but they fill up rapidly and must be changed more often.

Although most coated tape is slit into narrow widths, a reasonable percentage is cut into discs, belts, and flat sheets for special purposes, such as dictation records and floppy discs, for example.

Alternative Coating Techniques

With knife coaters, the deposited film thickness varies if there is any variation in base thickness, or imperfect contact between the base and its backing under the doctor blade due to waviness, air pockets, or debris. Techniques described below are sometimes used to avoid such imperfections.

Roller Coating (Gravure). Roller coaters deposit a film whose thickness does not depend on the above variations. As seen in Figure 3-18, a roller covered with an even layer of magnetic dispersion applies it to the base, transferring the coating in a manner analogous to roller painting, or to the inking rolls of a printing press. Not all of the dispersion is transferred, as is evident from the fact that the roller remains wet after parting from the base. The best results are obtained with a gravure roller whose surface is covered with a dotted pattern such as that used for half-tone illustrations in printing. The thinner the coating, the finer the pattern that can be used, so gravure roll coating is best suited for thin coatings. The main objections to gravure coating are the cost and maintenance of the patterned roller and the possibility that vestiges of the pattern will remain in the finished tape to give a periodic modulation of the recorded signal.

Reverse Roller Coating. Figure 3-19 shows a reverse roller coater in which the coating thickness is metered by the spacing between the reverse roll and its

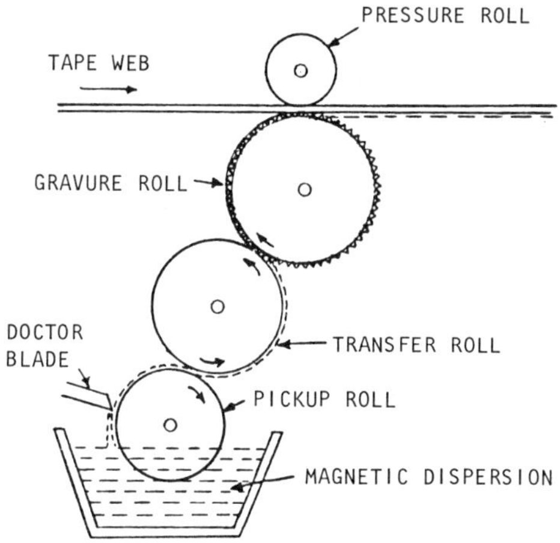

Figure 3-18. Gravure roller coater.

adjacent roll, although the viscosity of dispersion and coating speed are also important. As with the gravure roll coater, variations in web thickness do not affect the coating thickness.

Multiple Layer Coatings. The idea of tapes coated with two or more layers has fascinated experimenters for a long time. Possible variations include the following:

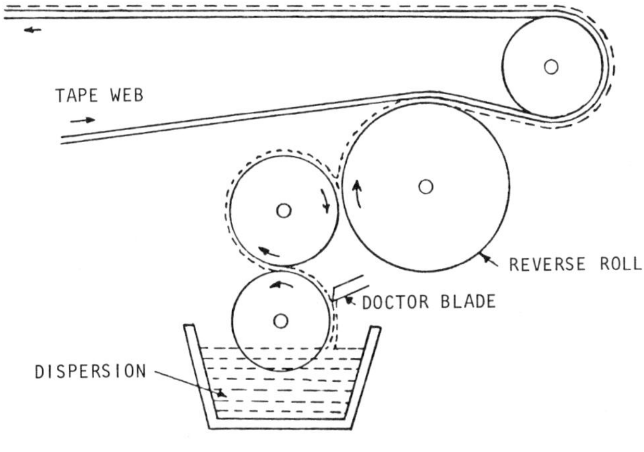

Figure 3-19. Reverse roller coater.

1. Both layers can be the same, but random irregularities are reduced by the overlapping.
2. The base layer can be of high-remanence, low-coercivity material and the top layer of high-coercivity material, matching the head field that is stronger at the tape surface and matching the wavelengths where short waves are recorded more efficiently at the high-coercivity surface (Camras 1965). Ferrichrome tapes have two coatings, with a base layer of ordinary γFe_2O_3 surmounted by a high coercivity layer of CrO_2.
3. The base layer can be of coarser, densely packed particles for high output at long wavelengths, whereas the top layer can be of very fine particles to register the shortest wavelengths.
4. The base layer can be of acicular particles oriented longitudinally, whereas the upper layer can be oriented perpendicular to its surface, enhancing the short wave response.
5. The base layer and top layer particles can be oriented at right angles to one another, longitudinally and laterally, for example, to allow two different distinguishable recordings be retained, one made with a longitudinal head, the other with a transverse head.
6. The base layer can be of magnetically soft material, not recordable by itself, but completing the magnetic circuit of a surface layer of high coercivity. The surface layer may be vertically oriented for high density recording.
7. After the base layer is finished, it may be saturated magnetically with a DC field to develop out the noisy areas due to small irregularities in the dispersion. The magnetized layer is then exposed to a suspension of fine magnetic particles that are preferentially attracted to the noisy regions, filling in the gaps to give a tape of very low noise level after this second layer is bonded in place.
8. The base layer can be oriented while still wet according to a special code, which after drying is permanently set and cannot be erased magnetically. The top layer is capable of conventional magnetic recording. Such a tape is useful for credit cards, for example, which are authenticated by a reading device that erases the top recording and senses the *magnetic watermark* coding of the underlayer.
9. A similar scheme has an underlayer of uniform orientation but extremely high coercivity; this underlayer can be coded individually by a super-high-intensity recorder but cannot be altered by ordinary recorders that operate on the top recording layer.

In practice many of the above schemes do not work out as anticipated, or do not give advantages which are worth the added cost, and therefore multiple layer tapes have not become as popular as might be expected.

COATED DISCS AND DRUMS

Magnetic particle dispersions are useful for the surface of rigid magnetic discs and drums in high-density computer memories. For this application, the coating does not have to be pliable as in tapes, and therefore a harder, more brittle binder is permissible. Also the heads do not ride on the magnetic surface but "fly" above it, so wear and abrasion are no problems (except for head-crashes). Cylindrical drums were most common in early computers, but have largely been replaced by discs that can be stacked to give a much greater storage capacity in given volume. The coating techniques are quite similar for discs and for drums; therefore the description for discs applies to both.

The base disc is usually of aluminum, 203 to 356 mm (8 to 14 in) in diameter, with a center hole 168.2mm (6.625 in) in diameter for the 356 mm disc, and a thickness of 1.27 mm (0.050 in), although giant discs 1219 mm (48 in) in diameter have been made. The aluminum surface may be anodized for extra hardness and durability. The disc is finished to great precision and flatness by grinding and lapping, with special care given to its surface to ensure absence of holes and pits. The coating dispersion in a hard grindable binder is applied in excessive thickness and, afterwards, ground down and lapped to finished thickness. The original thick rough coating may be applied by spraying of multiple coats, by spin coating using centrifugal force, by flow coating with a pool of dispersion, and even by electrophoresis.

DEPOSITED MAGNETIC FILMS

Continuous metal films on a nonmagnetic base are like solid metal tapes except that they are usually $\frac{1}{10}$ to $\frac{1}{100}$ as thick, and only one face of the metal is exposed. The metal film may be created by electroplating, by electroless (chemical) plating, by sputtering in a rarefied atmosphere, or by evaporation in a vacuum (Ransom and Zentner 1964). In deposited films we expect a high remanence because the magnetic material is not diluted as with the binder in coated tapes. We also expect a low noise level in a continuous film because there should be no granular particles or uneven dispersion. These expectations are not fulfilled completely because there is evidence that good permanent magnet metals have a structure where single domain crystals are suspended in a nonmagnetic matrix. Therefore even the continuous films have a granular magnetic nature, but possibly finer and more uniform than in particle coated tapes.

Electroplated Media

The first major use of electroplated media was a magnetic recording wire manufactured in the 1940s by the Brush Development Company. The wire had a bronze core 0.004 in. (100 μm) in diameter plated with a nickel-cobalt alloy 0.00035 in. (8.7 μm) thick (see Figure 2-14 and Table 2-3). When the need arose for drums and discs in computer memories, a similar plating was very successful for their magnetic surface (Bate 1965). Subsequently magnetic oxide particle coatings became popular and other nonplating techniques were also used, but plated discs have always been prominent (Wielinga et al. 1982).

Plated metal tapes were used in early Univac computers, where a bronze ribbon 0.001 inch (25 μm) thick and $\frac{1}{2}$ inch (12.7 mm) wide was plated with a magnetic surface of micron thickness. However such tapes were expensive, heavy, hard to handle, easily injured by creasing, and too stiff to conform to head surfaces. They were eventually replaced with oxide-coated tapes.

Platings deposited on plastic film bases have also been tried from time to time but have not been able to compete with ordinary oxide particle tapes (Allen et al. 1964). The platings on these tapes are very thin and easily eroded. On ordinary recorders, these tapes do not give outstanding performance, but require special heads, bias, and equalization to take advantage of their construction.

Plated discs and drums for computers have been very successful, probably because the heads do not contact the record and because the extremely thin coatings are less susceptible to "peak shift" troubles at high recording densities than are the thicker oxide coatings. Electroplating is done in an aqueous bath in which the disc is rotated slowly to give a uniform layer on its surfaces.

Electroplating Process. A nickel-cobalt alloy is extensively used for the electroplated deposits. An example of plating procedure is as follows:

- 101 parts nickel chloride hexahydrate ($NiCl_2 \cdot 6H_2O$)
- 101 parts cobalt chloride hexahydrate ($CoCl_2 \cdot 6H_2O$)
- 25 parts boric acid (H_3BO_3)
- 0.1 part sodium thiosulfate ($Na_2S_2O_3$)
- 1000 parts water

The plating bath is made by mixing the above ingredients. It is adjusted to a pH = 4 and heated to 70° C. A brass disc 150 mm in diameter, carefully polished and cleaned, is rotated slowly in the bath while attached to the negative DC terminal to act as a cathode. Two anodes are used, one of nickel and one of cobalt, each with a variable resistor and ammeter in series, as shown in Figure 3-20. Their relative currents are adjusted according to the plated proportions, which are about 20% nickel and 80% cobalt. The DC current density is 0.108 amperes/cm^2 of cathode area. Superimposed on the DC is an AC current, which can be adjusted with a variac set to give an AC/DC voltage ratio of 13/2. The plating is carried out to a thickness of 0.0004 inch (10 microns),

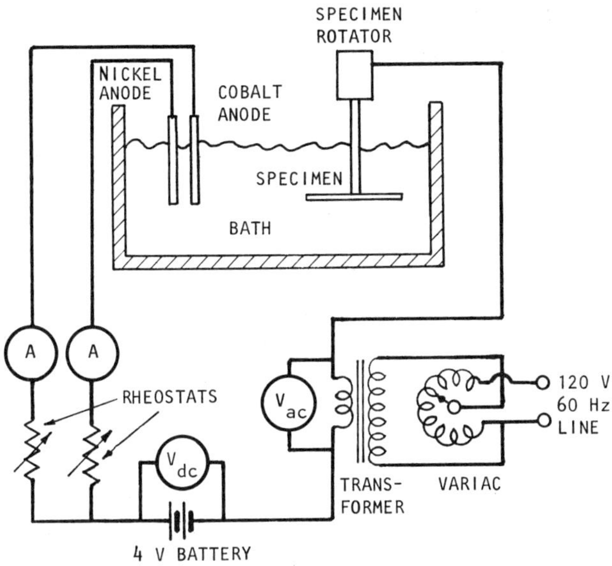

Figure 3-20. Electroplating process for nickel-cobalt alloy.

resulting in a plated surface of bright appearance with an H_c = 21.5 kA/m (270 Oe) and B_r = 0.8T (8000 Oe).

Electroless Plating (Autocatalytic Deposition)

Electroless plating, like electroplating, is carried out in chemical solutions, except that no electric current is used. (Allen et al. 1964). The process is analogous to silvering of mirrors by purely chemical means. It has the advantage that deposits are readily made on nonconductive substrates such as glass or polyester film. Also, the deposition is uniform and does not depend on shape and placement of anodes and cathodes. On the other hand, the chemical baths become depleted and must be monitored and readjusted.

Electroless plating gives high coercivity deposits of nickel-cobalt alloy, and also of pure cobalt and of cobalt-phosphorous, with the coercivity increasing as the phosphorous content increases from 1% to 6%. Extremely thin deposits can be made in the order of 1 to 10 microinches (0.025 to 0.25 μm). Electroless plating has been very successful for producing the surfaces of memory discs and drums. Magnetic tapes of polyester have also been tried, but the thin metal coatings lack durability when run in contact with heads. The thin films can also corrode or oxidize and change their magnetic properties (Judge et al. 1965).

Electroless Plating Procedure. An aqueous bath is prepared with the following concentrations of ingredients:

- 0.11M $CoSO_4$
- 0.19M NaH_2PO_2
- 0.50M $(NH_4)_2SO_4$
- 0.12M Na_3(citrate)

Temperature = 80° C; pH = 8.7 at 50° C, adjusted with NH_4OH. There is no agitation. The substrate surface of polyethylene terepthalate film is treated with a hot chromicsulfuric acid solution, followed by a hot NaOH solution. It is then sensitized with a stannous chloride solution and finally with a palladium chloride solution. The sensitized surface becomes plated when immersed in the main bath, with plating thickness depending on time of immersion. (Judge et al. 1965).

The coercivity varies with thickness of deposited film, reaching H_c = 96 kA/m (1200 Oe) when the film is 0.06 μm thick, and levelling down to H_c = 32 kA/m (400 Oe) for thicker films of 0.5 microns. A modified bath containing nickel gives a film that has a coercivity H_c = 103 kA/m (1300 Oe) independent of film thickness between 0.05 to 0.2 μm. This modified bath contains 76% cobalt sulfate and 24% nickel sulfate instead of 100% cobalt sulfate.

Evaporated Films

The most sophisticated of all coatings are evaporated films, where accuracies and finishes are truly of atomic dimensions (Bate 1965). Although alloys such as Ni-Co or Ni-Co-P can be evaporated, it has been found that pure cobalt or iron films have high coercivities of 24 to 80 kA/m (300 to 1000 Oe) if the deposited atoms arrive at a somewhat grazing angle to the substrate surface, as for example 70° to 80° from the normal. Nickel is less responsive to this method but gives coercivities to about 28 kA/m (350 Oe).

Evaporation Process. Deposition is carried out in a bell jar or in equipment that has been highly developed for the vacuum metallizing and semiconductor industries, for example under the following conditions.

- Substrate material: metal, plastic, or glass
- Precleaning of substrate: glow discharge bombardment
- Substrate temperature: 20° to 50° C
- Deposited metal: cobalt
- Heating method for evaporation: induction or electron beam

- Rate of deposition: 0.005 μm per second
- Angle between source and normal to substrate surface: 75°
- Size of crystallites: 0.1 μm
- Coercivity of film: 48 kA/m (600 Oe)

Sputtering Process Sputtering uses the same equipment as for vacuum deposition, except that the bell jar atmosphere is a low-pressure gas such as argon and a voltage may be applied between the work (cathode) and the source (anode).

PERPENDICULAR-ORIENTED RECORD MEDIA

Since 1975 Dr. Iwasaki and his coworkers have announced experiments with recording media having perpendicular-oriented crystal structure and giving linear recording densities of 6000 bits/mm (150 000 bits/in) or more (Iwasaki and Nakamura 1977; Iwasaki et al. 1980; Iwasaki 1980). Perpendicular orientation favors the shortest wavelengths because the perpendicular component is less affected by demagnetization as density increases, while the opposite is true of longitudinal recording. The vertical stacking of needlelike particles gives a much finer grain recording surface than if the same particles were aligned parallel to the surface. The short-wave magnetization mode is U-shaped so that deeper parts of the recorded layer contribute to the magnetization. These records also have the usual advantages of deposited metal films: high coercivity, high retentivity, and low thickness. Disadvantages include poor resistance to wear and corrosion.

Perpendicular recording films are especially suitable for magnetic memory discs, where digital recording utilizes their shortwave superiority and where the use of flying heads and protective environment minimizes erosion of the delicate record surface.

Cobalt-Chromium Films

The perpendicular-oriented film described by Iwasaki and Ouchi has crystalline anisotropy, a rodlike structure (Figure 3-21), and *M-H* curves (Figure 3-22) showing a perpendicular H_{ci} of about 80 kA/m (1000 Oe) and a rectangular hysteresis loop (Ohta et al. 1982; Ouchi and Iwasaki 1982). It is made by RF sputtering cobalt-chromium (83% Co, 17% Cr) to give a layer about 0.5 to 1.0 μm thick on a polyimide base film. An argon gas pressure of 0.01 Torr and an RF power density of 0.44 watt/cm^2 give a sputtering rate of 0.33 μm/hour, which is optimum for perpendicular anisotropy.

A special perpendicular head with stylii of different widths on opposite sides

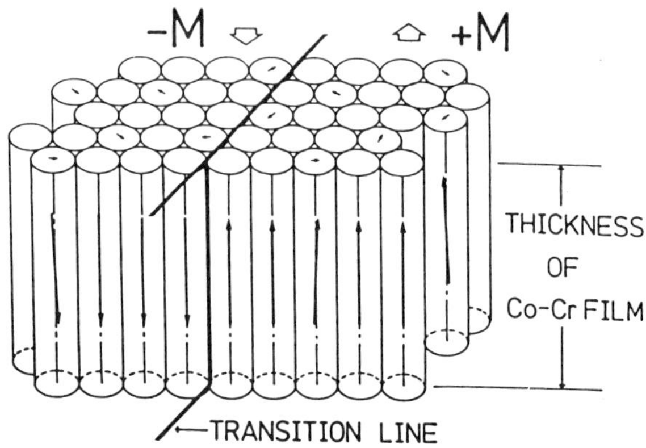

Figure 3-21. Crystalline structure of cobalt-chromium perpendicular-oriented film. (Ouchi and Iwasaki 1982) (IEEE Trans. MAG-18, 1982).

of the record has been designed for these media, as described in Chapter 5. For single-sided heads, the medium is backed with a layer of permalloy to provide a return path for the U-shaped magnetization (Iwasaki, 1980).

Other Perpendicular Films

Films of cobalt alloyed with metals other than chromium also have a perpendicular crystal structure. Sputtered cobalt-vanadium films with vanadium con-

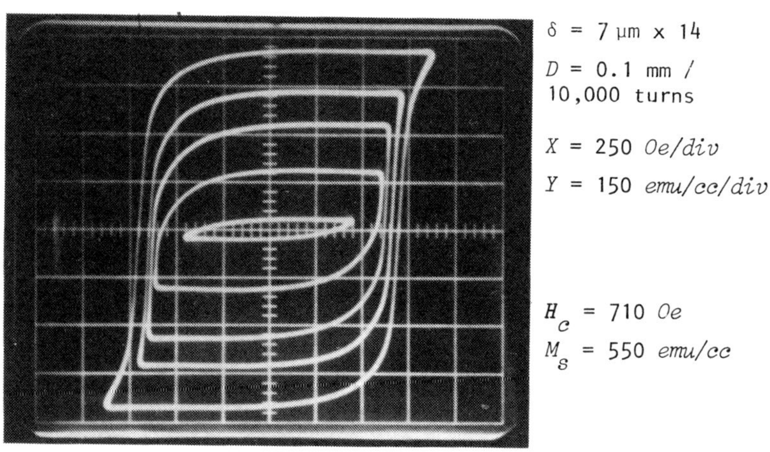

Figure 3-22. Family of intrinsic *M* vs. *H* loops of cobalt-chromium in perpendicular direction of sputtered film. (Ouchi and Iwasaki 1982) (IEEE Trans. MAG-18, 1982).

tent from 9.8% to 18.4% have a structure and magnetic behavior similar to cobalt-chromium films (Fukuda et al 1982). Possible advantages of other alloys are better frictional qualities, wear resistance, and flexibility.

A film of pure cobalt with isolated acicular particles normal to the film plane has been produced by electroplating from a 1 molar Co sulfamate solution of pH = 6.5 at a current density of about 7 mA/cm^2 at 35° C, followed by an anodizing treatment. The resulting film 1.0 μm thick has a coercivity of 80 to 160 kA/m (1000 to 2000 Oe), depending on plating current and temperature (Chen and Cavallotti 1982).

Barium Ferrite Perpendicular Media

Barium ferrites of perpendicular orientation have been prepared as sputtered films, and also as fine particles for coated media (Kubo et al. 1982) The particles are single domain thin hexagonal platelets with direction of easy magnetization perpendicular to their plane. Tapes are prepared by conventional coating techniques except for orientation. Recording densities of 3000 bits/mm (75000 bits/in) have been attained (Fujiwara et al. 1982).

TAPE TESTING

During tape manufacture many tests are made to insure that the finished product will be up to standard, including physical, chemical, magnetic, and recording-playback tests. The following is an abbreviated summary. Comprehensive standards may be obtained from the IEC (see Appendix G).

Physical Tests. Visual inspection by an experienced operator will detect off-standard raw materials, bad slitting, and so on. Microscopes disclose the condition of particles, dispersions, and surface finishes. Tensile strength testers, viscosimeters, and similar instruments are routinely used in laboratories of plastic film and coating industries. Table 3-8 lists various physical tests and appropriate methods of carrying them out. The numerical results of such tests have significance only when test conditions are also specified.

Since the base film is usually purchased, its specifications are spot checked by the tape manufacturer, and detailed tests are made only if there are signs of trouble. Coating, slitting, and winding are all critical, however, and must be monitored constantly. The ultimate test, of course, is the record-playback performance on a machine of high quality.

One useful test on a coated tape web is to observe the light transmitted through it from incandescent lamps on the other side. Gross variations in coating thickness or uniformity show up as darker and lighter patches on a translucent tan background. It is not well known that ordinary gamma oxide coatings are fairly

Table 3-8. Physical Tests of Magnetic Recording Media.

Characteristic Tested	Method or Instruments
1. Tensile strength	Weights suspended from clamped tape; commercial tensile test instruments.
2. Impact resistance	Pendulum impact tester
3. Resistance to tearing	Notch tester (Elmendorf tear tester)
4. Stretch	Force that gives permanent elongation of 5% (or other %) under specified conditions
5. Waviness	Visual observation; microscope observation during slow reeling
6. Bias	A cut tape one meter long is laid horizontally adjacent a straight edge
7. Edge smoothness	Microscope observation; observe edge of tape on wound reel
8. Coefficient of expansion a) with temperature b) with humidity	Dial indicator attached to a clamped tape length in controlled temperature or humidity environment
9. Surface finish	Profilometer; microscope; Scanning electron microscope (SEM)
10. Anchorage	Fold and unfold; pull coating from base with adhesive tape
11. Blocking	Observe interlayer adhesion of tape on reels; self-unwinding from a small mandrel
12. Cupping	Microscope measurement of angle of tape curvature in plane perpendicular to tape axis
13. Coefficient of friction	Calculated from difference of tape tension when pulled around a stationary or rotating drum
14. Head wear	Weight loss of a block of head-face material under specified conditions
15. Squealing	Audible; tone modulation of a steady recording
16. Flammability	Should self-extinguish when touched to a flame.
17. Ageing	Test after long storage; temperature and humidity cycling; fadeometer
18. Tape wear	A loop of tape is run until a defined deterioration occurs in performance
19. Particle size and shape	Electron Micrographs; X-ray Diffraction studies
20. Particle dispersion	Scanning electron micrographs of coating section; high-power optical microscope for rough check
21. Electrical resistance	Megohm-meter; volt-microammeter method with high voltage
22. Visible recordings	Develop out with suspension of fine magnetic particles on tape surface
23. Dimensions	
a. Total thickness	Micrometer, stage micrometer; comparator; microscopy of tape section; direct, or multiple thicknesses for accuracy and for averaging
b. Base thickness	Remove coating by dissolving or abrasion and measure as above
c. Coating thickness	By difference between *a* and *b* above
d. Back coating thickness	Same as *c*
e. Width	Micrometer, of several wound tape turns; traveling microscope
f. Length	Roller gauge for measuring textile length

transparent to near infrared light (800 nm wavelength). The particles are opaque to shorter wavelengths, while the plastic base stops longer waves of about 2 um.

Chemical Tests. Chemical tests such as the ones in Table 3-9, together with magnetic tests can trace faults in processing that lead to substandard material. Tests of the binder are used for quality control and for comparison with other formulations. Magnetic pigment weight (in the order of 20 mg per meter of half-inch tape) correlates with maximum output at long recorded wavelengths.

Magnetic Tests. Magnetic tests as in Table 3-10 on the recording material allow one to estimate the fields needed for biasing, recording and erasing, the playback output, and the frequency response. The usual specifications include retentivity (B_r), intrinsic coercivity (H_{ci}), and squareness ratio (B_r/B_{mi}), where B_{mi} is the saturation magnetization. For finished tape, a rating is often given of the *fluxivity*, which is the remanent flux per unit of tape width in nanowebers per meter (nW/m). Sometimes the fluxivity is expressed as maxwells per $\frac{1}{4}$ inch of tape width; multiply this figure by 1570 to obtain nW/m. When the fluxivity is known, the playback head output in millivolts can be calculated from $E = N\, d\Phi/dt$ (Chapter 4 gives more details of such calculations).

Most magnetic data are obtained by running a family of hysteresis loops on a test sample in a hysteresis loop tracer, or by use of a vibrating sample magnetometer as described in the section on magnetic testing at the end of this chapter. Anhysteretic data have also been suggested as a criterion for tapes used with AC biasing. Table 3-10 gives various magnetic tests for recording media, and suggests equipment or procedures.

Record-Playback Tests. Testing of the finished record should be under conditions of actual use, since properties of critical importance for digital re-

Table 3-9. Chemical Tests of Magnetic Record Media.

Characteristic Tested	Method or Instruments
1. Solubility of coating	Softening or dissolving of binder when subjected to solvents of different potency (ketones, esters, aromatic hydrocarbons, aliphatic hydrocarbons, etc.)
2. Degree of polymerization	Softening by heat or by solvents
3. Magnetic pigment loading	Quantitative analysis after removing organic material from a fixed area of tape
4. Pigment purity	a. Test for *ferrous iron*
	b. Test for *total iron*
	c. Test for *metallic iron*
	d. *Ferric iron* by difference, $d = b - c - a$

Table 3-10. Magnetic Tests of Magnetic Recording Media.

Characteristic Tested	Method or Instruments
1. Coercivity (H_c)	Hysteresis loop tracer; vibrating sample magnetometer; coercimeter.
2. Retentivity (B_r)	Hysteresis loop tracer; vibrating sample magnetometer
3. Saturation density (B_s)	Same
4. Squareness ratio (B_r/B_s)	Same
5. B_r vs. H curve	Same
6. Saturation hysteresis loop	Same
7. Hysteresis loop family	Same
8. Anysteretic B_r vs H	Modified B-H curve tracer
9. Preisach Diagram	Calculated from data taken with modified B-H curve tracer

cording may have very little value for audio cassettes and vice versa. Also the tape is handled quite differently in video recorders, instrumentation devices, or continuous loop drives with regards to tape tension, head pressure, reel storage, repeated cycling, and so on. Therefore each application has developed appropriate criteria for testing its media. Tables 3-11 and 3-12 summarize performance tests for different uses. We caution again that test results should indicate the conditions of testing unless they are for comparison only.

WEAR LIFE OF COATED TAPES

Ordinary tapes coated with particles in a binder have a useful life that is seldom determined by wearing-off of the coating, but rather by the coating becoming irregular with tiny embedded clumps and islands of debris and wear products. These include oxide and binder chunks that are "melted" into the coating, perhaps by localized heat and pressure during tape friction against the head. The high percentage of magnetic pigment loading in tape coatings contributes to this tendency. As debris accumulates, the background noise and dropouts increase, becoming troublesome and finally unsatisfactory. Tape cleaning and rejuvenating machines loosen the debris by chemical means and by scraping with sharp edges. Their effectiveness is controversial: they are apt to leach out binder ingredients such as plasticizers or lubricants, and the scraping may cause mechanical damage to the tape surface. For new tapes, there are surface lubricant treatments that increase tape life as mentioned in connection with tape coating.

Since tapes have a minimum life of hundreds or even thousands of plays, tape wear is of little concern in applications where recordings are mostly stored and seldom played repeatedly, as in home use, mastering, broadcasting, and many kinds of data recording. However, tape wear is of concern in endless-

Table 3-11. Performance Tests of Magnetic Record Media (Audio and Analog Applications).†

Property	Test*
1. Bias requirements	Record a small amplitude test signal, noting output level in dB vs bias currrent in dB
2. Sensitivity	At "optimum" bias, note output volts (or voltage level dB) vs signal current
3. Maximum output	At tape saturation, note maximum output volts (or voltage level dB)
4. Maximum "undistorted" output. MOL = maximum output level	At 3% third harmonic distortion (or other specified percentage) note the output voltage (or voltage level dB); run at different signal frequencies
5. Distortion	Percentage of harmonic distortion or intermodulation distortion in flat equalized system at specified fluxivity.
6. Frequency response unequalized	For constant recording current at frequencies in the operating spectrum, note the playback volts (or dB level) from a flat playback amplifier; see Chapter 6 for tests with a standardized reference tape
7. Freq. response (Eq.)	With specified record and play equalization, note the playback voltage level dB for constant input voltages over the operating spectrum
8. AC noise (Background Noise)	For a bulk erased tape, or an erased and biased tape, note the background noise voltage dB level
9. DC noise	For a tape after saturation with a dc field in recording direction, note the background noise dB
10. Modulation noise	Noise spectrum when an ac signal is present compared to noise with no signal applied. Measure the increased noise with a spectrum analyzer
11. Uniformity	Record a constant amplitude signal of long, medium, or short wavelength. Note fluctuations of playback output
12. Erasability	Vary the erase current in a specified erase head and note decrease in level of a saturated recording. An 80 kA/m (1000 Oe) field should give over 60 dB erasing of ordinary γFe_2O_3 tape coatings. Erasability depends greatly on bias current during recording, wavelength and erase head construction
13. Reversibility	Turn the tape end for end and note if there is a change in record-play level, especially at high frequencies
14. Print through	Record a short burst of signal on a reel of blank tape. Note spurious reproduction from adjacent turns for different signal levels, storage time and temperature. Use a specified procedure according to recommended standards

†Detailed procedure for such tests are found in ANSI, SMPTE, and other standards including IEC Publication 94, Parts 4 and 5.

*For specified conditions of tape speed, track width, head, etc.

Table 3-12. Performance Tests of Magnetic Record Media (Digital and Computer Applications).

Property	Test*
1. Recording current requirement	Minimum recording current that produces full amplitude pulses
2. Pulse output	Pulse output voltage for optimum recording current
3. Pulse packing density	Maximum (useful) bits or pulses per mm or per inch of tape
4. Pulse crowding	Error in time interval between a pair of adjacent pulses at high density of recording
5. Dropouts	Count playback pulses that are below 50% of amplitude of normal pulses
6. Erasability	Residual signal vs erase head current; erase current that reduces signal below a specified value
7. Noise level (drop-ins)	Pulse noise level; pulse noise 20 dB below useful signal pulses are called drop-ins

*For specified conditions of tape speed, track width, head, etc.

loop audio masters of high-speed duplicators, and in digital computers where reels of tape are in constant use, with information repeatedly erased and updated, and in data recorders where high tape tension is used. Tape wear is of greatest concern in video recorders where high-speed rotating heads require a tough tape coating, especially during *still-framing*, when heads scan the same tape area 60 times a second (216,000 times per hour).

Tape life is influenced greatly by friction at the head, by handling in the tape transport, and by stresses while wound on the reels. These depend, in turn, on tape surface lubrication, head surface material, aerodynamics at the head-tape interface, bending around tape guides, tape tension, stresses during starting and stopping, evenness of winding on the reel, irregularities that build up local stresses in wound tape, temperature, humidity, and incompatible expansion of tape reel hubs.

USE AND CARE OF MAGNETIC RECORDS

Magnetic records would never wear out if they did not have to make frictional contact with heads and guides, or be subject to flexing, stressing, and exposure to dust during winding and reeling. This ideal is approached in disc or drum recorders where out-of-contact heads are spaced a fraction of a micron from the magnetic surface in a clean atmosphere. Even with contact recording, 50,000,000 plays have been achieved without wearing out a plated drum record. But usually magnetic records have a finite life depending on use and care.

For use and storage of tape, normal comfortable room temperatures of about

20° C (68° F) and 40% to 50% relative humidity (RH) have been recommended. Even a range from 5° C to 32° C (40° F to 90° F) with RH from 20% to 80% have been mentioned as being satisfactory. Later studies have shown that polyurethane binders, most widely used for tapes, undergo hydrolysis rapidly (four to eight weeks) at RH approaching 100%, becoming sticky with high friction against heads and with interlayer adhesion (blocking) in wound reels. In mild cases, the binder is restorable by treating it at RH of 0 to 11%, since the hydrolysis is reversible. In extreme cases, the binder releases amines that attack the base of polyethylene teraphthalate, rendering it weak and brittle. All of these reactions are aggravated at high temperatures (75° C). The base material itself is otherwise practically immune to such environmental conditions. Conclusions from these studies are that for archival storage, tapes with polyurethane binders should be kept at 24% or lower relative humidity and at room temperature of 20° C or preferably lower to slow down any chemical changes. (Cuddihy 1980).

Winding and Storage

Some recommendations for tape users are as follows (Ampex Bulletin 1976; Minnesota Mining 1970b;):

1. Be sure that winding and rewinding tensions are correct. Manufacturers' recommendations vary with base thickness, and so on, but are in the order of 450 gm/in. of tape width for thick-base tapes. Storage at excessive tension imparts waviness. Loose winding causes cinching where loose turns are pulled together by sudden tension changes during starting and stopping. Loose windings will also telescope or fall away laterally from turn to turn. Unwinding tension is apt to be jerky.
2. Wind tapes under clean environmental conditions (no food or smoking in the same room); avoid touching their surfaces; handle reels at the hub and not the flanges. Heads, pinch rolls, capstans, and guides should be cleaned regularly, since abrasion of tape against built-up tape deposits is particularly harmful.
3. Heads should be properly demagnetized. Running the tape over a magnetized head during playback increases the noise level and partially erases short wavelengths.
4. Store tapes with reels vertical at room temperature of 5° C to 32° C (40° F to 90° F) and relative humidity of 20% to 60%. For long-time storage, a 20° C or lower temperature and 40% RH are recommended.
5. Avoid reels with hubs that are small, flimsy, or with threading slots or similar irregularities.
6. Some operators will store a tape without rewinding it after recording or

playback, because most machines wind the tape more evenly in the forward direction and because interlayer transfer is reduced if the tape is rewound immediately before playback. There are two minor difficulties: (1) labeling the takeup spool with the name of the program instead of the supply spool; and (2) not being able to play the record immediately, for you must rewind it.

7. If stored for a very long time, the tapes may be wound back and forth every few years to ensure that the turns do not set together in a particular way, and to reduce blocking (adhesion between layers), and print-through (Geller 1983).

MAGNETIC MEASURING INSTRUMENTS

Magnetic properties of tapes are measured most conveniently with a *hysteresis loop tracer*, while a *vibrating sample magnetometer* is often preferred for powders. Either instrument is capable of measuring both tapes and powders. The hysteresis loop tracer applies an AC field (as 60 Hz) to the sample, using it like the core of a transformer. The vibrating sample magnetometer applies a DC field and relies on movement of the magnetized sample to induce voltages in nearby pickup coils. Fields up to 400 kA/m (5000 Oe) are obtained with air core solenoids and up to 1750 kA/m (22,000 Oe) with ferromagnetic cores.

Hysteresis Loop Tracer

The instrument shown in Figure 3-23 displays the hysteresis curve on an oscilloscope together with calibration marks from which significant measurements are made of:

- H_m: maximum applied field
- B_m: maximum flux density (magnetic induction) or B_{mi} (intrinsic induction)
- B_r: retentivity
- H_c: coercivity or H_{ci} (intrinsic coercivity)
- B_r/B_m: squareness ratio
- μ: permeability

Alternatively, the hysteresis loop is traced on an *X-Y* plotter with microprocessor printout of the above readings. Additional data of interest to the experimenter are the B_r vs H curve and the H_{ci} vs H curve, which can readily be plotted from hysteresis loops taken at a series of applied H values.

The practical version of a loop tracer for magnetic recording tapes shown in Figure 3-23 departs in several ways from the basic unit described in Figure 2-4, mainly in the coils and integrating circuit. The recording tape loop tracer

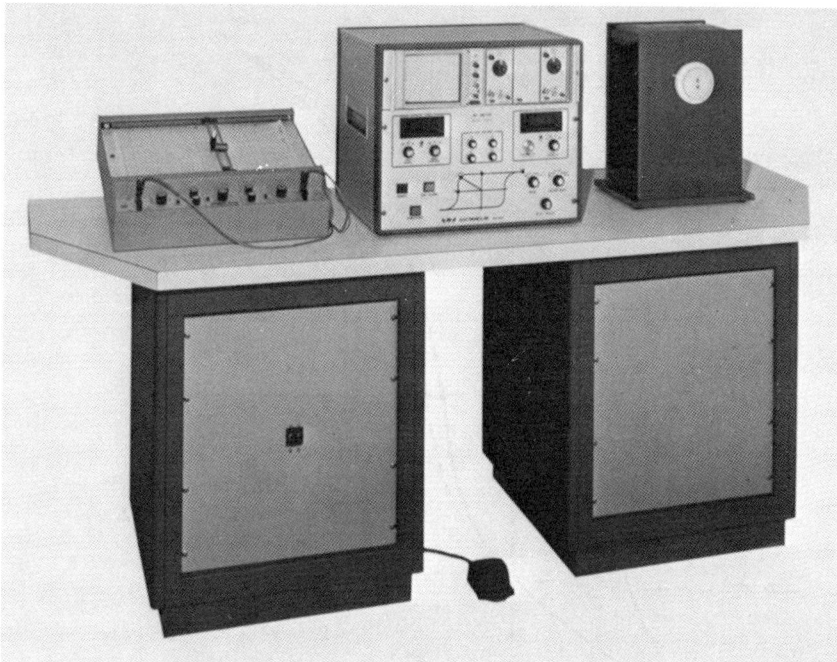

Figure 3-23. Hysteresis loop tracer. On the right side is a solenoid that receives the sample. In the center is a control unit with oscilloscope display. At the left is an X-Y plotter. Power supplies are under the table. (Courtesy LDJ Electronics, Inc.)

uses a long solenoid instead of a toroidal winding. Thus the sample can be a single piece of tape about 4 inches long cut from a reel of tape and inserted into the solenoid; or several pieces can be used in parallel for greater sensitivity if the magnetic coating is very thin. Inside the primary solenoid is a secondary winding with many turns of fine wire surrounding the tape sample.

Also necessary is a *bucking* coil in the secondary circuit to cancel the effect of air flux linking the secondary winding, since this air flux would overwhelm the contribution due to the thin magnetic layer on the tape sample. Depending on adjustment of cancellation, two useful kinds of hysteresis loops can be displayed:

1. An *intrinsic magnetization (B_i vs H) curve* (Figure 3-24 *A*) with flat-topped saturation B_{mi} (or J_m) and intrinsic coercivity H_{ci} where the loop intersects the H axis. Tape specifications are usually given in terms of intrinsic values, even if H_{ci} and B_{mi} are carelessly denoted as H_c and B_m, respectively. An intrinsic curve is obtained if the bucking coil is adjusted for complete cancellation when the magnetic sample is absent. It is really a plot of (B

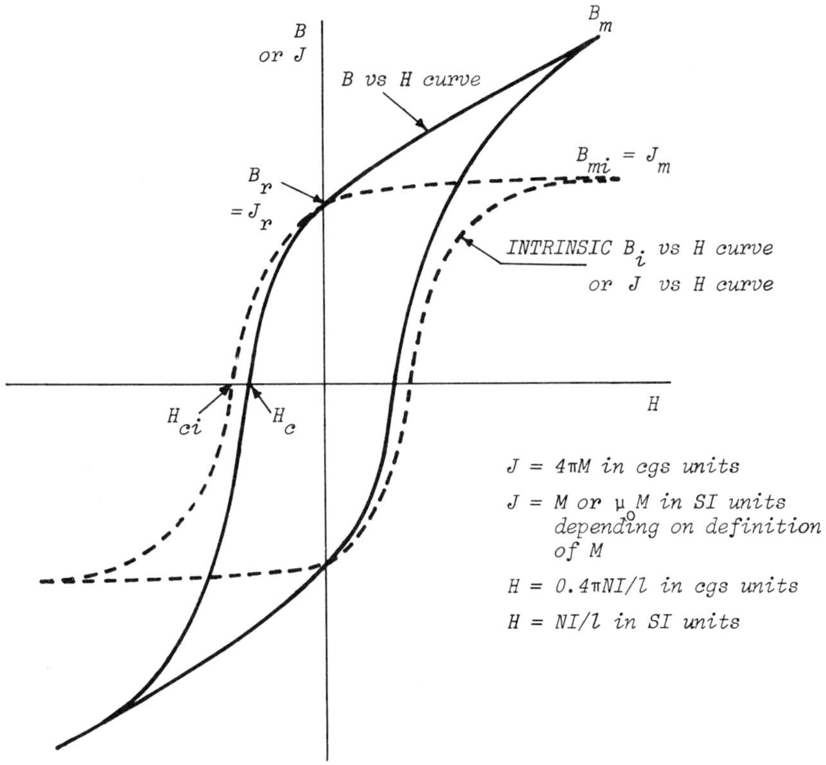

Figure 3-24. Intrinsic and "true" *B* vs. *H* curves.

$- \mu_o H$) vs H. If B_i is multiplied by $1/4\pi$, we have the M vs H curve when cgs units are used.

2. A *true B vs H curve* (Figure 3-24*B*), accounts for the total flux within the sample including the component of H field inside the sample. For a B vs H display the bucking coil action must fall short of complete cancellation, since it must not cancel the $\mu_o H$ flux carried by the cross section of the magnetic specimen. The actual cross section of the sample enters in any quantitative measurements, the calibration during setup taking the sample area into account. Figure 3-25 shows the coil arrangement and electronic circuits of a classical hysteresis loop tracer.

Vibrating Sample Magnetometer

Very high magnetic fields are obtained with modest power requirements by using an iron core structure that concentrates the field in an air gap, as shown

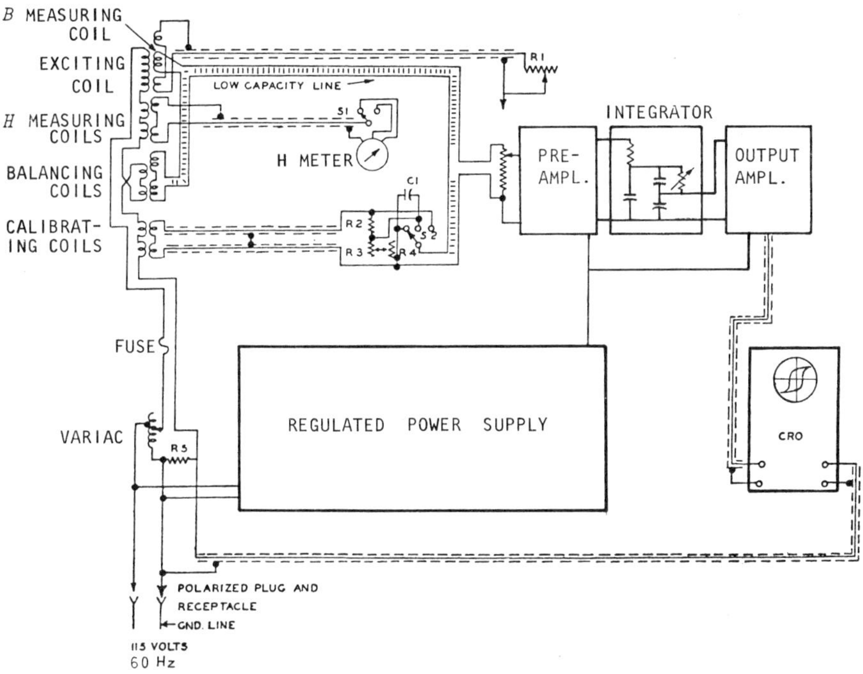

Figure 3-25. Coil arrangement and electronics of *B-H* tracer. (After Weigand & Hansen, AIEE Convention Paper 47-18, January 1948).

for the vibrating sample magnetometer in Figure 3.26. The sample is attached to a rod vibrated by a *voice coil* or by a motor driven eccentric. A reference signal is generated by a separate magnet at the top of the rod to standardize the amplitude of vibration. A phase-sensitive detector locked to the driving frequency improves the signal-to-noise ratio so that very small samples can be used.

The basic output voltage of a vibrating sample magnetometer is proportional to magnetic moment, from which the magnetic moment per unit volume or per unit mass or B, H, μ, are calculated. Samples are compared on a relative basis if they are of equal size. For this purpose a calibrated standard of magnetic moment is provided, as for example a small standardized sphere of pure nickel.

The pickup coils of a vibrating sample magnetometer must give an output that is sensitive to slight movement of the sample in its direction of vibration. This requirement leads to an elaborate and tricky arrangement of multiple coils balanced in opposition, with the sample between the coils at a saddle point in their "field." Two to eight pickup coils have been used. Relative to these coils, the sample must be positioned and aligned with great accuracy. The field func-

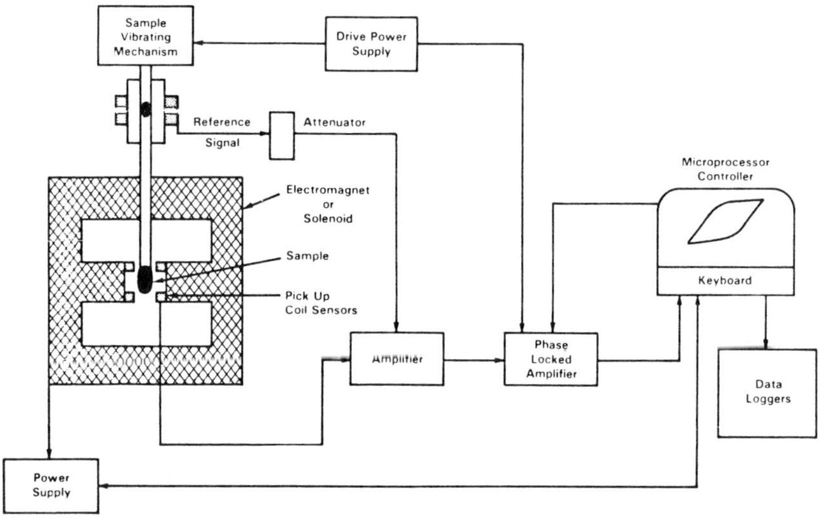

Figure 3-26. Vibrating sample magnetometer. (Courtesy LDJ Electronics, Inc.)

tion of these coils is so complex that it is impractical to calculate the output voltage. Therefore the calibration is made in terms of a standard sample of known magnetic moment, or with a small current-carrying coil of known dimensions.

To compensate for amplitude variations, it is advantageous to use the null method of Figure 3-26, where a reference voltage generated by a vibrating magnet at the top of the rod is attenuated enough to balance out the voltage from the test sample coils. The attenuator ratio that gives this null point is then proportional to the magnetic moment of the sample.

Comparison of Measuring Methods

For routine tests of magnetic tape samples, the hysteresis loop tracer is simplest to use and to calibrate. For research work, under special conditions a vibrating sample magnetometer may be required, although it is more difficult to adjust.

Because of its high sensitivity, the vibrating sample magnetometer is useful for very small or very dilute samples. The samples can be enclosed in a sealed jacket and subjected to extreme temperatures while readings are taken. Point-by-point characteristics are obtained in weak or strong magnetic fields, which can even exceed saturation of ferromagnetic polepieces if superconducting magnets are used.

4. Magnetic Heads

Recording, playback, and erase heads on magnetic recorders are called *transducer heads* because they convert energy from one form to another. Magnetic recording or erasing heads convert electrical energy into magnetic energy that reacts with the tape or other recording medium. Magnetic playback heads apparently convert magnetic energy of the recording into electrical energy at the output terminals. (More correctly, it is not the magnetic energy that is converted, but rather the mechanical energy that moves the tape magnets past the head).

Heads of different kinds will be described in Chapters 4 and 5. However of all transducer heads that have been used for recording, reproduction, or erasing of magnetic coated tapes, the closed-circuit (ring) head shown in Figure 4-1 and Figure 4-2 is the most important. It is efficient, easy to build and to use, and relatively inexpensive. Its universal use has encouraged extensive theoretical and experimental investigation of its properties. The head action is conveniently analyzed into (1) the external magnetic field at the gap(s) that interacts with the tape (giving rise to wavelength effects), and (2) all other parts of the magnetic circuit, including the electrical winding or other electromagnetic sensor (giving rise to frequency effects).

RING HEADS: THE EXTERNAL FIELD

The external gap field of heads such as the one shown in Figure 4-2 has been determined theoretically by conformal mapping and has been verified experimentally by probing (Eckstein and Gilbert 1947; Westmijze 1953; Karlquist 1954). Figure 4-3 shows a map of the field, from which some very useful relations can be observed. The lines represent direction of the field (H) and are semicircular arcs when far from the gap, becoming somewhat elliptical near the gap, and finally becoming straight lines within the gap. The field is strongest where the lines are concentrated together. Contours of equal field strength (regardless of direction) are given in Figure 4-4. Karlqvist has derived equations for flux density in the tape region (Karlquist 1954). His simplest relations are for a tape of relative permeability (μ_r) approximately unity:

$$B_x = -\frac{1}{\pi} B_o\left(\operatorname{arctg}\frac{N + x}{y} + \operatorname{arctg}\frac{N - x}{y}\right)$$

$$B_y = \frac{1}{2\pi} B_o \log\frac{y^2 + (N + x)^2}{y^2 + (N - x)^2},$$

Figure 4-1. Heads for magnetic recorders. Single or multiple heads have their gap surfaces exposed through a shield can or housing. Sometimes an erase head is built into the same shield can. The recording tape rides on the outer surface while it runs past the gap. (Courtesy Nortronics Corp.)

Where $N = \frac{1}{2}$ gap length, B_o = deep gap field, and x and y are distances from the origin at the gap center on the head surface. For tapes of higher μ_r, Karlqvist gives more complex expressions.

Because the field varies both in magnitude and in direction and because most tapes are isotropic, for analysis it is useful to divide the head field into hori-

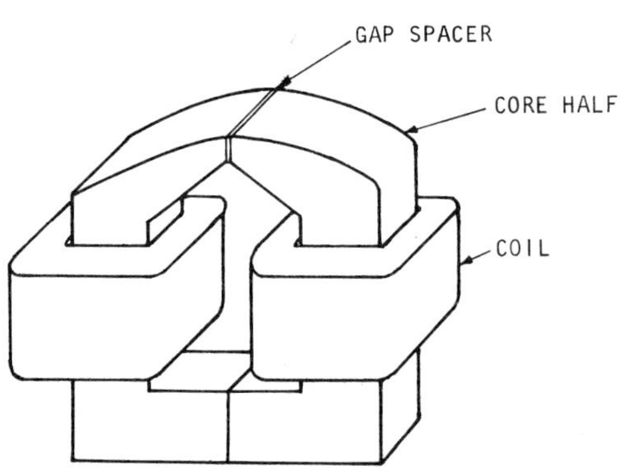

Figure 4-2. Internal construction of a closed circuit (ring) head.

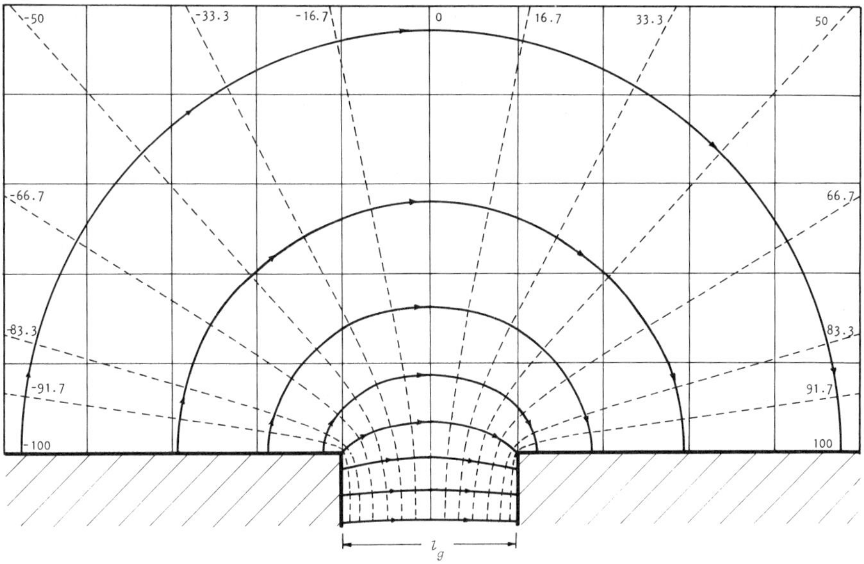

Figure 4-3. Map of magnetic flux density (or field) at a head gap. Flux density deep inside gap is 0.01T (100 gauss). Each band between flux lines carries a total flux of $1000/\pi$ = 318 nanowebers (31.8 maxwells). Equipotential lines are relative to magnetic poles at a potential of +100 and −100.

zontal and vertical components and to note how each component varies along straight line paths parallel to the direction of tape movement, thus simulating the field seen by a very small particle of tape as it approaches and recedes from the gap. An infinitely thin tape coating making perfect head contact would slide along line *A* of Figure 4-4. Thin coatings approximating these conditions have

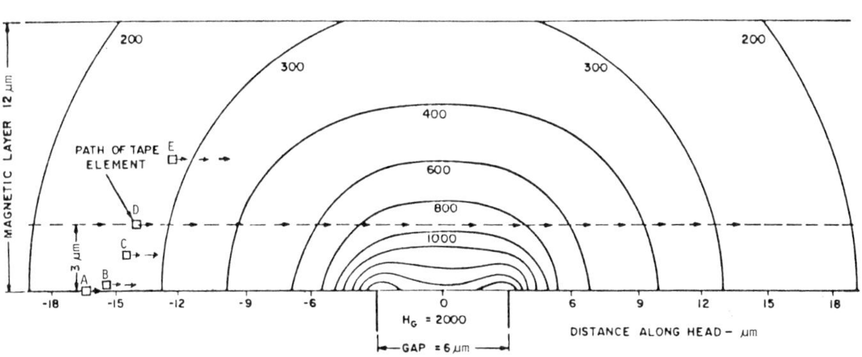

Figure 4-4. Contours of equal field strength at head gap. Element *D* spaced from the gap encounters an increase and decrease of field strength as it passes the gap. Other elements *A*, *B*, *C*, *E* encounter different field sequences as indicated in Figure 4-5.

been used and have certain advantages. However the tape coating is usually as thick as the gap length (ℓ_g) or even thicker, so that some magnetic particles in the tape move along a line *B* spaced from the gap, and some move along lines, *C, D,* or *E* spaced even farther. Figure 4-5*A* is a plot of the relative *horizontal field* component encountered by a particle as it moves past the head along line *A*, while Figures 4-5*B* to 4-5*E* give the *horizontal field* component along lines *B, C, D,* and *E*. Similarly, Figure 4-6 gives the *vertical field* component along lines *A, B, C, D,* or *E*. Figure 4-7 gives the total field at each point along the lines (vector sum of horizontal and vertical components, absolute value, regardless of direction). Directly above the center of the gap the field is purely horizontal. Figure 4-8 shows how this horizontal field decreases as we move vertically away from the gap. These curves are universal, giving the external field intensity relative to a field of 100 inside the gap. Thus if the inside gap field were 500 instead of 100, the values given in Figures 4-3 and 4-5 to 4-8 would be multiplied by five.

RING HEADS: THE MAGNETIC CIRCUIT

Before a head is built, its characteristics can be predicted from magnetic circuit calculations. An existing head can, of course, be measured, but the theoretical calculations can guide a designer toward improvements or toward modifying the head for a specific application.

Recording Heads. In recording heads, it is of interest to know what magnitude

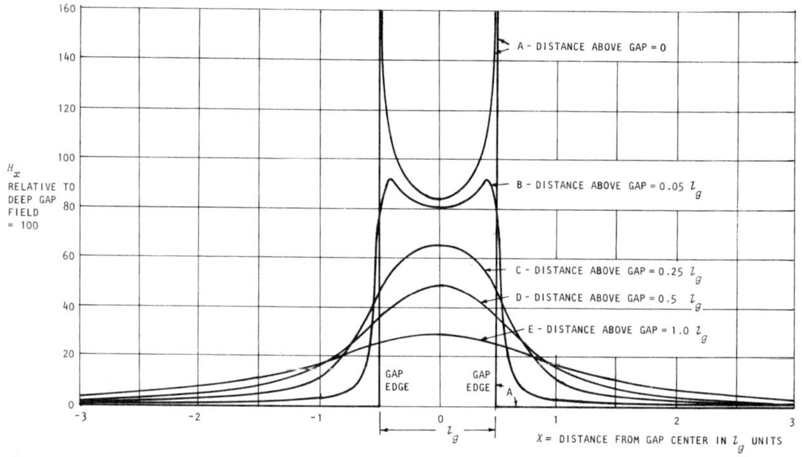

Figure 4-5. Horizontal field component along tape paths spaced from the head surface.

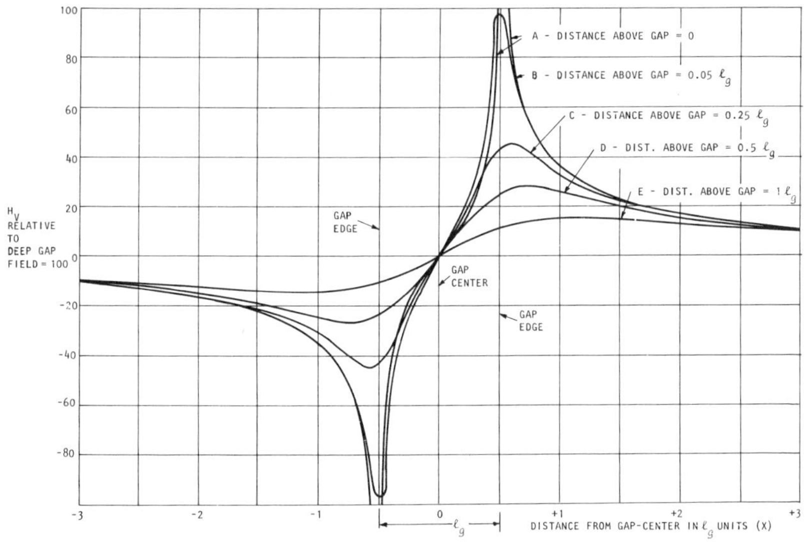

Figure 4-6. Vertical field component along tape paths spaced from the head surface.

of field will be expected at the gap as a function of current in the winding or, conversely, what current is necessary in the head winding to attain the required magnetic field at the head gap. It is also important to know the flux density at which the core operates, since when saturation is approached its recording characteristic becomes nonlinear. Then the core may not be able to supply sufficient

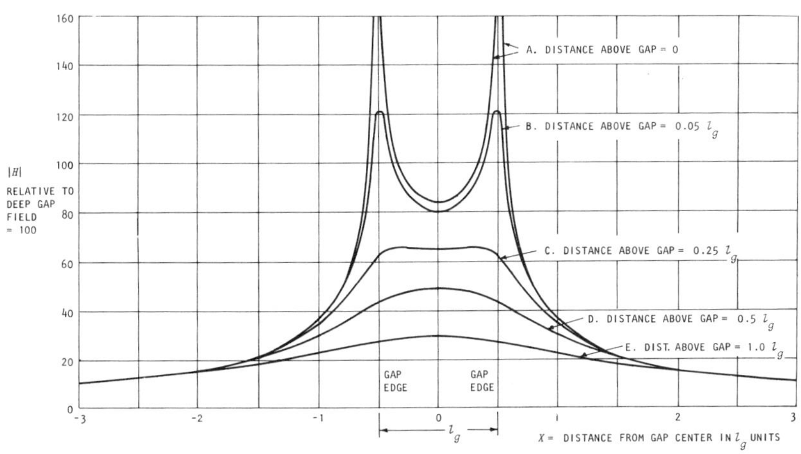

Figure 4-7. Total magnetic field along tape paths spaced from the head surface.

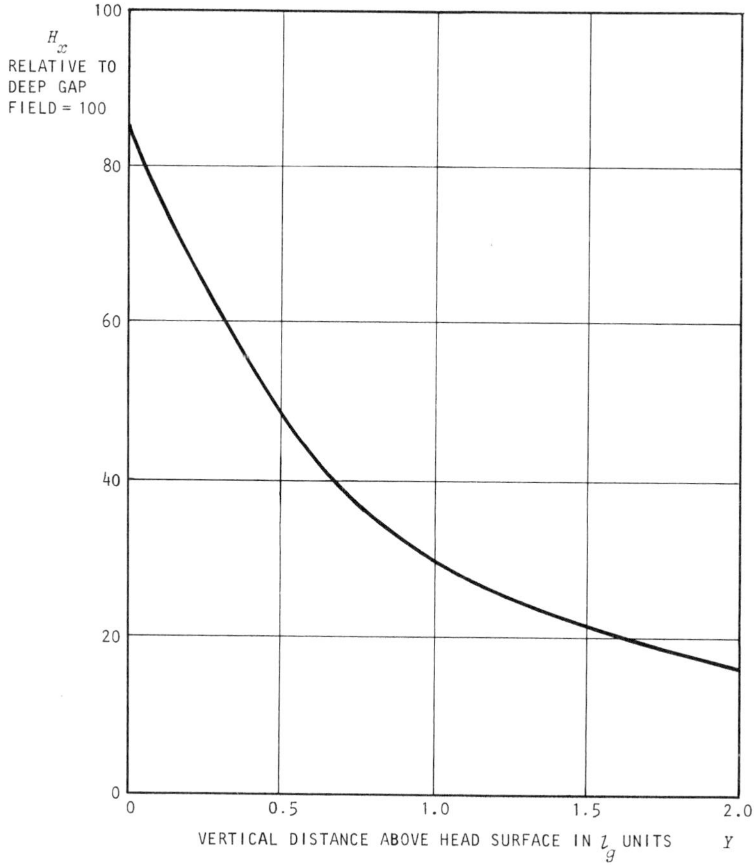

Figure 4-8. Decrease of magnetic field directly above head gap as spacing from head surface increases.

field to magnetize the tape to its full capability, and the head temperature may rise enough to harm the tape coating. The resistance and inductance of the head should be found in order to choose electronic circuits that will energize it properly.

Playback Heads. In playback heads, the most important considerations are the voltage that will be generated at the head terminals and the internal impedance of the head. Both of these are a function of frequency.

Erase Heads. In erase heads, the gap field, the impedance, and the power consumed are of importance. Other factors that determine the effectiveness of erase heads are considered later in this chapter.

MAGNETIC CIRCUIT CALCULATIONS

Magnetic calculations depend on core geometry and permeability. Magnetic circuits are the same for recording, playback, and erase heads, although each kind may have a different gap size. In erasing and recording heads, the source of mmf is the coil, and the utilization point is the gap-tape interface. In playback heads the roles are interchanged, but the efficiency, frequency response and impedance of a given head remains the same because of reciprocity, assuming that the permeability is constant.

Recording Head

Figure 4-9*B* shows a head core of rectangular shape with a tape against its top gap and an energizing winding at the bottom. Figure 4-9*C* is an electric circuit analogy of the magnetic circuit. To solve this circuit, we must find the reluctances (R) of the various flux paths. High accuracy is seldom required, so we are justified in making a few simplifying assumptions.

Simplifying Assumptions. The tape flux pattern is three dimensional, as shown by side fringing in the top view of Figure 4-9*A*, but this fringing may be ignored when the head width (W) is large compared to gap length (ℓ_g). Another simplification, Figure 4-9*D*, is to remove the tape whose permeability is quite low compared to head-core permeability. Removal of the tape decreases the flux density (B) above the gap, but increases the field intensity (H). However the semicircular shape of the field is almost unchanged, although its relative gradient increases somewhat when the tape is absent. The reluctance of the rear gap is usually considered negligible. The core in Figure 4-10*A* carries the main gap flux plus the exterior or stray flux. This exterior flux can be taken into account by substituting an equivalent deeper main gap in Figure 4-10*B* that carries the same total flux as in Figure 4-10*A*.

The simplified magnetic circuit shown in Figure 4-10*B* is like the electric circuit in Figure 4-10*C*, which is readily solved by Ohm's and Kirchoff's laws, using magnetic instead of electrical quantities. Ohm's law becomes $F = \Phi R$, where the magnetomotive force (F) is analogous to voltage, the flux (Φ) is analogous to current, and the reluctance (R) is analogous to resistance. When reluctances are in series, as in Figure 4-10*B*, each has a potential drop ($U = \Phi R$). The sum of these potential drops must be equal to F, the mmf due to ampere turns (NI) of the windings linking the magnetic circuit.

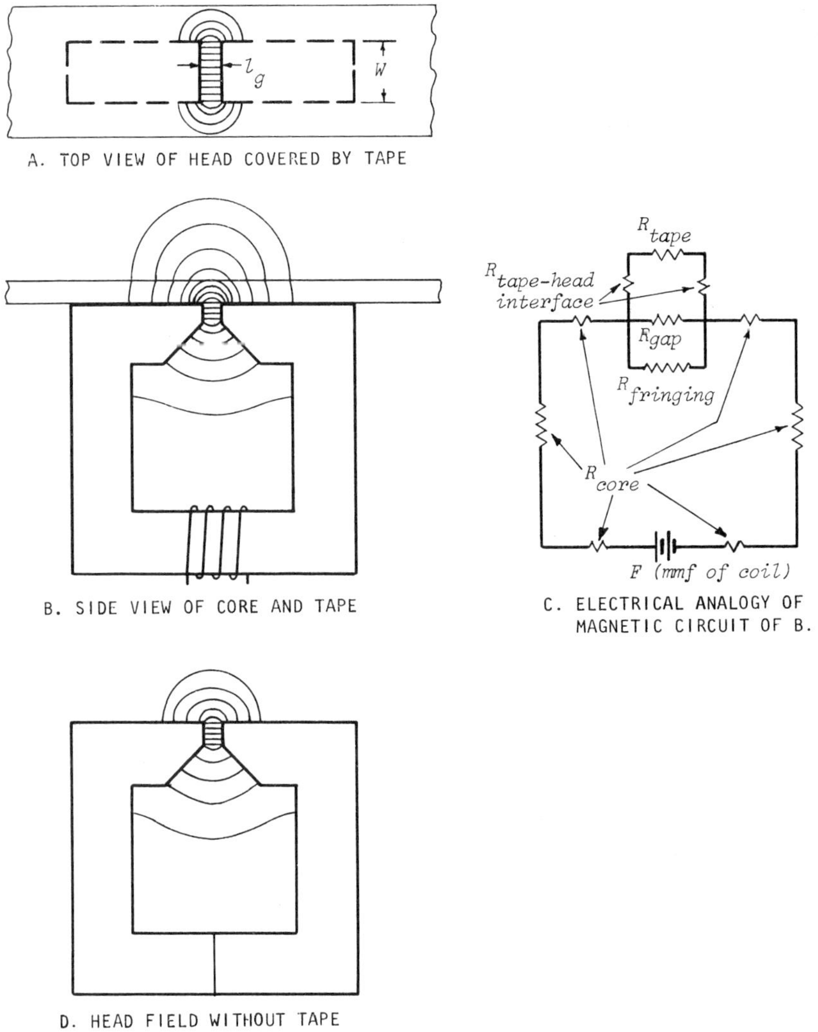

Figure 4-9. Flux pattern outside a head core and electrical analogy. A. Side fringing increases effective head width. B. Presence of tape increases B_g, decreases H_g, and "refracts" the air flux above the tape layer. A similar refraction occurs at the core-tape interface. C. Electrical analogy of magnetic circuit of B, where mmf is represented by emf, and reluctances of core, gap, and tape are represented by resistances. D. External head field when tape is absent.

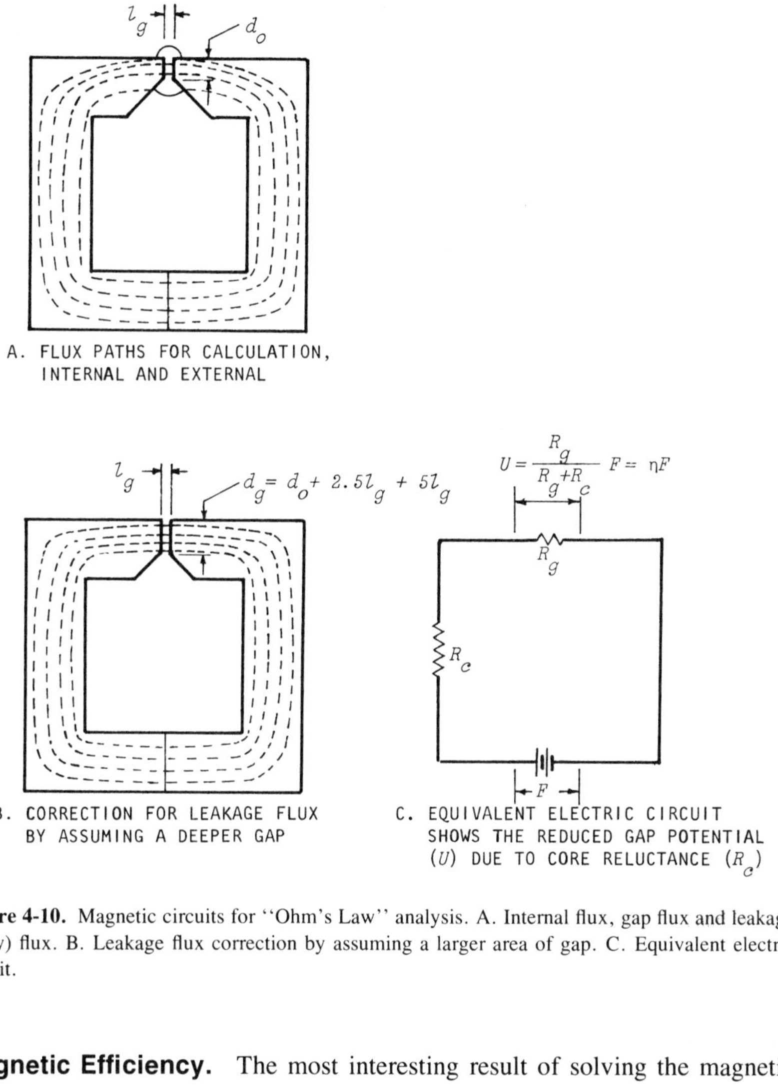

Figure 4-10. Magnetic circuits for "Ohm's Law" analysis. A. Internal flux, gap flux and leakage (stray) flux. B. Leakage flux correction by assuming a larger area of gap. C. Equivalent electric circuit.

Magnetic Efficiency. The most interesting result of solving the magnetic circuit is its efficiency, in other words the proportion of energizing mmf that reaches the recording gap. As seen from Figure 4-10*C*, magnetic efficiency (η) is simply the ratio of gap reluctance to total reluctance:

$$\eta = \frac{U_g}{F_t} = \frac{\Phi R_g}{\Phi R_t} = \frac{R_g}{R_t} = \frac{R_g}{R_g + R_c}. \qquad (4\text{-}1)$$

The deep-gap field is therefore:

$$H_g = \eta \frac{F}{\ell_g} = \frac{R_g}{R_g + R_c} \frac{F}{\ell_g}, \tag{4-2}$$

where ℓ_g is the gap length and $F = NI$ in the SI system, or $F = 0.4\pi NI$ in the cgs system. U_g is the F used by the gap. Equation (4-2) allows one to calculate the deep-gap field H_g if the coil current I is known, or conversely the current necessary to produce a required deep-gap field. For rough approximations, with reasonably efficient heads we can assume $R_c << R_g$. Then H_g becomes:

$$H_g = \frac{F}{\ell_g}. \tag{4-3}$$

Stray Flux. Although the fringing flux is the essential flux component that does the recording, its total amount is all that matters in calculating magnetic circuit efficiency by equation (4-1). The core reluctance is $R_c = \ell_c/\mu A_c$, where ℓ_c is total length of the core and A_c is its area. Similarly, the confronting faces of the gap have a reluctance $R_g = \ell_g/\mu_o A_g$. This gap face reluctance is good enough for most estimates where the gap is short and its area is considerable. Otherwise it could be reduced by the shunting effects of fringing reluctance, which can be calculated by various procedures (McKnight 1970), or from an equation (Unger and Fritzch 1970) (see Figure 4-11*B*):

$$R_f = \frac{\theta}{\mu_o W_c \, ln\,[(\theta L/\ell_g) + 1]} \text{ (in SI units).} \tag{4-4}$$

However the simplest approximation for fringing correction is to use a larger area for the main gap than its physical face area. Figure 4-11 shows typical stray fields at the edge of gaps in magnetic circuits that have different tapers of the confronting polepieces. The reluctance of each complex field can be replaced by an equal reluctance of a uniform field in the same gap by specifying the gap depth, which is shown by the equivalents at the right of each field map. Figure 4-10*B* indicates this correction. Gap width can be corrected similarly. Figure 4-12 shows the correction that is applied when a tape of permeability $\mu_r = 2$ is placed across the gap. Oxide tapes have a μ_r of about 1.5 to 3. In practice, it is tricky to calculate a permeability that simulates that of a moving tape while being recorded.

It should be understood that the equivalents of Figure 4-11 and Figure 4-12

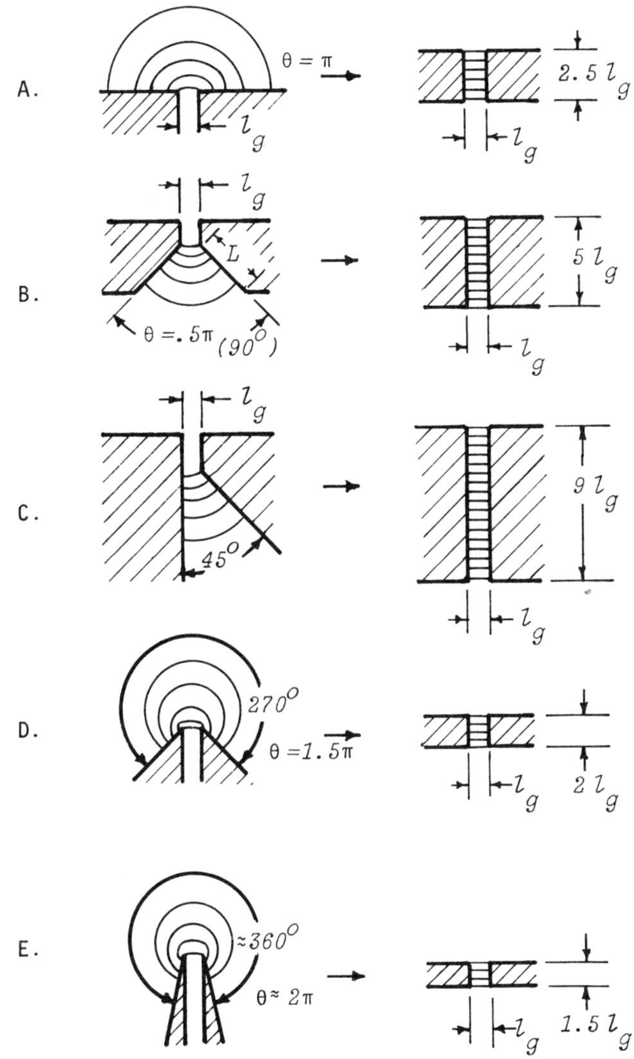

Figure 4-11. Leakage fields and their approximate equivalent gap corrections.

are approximate, depending on the length of taper and proportioning of head dimensions. In specific head designs, they can be made quite accurate by experimental measurements from which we can calculate an equivalent gap area that accounts for the measured flux. Figures 4-13 and 4-14 show a magnetic core and its dimensions that will be used in a sample calculation. The relative

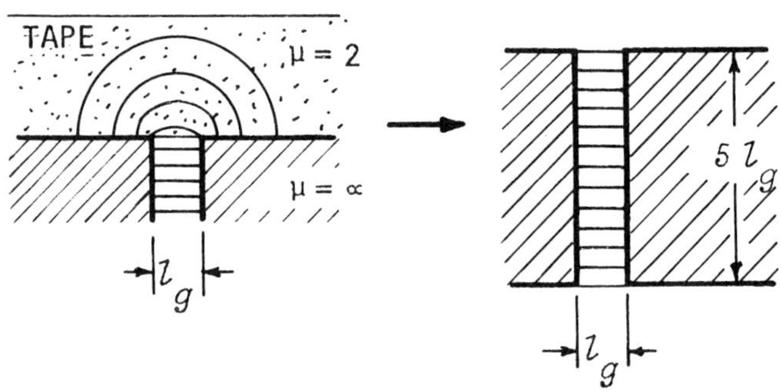

Figure 4-12. Field inside a permeable tape and approximate equivalent gap correction.

permeability of the core (μ_r) is obtained from tables or graphs of the chosen core material, as in Figure 4-15 or Figure 4-16. About ten factors go into the equation, making it look formidable, but it really is quite a simple matter of "plugging in" all of the core dimensions and so on, and working through the arithmetic. One derives great satisfaction when the answer turns out to be a

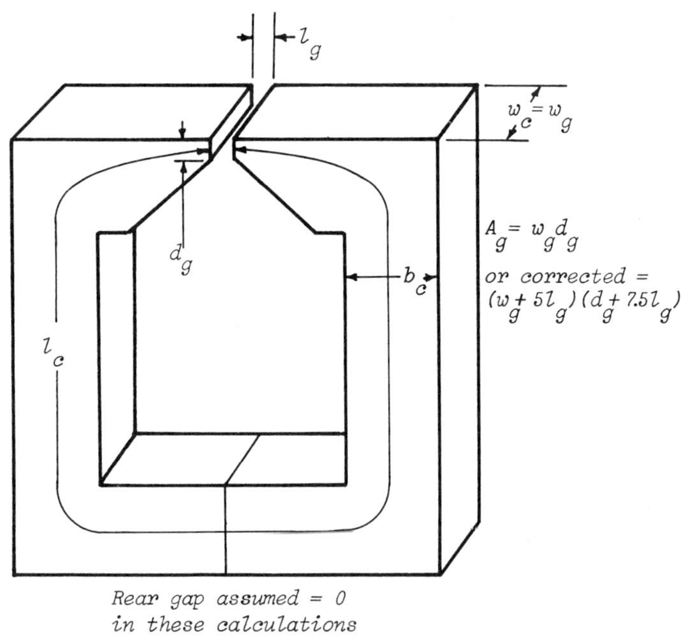

Figure 4-13. Core dimensions for magnetic circuit calculations.

QUANTITY (AND FORMULA)	SI UNITS	CGS UNITS
l_g = gap length	m (meters)	cm (centimeters)
l_c = core length	m	cm
A_c = core area = $w_c b_c$	m^2	cm^2
A_g = gap area = $w_g d_g$ or more accurately $(w_g + 5l_g)(d_g + 7.5l_g)$	m^2	cm^2
μ_r = relative permeability of core (relative to vacuum or air)	(pure number)	(pure number)
μ_o = mag. permeabilty of vacuum	$4\pi \times 10^{-7}$ (H/m)	(1)
μ = absolute permeability = $\mu_r\mu_o$	H/m	(pure number)
N = number of turns on winding	(pure number)	(pure number)
I = Electric current through winding	amperes	amperes
F = magnetomotive force	$A = NI$	$0.4\ \pi\ NI$ (gilberts)
U = magnetic potential = ΦR	A	(gilberts)
Φ = magnetic flux (total) = BA_c	webers (Wb)	maxwells (lines)
R = reluctance = $l/\mu_r\mu_o A$	H^{-1}	—
H = magnetic field intensity (magnetizing force) = U/l	$\frac{A}{m} = \frac{NI}{m}$	$0.4\pi NI/cm$ oersteds (Oe)
B = magnetic flux density = $\mu_c\mu_o H$ = ΦA	tesla (T)	gauss (G)

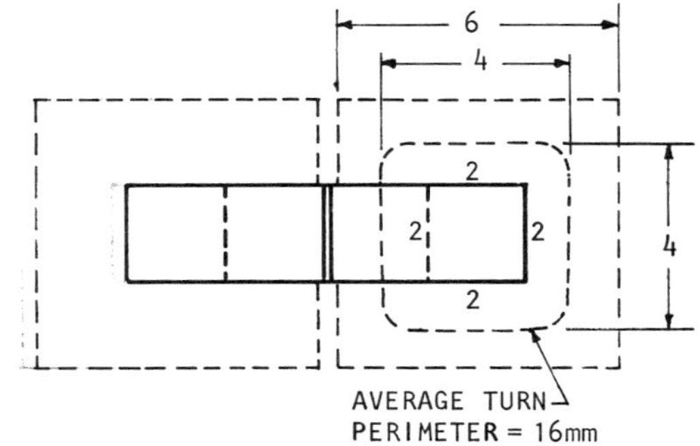

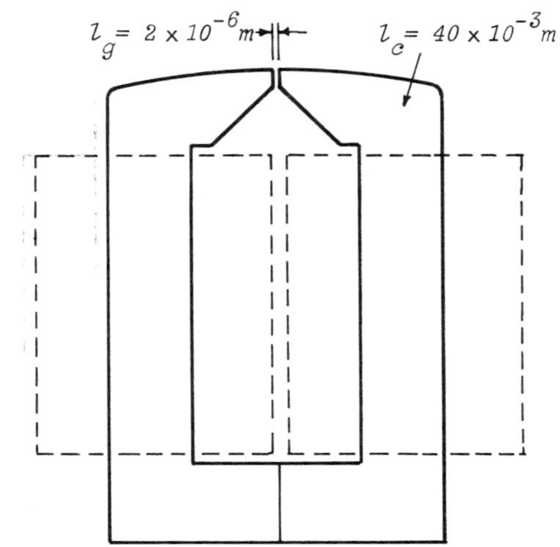

Figure 4-14. Symbols and core size for sample calculations.

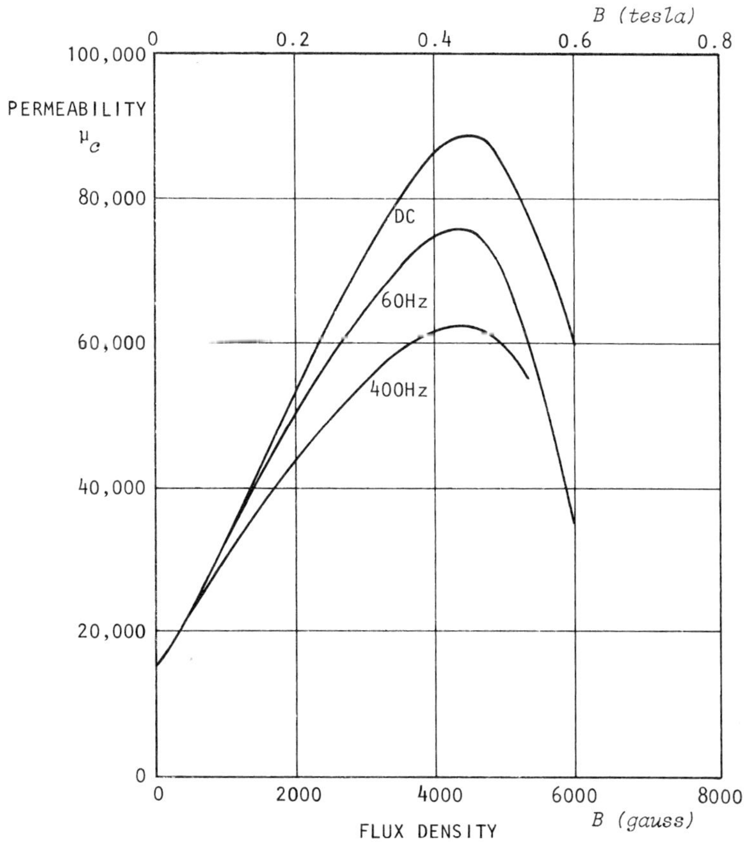

A. 4-79 PERMALLOY CORE MATERIAL. CORE THICKNESS IS 25 μm.

Figure 4-15. Permeability. A. 4-79 Permalloy core material. Core thickness is 25 μm (1 mil). B. Supermalloy core material. Lamination thickness is 100 μm (4 mils). The drastic reduction in permeability at higher frequencies is due to the thickness of laminations, but the permeability still reaches 100,000 at 400 Hz.

reasonable value, or when it checks closely with the measurements. High accuracy is not expected in magnetic circuit calculations because the initial data are uncertain. For example, the permeability specified for the core material is an average figure, while actually the value varies widely from batch to batch and with heat treatment, rolled thickness, frequency, flux density, mechanical stresses during fabrication, and so forth. Gap size and area are difficult to measure, and even the average length of the magnetic circuit (ℓ_c) is an approximation.

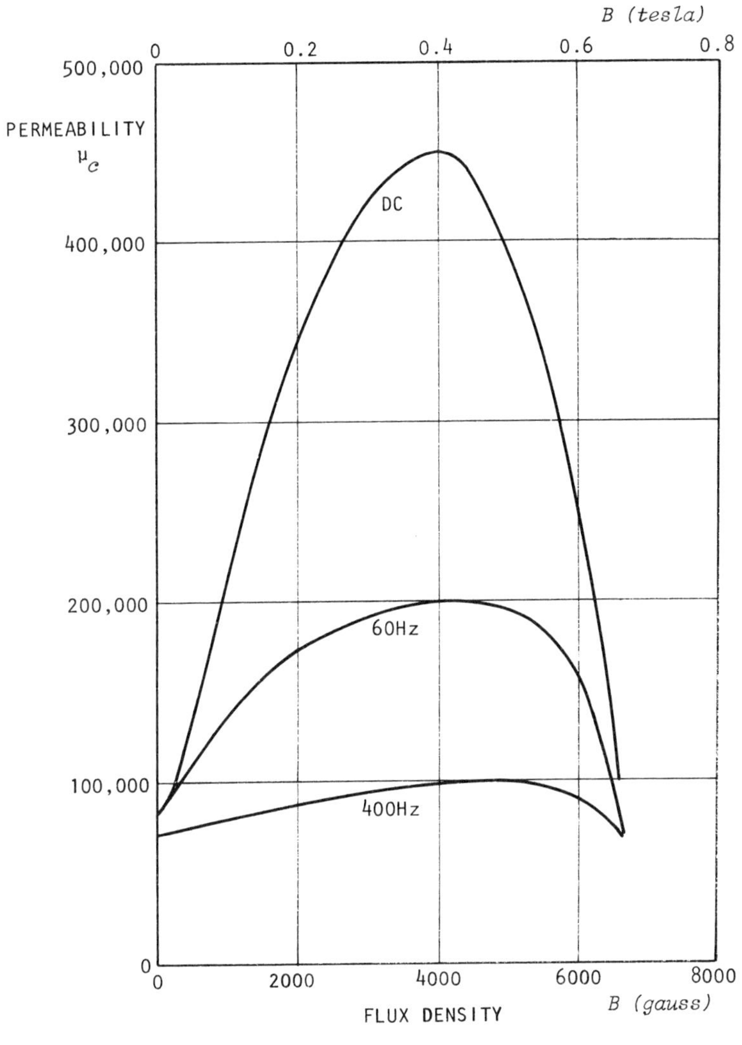

B. SUPERMALLOY CORE MATERIAL. LAMINATION THICKNESS IS 100 μm (4 MILS). *The drastic reduction in permeability at high frequencies is due to the thickness of laminations, but the permeability still reaches 100,000 at 400 Hz.*

Figure 4.15. (*Continued*)

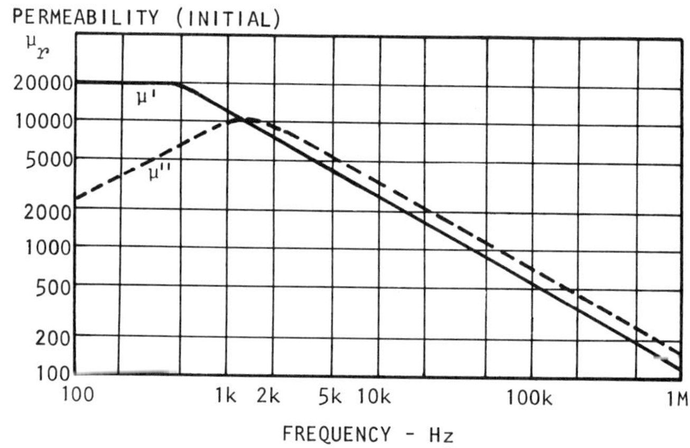

A. *Complex permeability of 79-permalloy 165 μm (6.5 mils) thick. Real (inductive) component =* μ'. *Imaginary (resistive) component =* μ''.

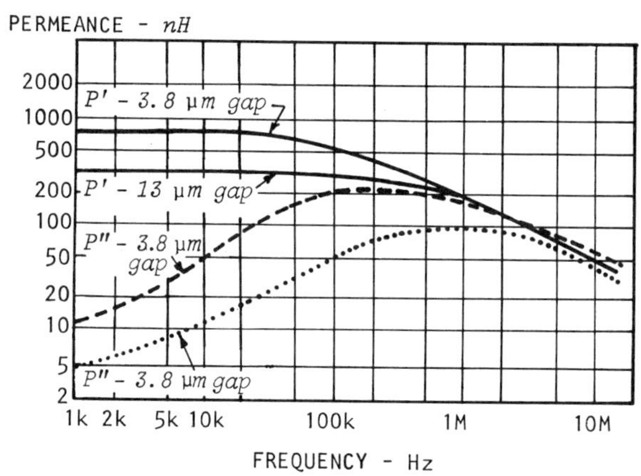

B. *Permeance (P) of head cores with two different lengths of gap spacer.* $P' = \mu' A_c / l_c$ *and* $P'' = \mu'' A_c / l_c$. *A longer gap spacer reduces permeance but increases the turnover frequency of* P''.

Figure 4-16. Complex permeability of core material and permeance of head cores. The general shape of these curves is typical of all core materials. (After McKnight 1977, 1979).

Once the deep-gap field is found, the intensity of all parts of the outside field map are also known, since the field maps are plotted relative to a deep-gap field of 100, and all parts on the maps will rise or fall proportionally to the deep-gap field. It is sufficient, then, that the following calculations are directed toward finding the deep-gap field for a given current in the head winding, or vice versa.

(The dimensions that enter into the equation are expressed in SI units as listed. However, many workers are accustomed to the cgs system, and write specifications and technical papers in cgs units. Therefore the results will also be given in cgs units.) Flux density in the head core:

$$B_c = \frac{\mu_r \mu_o A_g}{A_g \ell_c + \mu_r A_c \ell_g} F \tag{4-5}$$

Flux density in the head gap when pole pieces are tapered toward the gap, as is usual:

$$B_g = B_c \frac{A_c}{A_g} \tag{4-6}$$

Field at the gap when tape is not present:

$$H_g = \frac{B_g}{\mu_o} = 7.96 \times 10^5 B_g \text{ ampere-turns/meter (A/m in SI units)} \tag{4-7}$$

or

$$H_g = B_g \text{ oersteds in CGS units where } \mu_o = 1$$

Ampere turns (SI) or gilberts (cgs) to produce a specified core flux:

$$F = \frac{A_g \ell_c + \mu_r A_c \ell_g}{\mu_r \mu_o A_g} B_c \tag{4-8}$$

Ampere-turns (SI) or gilberts (cgs) to produce a specified gap field:

$$F = \frac{A_g \ell_c + \mu_r A_c \ell_g}{\mu_r A_c} H_g \tag{4-9}$$

Because the above equations do not account fully for all leakage flux, the flux density (B_c) in the core will usually be *higher than calculated*, while the gap field (H_g) will be *lower than calculated*. The equations are easily derived from the relation $F = \Phi R$, Ohm's law for magnetic circuits.

A sample calculation using SI units follows. The head drawn to scale is shown in Figure 4-14. Assume the following typical specifications

$$\mu_r = 50 \times 10^3$$

$$\mu_o = 4\pi \times 10^{-7}$$

$$w_c = b_c = w_g = 2 \times 10^{-3} \text{ meters}$$

$$A_c = w_c b_c = (2 \times 10^{-3})^2 = 4 \times 10^{-6} \text{ sq. meters}$$

$$\ell_g = 2 \times 10^{-6} \text{ meters}$$

$$\ell_c = 40 \times 10^{-3} \text{ meters}$$

$$d_g = 0.2 \times 10^{-3} \text{ meters}$$

$$A_g = w_g d_g = (2 \times 10^{-3})(0.2 \times 10^{-3}) = 0.4 \times 10^{-6}$$

(fringing corrections are neglected)

$$N = 100 \text{ turns}$$

$$I = 10^{-3} \text{ ampere}$$

The core flux given by Equation 4-5 is:

$$B_c = \frac{\mu_r \mu_o A_g}{A_g \ell_c + \mu_r A_c \ell_g} F$$

$$= \frac{(12.56 \times 10^{-7})(50 \times 10^3)(0.4 \times 10^{-6})}{(0.4 \times 10^{-6})(40 \times 10^{-3}) + (50 \times 10^3)(4 \times 10^{-6})(2 \times 10^{-6})} (100)(10^{-3})$$

$$= \frac{2512 \times 10^{-12}}{(16 \times 10^{-9}) + (400 \times 10^{-9})}$$

$$= \frac{2512}{416} \times 10^{-3} = 6.04 \times 10^{-3} \text{ T (tesla) (or 60.4 gauss).}$$

The gap flux is higher because of tapered polepieces. Equation 4-6 gives:

$$B_g = B_c \frac{A_c}{A_g} = 6.04 \times 10^{-3} \frac{4 \times 10^{-6}}{0.4 \times 10^{-6}}$$

$$= 60.4 \times 10^{-3} \text{ T (604 gauss).}$$

The gap field in air, from Equation 4-7, is:

$$H_g = 7.96 \times 10^5 B_g = (7.96 \times 10^5)(60.4 \times 10^{-3})$$
$$= 48.1 \times 10^3 \text{ A/m (604 oersteds).}$$

For a gap field of 79.6×10^3 A/m (1000 Oe) (the right magnitude for gamma oxide tape), the ampere turns required by Equation 4-9 are:

$$NI = F = \frac{A_g\ell_c + \mu_r A_c \ell_g}{\mu_r A_c} H_g$$

$$= \frac{(0.4 \times 10^{-6})(40 \times 10^{-3}) + (50 \times 10^3)(4 \times 10^{-6})(2 \times 10^{-6})}{(50 \times 10^3)(4 \times 10^{-6})} (79.6 \times 10^3)$$

$$= \frac{(16 \times 10^{-9}) + (400 \times 10^{-9})}{200 \times 10^{-3}} (79.6 \times 10^3)$$

$$= \frac{416 \times 10^{-9}}{200 \times 10^{-3}} = 166 \times 10^{-3} \text{ A (ampere-turns).}$$

The 100-turn winding therefore requires:

$$I = \frac{F}{N} = \frac{166 \times 10^{-3}}{10^2} = 1.66 \times 10^{-3} \text{ ampere.}$$

By rewriting the last equation for NI we can account for the different factors that use up the magnetomotive force supplied by the winding.

$$NI = H_g\ell_g + H_g \left[\frac{A_g\ell_c}{A_c\mu_r}\right] \tag{4-10}$$

The first term ($H_g\ell_g$) gives the ampere-turns that do the useful work in the head gap, while the second term gives the ampere turns required (and in a sense wasted) by the core. We notice that both H_g and ℓ_g are requirements imposed by the recording system and are not under control of the head designer. The second term, however, can be made small or negligible compared to the first term by minimizing the term in the brackets, making the core permeability (μ_r) large, and the core area A_c) large; while the core length (ℓ_c is made as small as possible, and the gap area (A_g) is minimized by tapering the core in one or two directions.

Simplified Equations. When the core permeability is very high and core length is short, working equations 4-5 to 4-9 may be simplified by neglecting the core magnetic potential drops. We then have

$$B_c = \frac{\mu_o A_g}{\ell_g A_c} F \tag{4-11}$$

$$B_g = \frac{\mu_o}{\ell_g} F \tag{4-12}$$

$$H_g = \frac{F}{\ell_g} \tag{4-13}$$

$$F = \frac{\ell_g A_c}{\mu_o A_g} B_c \tag{4-14}$$

$$F = \ell_g H_g \tag{4-15}$$

In the practical example worked out above, the simplified equations give answers that are within 5% of the more accurate ones. Experience with a specific head design will indicate whether the simple equations are appropriate.

Erase Head

The recording head equations apply also to erase heads. At high frequencies and high-flux densities usual for erase heads, their permeability is much lower. Their gap length (ℓ_g) is larger to give a high intensity field throughout the tape thickness. With increased gap length the fringing correction in calculating A_g becomes more important. Erase heads often have two gaps energized by a common coil on the center leg, as in Figure 4-17*B*. In such a head, the gaps are considered to be in parallel with each other, and therefore the center leg carries twice the usual flux. The fringing correction should be doubled when one pole-piece is straight instead of tapered, as indicated in Figure 4-11. Considerable power may be dissipated in both the core and the coil of an erase head, so that core losses and winding resistance should be minimized.

Playback Head

Playback heads are constructed almost identically with recording heads (except that they may use a smaller gap to advantage), and in fact most recorders use the same head, switching it from record to playback circuits. When the magnetic circuit of a playback head is analyzed as for the recording head, it is found

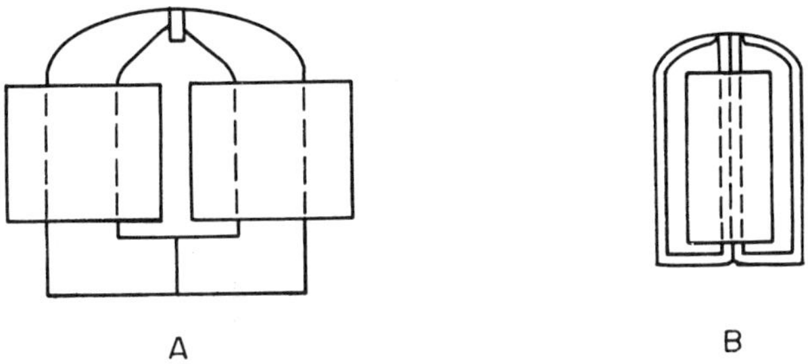

Figure 4-17. Erase heads. A. Single gap construction. B. Double gap construction. Erasing is more complete when gaps are separated farther and magnetic circuits are isolated.

that its "backwards" efficiency is the same; that is, if 33% of the coil mmf appears at the head gap during recording, then 33% of the tape flux at the gap will thread the coil during playback. However, the magnetic efficiency of a playback head is usually of much greater concern than of a recording head because in playback we are dealing with tiny magnets recorded on a tape and want them to generate the maximum possible electrical output at the head terminals to override noise in the sensitive playback amplifiers, while in recording there is usually ample power so we can waste some of it without harm as long as we do not saturate the core.

Operation of Playback Heads. As explained in Figure 2-15, the internal flux through any plane cutting a section through the recording on the tape is equal to the external flux through the same plane at the same section. When the tape is laid against the high permeability surface of a head core, most of the external flux completes its path through the core. As seen in Figure 4-18, the flux from a magnetic element of tape which straddles the head gap follows a magnetic circuit *B* through the core leg linked by an electric coil. A perfect playback head would absorb all of the external flux from the tape element and would channel it all through the coil. In an actual head, there are leakage paths that divert some of the flux and waste it, so that only a fraction of the total goes through the coil, depending on magnetic circuit efficiency.

ELECTRICAL PROPERTIES OF HEADS

When playback, recording or erasing heads are connected to electronic circuits, it is important to know their electrical properties, such as playback voltage, inductance, resistance, capacitance, and anomalous phase shifts. The high-permeability magnetic cores are nonlinear and lossy, causing the electrical prop-

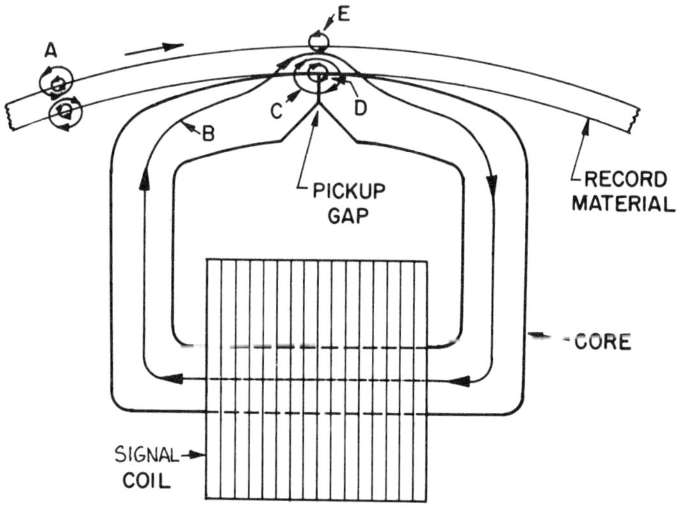

Figure 4-18. Flux paths in a magnetic head on playback.

erties to vary with amplitude and frequency. Heads with windings on such cores are assumed. Playback voltage can be calculated to a fair degree of accuracy in the midrange of head-operating frequencies as is described in the next section, and also the inductance and impedance can be approximated fairly well.

Playback voltage increases directly as the number of turns (N) on the playback winding. Usually we would like as much voltage as possible, but we are limited by head impedance which increases as N^2, much faster than the voltage. The inductive part of impedance reacts with circuit capacitance to spoil the high-frequency response. The resistive part creates noise during playback. In recording and erasing heads, the impedance should be suitable for their energizing circuits. The impedance of a head can be varied over wide limits by changing the number of winding turns.

PLAYBACK HEAD CALCULATIONS

Output Voltage of a Perfect Playback Head. When all the magnetic flux (assumed sinusoidal) of the record reacts with the winding, the voltage generated at the playback head terminals is:

$$E_{rms} = 2\pi f N\ (\Phi/\text{m})\ 10^{-9}\ W \tag{4-16}$$

where

E_{rms} = generated EMF in volts rms
N = No. of turns on winding
f = frequency in Hz
Φ/m = recorded magnetic tape fluxivity in nanowebers per meter (nWb/m, rms value)
W = head width in meters

All factors in this equation are known if the tape manufacturer specifies the recorded flux capabilities of the tape (fluxivity) in *nanowebers per meter*, which means simply that a recorded track *one meter wide* would supply the specified flux in nanowebers, or a recorded track *one millimeter wide*, which is more realistic, would supply 10^{-3} as many nanowebers. Some sources specify the recorded flux in *maxwells per $\frac{1}{4}$-inch tape width*, since $\frac{1}{4}$-inch tape is the most popular size. The conversion is:

$$1570 \text{ nWb/m} = 1 \text{ maxwell per } \tfrac{1}{4} \text{ inch.} \tag{4-17}$$

Note: Tape manufacturers often specify saturation fluxivity (peak) of their tapes. Calibration tapes use rms fluxivity.

Sample Calculation of Voltage from a Perfect Head. Figure 4-19 illustrates the dimensions for a sample calculation of a head that might be found in a compact cassette recorder. When the specified values are substituted into the above equation we obtain:

$$\begin{aligned} E_{rms} &= 6.28\ (10^3)(10^3)(320)(10^{-9})(0.5)(10^{-3}) \\ &= 1 \text{ millivolt} \end{aligned}$$

where

N = 1000 turns
f = 1000 Hz
Φ/m = 320 nW/m
W = 0.5 mm = 0.5×10^{-3} m

On an oscilloscope we would observe a peak-to-peak voltage of about 2.8 millivolts, which is a realistic rough estimate of what to expect from a head playing back a 1000 Hz recording. A general-purpose iron oxide tape retains about 1000 nWb/m peak value at saturation, corresponding to about 700 nWb/m rms. When

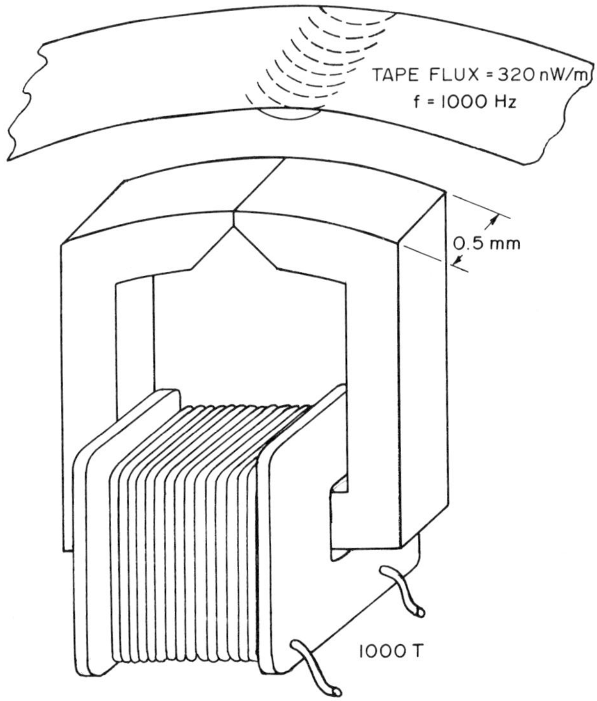

Figure 4-19. Head and tape parameters for playback output calculation.

recording level is monitored with a standard VU meter, the peak levels are about 10 dB above the meter reading due to meter averaging time. Therefore the zero dB reference deflection of the VU meter should be set 10 or 15 dB below peak tape fluxivity, at about 200 nWb/m.

Output Voltage of a Practical Head.—A real or practical head will always generate less voltage than the ideal head, and the actual playback voltage will be:

$$E_{rms} = 2\pi f\,(\Phi/\text{m} \times 10^{-9})\; W\,\eta\,N \qquad (4\text{-}18)$$

where η is an efficiency factor that takes into account the flux lost across the gap, and the other symbols are as listed previously. The factor η also includes other effects that reduce the playback voltage such as eddy currents, skin effect, hysteresis, and so on.

Jay McKnight suggests that ηN in the above equation is easy to measure and can be used as a calibration figure for any commercial head, although N itself and other internal conditions are hidden. This ηN figure is found from equation 4-18 by playing a known calibration tape and measuring the output voltage, frequency, and core width. The ηN figure represents an *equivalent number of turns*.

Playback Efficiency of an Imperfect Head. In Figure 4-18, most of flux B goes through the coil but a good part of C, D is shunted across the lower part of the gap, and some of it misses the head entirely. The shunting at C is an inevitable compromise where efficiency is sacrificed by making the gap length (ℓ_g) very small for short-wave response and the gap depth (d_g) large for long wear life. The component E is usually so small that it is disregarded. In fact the short-wave magnetization in the tape layer diminishes rapidly away from the surface that runs against the head.

Efficiency of the head shown in Figure 4-18 is expressed as a fraction less than unity:

$$\eta = \frac{\text{tape flux that threads the winding}}{\text{total flux available from tape at sensing gap}} \tag{4-19}$$

Since the tape flux divides itself inversely as the reluctance of each path B and C, the efficiency becomes

$$\eta = \frac{\Phi_c}{\Phi_c + \Phi_g} = \frac{R_g}{R_c + R_g} = \frac{1}{\dfrac{R_c}{R_g} + 1}, \tag{4-20}$$

which after substituting the core parameters can be written as:

$$\eta = \frac{1}{\left[\dfrac{A_g \ell_c}{\mu_r A_c \ell_g}\right] + 1} \tag{4-21}$$

When the core permeability (μ_r) is very large, the term in the brackets becomes negligible and the efficiency approaches unity, signifying a perfect playback head. When the term in the brackets is 1, the efficiency is 0.5: half of the flux threads the coil, the other half leaks across the gap. The bracketed term also tells us that efficiency is increased by minimizing A_g or ℓ_c and by maximizing A_c or ℓ_g.

The significant permeability μ_r for playback heads is the initial permeability μ_i, so that supermalloy with an initial permeability of about 70,000 (Figure

4-15*A*) is better than 4-79 permalloy with an initial permeability of 15,000 (Figure 4-15*B*), and both are better than ferrites with an initial permeability below 5000. In practical heads, the high permeabilities of the "super" materials are difficult to achieve, and their performance in heads may not be any better than of the less exotic materials. Some of the reasons are:

1. The very high μ_i of permalloys is achieved only by a critical annealing treatment with extreme cleanliness after all forming operations have been completed.
2. Mechanical stresses in the annealed material reduces the permeability, especially in the highest μ_i materials, and are difficult to avoid in cementing, mounting, grinding, lapping, and polishing operations when fabricating a head. Ultimately the tape running against the head stresses and erodes it to some extent.
3. Permeability falls quite rapidly with increased frequency mainly because of eddy currents, as may be seen in Figure 4-20. Even at high audio frequencies of 10 kHz, the initial permeability of supermalloy 100 μm thick is less than 10% of its 100 Hz value.

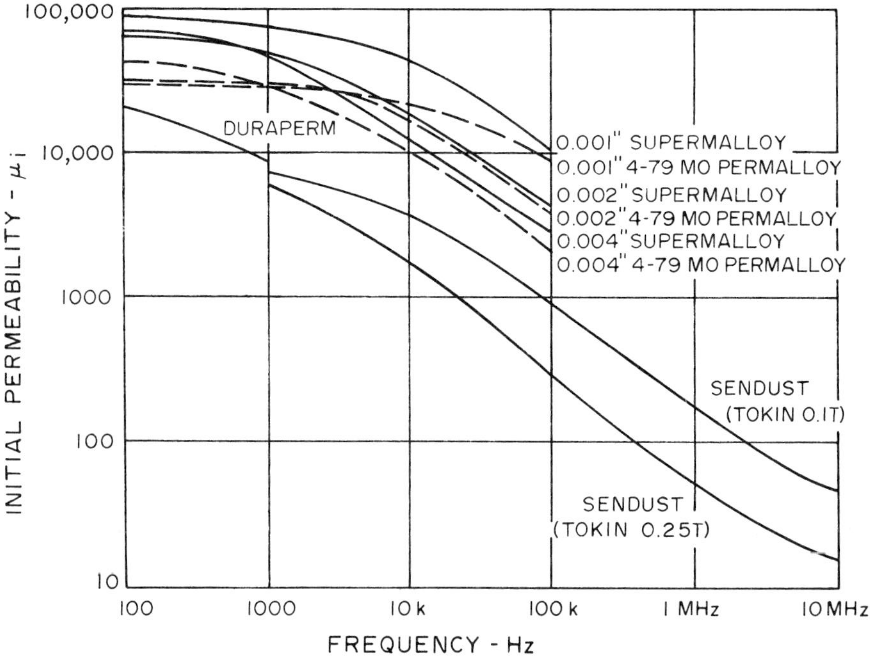

Figure 4-20. Decrease of initial permeability of core materials with frequency.

4. While thin laminations retain their permeability better at high frequencies, their theoretical advantage is cancelled by difficulties in keeping them free of stresses. The very thin gauges as manufactured also seem to be poorer in magnetic quality than thicker gauges because of surface peculiarities or effects of the rolling mills.
5. Additional gaps, such as the back gap in a two-piece head core, can decrease the effective permeability greatly. For example a one μm gap, not uncommon, in a core of $\mu_c = 50{,}000$, adds as much reluctance as a 50,000 μm length of core (50 mm or 2 inches).
6. The dimensions of the main head gap usually determines whether an increase in core permeability will give much benefit. Extremely small gaps generally benefit most from high μ_i. Application of the efficiency equation will show whether a higher permeability is justified.
7. Other losses in the head may decrease its magnetic circuit efficiency such as losses due to eddy currents, winding resistance, hysteresis, and electronic circuits that load the head unfavorably. If the other losses are high, then improving the magnetic leakage may hardly be noticeable

Example 1: Using the values of the previous example for the core parameters of Figure 4-14, $\mu_r = 50{,}000$, $A_c = 4 \times 10^{-6}$, $A_g = 4 \times 10^{-7}$, $\ell_c = 4 \times 10^{-2}$, $\ell_g = 2 \times 10^{-6}$, the efficiency equation 4-21 gives:

$$\text{Playback Efficiency} = \frac{1}{\dfrac{(4 \times 10^{-7})(4 \times 10^{-2})}{(50 \times 10^{3})(4 \times 10^{-6})(2 \times 10^{-6})} + 1}$$

$$= \frac{1}{0.04 + 1} = \frac{1}{1.04} = 0.96$$

Here the result is so near to perfection that there would be no justification in trying to increase μ_i to 100,000, since 98% efficiency would not be noticed compared to 96% efficiency.

Example 2: Using the not uncommon figures of $\mu_r = 2000$, $A_c = 1 \times 10^{-6}$, $A_g = 3 \times 10^{-7}$, $\ell_c = 1 \times 10^{-2}$, $\ell_g = 1 \times 10^{-6}$, the efficiency becomes:

$$\text{Playback Efficiency} = \frac{1}{\dfrac{(3 \times 10^{-7})(10^{-2})}{(2 \times 10^{3})(10^{-6})(10^{-6})} + 1}$$

$$= \frac{1}{1.5 + 1} = \frac{1}{2.5} = 0.40$$

In this case, bringing μ_r up to 6000 instead of 2000 would increase the efficiency to 67% instead of 40%, a worthwhile improvement.

IMPEDANCE CALCULATIONS

Playback, record, or erase head windings behave as lossy inductances with an impedance

$$Z = R_{eq} + jX_L = R_{eq} + j\omega L \tag{4-22}$$

$$|Z| = \sqrt{R_{eq}^2 + X_L^2} \tag{4-23}$$

where

Z = complex impedance in ohms
$|Z|$ = absolute magnitude of Z (modulus)
R_{eq} = equivalent resistance which includes winding resistance and core losses
$j = \sqrt{-1}$
X_L = inductive reactance in ohms
$\omega = 2\pi f$, where f is frequency
L = inductance in henrys

At lowest frequencies, the impedance is mainly head-winding resistance, but at medium and high frequencies, the impedance is dominated by the inductive reactance $j\omega L$ and by the resistive component reflected into the winding because of core losses.

Core Losses

Core losses in metal core heads are due mainly to eddy currents. But even in ferrite cores of extremely high resistivity, there is an inhibition of rapid domain wall motion at high frequencies that lowers permeability and increases losses. This so-called *spin relaxation frequency* limits the useful high-frequency capabilities of core materials. Typical relaxation frequencies are about 0.1 to 3 mHz, the higher frequencies being associated with lower permeability materials.

One way of accounting for core losses, regardless of their cause, is to treat the core permeability as a complex number, $\mu = \mu' + j\mu''$, where the real part (μ') gives the winding its inductance L, while the imaginary part (μ'') reflects resistance into the winding to represent core losses. Figure 4-16*A* shows how complex permeability varies with frequency for a typical core material (McKnight 1977). The variation with frequency of μ' and μ'' is influenced by the skin-effect where the magnetic flux is carried mainly by the outer layers of the laminations. The equation for skin-depth is:

$$\delta = 0.564 \sqrt{\frac{\rho}{\mu_o \mu_r f}}, \tag{4-24}$$

where δ is the skin-depth, ρ the resistivity (ohm-m^3), f the frequency and μ_r the permeability, all in SI units. ($\delta = 5030 \, (\rho/\mu f)^{1/2}$ in cgs). In permalloy, this depth is already 80 μm at 1000 Hz, which is approximately the transition frequency in 160 μm laminations (two skin-depths).

We now can explain the trends in the permeability and permeance curves of Figure 4-16. The μ' in Figure 4-16*A* is constant at low frequencies, giving a constant head inductance L. Above the transition frequency μ' falls at about 3 dB/octave (10 dB/decade) as predicted by the inverse $\sqrt{f}$ factor in the skin-effect equation. The imaginary permeability μ'' in Figure 4-16*A* rises steeply at low frequencies since eddy current losses are proportional to f^2, but the rise is checked at frequencies above the transition, where the effective lamination thickness is reduced by skin-effect. Finally, at high frequencies μ'' follows μ' in the 3 dB/octave downtrend. (Nevertheless the series impedance and resistance measured at the head winding continue to rise with frequency because of the $2\pi f$ factor in their equations).

A head made with core material of complex permeability has gaps and physical dimensions, which interact with the core permeability to give a complex permeance:

$$P = P' + jP'', \tag{4-25}$$

where $P = \mu_r \mu_o A_c / \ell_c$, which is the reciprocal of reluctance $R = \ell_c / \mu_r \mu_o A_c$. When the permeance of a specific head core is known, as in Figure 4-16*B*, the head impedance is easily calculated for a winding of N turns as explained in the following section. The slope of the permeance curves of Figure 4-16*B* at high frequencies also shows the $1/\sqrt{f}$ drop-off.

Inductance, Equivalent Resistance, and Impedance

In SI units the inductance of the winding linking a magnetic circuit is:

$$L = N^2 P', \tag{4-26}$$

where L is the inductance in henries, N is the number of winding turns, and P' is the real component of permeance of the magnetic circuit. The inductive reactance is $X_L = 2\pi fL$. Therefore,

$$X_L = 2\pi f N^2 P' \tag{4-27}$$

In the same way, the imaginary part P'' gives the reflected resistance:

$$R = 2\pi f N^2 P'' \tag{4-28}$$

We can find the permeances P' and P'' for any head at a given frequency by measuring its L and R with a bridge or impedance meter and substituting X_L and R into equations 4-27 and 4-28. Figure 4-16*B* shows how P' and P'' vary with frequency for a typical head core. The ohmic resistance of the winding itself is added directly to the reflected resistance to give total equivalent resistance:

$$R_{eq} = R_w + R = R_w + 2\pi f N^2 P'' \tag{4-29}$$

Conversely if we have a plot of P' and P'', we can calculate the head reactance and resistance from equations 4-27 and 4-28. The absolute head impedance is then

$$|Z| = \sqrt{R_{eq}^2 + X_L^2} = 2\pi f N^2 \sqrt{(P'')^2 + (P')^2} \tag{4-30}$$

The winding resistance usually has negligible effect except at lowest frequencies.

In designing a head we can calculate its permeance from core dimensions, core material, and permeability, as in Figure 4-16*A*. Thus we can predict impedance, inductance, losses, and efficiency before we build the head.

Inductance of a Head.—The head inductance equation $L = N^2P'$ is simple, but the calculation of $P = P' + jP''$ is not. For a core with a gap we have

$$\frac{1}{P} = \frac{1}{P_c} + \frac{1}{P_g} \tag{4-31}$$

Core permeance $P_c = \mu_o\mu_r A_c/\ell_c$ is a complex number because $\mu_r = \mu_r' + j\mu_r''$. Gap permeance $P_g = \mu_o A_g/\ell_g$ is a real number. Substituting these into equation 4-31 gives a lengthy expression in complex numbers, which must be reduced to the form $P = P' + jP''$. The calculation becomes even worse if we account for fringing, conductive gap spacers (Otto 1972), and degradation of lamination surfaces (Peterson and Wrathall 1977). McKnight (1979) gives a procedure and computer program for this arithmetic.

Most often we need only some reasonable estimates. If the head is efficient and there is no rear gap, as is usually the case, we can omit the P_c term so that $P = P_g = P'$, making the equation very simple again:

$$L = N^2P = N^2\,\mu_o A_g/\ell_g. \tag{4-32}$$

Sample Calculation of Inductance. Using the simplified equation 4-32 and the physical data of the previous examples:

$$L = N^2\,\frac{\mu_o A_g}{\ell_g} = \frac{(4\pi \times 10^{-7})(4 \times 10^{-7})}{2 \times 10^{-6}}\,(10^4)$$

$$= \frac{50.2 \times 10^{-10}}{2 \times 10^{-6}} = 25.1 \times 10^{-4}\text{ henry} = 2.51\text{ millihenry}$$

where

$N = 100$ turns
$\mu_o = 4\pi \times 10^{-7}$
$A_g = 4 \times 10^{-7}\,m^2$
$\ell_g = 2 \times 10^{-6}\,m$

A more accurate calculation gave 2.41 millihenry, but greater accuracy is not justified in this head design. Typical audio heads have an inductance of 2 to 800 mH. Digital head inductance is about .03 to 3 mH. Record-only heads are usually at the lower inductances.

Calculation of Reactance and Approximate Impedance. Considering only the head inductance and neglecting its winding resistance (which is later shown to be justified), the audio midband impedance at 1000 Hz is:

$$Z_{1000} = \omega L = 2\pi(1000)(2.51 \times 10^{-3}) = 15.2\text{ ohms}$$

This figure expectedly is too low because the neglected factors can only increase the impedance.

Sample Calculation of DC Resistance of Winding. In Figure 4-14, the window of the sample core measures approximately 4 × 8 mm for a winding area of 32 mm^2, shared by the two side windings. Since 100 turns are specified, this comes to 100/32 = 3.12 turns per mm^2. From Table 4-4, in the column for 25% winding efficiency, we find that No. 30 AWG wire will fit the space with a rating of 3.84 turns per mm^2. We can now estimate the total length of wire needed by noticing from the top view in Figure 4-14 that an average turn is about 16 mm long. A hundred turns uses 100 × 16 = 1600 mm= 1.6 meters of No. 30 AWG wire. Table 4-4 shows 0.339 ohms per meter, so the total winding resistance is R = 0.339 × 1.6 = 0.541 ohms. Typical audio head windings have a DC resistance of 4 to 800 ohms; digital heads from 0.1 to 20 ohms.

Equivalent Resistance, Impedance, and Q-Factor. If the winding resistance were the only resistive component of the head impedance, then the impedance equation 4-23 would still give Z_{1000} = 15.2 ohms as before, since

$$Z_{1000} = \sqrt{R^2 + X^2} = \sqrt{(0.541)^2 + (15.2)^2} \cong 15.2$$

Actually, at frequencies of 1000 Hz the resistive component at the head terminals is several times as high as the DC resistance of its wire, mainly because of core losses that are reflected into the winding. The resistance due to reflected core losses can be estimated by noticing that for heads with short gaps $P'' \cong P'$ at higher frequencies. The reflected resistance using equation 4-28 then approximates the inductive reactance. The overall impedance magnitude at high frequencies becomes

$$|Z| = \sqrt{2}\, X_L = 2\sqrt{2}\pi f N^2 P = 8.88\, fN^2\ \mu_o A_g/\ell_g. \qquad (4\text{-}33)$$

Skin effect will diminish the R, L, Z approximations of equations 4-32 and 4-33 by a factor of 0.707 per octave ($\sqrt{f_t/f}$) above the turnover frequency f_t.

The true R_{eq} may be found experimentally by measuring the Q factor (X/R) and inductance of the head at currents and voltages of actual operating conditions.

Since

$$Q = \frac{X_L}{R_{eq}},$$

Then

$$R_{eq} = \frac{X_L}{Q} = \frac{2\pi f L}{Q}, \qquad (4\text{-}34)$$

where R_{eq} is the equivalent resistance of the head including the winding resistance and all other losses. For example, a bridge measurement of the head might give a reading of $L = 3 \times 10^{-3}$ Hy and $Q = 12$ at 1000Hz. Then

$$R_{eq} = \frac{6.28(10^3)(3 \times 10^{-3})}{12} = 1.57 \text{ ohms}$$

or roughly three times as much as the winding resistance, and rising rapidly with frequency.

HEAD DESIGN: THE HEAD CORE

The theory and calculations above guide an engineer towards effective head designs for specific applications. A most important component of the head is its core. Practical core design takes into account physical size and shape, in other words its architecture, as well as the magnetic and electric qualities that are invisible but also important.

Core Architecture

Designing a new core shape or a new manufacturing procedure gives great satisfaction to an engineer who while exercising creative facilities imagines that he or she will achieve a breakthrough in frequency response, recording density, output, efficiency, signal-to-noise ratio, long life, low cost or perhaps all of these at the same time. The real world is often cruel to these fantasies; yet it is difficult to design a head that is utterly inoperative, although some designers have come close.

A few practical core shapes are shown in Figure 4-21; more will be described in Chapter 5. Core architecture may in general be directed toward core size or core shape, with operationally valid reasons for particular choices.

Core Size

It appears that a fundamental objective is to make the magnetic circuit as short as possible, which usually signifies a miniaturized core, although not necessarily since the turn-in-gap design shown Figure 4-21*K* allows a very minimal magnetic circuit in an otherwise massive core. After all, the head reacts with regions on the tape of micron size. During recording it is wasteful to energize a large body of ferromagnetic material when all that is used is the tiny field at the small gap. A shorter magnetic path takes fewer ampere turns to excite it and has a lower energy loss and lower remanent magnetization.

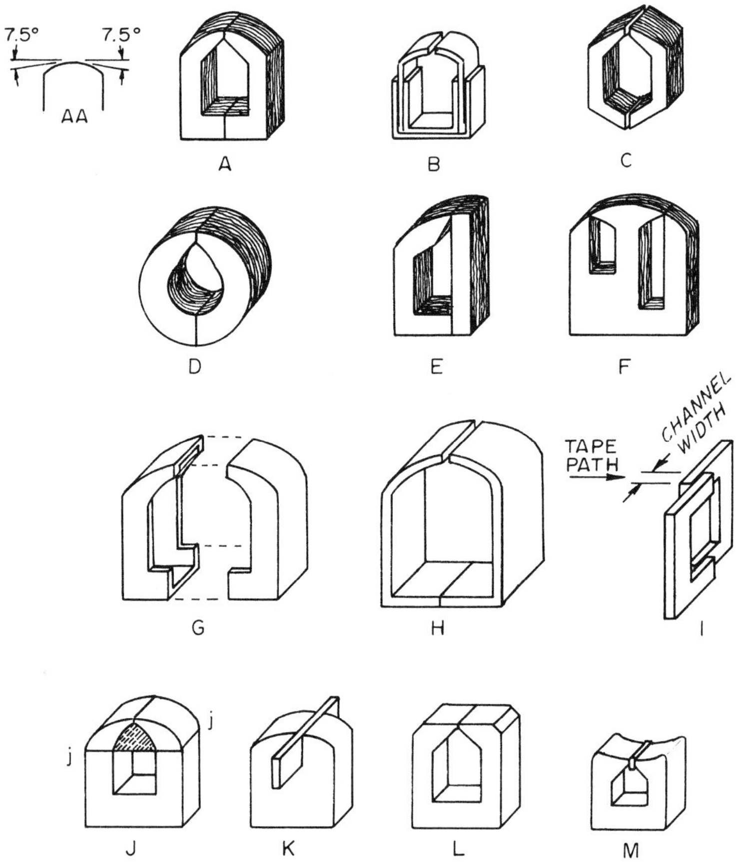

Figure 4-21. Variations in head core designs. A. Most popular core shape (Fig. 4-1 is a modified version) of stacked laminations or solid ferrite. B. Nested polepiece assembly lifts out of its yoke for replacement when worn. Gap assembly is reinforced underneath with a non-magnetic bar. C. Symmetrical gap structure at top and bottom is hum-bucking when equal coils are on each leg. D. Ring head of old German Magnetophones. E. Unsymmetrical C and I cores are compact and use one coil. F. Erase-record-playback combination head for wire recorders. G. Channel core head eliminates laminations and glue lines. H. Bent metal core for inexpensive heads. I. Side-gap head. J. Solid core head of ferrite with replaceable pole tips. K. Turn-in-gap head. L. Flat faced head used with pressure pad. M. Concave face head used against drum.

During playback the flux stored in microscopic areas of the tape is so small that we can well imagine that it could become lost in a large tape core. Yet there are good reasons for larger cores: (1) When a head must respond to long recorded wavelengths, a small core covers only a fraction of the wavelength and intercepts only a portion of the flux. Its low frequency response then drops an extra 6 dB per octave for wavelengths longer than the head dimension, as explained in Chapter 2. (2) At wavelengths about the same size as the head, there is pickup by other parts of the head, which enhances or decreases the main gap pickup, to give wavelength-dependent peaks and valleys in the frequency response. A small head in an audio recorder places these so-called *contour effects* in a very audible region of perhaps 200 to 2000 Hz, where otherwise they would be in a less sensitive part of the spectrum below 200 Hz. (3) A small core has only a small winding window, whose area decreases as the square of the shrinkage in linear dimensions. The winding therefore has a relatively high resistance for a given number of turns. The small assembly then has a lower Q factor than a larger head and might generate more Johnson noise at the input of an amplifier. A tiny coil is more expensive and less reliable than a large one. (4) Reasonable wear for a contacting type head requires a certain thickness of core substance, which is out of proportion in miniature heads. (5) Very small heads are more expensive to manufacture, at least by standard discrete methods, although this might not hold true for deposition techniques.

Core Shape and Construction

Each core shown in Figure 4-21 and elsewhere has a special shape. In every case there is a good reason for that shape. The top of core *A* on which the tape slides is usually rounded, so that tape can feed and depart at an angle where the tape tension forces the tape to press against the gap to insure good contact as in *AA*.

Nose Radius An easy radius at the top of the core, as in Figure 4-21*B, H,* or *J* gives a substantial area of tape contact and minimizes head wear, but irregularities in the tape, such as dimples, do not contact the gap closely and result in dropouts. Increasing tape tension is a partial remedy but decreases head life. The angular nose shown in Figure 4-21*C*, with only a slight rounding at the gap (small radius), gives better tape contact and fewer dropouts. As shown at *AA*, the recommended tape angle is 7.5° on approach and departure.

Flat Top. A core with a straight top, such as the one shown in Figure 4-21*L*, is practical where a pressure pad forces the tape against the head gap. Even here a new head top is usually convex although it wears flat eventually.

Negative Curvature. The top may even have a negative radius, as in Figure 4-21*M*, which fits the contour of a magnetic surfaced drum, or a drum over which the tape rides.

Symmetry and Hum Balancing. The core of Figure 4-21*C* has top and bottom symmetry, which leads to a humbucking construction if identical coils are wound on each side leg and if the bottom gap has the same size spacer as the top working gap. The coils are connected in series, aiding with respect to the internal signal flux that circulates through the core. However, a distant external stray field, as from a motor, gives the same flux in each core half, which induces opposite voltage in each coil, so the effect of the stray field cancels out. In practice, the interfering field is never distant enough to give satisfactory balance without shielding the head. When a shield that is closed at the bottom but open at the top is applied to expose the signal gap, then the balance is disturbed and the symmetry is no longer advantageous over a similar head where the cores are not symmetrical. The bottom taper and the rear gap reduce the effective permeability of the core; but since they no longer have a beneficial effect on reducing stray pickup, they might as well be changed to the more efficient structure of core *A*. Another suggestion has been to mount a symmetrical head such as the one shown in Figure 4-21*C* so that it can be inverted and the bottom gap put to use when the top gap wears out. This seems like a good idea, but is not economical considering the cost of finishing the rear head surface, the complex mounting, the lower efficiency, and the shielding problem.

Single-Gap Magnetic Circuit. A continuous magnetic circuit as in Figure 4-21*L* or *M* is efficient magnetically because it has no auxiliary gap but has two disadvantages: (1) The coil must be wound by threading the wire through the window turn-by-turn, usually by hand; and (2) it is difficult to make the main gap small, straight, and precise.

One-Piece Core. Some manufacturing procedures, such as glass bonding, yield one-piece cores that are permanently bonded and cannot be opened to insert a wound coil although they have two or more gaps, which incidentally can be very small and precise. These cores are wound as with the single-gap magnetic circuit discussed above. An example is the tiny ferrite rotary-scan head of a video recorder.

Two-Piece Core. Split core halves, such as those shown in Figure 4-21*A, C, D, G, H,* and *J*, allow the gap faces to be lapped and polished very precisely but introduce an auxiliary gap in the magnetic circuit. If coils are to be on the sides of the cores, the windings cannot be manufactured separately, but must be wound over the core itself, usually on a split nylon bobbin for insulation, as shown in Figure 4-24*A, B,* and *C*.

Cores for Prewound Coils. Prewound coils that can be slipped over the core are economical to manufacture. Therefore some core designs (Figure 4-21*B, E,* and *J*) have straight sidelegs that accept coils such as the one shown in Figure

4-24*E*. If a single coil is to be slipped on to the bottom leg only, then the designs *A, D, G, H,* and *I* of Figure 4-21 are suitable.

Replaceable Pole Tips. In the design shown in Figure 4-21*B*, the top polepiece assembly can be replaced when worn out simply by lifting out the top inverted U and sliding a new one into the bottom nest. This head was used in Magnecord and in Webcor tape recorders. A similar idea is shown in Figure 4-21 *J*, where a glass-bonded ferrite top is joined to a U-shaped ferrite yoke along a lapped plane (*jj*). The idea seems attractive technically, but the cost of replacement pole tips plus the system for fastening them accurately may be more than the cost of a complete conventional head. Replaceable tips should be economic for data recorders and computers, where heads are worn out by constant use and are very expensive to replace.

Laminated Cores. Reduction of eddy current losses and skin effects by use of laminated cores is well known in transformer and motor design. This principle is used in Figure 4-21*A, C, D,* and *E* where the core is built up of thin metallic laminations. The laminations are cemented together after they have been heat treated. Lamination thicknesses of 150 μm, 100μm and 50μm (6 mils, 4 mils, and 2 mils) are common, with a range from 25 to 500μm. The thicker gauges are found in audio heads; the thinner ones in video and computer heads. One procedure is to coat the laminations with thermoplastic material in a solution that dries by evaporation. The laminations are then stacked in a fixture and heated until the thermoplastic coating melts and fastens the stack together.

Laminated cores have certain disadvantages: (1) The magnetic core is likely to be stressed in assembly when it is squeezed during cementing, and after assembly when the cement expands or contracts at a different rate from the metal. Permalloys are very sensitive to stresses, losing their high μ_i and gaining a tendency to retain residual magnetism. (2) The layers of cement between laminations, although very thin, contact the recording tape and have an affinity for the tape coating that can shed off and build up on the head, eventually lifting the tape from the head gap. (3) Construction is expensive, especially when very thin materials are used. In spite of these disadvantages, laminated heads have been most popular in high-grade audio recorders and in computers.

Solid Metallic Cores. Solid (nonlaminated) cores of magnetic metal alloys have been useful in magnetic recording since the earliest days, where economy was the main consideration rather than maximum performance, and where a fixed core width was used in large quantities. Such cores are usually made of sheet metal about 350 μm thick (14 mils) in shapes, such as the ones shown in Figure 4-21*B, H,* and *I*, or of flat metal of similar thickness, (*F, I,* and *K*), or of channel shape with thinner 100μm (4 mil) material (*G*)., These nonlaminated cores perform surprisingly well, much better than expected from simple calculations, probably due to skin effects (see page 177).

Ferrite Cores. Magnetic ferrites, developed since 1950, have magnetic properties advantageous for head cores and their resistivity is so high that they do not have to be laminated. Solid cores of ferrite can then be shaped as in Figure 4-21*J, K, L,* or *M*. Ferrites are physically very hard and wear-resistant, resembling ceramics and glasses in these respects, so that ferrite cores wear longer than cores of any other material, although their gaps can fail prematurely by chipping, as explained on page 192.

Composite Cores. Composite cores are built up of two or more different magnetic materials, using each one to advantage for a better overall result than could be achieved with a single core material. Figure 4-22*A* shows a head for video recorders; it has a sendust pole tip bonded to a ferrite body. Figure 4-22*B* is a head of different construction made of similar materials for a wider track on a high-density, wide-band data recorder. (See also Figure 4-30). Sendust is a hard metallic material that retains sharp microgap edges that do not fracture and can carry about three times as much flux as ferrites. Hence sendust pole tips magnetize the tape to higher levels than ferrite during recording. They also seem to give more output on playback, possibly because they do not exhibit the *dead layer* effect of ferrites noted on page 193. The main head body is made of ferrite with ample cross section to carry the required flux, and with far lower loss at high frequencies than is possible with a main body of sendust or other metals.

Core Material Requirements

Head cores are made of so-called magnetically soft materials such as permalloys or ferrites. When choosing a core material, the properties to be considered depend on whether the head is used for playback, recording or erasing of audio,

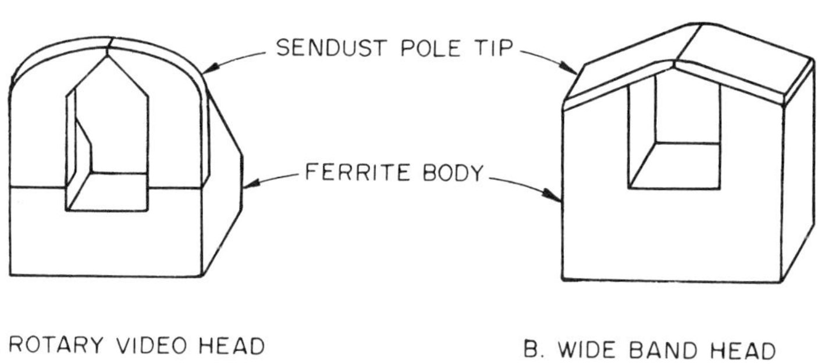

Figure 4-22. Heads with composite cores.

video, instrumentation, or computer data; and whether the head is in contact with the record surface, or spaced from it.

Playback Head Cores. Cores for playback heads should have:

1. *High initial permeability*, since they operate at very low flux densities of 1 millitesla (10 gauss) or less
2. *Low coercivity* so as not to retain any DC magnetization; a magnetized head can permanently degrade a tape by erasing it partially (short wavelengths are most susceptible), and will usually increase the noise level
3. *High electrical resistivity* for lower eddy current losses and better high-frequency response
4. *Thin laminations* in metal cores to reduce eddy currents (but problems arise if excessively thin)
5. *Low magnetostriction* to prevent noises caused by rubbing against the core as the tape moves over the head
6. *Uniformity of grain structure* to sharply define the gap edges, especially with gaps smaller than 100 microinches (2.5 μm)
7. *High wear resistance* for long life
8. *Workability* for economical and uniform production

Recording Head Cores. Recording head cores should have:

1. *High permeability at medium and high flux densities* at which the head operates
2. *High magnetic saturation* density so that the core does not saturate before it can magnetize high coercivity tape to its fullest capability
3. *High electrical resistivity* to reduce eddy currents that cause skin effects, high-frequency loss, and heating of the core, especially at HF bias frequencies
4. *Thin laminations* in metal cores to reduce eddy currents, with benefits as for high resistivity
5. *Uniformity of grain structure, high wear resistance,* and *workability* for the same reasons as in the playback head

When the same head is used for both recording and playback, a compromise is made, usually in favor of playback head characteristics.

Erase Head Cores. Erase head cores should have:

1. *High permeability at high flux densities*
2. *High magnetic saturation density* for maximum erase fields

3. *High electrical resistivity* to reduce losses and heating since erase heads take considerable high-frequency power
4. *Thin laminations in metal cores* to reduce eddy currents, with benefits as for high resistivity
5. *High wear resistance* and *workability* for long life and economy.

Remanent Magnetization. Although low retentivity is of lesser importance in the functioning of recording and erasing heads than in playback heads, it should be remembered that in most recorders the tape passes over all the heads even when they are turned off during playback or rewind. Remanent magnetization in such inoperative cores can harm the recording just as with a magnetized playback head. However the relatively larger gaps of erase heads and recording heads reduce their tendency to retain magnetization. Additional rear gaps are sometimes introduced for further reduction. The possibility of remanent magnetization in an inactive head should not be overlooked in good designs. It is customary to demagnetize heads periodically with the AC field of a head degausser, especially in maintenance of professional recorders.

MAGNETIC MATERIALS FOR HEAD CORES

Materials that meet the above requirements and have been used or proposed for magnetic heads are listed in Tables 4-1 and 4-2. Different sources of a named material may depart from specified values as compositions and treatments vary. A discussion of each kind of material follows.

Permalloy. The favorite material for head cores has been permalloy, such as 79 permalloy or 4-79 permalloy (80% nickel, 16% iron, and 4% molybdenum), or similar alloys with trade names such as Mumetal, Hy-mu 80, and so on, as listed in Table 4-1. The 79 permalloy is easily shaped into head cores by stamping, etching, or bending, and has high initial permeability, good saturation flux density of about 0.8T (8000 gauss), and good corrosion resistance. In its annealed state, it is soft and malleable, and its machinability resembles that of copper. Its wear-resistance is good, but not outstanding. For severe applications other materials surpass permalloy in wear resistance, in high-frequency performance, or in saturation density, but are usually expensive and difficult to fabricate.

Supermalloy. Supermalloy (79% nickel, 16% iron, 5% molybdenum) is a superior grade of permalloy that has approximately the same composition as 4-79 permalloy but is carefully controlled during manufacture. Its initial permability can reach 120,000, which is the highest for any commercially available material. Its coercivity of 0.4 A/m (0.005 Oe) is probably the lowest for any

Table 4-1. Magnetic Materials for Head Cores

Material	Composition	μ_r init	μ_r max	B sat mT (G)	H_c A/m (Oe)	Resistivity $\mu\Omega$-m^3 ($\mu\Omega$-cm^3)	Expansion per °C $\times 10^{-6}$	Specific Gravity	Hardness Rockwell (Vickers)	Curie Temp °C	Remarks
79 Permalloy	80Ni, 16Fe, 4Mo	20 000	100 000	850 (8500)	4 (0.05)	0.58 (58)		8.7	62B (130)	460	Arnold Engineering Carpenter Technology (Hy-Mu 80)
Mumetal	77Ni, 15Fe, 5Cu, 1.5Cr	20 000	100 000	650 (6500)	4 (0.05)	0.62 (62)	13	8.6	35B	750	
Wear resistant Hy-Mu 800	80Ni, 14Fe, 5Mo + 1% Hard Particles	11 000		730 (7300)	24 (0.3)	0.6 (60)	13	8.7		460	(Carpenter Technology) (Nortronics) Duracore®
Hardenable Permalloy	77Ni, 15Fe, +Cu, Mo, Ti, Nb		40 000	500 (5000)		0.9 (90)			(240)		
Supermalloy	79Ni, 16Fe, 5Mo	100 000	500 000	800 (8000)	0.16 (0.002)	0.65 (65)		8.8		400	Arnold Engineering
45 Permalloy	55Fe, 45Ni	2500	25 000	1600 (16000)	24 (0.3)	0.45 (45)		8.2		400	Similar alloys are 4750, Carpenter 49, Hypernik
MnZn Ferrite	$Mn_xZn_{(1-x)}$ Fe_2O_4	3600	5000	500 (5000)	15 (0.2)	10^{10} (10^{12})	12	5	54C (500)	130	
Single-crystal ferrite	19MnO,11ZnO, 70 Fe_2O_3	4000		550 (5500)	4 (0.05)	10^4 (10^6)	10	5.1	(640)	180	
Hot-pressed ferrite	15MnO, 15ZnO, 70 Fe_2O_3	3000	7000	400 (4000)	8 (0.1)	10^5 (10^7)	10	5.1	(640)	130	
Sendust	85Fe, 9.6Si 5.4Al	6500	30 000	1100 (11000)	4 (0.05)	0.9 (90)		8.8	48C (530)	500	Alfesil, Spinalloy, Duraperm Hamilton Prec. Metals
Amorphous alloy	72Co, 3Fe, 14Si 10B, Mn	50 000	600 000	900 (9000)	0.4 (0.005)	1.35 (135)			(900)	340	Low magnetostriction alloy (Matsushita)

Table 4-2. Special-Purpose Magnetic Materials for Head Cores

Material	Composition	μ_r init	μ_r max	B sat mT (G)	H_c A/m (Oe)	Resistivity $\mu\Omega$-m^3 ($\mu\Omega$-cm^3)	Expansion per °C × 10^6	Specific gravity	Hardness Rockwell (Vickers)	Curie Temp °C	Remarks
NiZn Ferrite	$Ni_xZn_{(1-x)}Fe_2O_4$	1500	4000	300 (3000)	12 (0.15)	10^{10} 10^{12}	8	5.3	(600)	150	
16 Alfenol	84Fe, 16Al	5000	100 000	850 (8500)	4 (0.05)	1.5 (150)	15	6.5	29C (280)	450	Alperm
Supermendur	49Co, 49Fe, 2V	800	4500	2400 (24000)	160 (2.0)	0.26 (26)		8.2		980	Vanadium Permendur (Arnold Engineering)
3% Si Iron	97Fe, 3Si	1500	40 000	2000 (20000)	12 (0.15)	0.47 (47)	12	7.7	79B	745	
6.5 Si Iron	93.5Fe, 6.5Si	8000	220 000	1800 (18000)	8 (0.1)	0.8 (80)	12	7.5			(Westinghouse)
Sinimax	54Fe, 43Ni, 3Si	5000	35 000	1100 (11000)	8 (0.1)	0.55 (55)		8.0	76B		(Allegheny Ludlum)
Deltamax	50Fe, 50Ni	1700	250 000	1550 (15500)	24 (0.3)	0.45 (45)		8.2	(500)		
Pure Iron	99.95Fe	10 000	200 000	2150 (21500)	4 (0.05)	0.1 (10)	12	7.9	30B	770	
Amorphous alloy	78Fe, 13B, 9Si	10 000	50 000	1560 (15600)	2.4 (0.03)	1.3 (130)				415	Excellent magnetic properties but high magnetostriction. (Allied Chemical Metglas®)

available material, and its resistivity is about 10% higher than that of 4-79 permalloy.

Wear-Resistant Hy-Mu 800. Wear-resistant Hy-Mu 800 (or Duracore) is a 79-type permalloy containing about 1% hard particles suspended in the alloy to improve wear resistance (Bendson 1976). Workability and hardness are nearly the same as for ordinary Hy-Mu 80, but head life is about ten times longer. Head core permeability is degraded slightly and may be compensated for by using more winding turns.

49 Permalloy. Where higher saturation density of about 1.5T (15000 gauss) is desired, another class of permalloys are available, having a composition of about 50% nickel and 50% iron, and are sold under such trade names as 45 Permalloy, 4750 Alloy, Carpenter 49, Monimax, and Hypernik. The 49 permalloys have much higher saturation induction but do not have as high an initial permeability as the 79 permalloys. However, the magnetic circuit design is usually such that an increase in permeability has negligible effect on the overall performance. This is true when the permeability is already quite high; in other words, if the permeability is 50,000, then raising it to 200,000 may hardly be noticeable in a magnetic circuit containing air gaps. The 49 permalloys contain less nickel and are therefore less expensive than the 79 permalloys.

One should be warned that permalloys are very sensitive to mechanical stresses, to annealing, and to extremely small percentages of impurities. A finished head usually falls short of ideal magnetic specifications because of possible mistreatment at many stages of its manufacture. The greater the permeability, the more sensitive a material is to degradation.

Heat Treatment of Permalloys. Permalloys, regardless of composition, are usually supplied "as rolled" because the half-hardness imparted by cold rolling makes it easier to stamp, machine, or form the permalloy. In this original condition the magnetic properties are quite poor. A special annealing treatment as described here is required to bring out the magnetic qualities of permalloys, and this treatment should be applied only after all mechanical fabricating steps have been completed. Before annealing, the fabricated head parts must be scrupulously cleaned with degreasing solvents, acetone, and distilled water, since the slightest traces of carbon from organic matter may be absorbed by the metal. Heat treatment is done in an atmosphere of dry hydrogen, purified to remove water and especially oxygen, which is harmful to permalloys at elevated temperatures. A suitable temperature schedule is to bring the temperature up to 1088° C (2000° F) at any desired rate, hold at 1088° C for two hours, and cool slowly in the oven for several hours by turning off the oven. The important range for slow cooling is between temperatures of abut 500° C down to about

100° C (the Curie temperature is 400° to 500° C). More rapid cooling may be done outside this range. The hydrogen atmosphere should be retained until the sample has cooled to 100° C. The basic treatment described above is sometimes modified as to time or temperature in order to save time or to give somewhat better properties, depending on the alloy, its thickness, and the heat treating equipment.

Ferrites. After permalloys, ferrites are the most widely used for head cores. With the advent of video recorders, ferrite heads have become popular because they are efficient at high frequencies and because they wear about 20 times as long as permalloys. Even before the video recording era, ferrite heads were used in computers, where they could handle megahertz bit rates better than metal core heads. Ferrite heads are also used in some audio recorders for extra long head life, but, as previously noted, they are vulnerable to gap erosion before they actually wear out.

Magnetic ferrites are modified iron oxides with nonmetallic properties, including ferrimagnetism, and very high electrical resistivity in contrast to ferromagnetic metals. Therefore ferrites are efficient at high frequencies in the megahertz range where ordinary magnetic materials fail because of eddy current losses and skin effects. Various compositions of ferrites are available with a range of permeabilities and effective operating frequencies. Manganese-zinc ferrites (for example; $Mn_{0.7}$ $Zn_{0.3}$ Fe_2O_4) are preferred for most heads because they have high initial permeability of about 5000 at frequencies common in magnetic recording to about 5 MHz. Nickel-zinc ferrites (for example $Ni_{0.3}Zn_{0.7}$ Fe_2O_4) have much lower permeability and higher resistivity, but they can handle ultra-high frequencies.

The high hardness of ferrites requires fabrication by sintering, grinding, and lapping rather than by machining methods. A very serious problem with ferrite heads is that tape abrasion erodes their gap edges by pitting and crystal pullout, causing premature failure by spoiling the short wavelength response. The problem has been mostly solved by glass bonding. Glass is melted into the gap space where it bonds with the ferrite pole faces to give a continuous solid path from pole to pole with no unsupported corners that can break off. Where extremely short gaps are required, as in video recording and high-density digital recording, the sintered ferrites are too porous to define a sharp gap edge. Special hot-pressed ferrites (Sugaya 1968) have been developed with low porosity, but the ultimate material is single-crystal ferrite, which is poreless. Glass-bonded and single-crystal ferrites have the best immunity to gap erosion and are widely used in video recorders and computers.

Ferrite heads are most successful where high surface velocities cause a film of air to separate them slightly from the record. In more ordinary applications, many advertised "long-life ferrite heads" had to be recalled because of gap failures and were replaced with conventional metal heads by embarassed manufacturers.

Because they are ferrimagnetic, ferrites inherently have a lower saturation flux density than ferromagnetic metals by at least a factor of 2. There is more possibility than in other core materials of saturating ferrite poletips or even the main core at high recording levels. Some workers have also reported a so-called *dead layer* at the surface of ferrite heads, which causes these to behave as if a film of air separated the head from the tape by about 0.2 to 0.5 μm, the effect being worse for nickel-zinc ferrite than for maganese-zinc ferrite. The dead layer (really a lower μ layer) is caused by stresses in the ferrite surface due to the grinding and polishing operations and may be removed by etching away the stressed layer.

Despite the success of ferrite heads, their low-saturation density makes their performance somewhat below par even for gamma iron oxide tapes of ordinary magnetic properties. With the advent of chromium dioxide, metal particle, and deposited magnetic tapes having coercivities of about 1000, the ferrite heads have become even more inadequate, although successful designs have been reported (Ketori et al. 1979). The low-saturation density is a fundamental limitation of ferrimagnetism, and there is no prospect that it can be improved.

Sendust. In the search over several decades for long-wearing head materials of high saturation and good high-frequency performance, the best metal alloy found to date has been sendust, an alloy containing 85% iron, 9.6% silicon, and 5.4% aluminum, (magnetic properties given in Table 4-1 and Figure 4-20). Its peculiar name signifies the purpose for its original development by Masumoto in 1936. Sendust was an alloy so brittle and friable that it could be crushed into a fine dust to be used in powder core toroids for telephone repeaters. Its extreme hardness (in uncrushed state) as well as its excellent magnetic properties make is suitable for head cores, but it cannot be machined and must be shaped by special methods. An added complication is that, unlike ferrites, it is not efficient in thick sections, since its resistivity is low, being a metal rather than an insulator. Sendust must therefore be reduced to thin plates, used as overlays in composite heads (Figure 4-22), or shaped into cores for narrow-track heads. A process for making sendust in thin ribbons (Tsukagoshi et al. 1981) is important for solving these fabrication problems. Sendust-type material is also known by trade names such as Alfesil, Duraperm, and Spinalloy. Sendust has a wear life about three to four times as long as 79 permalloy and a saturation density of 0.9 to 1.0 T, about 20% to 50% higher than of permalloys. It has been used in video heads for quadruplex recorders, in wide-band data recorders, and in other high-usage professional recorders where head wear is a major problem and where the expense of sendust heads is justified (Yasuda et al. 1980). In home recorders with gamma oxide tapes, neither wear nor saturation have been problems with permalloy heads in the past. But the introduction of high-coercivity tapes (metal particle or cobalt-adsorbed) have made sendust heads necessary even in higher grades of home recorders.

Alfenol. Alfenol, a magnetic alloy of iron and aluminum, is a substitute for sendust but is not as brittle and can be rolled into sheets (Adams 1962). It is therefore less expensive and more workable than sendust. Its name is composed of its ingredient symbols Al, Fe, and NOL, an acronym for *N*aval *O*rdnance *L*ab (USA) where it was developed. The 16 Alfenol is harder than 12 Alfenol, with the former containing 16% and the latter 12% aluminum. Both are too brittle to be formed by bending, but they can be stamped into laminations. The thin gauges of about 50 to 100μm (2 to 4 mils) necessary for efficient laminated heads apparently have been impractical to produce; and the thicker gauges of about 350 μm have not become popular. Magnetic properties are given in Table 4-2.

Permendur and Supermendur. The highest saturation density that has ever been obtained in any magnetic material is 2.45T (24500 gauss), and the material is a 50% iron 50% cobalt alloy named Permendur because its permeability endures to this very high density. As a head core, permendur has undesirably low resistivity, low permeability, and excessive coercivity. The undesirable characteristics are alleviated by adding vanadium to the alloy. Supermendur contains 49% iron, 49% cobalt, and 2% vanadium, and has the magnetic properties specified in Table 4-2, which include a saturation of 2.3T (about three times as high as 79 permalloy and six times as high as nickel-zinc ferrite), together with a maximum permeability of 4500 and reasonable resistivity and coercivity for a recording or erasing head.

Silicon Iron. Ordinary silicon transformer iron is a good example of what occurs frequently in magnetic recording: Something that should work doesn't. Silicon iron is produced in vast tonnages for transformers, motors, generators, relays—in fact for almost any use where a soft magnetic material is indicated. By analogy, one would expect it to make good heads for tape recording, but experiments with silicon iron cores give a discouraging lack of output, plus a tendency toward remanence that erases what little output there is. The performance is traceable to a dramatic loss of magnetic qualities at very low flux densities where magnetic heads operate. The initial permeability is only 500, and the coercivity is 0.5 oersted. Grain-oriented silicon iron is better, but head cores usually have some portions oriented against the grain and cannot make full use of the orientation.

An experimental program to adapt silicon iron has resulted in heads with sharp microgaps that outwear the permalloys, yet their cores are not brittle like those of sendust. To take advantage of grain orientation, the core may be bent, as in Figure 4-21 *B*, *G*, or *H*, with the grain running in the direction of core flux. Such a core may be improved further by annealing while it is magnetized with a nonmagnetic gap spacer in place to create a circuital flux in the direction of ultimate use. A preferred material for silicon-iron cores is an alloy containing 6% silicon instead of the usual 3%, the high silicon material having longer wear life, higher resistivity, and better magnetic permeability at low flux densities, with a slight sacrifice of saturation density (Camras 1971)

Amorphous Alloys. A glasslike ribbon of high magnetic permeability can be produced by directing a thin stream of molten nickel-iron-boron alloy against the surface of a spinning cold-metal roll. The ultra-rapid cooling plus the appreciable content of such elements as boron, silicon, or phosphorous gives an amorphous instead of crystalline structure. Of the many possible alloys, the most successful for heads have been the low magnetostriction alloys of about 72% Co, 3% to 5% Fe, plus Si and B (Meguro et al. 1981). Amorphous alloys (Metglas®) have excellent wear-resistance.

HEAD GAPS

The part of the head that does the work is its gap, literally something that is not there. A gap interrupts the high permeability core material that otherwise forms a closed magnetic circuit. Such a discontinuity creates the external field that interacts with the tape. The gap is usually filled with a spacer of low magnetic permeability that (1) maintains an accurate space between the core pieces; (2) provides mechanical support for the core edges, which prevents erosion, fracturing, or mushrooming; (3) prevents filling of the gap with debris from tape and head wear; and (4) may enhance the head field by current conduction as shown in Figure 4-23 (conductive gap spacer) (McKnight 1978).

Gap Material. The following materials have been used for gap spacers: mica, silicon monoxide (SiO), silicon dioxide (SiO_2), glass of various compositions and expansion coefficients, epoxy resins, fish paper, air, copper, silver, gold, beryllium copper, platinum, brass, and solder (hard and soft). In fact, any nonmagnetic or slightly magnetic material is operative. The choice is often made as a matter of convenience. For example, mica, silver, gold, and platinum are commercially available in thin sheets or foils, while silicon monoxide can be evaporated in accurately calibrated thin films. Beryllium copper is a preferred spacer material because its hardness prevents erosion of the gap edges. Ferrite cores can have their gap filled with molten glass that bonds to the cores to give a solid unbroken path across the head. Vacuum deposition techniques are used for film thickness down to Angstrom size (1 Angstrom = 0.0001 μm).

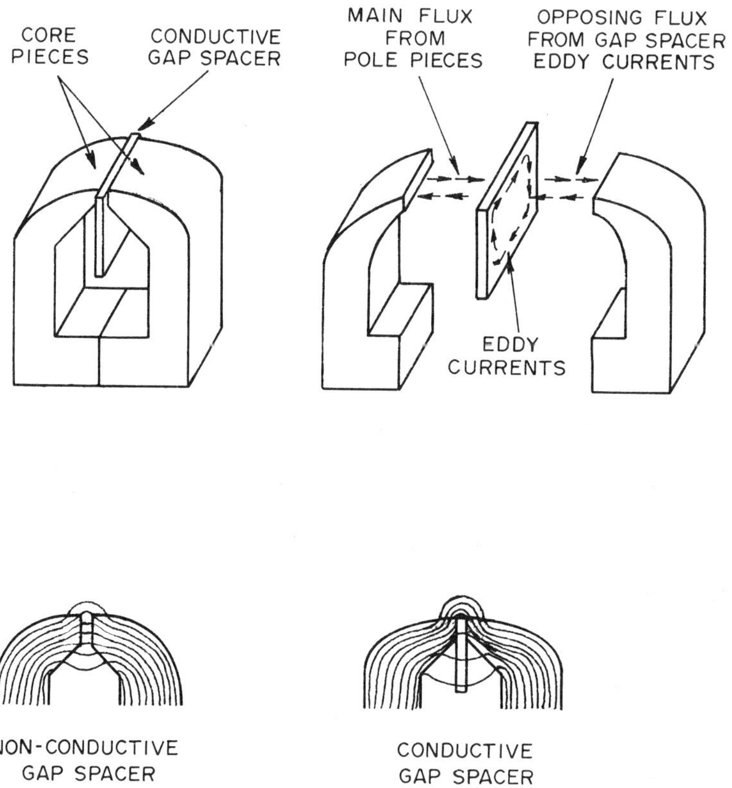

Figure 4-23. Conductive gap spacer for eddy-current shielding or ''focussed gap'' effect.

''Focused Gaps'' and Eddy Current Shields. A gap spacer of high-conductivity material such as copper or silver will generate internal eddy currents due to the AC gap field across it, as shown in Figure 4-23. The eddy currents tend to reduce the internal ''deep'' gap field and to increase the field that acts on the tape just outside of the gap. Both the efficiency and the resolution of a head are improved (Otto 1972).

The greatest improvement occurs for high-frequency recording where eddy currents are highest in the spacer. This improvement is especially welcome for metal core heads that have high losses at high frequencies. Video and data recorders often operate at 10 MHz or more. Even audio heads may carry high frequency bias flux of 0.1 to 1.0 MHz. Playback efficiency also benefits from conductive spacers that reduce the AC playback flux shunted across the gap and send more of it through the main magnetic circuit.

The benefits of conductive spacers were called *eddy current shielding* in earlier heads, indicating that the spacer intercepted undesirable magnetic flux. The gap spacer was often extended well below the gap itself to further reduce leakage flux between the gap poles. Later the term " focussed gap" was used, implying that a sharper field is obtained. However it is doubtful that the field shape above the gap in a high permeability core is affected directly by an eddy current spacer, although the spacer may give indirect improvements by reducing saturation of gap corners and so on.

Application. The gap spacer material may be cut from sheet material or foil and then adhered to the pole faces with cement; or it may be bonded, for example, by melting of glass, soldering or welding; or it may be evaporated or sputtered on to the pole face, which is a preferred method for very short gaps.

Gap Length. Gap length is determined by the thickness of spacer in the direction that interrupts the magnetic circuit. Gaps of 0.4 mm (0.015 in) and larger have been used in erase heads, while gaps down to a few microinches (0.1 μm) have been made for high-density video and data recorders. Typical gap lengths are listed in Table 4-3.

In playback heads, the gap lengths should be minimized for best resolution because the gap sensitivity is symmetrical in both forward and backward directions. Gaps of longer lengths are advantageous in recording heads because the recording process is unsymmetrical and occurs at the trailing edge of the gap. Sharpness, straightness and good tape contact at the edge of the trailing polepiece are important. Wavelengths smaller than the gap length can be recorded. The larger gap improves head efficiency, allowing the core to operate at lower flux density for the required gap field.

Heads with larger gaps record a somewhat different flux pattern on a tape than short gap heads, and some problems of interchangeability among different machines have been attributed to this effect, which is called *crossplay* (see Chapter 13).

Table 4-3. Gap Lengths of Magnetic Heads

Heads	Gap Lengths
Erase heads	125 to 25 μm (0.005 to 0.001 in.)
Audio recording heads	12 to 2.5 μm (0.0005 to 0.0001 in.)
Audio playback heads	6 to 1.5 μm (0.00025 to 0.00006 in.)
Audio combined record-play heads	6 to 1.5 μm (0.00025 to 0.00006 in.)
Video recording heads	1 to 0.5 μm (0.00004 to 0.00002 in.)
Video playback heads	0.75 to 0.3 μm (0.00003 to 0.000012 in.)
Combined video record-play heads	1 to 0.3 μm (0.00004 to 0.000012 in.)

Microgaps. As gap length is decreased in an attempt to achieve higher recording densities, an optimum gap length is found. With even smaller gaps the recording density no longer improves for a specific tape, and head efficiency is reduced greatly. Too small a gap in a recording head causes the head to saturate before it can fully magnetize the tape layer. Too small a gap in a playback head gives lower output and poorer signal-to-noise ratio. This is not to say that a more efficient design cannot utilize a smaller gap, but only that it too has a limit where a smaller gap makes it operate worse instead of better.

As the quality of tape has improved in surface smoothness and in magnetic uniformity, heads with gaps of fractional micrometre size have become feasible, especially where only short wave-lengths are recorded as in video, FM instrumentation, and digital recorders.

HEAD WINDINGS

The winding that surrounds the head core is probably the least critical component of the head. Any construction with the specified wire size and turns will do the job, but a few precautions are noted below. Copper wire insulated with a synthetic enamel such as formvar is the most common, although a flat ribbon, spirally wound, has been used. Windings can be designed to minimize self-capacitance, but this consideration is usually neglected in magnetic recorders.

Winding Details. Some core designs allow for a separate winding that can be slipped over the core legs during assembly. A molded nylon bobbin (Figure 4-24*A*) is convenient. Fine wire is generally random wound, but thicker wire, when layer wound, makes more efficient use of the space. Alternatively, a tube of stiff insulation without end pieces (Figure 4-24*D*) may be wound with wire, the layers being coated with fast-drying cement to hold them in place for a finished coil (Figure 4-24*E*). It is a good idea to check whether the insulating material will react chemically to corrode the copper winding, especially if fine wire sizes are used, or if there is a DC voltage between the winding and core, or if high voltages are present across the winding. (High-frequency bias can reach 100 volts or more.) Where the winding cannot be slipped over the core leg, a split bobbin (Figure 4-24*A*) is assembled over the core (*B*), and the coil is wound over the assembly (*C*). Cores requiring very few turns are often insulated and then wound by threading the wire through the window turn by turn (Figure 4-24*F*). Such a head, used in rotating head video recorders, may be only 1 or 2 millimeters high.

Wire Sizes. Usually the number of turns and window area of a head are

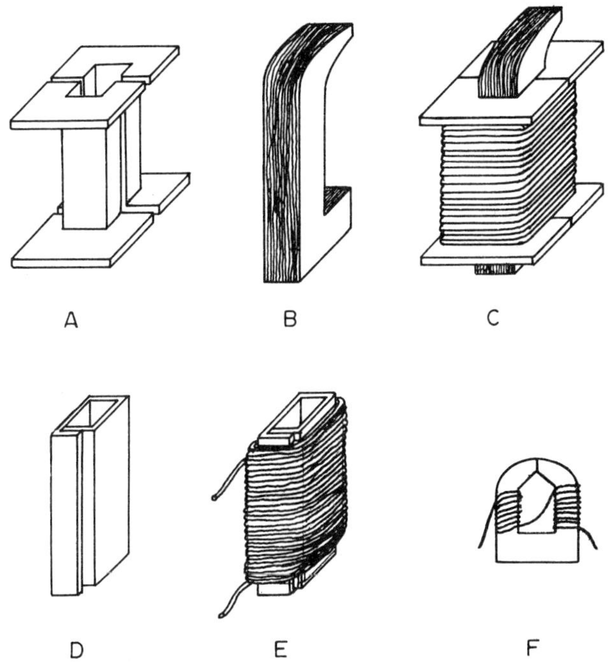

Figure 4-24. Head windings. A. Split bobbin of molded nylon. B. Core half of glued metal laminations that receives the split bobbin. C. Core half after winding. D. Insulating sleeve for self supporting coil. E. Completed self supporting coil. F. A continuous core requires turns to be wound through its window.

known, in which case the wire gauge can be found in Tables 4-4 or 4-5 which give the diameter, resistance, and window area needed for small wire sizes. Even numbered sizes are listed in AWG (American Wire Gauge, which is the same as B&S gauge). Odd-numbered sizes are generally not available except on special order. Theoretical number of turns in an area does not allow for bobbins, insulation, or randomness of winding. The 25% winding efficiency column is a good one to use in first designs of small heads. Experience will then indicate whether a higher winding efficiency can be achieved.

Number of Turns. The number of turns on a head winding determines its impedance. The impedance can therefore be adjusted to any amount that is optimum for the external circuitry to which the head is connected. Head impedance varies as the square of the number of turns, so if the impedance and turns of a head are known it can be rewound for a different impedance according to the equation:

Table 4-4. Wire Sizes, Turns in a Window Area, and Resistance. (Metric Measurements)

Gauge A.W.G.	Diam. mm	Theo. Turns Per mm	Theo. Turns Per mm^2	75% Eff. Turns Per mm^2	50% Eff. Turns Per mm^2	25% Eff. Turns Per mm^2	OHMS Per Meter
24	.510	1.96	3.84	2.88	1.92	.96	.084
26	.404	2.48	6.13	4.60	3.06	1.53	.134
28	.321	3.12	9.70	7.28	4.85	2.43	.213
30	.255	3.92	15.4	11.5	7.69	3.84	.339
32	.202	4.95	24.5	12.2	18.4	6.13	.538
34	.160	6.25	39.1	29.3	19.5	9.76	.856
36	.127	7.87	62.0	46.5	31.0	15.5	1.36
38	.101	9.90	98.0	73.5	49.0	24.5	2.16
40	.080	12.5	156	117	78	39	3.44
42	.064	15.8	248	186	124	62	5.51
44	.051	19.7	388	291	194	97	8.66
46	.040	25	628	471	314	157	14.2
48	.032	32	995	746	498	249	22.4
50	.025	39	1550	1162	775	388	33.9
52	.020	51	2550	1913	1275	638	55.1
54	.015	66	4328	3246	2164	1082	90.6
56	.013	79	6200	4650	3100	1550	134

Table 4-5. Wire Sizes, Turns in a Window Area, and Resistance. (Inch Measurements)

Gauge A.W.G.	Diam. Mils	Theo. Turns Per Inch	Theo. Turns Per $Inch^2$	75% Eff. Turns Per $Inch^2$	50% Eff. Turns Per $Inch^2$	25% Eff. Turns Per $Inch^2$	OHMS Per Inch
24	20.1	49.8	2.48×10^3	1.86×10^3	1.24×10^3	0.62×10^3	.0021
26	15.9	62.9	3.96	2.97	1.98	0.99	.0034
28	12.6	79.4	6.30	4.72	3.15	1.58	.0054
30	10.0	100	10.0	7.5	5.0	2.5	.0086
32	7.95	126	15.8	11.9	7.91	3.96	.014
34	6.30	159	25.2	18.9	12.6	6.3	.022
36	5.00	200	40.0	30.0	20	10.0	.035
38	3.96	253	63.8	47.8	32	15.9	.055
40	3.14	318	101	76	51	25.4	.087
42	2.5	400	160	120	80	40.0	.14
44	2.0	500	250	188	125	62.5	.22
46	1.57	637	406	304	203	101	.36
48	1.25	800	640	480	320	160	.57
50	1.0	1000	1000	750	500	250	.86
52	0.78	1280	1640	1233	822	411	1.4
54	0.6	1667	2800	2083	1389	694	2.3
56	0.5	2000	4000	3000	2000	1000	3.4

$$N_2 = N_1 \sqrt{\frac{Z_2}{Z_1}} \tag{4-35}$$

where

N_2 = number of turns required for new impedance
N_1 = number of turns on original winding
Z_2 = new impedance desired
Z_1 = original impedance

The core and gap, as well as the number of turns, determine the head impedance, which can be calculated from equations 4-22 to 4-34.

POSITION OF WINDING ON THE CORE

The equations we have used for magnetic circuit calculations are the same regardless of where the coil is positioned on the core, and therefore from the idealized model we might expect that positioning as is shown in Figure 4-25 makes no difference. Surprisingly enough, the coil position does make a difference, both in efficiency and in flatness of response. The absolute efficiency, however, can still be very high and quite adequate even in the least efficient designs if other conditions are favorable.

Efficiency. Of the ordinary designs of heads, the most efficient is the arrangement found in Figure 4-25*A*, with two side coils; the least efficient is *B*, with a single coil on the bottom leg; and intermediate is *C* or *D*, with a single side coil. In *A*, only the top portions of the head are at opposite magnetic potential with respect to each other, (+) representing north polarity, (−) south. Leakage flux of *A* follows various paths in the three-dimensional space between the top pole tips somewhat as pictured by the dotted lines. But *B* has much larger exposed pole portions of opposite polarity, including the sides and bottoms of poles inside and outside the winding window and therefore the three-dimensional leakage flux is much greater than in *A*. To make matters worse, if a specified potential difference is required across the head gap, the arrangement in *B* will need an even larger average potential between its poles than *A* does because of the potential drop in the side legs of *B*. The extra leakage flux carried by the bottom yoke of *B* increases its losses during recording or erasing and may cause it to saturate.

Arrangements found in Figure 4-25 *C* and *D* are intermediate in efficiency between *A* and *B*. Arrangement *E* makes the opposite polarity areas very small by concentrating the coils very close to the gap. All of the above arrangements

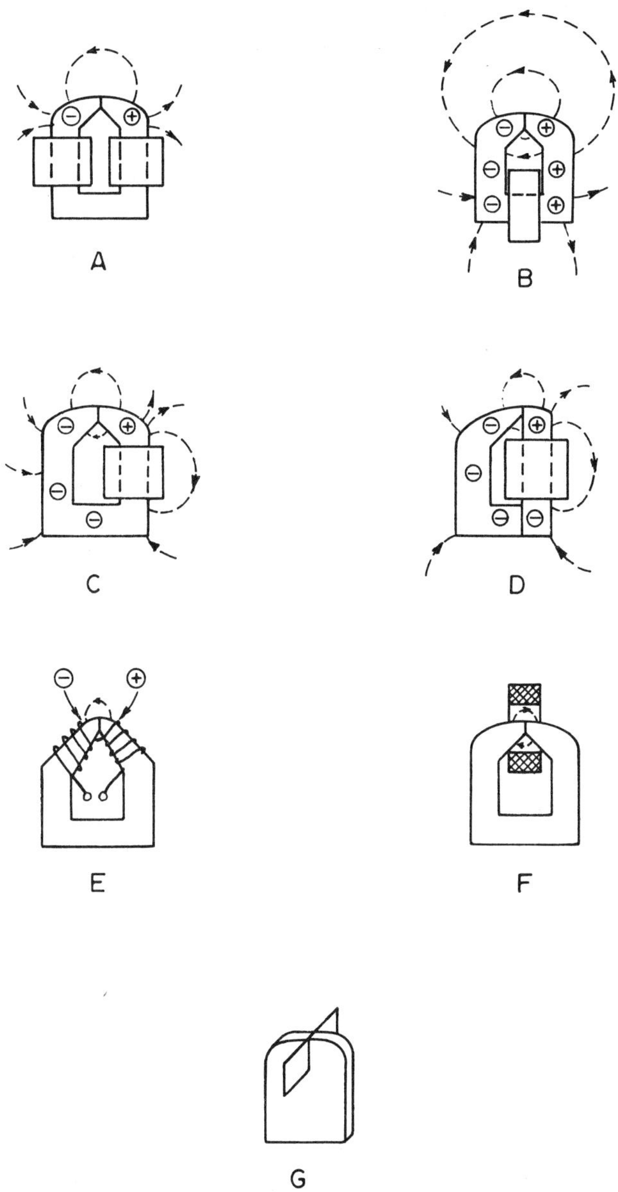

Figure 4-25. Effect on head field of position of winding on the core. Magnetic potential is indicated in the circles.

are worsened if the head is enclosed in or adjacent to a shield that provides a shorter path for leakage flux. Arrangement *F*, the so called *coil-over-gap* or *closed-head*, is the most efficient of any since the potential difference of the core is greatest across the gap faces and decreases to zero at the coil boundaries. This head has other remarkable properties, described in Chapter 5. The turn-in-gap head *G*, uses the gap spacer as a one-turn energizing winding and thus generates the magnetic field in the same region where the field is utilized. The magnetizing field (*H*) due to current in the spacer adds to the flux density (*B*) of the magnetic pole tips of the core, giving a stronger field above the gap than if the coil is elsewhere. The (*H*) component from the winding is not affected by core saturation and can be very helpful at high densities. Leakage flux in head *G* depends on whether the external electric circuit that energizes the gap spacer is completed above, below, or at the side of the magnetic core; the most efficient arrangement is with the coil over the gap (Camras 1944).

Flatness of Response and Contour Effects. At medium and short wavelengths, all the heads shown in Figure 4-25 have a relatively uniform integrated response, each at a level according to its own efficiency. At longer wavelengths comparable to the core dimensions, there are additional flux paths linking the winding, other than through the main gap (Westmijze 1953; Fritch 1968). The additional flux adds to the gap flux at some wave-lengths and subtracts at others. As shown in Figure 4-26, side coils will have dips in the output when p (the half-span of the core) is a whole number of wavelengths on the tape ($p = \lambda, 2\lambda, 3\lambda$, etc.); while peaks in the output will occur where p is an odd number of half wavelengths ($p = \frac{1}{2}\lambda, \frac{3}{2}\lambda, \frac{5}{2}\lambda$, etc.). These peaks and valleys are called *contour effects*. The effective pole half-span p is usually not exactly equal to the physical distance, but may be inferred from the response curves. The peaks and valleys die out as wavelengths decrease because the shorter flux paths have less effective coupling with the coil. Superposed on these undulations is a slight steady long-wavelength rise due to a "secondary-gap effect" (Fritzsch 1968); but as wave-lengths become very long compared to head dimensions, less and less tape flux is intercepted by the head and the output falls toward zero.

Contour effects occur almost entirely during playback, which is a linear process, and not during recording because the threshold at which recording takes place is reached only at the gap and not at the external pole edges, especially where high-frequency bias is used. However, in exceptional cases the recording can also be affected: if there are very high levels of recording, or if the main gap is very small, or if the coils are very close to the tape, or if the outer edges of the poles can concentrate leakage flux on the tape.

Eliminating Contour Effects. If only one main dip in response is to be cor-

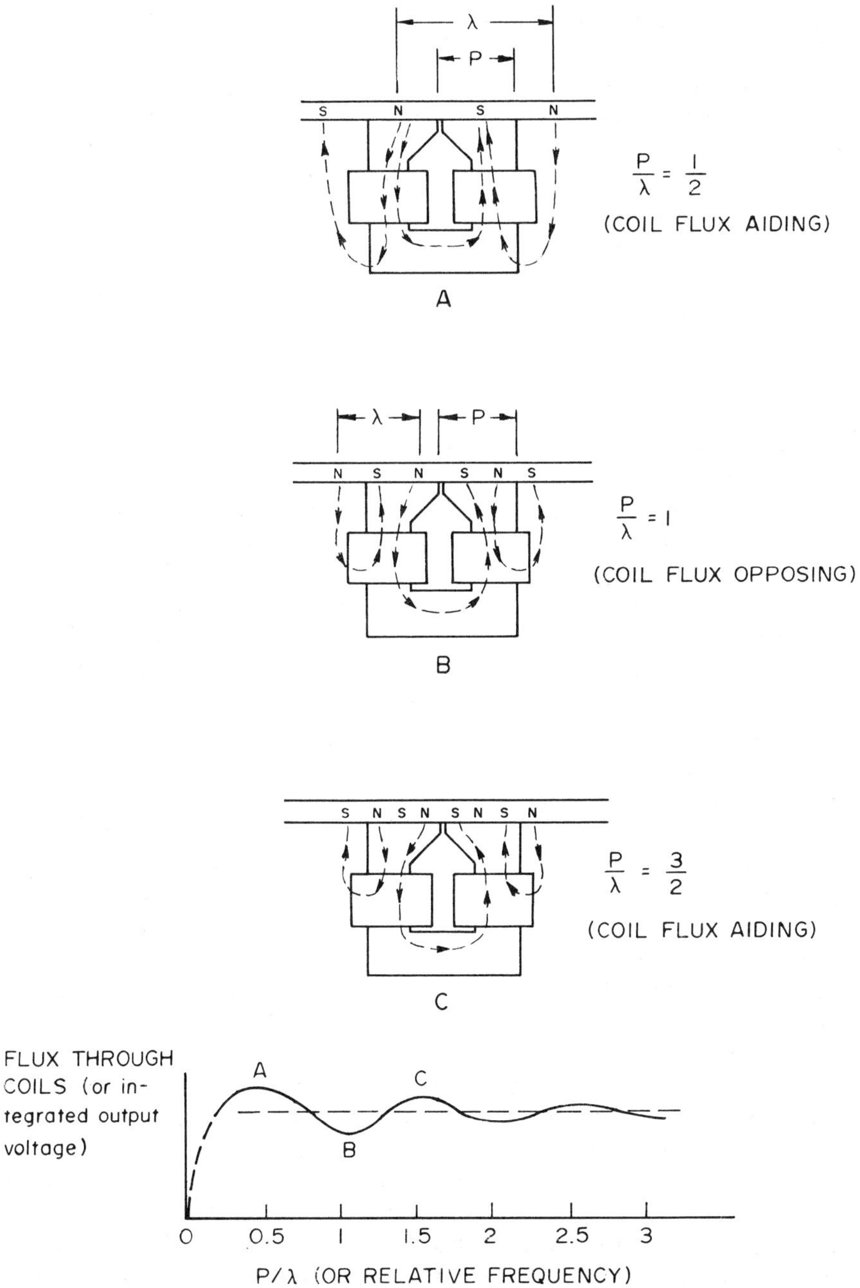

Figure 4-26. Head contour effects from pickup outside of head gap.

rected, an equalizer can be used, but is limited to one tape speed. Contour effects are reduced considerably by rounding the pole faces at the regions where the tape first touches the head at entry and departure, by avoiding sharp edges that concentrate the flux, and by spacing the shield away from tape entry and departure regions. Unequal lengths of core halves reduce the amplitude of undulations. Coil position on the core is important. As seen in Figure 4-26, hardly any of the spurious flux threads the bottom leg; a bottom coil position, although inefficient, reduces contour effects. The closed head shown in Figure 4-25*F* has no contour effects because the coil does not link the spurious flux. A special core design that balances out contour effects is described in Chapter 5.

HEAD SHIELDING

Playback heads that give useful output over a range of 40 to 80 db are sensitive to magnetic fields $\frac{1}{100}$ to $\frac{1}{10000}$ weaker than the maximum. Hence they will be disturbed by fields from nearby motors, transformers, and loudspeakers. They will also pick up recordings from adjacent tracks on tapes that have multiple tracks, such pickup being greater at longer recorded wavelengths.

If the playback head is near an energized recording head or erase head, any magnetic or capacitive coupling will induce unwanted recording signals and high-frequency bias into the sensitive playback circuits. This occurs when, for example, simultaneous recording and playback is arranged for monitoring the recording shortly after it is made.

Playback Head Shielding. The simplest shield for the playback head shown in Figure 4-27*A* is a mumetal box of rectangular shape, open at the top to expose the gapped pole faces that contact the tape. The effectiveness of a shield depends on its ability to intercept and bypass unwanted external flux, preventing it from threading the head core and pickup coils. A simple shield is usually adequate in low-cost recorders, where low-frequency response and dynamic range are limited and where hum sources are oriented for minimum interference.

Effectiveness of the head shield is improved greatly if the top of the head is also shielded. The shield then becomes an almost closed box (Figure 4-27*B*) with only the tips of the polepieces exposed to contact the tape. However the head efficiency is reduced when the shield is closely adjacent to poles that are at high magnetic potential, so a certain amount of spacing is required between polepieces and shield. If a multi-track tape runs over the head, the shield can be designed to act as a "keeper" that short-circuits the flux of tracks adjacent to the track that is being played and thus prevents unwanted interference from adjacent tracks (Figure 4-27*B*). The side spacing between head and shield to

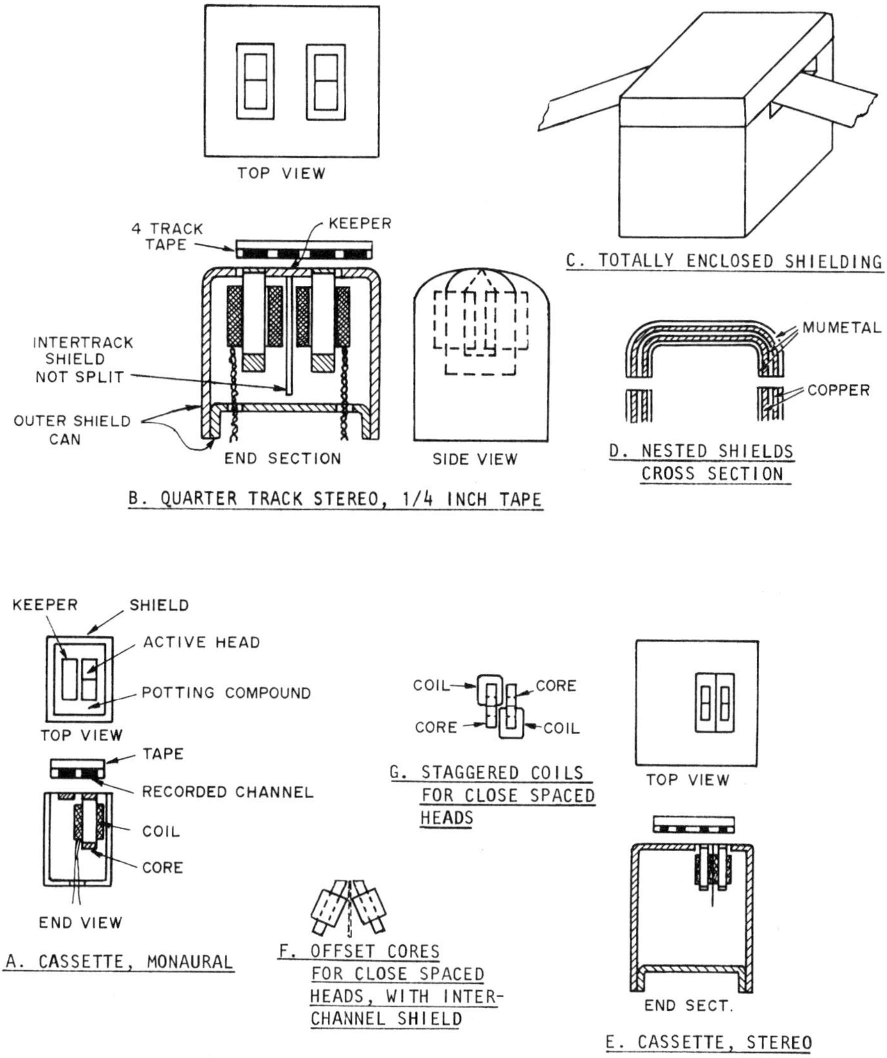

Figure 4-27. Heads and shields for tape recorders.

give this keeper effect is less than would be chosen for hum shielding alone, so the keeper advantages are at the expense of reduced head efficiency.

The ultimate in head shielding is to enclose the entire head in a shield, including the gap, and to bring the tape in and out of the shield through side openings (Figure 4-27*C*). To allow convenient tape threading, the shield top has a lapped joint that allows it to open. Such a shield is used in professional recorders for recording and broadcast studios.

Shield Construction. Mumetal is a favorite material for head shields, being the least expensive of permalloys with high initial permeability essential for effective shielding. Thin gauges down to 25μm (1 mil) are operative, but thicker gauges of 14 mils, strong enough to be self-supporting and to be die formed into seamless cans, are usually chosen. Typical commercially available shielded heads are shown in Figure 4-1. The shielding effectiveness increases slowly as the shield is made thicker. Nested shields are much better, consisting of shields within shields, each separated from its neighbors by a non-magnetic spacer. For AC shielding, the nonmagnetic spacers can be solid copper. Eddy currents in the copper oppose the penetration of AC fields and are most effective at high frequencies. The best shields then are made of alternately nested boxes of permalloy and copper (Figure 4-27*D*), and these are also the most expensive. Permalloy shields must be annealed after all mechanical working is completed. The hydrogen anneal schedule described previously for cores is recommended.

RECORDING EFFECTS

If one looks for recorded spillover on tape beyond the side edges of the recording head one will find little or none. Fine gaps have a negligible distance of effective fringing. Also, the recording process has a threshold below which nothing is recorded; high-frequency bias and/or strong signals are needed to overcome the threshold. Recording interference does occur, however, but by more subtle mechanisms.

In multitrack heads, there is an unavoidable amount of magnetic coupling between adjacent heads while recording. Even though the heads are carefully shielded from one another, their pole tips must all be exposed to the tape surface, so magnetic coupling inevitably takes place in these exposed areas. Core 1 and core 3 then carry a small amount of magnetic signal flux that was meant for core 2 alone. Now if all the cores are high-frequency biased for recording, then core 1 and core 3 will record the magnetic coupled signals they picked up from core 2 along with their own signals, if any. Playback of tracks 1 and 3 will always give some track 2 signals and these may be attributed wrongly to recorded spillover or to pickup from track 2 during playback. This theory is easily disproved by removing the bias from heads 1 and 3, thus desensitizing them while recording track 2, in which case no interference is detected on adjacent tracks.

Eliminating the Recording Effects. Magnetic coupling opposite to the natural coupling in between head cores can be introduced in the construction of multichannel heads to reduce interference during recording. Or currents from adjacent heads can be mixed in each head in direction and magnitude to balance out the stray flux. Or small auxiliary windings on heads 1 and 3 can be connected to the main winding of head 2, and so on. The cancellation of interference already recorded on the wrong channels can be done during playback by mixing their output with opposite output from the adjacent track on which the interference originated. None of these schemes is completely effective over a wide-frequency band, because couplings and phase shifts are determined by nonlinear magnetic elements.

SIMULTANEOUS RECORD AND PLAYBACK

Isolation of the pickup head is a challenge when a high-level recording head is operating close by. Fortunately the isolation experiments can be carried out when the tape is not running. Also, the quality during simultaneous playback usually does not have to be as perfect as the quality on subsequent playback when the recording function is turned off. The following suggestions may be helpful:

1. Separate shielding of both the recording and the playback heads is in order. Possibility of pickup from the erase head or its circuits should not be overlooked.
2. A low pass filter or bias trap in the playback circuit prior to the first amplifier will eliminate bias and erase pickup before it can overload the input circuits.
3. It is very likely that coupling between the electronic circuits of the record and playback amplifiers exists through ground loops, common power supplies, and so forth. A resistor substituted for the playback head should give no feed-through of the recording or bias signals if the electronics are isolated properly. Two-wire shielded cables may be beneficial, with the shield not carrying any signal current.

MULTICHANNEL HEADS

Because many parallel channels can be recorded on the surface of a wide tape, multitrack records are almost universal, and single track tapes are the exception. The multiple tracks do not have to be used simultaneously in all cases. In video recorders, one track is read at a time. In stereo machines, only two of the multiple tracks are read at once. But in digital and data recorders, all tracks usually must be read at the same time, and a complex multitrack head is then required. Multitrack head dimensions, mostly for audio, are given in Figure 4-28. Heads for other applications are detailed in later chapters.

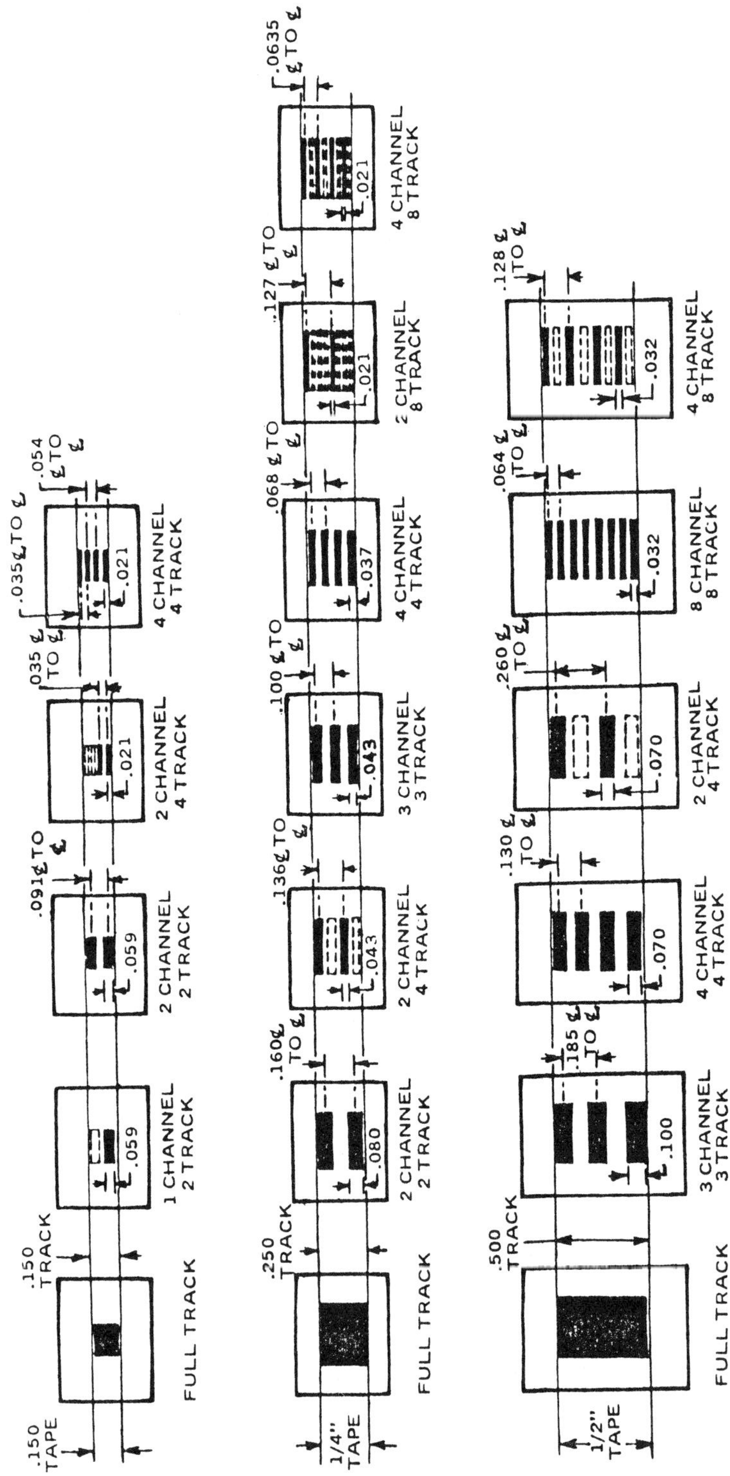

.150 TAPE
.150 TRACK
FULL TRACK
1 CHANNEL 2 TRACK
.059
2 CHANNEL 2 TRACK
.059
.091℄ TO ℄
2 CHANNEL 4 TRACK
.021
.035 ℄ TO ℄
4 CHANNEL 4 TRACK
.021
.035℄ TO ℄
.054 ℄ TO ℄
1/4" TAPE
.250 TRACK
FULL TRACK
2 CHANNEL 2 TRACK
.080
.160℄ TO ℄
2 CHANNEL 4 TRACK
.043
.136℄ TO ℄
3 CHANNEL 3 TRACK
.043
.100 ℄ TO ℄
4 CHANNEL 4 TRACK
.037
.068 ℄ TO ℄
2 CHANNEL 8 TRACK
.021
.127 ℄ TO ℄
4 CHANNEL 8 TRACK
.021
.0635 ℄ TO ℄
1/2" TAPE
.500 TRACK
FULL TRACK
3 CHANNEL 3 TRACK
.100
.185 ℄ TO ℄
4 CHANNEL 4 TRACK
.070
.130 ℄ TO ℄
2 CHANNEL 4 TRACK
.070
.260 ℄ TO ℄
8 CHANNEL 8 TRACK
.032
.064 ℄ TO ℄
4 CHANNEL 8 TRACK
.032
.128 ℄ TO ℄

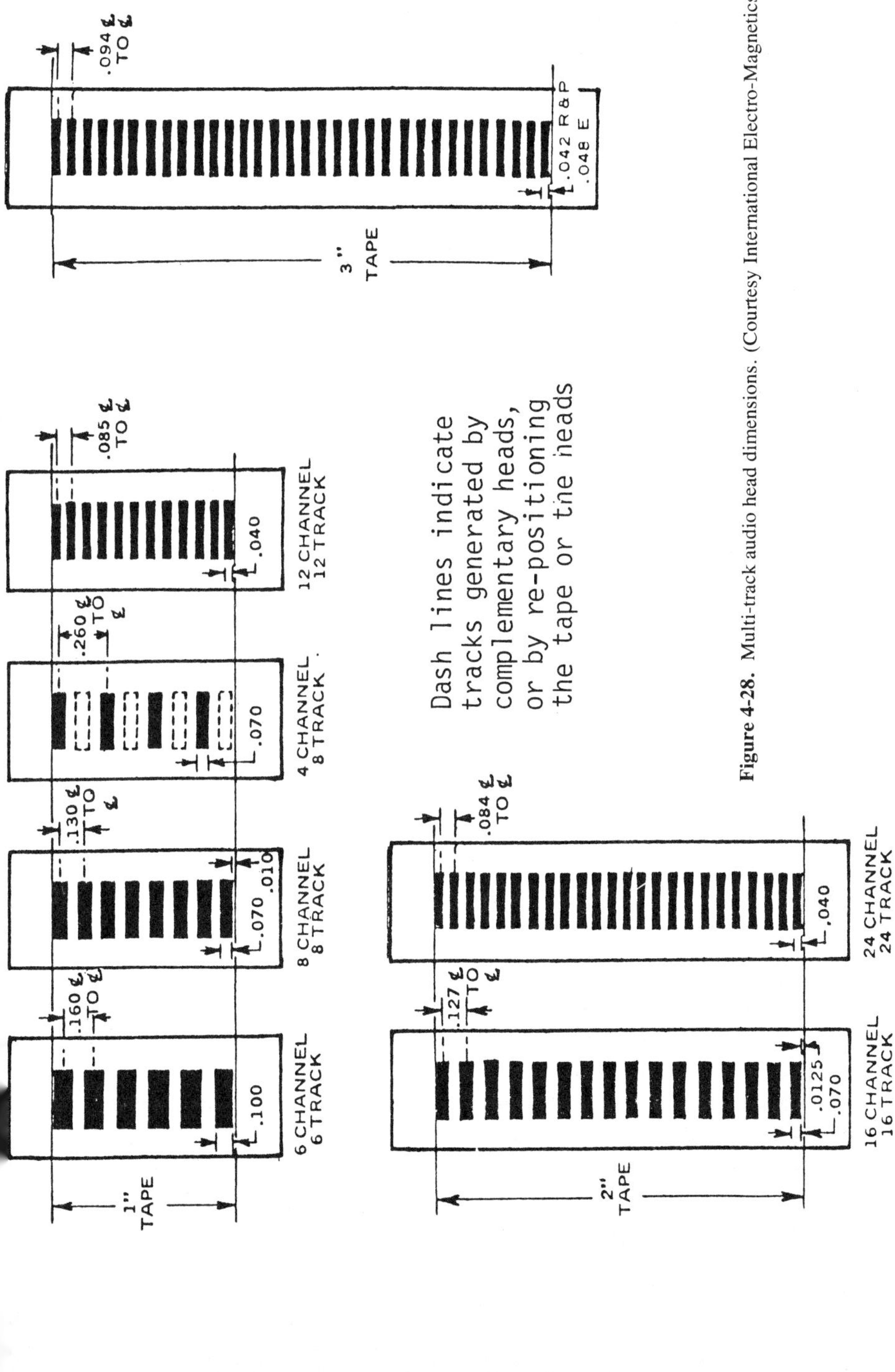

Figure 4-28. Multi-track audio head dimensions. (Courtesy International Electro-Magnetics, Inc.)

Single Head for Two-Track Tape. The elementary head shown in Figure 4-27*A* is for reading one track at a time of a two-track tape, as in a compact cassette recorder. This head is encased in an open-top shield-can filled with casting resin. A blank strip of permalloy is fixed adjacent to the active head poles in order to contact the unused track and to short-circuit its flux to prevent interfering pickup. The added strip is called a *keeper*, by analogy to the bar that is placed across a permanent magnet to shunt its flux when the magnet is not in use.

Two Channel Heads for Four-Track and Eight-Track Tapes. Standards for quarter-inch reel-to-reel tapes call for two interleaved stereo pairs, four tracks altogether, the tape utilizing the second stereo pair with the same heads when the reels are interchanged and turned over. A head for such a recorder is shown in Figure 4-27*B*. The top shield can also act as a keeper. Heads for eight-track stereo cartridges that use quarter inch tape have the same two channel construction, but the heads are narrower and spaced farther apart to contact tracks 1 and 5 for the first stereo pair. The heads then shift to tracks 2 and 6 for the second pair, then to 3 and 7, and finally to tracks 4 and 8, the endless loop of tape always moving in the same direction.

Two-Channel Head for Stereo Cassettes. Compact cassette stereo recorders use four tracks on a narrower tape than the audio recorders, 0.15 inch (3.81 mm) wide, and the stereo paired tracks are immediately adjacent to one another instead of interleaved. This calls for very narrow heads crowded in the side of the head casing devoted to a stereo pair, as shown in Figure 4-27*E*. Isolation between such closely adjacent tracks is difficult, but fortunately the stereo information on both tracks is coherent and slight cross-talk is tolerable. A modified construction that gives more room for the head windings is shown in Figure 4-27*F*, where the cores bodies are offset at an angle. Another possibility is shown at *G*, where a coil is wound on only one side leg of each core, with the coil sides staggered.

COMPUTER AND INSTRUMENTATION HEADS

Computers have standardized on heads for nine-track ASCII format, although the older seven-track format is still in use. Instrumentation recorders are more diverse, ranging from one track to more than fifty with heads from 6.3 mm ($\frac{1}{4}$ in) to 76 mm (3 in) or even wider. IRIG standards (Figures 13-2, 13-3, 13-4) have largely been adopted, bringing order into a chaotic field where each agency would design its own customized format. (IRIG is the Inter-Range Instrumentation Group, a U.S. agency).

Nine-Track Head for Computer Tapes. Multichannel heads for computers are more precise, more durable, and far more expensive than audio heads. Computer tapes are usually 12.7 mm ($\frac{1}{2}$ in) wide and use nine tracks 1.09 mm (43 mils) wide. See Figure 14-2 for details of the nine-track format and Figure 13-2 for details of the seven-track format. Figure 4-29 shows a read-after-write head for computers.

Wide-Band Data Recorder Heads. Very high performance is required in wide-band data recorders. The most popular of these has used the IRIG standards of seven tracks on 12.7 mm ($\frac{1}{2}$ in) tape, with reliable output from 400 Hz to 2 MHz at a tape speed of 3048 mm/s (120 in/s), and with proportional re-

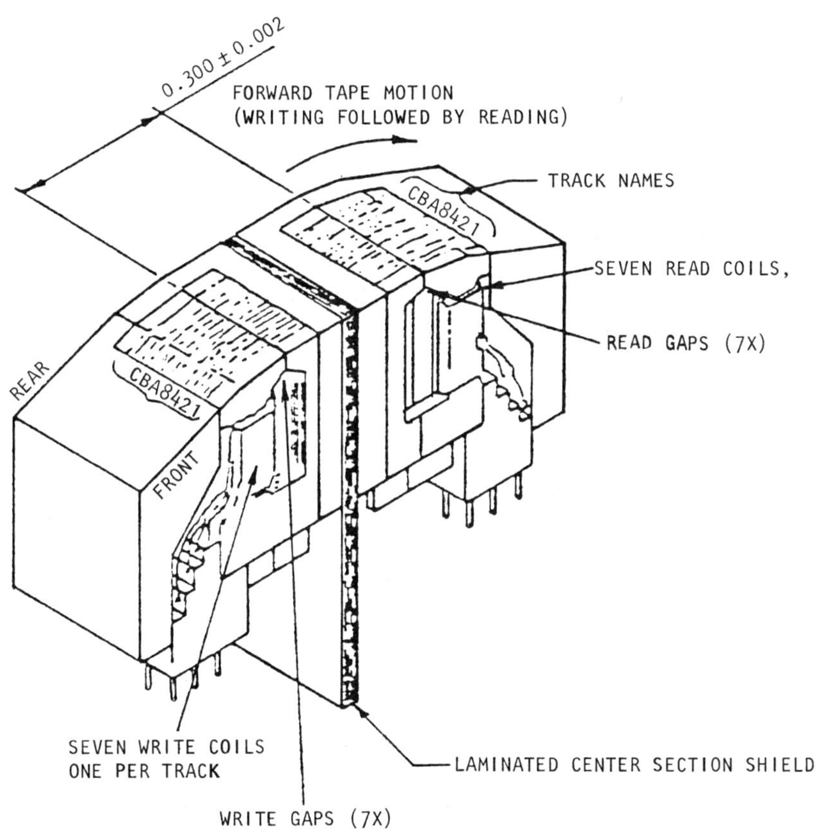

Figure 4-29. Read-after-write head used on the IBM 729 digital computer (IBM J. Res. Develop., Sept. 1981).

sponse at lower speeds. Improved tapes and heads have allowed the upper limit to reach 4 MHz.

Nowadays, more sophisticated recorders use wider tapes (25.4 mm or 1 in wide), and higher track densities (IRIG 42 tracks/in). A successful head construction for this application uses sendust inlays for the gap defining poletips, with ferrite for the rest of the body, as shown in Figure 4-22. Competitive heads by other manufacturers use all-ferrite construction with glass-bonded poletips. Figure 4-30 shows some components and assembly of a sendust inlay head.

Special Heads. Heads made by special techniques or for specialized applications are described in Chapter 5 and in later chapters. These include flying heads, thin-film heads, and stylus heads, among others. Basic principles of this chapter also apply to these special heads, but their adaptation is better understood in context with their uses.

Additional Design Considerations. Where multitrack tapes are to be interchangeable among several machines, the width and spacing of heads in the multihead assembly must be maintained to close tolerances. The most successful multihead mounts are solid blocks with milled slots that locate the cores; and with associated cavities that accommodate the coils and intertrack shields. Two similar blocks are used for half-assemblies, the head being divided along a plane through the gap faces (Figure 4-30*B*). Each half is polished along the dividing plane to nearly optical flatness. A thin film is then deposited over the gap faces to serve as a gap spacer. The halves are assembled, and, as a final step, the surfaces that accommodate the tape are finished and polished.

The above construction is by no means the only possibility. Other designs are based on modular construction, where the multitrack head is built as a stack of thin individual head sections. A modular head has advantages of economy, versatility, and the possible replacement of defective sections. But its dimensional accuracy is much less than with one-piece multislot head blocks. Another variation is the common pole head shown in Figure 4-31: One of the core half-assemblies is replaced by a solid piece that acts as a common yoke for all channels (Anderson et al. 1960).

Heads that are placed very close together do not allow much room for the winding. Often a single-layer winding is sufficient. The construction shown in Figure 4-27*G* saves room by staggering the windings on alternate legs.

Gap Scatter. In some applications timing errors or phase differences between parallel channels is critical, in which case gap scatter becomes important. Gap scatter is the deviation from a straight line of the magnetically effective gaps. When measured in terms of signal time-delay, fractions of a μm gap scatter are detectable even in the most carefully made heads, since the gap edges

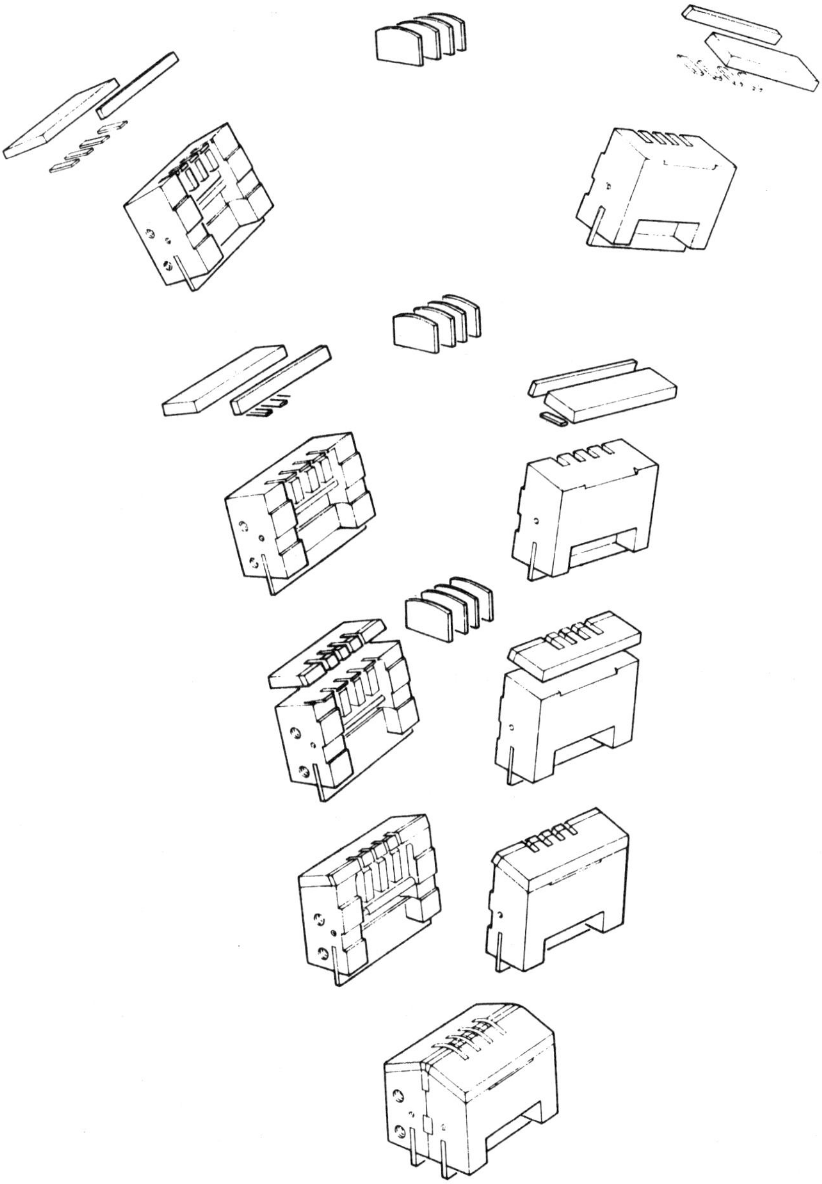

Figure 4-30. Some component parts and assembly of a sendust-inlay head. (Courtesy Ampex, Mersing 1982).

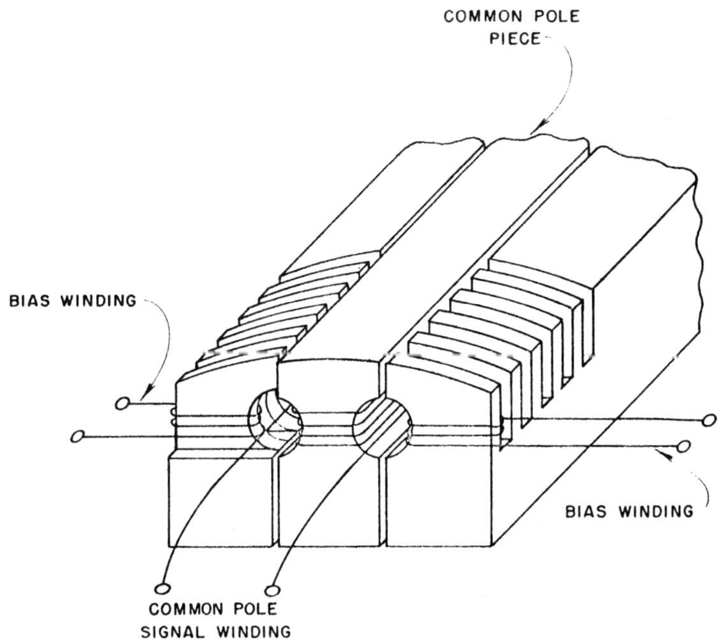

Figure 4-31. Common pole head. The unusual windings are for operating the head in a video scanning mode. See Figure 5-26. (Anderson 1960).

are eroded randomly by passage of the tape, and tape-contact and bias may vary from head to head, shifting their recording points. The single-block design described above will minimize gap scatter caused by constructional variations.

Tunnel Erase Head. In digital NRZ saturation recording, the same head blots out the old record whenever it writes new data. In such recording, the head must track the old record with great accuracy; otherwise it will not obliterate an edge of the old recording, and during playback this old fringe will generate clutter and noise. Accurate tracking is very difficult if the recording is interchanged among several machines. The tunnel-erase head (Figure 14-21) solves this problem by erasing portions of the track at both edges of the write head, trimming off any recording from narrow stripes adjacent the head and leaving a clean new track after every write operation.

HEAD WEAR

A tape stretched between a pair of guides and running over a head in the usual manner does not exert uniform pressure across the head, but exerts maximum pressure according to an anticlastic function, as in Figure 4-32*A* (Arnold et al.

1972). After considerable wear, the head should therefore be expected to have a concave wear pattern, as shown in Figure 4-32*B*. In practice this does not seem to happen, and in fact the groove in worn heads usually has a convex cross section except for rounding at the very edges.

There are many other paradoxes in head wear. For example, wear resistance does not always increase with hardness of the wearing surface, and wear rate

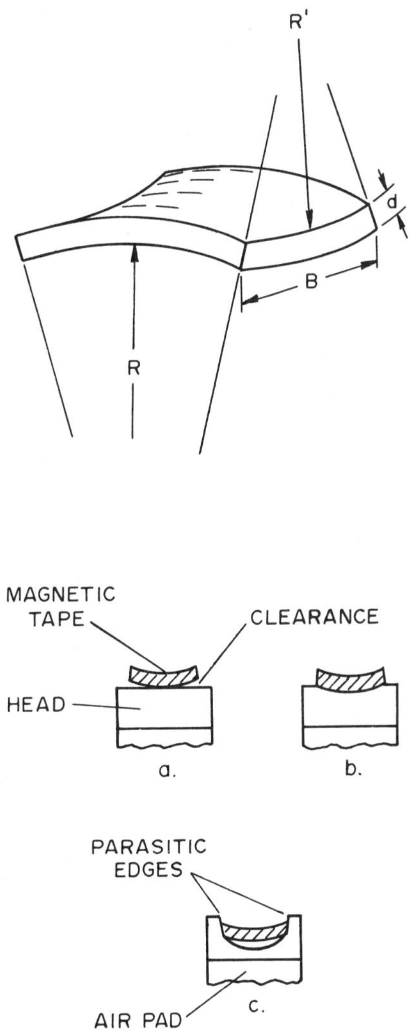

Figure 4-32. Pressure variations across tape width due to anticlastic curvature (Arnold, jAES, July/Aug 1972).

often decreases when tape speed increases. Head wear is therefore best studied under actual operating conditions. Accelerated or simulated studies are usually misleading. In the metallurgical and glass-polishing industries, it is recognized after centuries of experience that the polishing action of different agents is still an empirical art. With these caveats, we list some factors that affect head wear:

1. Abrasiveness of tape coating (which decreases as tape wears)
2. Pressure between tape and head
3. Hardness of polepieces (with exceptions)
4. Hardness of head block (polepiece support)
5. Lubricant film between tape and head
6. Accumulation of grit, debris, and brown stain
7. Relative speed between tape and head
8. Aerodynamic separation of tape from head
9. Chemical affinities at tape-head interface
10. Humidity above 50% (as RH increases from 50% to 80%, wear may increase 10×).

As a rough estimate of head-wear experience, tests on 79 permalloy gave wear rates ranging from 30 to 1500 μm per 1000 hours, with 300 being typical. Wear-resistant Hy-Mu 800 indicated wear of 10 to 50 μm per 1000 hours. Sendust wore at a rate of 2 μm per 1000 hours. A test of relative wear rates, all under the same conditions, gave permalloy 30, sendust 10, amorphous alloy 5, and ferrite 2. Another comparison gave hot-pressed ferrite a 5:1 to 10:1 advantage over sendust.

The head block material should preferably be chosen to wear at a slightly faster rate than the polepieces themselves, allowing the tape to bear most firmly on the polepieces at the gap. Figure 4-33 gives comparative wear results of head block and core materials used in instrumentation heads.

Brown Stain. With metal-faced heads and relative humidities of less than 35%, a condition occurs that is the opposite of head wear: a hard brown film forms on the head surface. Brown stain of thickness from 0.2 to 5μm has been measured, reducing short wavelength response because it separates the tape from the head. Brown stain resists solvents and must be removed with an abrasive polishing tape such as green tape or gray tape made for head-lapping purposes. Some recording tapes contain a small percentage of abrasive to alleviate the above problem.

Hard Inserts. An old favorite idea of head designers who want to extend head life is to incorporate inserts of hard or wear-resistant material into or adjacent to the polepieces where the tape bears against them. The hard inserts then

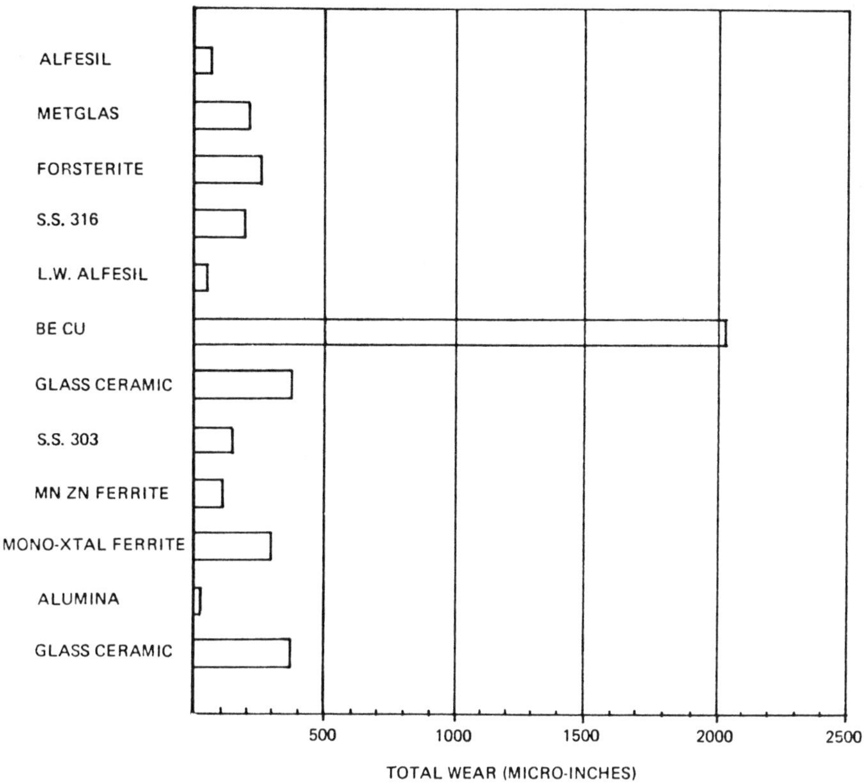

TRANSPORT – FR3030
TAPE – AMPEX 797, 1 INCH
SPEED – 120 IPS
TENSION – 11 OZ

RUN TIME – 113 HOURS
HUMIDITY – 55%
RADIUS – .118"
WRAP – 7°

Figure 4-33. Wear results of head block and core materials used in instrumentation heads. S.S. = stainless steel; L.W. = low wear. (Courtesy Ampex Corp., Mersing 1982)

carry the brunt of wear from the tape. Such "inserts" can be a chromium plating or a ceramic facing that increases head life 5 to 50 times (Nortronics Co. 1976). In practice, the tape may scoop out pockets in the softer core material and prevent ideal contact, thereby giving reduced and erratic response at short wave-lengths.

LUBRICANTS

Lubricants operate at the head-tape interface and can be affiliated with either the head or the tape.

Air Films. The best lubricant is probably an air film that prevents contact between tape and head, as in the flying heads used with magnetic discs and drums. Air-film separation also occurs when tapes are run at very high speeds in ordinary configurations, giving rise to the paradox that head wear decreases as tape speed is increased. Proper aerodynamic design allows separations down to a small fraction of a micron, giving excellent short wave response.

Waxy Lubricants. Waxy or semiliquid lubricants have been an ingredient of tape coating since the earliest days. Silicones in tapes were much advertised when the name implied a miracle product of atomic age chemistry. Lubricants in tapes are not an unmixed blessing because they can cause stickiness to the head poles, or to head imbedding resins, or even between layers of the tape itself. They may weaken the tape coating, ooze out from its surface, or make the tape unstable with respect to temperature, pressure, and time.

Solid Lubricants. Solid lubricants such as graphite, molybdenum disulfide, and Teflon® are also used as tape lubricants. In endless loop tape cartridges, a lubricating coat is applied to the back side of the tape, enabling the adjacent turns of tape to slip relative to each other. Some of the lubricant is inevitably transferred to the magnetic-coated side, reducing friction against the head and increasing head-to-tape spacing losses somewhat.

Lubricating Guides. Guideposts for the tape are sometimes made of graphite or other materials that meter out lubricant to the tape as it passes over the guides. The benefits of this method are often marred by accumulations of the lubricant and agglomerated debris in unwanted areas.

HEAD ADJUSTMENTS

Azimuth adjustment refers to the angle or tilt of record or play gaps, which should be perpendicular to the direction of tape motion. When the playback gap angle differs from the recording angle, there is a loss of output. In fact the output goes through a null when the angular error (θ) gives a difference of one wavelength across the head width span ($\lambda = w \tan \theta$). The azimuth adjustment is critical for short wavelengths and wide tracks. It is most readily corrected to the standard 90° angle by playing a calibration tape, but can also be set optically or by turning over a tape and playing it though its backing layer. In the turnover method, the head angle is set at an average between settings for maximum output on frontside and back-side playback. A new recording is then made and the process repeated until the maxima coincide.

Zenith or forward-backward tilt is usually aligned mechanically but can be corrected by painting the head with blueing and observing the wear after a run of tape.

Head polarity or phasing is checked with a small compass needle near the pole tips while energizing the winding with a few milliamperes of DC current. After this procedure, the head should be degaussed thoroughly.

5. Magnetic Heads: Special Designs

As magnetic recording has developed, many different kinds of heads have been used or proposed. Some are quite different from standard designs and have worthwhile advantages. Why weren't they used more widely? Some were forgotten and some never publicized. Some required materials or techniques that weren't available. Some didn't have enough advantages; some had advantages that weren't needed and some were too costly. Recent concepts are still under development, with problems yet unsolved. However, all of these are useful in stimulating thoughts for new designs or unusual applications.

HIGH-DENSITY HEADS

Heads that will increase recorded density are of great interest in magnetic recording. Although the standard designs of heads achieve extremely high densities when scaled down to use submicroscopic gaps or stylii, there are special designs that give sharper gradients.

Cross-Field Heads

The cross-field head (or X-field head, as it is sometimes called) modifies the generally semicircular field of an ordinary head by superimposing on it a vertical field, as shown in Figure 5-1. The components combine vectorially to give a higher field gradient at the trailing edge of the head. A high gradient at the trailing edge increases the recording resolution because short recorded wavelengths pass out of the influence of the gap before they can be altered by the next cycle of recording or bias (Camras 1952).

Overhanging Pole Heads. The cross field can be supplied by a polepiece that overhangs the head gap (Figure 5-2), the polepiece being offset or shaped to favor the recording edge of the head gap. With the coil on the cross-field polepiece connected in series with the main head coil, the field at the trailing edge of the gap is neutralized at some region q in Figure 5-1C becoming zero and reversing in polarity at the right-hand side of this region. Because the trailing-edge field decreases rapidly to zero, its gradient or rate of change is higher than of the semicircular field of a conventional head. If the relative strength of the crossfield is increased, the zero field region moves closer to the gap and the gradient becomes higher, but the cross field should not be increased to the point where the reverse polarity region will affect the recording. The relative cross-

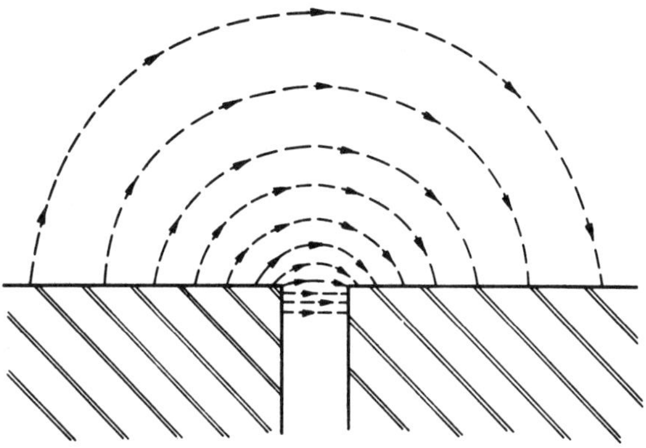

A. MAGNETIC FIELD PRODUCED BY GAP

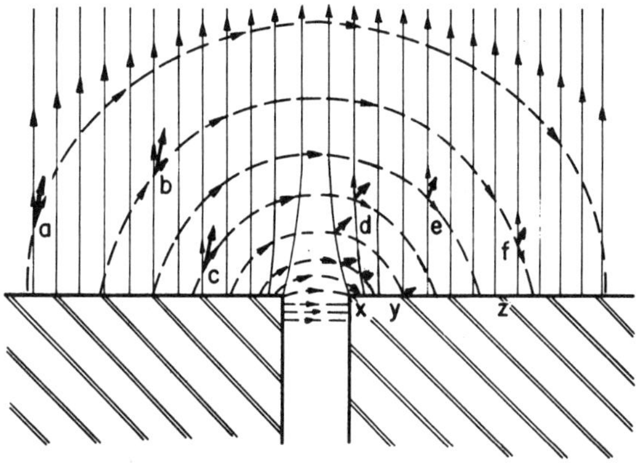

B. SUPERPOSED CROSS FIELD ADDS VECTORIALLY TO THE GAP FIELD

Figure 5-1. Field components and resultant field of an X-field head.

field strength can be adjusted by varying the number of turns on the winding of coil, by varying the distance between the auxiliary polepiece and the head, or by changing the offset of the polepiece from the gap.

Where high-frequency bias is mixed with the recording signal, both are usually applied to the cross field winding as well as to the main winding; but various combinations of separate or mixed signals to each winding will give different effects. For example, the record should retain a higher degree of magnetization if the HF bias (high frequency bias) field fades out in the direction of tape movement while the signal field is still present. This condition is obtained if the cross-field effect with its higher gradient is used for the HF bias

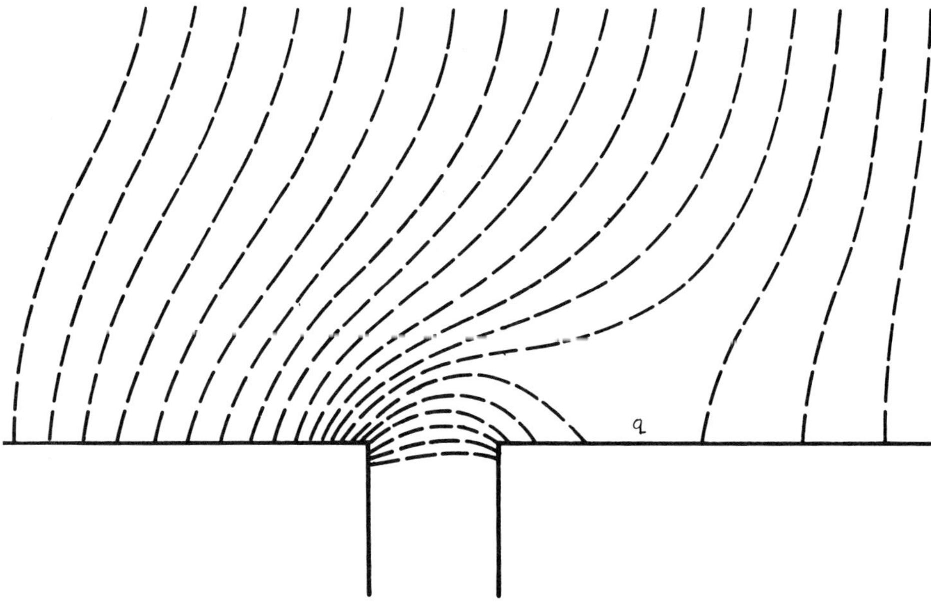

C. RESULTANT OF GAP-FIELD AND X-FIELD

Figure 5-1. *(Continued)*

alone (both coils energized), while the signal is applied only to the main gap coil, or only to the auxiliary coil if vertical recording is to be emphasized.

The cross-field principle is beneficial where the layer of recording material on the tape is reasonably thick compared to the shortest wavelengths to be recorded and where longer wavelengths, which make use of the thickness of the layer, are also recorded. Figure 5-3 shows that the cross-field head gives a more uniform distribution of recording field through the thickness of the tape layer than an ordinary head does. A weaker field can therefore be used than would otherwise be necessary for recording throughout the tape layer thickness. Optimum bias for long wavelengths can be set at a lower level, which does not erase the short wavelengths, as indicated in Table 5-1.

Single-Sided Cross-Field Heads. An overhanging polepiece is objectionable because it makes threading of the tape more difficult and is liable to catch tape splices and to accumulate debris. The cross-field pole can be lifted away during threading, but a precise mechanism is required to relocate it always with an accuracy of a few microns. Single-sided cross-field heads have been devel-

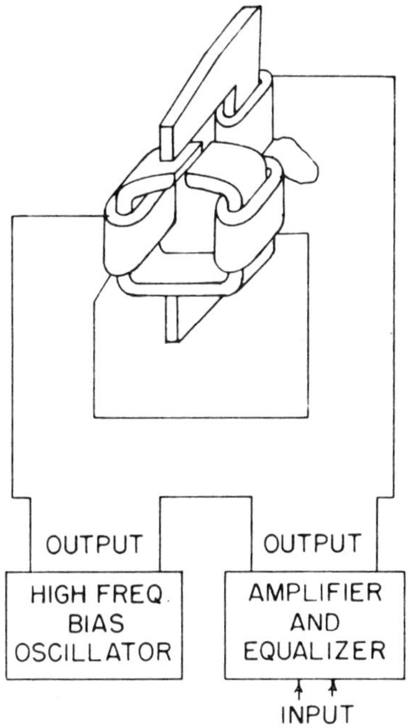

Figure 5-2. X-field head with overhanging polepiece. Coils are connected in series as shown (Camras 1952).

oped to overcome these objections (Camras 1964). The head shown in Figure 5-4 has a large erase gap ahead of the recording gap. The erasing field strength and its distance from the recording gap is such that it sets up a desirable cross-field component of bias at the recording gap. Where very high frequency bias is used (up to 100 MHz or more), the cross-field source can be a flattened wire imbedded in the core (Figure 5-5), the wire being energized with high enough current to give the required field.

Double Pole Opposed Head

A close relative of the cross-field head described above is the arrangement shown Figure 5-6, where a second head is placed above the back of the tape overlaying the main head. The field of the second head is usually depicted as shown in Figure 5-6*A*, but if it is spaced 25 μm or more from the tape coating (as it must be by the base layer of the tape), the direct influence of the opposed head on the tape is quite minimal, though it induces some flux into the main head core and thus influences the tape indirectly through the main head gap (Figure 5-6*B*). Sometimes the opposed head is energized with high-frequency bias only,

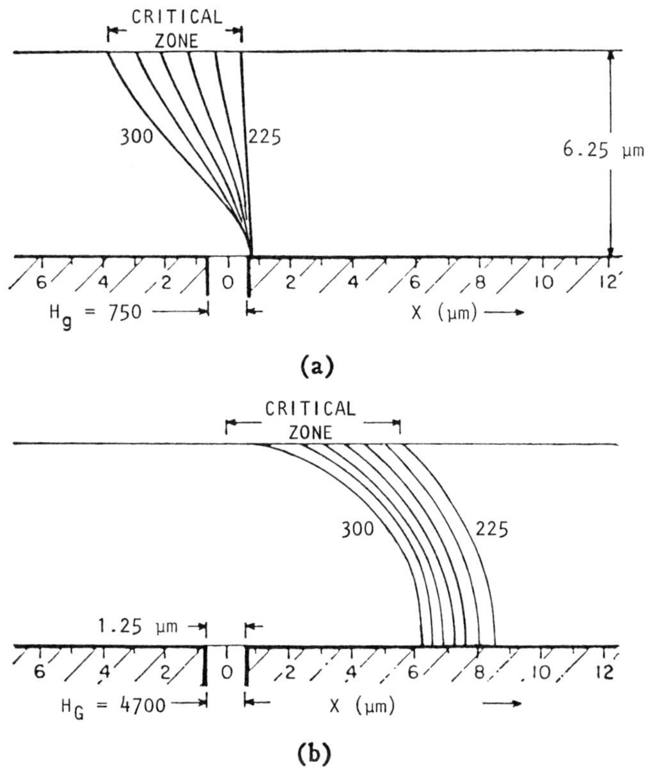

Figure 5-3. Effective recording field. (a) Critical zone for an X-field head. (b) Critical zone for same head used without X-field connection (Camras 1964).

Table 5-1 X-Field Head Connection Reduces Loss of High Frequency Signal, Compared with Same Head Without X-Field.

Connection	Bias required for undistorted 100-Hz output	Bias for max. 10-kHz output	Loss at 10kHz due to increased bias
Standard	1000 mA	600 mA	7 dB
X-field	650 mA	560 mA	0 dB
X-field	(Purposely overbiased to 1000 mA)		1.5 dB

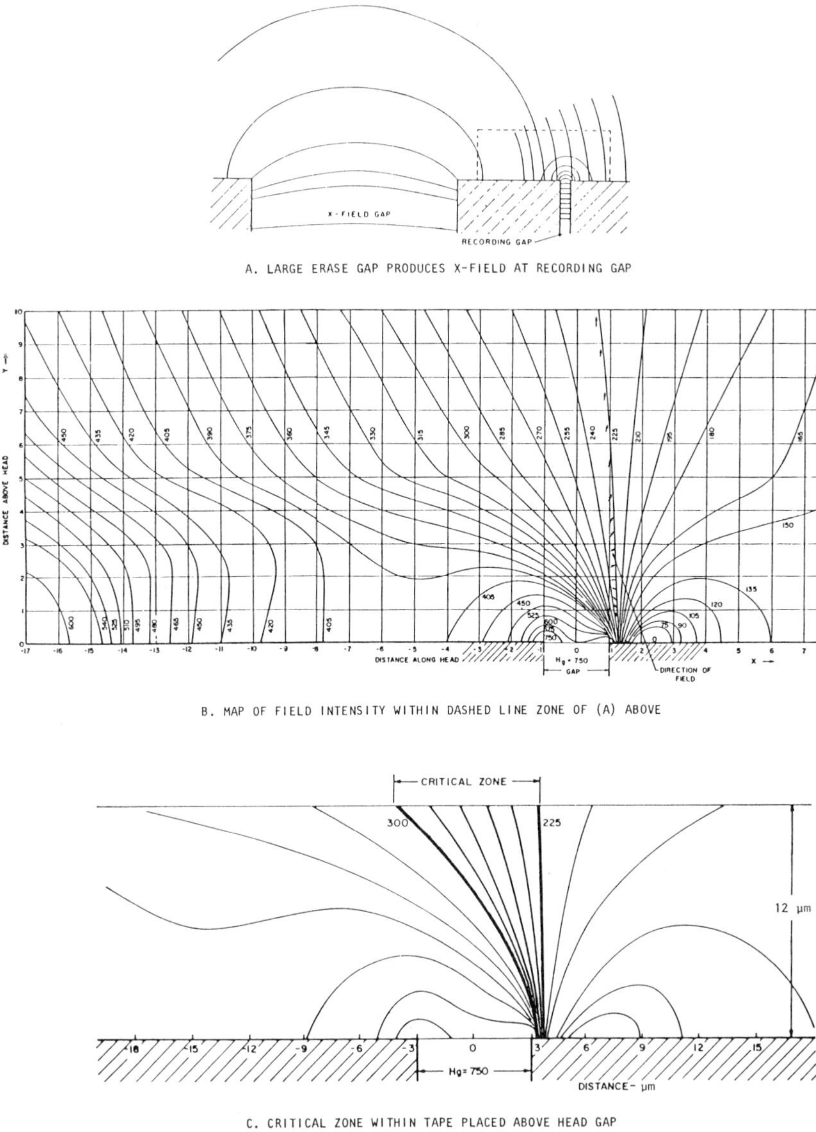

Figure 5-4. Single sided cross-field head (Camras 1964).

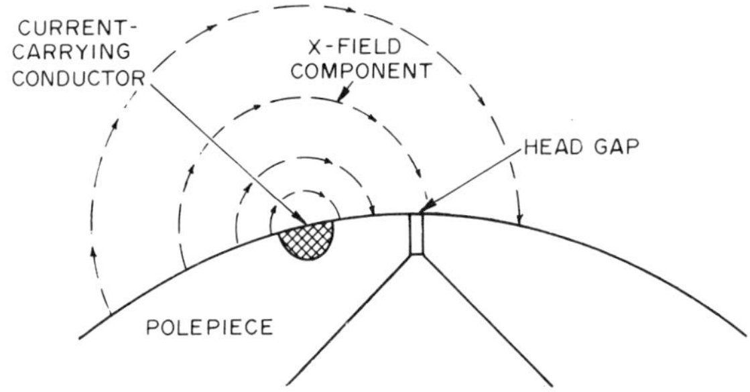

Figure 5-5. Cross-field component supplied by adjacent current carrying conductor.

while the main head is energized with signal only. In this case the main head gap supplies the signal flux and also supplies a significant part of the high-frequency bias by induction from the opposed head. The opposed head gap can be offset slightly upstream or downstream from the main gap to give various cross-field effects. It can also be reversed in polarity compared to the condition shown in Figure 5-6*A*, in which case the horizontal components of the recording field are reduced while the vertical components are enhanced (Figure 5-6*C*).

Outside-Coil Head

Another cousin of the cross-field head is the outside-coil head (Figure 5-7), where the main head gap and its associated magnetic structure are modified to take further advantage of the magnetization induced by a structure on the opposite (base) side of the tape (Camras and Sears 1961). The outside coil head is most efficient at high frequencies, and alleviates one of the dilemmas of head design where ordinarily the gap depth (d) not only should be minimized for high efficiency but also should be maximized for long wear. In the outside-coil head, the gap and the polepieces are relatively deep, but only the thin outer skin is utilized because the flux does not penetrate to the interior at high frequencies. Thus, the effective depth of the polepieces is calculated to be only about 1.8 μm at 1.0 MHz. The magnetic circuit of the main gap is not completed by a yoke, since this would bypass some of the flux.

Short circuited turns or a mass of copper is advantageous in the return circuit area below the gap. The energizing winding; or "outside-coil" is on the opposite side of the tape, and can be a coreless coil; but a ferrite core in the outside coil (Figure 5-7) improves the magnetic efficiency. Energizing flux ϕ enters and leaves the polepieces P,P over such a wide area that it is too low in density to affect the recording except where concentrated at the gap. As the head wears,

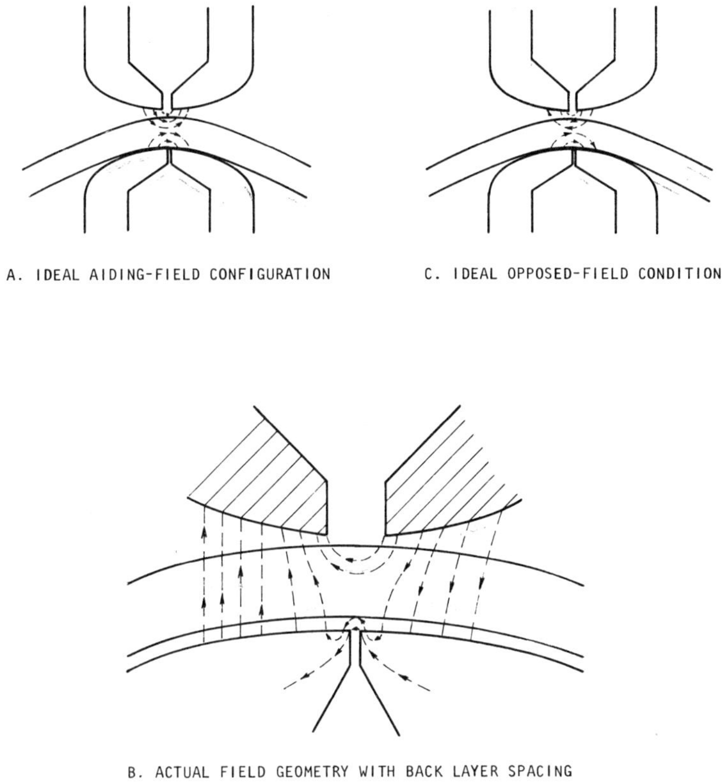

Figure 5-6. Double pole opposed heads.

a new outer skin becomes effective so that high efficiency is maintained over the life of the head. The outside coil principle is useful for both recording and playback.

HIGH FREQUENCY–HIGH FIELD HEADS

Turn-In-Gap (TIG) Heads

Recording and playback with a single conductor has always been an intriguing possibility. The basic idea is shown in Figure 5-8*A*, where one edge of a copper wire loop traverses the recording tape surface. For recording, a current through the wire sets up a magnetic field that encircles the wire and magnetizes the record. For playback, the field of the recorded tape links the wire and induces

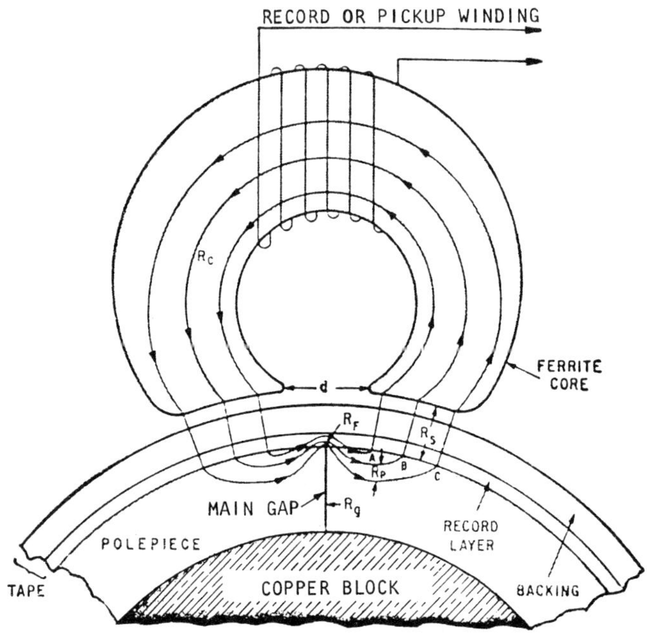

Figure 5-7. Outside-coil head. (Camras and Sears 1961).

a voltage in it. With an open loop of this kind, both the recording and playback are inefficient and of poor resolution because most of the magnetic circuit is through air.

Core Type TIG Heads. The shortcomings of open loop heads are cured by adding a core, as in Figure 5-8*B*. It then becomes advantageous to flatten the wire into a narrow strip, making the gap as short as possible, while providing enough cross section and surface area of the conductor to handle the current and to carry away the heat during recording (Figure 5-8*C*). The gap conductor now resembles the gap spacer found in conventional heads, except that it has an electrical connection at each edge. The core may be of ferrite or of any other high-permeability head material. It can be large or small; in fact, it can be fabricated by microcircuit techniques. Core loss is very low because a minimal volume of core carries the flux. High-frequency response is maximized because some of the tape flux is coupled directly to the conductor, while the rest traverses a short path through the core (Camras 1949; Wiegand and Zenner 1948).

Impedance Matching. The head winding is a single turn, making the head impedance so low that it is difficult to match it directly to drivers or amplifiers.

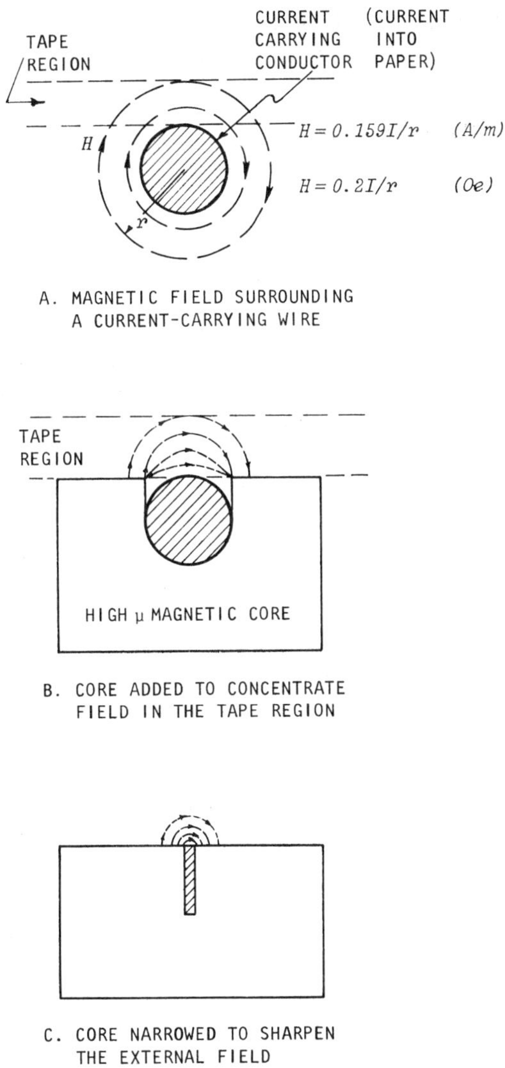

Figure 5-8. Turn-in-gap heads.

A transformer is usually necessary both for recording and for playback. The transformer may have a single turn on its primary and 20 to 2,000 turns on its secondary, as required for frequency range, voltage level, and impedance level. The connections between head and transformer must be very short and close together to avoid losses and hum pickup, making it advantageous to build the head and transformer as a unit (Figure 5-9*A*). The hybrid head design shown in *B* of the same figure uses the head core itself as a transformer.

Calculation of TIG Recording Field. The external magnetic field in the air surrounding the current-carrying conductor of Figure 5-8*A* is uniformly

$$H = \frac{0.159\, I}{r}, \tag{5-1}$$

Where H = magnetic field in A/m, I = current in amperes, and r = distance from the center of the wire in meters. (The corresponding cgs relation is $H = 0.2I/r$ with r in cm). When a tape of relative permeability μ_r is held against this wire it changes the field distribution, decreasing the H inside the tape to approximately $H_t = H/\mu_r$ while correspondingly increasing H of the remaining field in air. As an estimate, a field of at least H_t = 24 kA/m (300 oersted) including bias is required to record on ordinary magnetic oxide tape. Taking $r = 12.5 \times 10^{-6}$m, and $\mu_r = 10$, the current required to record on a tape with a single conductor 25 microns in diameter is

$$I = \frac{Hr}{0.159} = \frac{\mu_r H_t r}{0.159} = \frac{(10)(24000)(12.5 \times 10^{-6})}{0.159} = 18.8 \text{ amperes} \tag{5-2}$$

This is a formidable current, and the several watts to be dissipated or heat-sinked would cause the recording filament to run rather hot, to say the least.

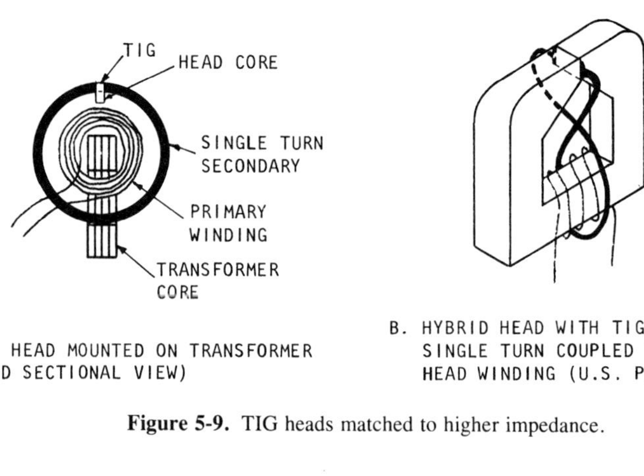

A. TIG HEAD MOUNTED ON TRANSFORMER (END SECTIONAL VIEW)

B. HYBRID HEAD WITH TIG ENERGIZED BY SINGLE TURN COUPLED TO MAIN HEAD WINDING (U.S. PAT. 2,651,097)

Figure 5-9. TIG heads matched to higher impedance.

The situation is improved by several orders of magnitude when the TIG is a ribbon imbedded in a high-permeability core. The gap field now becomes

$$H_g = \frac{I}{\ell} \tag{5-3}$$

Where ℓ is the gap length in meters. The tape permeability doesn't reduce the field appreciably, so a small current is enough to obtain a field of $H = 24000$ A/m in a gap 12.5 μm long:

$$I = H_g\ell = 24000(12.5 \times 10^{-6}) = 0.30 \text{ ampere} \tag{5-4}$$

The depth of the ribbon decreases the electrical resistance of the TIG, and the ribbon face in the contact with the core allows the ribbon to dissipate the several milliwatts of generated heat. The ribbon can be of micron or submicron thickness, deposited on the face of the polepiece; and can be of sufficient depth to take care of wearing problems if the tape rides in contact with the head.

Advantages and Disadvantages of TIG Heads. TIG head designs can reduce or overcome limitations of ferromagnetic head cores, they can operate at gigahertz frequencies and at flux densities above 2 telsa. Their chief disadvantages are their low impedance, which usually requires a matching transformer, their limited ability to dissipate heat, and problems of fabrication and of making adequate electrical connections.

CONTOUR-FREE HEADS

Closed Heads

The closed head, where the coil surrounds the gap (Figure 4-25*F*), overcomes two difficulties of ordinary heads: the rapid fall in response at very long wavelengths, and the contour effects that give irregular output at recorded wavelengths near the head dimensions (Camras 1944). The closed head is insensitive to external hum fields, which are bypassed through coil-less legs. It also enhances high-frequency response by minimizing the effect of core losses.

Long Wavelength Response. When the head coil surrounds the magnetic recording, it is linked with the *internal flux* of the record. The internal remanent flux, which is the retained magnetization in the record, does not diminish as the wavelength becomes longer; in fact, it increases slightly. Thus, the maximum flux linking the pickup coil remains constant even at an infinite wavelength (DC recording). A conventional head, in contrast, picks up a fraction of the *external* *flux*, and as the wavelength grows longer, links less and less of this external flux. The effect on low-frequency response is shown in Figure 5-10, where the closed head response follows the theoretical 6 db/octave slope, while the conventional head falls off at a steeper rate of 12 db per octave when the recorded wavelength exceeds the playback head dimensions.

Contour Effects. The coil of a closed head provides a magnetic potential difference at the gap edges for recording and a reciprocal sensitivity on playback, but the rest of the head core is at zero magnetic potential, including the regions where the tape approaches and recedes from the head. This is apparent in Figure 4-25*F*. There is no spurious pickup in these regions, so the contour effects explained previously are absent in closed-head designs.

Disadvantages. The main objection to a closed head is that the tape must pass through the coil. The beginning of a tape on a reel must be threaded through a closed loop, and while the tape is on both reels it cannot be removed from the machine, but must first be wound on to one reel or the other, or else cut for later splicing. This objection can be overcome by coils that open up for tape insertion; such coils can have a single turn or multiple turns.

Another difficulty is that closed heads are not readily adaptable for multiple tracks on a tape, difficulties occurring when several tracks are used simultaneously, and even when only one track is used at a time. Hundreds of tracks, closely packed together, are quite common on magnetic discs, drums, and wide tapes.

Contour-Free Open Head

An open head that minimizes irregularities in response due to contour effect is shown in Figure 5-11 (Zenner 1947). The coils are wound in opposition or

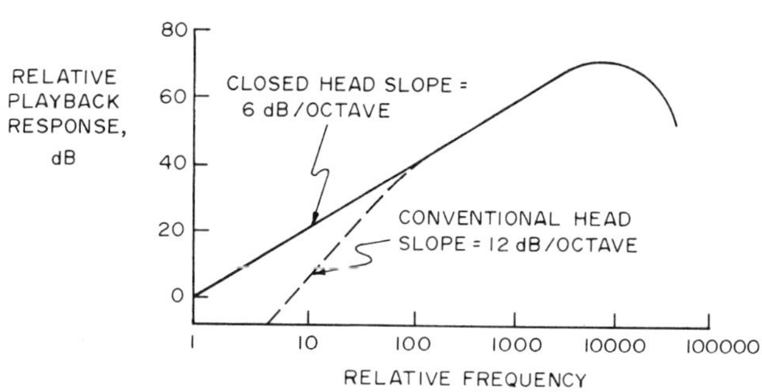

Figure 5-10. Long wavelength response of a closed head.

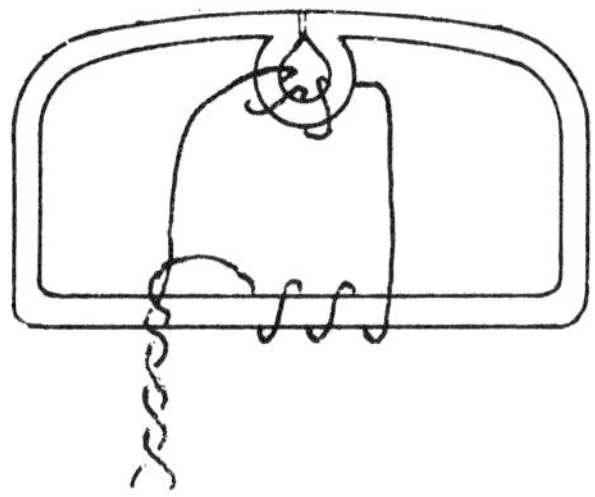

Figure 5-11. Contour-free open head.

bucking relation with respect to the large outside core, thus cancelling out the sensitivity to pickup of long wavelengths by the external core. The upper winding is effectively coupled to flux at the main gap, while the lower opposite winding is less effectively coupled because of its long core path.

PERMEABLE GAP HEAD

It is axiomatic that a head gap should have lowest possible magnetic permeability to maximize the fringing field that interacts with the tape. A gap spacer having the same (high) permeability as the core produces no external field. But if such a gap spacer carries a current, it becomes a source of magnetic field that can affect the tape. The gap current also sets up a magnetic flux circulating within the gap spacer; this may saturate it and cause it to act as a low-permeability gap. We can thus achieve a dynamic variation in effective gap dimension by varying the current through a magnetic gap spacer.

One application is a head with a very small gap for best playback resolution, the effective gap being enlarged during recording for optimum penetration through the recording layer. Another is a recording or erasing head whose gap "disappears" when the head is turned off, so a tape riding over it is free from possible degradation due to residual magnetization in head cores. This feature is also valuable in single-sided crossfield heads. U.S. Patent 3,686,467 (Camras) elaborates on construction and applications of permeable gap heads.

FLUX SENSITIVE HEADS

Playback heads with windings that link the recorded magnetic flux ϕ give an output voltage proportional to $d\phi/dt$. When the recorded flux has a constant maximum value ϕ_m and varies sinusoidally, $\phi = \phi_m \sin 2\pi ft$. Therefore, the output is $E = K(d\phi/dt) = K\phi_m 2\pi f \cos 2\pi ft$. Since K, ϕ_m, and 2π are constant, the output voltage rises in direct proportion to the frequency f, giving the familiar rising response curve (Figure 5-12*A*), which must be equalized rather

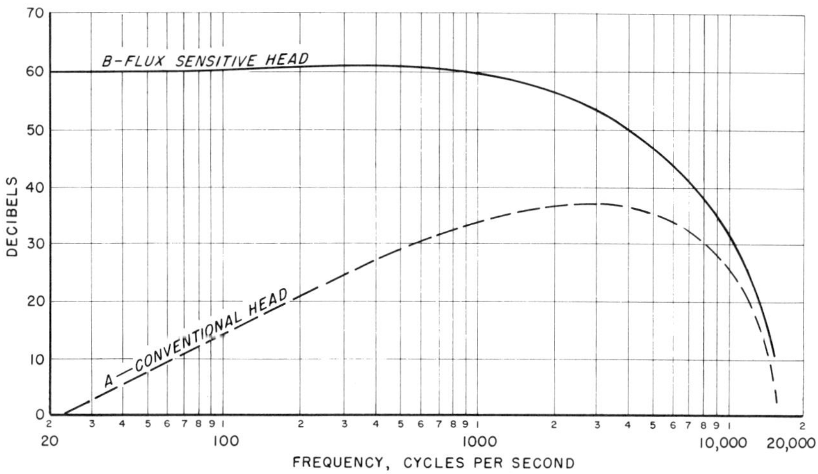

Figure 5-12. Unequalized playback response of flux-sensitive vs. conventional head for a constant current recording.

drastically to give a flat response of playback voltage corresponding to a constant recording current. Since "perfect" equalization is difficult to achieve over a wide range of frequencies, and since ordinary equalization (minimum phase) does not correct for certain errors in magnetic recording, a great deal of effort has been spent in developing a flux-sensitive head whose output is proportional to the recorded flux (Figure 5-12*B*) rather than to rate of change of flux.

Magnetic Modulator Heads

The earliest flux-sensitive head that gave high-output voltage and had a wide dynamic range used the magnetic modulator principle (Figure 5-13) (Wiegand 1952). The recorded flux is not sensed directly; rather, it unbalances a delicately balanced magnetic circuit that operates at a carrier frequency several times higher than the highest frequency to be played back. The balanced circuit, termed the *magnetic modulator*, is made of thin strips of permalloy, of dimensions approximately 0.025 × 1.0 × 22 mm, wound with carrier frequency energizing windings 1,2, which aid each other with respect to the local magnetic circuit *a b c d.* Pickup windings 3,4 each have a strong signal induced from the circulating flux through *a b c d*, but since they are connected in opposition, the induced voltages will cancel each other when the magnetic circuit is balanced. However, the voltages do not cancel if the magnetic circuit becomes unbalanced because of recorded flux from the tape channeled through it by pickup pole-

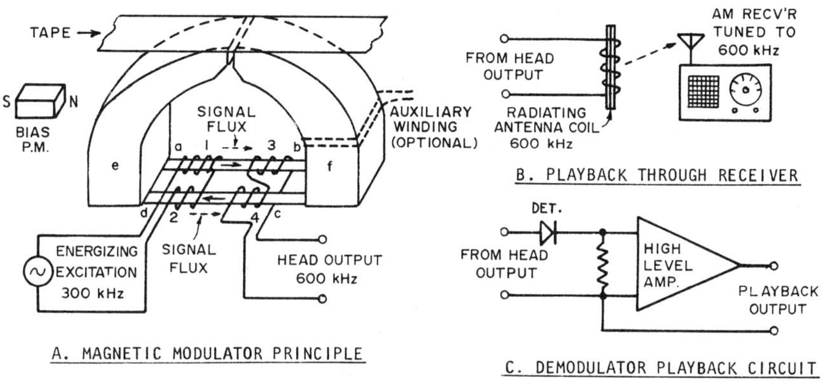

Figure 5-13. Magnetic modulator head.

pieces *e f*. High-frequency signals then appear at the output terminals, modulated by the variations in signal flux.

Because of symmetry, the positive and the negative cycles of energizing current both modulate the output in the same direction. Therefore the output carrier is twice the frequency of the energizing current. For example, if the energizing current in coils 1,2 is 300 kHz, the output from coils 3,4 will be at 600 kHz. A similar effect occurs for the modulation corresponding to the signal flux. Modulation theory indicates that this action generates a suppressed carrier double sideband modulation. The original signal cannot be recovered from suppressed carrier modulation by an ordinary diode detector unless the main magnetic circuit is offset by a DC component. The offset is conveniently supplied by a small nearby permanent magnet or by current in an auxiliary winding linking the signal magnetic circuit. The auxiliary winding on the main core may serve as an ordinary recording winding if the same head is to be used both for recording and for playback.

Noise Considerations. The magnetic flux in the thin permalloy strips generates Barkhausen noise. There is an optimum cross-sectional area for the strips that gives maximum signal to noise ratio, and this has been found to be 0.8 × 0.025 mm (0.031 × 0.001 in) for a full width recorded channel on 6.25 mm ($\frac{1}{4}$ in) audio tape. Under these conditions, signal-to-noise ratios greater than 70dB are obtained. Strips of other dimensions having the same cross-sectional area give equivalent results.

Output Voltage. The magnetic modulator acts as a magnetic amplifier. Strong signals in the order of a volt or more are obtained directly from the pickup windings on the head, without further amplification. These windings can be connected to an antenna loop (Figure 5-13*B*), which will radiate the 600 kHz signals, for example, to nearby AM radios, and the radios will then reproduce the tape recordings just as if they were tuned to a broadcast station. Or the head output can be demodulated (Figure 5-13*C*) and the high output can be utilized with only a moderate amount of amplification, much less than the amplification required with an ordinary head. Although their frequency response is inherently flat, modulator heads can use some high-frequency equalization to compensate for tape and core losses and may need low-frequency boost when recorded wavelengths exceed head core dimensions.

Alternative Forms of Magnetic Modulators. Figure 5-14*A* shows a simplified design of modulator head requiring only a single strip of permalloy surrounded by oppositely wound energizing coils and by a single pickup coil. The strip is locally magnetized in two segments, its operation being analogous to the two-strip modulator. Another improvement is the synchronous detector circuit (Figure 5-14*A*), which uses a frequency doubler to supply the missing carrier, eliminating the need for permanent magnetic bias and doubling the signal voltage swing for a given flux capability (and noise generation) of the core. It is possible also to use an ordinary head as a modulator head by drilling a pair of holes through the polepieces and threading a winding through the holes, as shown in Figure 5-14*B*. Such a head is less efficient than the previous types, but it is much easier to build.

Advantages of the Magnetic Modulator Head. Without equalization, the magnetic modulator head gives a flat frequency response (except for longest and shortest wavelengths) and reproduces waveforms of recorded flux with good accuracy. Linearity is excellent and distortion is low. The output voltage is high (a volt or more) and is constant at all speeds, down to zero speed, for a given wavelength of recording. Signal-to-noise ratio, above 70 dB, exceeds the capability of most tapes. Modulator heads are stable and reliable.

Disadvantages. Modulator heads are more expensive to build than ordinary heads, and they require special circuity: an oscillator for energization and a detector for demodulation. They may require adjustment, as for example of the biasing magnet or resonant circuits. The high-frequency energization and/or DC biasing may set up a field in the reading gap that affects the record, although tests have indicated negligible effects. The high frequency carrier may be picked up in nearby electronic circuits.

Despite these disadvantages, the modulator head has been the preferred choice for a flux-sensitive head. Improvements in Hall-effect heads and in Magneto-

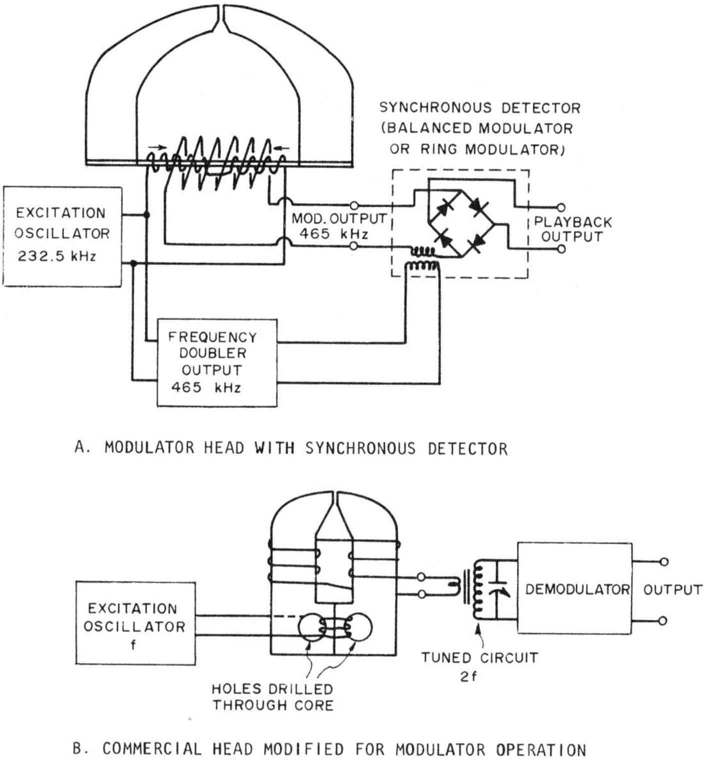

Figure 5-14. Alternative forms of modulator heads.

resistive heads may challenge the modulator head, but these other types have also had problems in the past.

Electron Beam Heads

An electron beam passing through a weak magnetic field is deflected through a small angle proportional to the field. If the magnetic field of a recorded tape is channeled so that it can deflect the beam, and if the beam strikes a target shaped so that its current changes in proportion to the beam deflection, then the target will given an electrical output signal proportional to the recorded flux. As shown in Figure 5-15*A*, the electron beam tube is built like a small cathode ray tube but with magnetic deflection polepieces that are coupled to magnetic flux picked up from the tape (Skellett et al. 1953). The beam strikes a split target, such that

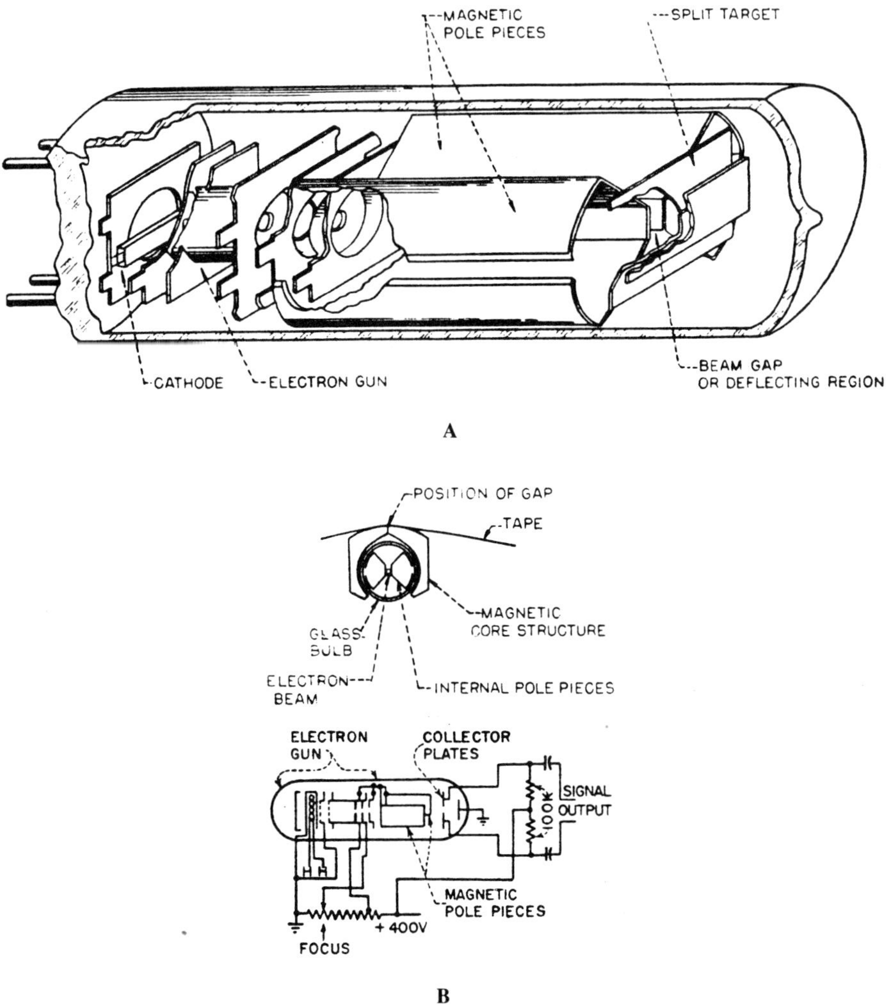

Figure 5-15. Electron beam head (Skellett 1953) (Electronics, Oct 1953).

any beam deflection perpendicular to that split unbalances a push-pull output circuit, as shown in Figure 5-15*B*. The beam deflecting field available from a full-width recording is about 3.18 A/m (0.04G) or $\frac{1}{10}$ as much as the earth's magnetic field; but the beam deflection in this small field is enough to give 0.6 volts peak output between the two collector plates in the circuit shown in Figure 5-15*B*. Over the working frequency range at normal recorded levels, the head gives 0.2 volt output.

The high sensitivity to weak fields requires careful magnetic shielding of the head. One successful arrangement uses three nested shield boxes; the electron beam tube is mounted inside the innermost shield with slots cut in the boxes to allow the recorded tape to pass through.

Although the electron beam playback principle was announced as early as 1953, it has not been applied to any commercial devices, probably because the tube itself is expensive and requires elaborate shielding, a regulated high voltage power supply with beam control electronics, and a skilled operator to keep the beam in adjustment. Signal-to-noise ratio is poor in comparison to the performance of ordinary heads, and since there are other ways of obtaining a flux-sensitive output, it has not been competitive.

Electron Cloud Heads

To overcome disadvantages of the electron beam head, a smaller and simpler version which resembled a seven-pin miniature vacuum tube was designed. The somewhat diffuse low-voltage beam (electron cloud) is directed into a constricted neck and then allowed to fall back on to a split target electrode at the front end of the electron gun. While in the constricted neck, the beam is acted on by the field between external polepieces of a head coupled to the magnetic tape (Camras 1963a).

Transistor Sensors

Magnetically sensitive transistors have been announced from time to time. Such transistors could be solid state substitutes for the electron beam tape sensors. Early devices have not been developed commercially because they did not promise to have advantages over available solid-state sensors such as Hall-effect or magnetoresistive devices (Lopez 1982).

Hall-Effect Heads

The simplest of all flux-sensitive heads makes use of the Hall-effect principle, shown in Figure 5-16, where a magnetic field unbalances the current flow I in a semiconductor and gives an output voltage across the side tabs equal to

$$E_H = \frac{RIH}{T} \tag{5-5}$$

the units being in any self-consistent system (Camras 1962). For high sensitivity, a semiconductor with high carrier mobility (μ) is chosen. Indium-antimonide in single crystals or deposited by special techniques is a preferred material

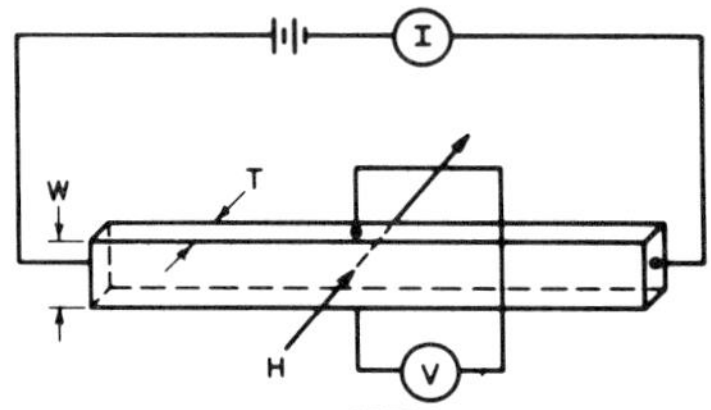

$$\text{HALL VOLTAGE, } V_H = \frac{RIH}{T}$$

I = CURRENT
H = MAGNETIC FIELD
T = THICKNESS
R = HALL COEFFICIENT $= \frac{1}{ne}$
n = NO. OF CARRIERS
e = CHARGE PER CARRIER
μ = MOBILITY OF CARRIERS
ρ = RESISTIVITY $= \frac{1}{ne\mu}$
$\frac{R}{\rho} = \mu$
μ = 2000 FOR GERMANIUM
μ = 30000 FOR InSb

Figure 5-16. The Hall effect.

(Kotera 1978). The Hall effect is responsive to frequencies as high as 10,000 MHz.

In Figure 5-17 the sensitive Hall element is shown conveniently placed in the back gap of the polepiece structure. The head pictured here is for playback only, and yields a flat playback response, as in Figure 5-12*B*. If the same head is to be used for recording, a coil can be added.

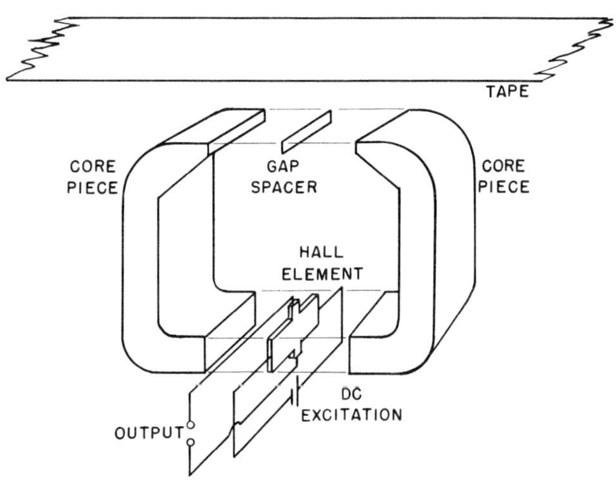

Figure 5-17. Elements of a back-gap Hall effect head.

Making the Hall element very thin improves the sensitivity of and minimizes gaps in the magnetic circuit of the core. Such thin elements lend themselves for use as a main gap spacer (Figure 5-18), eliminating the back gap and increasing the high-frequency response. When the Hall element is in the main gap, the upper tab may be eliminated by a phantom equipotential tap, a push-pull arrangement, or a K-shaped element, as shown in Figure 5-19.

For maximum output, the exciting current (*I*) is set at the highest value that will not cause overheating. Signal-to-noise ratio improves as current increases, up to the point where self-generated noise in the Hall element predominates over the tape and amplifier noises. Beyond this point, signal and noise rise together in the same ratio; ratios greater than 60 db have been attained.

High-frequency AC can be used instead of DC for exciting the Hall element, in which case the output will be a modulated carrier that can be amplified and filtered more readily than a direct signal.

Magnetoresistive (MR) Heads

Although it has been known since 1857 that the electrical resistivity of magnetic materials changes when magnetized, this principle was not seriously considered

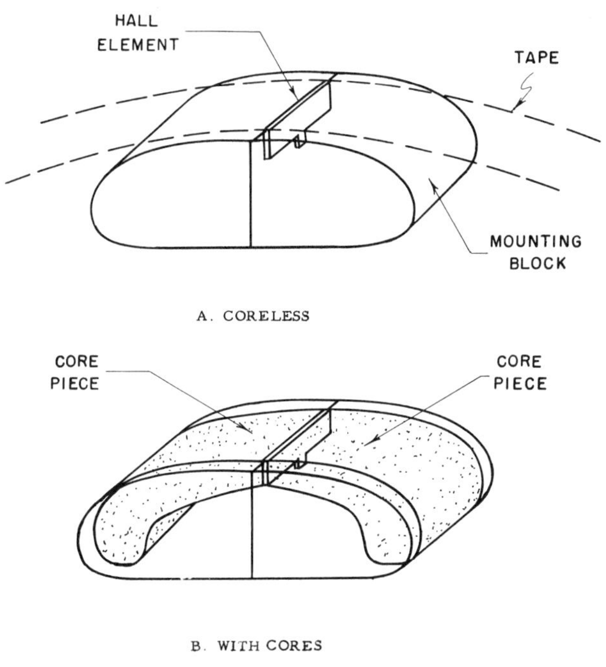

Figure 5-18. Front gap Hall-effect heads.

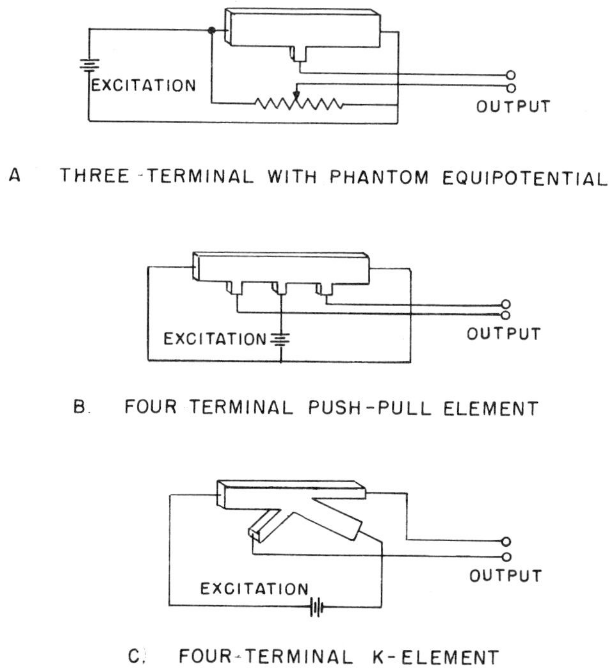

Figure 5-19. Circuits and elements for front gap heads.

for the flux sensitive playback heads until about 1971 (Hunt et al. 1971; Thompson 1957). Since that time, considerable effort has been made to adapt the unusual properties of magnetoresistive elements to make a useful magnetoresistive head (MRH). The classical configuration for measuring magnetoresistance is shown in Figure 5-20*A*, where the applied field of the magnet (*N*,*S*) is in the same direction as the current through the specimen, which is for example, a thin film of permalloy. The increase in resistance R of a demagnetized specimen when a saturating field is applied is ΔR, amounting to about 2% of the initial resistance for ordinary 79-permalloy at room temperature, that is

$$\frac{\Delta\rho}{\rho} = \frac{\Delta R}{R} \cong 0.020, \tag{5-6}$$

where ρ is the resistivity of the material. The same change in resistivity $\Delta\rho$ occurs regardless of magnetic polarity. The magnetoresistivity of some ferromagnetic materials is given in Table 5-2.

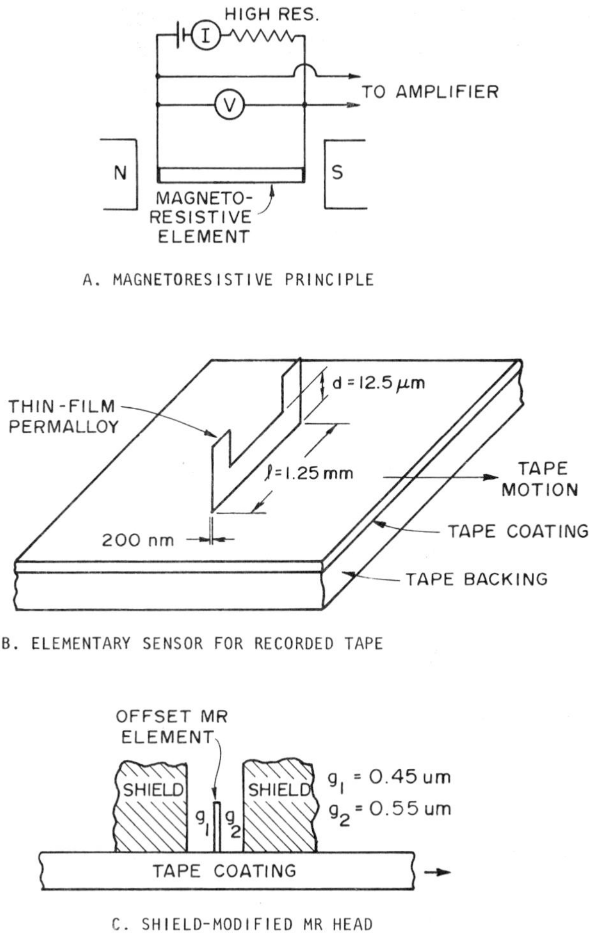

Figure 5-20. Magnetoresistivity and magnetoresistive heads.

In general, the magnetoresistivity coefficient increases at low temperatures, a figure of 0.16 having been attained at −195° C. The change in resistivity is sensitive to the angle of the saturating field, becoming negative for nickel in a perpendicular field; it is also sensitive to magnetostriction, so that tensioning or compressing the specimen can aid or counteract the effect of the field. Theory explains that domain orientation influences the passage of conduction electrons and that domain orientation is affected by applying a magnetic field or, similarly, by applying stresses to the material. If the domains are initially distributed

TABLE 5-2. Approximate Magnetoresistivity of Some Ferromagnetic Materials When Saturated Longitudinally.

Material	Temperature (° C)	$\Delta\rho/\rho$
79 Permalloy	20° C	0.020
69 Permalloy	—	0.016
89 Permalloy	20° C	0.040
89 Permalloy	−95° C	0.073
89 Permalloy	200° C	0.014
Nickel	—	0.016
Iron	—	0.003
50 Cobalt Iron	0° C	0.010
30 Cobalt Nickel	0° C	0.060
30 Cobalt Nickel	−195° C	0.163
7.5 Copper Nickel	0° C	0.024
3 Silicon Iron	0° C	0.005

uniformly in all directions and are subjected to a magnetic field at an angle θ with respect to the direction of current, then

$$\frac{\Delta\rho}{\Delta\rho_{\max}} = 3/2\ (\cos^2\theta - 1/3), \tag{5-7}$$

indicating that $\Delta\rho$ is maximum when $\theta = 0$, becomes zero when $\theta \cong 55°$, and becomes negative for θ between 55° and 90°. At 90°, the sensitivity $\Delta\rho$ is half as much as at 0°.

Elementary MR Head. An elementary sensor for the recorded magnetic tape shown in Figure 5-20*B* consists of an evaporated film of permalloy 200 nm thick on a glass substrate. The film is trimmed to a depth of 12.5 μm (0.5 mil) and a length (ℓ) of 1.25 mm (50 mils), corresponding to the recorded track width. The element has a resistance of about 500 ohms and is usable with a current of 10mA. When it is held edgewise against a moving tape that has previously been recorded, it gives a peak output of 25mV, with discernible output to wavelengths as short as 1 μm and a maximum signal-to-noise ratio of 67 db. This elementary sensor is a miniature version of the single polepiece head shown in Figure 5-23*A*, sensing the vertical component of tape field. It is very simple in construction compared to ordinary heads that have coils, pole-piece pairs, and gaps; but it can be used for playback only. Its midrange frequency response has a 6 db per octave rising characteristic, like conventional rate-of-change-of flux heads rather than like flux-sensitive heads (see Figure 5-12). This is attributable to its small dimensions compared to the longer recorded wavelengths. Unlike rate-of-change-of-flux heads, the output at a given wavelength remains constant, regardless of tape velocity, giving the same output at near zero speed as at high speeds.

Advantages of MR Heads. Great interest has been shown in magnetoresistive heads because they have:

1. High sensitivity
2. High resolution
3. Good response at low tape speeds
4. Output independent of tape speed
5. Capabilty of multitrack heads on one substrate
6. Capability of being fabricated by microcircuit deposition techniques

Limitations of MR Heads. Magnetoresistive heads have the limitations of most other no-winding heads: They are for playback only and cannot record or erase; their signal-to-noise ratio is generally not as good as for the best conventional heads; and they have special problems in practical applications, such as stability, uniformity, wear resistance, and so on.

Magnetoresistive heads have, in addition, some specific problems of their own:

1. They require both current biasing and magnetic biasing.
2. Their magnetic biasing is of a high order, approaching element saturation for best sensitivity.
3. Their output is extremely nonlinear, introducing considerable distortion.
4. Thin film elements of Angstrom dimensions are used, which are subject to deterioration and corrosion.
5. Thermal sensitivity generates noise due to frictional heating when tape runs adjacent the element.
6. Barkhausen noise is prominent.
7. Thin magnetic materials generate a strong demagnetizing field perpendicular to their surface.

Some of the above problems are alleviated by MR heads of special designs, which also may improve the recording density as described below.

Special Designs

Shield-Modified MR Head. The elementary head shown in Figure 5-20*B* may be modified by adding a pair of side shields, as in Figure 5-20*C*, to increase the resolving power (Schwarz and Decker 1979). The side shields should cause interference at wavelengths comparable to gap dimensions, but this effect is alleviated by locating the element off-center. A head of dimensions indicated in Figure 5-20*B*, has been able to read a density of 5600 flux changes per mm (140,000 bits/in).

Biased Film Heads. Magnetic bias can be supplied conveniently by a conductor, as in Figure 5-21*A*. An additional film of titanium is deposited adjacent to the MR film. The titanium film carries a current that magnetically biases the MR film.

Super-Narrow MR Head. In Figure 5-21*B*, the head core is like the core of a conventional head with polepieces (*P,P*) defining a gap (*G*), which rides against

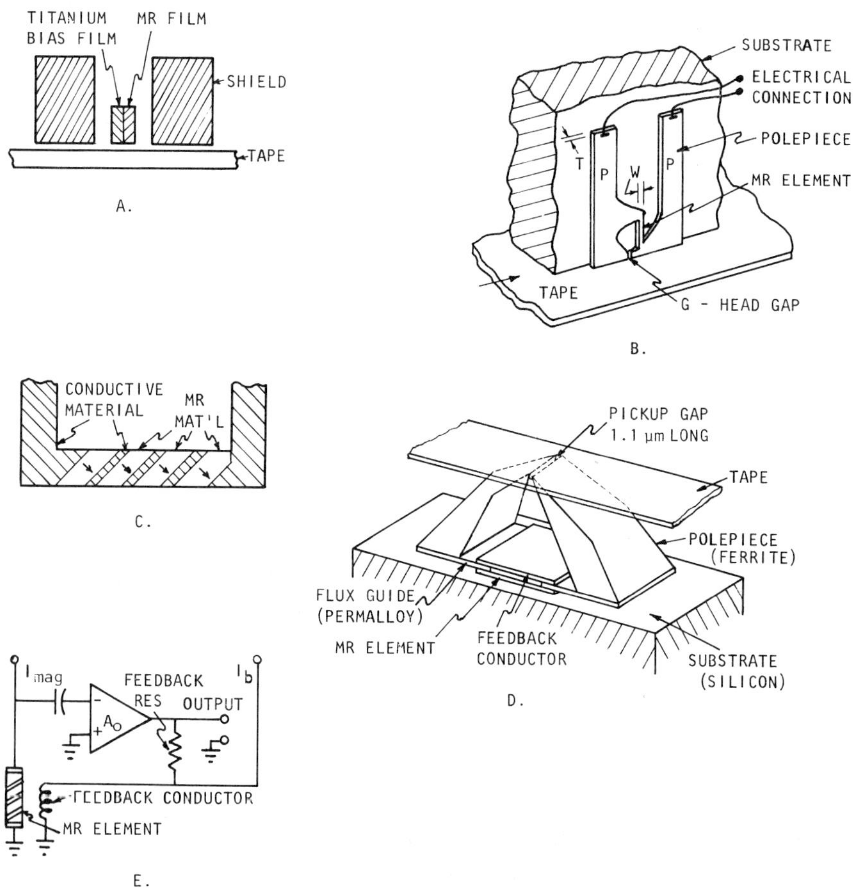

Figure 5-21. Alternative magnetoresistive heads.

the tape surface. The MR element is a constricted portion of width (w), to which the electrical connections are made by polepiece extensions (P,P). The entire head is less than 1 mm in size, made by deposition techniques with a thickness T of 2μm or smaller, to operate on a super-narrow track.

Barber Pole MR Head. Although MR heads have a nonlinear response, the best approximation to linearity is achieved if their sensor element is magnetically biased at an angle approximately 45° from the direction of current flow. One way to achieve this relation is to direct the current flow to a 45° angle by coating the MR film with conductive stripes (Figure 5-21*C*). Current flowing through the uncoated MR portions then takes the shortest direction between conductors, as shown by the arrows, and provides both the magnetic biasing and current biasing (Kuijk et al. 1975).

Feedback MR Head. Magnetic feedback directly to the MR element has been used to reduce its nonlinearity for application to audio recorders. In Figure 5-21*D*, the MR element is in the back gap of a set of polepieces. Closely adjacent is a negative feedback conductor, energized from the amplifier output in a direction to oppose the signal field at the MR element. Except for the main polepieces that contact the tape, the entire head is made by depositing thin films on a substrate. The basic amplifier with feedback circuit is shown in Figure 5-21*E*. It was found that the feedback reduced Barkhausen noise in addition to reducing distortion. One experimental feedback MR head for a cassette recorder has an inherent signal-to-noise ratio of 70 db (limited by tape noise to 60 db), with negligible distortion and Barkhausen noise, and noncritical setting of the bias point (deNiet and Vreeken 1979).

THIN FILM HEADS

Microcircuit techniques lend themselves admirably to the fabrication of thin film heads. Highly developed for integrated circuits, these techniques enable one to deposit magnetic cores, electrical conductors, and insulators, which are the three basic elements of a magnetic head. As with integrated circuits, hundreds of devices can be made simultaneously by vacuum deposition, photo masking, etching, and laser-trimming techniques.

Some features of thin film heads are beneficial, but others present great difficulties if such heads are to be adapted as substitutes for ordinary heads. Problem areas are the fabrication of multiturn windings of reasonably low resistance, deposition and adherence of magnetic film cores of sufficient thickness to carry the required flux, and mounting the head in a wear-resistant assembly for accurate interfacing with the tape.

The first commercial use of thin film heads was as flying heads for data discs.

Thin film heads are especially suitable for this application, where wear is not a problem, where microscopic size and low mass are great advantages in following the tracks, and where heads 25 μm wide or less are easily fabricated. Figure 5-22 shows a commercial thin film head for magnetic discs. Further details of its application are found in Chapter 14.

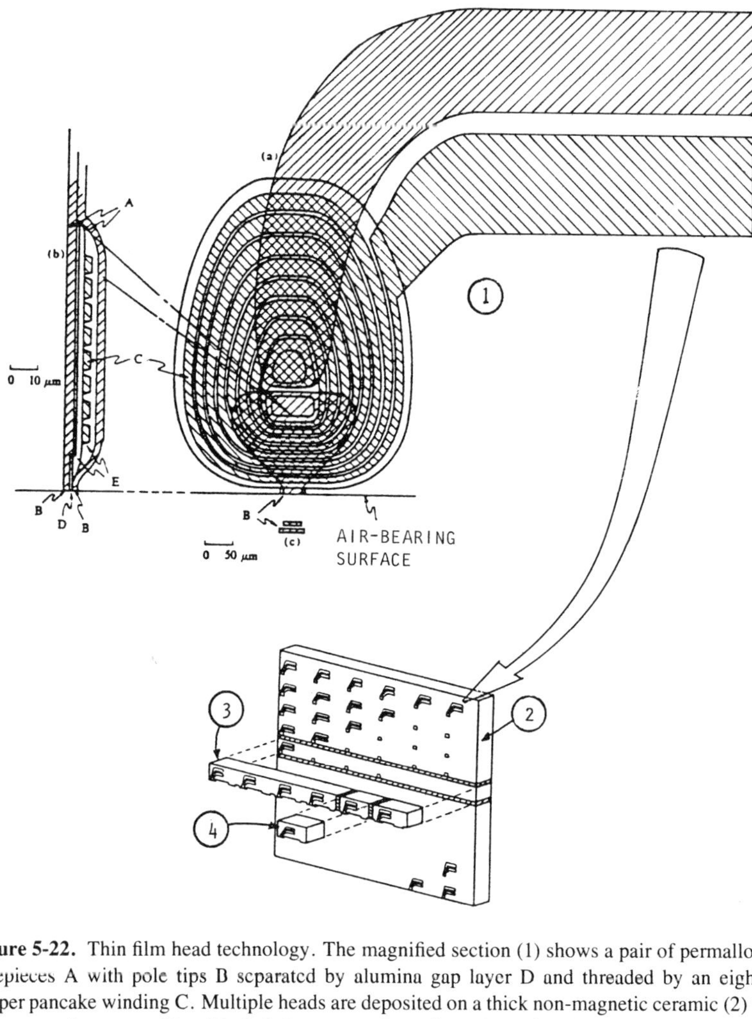

Figure 5-22. Thin film head technology. The magnified section (1) shows a pair of permalloy film polepieces A with pole tips B separated by alumina gap layer D and threaded by an eight turn copper pancake winding C. Multiple heads are deposited on a thick non-magnetic ceramic (2) which is later cut into rows (3) with their bottom surfaces lapped to give the head-throat height, and shaped to form the sliders or skis that will float on the high speed magnetic disc surfaces. The rows are then sliced to give individually mounted heads (4). (Harker et. al., IBM J. Res. Develop., Sept. 1981).

STYLUS HEADS

A small gap between magnetic poles is commonly used to concentrate a magnetic field on a small region of the record but a sharp magnetic stylus is also practical and was widely used in early recorders. The field of a single stylus head in the tape region (Figure 5-23*F*) resembles the dual or orthogonal function

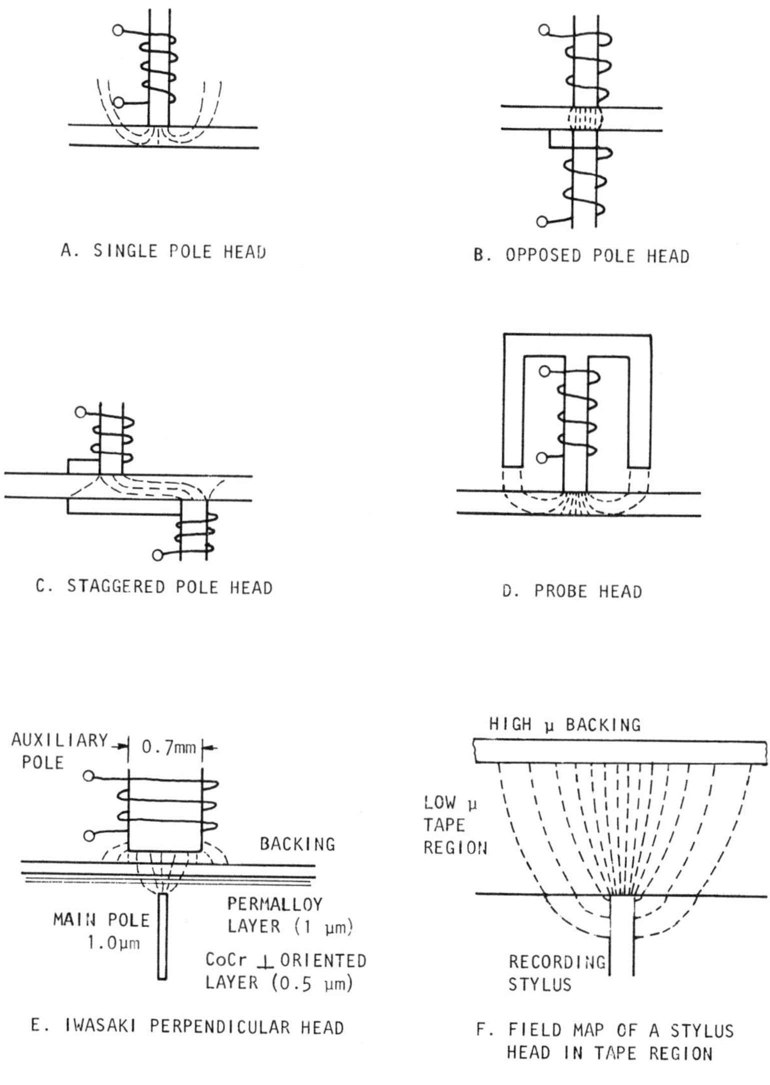

Figure 5-23. Stylus type heads.

of the gap fields that have been described in detail in Figures 4-3 to 4-8. Although stylus heads have been considered obsolete since about 1940, there was a revival of interest in the 1950s when probe heads were announced, and again in the late 1970s when very high recording densities were obtained with a stylus head on perpendicularly oriented tape coatings (Iwasaki and Nakamura 1978).

Early Stylus Heads Telegraphone machines invariably used stylus type heads consisting of long straight polepieces (Figure 5-23), perpendicular to the record axis, with their tips sliding along the wire or tape (see also Figure 2-14). Although a single polepiece (Figure 5-23*A*) was effective, two opposed polepieces (Figure 5-23*B*) gave more output and better resolution. The opposed polepiece head magnetized the tape essentially perpendicular to its surface. Still more output at a sacrifice of resolution was obtained by staggering the polepieces (Figure 5-23*C*). Staggered polepieces gave somewhat longitudinal magnetization of the record, which, in the case of wire recordings, minimized changes in output due to the twisting of the wire between recording and playback operations.

Stylus Wear. In stylus heads, when straight polepieces of substantial thickness were used, they were movable within the coil, being spring pressed against the steel tape or wire; therefore as they wore against tape their faces were renewed, and when worn out they could be replaced in much the same way as the brushes on a motor commutator. Thinner polepieces, too fragile to be self-supporting, were imbedded in a nonmagnetic bar with only the magnetic tip exposed and then used as above.

Disadvantages. Early stylus heads used relatively thick polepieces and thick record media that ran at high speeds because resolving power was poor. Their magnetic efficiency was low and they were sensitive to hum pickup from external fields. The two pole version was difficult to thread with tape. Thin polepieces tended to saturate magnetically. Contour effects occurred at wavelengths in the mid-band of the operating range rather than at the low-frequency end, where they were less troublesome.

Probe Heads. The single-pole head shown in Figure 5-23*D*, called the *probe head*, was developed in the 1950s for digital recording and was capable of high recording densities. This head took advantage of the fact that in ultra-short wavelength recordings, the vertical component of magnetization is prominent and contributes most importantly to resolving power and high bit density.

Thin Film Stylus with Perpendicularly Oriented Tape Coating. A later version of stylus recording has aroused great interest. With a thin film recording-stylus and a perpendicularly oriented tape coating of cobalt-chromium (CoCr) (Figure 5-23*E*), a recording density of 4000 bits/mm (100,000 bits/in) was reported (Iwasaki et al. 1979). The playback voltage output was doubled without affecting the resolution by adding an underlayer of permalloy to complete the back magnetic circuit through the vertical particles. The energizing winding was on an auxiliary polepiece to reduce saturation effects in the thin-film main polepiece. For playback of the stylus-made recordings, a ring head (conventional gap type) gave more output and a higher null-frequency than the stylus head did.

Advantages. The revival of perpendicular recording and stylus heads may be attributed to thin film technology that makes special tapes and a very fine stylus possible. Also the response at medium and long wavelengths is no longer important for digital and FM applications. It is a fact that as recording density increases (short wavelengths), the demagnetization increases for the horizontal recorded component, while demagnetization decreases for the vertical recorded component. Therefore, at extremely high densities, special attention is given to the vertical component.

BOUNDARY DISPLACEMENT HEAD

Boundary displacement recording, which is a variable-area magnetic recording, requires an unusual recording head (Daniels 1952). As seen in Figure 5-24*A*, the right side of the head has a lamination stack biased by a permanent magnet, so that each lamination differs from its neighbor in magnetic potential, ranging from full north on one side, to zero at the center, to full south on the other side. Figure 5-24*B* and *C* show a U-shaped signal core that completes the magnetic circuit, with a recording gap at the bottom. When there is no signal in the winding, the signal core is at a neutral, or zero, magnetic potential, matching the center lamination, but all other laminations are at a strong north potential if they are on the left side or a strong south potential if they are on the right side. A tape running past the head gap is then recorded as in the no-signal condition of Figure 5-24*D1*, with half of the recorded track saturated in one direction and the other half saturated in the opposite direction. With a signal in the winding, the signal core momentarily assumes a potential that is north or south of neutral and therefore matches some other lamination to the right or left of center, moving the non-saturated zone across the tape, as in the sinusoidal signal example of Figure 5-24*D2*.

Boundary displacement recordings are played back with a conventional head, which averages the opposing fluxes separated by the boundary. The recording on the tape can be made visible by allowing the tape surface to attract magnetic powders that can be fixed to the tape or transferred to a more suitable base. The

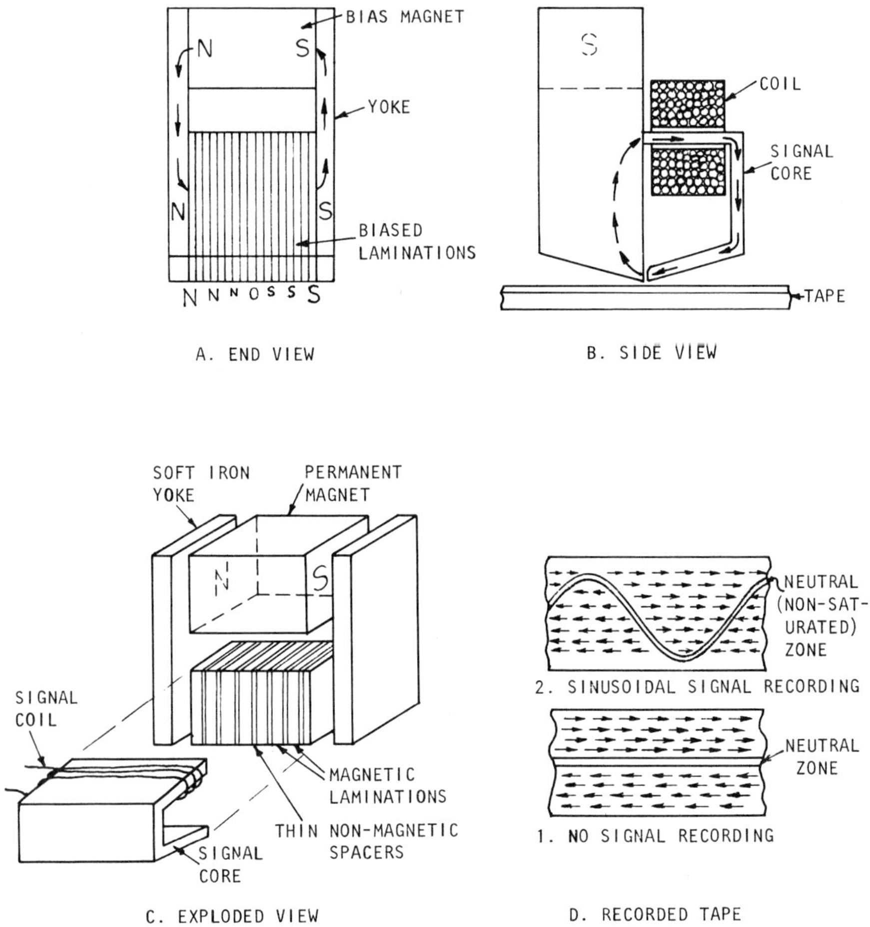

Figure 5-24. Boundary displacement head.

pattern is an oscillogram of signal waveform. Wide heads and tapes of 50 mm (2 inches) width or more yield oscillograms of a useful size.

Advantages of boundary displacement recordings include the following:

1. Bias (DC or high frequency) is not required.
2. The recording is insensitive to the variations in head-to-tape contact that cause dropouts in conventional recording.
3. The recorded waveform can be made visible by magnetic particles.

Disadvantages of boundary displacement recording are that:

1. It gives a higher noise level than conventional recording, corresponding to *DC* saturated tape rather than to erased tape, although signal-to-noise ratios in excess of 40 db have been achieved.
2. It has limited short wavelength response, due to finite width of the boundary.

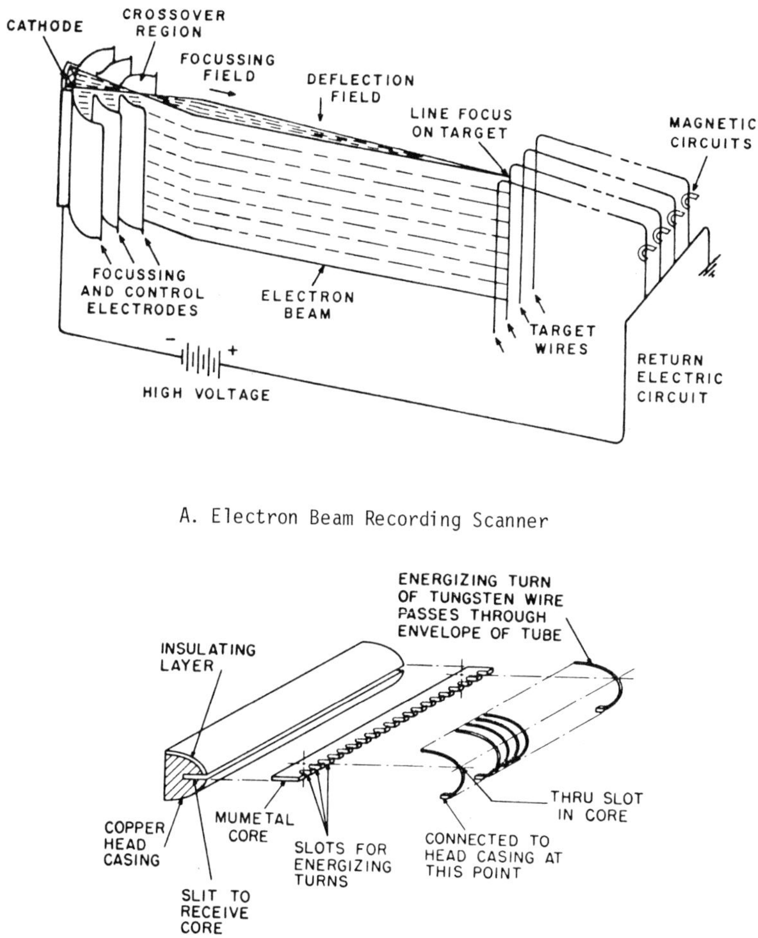

A. Electron Beam Recording Scanner

B. Detail of Recording Head

Figure 5-25. Electron scan recording system (Camras 1963).

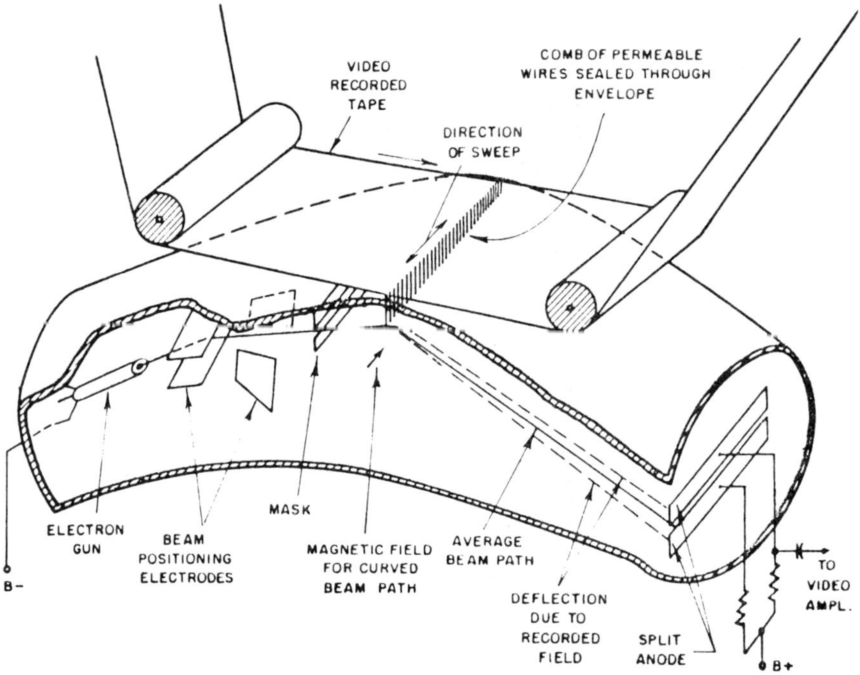

C. Permalloy-Comb Playback Tube

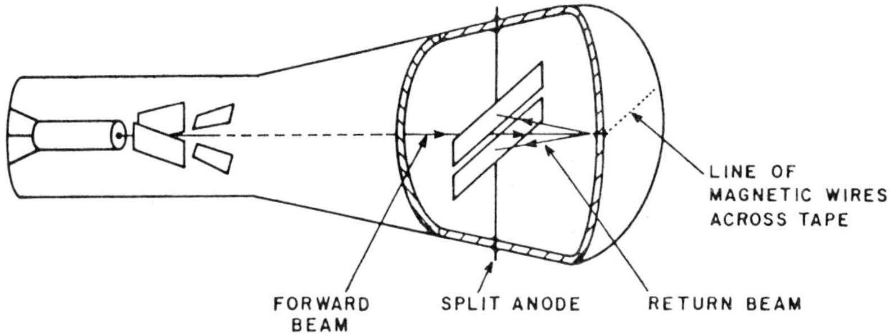

D. Reflection Type Playback Tube

Figure 5-25. (*Continued*)

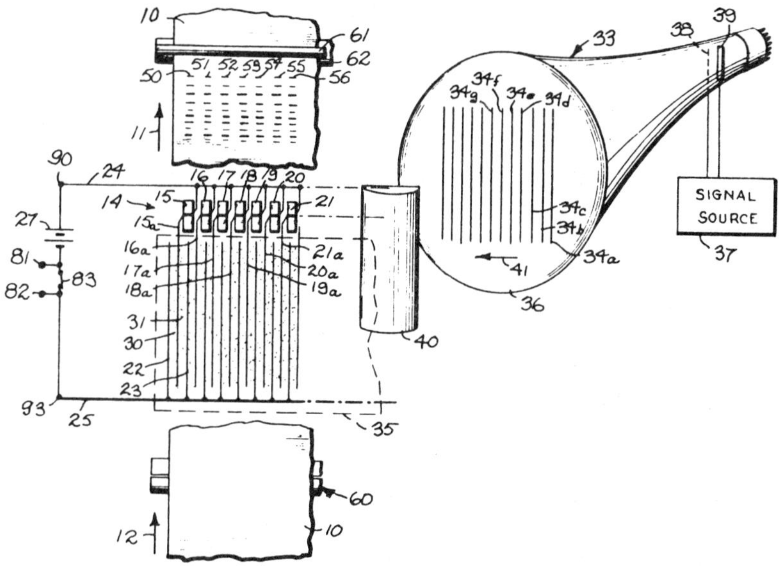

E SOLID STATE SCANNING HEAD

Figure 5-25. *(Continued)*

3. A separate head must be provided for playback, and the recording head must be taken out of contact with the tape at all times when a recording is not being made.
4. The head is more expensive than conventional kinds.

SCANNING HEADS

The concept of a video tape recorder where the high-speed scanning is accomplished without a rotating head is very attractive. Stationary head scanners may use electron beams, solid-state switching, or magnetic scanning.

Electron Beam Video Heads

A stationary scanner can use an electron beam for high-speed scanning of recording head elements while the tape advances forward slowly across the head at audio speeds. For recording, a high-current beam is needed to give sufficient magnetic field strength, especially if simple, closely packed heads with few winding turns are used. The experimental recording tube shown in Figure

5-25*A* uses a sheet-shaped beam of cross section 0.1 mm × 40 mm that sweeps across an array of tungsten wires on a target attached to the face of a cathode ray tube (Camras 1963b). One end of each tungsten wire passes through a seal in the face of the tube and over a permalloy head piece with a slot or gap for each wire, ending in a common terminal for all wires (Figure 5-25*B*). As the beam sweeps across the target at a video scan rate, it energizes the head gaps in sequence to record one video line at a time across a slowly moving magnetic tape.

To play back the above video recording, a low-current electron beam is required for sensitivity to the weak magnetic flux from the recorded tape; and the beam must be closely coupled to the recorded magnetization to avoid loss of resolution. The playback tube shown in Figure 5-25*C* uses a beam that passes through a comb of high-permeability wires sealed into its side. The wires bring into the tube the magnetic field from the external tape that runs over the wires. The beam is bent downward toward the target as it passes the wires, the amount of bending being modulated by the signal fields as the beam is swept sidewise for scanning. Figure 5-25*D* shows a modification where the beam is reflected. An array of magnetic wires bring the recorded field to the electron-reflecting

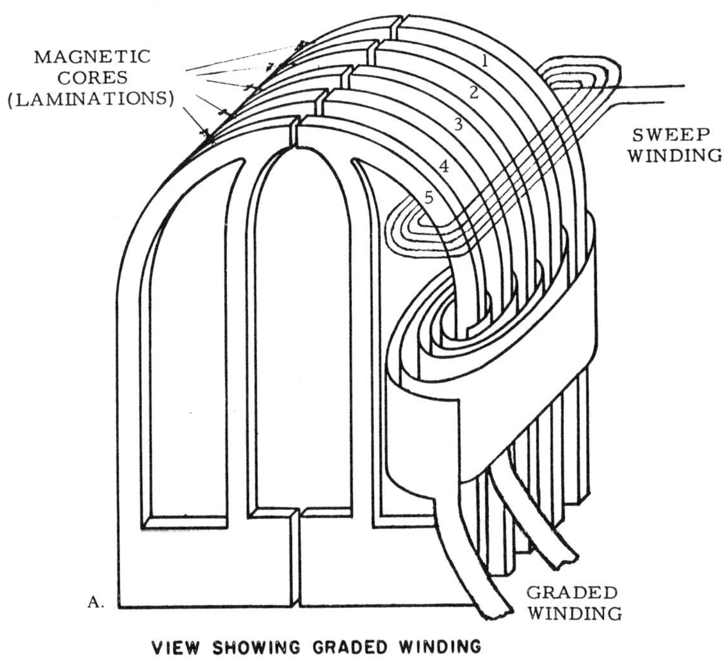

Figure 5-26. Magnetic scan head (Camras 1967).

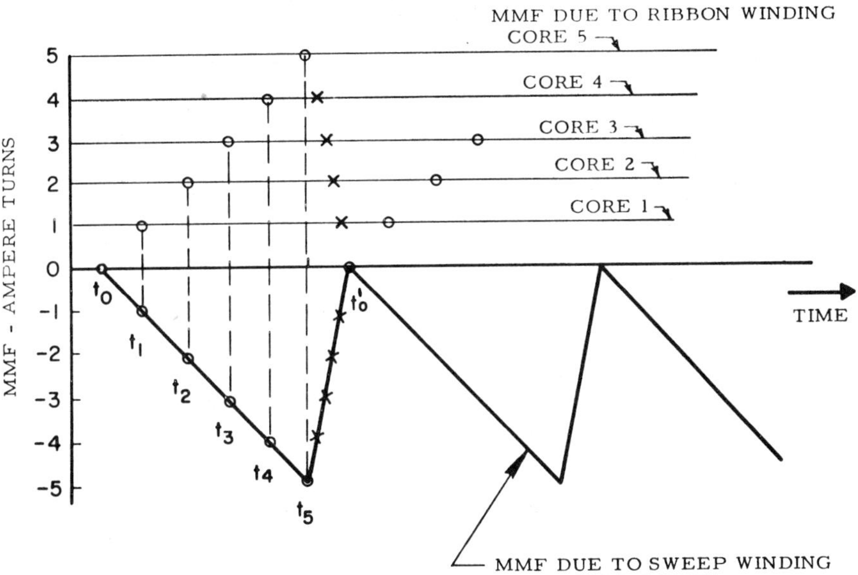

B.

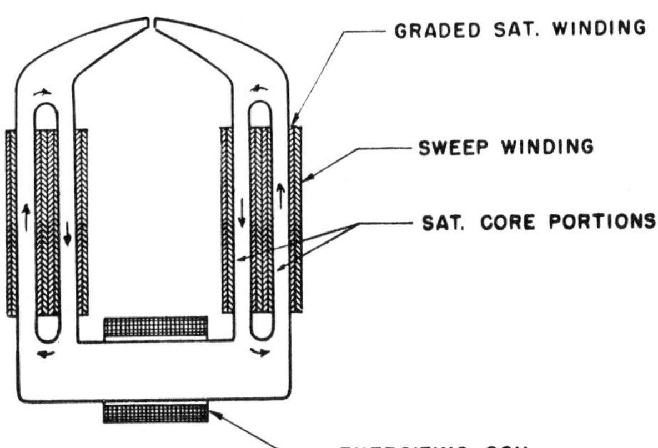

C.

Figure 5-26. (*Continued*)

surface. Alternatively, a very thin sealed window can allow the tape field to interact with the electron beam.

Solid-State Video Scan Head

A solid-state alternative to the electron beam recording tube is shown in Figure 5-25*E*. A light-activated solid-state switching grid passes the high currents re-

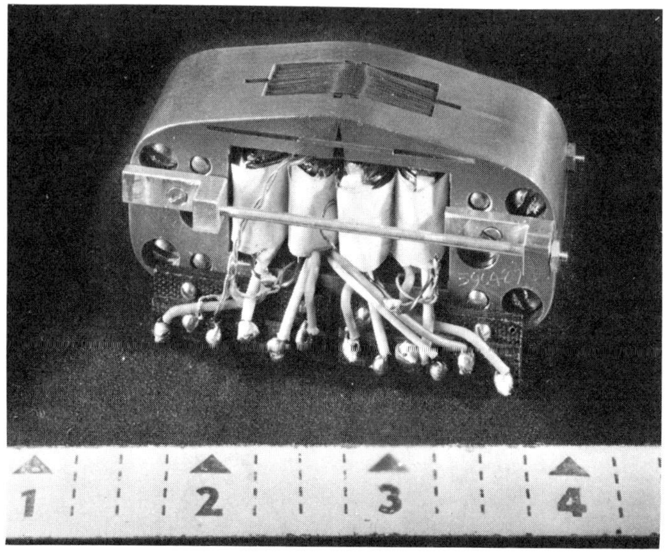

D.

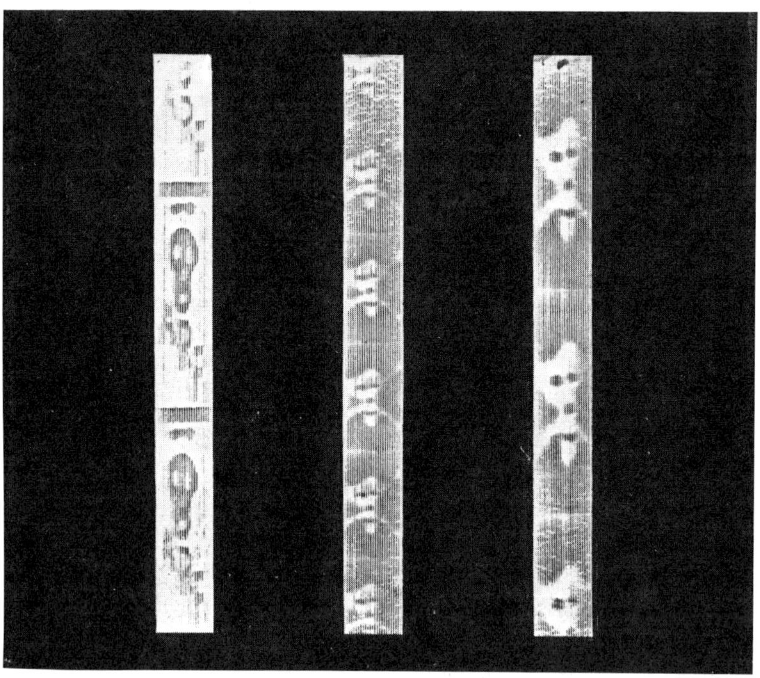

E.

Figure 5-26. (*Continued*)

quired by the scanned recording heads, and the same arrangement is also useful in playback scanning.

Magnetic Scan Head

The magnetic scanning head has a stack of laminations in which only one lamination at a time is operative, while the others are inactivated. Its principle of operation is shown in Figure 5-26*A*, where the side legs of a head are saturated by a graded ribbon winding that encircles each successive lamination with more turns than the previous one (Camras 1967). Thus, if the winding carries a steady DC current of one ampere, the first lamination is saturated by an mmf of one ampere-turn, the second by two ampere-turns, and so on. A sweep winding encircling all laminations carried a sawtooth wave of current (Figure 5-26*B*). As the sweep current grows, it cancels the effect of the ribbon-winding current in each lamination in succession, allowing that lamination to transmit video from the energizing coil (Figure 5-26*C*) to the top gap, or vice versa. Figure 5-26*D*, shows a 70-channel experimental head for real time magnetic recording of video, and Figure 5-26*E* shows test patterns recorded with this head.

6. Electronic Systems

Without electronics, would magnetic recording be impossible? Poulsen's telegraphone had no amplifiers. One can even conceive of magnetic recording without electricity, by recording with a vibrating magnet attached to an acoustical diaphragm. Playback could use the varying attraction to a small piece of iron as the magnetized record was drawn past it by a spring-wound motor. However magnetic recording, as we know it, depends strongly on electronics.

Electronic systems for *recording* supply signal and bias currents to recording heads capable of magnetizing the record to its maximum extent over the recording spectrum. For *playback*, they convert the transducer head voltage into desired output such as sound, visual displays, or digital pulses, adding minimum distortion or noise. For *erasing*, they supply enough power to remove the recorded information. For *tape transport*, electronic systems control tape tension, reduce flutter, and provide convenient indexing.

Electronics for recording are quite different in audio, video, and digital systems, and must be coordinated with their own playback and control electronics. For this reason we will describe the overall electronics of different kinds in the chapter devoted to each recording specialty (Chapters 9 through 14). The present chapter deals with elements and circuits that are the building blocks common to all systems. *Amplifiers* are a basic element of all electronic systems, and are discussed first. *Analog systems*, next discussed, are the oldest and most developed art for audio, instrumentation, and general purpose recording. All analog systems require equalization, preemphasis, and calibration according to agreed standards which are detailed. *FM systems* and *digital systems* are quite different from analog. Basic circuits are introduced whose uses are expanded in Chapters 11 to 14. *Erasing* and *biasing* are found in many recorders, but absent in others. Electronics for these are presented. *Control* functions, though they use amplifiers, are related to drive systems and are described in Chapter 7.

AMPLIFIERS

In the vacuum tube era, amplifiers for intensifying the signal were bulky, heavy, and fragile, with consumable tubes that were microphonic and power hungry. Today a linear integrated amplifier on a tiny silicon chip has enough amplifier stages for almost any application, as well as a power output stage that drives the recording head. A more elaborate but still negligibly small hybrid amplifier combines opamps (multistage linear integrated voltage amplifiers) with a separate output stage. Another option is an amplifier with discrete (single) transistors, which allows great flexibility in choice of circuits for equalization, low noise level, and low distortion. However, integrated circuit devices have become so excellent that they are now preferred for most preamplifiers.

Integrated Circuit Amplifiers. The simplest and least expensive low-level voltage gain amplifiers are opamps, as shown in Figure 6-1. An *opamp* is an integrated circuit having 20 or more transistors on a tiny silicon chip connected to give a high gain of about 100,000 if no feedback is used. Feedback is invariably used for stability and flat response. The effective gain of an opamp is therefore controlled by its feedback resistor and is essentially:

$$G = -\frac{R_F}{R_I} \tag{6-1}$$

where

G is the voltage gain of the opamp
R_F its feedback resistor
R_I its input resistor;

and the negative sign signifies an inversion of polarity between output and input. Opamps can also be connected so their outputs are not inverted, as shown in Figure 6-2*C*. With negative feedback a gain of 100 to 10,000 is readily

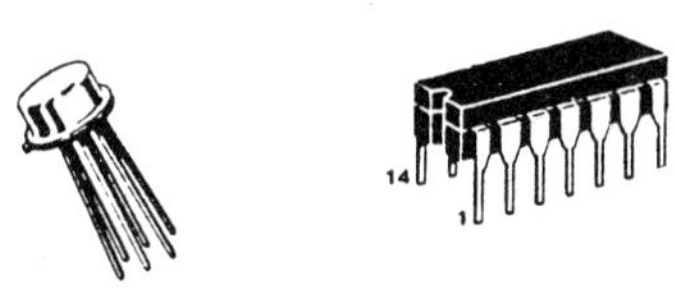

A. INTEGRATED CIRCUIT PACKAGES

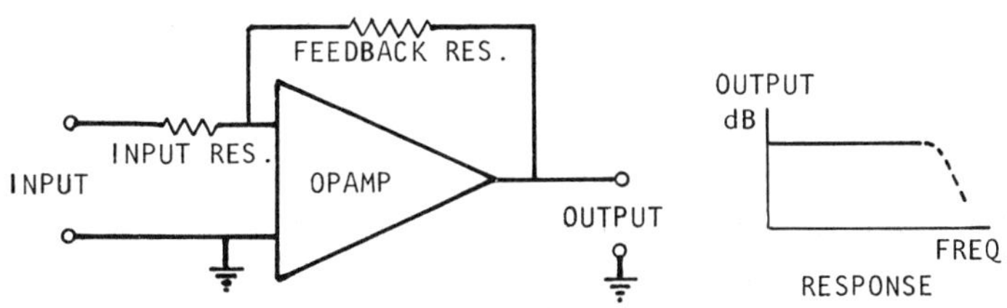

B. BASIC OPAMP CIRCUIT WITH FLAT RESPONSE

Figure 6-1. Integrated Circuit Amplifiers.

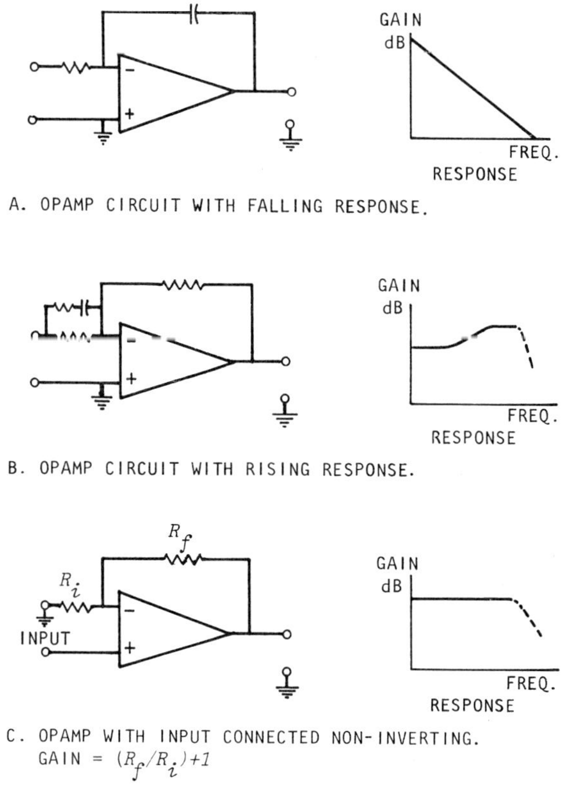

Figure 6-2. Opamp circuits with frequency response modified by RC feedback. A. The feedback capacitor gives a sloping response of an integrating circuit, useful for playback. B. The rising hf response is useful for pre-emphasis. C. The non-inverting connection has higher input impedance and can be used with any of the above frequency shaping networks.

achieved, depending on requirements of band width, gain stability, and slew rate. Slew rate can contribute to transient distortion if the gain is too high. A frequency corrective network can be used in place of R_F or R_I for convenient equalization, as shown in Figure 6-2*A* or *B*.

Discrete Transistor Amplifiers. Voltage amplifiers using a series of stages with separate (discrete) transistors are an alternative to opamps, but in recent years integrated circuits have been developed that equal or surpass the performance of discrete devices for most applications. A typical two-stage amplifier is shown in Figure 6-3.

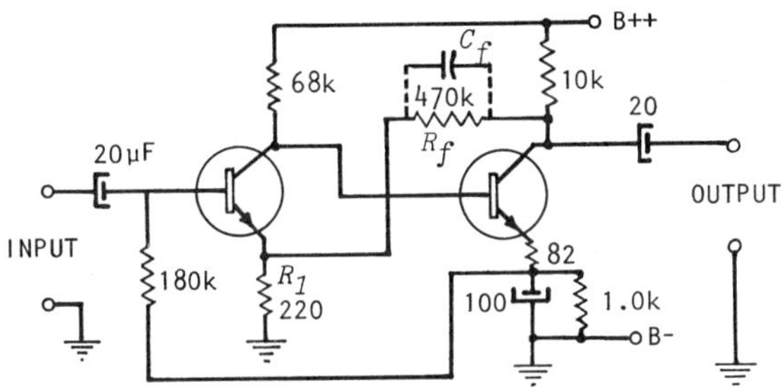

Figure 6-3. Two-stage transistor amplifier. Discrete NPN bipolar junction transistors are stabilized with negative feedback through R_f and R_l for flat response if C_f is not present. The frequency response can be modified by R_fC_f or other networks in the feedback circuit. Response shaping networks can also be included at the input or output terminals.

In a discrete amplifier, for the critical first stage one can choose special low noise transistors, and by using plug-in sockets one can even select individual transistors that give best results. For lowest possible transistor noise a cascode circuit is sometimes used for the first stage (Figure 6-4). Discrete transistor circuits also allow the most advantageous placement of equalizer and signal processing networks between their stages (Mochtenbacher and Fitchen 1973; Doyle 1967).

In the early stages, amplifiers for microphones or playback heads need to handle only millivolt and milliwatt levels. In later stages, the voltages reach a level of 1 to 100 volts and appreciable power must be handled. These high-

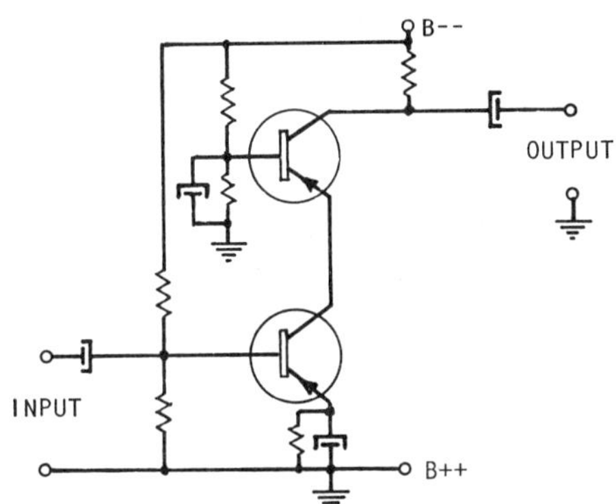

Figure 6-4. Cascode circuit for a low noise input stage. The cascode circuit uses two transistors with the emitter of the second transistor direct coupled to the collector of the first transistor for lowest noise. This example uses a pair of low noise PNP transistors.

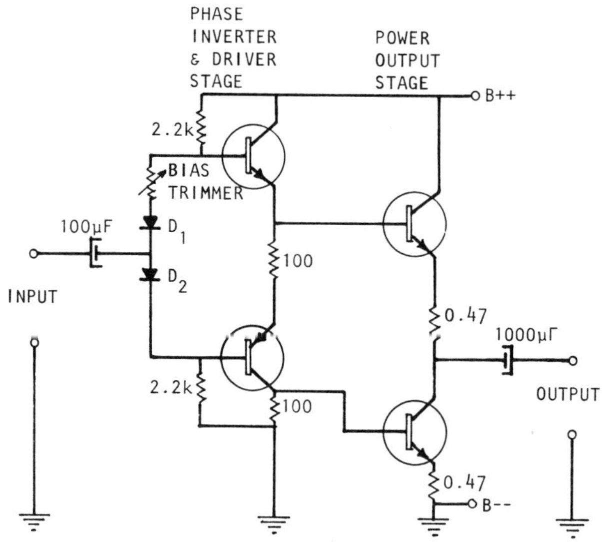

Figure 6-5. Push-pull power amplifier with complementary-symmetry driver.

level stages customarily use discrete transistors or hybrid amplifiers mounted on heat sinks to carry away their power losses. Figure 6-5 shows a typical circuit for a power amplifier using a complementary-symmetry driver stage. In power circuits, as elsewhere, the integrated circuit designers are offering economical alternatives to discrete devices. For mass production it is often feasible to custom design special-purpose integrated circuits for the application, rather than to compromise with off-the-shelf devices. Table 6-1 gives data on solid-state low-noise preamplifiers and power amplifiers.

Vacuum Tube Amplifiers. Vacuum tube amplifiers are still found in old work-horse tape recorders, operating faithfully over many decades. Figure 6-6 shows a vacuum tube amplifier stage compared with an analogous field-effect transistor stage. One advantage of vacuum tubes is that they can withstand heavy overloads and large spikes of voltage that would burn out transistors. Latest designs of transistors and their circuits are protected against these problems.

There has been a perennial argument that for power output stages vacuum tubes and especially triodes are still unexcelled. Even though tests show almost zero distortion for competitive transistor amplifiers, the vacuum tube advocates claim an audible difference of clean sound that cannot be matched by transistors. Their theory is that tubes overload gradually and gracefully just as the human ear does, while transistors give sharp ''break-up'' transients at overload, which are unnatural to our hearing although so slight that they hardly register

Table 6-1. Preamplifiers and Power Amplifiers

Type	Function	Input Noise $nV/\sqrt{Hz}$ at 1 kHz	Broadband Noise Figure dB	Gain (Max)	Gain Bandwidth MHz	Power Bandwidth	Transient Response ns	Slew Rate $V/\mu s$	Output Res. Ohms	Remarks
NE5533AN	Dual low-noise opamp	4	0.9	100 000	10		20	13	300	Signetics
OP-27	Ultra low-noise opamp	3		1 000 000	8			2.8	70	Precision Monolithics, Inc. (PMI)
LM194	Supermatch pair (discrete)	1.8	0.2 (1 kHz)	$h_{FE}* = 700$						National Semiconductor
LM381	Dual low-noise preamplifier	0.5 μV (10 kHz band)		300 000	15	0.075			150	National Semiconductor
2N4250	Low-noise amplifier (discrete)		0.7	$h_{FE} = 350$						Fairchild
LH0101	Power opamp (5A peak output)	25		200 000	5	0.3	200	10	10	National Semiconductor
LM2878	Dual 5-watt power amplifier	0.25 μV (20 kHz band)				0.065		2	8	National Semiconductor

*h_{FE} = current gain

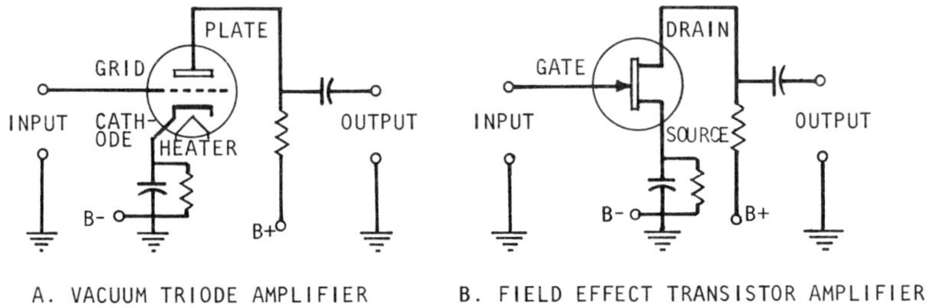

Figure 6-6. Vacuum tube and field effect transistor amplifiers. The circuits are almost identical.

on measuring instruments. Listening to transistor amplifiers over a period of time is said to fatigue the hearing even when distortion is otherwise undetectable. Transistor supporters hold that these theories are imaginary.

Audio Transformers

In the days of vacuum tubes, every studio used a great abundance of audio transformers. They had almost no substitute for:

1. Impedance matching. Vacuum tube amplifiers operate at such high impedance levels (thousands to hundreds-of-thousands of ohms) that an input transformer and an output transformer are necessary to match them to the 600 ohm line impedances.
2. Isolation, insulation, and common mode rejection. Since the primary and secondary windings are completely insulated from one another and may be electrostatically shielded if necessary, their inputs and outputs may be grounded to different independent points, easily solving knotty problems of ground currents (ground loops) where remote sources and separated independent amplifiers are links in a long chain. By means of a "phantom" circuit, an insulated primary winding can transmit DC power to a microphone through the audio wires of the microphone cable, energizing its built-in amplifier while simultaneously receiving the microphone audio. An output transformer removes the DC power supply component from the audio, which otherwise would require a large blocking capacitor.
3. Hybrid coupling. Traditional push-pull amplifiers use an output transformer as the most efficient method of coupling oppositely phased amplifiers to a single-ended low-impedance load such as a loudspeaker. Transformers handle in the most straight-forward way all manners of balanced to unbalanced interchanges, single ended to push-pull, and impedance matching.

Transistors have changed this picture to a great extent because their operating impedances can be matched directly to the lines. Transistors are compact, energy efficient, and inexpensive, making them useful in complex circuits such as differential amplifiers and emitter-followers, as well as in complementary symmetry modes that were not practical or possible with vacuum tube technology.

Although transformerless amplifiers are the rule, some installations have problems that are best solved with transformers. Transformers have persisted because they are reliable and durable, and give inputs (or outputs) that are electrically insulated from the amplifier, withstanding hundreds of volts of common mode interference that would disable direct inputs to amplifiers. Disadvantages of transformers are cost, possibility of hum pickup, possibility of adding high-frequency noise in low-level circuits, and possible distortion and frequency limitations, which are quite low at medium impedance levels used in studios, but add to transient and phase shift distortion. Table 6-2 gives characteristics of some standard audio transformers useful for magnetic tape amplifiers. Transformers that couple magnetic heads to amplifiers are usually designed specially for the purpose, since heads are highly inductive and are apt to resonate with circuit capacitances. Damping resistance is usually needed to flatten the wideband response. Transformers are often used with video heads that have very low impedance windings.

Transformer Design and Characteristics. Ideal transformer equations give an excellent approximation to actual transformers.

$$\text{transformer ratio, } \frac{E_s}{E_p} = \frac{I_p}{I_s} = \frac{N_s}{N_p} = a \tag{6-2}$$

$$\text{Impedance ratio, } \frac{X_s}{X_p} = \frac{R_s}{R_p} = \left(\frac{N_s}{N_p}\right)^2 = a^2 \tag{6-3}$$

$$\text{Flux density in core, } B_{\text{peak}} = \frac{E_{rms}}{4.44\, fNA_c} \tag{6-4}$$

where *E, I, X, and R,* = the voltages, currents, impedances, and resistances of primary windings p, or secondary windings, s; N = the number of turns on the winding; B_{peak} is maximum flux density (tesla) in transformer core (keep well below maximum rating of chosen core material); A_c = core area (m^2); (or if B and A_c are in cgs units, G and cm^2, multiply the right-hand side of the equation by 10^8); f = frequency, Hz.

Table 6-2. Audio Transformers

INPUT AND MATCHING TRANSFORMERS*

Type No.	Application	Primary Imp. Ohms	Unbal DC	Secondary Imp. Ohms	Response ±2 dB (Hz)	Pri. Res. Ohms	Max. Level dbm	Max. Level mw
A-10	Low impedance to grid	50, 125/150,200/250, 333, 500/600	0	50,000 split	20–20,000	59	+15	30
A-20	Mixing, matching	50, 125/150,200/250, 333, 500/600	0	50, 125/150, 200/250, 333, 500/600	10–50,000	64	+15	30
A-39*	Line to transistor	600/150 split	0	2000/500 split	20–20,000	70	+10	10

INTERSTAGE AND OUTPUT TRANSFORMERS*

Type No.	Application	Primary Imp. Ohms	Unbal DC	Secondary Imp. Ohms	Response ±2 dB (Hz)	Prim Res. Ohms	Max. Level dbm	Max. Level mw
A-15	Transistor Interstage	10,000/2500 split	8 mA	2000/500 split	40–20,000	600	—	1w
A-18	Single or PP plates to PP grids	15,000 split	0	80,000 split	20–20,000	1040	+15	30
A-23	Tr. Intstg. or output	500 CT	20 mA	16/4 split	40–20,000	36	—	1w
A-24	Single or PP plates to line	15,000 split	0	50, 125/150, 200/250, 333, 500/600	20–40,000	1430	+15	30
A-44	Tr. Intstg. or output	4K/1K split	12 mA	600/150 split	30–20,000	310	—	1w

*TRW Inductive Products, New York, N.Y. 10013

ANALOG RECORDING, AND EQUALIZATION OF FREQUENCY RESPONSE

For linear recording (*analog* in contrast to *digital* or *FM* recording), it is assumed that high-frequency bias will be incorporated as indicated at the end of this chapter. In the following discussion only the signal itself, apart from the bias, is treated.

The recording amplifier (Figure 6-7) is usually of straightforward design, with a flat voltage response over the operating frequency range. As a first approximation, a constant voltage input to the amplifier should energize the recording head with equal currents at all frequencies (constant current response). A lossless head would then apply to the tape an equal intensity of magnetizing (fringing) field. This first approximation is later modified to take care of losses.

Constant Current Recording. Since the head winding is inductive, its reactance ($X = 2\pi fL$) increases with rising frequency, and its current therefore decreases if driven by an amplifier that has flat voltage response. The usual remedy is to insert in series with the head of Figure 6-7*A* a resistance R_s high enough to swamp out the inductive reactance of the head and therefore to approximate the constant current condition. The amplifier supplies enough voltage to take care of losses in the series resistor, usually many times greater than in the head alone. Alternatively, where efficiency is important, the amplifier itself

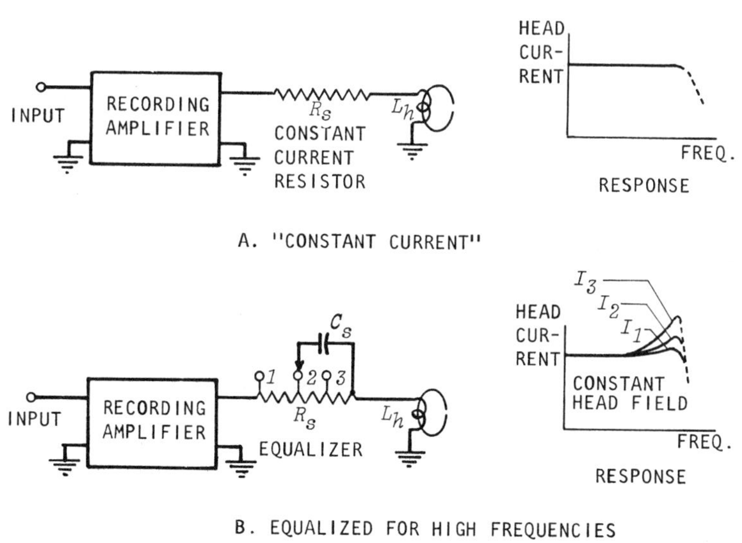

Figure 6-7. Head driver amplifier with basic recording equalizer.

may be of the constant current type to drive the head directly over a limited number of octaves.

Basic Recording Equalizer. Although driven by constant current, the recording head gap field and fringing flux decrease somewhat at high frequencies because of head core losses, and therefore the driving current should rise enough to compensate for these losses. This rise is readily achieved by a series capacitor (C_s) across all or part of R_s, as shown in Figure 6-7*B*. The correction needed to give a constant head-field is usually small, so that the term *constant current recording* is often used to describe *constant head-field recording*. In our present discussion we are concerned only with frequency-dependent losses, but we should remember that a constant head field does not give constant remanent tape flux at short wavelengths because of wavelength dependent losses.

At a high frequency the capacitor C_s resonates with the head inductance to give a peak in head current. By judicious choice of C_s and the proportion of R_s that it bridges, the recording response can be shaped advantageously by the simple network shown in Figure 6-7*B*.

ELO Equalization. If we were to use our equipment only to record test signals of equal intensity at all frequencies, the recording current response of Figure 6-7*B* giving a constant head field at all frequencies would be an ideal first approximation to our recording equalization. However, in recording speech and music the input signal may not have a flat spectrum, and therefore some frequencies reach overload levels while others are far below, in effect wasting the potential capability of the system. Where the spectrum is known, it is logical to emphasize the weak frequencies before recording them, flattening the recorded spectrum such that all frequencies are *equally likely to overload* (ELO). Such ELO equalization, added to the constant flux equalization prior to recording is a form of preemphasis and should be balanced out during playback with a opposite amount of postemphasis (deemphasis). When used properly, this pre- and postequalization improves the signal-to-noise (*S/N*) ratio of the system. The better *S/N* ratio justifies such equalization even though the *ELO* theory for its use has been questioned.

High-Frequency Preemphasis. In 1929 Sivian, Dunn, and White of Bell Laboratories published a classic paper on the spectrum of musical instruments and orchestras (Sivian, Dunn, and White 1959). Figure 6-8*A*, their spectral response of a 75-piece orchestra, shows that the average amplitude falls at frequencies above 300 Hz. In recording this spectrum with a constant current recorder set so that 300 Hz does not overload, it appears that frequencies above 4 kHz have about 14 dB of head room and that the tape recorder capability is under-utilized at these high frequencies. Therefore according to ELO theory it

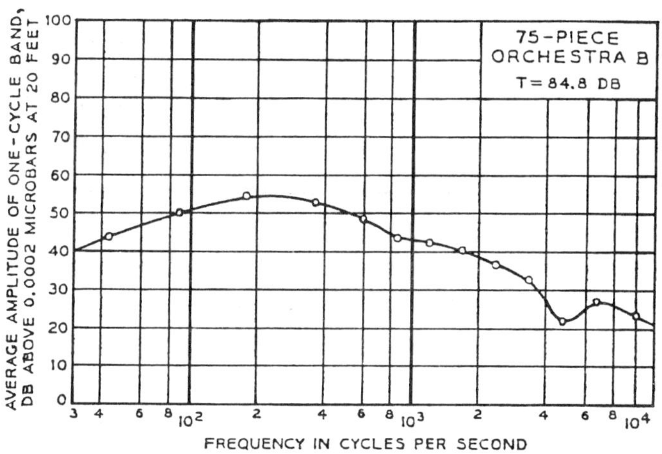

A. TRENDS OF SPECTRAL RESPONSE CURVES ARE USED TO DERIVE ALLOWABLE PRE-EMPHASIS

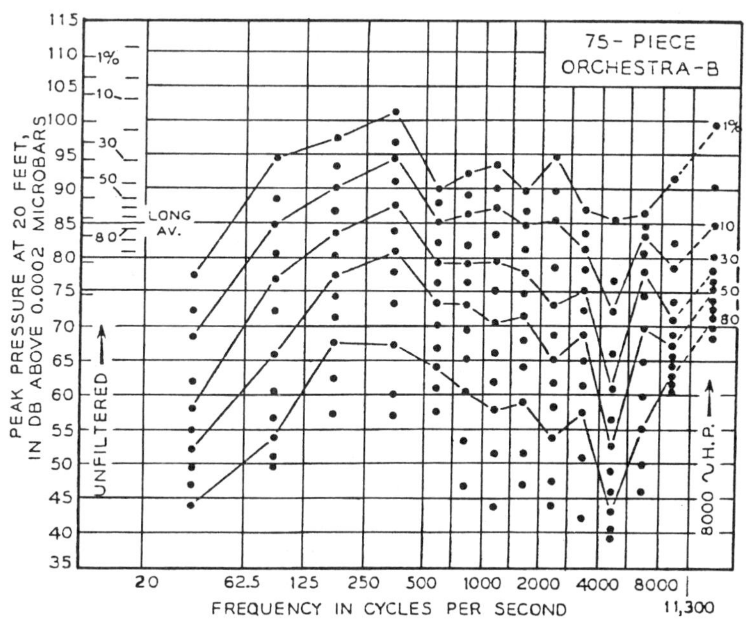

B. PEAK SOUND PRESSURE EXCEEDS THE TOP CURVE 1% OF THE TIME, AND OTHER CURVES BY THE INDICATED PERCENTAGES

Figure 6-8. Spectral response of a 75-piece orchestra (Sivian, Dunn, and White 1959).

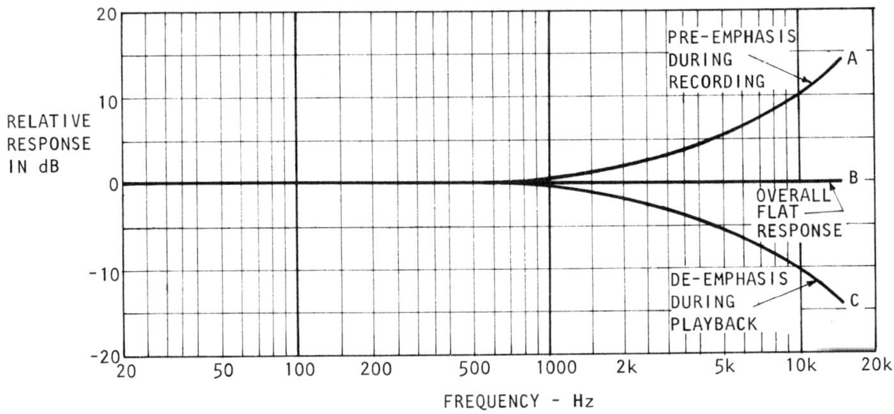

Figure 6-9. Preemphasis during recording and deemphasis during playback give a flat overall response with reduction of noise. Frequency response of recording head field of Figure 6-7*B* is shown by curve *A*.

would be permissible to emphasize the highest frequencies as much as 14 dB without danger of overload (Figure 6-9*A*). Such preemphasis on recording is accompanied by the same amount of deemphasis during playback to give a flat overall response, as shown in Figure 6-9*B*. The advantage of playing back with deemphasis (Figure 6-9*C*) is that the falling response at high frequencies reduces tape hiss and noise, which would otherwise be most audible and most annoying in the deemphasized band. Although deemphasized, the high frequencies in the music come through normally, standing out above the noise since the music has been preemphasized far above the tape noise level.

Using a response such as the one shown in Figure 6-8*A* to justify preemphasis is controversial because a peak power spectrum of the same orchestra in Figure 6-8*B* shows a rise at highest frequencies, which tends to level out the response curve. A better explanation is that the chosen equalization trades off an occasional brief distortion transient that is quite tolerable, for a continuously lower background noise that is a welcome improvement. Even more welcome is additional noise reduction obtained with dynamic compression, as will be described in Chapters 9 and 10 (Figures 9-8 and 10-8).

After all, a sound recording is designed ultimately to please the ear. A musical waveform has a complex shape, the resultant of many components. Statistically the peaks of the components will at times add in phase to give an enormous peak, which cannot be reproduced by the most conservative system, so even a perfectionist must accept overloading for a small percentage of the time. Fortunately the ear tolerates this distortion. We can thus explain the chosen equalization as an empirical art rather than a science, derived from experience to give the most pleasing results from an imperfect medium. And this is true of all art.

Low Frequency Preemphasis. A similar argument with respect to Figure 6-8*A* is that low frequencies also could be boosted up to 7 dB at 30 Hz to achieve ELO, with benefits of reduced hum and low frequency noise on playback. A low frequency boost has indeed been incorporated in NAB standards for 7.5 and 15 in/s professional recorders, and in home-use recorders to reduce hum problems. The NAB standard is used mostly in the USA and is considered less favorable than the European IEC (flat LF) standard for reasons given below. Many recorders allow either one to be selected.

Limitations of Preemphasis. For classical orchestral music the preemphasis based on Figure 6-8 has been found advantageous. But for organ music the low frequencies are prominent and must not be preemphasized, while the extreme highs could use more emphasis. Although a special recording curve might be devised for organ music, it would entail a special compensating curve on playback. Popular music, rich in both highs and lows, would be better served by a flat curve. *Close mike technique* causes the spectral balance to change radically when the microphone position changes a few inches. Here also a flat response would be a good policy. Although a standard preemphasis is built into recorders, the user is not helpless because the recording console has elaborate equalizers that can modify the effective equalization and allow the user to record at a higher level than with a flat input. In principle, the playback response should be adjusted to compensate, and this may not be known to others who play a nonstandard recording. In practice, the playback is usually adjusted by ear to give a satisfactory result. There is a definite trend to use less preemphasis as quieter tapes become available and better dynamic noise suppressors take care of the noise problem.

Standardization. Standardizing the *recording equalization* was suggested in early days of magnetic recording, but this suggestion proved totally impractical. Not only are the program spectra variable, but other factors will also vary the recorded response, including tape speed, biasing levels, tape coating thickness, type and concentration of tape coating, head gap size and depth, and core losses, to name a few.

An ingenious solution for these problems of standardization will be described in the next sections on playback. In essence it involves a calibration tape, recorded with signals of definite intensity over the entire range, obtainable from standards laboratories that guarantee its accuracy. The playback equalization is adjusted to give a flat response when playing the calibration tape. Next the equalization of the recorder is adjusted such that its recordings will give the same response as the calibration tape on the adjusted playback machine. When these steps are carried out both the recorder and player are standardized. Recordings will then be interchangeable among different machines, provided, of course, that mechanical considerations such as tape speed and track alignment are observed. Unrecorded tapes of standardized properties (reference tapes) are also available for comparing blank tapes.

Equalizer Considerations. The advantage of an RLC equalizer directly in the head circuit, after all the amplifiers, is that its differentiator action reduces hum or $1/f$ noise that might be superposed on the weak head current. A disadvantage is that more driving power is required, mostly wasted in the equalizer network, but even so the power is only milliwatts and of importance mainly in battery-operated portables. Another disadvantage is that the head inductance is part of the network and limits the design flexibility.

Recording equalizers may be inserted between stages of the recording amplifier for increased head-driving efficiency and for equalizer design versatility. This arrangement is practical if the head driver stage is quiet, especially at low frequencies to which the head is most sensitive. Some basic equalizer networks and their responses are shown in Figure 6-10.

It is usually desirable to limit the recording amplifier response to the frequency band of interest. Cutoff at the high-frequency end prevents intermodulation and spurious beat notes between the bias oscillator and high-frequency signals that are of themselves above the recording range. Cutoff at low frequencies reduces hum, rumble, and transients that might be present on the signal, giving a "cleaner" recording. Theoretically the frequency response cutoffs are accompanied by phase shifts near the band limits. In audio work these have not been of major concern, although many enthusiasts will argue about phase perception. In data and video recording where phase is important the bandwidth should be extended or the phase corrected.

PLAYBACK OF ANALOG RECORDINGS

Analog playback requires sensitive, noise-free amplification and special correction and equalization. Playback heads usually give feeble output voltages of about 0.1 to 10 millivolts, requiring voltage amplifiers of the same kind as described for recording. A major difference is the need to correct for the $d\phi/dt$ character of induced voltages in coil-wound heads. This correction is made with an *integrating network* or its equivalent. Capacitive feedback networks in high-gain preamplifiers give such response (Figure 6-2*A*) and are commonly used.

Playback Amplification and Equalization. Figure 6-11*A* shows the playback response of a constant current recording when played back with an ordi-

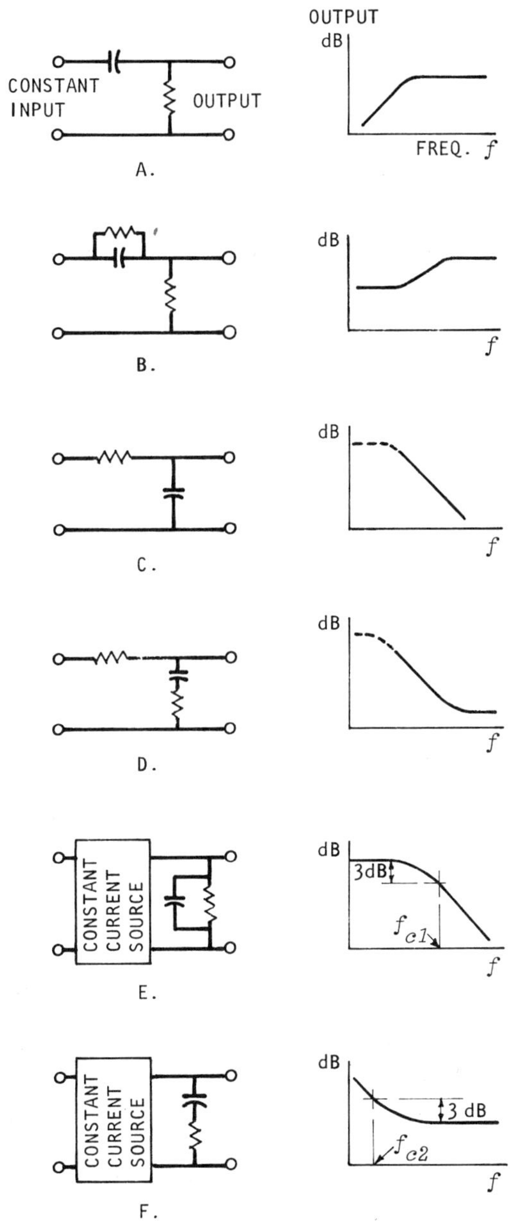

Figure 6-10. Equalizer networks and their responses. Network B is useful in high frequency pre-emphasis. Network C is an integrating network for playback equalization. Network D approximates closely the overall playback equalization required. Networks E and F define the standards of recorded tape flux. Networks like A give the inverse of their response when used in negative feedback circuits.

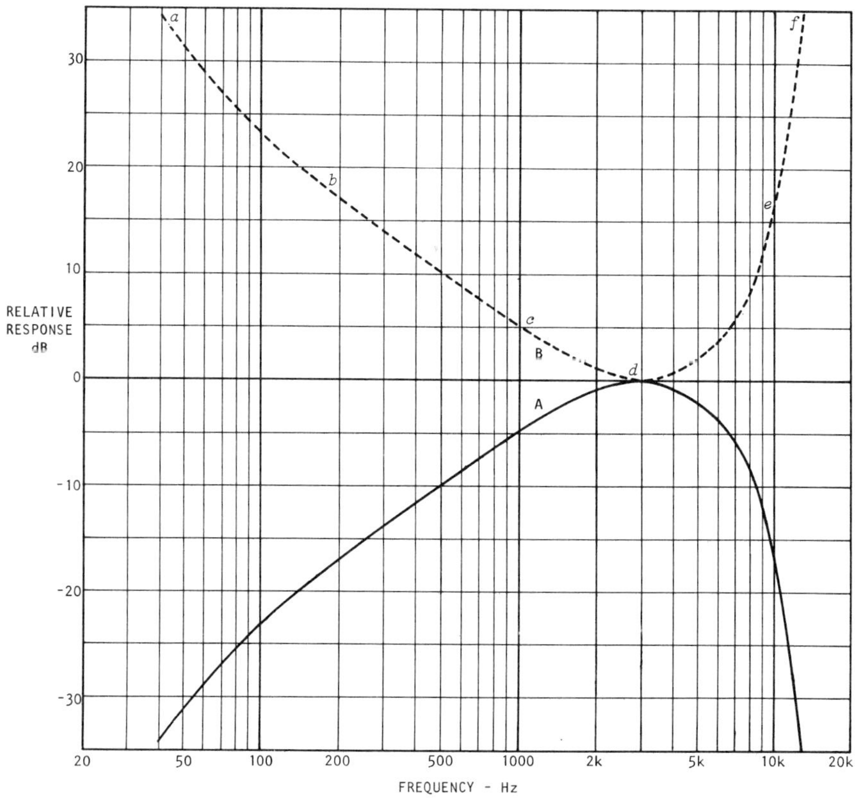

Figure 6-11. Playback response and equalization for a constant current recording. The general shape of *A*, typical in most recorders, is due to $d\phi/dt$ and to frequency and wavelength effects explained in Chapter 2.

nary (coil-wound) head, as explained in Figure 2-48. Correcting this response to a flat characteristic requires that networks of complementary overall response (Figure 6-11*B*) be included somewhere in the playback amplifier. Also an amplifier with an overall voltage gain of perhaps 10,000 to 100,000 (80 to 100 db) is needed to amplify the weak output of 0.1 millivolt at 50 Hz up a usable level of 1 to 10 volts.

The complementary response along the sloping portion *ab* in correction curve shown in Figure 6-11*B* is supplied by the integrating network of Figure 6-10*C*. At shorter wavelengths and higher frequencies, the integrating function is eased off into a flat response at portion *cd*, which really amounts to boosting the output from the recorded short wavelengths that are deficient because of losses in the recording process described in Figure 2-54. At *de*, where the correction curve takes a positive slope, the playback amplifier would require a network having a rising response. Such correction is generally avoided by adding preemphasis while recording, so that the preemphasized playback curve will not turn down until the maximum upper band limit is reached.

At the lowest frequencies, represented by *ab* of Figure 6-11, the curve shows a deficiency because recorded wavelengths become longer than their contact with the playback head. Standards for professional machines mostly call for a flat recording of tape flux at low frequencies down to 16 Hz, and therefore any deficiencies due to head size, for example, must be corrected by low-frequency boost in the playback amplifier. However, the standards for home-use recorders, especially cassette and cartridge machines, recommend a low-frequency preemphasis below 315 Hz, reaching 9 dB at 20 Hz. This low-frequency preemphasis eases the head-shielding and hum reduction problems in compact economical recorders.

Tape Speed Effects. Tape speed is usually the most important variable that determines response and equalization. Figure 6-12 shows the variation in response with tape speed. It is interesting that when the same recording is played at half speed, all of its frequencies are reduced to half their former values, and their corresponding induced voltages in a coil head are cut in half since $d\phi/dt$ is half as much. On the semilog plot of Figure 6-12, this amounts to shifting the original playback curve *B* an octave to the left and 6 dB downwards, in effect sliding it along the 6 dB-per-octave line to give curve *C*. The straight part of the new curve mostly coincides with the old one except that the high end

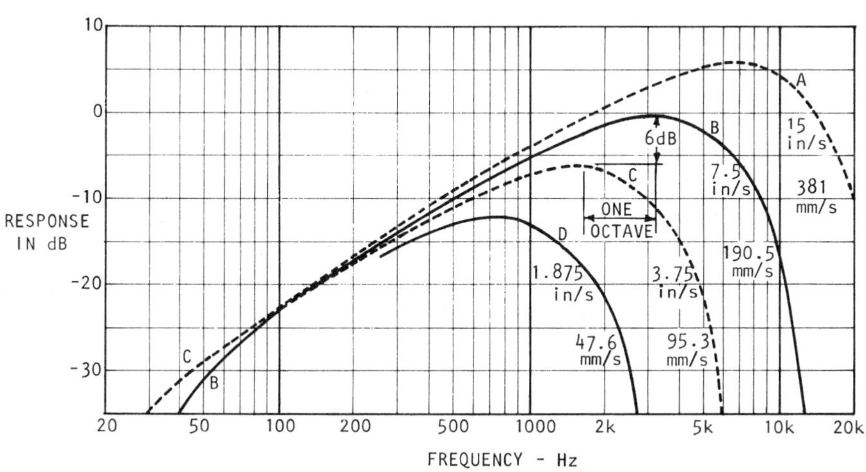

Figure 6-12. Effect of tape speed on frequency response.

turns downwards an octave sooner, and the low end continues an octave lower before it begins to drop more steeply.

Although the above analysis was made for a given recording on the tape, the same reasoning would apply if the tape were *recorded* as well as played back at the new speed. This is true because a constant current recording will magnetize the tape to nearly the same level at the same wavelength regardless of the speed at which the recording takes place (using the same head, of course, and neglecting minor frequency dependent losses).

EQUALIZATION STANDARDS FOR ANALOG SYSTEMS

Standards are most important for proper playback of tapes made on a different recorder, as when professionally taped programs are exchanged, or when recorded tapes are purchased for home-use tape players. Recorders are prone to have widely different response curves even when they are of the same make, depending on tapes, bias, heads, head wear, etc.; and differences are even greater among different makes, although in most cases the response has been adjusted so that a tape plays correctly on its own recorder. When machines are standardized as described below, their recordings are freely interchangeable within their own classification.

As seen in Figure 6-12, the range of tape speeds in common use gives such an enormous difference of high-end response that a single equalization standard is not practical. Therefore a family of equalization standards has been designed, with one appropriate for each chosen speed, head gap, and application.

Designation of Equalization Standards

The plotted response curve, as shown in Figure 6-13*A* and *D* of the standard tape describes completely the shape of its frequency response. But it is convenient to have a shorthand label for the response so we do not have to display the actual curve every time we want to talk about it. We can do this by specifying the time constant, $t_1 = RC$ (microseconds), of the parallel network in Figure 6-10*E*, which has the required response. If the product of $R \times C = 70$, where R is the resistance in ohms and C is the capacitance in microfarads, then a flat input to the network will give an output response (Figure 6-13*A* and *D*) with mathematical exactness. It does not matter what specific values are chosen for R or C as long as their product is 70 μs; thus we could combine $R = 1000$ ohms with $C = 0.07$ uF, or $R = 50$ ohms with $C = 1.4$ uF, and so on. The time constant specification also helps in designing equalizers that conform to the standards. Figure 6-14 gives additional standard curves for commercial and home-use recorders.

If there is no low- frequency correction the single time constant (t_1) suffices

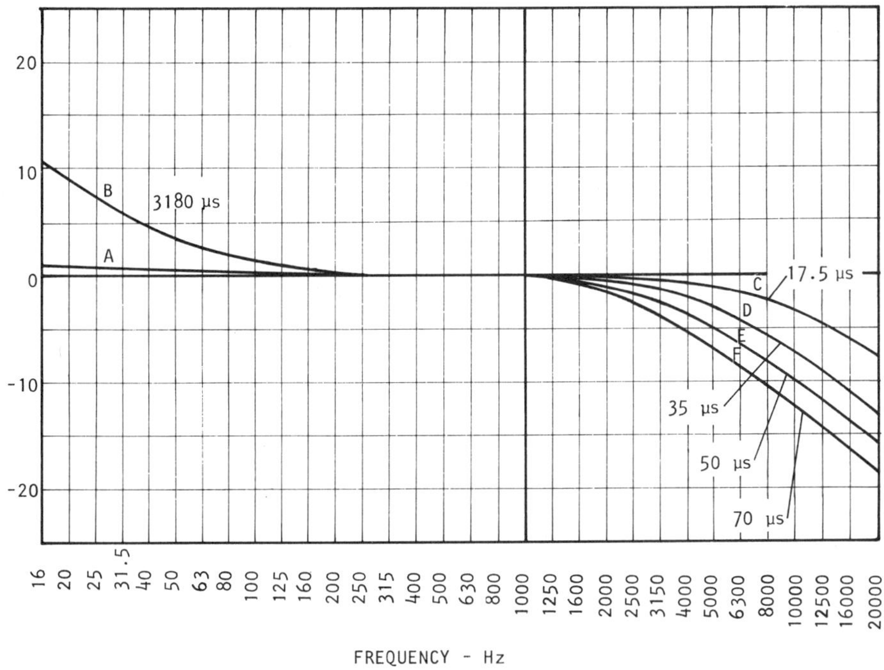

Figure 6-13. Standard response curves for professional tapes. These curves represent relative short circuit flux of standard tapes in Table 6-1 for highest quality audio.

for depicting the entire response. Where there is low-frequency correction, a second time constant (t_2) is necessary to shape the low-frequency rise, as in Figure 6-13*B*, which shows the response when $t_2 = RC = 3180$ microseconds, for the series network of Figure 6-10*F*. At low frequencies there is not as much variation in response at different speeds, as is evident from Figure 6-12, and therefore it is not necessary to standardize on many different values of t_2. Where the low-frequency response is flat, $t_2 = \infty$. This is assumed when t_2 is not specified. The overall response with correction at both high and low frequencies is obtained by combining in tandem the networks *E* and *F* of Figure 6-10. In practical equalizers, a single network can be devised to give this response.

Mathematically the equation of the response curves can be derived from the impedance of the networks in Figure 6-10*E* and *F* by letting $R = 1$, in which case $C_1 = t_1$ and $C_2 = t_2$. Any curve in Figure 6-13 or 6-14, or any curve having other time constants, can be plotted from the equation:

$$N\text{ (dB)} = 10 \log \left[1 + \frac{10^{12}}{(2\pi f t_2)^2}\right] - 10 \log \left[1 + 10^{-12}(2\pi f t_1)^2\right] + K \quad (6\text{-}5)$$

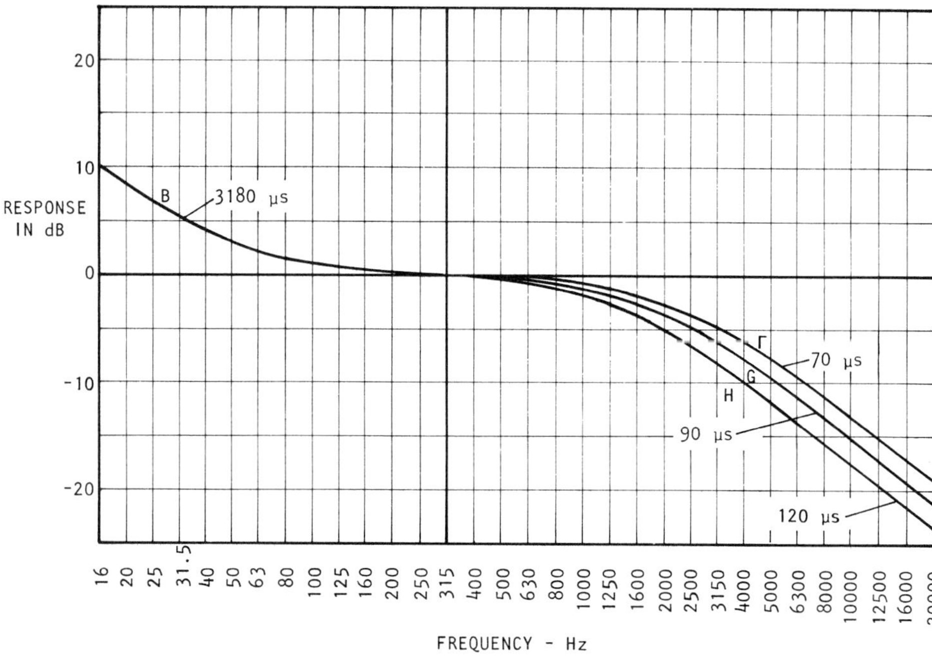

Figure 6-14. Standard response curves for commercial and home use tapes. These curves represent relative short circuit flux of standard tapes in Table 6-1 for various slower speed uses.

where t_1 and t_2 are in microseconds. The constant K can be chosen to shift the zero dB reference level to 1000 Hz or 315 Hz, etc., without changing the shape of the curve. The output N at frequency f is usually referenced to a chosen tape flux level, as for example zero dB = 110 nW/m.

Each response curve departs 3 dB from a flat level at its so-called cutoff frequency (or transition frequency, or corner frequency). The cutoff frequency (f_{c1}) is related to the time constant (t_1) by the equation:

$$f_{c1} = \frac{10^6}{2\pi t_1} \tag{6-6}$$

and similarly for f_{c2}. For example if $t_1 = 35\ \mu s$, its cut off frequency is f_{c1} = 4500 Hz. When $t_2 = 3180\ \mu s$, its cutoff frequency is $f_{c2} = 50$ Hz.

Those who like mathematical exercises can transform equation (6-5) into an equivalent equation by replacing t_1 and t_2 with equivalent f_c using equation (6-6). The result is:

Table 6-3. Recommended Response Time-Constants*

Application	t_1 (μs)	t_2 (μs)
Professional—Reel-to-Reel		
760 mm/s (30 in/s) IEC 1 (1968)(Obsolete)	35	∞
760 mm/s (30 in/s) IEC 2 (Same as AES 1971)	17.5	∞
380 mm/s (15 in/s) IEC 1 (1968)	35	∞
380 mm/s (15 in/s) IEC 2	50	3180***
190 mm/s (7.5 in/s) IEC 1 (1968)	70	∞
190 mm/s (7.5 in/s) IEC 2	50	3180
380 mm/s (15 in/s) ANSC Sound track on 2″ quadruplex videotape	35	2000
190 mm/s (7.5 in/s) ANSC Sound track on 2″ quadruplex videotape	35	2000
Professional—Cartridge		
190 mm/s (7.5 in/s) (NAB 1975)	50	∞
Commercial and Domestic (Home-Use) Reel-to-Reel		
380 mm/s (15 in/s) (NAB 1965)	50	3180
190 mm/s (7.5 in/s) (NAB 1965)	50	3180
95 mm/s (3.75 in/s) (NAB 1965; Same as IEC 1968)	90	3180
47.6 mm/s (1.875 in/s)	120	3180
Commercial and Domestic—Cartridge		
95 mm/s (3.75 in/s)	90	3180
Commercial and Domestic—Cassette		
47.6 mm/s (1.875 in/s)	120	3180
47.6 mm/s (1.875 in/s) (High resolution tape)**	70	3180
Commercial and Domestic-Betamax and VHS videotape		

* Refer to Figures 13 and 14 for response curves corresponding to t_1 and t_2.
** Sensing holes in cassette body may switch bias and equalization.
*** Sometimes the 3180 designations are rounded to 3150 μs.

$$N\ (\mathrm{dB}) = 10 \log \left[1 + \left(\frac{f_{c1}}{f}\right)^2\right] - 10 \log \left[1 + \left(\frac{f}{f_{c2}}\right)^2\right] + K \qquad (6\text{-}7)$$

The recommended frequency response standards are listed in Table 6-3, based on practical tapes, heads, and speeds that are in common use. The standards specify that the short-circuit magnetic tape flux recorded on the tape shall follow the t_1 and t_2 response curves when a sine wave of constant amplitude is applied to the input of the recorder. Short-circuit magnetic tape flux is the flux supplied by the tape to a high-permeability head core with which it is in contact. (Formerly the open-circuit tape flux measured with a coreless pickup was considered a standard, but the short-circuit flux is more realistic as it accounts for the keeper effect that increases the flux when a head core is against the tape.)

Standardizing a Playback System

A playback system is considered to be standardized if it gives a flat frequency response when playing back a tape whose recorded flux follows the specified standard. For example, a calibration tape is available with a recorded flux following a 50 to 3180 μs curve at 7.5 IPS (19.05 cm/s). The procedure is to first ascertain that the playback machine is in proper mechanical adjustment of head azimuth, track alignment, and tape speed. The head should be demagnetized so that it will not accidentally decrease the recorded level of the precious calibration tape to be played on it. The calibration tape is then played back and deviations noted from a flat output response. Playback equalization can then be modified and the calibration tape replayed, repeating the equalization adjustments until the response is within specifications. Tolerances are specified for response flatness, with extreme limits of 4 dB for professional programs and 12 dB for home-use recordings. Figure 6-15 and Figure 6-16 are graphs of the equalization that must be supplied by the playback system for recordings made by the standards given in Table 6-3 in order to yield a flat output (neglecting playback head loss).

Standardizing a Recording System

A recording system is considered standard if it records a tape to a short-circuit flux of the chosen curve shown in Figure 6-13 or 6-14, 50 μs to 3180 μs, for example, when a constant amplitude sine wave is applied to the recorder input. Very few users have the facilities to measure tape flux accurately; almost everyone relies on calibration tapes that are issued by laboratories equipped to make and calibrate them. The procedure is to make a recording with the recorder you are standardizing, and to note its response when played through a playback system, preferably one that has been standardized as above. The laboratory calibrated recording is then played under the same conditions and its response is also noted. The difference between response of our recording and of the lab standard tells you how much the recording equalization must be changed to bring your recording up to standard. The procedure is repeated until the recorder meets the specification.

INSTRUMENTATION: DIRECT RECORDING

Direct or analog instrumentation recorders are required to cover a spectrum ranging in some cases from 400 Hz to above 2 MHz, or more than 12 octaves,

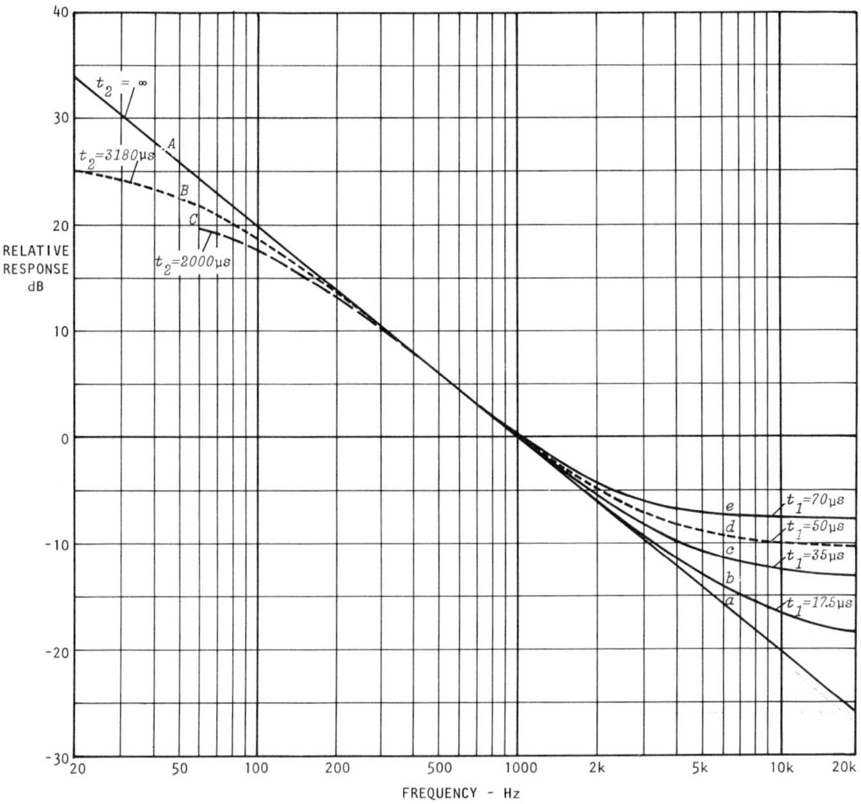

Figure 6-15. Basic playback equalization - professional. A tape recorded to the 17.5 μs standard needs the playback response of curve Ab to give a flat overall response if playback head loss is negligible. Other standards require playback responses as follows: 35 μs tape = curve Ac
50 μs tape = curve Ad
70 μs tape = curve Ae
50 μs − 3180 μs tape = curve Bd
35 μs − 2000 μs tape = curve Cc

and hence are known as *wide-band recorders.* Audio recorders usually cover about 9 octaves. Electronic systems for wide-band recorders are quite analogous to those used in the finest professional audio recorders, except that they do not have to respond to the lowest audio frequencies unless there is a provision for running the tape at audio speeds. In common with audio recorders, the greatest care is required in designing the low-level stages of playback amplifiers for least possible noise.

Several important differences between audio and wide-band electronics in-

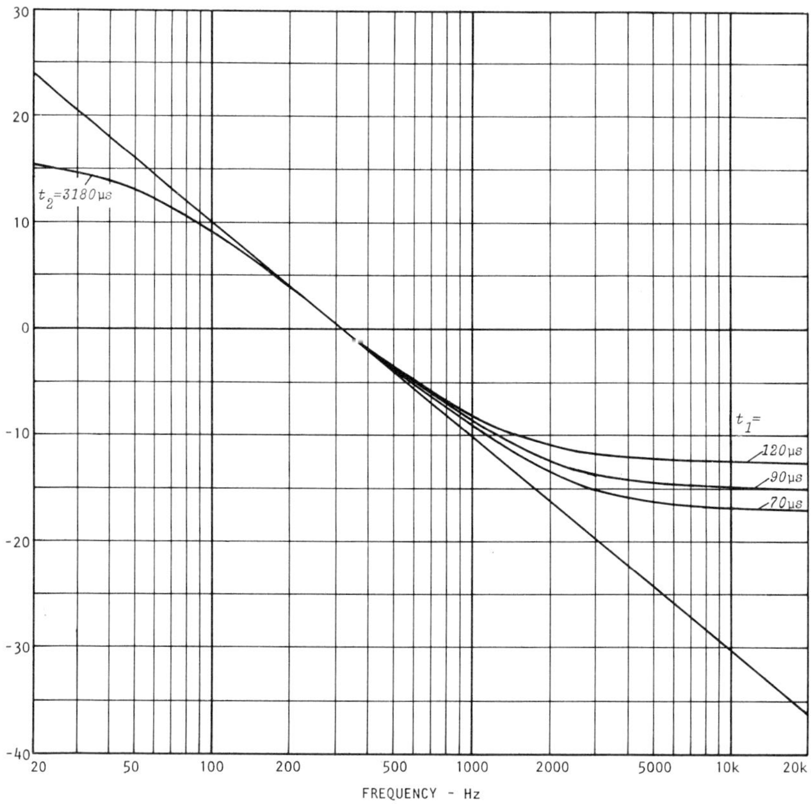

Figure 6-16. Basic playback equalization – commercial and home use. These represent the playback equalization (neglecting head loss) to give flat response from tapes of the designated t_1 and t_2 standard. Zero dB is at 315 Hz instead of 1000 Hz as in Fig. 6-15.

volve noise, preemphasis, and phase. In going to frequencies 100 times above the audio high limit, video amplifier techniques are used, with more amplifier stages and less gain per stage. The overall band width is 100 times as high, and the noise band width is in proportion, or about 40 dB greater. However the higher tape speeds to 120 in/s also give more signal voltage, perhaps 20 dB more, so the net loss of signal-to-noise ratio is about 20 dB. In principle, preemphasis can be used only to take care of recording head losses, since the input data is likely to occupy any part of the spectrum. In practice, some additional preemphasis may be introduced for flatter response at the expense of dynamic range. Phase equalization is usually neglected in audio recording but is important for instrumentation where recorded wave shapes must be reproduced accurately. The special kinds of phase equalization needed and other problems are discussed in Chapter 13.

INSTRUMENTATION AND VIDEO: FM SYSTEMS

Frequency-modulated (FM) recordings are made with a constant recording level at tape saturation or near it. The FM carrier is chosen close to the upper frequency limit of the record-play system while still leaving room for the upper sidebands. (Sometime the upper sidebands are cut off). The recording contains mainly high frequencies (short wavelengths) and little or no low frequencies; and these high frequencies encompass only a limited band, often less than an octave, compared to the 9 or 10 octaves required for audio.

The equalization needed for FM work is therefore minimal. Precautions should be taken against too strong a recording field at highest frequencies. At high frequencies the dynamic range is quite limited, especially if no bias is used, and when the optimum head current is exceeded the output falls instead of increases. Playback equalization is useful for reducing noise, but in any case the playback signal undergoes severe limiting (clipping) to remove amplitude variations prior to demodulation.

As shown in Figure 6-17*A*, the FM recording head current is generated by an oscillator whose frequency varies with the signal voltage. The playback FM signal is successively amplified and limited and finally demodulated, as in Figure 6-17*B*. Further details on FM recording are given in Chapter 11.

DIGITAL SYSTEMS

Because only two states, ones and zeros, are recognized in the digital world, the electronic circuits contain gates, flip-flops, multivibrators, Schmidt triggers, and other all-or-nothing devices that are not found in analog recorders. But it is an annoying fact of life that analog effects are very prominent in the head-tape interface and must be dealt with, especially at high recording densities. Some of these are dropouts, dropins, pulse crowding, contour effects, tape skew, and gap scatter.

In digital recording and in FM recording, the bias is usually omitted since linearity is not necessary when the recording is always saturated either in one direction or the other. However the use of high-frequency bias even in these applications may improve recording density and signal-to-noise ratio.

Most digital recorders use a variation of NRZ (non-return to zero) coding such as NRZ-M, NRZ-I, or NRZ-C. Other codes are in the minority, including RZ (return to zero), RB (return to bias), teletype, and double pulse. Basic dig-

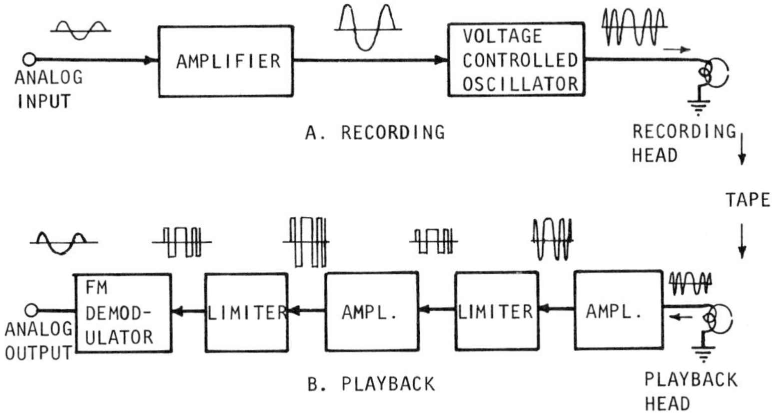

Figure 6-17. Frequency modulation (FM) recording and playback system.

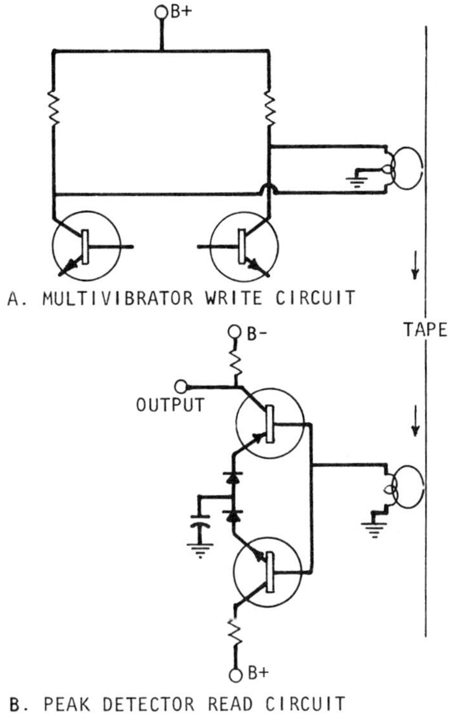

Figure 6-18. Write and read circuits for digital recording.

ital recording circuits merely activate the write-head coil on one side or the other of its center tap for positive or negative saturation of the tape (Figure 6-18*A*). Basic digital readout circuits sense some feature of the signal from the reading head, such as a local peak or a zero crossing or the lack of any output during a time-slot interval, converting it as soon as possible into the digital realm (Figure 6-18*B*). Some digital recorders use signal processing, analogous to equalization in analog recording. More details are given in Chapters 13 and 14 on digital recording.

BIASING AND ERASING

Circuits and specifications for generating and using high frequency bias and erase currents are discussed in this section. Analog recorders that use high-frequency bias require an oscillator at least for recording-bias and usually also for erasing. The bias frequency should be as high as feasible to prevent bias beats and to give a low background noise. A rule of thumb is to choose a bias frequency about five times as high as the highest signal frequency; therefore the bias in audio recorders is set at about 75 kHz to 100 kHz, although early recorders went as low as 16 kHz and some professional recorders run at 200 kHz or higher. Wide-band instrumentation recorders are biased at 10 MHz, and even higher frequencies have been used.

Bias Current

Amplitude of the bias flux is in the order of ten times the maximum signal flux, depending on coercivity of the record material and on steepness of the sides of its hysteresis loop. The bias current is usually set by experiment, noting the playback-output versus recording-bias of the recorder under actual conditions of use, as specified by the manufacturer. In audio work, a low-level test signal of about 300 to 1000 Hz is chosen. Bias is optimized at a current somewhat higher than the current that gives peak output, usually enough to reduce the output about one dB below peak. More complex procedures are sometimes specified, such as measuring distortion levels and using both high and low-frequency test signals. Calculating the bias current would otherwise be difficult since head losses and tape interaction are uncertain at high-bias frequencies.

Bias Waveform

Although a sinusoidal bias waveform is considered ideal, a distorted wave can be tolerated if it is symmetrical, with positive and negative cycles being of the same shape. An unsymmetrical wave is undesirable because it can offset the recording in the same way as adding a DC component to the signal, increasing noise and possible distortion. In fact the remedy for small amounts of unavoidable asymmetry is to introduce an opposite DC current in the recording head, adjusting it for minimum playback noise.

The best way to generate a symmetrical bias waveform is with a push-pull oscillator, as with the circuit shown in Figure 6-19. A high Q resonant "tank" circuit, lightly loaded, improves the waveform purity. Oscillator loading can be made negligible by isolating it from the load with an amplifier or buffer stage.

Coupling or Mixing the Bias

Several ways of adding high-frequency bias to the signal flux are shown in Figure 6-20. All are effective, and the choice depends on convenience and economy. Although the signal and bias are well-mixed at the head gap, their mixture elsewhere may be troublesome.

Stray bias frequencies in the signal circuits may overload them. Stray bias frequency output from signal circuits can also mix in the head with the regular bias, adding or subtracting from it. The total bias may then change whenever the gain control is varied, causing underbias or overbias conditions even though the recorder has been set up according to specifications. Similar variations can occur when plug connections are changed. These possibilities should not be overlooked. Bias traps (wavetraps) are useful in confining the bias currents, as are also signal amplifiers that cut off above the useful band.

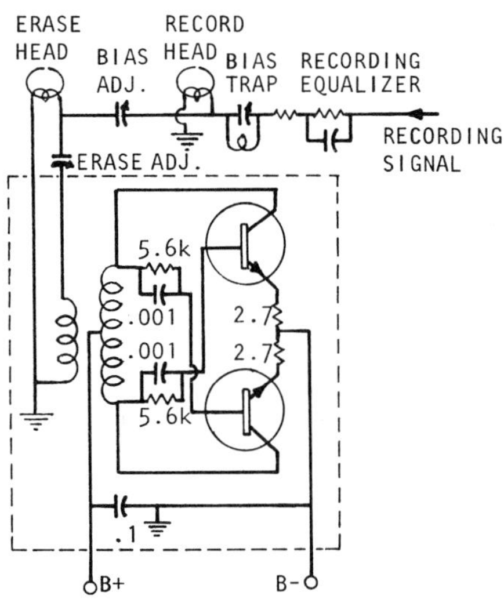

Figure 6-19. Push-pull oscillator for biasing and erasing.

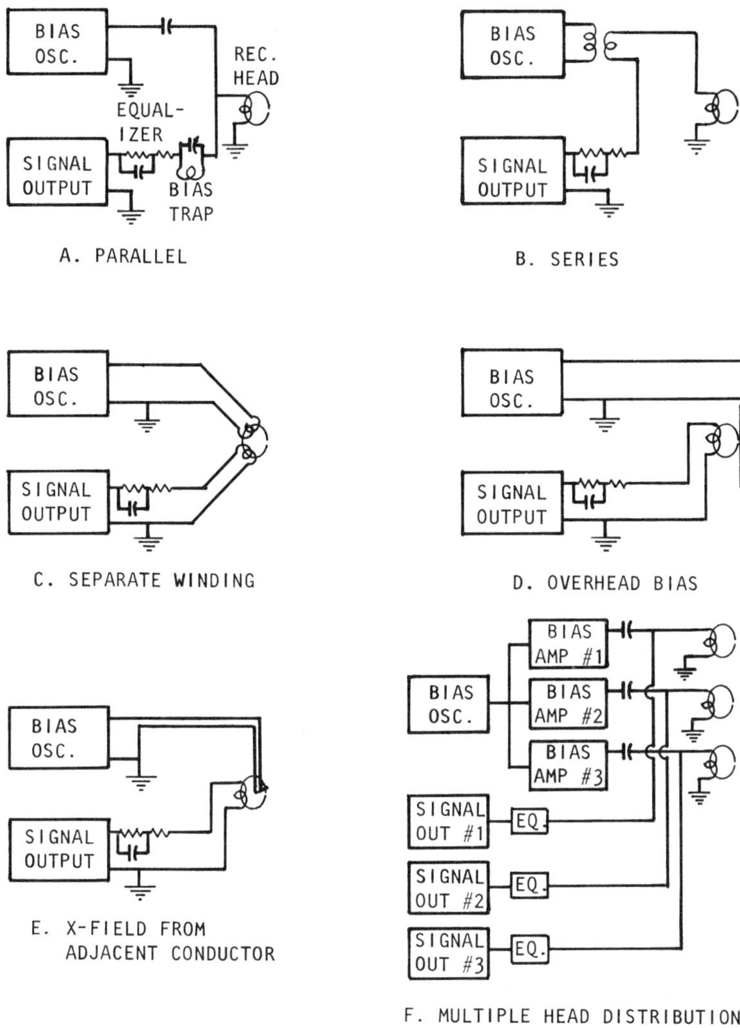

Figure 6-20. Bias coupling circuits.

Signal currents in the bias source, if coupled into the erase circuits, add unwanted background interference to the signal. In multitrack recorders, several heads are often fed from a common bias source, which can cause cross-coupling among heads. To avoid cross-coupling, the internal impedance of the common bias source must be so low that whatever signals appear across it are below the level of interest. An excellent isolation scheme is to use a common oscillator with a separate buffer for each head (Figure 6-20*F*), which also allows independent adjustment of bias for each head. Separate oscillators for each channel will beat against each other unless locked to a common source.

DC bias is explained in Figure 2-22. It is almost never used in modern recorders.

Erasing Considerations

In most recorders the same oscillator used for biasing also energizes the erasing head (Figure 6-19). The oscillator power output capability is about 0.5 to 5 watts, with most of the capacity not utilized, or wasted in coupling circuits. A convenient arrangement is loose inductive coupling to the tank coil with a series resonant head circuit as shown in the figure.

Many variations have been used:

1. A low power oscillator fed into a power output amplifier gives better stability and waveform.
2. The erase frequency can be a submultiple of the bias frequency (as $\frac{1}{2}$, $\frac{1}{3}$, $\frac{1}{4}$). Core losses in the erase head are reduced greatly at the lower frequency, giving a stronger erase field with less oscillator power. At the same time the bias is at an advantageous higher frequency, too high for efficient erasing.
3. The head winding is used as the tank coil of the oscillator to save the cost of an extra coil. If the coil carries DC, such a head may give noisier erasing.
4. The erase oscillator is eliminated, the tape being erased by a set of multipole magnets (Figure 2-15*C*). A bias oscillator is still required but is of much lower power.
5. No erasing is provided on the recorder, to ensure that data cannot be erased accidentally. Reels are bulk erased if necessary (Figure 2-55).
6. The recording head alone gives reasonable erasing while recording if it is overbiased and its gap is not too small. This arrangement is used in belt or drum dictation machines where a separate erase head would give an undesirable lapse during corrections, and where loss of recording resolution is tolerable.
7. Recording heads in digital recorders using NRZ coding are also self-erasing, since they always give a saturating field in one direction or the other.
8. DC erasing saturates the tape prior to recording with DC bias. This method is mainly of historical interest.

7. Tape Transports

The simple spool drives shown in Figure 7-1*A* and *B* gave satisfactory results for voice recordings in the earliest recorders and are even found in modern-day dictation machines. Indeed the drive stability was largely ignored while designers' attention was concentrated on shuttling wire or tape back and forth conveniently and rapidly, together with indexing, backspacing, level winding, and other niceties.

When tape machines were seriously considered for music recording in the 1930s, they were equipped with a capstan for driving or "metering out" the tape at a constant rate (Figure 7-1*C*) while the supply and takeup drives merely tensioned the tape as it was pulled or fed by the capstan. A flywheel added on the capstan shaft gave the basic drive still found in most home-use recorders. If even better stability was needed, as in professional and high fidelity recorders, the next step was to isolate the heads from the supply reel or to use more sophisticated drives.

Experience has shown that for voice recording a drive stability within about 0.25% wow and flutter content is satisfactory. For music a stability of 0.15% or better is specified. These guidelines are subject to variations and interpretations depending on the nature of the disturbances, method of measurement, program material, and listening conditions. For perfectionists, even the best drives are not ideal since the human ear can detect instabilities of 0.005%, far beyond the capability of even the best machines. Hence there is continued challenge and competition to improve audio transports by precision tolerances to fractions of a μm, by perfecting the mechanical filtering, by servomechanisms, and so on.

Tape drives for digital computers do not require the steady tape motion of audio recorders, but must start, stop, reverse, and repeat at super-rapid millisecond rates. Hence they use vacuum bins and special capstans. Instrumentation and FM recorders (Chapter 13), however, do require very steady drives, as well as a wide selection of accurate tape speeds. These machines feature the most precise, elaborate, and expensive drives that the state of the art can provide.

Magnetic records also take the form of discs, drums, belts, endless loop windings, and helixes, among others, for which appropriate drives must be designed. Audio and video cassette drives are treated in Chapters 10 and 12. Computer and disc drives are additionally detailed in Chapter 14.

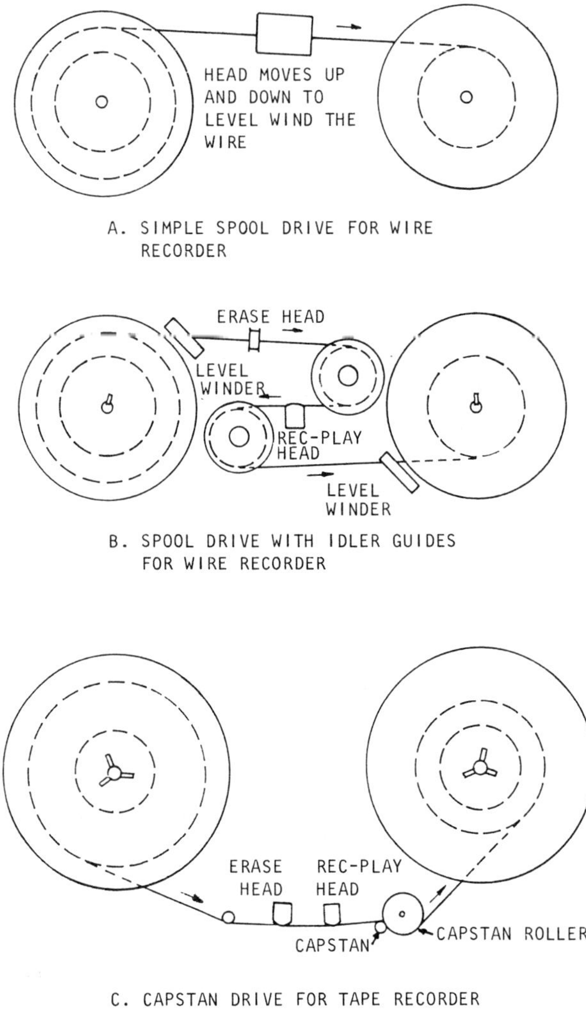

Figure 7-1. Elementary drives for wire and tape recorders.

GENERAL-PURPOSE AND AUDIO TAPE DRIVES

In the late 1940s and in the 1950s, low-priced reel-to-reel recorders were a flourishing new consumer industry. Old-line record-changer and electronics companies added recorders to their product line and new companies sprang up to meet demands of the mass market. Most designs were variations of the basic drive shown in Figure 7-2*A*, which gave satisfactory results for ordinary appli-

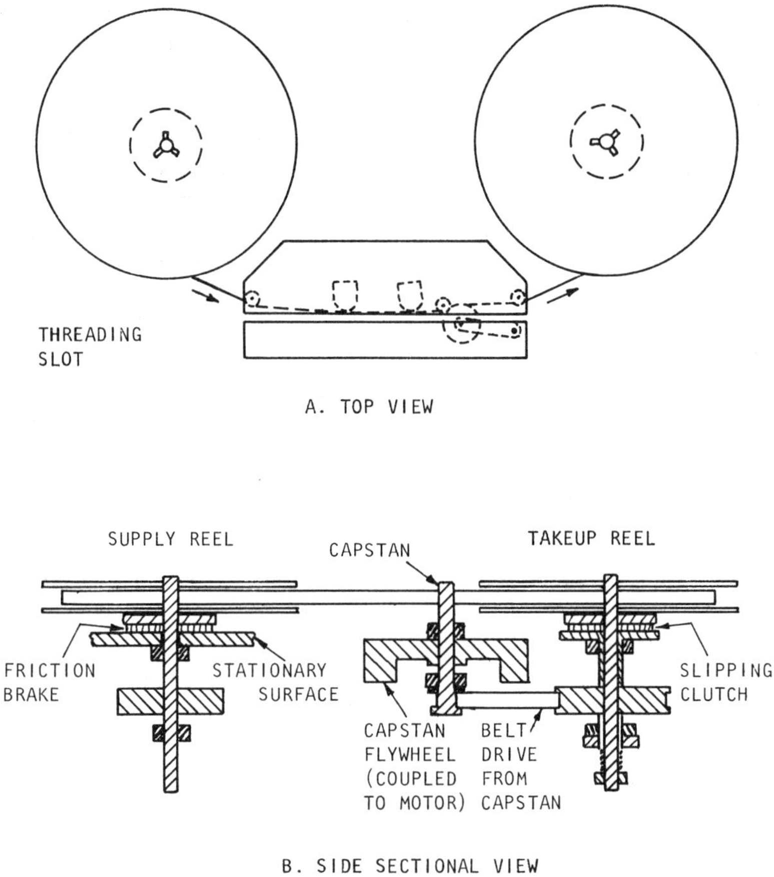

Figure 7-2. General purpose tape transport.

cations and were the logical no-nonsense approach that appealed to mechanical designers.

Typical Audio Drive

The popular reel-to-reel drive shown in Figure 7-2*A* features *straight line threading*, in which the tape is dropped into a slot defined by a cover for the heads and capstan. As shown in Figure 7-2*B*, the tape is pulled off the supply reel against the action of a friction brake that tensions the tape to about 1.11 N (4 oz) per 6.3 mm ($\frac{1}{4}$ in.) tape width. The pull is supplied by a capstan with a rubber pinch roller that bears against it with a fairly heavy pressure usually amounting to 1 or 2 kg (2 to 4 lb) or even more. The usual capstan is about 6.35 mm ($\frac{1}{4}$ in.) in diameter, but may range from $\frac{1}{16}$ to $\frac{1}{2}$ in. or more with a flywheel on the capstan shaft to give it a smooth, steady motion. As the tape feeds out of the capstan drive, the takeup reel must receive it with a tension approximately 1.11 N (4 oz) per 6.3 mm ($\frac{1}{4}$ in) tape width as from the supply reel. The takeup reel is overdriven through a slipping clutch and its turning speed depends on how fast the tape feeds to it and on the tape winding diameter. Finally the tape must be rewound at high speed while being decoupled from the capstan, and similar fast-forward speed is needed for rapid indexing.

Sources of Mechanical Power

Electric motor drives are so universal for tape transports that anything else is almost unthinkable. Yet spring-wound motors have been used in the past, and for special devices there are possibilities of gravity (descending weights), hydraulics, and even hand cranking. Electric motors offer a bewildering variety of characteristics, and the choice is a compromise between many factors, as considered below.

Cost. In lowest-cost recorders, where performance, efficiency, and long life are sacrificed, the motor may be of shell construction with a minimum of low-cost parts. AC motors are usually of the shaded-pole type; DC motors have permanent magnet fields. Expensive recorders use better motors.

Efficiency. In line-operated recorders energy efficiency is usually ignored, but in portable recorders efficient motors can mean smaller, lighter batteries and longer battery life. Well-designed small motors are about 80% efficient, so there is little to gain by improving the motor when much more can be done elsewhere in the drive train. Efficient DC motors use permanent magnets for their fields, while efficient AC motors include capacitor-run types with hysteresis or induction rotors. A very small air gap between rotor and field poles increases efficiency at the expense of close mechanical tolerances.

Size and Weight. Other things being equal, the horsepower delivered by a motor is proportional to its speed, so a high RPM motor is advantageous for reducing size and weight. Two-pole motors run the fastest. AC motors run faster at higher frequencies, and therefore 400 and 800 Hz are common in lightweight aircraft equipment. DC motors run faster with fewer armature turns and weaker fields. High-speed motors are seldom used in magnetic recorders because they introduce many problems. They are noisy, mechanically and electrically, have a shorter life, and require a gear box to reduce their speed to working range.

Torque Characteristics. Capstan motors usually run at full speed and moderate torque, but must give the steadiest possible tape drive in audio and analog recorders. Reel motors run much slower and deliver much higher torque. Reel torque ideally should vary inversely with speed to keep tape tension constant whether the reels are full or empty.

For a capstan drive the smoothest, quietest motor that can be chosen is the AC hysteresis synchronous type, which has a rotor of cobalt alloy steel, without slots, bars, or windings of any kind. The hysteresis sync motor runs at a synchronous speed of RPM = 120 × line-frequency/number-of-poles, so that a 60 Hz four-pole motor would run at 1800 RPM. Of course the tape does not necessarily run synchronously even with a sync motor because capstan slippage can vary. Less costly drives use a capacitor-run squirrel cage induction motor, or even a phonograph type of shaded pole motor, both running at somewhat less than synchronous speed according to loading. In all AC motors the stator generates a rotating field that drives the rotor, but this field itself fluctuates somewhat as it rotates because of discontinuous winding distribution and imperfect phase conditions in the quadrature windings. Moreover, the driving field is linked elastically to the rotor, so the rotor position lags behind whenever the motor load increases. DC motors for constant speed are of the permanent magnet or shunt type, with governor control for speed regulation. As with AC motors, the torque within each revolution can be uneven. It is apparent that even the best capstan motors are imperfect, but they can be improved by flywheels or servo control.

For reel drives DC motors with series windings come closest to giving an ideal torque-speed characteristic but have been avoided in the past because they use brushes and commutators, which generate electromagnetic interference and require maintenance. AC motors were usually adapted for reel drives, although they were inherently unsuited for variable low speeds at high torque, and hence were oversized or heated considerably or gave less than ideal tape tensioning.

In recent years new kinds of motors have been very successful for tape transports, including printed circuit motors and brushless DC motors. Brush-type motors have also become popular because their interference problems have been solved. Most important of all, servo control has become practical with the advent of low-cost microcircuits. Servo control allows the designer to regulate tape tension, drive steadiness, and rapid starting and stopping to a degree heretofore not attainable.

Capstan Drives

The capstan itself is a tape driving shaft of high precision, preferably with a matte finish to give a nonslip grip on the tape. A rubber pinch roller holds the tape against the capstan with a force that may amount to several pounds or

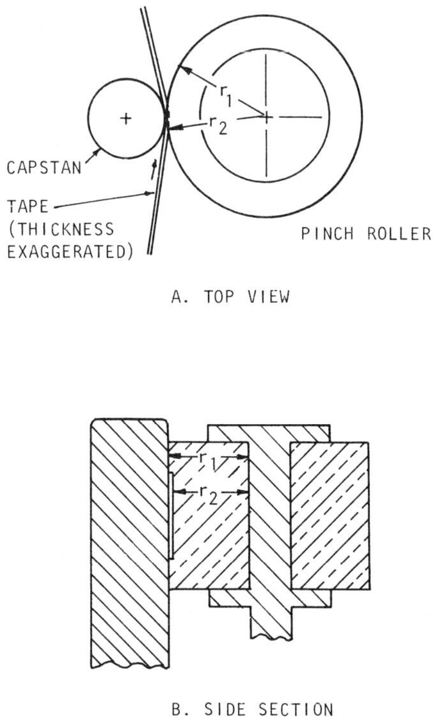

Figure 7-3. Basic capstan configuration.

kilograms. As shown in Figure 7-3, the pinch roller is usually wide enough to mushroom around the tape and to contact the capstan above and below it. Since the friction is greatest between the rubber and any surface it touches, we can infer that the capstan drives the roller at its top and bottom, while the roller thus driven, drives the tape from its back side. Because the tape has finite thickness, its back side has slightly higher surface velocity than its front side, which wraps against the capstan. Hence there may be slippage between front tape surface and capstan because friction is least at the front tape surface. It might appear that the roller drives the back of the tape at reduced velocity because r_2 is less than r_1 (Figure 7-3*A*), but another effect counteracts this one. The area at cross section *2* is less than at cross section *1* (Figure 7-3*B*), but as much rubber must move through the smaller area as it does through the larger one in going past the pinched region, and hence the rubber material moves faster where it is pinched more, analogous to Bernoulli's well-known principle. The extent to which this theory holds in practice depends on the type and degree of mushrooming, capstan and roller diameters, and so on. There is, however, positive evidence of slippage between tape and capstan since considerable wearing of the capstan is noticeable where it rides against the tape.

One way to avoid the above roller-edge effect is to make the pinch roll width equal to tape width or slightly less (Figure 7-4*A*). Then the capstan must necessarily drive the tape. This arrangement requires much more pressure of the

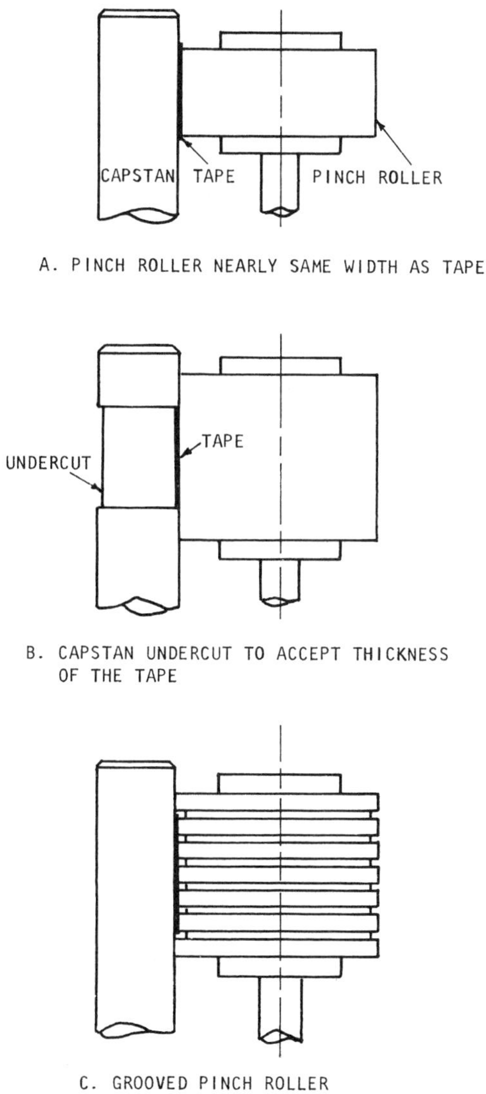

Figure 7-4. Modified capstan configurations.

pinch roll to give dependable drive, often to the extent where tape, roller, or bearings become marginally damaged. A better solution, shown in Figure 7-4*B*, is to undercut the capstan with a groove equal to the tape thickness; now all parts of the overlapping pinch roller move at the velocity of the outer tape surface. Another refinement is a grooved pinch roller (Figure 7-4*C*), where each protruding portion can expand endwise when under pressure, which is especially helpful when driving a wide tape.

Capstan Design

The same tape speed can be attained with a large-diameter capstan turning slowly, or with a small diameter capstan turning rapidly. A large capstan has better relative accuracy than a small one when they have the same dimensional error. For example, a runout (eccentricity) of 0.001 in. on a 1-in. diameter gives a 0.1% error, whereas the same runout on a 0.1-in. capstan gives a 1% error. A large capstan also minimizes stress in the tape since the pinch roll pressure acts over a larger area and the tape is not bent over a small radius. Another advantage is physical strength, making a large capstan less likely to bend under normal, or abnormal, conditions.

The great advantage to using a small high-speed capstan, however, is that it can store much more energy in a flywheel of reasonable size and weight, since the energy storage increases as the square of rotational velocity. Small-diameter capstans are therefore found on compact machines and where economy is important. Cassette recorders sometimes have capstans about $\frac{1}{16}$ in (1.5 mm) in diameter, and their capstan accuracy must be held to submicron dimensions to achieve low flutter for music reproduction.

A capstan surface with a matte finish to give good frictional driving action for the tape is obtainable with a dull chromium plating. A capstan made of ceramic material also has excellent frictional qualities and long life. Sleeve bearings or precision ball bearings are generally used for capstan and flywheel shafts. Although ball bearings are found in the finest commercial machines, some experts insist that ultimate precision is attained only with sleeve bearings. Where audible noise is a problem, sleeve bearings run more quietly, especially at high speeds.

When a pinch roller is used, it is good practice to offset it, as shown in Figure 7-5, so that the tape coming from the head wraps slightly around the capstan before reaching the pinch roll. Otherwise the less precise free surface of the rubber roll might disturb the tape motion.

One Motor or Three?

Low-cost transports use a single motor for all functions: capstan drive, takeup reel, rewind, and fast forward. Each mode is selected by cams and levers that

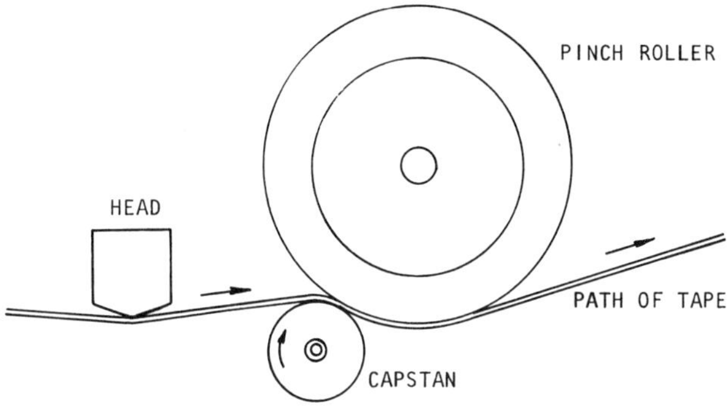

Figure 7-5. Offset pinch roller.

shift belts and rubber drive wheels. Of the many possibilities, one successful arrangement used in a cassette recorder is detailed in Figures 10-2 and 10-3. The single-motor drive transmits to the capstan some irregularities from the takeup spool system, which it drives together with the capstan. As shown in Figure 7-2*B*, the mechanical designer finds it most convenient to drive the takeup reel from the capstan flywheel, even though this is not ideal for minimizing flutter.

A separate motor for each reel and for the capstan is a neat solution for the designer, giving a steadier drive and eliminating most parts that wear out or require adjustment and maintenance. A three-motor drive is shown in Figure 7-6. The additional expense is not as great as might be expected, since reel motors are of less expensive construction and can eliminate bearings and pulleys otherwise required for the reel shafts. Of course, one could use only two motors, one of them having a double function including high-speed rewind, but the savings usually do not justify the added complexity.

With three motors the drive functions are switched electrically, giving the immediate possibilities of remote control and automation. Supply and takeup tensions are adjusted electrically and will maintain their settings. Two difficulties with reel motors soon become apparent: (1) An ordinary motor for directly tensioning a tape is not light in weight, since it must supply about 0.07 Nm (10 in.-oz.) of torque to a 7-in reel of $\frac{1}{4}$-in. tape, and proportionately more for wider or longer tapes; hence the tape deck becomes quite heavy. (2) Reel motors of the AC induction type operate at nearly 100% slip, making them very inefficient, especially if they are of the shaded pole variety, so they generate a great amount of heat, in some instances enough to melt the hub of plastic reels attached to their rotor shaft.

Cooling fans on such motors are useless because they turn so slowly. Forced cooling is possible, but adds noise and complexity. Capacitor type motors are

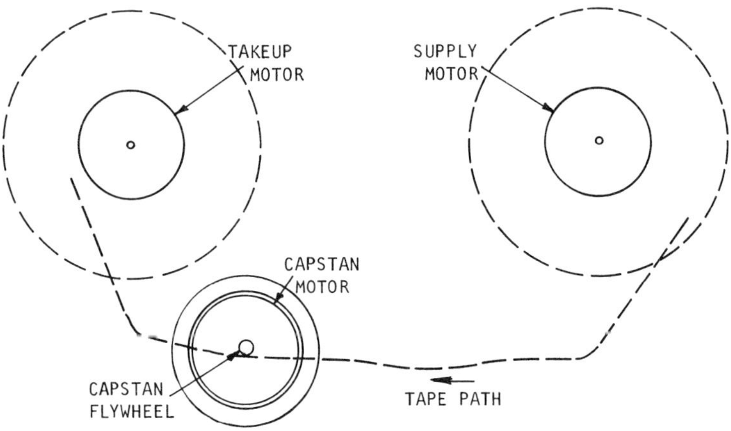

UNDERSIDE OF TAPE DECK

FUNCTION	TAKEUP MOTOR	SUPPLY MOTOR	CAPSTAN MOTOR
Off	*Off*	*Off*	*Off*
Standby	*Off*	*Off*	*On, not engaged*
Record-Play	*Reduced torque*	*Reduced torque*	*Engaged*
Rewind	*Reduced torque*	*Full torque*	*On, not engaged*
Fast-forward	*Full torque*	*Reduced torque*	*On, not engaged*

Figure 7-6. Three motor tape deck.

more efficient but more expensive. A DC or universal motor is even more efficient for low-speed, high-torque service, but electrical interference from its commutator must be filtered out. These challenges to designers have been solved in professional and semiprofessional models, and in the best-quality "hi-fi" machines.

Power Transmission

Gears, worms, sprockets, chains, and other positive drives found in most machinery are conspicuously absent in tape transports, although almost every novice designer in this field reintroduces them with confidence and withdraws them with chagrin. Belts and rubber-faced friction wheels are used almost exclusively in tape drives for speed changing and transmission, giving smooth, silent, ripple-free motion that is difficult to achieve with any drive having toothed wheels of reasonable cost.

Belts and Friction wheels. Our hearing is exceptionally sensitive to short-term ripples of gears but ignores long-term slippage of belts or friction wheels, so there is a human-engineering factor that favors the latter. But neither belts nor friction wheels are durable and they are the weakest link of drive systems that use them, requiring special precautions, maintenance, and replacement.

The rubber (or synthetic substitute) facing on a wheel develops a dent in its surface where it is pressed against another wheel when stationary. Hence the friction drive should be unmeshed when not in use. Surface wear of rubber is not uniform, so after a while a perfectly ground wheel loses its accuracy and must be reground or replaced. The rubbery material is not perfectly uniform in hardness (durometer) from point to point so that even a new wheel does not give perfect drive. The rubber surface deteriorates with time, becoming glazed, hardened, cracked or dirty and losing its frictional qualities.

Rubber belts may be of homogeneous elastomer or may be reinforced with cotton or synthetic fabric. The homogeneous types have disadvantages as described above for friction wheels. It is usually impractical to release their tensioning when not in use but their "set" is not as serious as in drive wheels and works out after running a while. On the other hand, such belts can break suddenly at a critical time and can stretch gradually until they are useless and must be replaced. Reinforced belts overcome these objections, but since they are not very elastic they require an idler or equivalent to take up slack. Vee belts in small-diameter V pulleys do not have good constancy of speed since they tend to ride up and down the V groove, changing the drive ratio. Elastic belts of square cross section used in cassette drives perform better, but have to be replaced frequently.

Direct Drives. The capstan motor shaft can itself be the tape drive capstan if its driving diameter and the motor speed are correct. Table 7-1 gives the ideal capstan diameters for commonly used tape speeds and motor speeds. Practical diameters are slightly larger to compensate for tape slippage. Direct drives are simple and reliable over long periods of time. They are most convenient for high-speed drives when two-pole or four-pole motors and reasonably large capstan diameters are usable. For low speeds the capstan diameter becomes too small, or a large number of poles must be used in the motor, making it somewhat special. Multispeed AC motors are available with electrical switching that changes the number of poles, allowing synchronous speed selection in a 2:1 ratio for example.

PROFESSIONAL TAPE TRANSPORTS

The simple tape transport shown in Figure 7-2 can be improved further by stabilizing the tension from the supply reel. Variations in supply tension stretch the tape more or less in the region between capstan and head. In this critical

Table 7-1. Direct-Drive Capstan Diameters for Standard Tape Speeds

Tape Speed		Capstan			Capstan			Capstan			Capstan			Capstan		
			DIAM			DIAM			DIAM			DIAM			DIAM	
in./sec	(mm/sec)	RPM	in.	(mm)	RPM	in.	(mm)	RPM	in.	(mm)	RPM	in.	(mm)	RPM	in.	(mm)
120	(3048)	3600	.637	16.18	3000	.764	19.41	1800	1.273	32.33	1500	1.528	38.81	1200	1.91	48.51
60	(1524)	3600	.318	8.08	3000	.382	9.70	1800	.637	16.18	1500	.764	19.41	1200	.955	24.26
30	(762)	1800	.318	8.08	1500	.382	9.70	1200	.477	12.12	1000	.573	14.55	900	.637	16.18
15	(381)	1800	.159	4.04	1500	.191	4.85	1200	.239	6.07	1000	.286	7.26	900	.318	8.08
7.5	(190.5)	1200	.119	3.02	1000	.143	3.63	900	.159	4.04	750	.191	4.85	600	.239	6.07

region even a microscopic stretch or shrinkage produces a high percentage of flutter if the change is sudden enough.

Open Loop Professional Transports

An effective, although expensive, stabilizer of supply reel tension is a tape-driven flywheel between supply reel and head assembly, giving the basic open loop stabilized drive of Figure 7-7. This drive has been more widely used in professional tape transports than any other because it is very reliable and easy to maintain. The stabilizing roll has an outer diameter of about 1.5 in. (38 mm) and ball bearings, so the tape can drive it by wraparound alone, aided by the spring-tensioned wraparound guide as shown. Its flywheel is about 4.5 in. in diameter and 1.5 in. high in professional recorders, but is scaled down in smaller decks. Internal damping is advantageous (as described in Figure 7-26*A*).

Dual Capstan Drives

Closely related to the open-loop-with-stabilizer concept is the dual capstan drive shown in Figure 7-8. The capstans isolate the tape at the heads from the reels on each side. To maintain proper tape tension, the trailing capstan must tend to

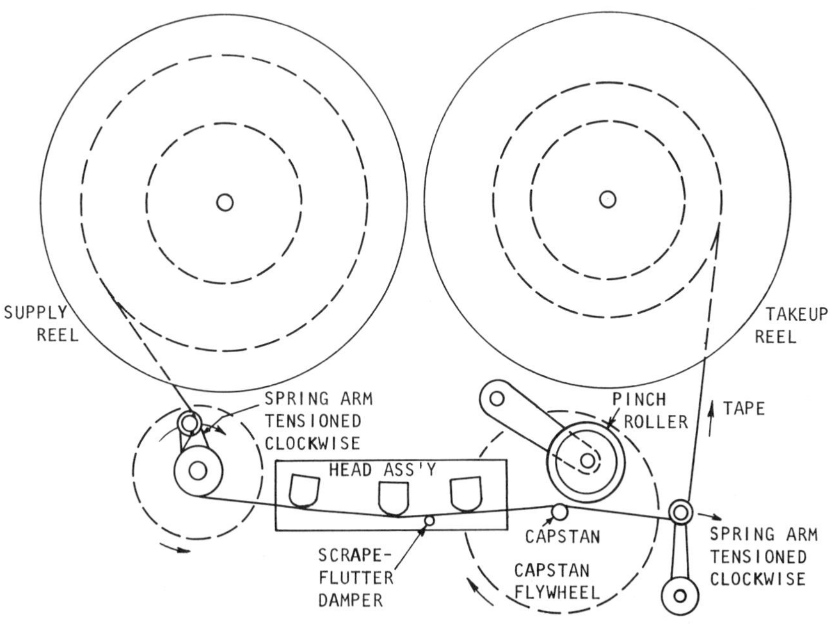

Figure 7-7. Open loop stabilized drive.

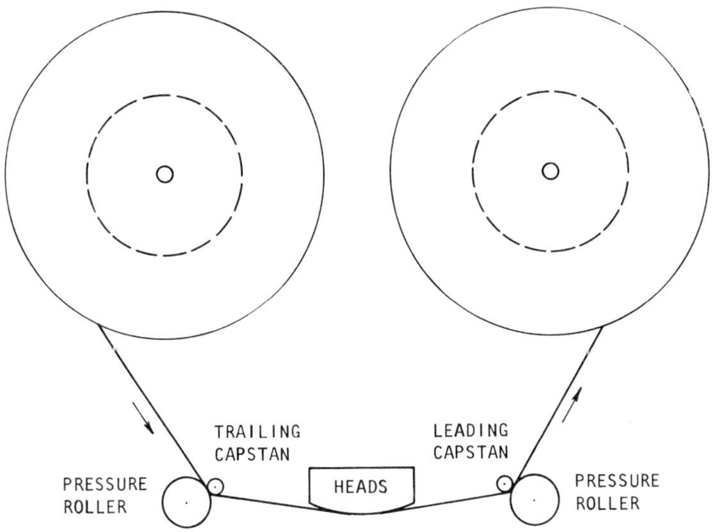

Figure 7-8. Dual capstan drive.

go slower than the leading capstan; in other words, the tape must pull the trailing capstan instead of being driven by it. Since we want positive drive at the main leading capstan, there must be some "give" or slippage associated with the trailing capstan. Otherwise in a friction drive the smallest cumulative error would eventually release the tape tension between capstans or would increase it to the breaking point. The smoothest slippage is electromagnetic, as in the rotor of a motor. Some transports rely on the slight tape slippage at the capstan, which is inevitable no matter how carefully avoided and which is put to beneficial use in this case. Where belts are used between capstan and motor, the slipping can take place at a belt. Two-capstan drives have been designed to run the tape in either direction for convenient playing of quarter track tapes, usually on subprofessional machines.

If both capstans and flywheels are practically identical and mechanical damping is low, there may be an interchange of resonant energy that aggravates flutter whenever the mechanical system is disturbed by a transient impulse.

Closed Loop Drives

In Figure 7-9, the same capstan acts on the tape both as it enters and as it leaves the heads, thus isolating a loop of tape from the supply and takeup reels with a single capstan-flywheel. To keep the isolated loop taut at the heads, the capstan has two different diameters, with an aligned pinch roller for each diameter. The

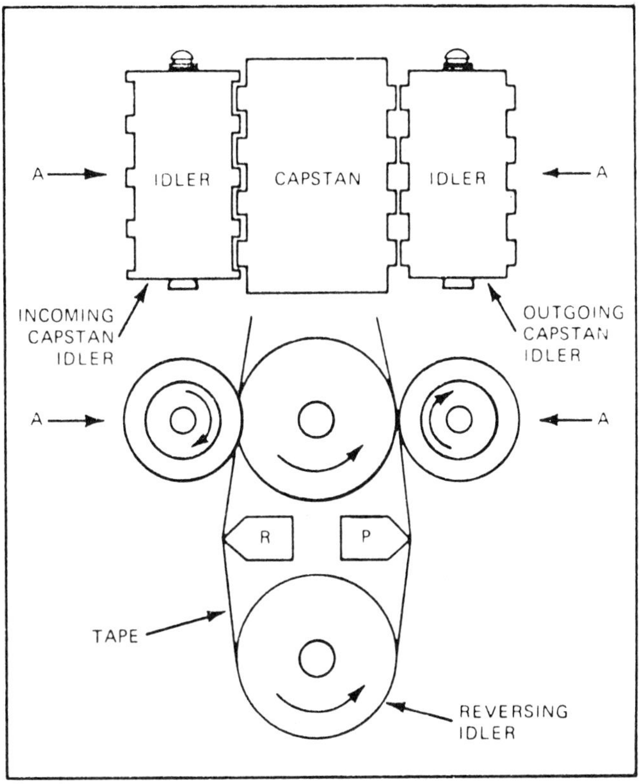

Figure 7-9. Closed loop drive (Courtesy Mincom Division, 3M Co.).

outer, faster drive is on the takeup side whereas the inner, slower drive handles the supply, giving a differential driving effect, with inevitable slippage preferably on the supply side.

The economy of a single flywheel is diminished somewhat by the requirement of a precision turnaround roller and bearings. The elegant concept of double duty from a single capstan is also diminished by the fact that tape tension depends on frictional slippage, which can vary with the tape and capstan surfaces, pinch roll pressure, and other factors.

Zero Loop Drives

In principle the ultimate capstan drive is one that has no loop at all, as in Figure 7-10, where the head rides on tape that is on the capstan itself and the tape motion therefore assumes the perfection of a precision capstan.

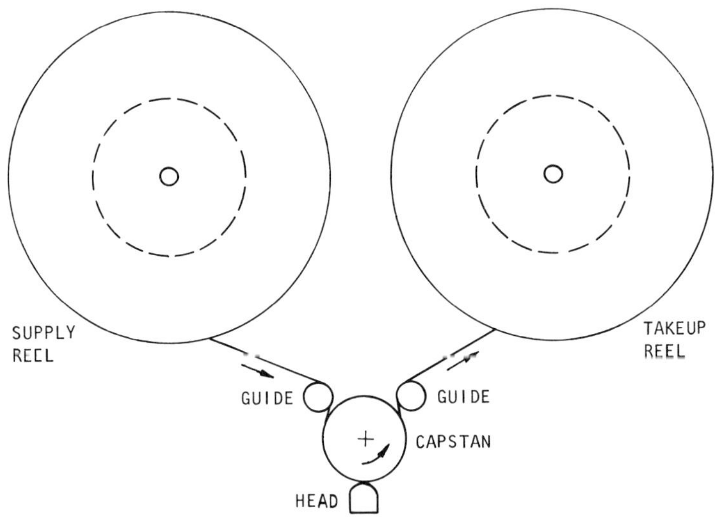

Figure 7-10. Zero loop drive.

Several difficulties become apparent when putting this ideal concept into practice:

1. A metal capstan is too hard a backing for the tape when the head presses against it, giving excessive head wear and rubbing noise. Also there is usually a magnetic noise due to a minute percentage of magnetic impurities in the capstan or its plating, even though made of nonmagnetic materials.
2. A rubber or plastic surfaced capstan is softer, but does not retain its accuracy and tends to become loaded with debris. Some plastics have a memory or time delay between pressure and deflection and therefore act as if they were brittle when rotating rapidly. The larger diameters required for soft rollers need more speed reduction and heavier flywheels.
3. The head wears to a hollow curvature that fits the capstan but is sensitive to the slightest longitudinal shifts, causing loss of contact between head gap and tape.

One solution to these difficulties is a grooved capstan such as the one shown in Figure 7-11*A*, where the head rides on the tape above the groove, thus removing the hard backing of a metal capstan. Another alternative is the modified zero loop drive of Figure 7-11*B*, where a guide lifts the tape from the capstan to allow heads such as H_1 and H_2 to contact the tape at the lifted portions. Still

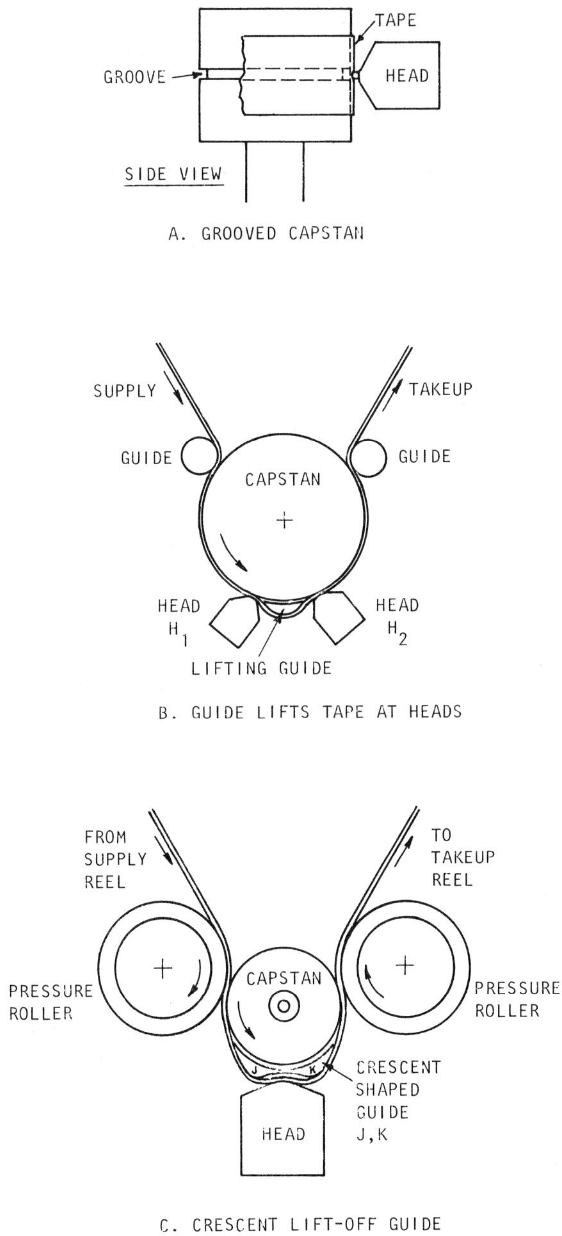

Figure 7-11. Modified zero loop drives.

another is the crescent guide of Figure 7-11*C*, which supports the tape at *J*, *K* where lifted from the capstan, allowing the head to ride at *L*.

Vacuum Capstans

A vacuum capstan grips the tape to give positive driving action with a metal capstan and eliminates the need for pressure rollers or rubber-faced capstans. Figure 7-12*A* shows one version with an inner manifold and valving arrange-

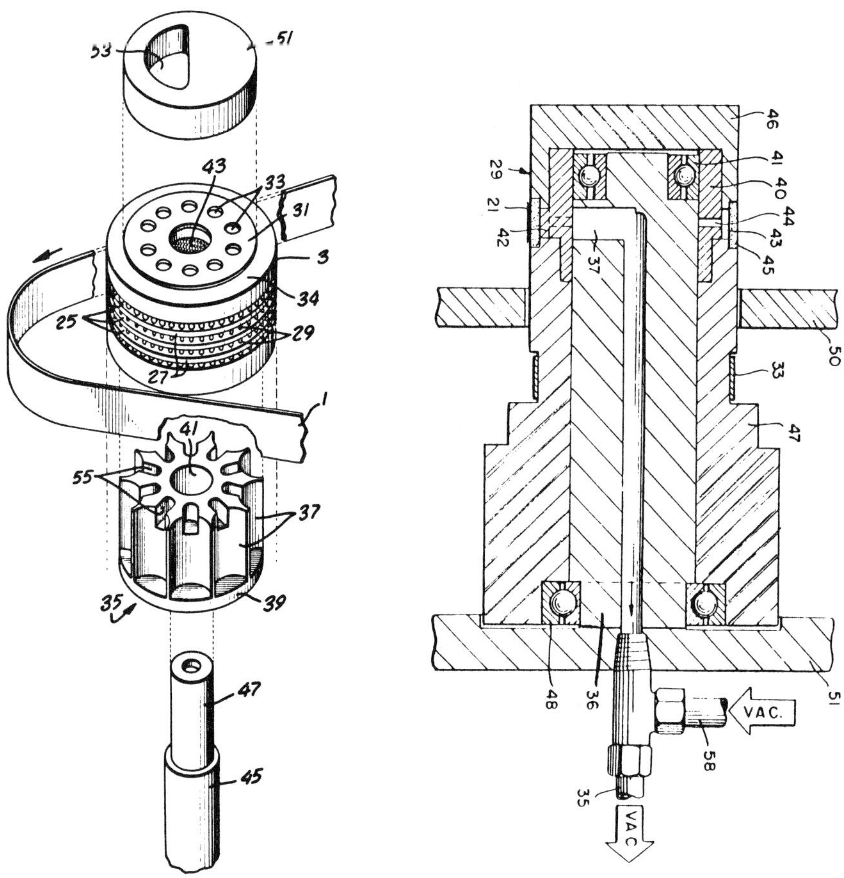

A. MANIFOLD TYPE

B. POROUS INSERT TYPE

Figure 7-12. Vacuum capstans. A. The manifold type has sliding valves that couple the vacuum to holes that hold and release the tape. From U.S. Pat. 2,747,025 to W. T. Selsted, May 22, 1956. B. A porous ring *45* pulls tape *21* against the capstan, aided by vacuum through duct *37*. From U.S. Pat. 3,441,187 to R. S. Tveter and S. Himmelstein, Apr. 29, 1969.

ment that shuts off the vacuum where the tape enters and leaves the capstan. Figure 7-12*B* shows another version with a porous insert that allows the vacuum to function with a continuous surface that has no drilled holes.

Cartridges, Cassettes, and Discs

Where state-of-the-art perfection is not required, inroads have been made by cartridges, cassettes, and discs. Higher quality items are available for the professional market than are for the mass market and are not necessarily interchangeable with the mass standardized products. An example is the endless loop NAB cartridge with a more precise capstan drive, specially developed for the broadcast industry. The mechanical drives for professional cassettes and cartridges are quite similar to the ones described for ordinary cassettes and cartridges, while magnetic disc drives are like the ones used for phonograph records.

COMPUTER AND INSTRUMENTATION DRIVES

Computers rely on magnetic recording for mass memory. The unusual drives in their realm fulfill special needs for rapid access, fast readout, repetition, random access, and extreme reliability. Wow, flutter, and steadiness of motion are of lesser importance than in other fields of tape recording. Instrumentation drives, on the other hand, do not need rapid tape maneuvering, but otherwise have the greatest drive precision, highest recording density, and lowest noise and distortion that can be achieved by state of the art.

Reel Isolation

Tape reels in computers require an exceptionally high degree of isolation from capstans and heads. Here the tape must shuttle back and forth across the heads, reaching full speed, stop, or reverse in a millisecond or so. It is impractical to accelerate the heavy reels at this rate so, as shown in Figure 7-13, there are sizeable U-shaped loops of tape stored in vacuum chambers on each side of the tape drive. The drive needs to accelerate only a short, lightweight strip of tape, lifting it out of the vacuum chamber that tensions the strip. A sensor in the vacuum chamber starts the reel motors to replenish or take up from the bins as they become depleted or filled.

An alternative transport for the same purpose, shown in Figure 7-14, uses multiple spring arm guides from each reel. No vacuum system is needed, but performance is not as good and threading is more complicated.

The more complete reel isolation may, of course, be used in other applications, such as instrumentation and audio, but is usually not justified where fast starts and stops are not needed.

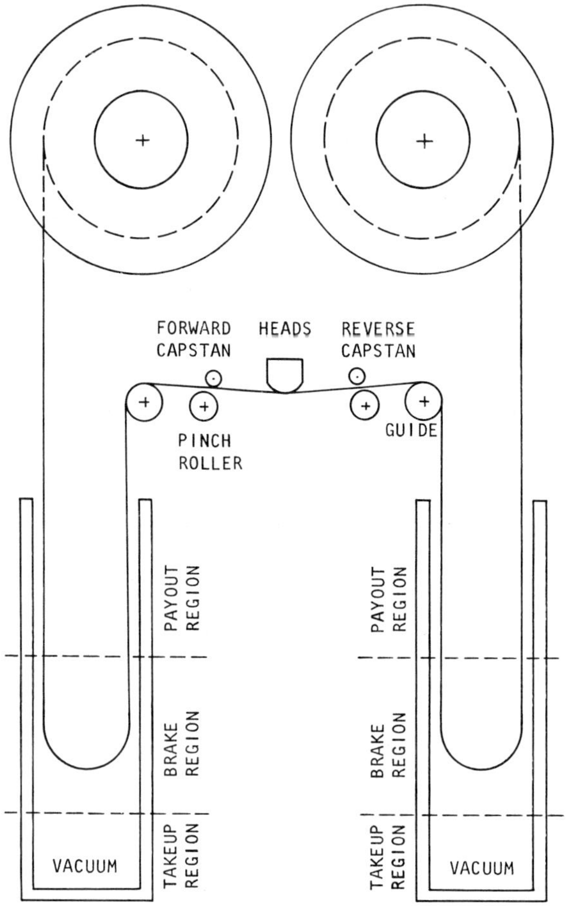

Figure 7-13. Vacuum bin tape transport.

Discs and Drums

Disc memories for computers do not start, stop, or reverse during operation, but spin along continuously while indexable or fixed heads read and write the data as they fly over the magnetic surface. The drive itself is simply a motor, preferably synchronous and usually connected directly to the disc shaft, with means for bringing it up to speed or stopping it smoothly on occasions when the disc file is to be changed.

Rigid discs spaced and stacked in layers are the intermediate memory backbone of medium and large computers. Flying heads are sandwiched between layers to ride on magnetic surfaces on both sides of every disc. The important

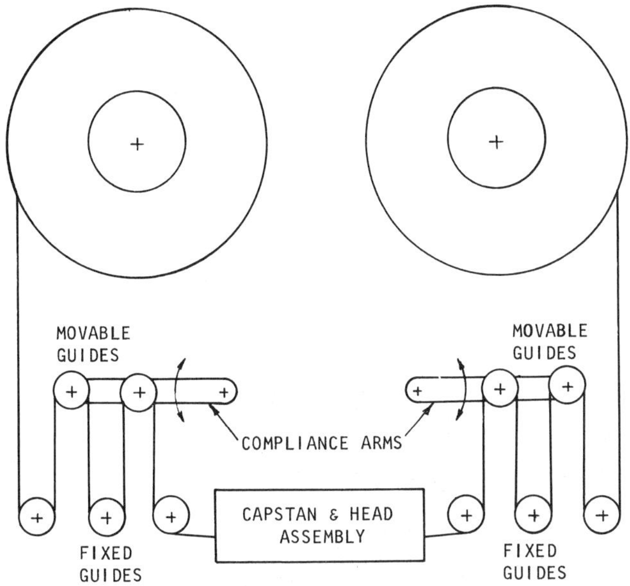

Figure 7-14. Computer transport with spring arms.

mechanics of disc files is in their head indexing, described in more detail in Chapter 14 on computer applications.

A floppy disc is a small, thin, inexpensive plastic disc, 8 in. (203 mm) or 5.25 in. (133 mm), or smaller, in diameter, magnetically coated, and permanently enclosed in a paper jacket. The jacket remains stationary while the inner disc is driven and heads make contact through windows in the jacket. Although its memory capacity is more limited and less reliable than the rigid disc memory, the floppy disc is a favorite in small computers because it is easy to handle and to file. The drive for the floppy disc shown in Figure 7-15 is quite straightforward and low priced.

Magnetic coated drum memories (Figure A-18*C*, Appendix) have become obsolete by the development of multiple disc memories. Their drive is most elementary, the drum usually being coupled directly to the shaft of a high-speed motor.

Instrumentation Drives

Instrumentation drives emphasize versatility in selection of different tape speeds, accuracy of tape motion, and high-density recording. Quick starts and reversals are of little or no importance, and therefore instrumentation drives resemble professional audio transports rather than computer drives. Instrumentation cov-

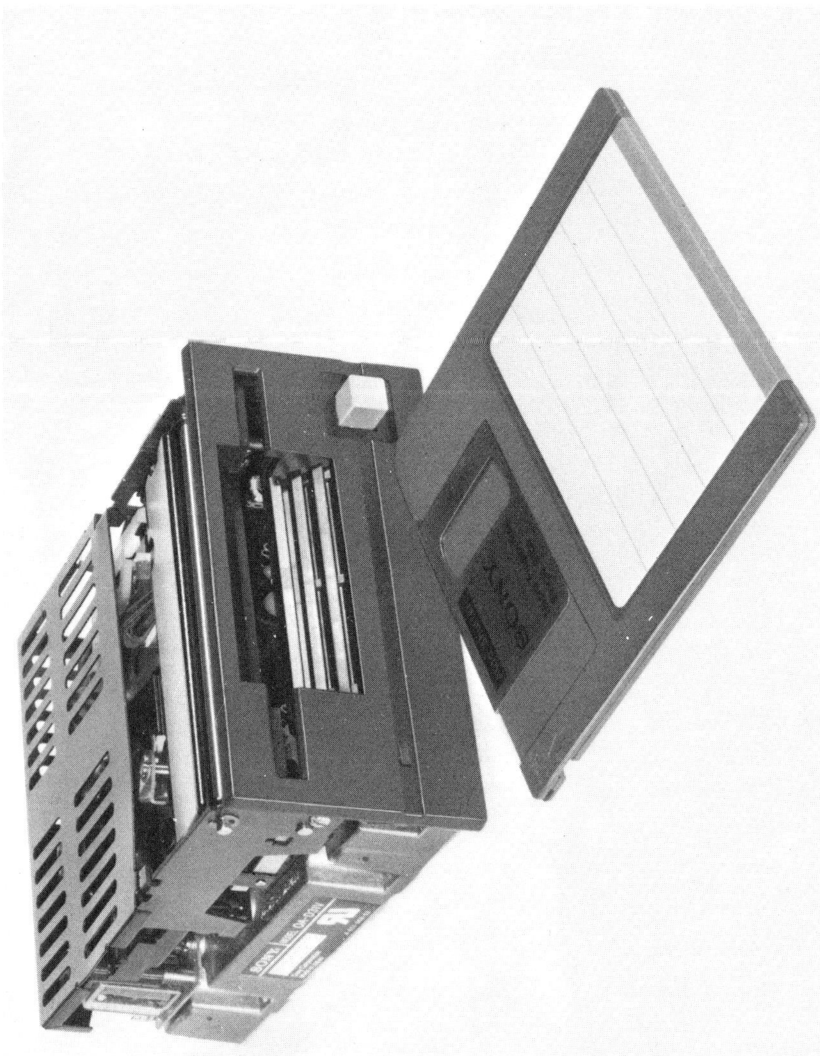

Figure 7-15. Drive for floppy disc. This $3\frac{1}{2}$ inch microfloppy by Sony holds 437.5k bytes at 135 tracks/in and 7610 bits/in density.

Figure 7-16. Instrumentation recorder, Honeywell model 101 portable.

ers a frequency range to 2 MHz or more, and thus requires tape speeds to 3048 mm/s (120 in/s) and higher, plus heads of high-frequency and high-density capability. Figure 7-16 shows one form of instrumentation drive.

SPECIAL DRIVES

Magnetic recording holds a special fascination for inventors of ingenious and unusual drives. Some of these Rube Goldberg devices have become commercially feasible, as for example rotating-head video recorders, endless loop cartridges, and chopper-type speech compressors. Others remain on the sidelines, waiting for a new application that fits them, or are lost in the competitive scramble.

Energy-Saving Drives

A fundamental calculation, seldom considered, is the amount of energy actually used for moving the tape at recommended tensions. A tape moving at 191 mm/s (7.5 in/s) with 1.11 N (4 oz) tension uses 0.212 Nm/s (0.156 ft-lb/s) of power, equivalent to 0.000284 horsepower or 0.212 watts of power. Even with additional losses and inefficiencies, this is not a great deal for a line-operated recorder, but is sizable for solar cells in satellite recorders or even for flashlight battery operation. In proportion, a 120-in/s recorder with 1-in. tape uses 0.0182 horsepower or 13.6 watts net driving power, while a cassette tape at 1.875 in/s (47.6 mm/s) uses only 0.0000416 horsepower or 0.032 watt.

The tensioning energy need not be wasted as frictional loss in a supply-side brake but can be fed back to the takeup side or to the capstan, storing up any difference in a spring if desired. One such scheme is the *negator spring* system described in more detail in Figure 13-17 in connection with satellite recorders. Here the tensioning energy from the tape is captured by the supply reel that transmits it to the takeup reel. Since the reels turn at somewhat different speeds as they fill and empty, the connection between them is through a long spiral *clockspring* that takes up the temporary differences in rotation.

The Newell Drive

A tape transport having many novel features is shown in Figure 7-17. Reels and takeup drives are not used; instead, the tape winds into and out of self-

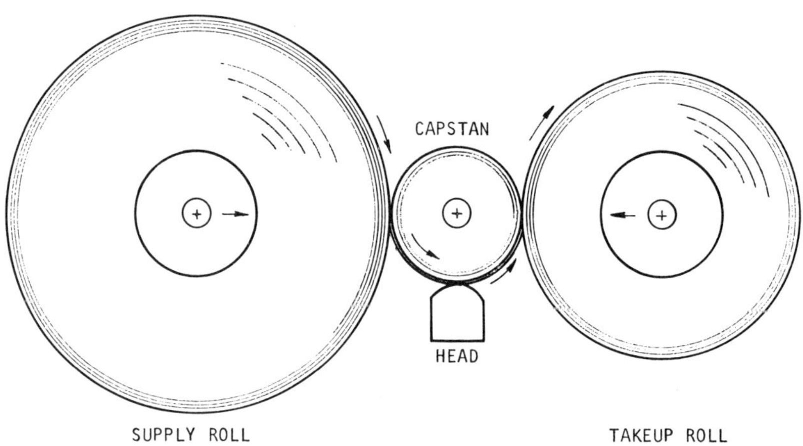

Figure 7-17. The Newell drive.

supporting rolls that are pressed against a resilient surfaced capstan of rather large diameter, the only motor-driven element. Heads ride on the tape while it runs over the capstan; or other zero-loop arrangements can be used. This drive has been suggested for speeds as high as 38.1 m/s (1500 in/s). It is readily reversible merely by reversing the capstan motor.

These ideal features are not without problems. Those of zero-loop drives have already been discussed (Figure 7-10). The self-supporting rolls are at the expense of a tight winding, which may warp the tape during long storage or environmental changes. The movable takeup hubs must run precisely parallel to the capstan or else the winding will be defective.

Endless Helical Drive

Somewhat analogous to the endless spiral tape cartridge is the endless helix tape arrangement developed in France (Figure 7-18). The helix does not have slippage between tape layers, which has been the nemesis of spiral windings. It therefore does not require back-lubrication or limited-size windings. The helix can be directly driven from its center like a phonograph turntable, at synchronous speeds if desired, without a capstan or other friction element. Multiple tracks are handled as with other endless loops.

Difficulties with the endless helix are that the tape itself is expensive because it must be specially manufactured to its unusual shape. The gearlike drive from the center may introduce a ripple into the tape motion; however, a friction drive could be used instead. The tape cartridge is bulky and has an unusual figure-eight shape for smoothly transferring the top turn to the bottom.

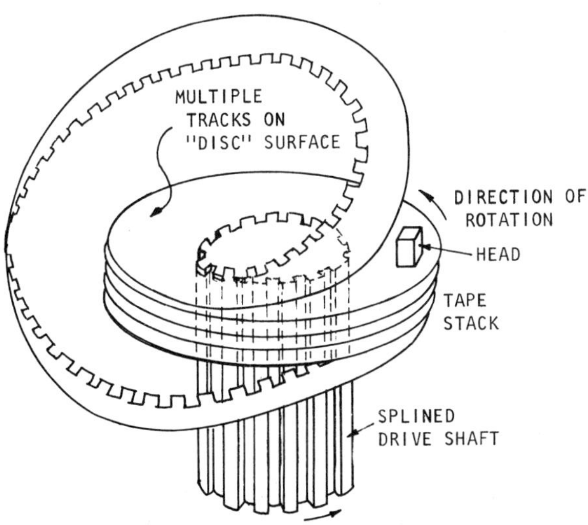

Figure 7-18. Endless helix drive.

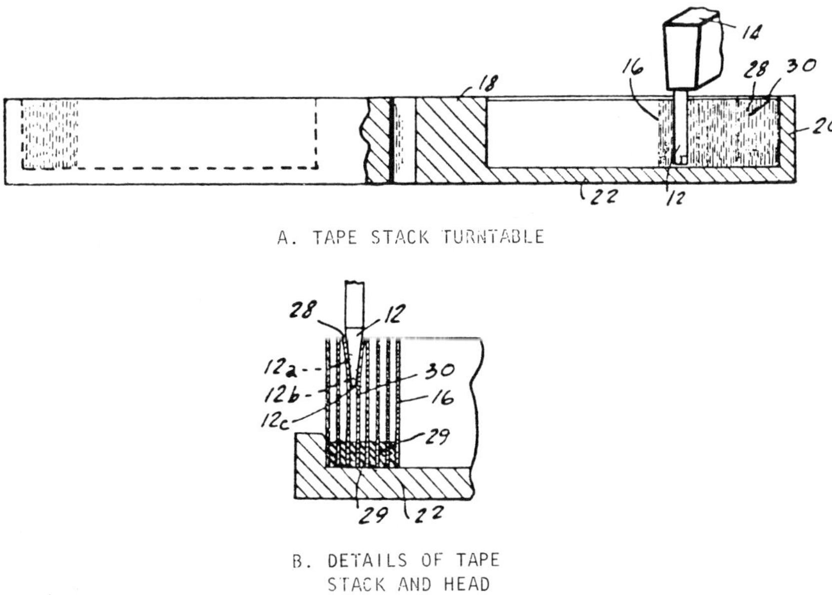

Figure 7-19. Tape stack turntable (From U.S. Patent 3,534,966 to M. Camras, Oct. 20, 1970).

Tape Turntable

An arrangement combining the advantages of reeled tapes and of discs is shown in Figure 7-19. Reeled tapes store an enormous amount of information and can contain many parallel tracks, but must be wound and rewound to get at a particular selection. Discs allow immediate random access, can be replayed without rewinding, and do not have to be threaded, but carry limited information. In Figure 7-19, the tape is wound loosely so that a thin head can fit between its turns. One edge of the tape winding is fastened to a disc, which is placed on a phonograph-type turntable. The head is mounted on a *tone arm* that follows the tape spiral or can be set down anywhere within the tape platter. The penetration depth of the head can be adjusted to various levels for playing different tracks. An automatic changer can bring the tone arm to the beginning after each selection and then play another track, either in succession or according to a chosen plan.

Miscellaneous Drives

Drives in common usage are described in the chapters appropriate to their application. Such drives include endless cartridges, tape bins, cassettes, side scan and helical scan video, sprocket films, cards, and wire recorders, among others.

SERVOMECHANISMS AND ELECTRIC CONTROLS

As purely mechanical drives approached perfection, the remaining frontiers of progress in drives have been in electronic devices. These have revolutionized the more advanced mechanisms not only in performance but also in conveniences such as fast, smooth starting and stopping, speed changing, aids to editing, and such indirect benefits as increased tape life by optimum tensioning. Electronic aids such as servomechanisms and tension regulators became economically feasible when transistors replaced vacuum tubes to give tiny, reliable, cool-operating devices, and became more compelling as discrete components progressed to gates and opamps, to larger-scale integrated circuits, and to microprocessors of unbelievable capability and low cost.

Servo Controls

Figure 7-20*A* shows the basic concepts of a servo control, which is a closed loop negative feedback hookup that senses the output of a system, compares it with a desired standard, and transmits an error signal, plus or minus, that corrects the system to bring it to the standard. The greater the gain of feedback loop, the more sensitive and faster the response to an error, but also the more likely the servo will overshoot or become unstable. Detailed behavior and optimization of servos is a specialty in itself.

An early example of a servo applied to tape recording was the Dingley flutter corrector (Figure 7-21). Dingley's idea was not to worry too much about the drive itself, but to correct its imperfections by fluttering the pickup head in a direction opposite to the tape-speed errors, Although attractive in principle, this device has not become commercially useful because of difficulties in accurate high-speed control of a sensitive shielded head, and because a compliant flimsy filament of tape reacts to head movement.

On the other hand, servocontrolled capstans for average long-term tape speed are a great success, as they are for medium-term control in video recorders where they correct the forward tape speed to lock in the picture signals at the vertical rate of 50 or 60 per second. Instead of relying on a heavy flywheel to maintain the capstan at a steady speed, the capstan wheel, shown in Figure 7-20*B*, has a series of *gear teeth* (or equivalent) that modulate a light beam or magnetic sensor to generate an output voltage of frequency proportional to the capstan speed. The generated frequency is compared to an accurate reference frequency to give a $\pm$ error signal. The error signal regulates a brake on the capstan wheel, keeping the capstan speed constant. Alternatively or concurrently, the capstan motor speed may be changed. Servo control remains effective at very low flutter rates where flywheels become almost useless. At the highest flutter rates that can not be handled by servos, the inertia of even a lightweight capstan wheel is sufficient to correct the flutter.

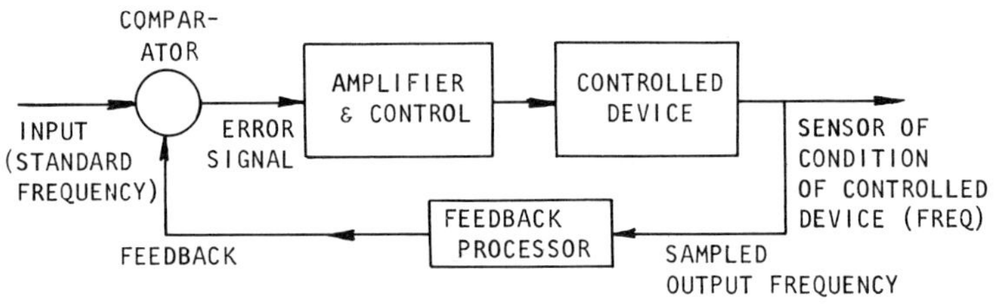

A. ELEMENTS IN A SERVO CONTROLLER

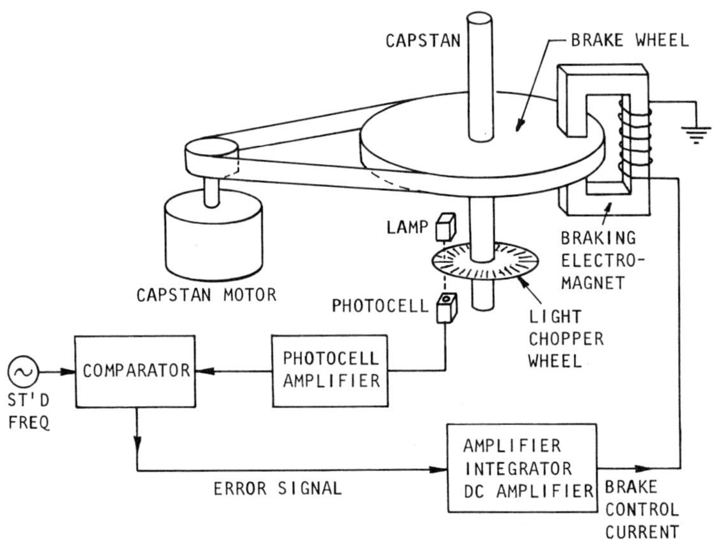

B. SERVO CONTROLLED CAPSTAN

Figure 7-20. Servo control.

A very useful feature of servo capstans is the ease of changing the tape speed by setting the reference frequency to a different rate. Moderate changes are simple. Changes over a gamut of speeds, as for example 120, 60, 30, 15, 7.5, 3.75, 1.875, 0.938 in./s, are generally accompanied by changes in drive motor speeds or stepdown ratios.

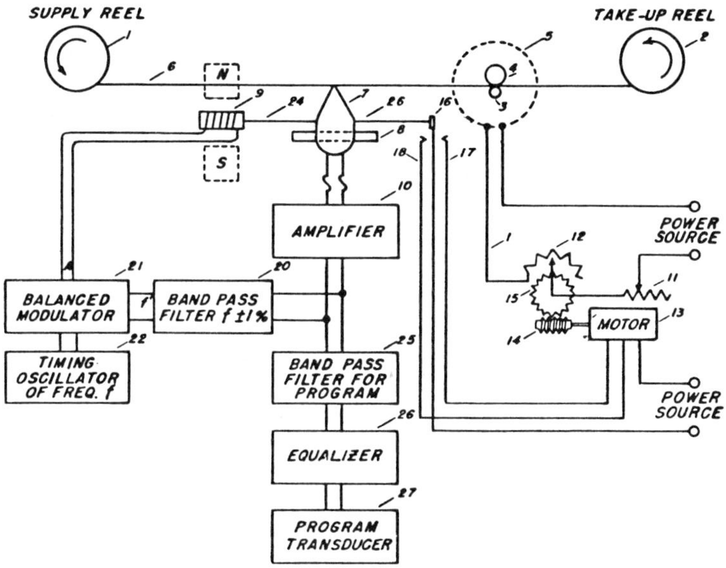

Figure 7-21. Dingley flutter corrector. Head *7* is moved by solenoid *9* along *24, 26* to correct short term errors in tape speed. Long term errors are corrected by contacts 16, 17, 18 which readjust the capstan motor. (From U.S. Re 24,182 to E. N. Dingley, July, 17, 1956).

Tape Tension Control

In low-cost decks where the spool torque remains nearly constant, the tape tension may vary by a 4:1 ratio as a tape reel goes from full to empty. A tolerable tension for a full reel then overtensions the tape near the hub, almost certain to injure the tape during long-term storage and possibly causing capstan slippage and other problems in drive steadiness.

A simple remedy used in early home recorders was to have the weight of the reel rest on its brake or clutch, thus increasing the torque as the reel became filled with tape, as seen in Figure 7-2*B*. A different scheme, used on professional tape decks, was a delicately balanced sensor arm to detect the fullness of each reel. As the sensor arm changed in position, it varied a resistor that was tapered to adjust the reel motor for constant tension whether the reel was full or empty.

A photosensitive detector gives more delicacy without tape contact but requires electronic circuits. Another solution is to sense the turning rate of the reel shaft, which turns faster as the reel becomes emptied. A suitable sensor is a photoelectric detector of segments on a wheel carried by the reel shaft, similar to the capstan sensor described in Figure 7-20. Electronic circuits then adjust the reel motor torque to maintain constant tape tension according to readout frequency from the sensor.

Convenience Electronics

Convenience electronics includes remote control, indexing, automatic stop, channel selection, and editing facilities. Remote control was possible even on the oldest professional machines where all mechanical shifting (record, playback, rewind, fast forward) was done by solenoids.

Indexers of selections on a reel often use a revolution counter on the supply reel. More sophisticated indexers are calibrated in minutes and seconds. Some are even locked to time code signals recorded on the tape to mark the exact time any part of a recording was made (see Chapter 11). Electronic timing can be read remotely, of course. Advanced indexers with cuers will wind the reels at high speed to any desired selection and play it at will. They can stop, reverse, and change to different tape tracks or to different machines. Microprocessors and memory chips allow complex programming for cueing and playback of selected program material in prearranged sequence. In automatic broadcast stations the control system intersperses commercials with music selections. Home video recorders have timers that can be programmed to turn on the recorder at preselected intervals to preselected stations for a week or more in advance. Automatic stop, automatic reverse, and channel changing are also incorporated in many home-type audio recorders.

Editing facilities, which shuttle the tape slowly past the heads to locate words and syllables accurately are featured in some machines. Some have rotating heads that read a length of tape repeatedly, even when the tape itself is not moving.

Tape Transport Maintenance

Freon TF in bottles is recommended as a solvent for cleaning tape guides, etc. Isopropyl alcohol or methyl alcohol are not as effective and are flammable. Xylene damages polycarbonates (some drive rollers) and is flammable and toxic. Spray cans should not be used because they contain metal particles.

ELONGATION, BENDING, AND VIBRATION OF TAPES

Three interesting characteristics result from elasticity and mass of the tape itself. (1) Thinness and elastic modulus allow stretch of the tape, which is appreciable at normal running tension. If tape tension is uneven, the variation in stretching is an important factor in flutter content of tape transports (Werner

1952). (2) Limpness is a desirable mechanical quality of tape that allows it to conform to the head. Tape limpness depends both on its thinness and elastic modulus. (3) Unsupported lengths of tape between guides will tend to vibrate like a violin string at frequencies determined by length, tension, and density, producing flutter at kilohertz rates.

Elongation

Tape elongation or stretch (Figure 7-22*A*) is calculated from the following relation:

$$\Delta \ell = \frac{T\ell}{Ewt} \tag{7-1}$$

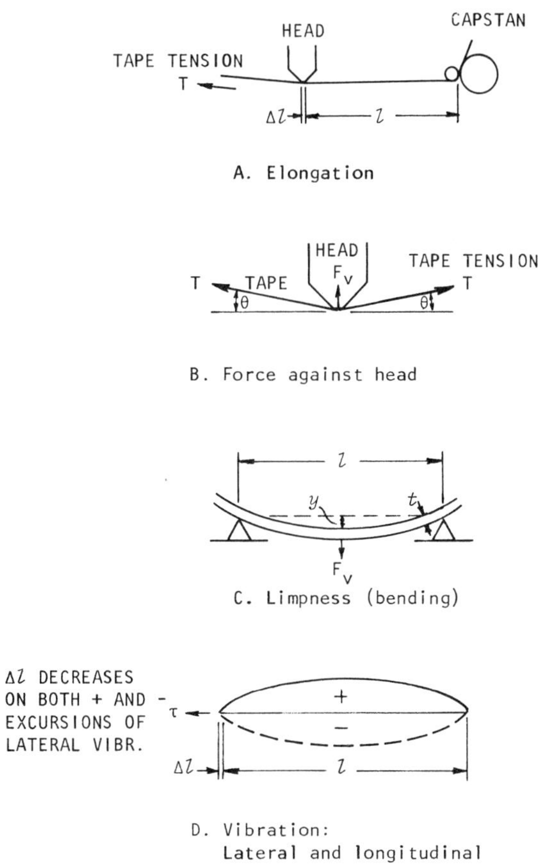

Figure 7-22. Elongation, bending and vibration of tapes.

where

$\Delta\ell$ = stretch (elongation) in meters (in.)
ℓ, w, t = length, width, thickness of tape in meters (in.)
T = tape tension in newtons (lb)
E = elastic (Young's) modulus in pascals (lb/in^2)

The example that follows measures the stretch of Mylar® tape 25μm (1 mil) thick in the length of 127 mm (5 in) distance measured between capstan and head of a professional audio recorder. With a recommended tension of 4.448 N per 25 mm tape width (1.0 lb/in.), and E = 3790 MPa (550,000 lb/in^2) from Table 3-4, we have, in SI units:

$$\Delta\ell = \frac{(4.448)(127 \times 10^{-3})}{(3790 \times 10^{6})(25 \times 10^{-3})(25 \times 10^{-6})} = 0.000238 \text{ m} = 0.238 \text{ mm}$$

or about 9 mils.

From this calculation we can make the following observations:

1. Tape stretches about 0.2% at typical operating tensions.
2. Large amounts of flutter can result from this percentage change taking place in a fraction of a second.
3. In a professional recorder at 190 mm/s (7.5 in/s) tape speed the stretch is more than a wavelength at 1 kHz and more than 10 wavelengths at 10 kHz.
4. Percentage of stretch remains approximately constant for different tape widths, varies directly with tension for a given tape width, and varies inversely with base thickness (coating contributes negligibly to tape strength).

If tape tension in Figure 7-22*A* is consistently greater in the region ℓ than at the reels, the tape velocity will be higher in this region ℓ (McKnight 1968).

Longitudinal Springiness, Compliance, or Spring-Constant. Where the longitudinal springiness of tape is of interest (as for electromechanical analogies) it may be calculated from:

$$k = \frac{\ell}{Ewt} \tag{7-2}$$

where k = spring-constant in meters/newton (in./lb) and other quantities are as in equation 7-1.

Limpness or Lateral Stiffness of Tape (Bending)

When the tape path makes an angle with the head, as in Figure 7-22*B*, the vertical force of the tape against the gap region is:

$$F_v = 2T \sin \theta \tag{7-3}$$

where

T = tape tension in newtons (lb)
F_v = vertical force in newtons (lb)
θ = angle of approach or departure from head.

A measure of the conformity of tape against head is how easily the tape will bend in response to a lateral force F_v. The deflection y in Figure 7-21*C* is given by the *beam* equation:

$$y = \frac{F_v \ell^3}{4\,Ewt^3} \tag{7-4}$$

where

F_v = force at the center of the span in newtons (lb)
ℓ, w, t = length, width, thickness in meters (in.)
E = elastic (Young's) modulus in pascals (lb/in^2)

For comparative purposes, vary the thickness t, keeping all other quantities constant on the right-hand side of equation 7-4. Note that the deflection y will vary inversely as t^3. If we select 100% as the relative limpness for 25 μm (1 mil) base Mylar® then the limpness of other bases will change rapidly with thickness, as in Table 7-2.

Table 7-2. Limpness of Tapes

Mylar Tape Base Thickness		Relative Limpness
μm	(mils)	%
38	1.5	30
25	1.0	100
16	0.67	338
12	0.5	800

Natural Vibration Frequency of Tape

The violin string equation may be applied to a stretched tape vibrating laterally perpendicular to its surface, as in Figure 7-22*D*. The fundamental frequency is:

$$f = \frac{1}{2\ell}\sqrt{\frac{T}{m}} \tag{7-5}$$

where

f = natural frequency of vibration in hertz
ℓ = length of the string in meters (in.)
T = tension of the string in newtons (lb × 384)
m = mass per unit length in kilograms/meter (lb/in.)

The frequency calculation is the same for narrow or wide tapes when the tensioning is increased in proportion to tape width, as is usual in tape drives. It is of interest that the longitudinal flutter generated by the lateral vibration is at twice the lateral frequency since the tape is accelerated longitudinally twice in a lateral cycle: once when the tape moves downwards and again when the tape moves upwards, as in Figure 7- 22*D*

As an example, a typical professional audio recorder may have a 127 mm span of 25 μm Mylar® between capstan and head, a tension of 4.448 N per 25 mm tape width, and a density of 1.395 for Mylar® (Table 3-4), giving an *m* value of 0.0009 kg/m for a tape 25 mm wide. The lateral vibration frequency according to equation 7-5 is:

$$f = \frac{1}{2\ (0.127)}\sqrt{\frac{4.448}{0.0009}} = 277\ \text{Hz}$$

The longitudinal flutter is twice this amount, or 554 Hz. Table 7-3 gives flutter frequencies for other conditions. Higher harmonic modes are possible at integral multiples of the listed fundamental frequencies. The forward movement of the tape is assumed negligible during a cycle of vibration. Otherwise the analysis should be modified to account for doppler effects.

Flutter in the range of hundreds or thousands of vibrations per second is perceived as a modulation-noise rather than a flutter. It may be reduced or eliminated by a damper roll, lightly touching the tape where the tape's amplitude would be maximum, or by alleviating stick-slip conditions between tape and fixed surfaces such as heads. *Stick-slip, stiction, scrape flutter,* and *violin-stringing* are terms used to describe the vibration, chattering, or squealing caused by high static friction alternated with lower dynamic friction between surfaces,

Table 7-3. Lateral Vibration of Tape - Violin String mode*

Free Length of Tape	Location	Lateral Vibr. Frequency Fundamental (Hz)	Longitudinal Flutter Frequency (Hz)
12 mm (0.5 in)	Between heads	2770	5540
25 mm (1.0 in)	Head to capstan or between heads	1385	2770
50 mm (2.0 in)	Head to capstan or between heads	692	1384
76 mm (3.0 in)	Head to capstan or to guide	462	923
127 mm (5.0 in)	Head to capstan or to guide	277	554
203 mm (8.0 in)	Reel to guide	173	346
305 mm (12 in)	Large reel to guide	115	230

*Mylar® tape 25 μm (1 mil) thick, tension = 4.448N/(25 mm width). For 38 μm (1.5 mil) Mylar® multiply above frequencies by 0.82. For 12 μm (0.5 mil) Mylar® multiply above frequencies by 1.4.

which is the basis for violin bowing. Stick-slip is apt to occur with very smooth tape and head surfaces, ''sticky'' tape coatings, high tape-head pressures, and low tape speeds.

THEORY OF TAPE TRANSPORTS

Designers of tape drives usually work by instinct and experience starting with a known design that has been successful and improving it perhaps with a better motor, heavier flywheel, more accuracy, more rigidity, and more convenience and reliability of operation. Common sense tells the designer that a heavy flywheel at the capstan will resist speed changes due to irregularities in tape tension, and that elasticity of the tape or springy guides will absorb tensioning irregularities when used with the flywheel. Experience shows that although the irregularities themselves are smoothed out, their impulses can excite the elasticities and masses into back-and-forth oscillations that are more annoying than the original disturbances. Eliminating these unexpected effects requires insights that are usually beyond instinctive brute-force approaches. As described previously in this chapter, the newest tape drives use servomechanisms instead of massive flywheels to reduce tape drive flutter. However these also are limited by residual masses, compliances, and frictions that must be considered in servo design.

Mechanical Filter Theory

The interactions of mechanical elements having compliance, mass, and friction are described by second order differential equations that have the same form as the equations for electric circuits containing inductance, capacitance, and resistance. In early days, when electricity was somewhat of a mystery, it was fashionable to explain electrical behavior by mechanical analogies. With the universal spread of AC power and radio communication theory, however, the tables have been turned, and mechanical behavior is conveniently explained by electrical analogies.

Electrical Analogies

Electrical analogies give a satisfying qualitative understanding of mechanical systems. Electrical components may also be connected in a quantitative circuit diagram that represents the elements of an existing mechanical system such as a tape transport. The circuit is solved for its behavior with respect to time or frequency by standard methods, using a computer if desired. Or an actual electrical circuit can be put together, since all parts are easy to obtain. Electrical measurements then are made on the circuit, which can be modified until it is optimum. This procedure is much simpler and quicker than obtaining and machining the mechanical components.

The mobility analogy is easier to set up and gives better insight than the traditional impedance analogy because the mobility *circuit diagram* resembles the mechanical connections. Figure 7-23*A* shows a layout representing a widely used transport, and Figure 7-23*B* is its corresponding electrical analogy. Analogs are shown in Table 7-4. Most elements in the transport are rotational; so their angular velocity, inertia, and torsional compliances must first be converted to linear equivalents, as in Table 7-5, then combined with linear quantities, and finally converted to electrical analogs, as in Table 7-6, giving quantitative values in the analog electrical circuit of Figure 7-23*B* (Wolf 1966).

The completely specified electrical analog circuit is easily hooked up. The scaling factors *m* and *n* in Table 7-6 are chosen to give available *R, L, C,* and *f* values, as in Figure 7-23*B*. An oscilloscope is then connected to head point (*H*) while test voltages (representing mechanical disturbances) are inserted at points of interest such as *a* to *j*. The more effective the filtering, the less the voltage that will appear at head point (*H*). Test voltages can be sine waves of chosen frequencies, or the frequency can be swept across the band to find expected and unexpected resonances. Test voltages can be impulses or step functions to bring out ringing effects due to sudden changes. Circuit components are easy to change, and the results are checked quickly. When optimized, the electrical values are reconverted to mechanical ones, saving much cut-and-try in the machine shop.

The steady driving motion of the motor is represented by a DC supply voltage (battery) at *g*, but for our purpose we can ignore the DC component since we are interested mainly in AC and transient disturbances inserted at points *a b c*

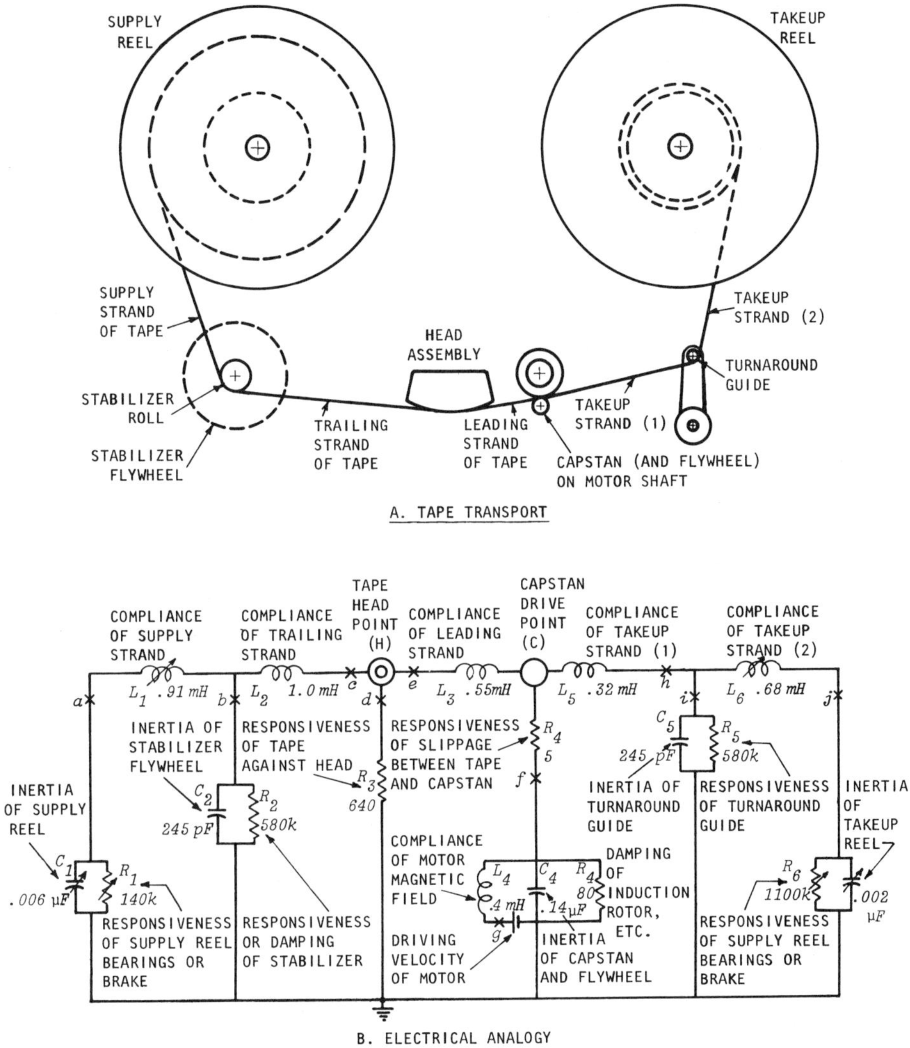

Figure 7-23. Professional tape transport (simplified) and electrical analogy. Typical electrical analogous quantities are from W. Wolf 1966.

d e f g h i j, and in what effect they will have on tape velocity at the head. The deviations being very small, we can assume a linear response and test the effects of disturbance at each point, taking the points one at a time. The results can be superimposed if desired, but it is most helpful to consider them separately.

Referring to Figure 7-23*B*, tests are made by applying an electrical distur-

Table 7-4. Electrical Analogs of Mechanical Quantities

Mechanical Quantity	Electrical Analog
Compliance	Inductance (L)
Mass	Capacitance (C)
Responsiveness = $(\text{Friction})^{-1}$	Resistance (R)
Velocity (between ends of an element, or to reference ground which is the frame of the machine)	Voltage (V)
Force (through a mechanical element)	Current (I)
Frequency of vibration	Frequency (f)

bance at each of the lettered points *a b c d e f g h i j* in turn, and noting the effects at the tape head point H with an oscilloscope:

a. Here we insert an AC voltage at the scaled frequency of rotation of the supply reel, representing off-center effects in the reel or cumulative errors in the tape wound on it. As the tape supply is emptied, the supply reel rotates faster, its frequency increases, its inertia decreases, and its responsiveness (friction^{-1}) also varies. Depending on reel size and tape speed, the frequency of supply reel eccentricities is in the order of 0.25 to 5 Hz. Higher frequencies may also be present from hum or cogging of the reel motor or errors in belts, pulleys, and brakes if they are used.

Transient irregularities such as sticky tape splices or brief rubbing of tape against the rim are investigated by inserting voltage pulses at point

Table 7-5. Conversion of Rotational Mechanical Quantities to Equivalent Linear Mechanical Quantities* (SI Units)

Equation of Conversion	Symbols Linear quantities on first line of each group Rotational quantities on other lines
$M = \frac{I}{r^2}$	M = equivalent mass, in kilograms I = moment of inertia, in kilogram-meters^2 r = radius where force is applied, in meters
$K = kr^2$	K = equivalent compliance, in meter/newton k = rotational compliance, in radians/newton-meter r = radius where force is applied, in meters
$R = \rho r^2$	R = equivalent responsiveness, in $\frac{\text{meters/sec}}{\text{newton}}$ ρ = rotational responsiveness, in $\frac{\text{radians/sec}}{\text{newton-meter}}$ r = radius where force is applied, in meters

*The converted values may be used in the relations of Table 7-4.

Table 7-6. Conversion of Linear Mechanical Quantities to Analogous Electrical Quantities (SI Units).*

Equation of Conversion	Symbols Electrical quantity on first line of each group Mechanical quantities on other lines
$C = Mm$	C = analogous capacitance, in farads M = equivalent mass, in kilograms m = scale factor, chosen for convenience
$L = Kn$	L = analogous inductance, in henries K = equivalent compliance, in meters/newton n = scale factor, chosen for convenience
$R = \mathcal{R}\sqrt{\frac{m}{n}}$	R = analogous resistance, in ohms $\mathcal{R}$ = equivalent responsiveness, in $\frac{\text{meters/sec}}{\text{newton}}$ m,n = scale factors, same as above
$f = f_M \frac{1}{\sqrt{mn}}$	f = analogous electrical frequency, in Hz f_M = actual mechanical frequency, in Hz m,n = scale factors, same as above
$V = v_m{}^4\sqrt{\frac{n}{m}}$	V = analogous elecrical emf, in volts v_m = mechanical velocity in meters/sec m,n = scale factors, same as above
$I = F_m{}^4\sqrt{\frac{m}{n}}$	I = analogous electrical current, in amperes F_m = mechanical force, in newtons m,n = scale factors, same as above

*In the above relations choose m and n to give convenient sizes of L, C, f that can be used in building the electrical analog.

a instead of a steady AC voltage. Transient behavior can also be derived from a frequency response plot over a wide range of frequencies taken with a constant input at point *a*.

The voltage at the head point (H) corresponding to a disturbance at *a* is then determined from response of the electric circuit. We would expect L_1C_2 to act as a low-pass filter reducing AC transmission from *a*, with further reduction by the L_2R_3 filter to H. On the other hand, L_1 $(C_1 + C_2)$ is a series-resonant circuit that at certain frequencies could increase the voltage to H, and for an impulse at *a* it could ring at resonance, causing many cycles of fluctuation at H.

b. Irregularities at the stabilizer roll and flywheel are represented by inserting at point *b* a voltage having a scaled frequency of about 1.0 to 6.0 Hz.

With high-precision parts, the disturbances from the flywheel are a minor part of the overall flutter.

c. Although the resultant of previously mentioned AC components passes through point *c*, the unique contribution here is from tape vibration (violin stringing or scrape-flutter) caused by lateral and longitudinal vibration of the free tape marked *trailing-strand*. The frequency may be in the squeal range of several kHz, far removed from pulley and reel frequencies.

d. Noise voltage at point *d* represents the friction of the tape's rubbing against head polepieces. High friction (low resistance) here would shunt out the flutter transmitted to the head, but would also give intolerable head wear and tape wear.

e. Symmetrical with point *c*, there is tape vibration of the leading strand plus transmitted AC that gets through the capstan drive point (*C*).

f. Voltage at point *f* has an AC component at capstan rotational frequency which ranges from 10 to 30 Hz when the capstan shaft is directly driven by a conventional motor, or possibly 2 to 20 Hz with a reduction train. There is possibility of hum and motor torque pulses at power line frequency or its harmonics, as 60 Hz and 120 Hz.

g. Imperfections in motor balance, hum, and so on, are represented by a voltage at line frequency as 60 Hz. Single-phase motors also give a substantial component at twice the line frequency as 120 Hz. Higher harmonics are possible from motor tooth ripple. When a reduction train from the motor is used, additional coupled elements and frequencies are introduced.

h. Conditions at the takeup strand (*1*) *h* are similar to those at the trailing strand *c*, except that they are isolated from the head by intervening capstan point (*c*).

i. The turnaround guide at *i* has considerable friction but negligible inertia compared to conditions at C_2R_2.

j. Fluctuations at takeup strand (2) *j* are like the ones at supply strand *a*.

Sinusoidal voltages of appropriate frequency introduced at test points *a* to *j* in Figure 7-23*B* produce voltages of the same frequency at tape head point (*H*), representing changes from a steady tape velocity at the head. But pulse voltages at the test points set up damped oscillations (ringing) of the tuned circuits, giving additional disturbances near their natural frequencies. Unfortunately the inherent mechanical damping of tape drives is very low because bearing losses and tape friction are negligible compared with the energy stored in massive flywheels, giving high *Q* resonant circuits with a strong ringing tendency. Figure 7-24 shows the response to an impulse of resonant circuits with different *Q* factors.

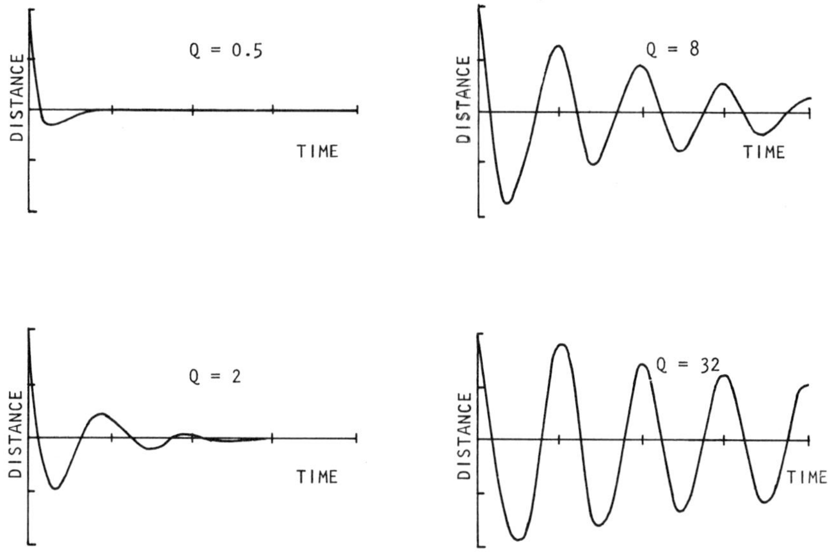

Figure 7-24. Impulse (ringing) response of resonant circuits as damping is decreased (Q increased).

Damping Considerations

To damp out the ringing oscillations, friction or other lossy elements must be added to the resonant components. Such friction must increase in proportion to velocity (viscous friction); steady friction independent of velocity merely acts as a drag or tension and does not damp the oscillations. Damping friction may be added in two ways:

1. Lossy elements may be included in the steady-state drive. Two examples are shown in Figure 7-25. Both waste considerable amounts of steady-state power especially at high tape speeds, requiring stronger motors and giving undesirable effects such as tape stretch, head wear, or poor steady-state speed regulation.
2. Lossy elements may be inserted within the tuned circuits, responding to AC only, but isolated from the steady-state drive. This second option is more desirable but somewhat tricky to put into practice. The most successful is the oil-damped flywheel of Figure 7-26*A*, where the outer flywheel shell contains a *floating* inner flywheel centered on ball bearings but not connected to the outer shell except through a viscous fluid (silicone oil) that fills the sealed inner space between flywheels. As long as the

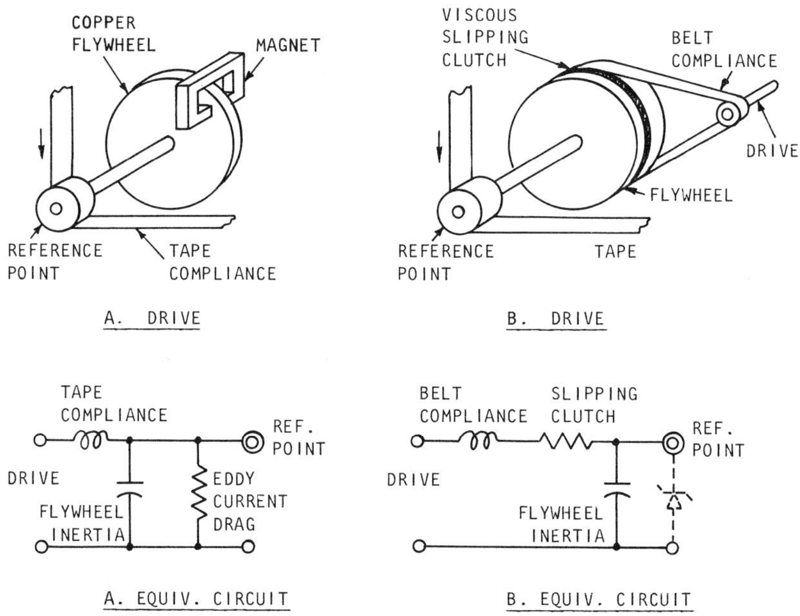

Figure 7-25. Lossy elements included in steady-state drive. A. System useful in supply side stabilizer. B. System useful with servo on output side, represented by dashed-line zener regulator.

main drive shaft connected to the shell flywheel maintains a constant velocity, both flywheels turn together. But if the outer shell fluctuates in angular velocity while the inner flywheel coasts along at steady velocity, a shearing damping action takes place in the coupling fluid. Figure 7-26*B* shows a different arrangement, where a dashpot damps the tape compliance. The dashpot must be designed very carefully because small forces may not actuate it where there is lost motion, side play, or static friction, among other factors.

FLUTTER AND WOW

Even the best tape drives give imperfections in the steadiness of tape motion amounting to a small fraction of a percent. Flutter and wow are audible when sounds are reproduced at an unsteady speed, but unsteadiness is also troublesome in video or instrumentation, for example. In a critical environment, listeners can discern a flutter of 0.005%; however, a transport with 0.1% flutter is considered excellent for music reproduction while one with flutter content of 0.25% is adequate for reproduction of speech. An arbitrary classification for the wow range is 0.1 to 10 Hz; for the flutter range 0.1 to 200 Hz.

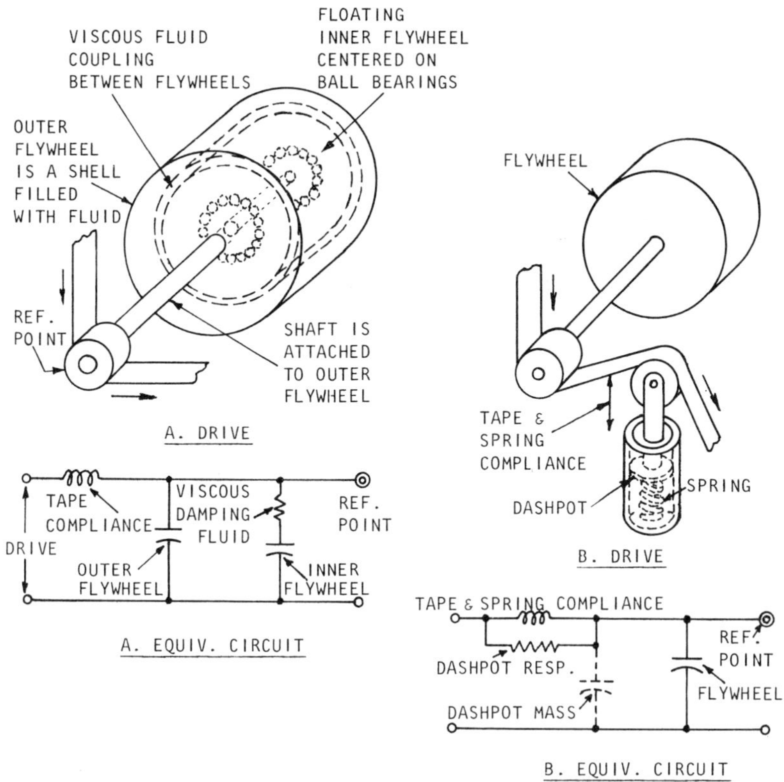

Figure 7-26. Oil damped flywheel, and dashpot damped compliance.

Flutter is more noticeable with some kinds of sound than with others, piano music being especially critical (Comerci 1955; Stott and Axon 1955). Flutter rate is as important as flutter magnitude since our hearing is most sensitive to a rate of about 5 per second (the "wow" range). If we insist on a single number for designating the flutter quality of a system, then we must apply a weighting curve according to the flutter frequency to make the flutter percentage rating more meaningful (IEEE Standard 1971, McKnight 1972). Flutter content has been expressed as RMS, peak, and peak-to-peak deviations. A given transport with infrequent spikes of flutter can be given an impressively low RMS flutter percentage rating while sounding quite unimpressive. A fair rating must therefore be based on the same standardized weighted procedure, such as the DIN 45-405. As a sidelight it might be mentioned that such standards are adopted after many years of discussion by industry representatives who like a standard that gives optimistic absolute numbers for everybody's equipment.

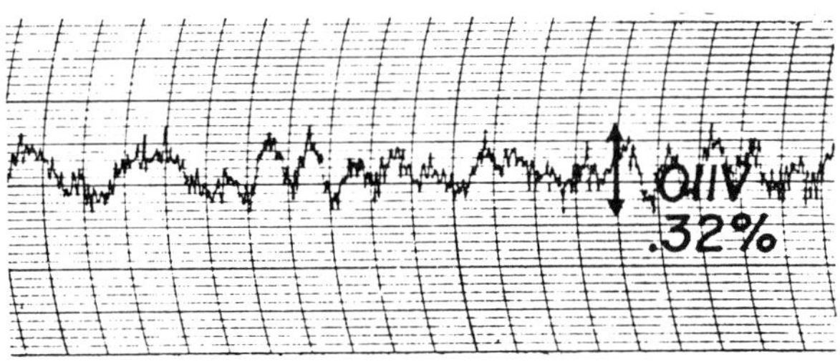

Figure 7-27. Strip chart of flutter. (From Balducci 1963).

Some Aspects of Flutter

From a purely scientific standpoint, the entire flutter story is told by a strip chart recording of instantaneous speed versus time (Figure 7-27), which may be taken by playing a perfectly recorded constant frequency through a discriminator whose output is proportional to the slightest deviation from constancy. Such a chart tells how much and how often the errors occur, and their regularity gives clues to mechanical causes that need correction.

An age-old problem is how to obtain a perfect recording in the first place so that the results truly express the imperfections of the playback transport alone, rather than being a hodgepodge of both recording and playback flutter. If the transport under test is much poorer than the equipment that made the standard tape, say by a factor of 5:1, then the error is negligible. When testing the best state-of-the-art transports, considerable ingenuity must be exercised. One can resort to the relation

$$F_T = \sqrt{F_1^2 + F_2^2 + F_3^2 + F_4^2 \ldots} \tag{7-6}$$

between total percentage flutter F_T, and its components, F_1, F_2, etc, where for example F_1 is the flutter due to the recorder and F_2 the flutter due to the playback machine.

The above equation assumes a random or constantly changing phasing between the various components; this is achieved automatically if the record and playback drives are different, or if the steady playback speed differs however slightly from the recording speed. When the recording and playback are both made on the same transport, an interesting phenomenon sometimes occurs where, at times, the playback flutter cancels the recorded flutter, and this can happen even for complex flutter patterns. At other times, a maximum flutter condition occurs where the playback flutter reinforces the recorded flutter to give twice the amplitude but the same waveform. When the phase of recorded and playback flutter are between these extreme conditions, they will generate double frequency components that give confusing results on the graph, as in

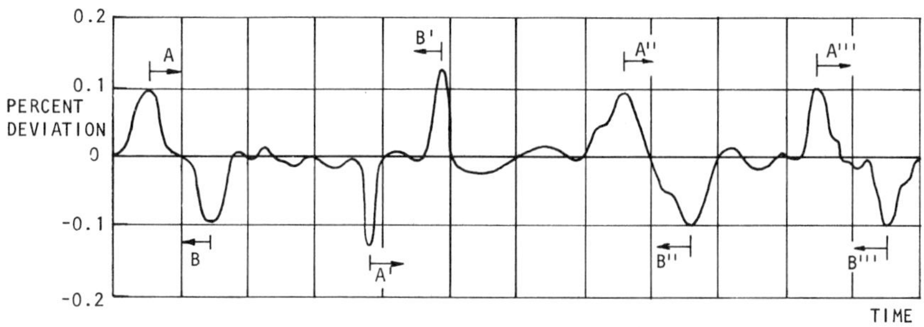

Figure 7-28. Combined flutter in record and playback (idealized). Chart is the resultant of flutter deviations A, A′, A″, A‴, during recording, and the same deviations B, B′, B″, B‴ during playback which are of opposite phase but displaced in time by amount indicated by arrows. Flutters cancel when time displacements become zero, except for smaller random errors.

Figure 7-28. One can distinguish the double effects of record-playback flutter by studying the graph as the playback gradually creeps in and out of cancellation, or, qualitatively, one can observe the overall effects of the gradually changing phase.

Amplitude flutter is also generated in recordings, either by itself or as a consequence of frequency flutter where output is proportional to $d\phi/dt$. Theoretical analysis shows that amplitude modulation is necessarily accompanied by frequency changes (sidebands of sum and difference frequencies) according to:

$$V = V \cos \omega_c t + 0.5m\, V \cos (\omega_c + \omega_m)t + 0.5m\, V \cos (\omega_c - \omega_m)t \qquad (7\text{-}7)$$

where ω_c is the steady (angular) frequency, ω_m is the amplitude flutter frequency, V is the amplitude, t is time, and m is the degree of modulation. Similarly, frequency modulation generates amplitude modulation, so either type of modulation ordinarily implies the presence of the other kind and this is borne out by subjective listening tests.

Subjective Effects of Flutter

Very slow and small percentage changes in tape speed, at say 0.01 to 0.1 Hz, may be noticed as slight departures from perfect musical pitch by musically trained listeners. Variations from 1 to 10 Hz are heard as sour notes, where steady tones may sound like the word *wowowow*; hence the name *wow*. And yet a vibrato purposely and artistically applied by the performer to violin and vocal music at about 8 Hz enriches the music and is very pleasing. Higher flutter rates of 20 to 50 Hz sound like a warble, which changes to a buzz or roughness at still higher frequencies, finally being perceived more as a noise or harshness than as a tone at rates of 1000 to 10,000 Hz.

Flutter Thresholds

The threshold of perception of wow and flutter is important for setting minimum standards for recording equipment and has been determined by listening tests using appropriate audiences. The threshold depends on the type of sound, the listening room, and the listener. Steady tones, reverberant rooms, and female listeners gave the lowest thresholds in one series of studies (Stott and Axon 1955). Under the most critical conditions of a steady 3000 Hz tone and reverberant room, 0.005% peak flutter is perceptible. However, in order of magnitude, more flutter is tolerable for piano music, still more for other instruments, and even more for orchestra or voice. Since the piano was considered the "worst case" in practical music, its threshold where flutter could be perceived by an average listening audience was chosen as a practical guide, with results shown in Figure 7-29. Note that a threshold of 1% peak flutter on this graph would correspond to only 0.707% rms flutter (sine wave) or as much as 2% peak-to-peak flutter, so one must be very careful to notice what scale is specified.

Weighted Flutter Rating

A single number is usually demanded as a figure of merit specification of flutter. A weighted index comes closest to expressing the annoyance factor in audio and is based on applying a curve, such as in Figure 7-30, to the flutter frequen-

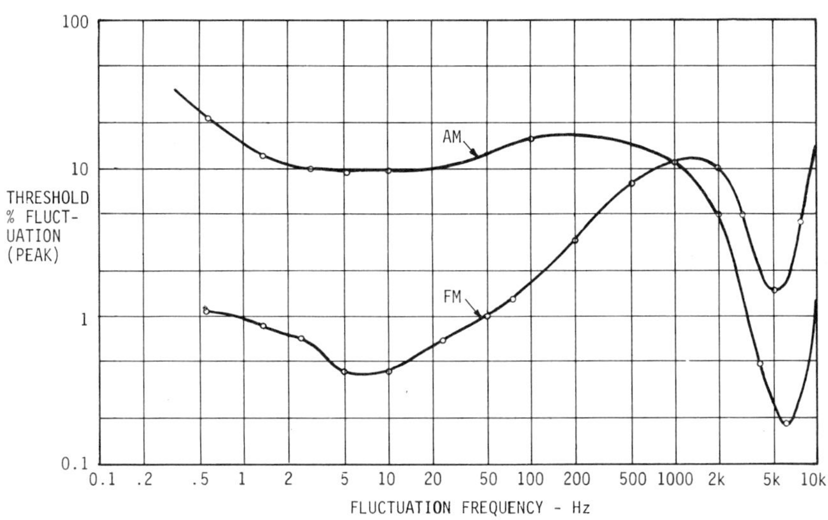

Figure 7-29. Threshold sensitivity to FM and AM flutter in piano music (After Stott and Axon 1955).

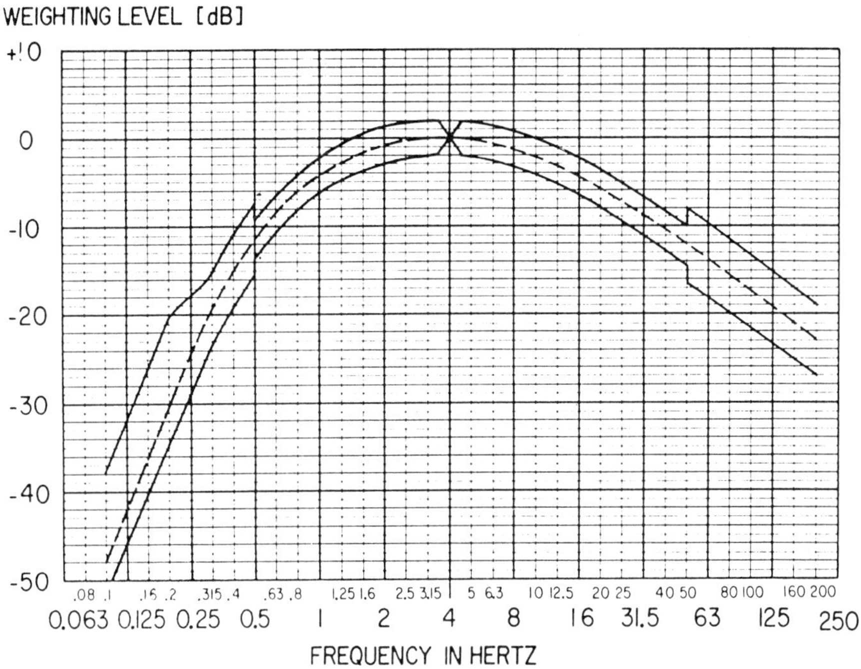

Figure 7-30. Weighting curve for IEEE 193-1971 flutter index (IEEE Standard 1971).

cies present. An instrument that reads such weighted flutter gives the most weight to components that are most perceptible to listeners, but does not, of course, tell the whole story. A testing frequency of 3150 Hz is recommended. Accepted standards are the *weighted RMS,* abbreviated WRMS, according to DIN 45-405, or IEEE 193-1971 or ANSI SA.3, which are all basically similar. The WRMS rating gives a lower number than if peak values were specified. A typical value claimed for high-grade cassette decks is 0.07% WRMS or 0.12% RMS. Some professional recorders specify 0.05% WRMS. Portable cassette decks specify 0.3% WRMS.

8. High-Density Recording and Noise Limits

After he had listened to his voice on the wonderful wire recorder of the late 1930s, Dr. Vagtborg, director of Armour Research Foundation asked, "Why must the wire move so fast? Why can't we record slower and longer on the same record?" In other words, why not increase the recording density?

In answering these questions the most obvious considerations are purely mechanical ones: How small can we made a head gap? How smooth can we make a tape surface? How narrow a channel can we track reliably? In 1945, the Model 50 head gap of 0.002 in. (51 μm) was considered too small by potential manufacturers, who said they would need a microscope even to see it. In 1981 gaps of 0.000020 in. (0.5 μm) were produced routinely. Similarly, the tracks can now be two orders of magnitude narrower and tape smoothness has improved likewise.

A more subtle consideration is the noise level. As smaller and smaller areas of tape are scanned, the heads give less electrical output, until, finally, the pickup voltage is lost in the noise of the electric circuits. Aside from electric circuit noise, the small tape areas are themselves irregular magnetically because permanent magnetism depends on arrays of domains, whose granularity becomes noticeable when our scanning is fine enough to distinguish domains.

What are the ultimate limits of recording density? How close are we to these already? What are the most feasible and economical directions to take in reaching for the ultimate? These are subjects of great interest which have been explored recently and predictions made for the future. A word of caution, however, about negative predictions. In the early 1920s, Austin and Cohen proved rigorously that frequencies above 1.5 Mhz would be useless for long-distance radio communication because ground waves could not follow the earth's curvature. On the basis of the famous Austin-Cohen formula, the worthless higher frequencies were given away to amateurs, who promptly discovered they could converse around the world with a few watts of power by reflections from the then unknown Kennelly-Heaviside layer (ionosphere).

RECORDING DENSITIES

High-density recording has always been a favorite topic of conversation among recording engineers. In 1940 they were impressed when recorded wavelengths figured out to be $\frac{1}{100}$ of an inch long. In 1950 they were recording 1000 waves

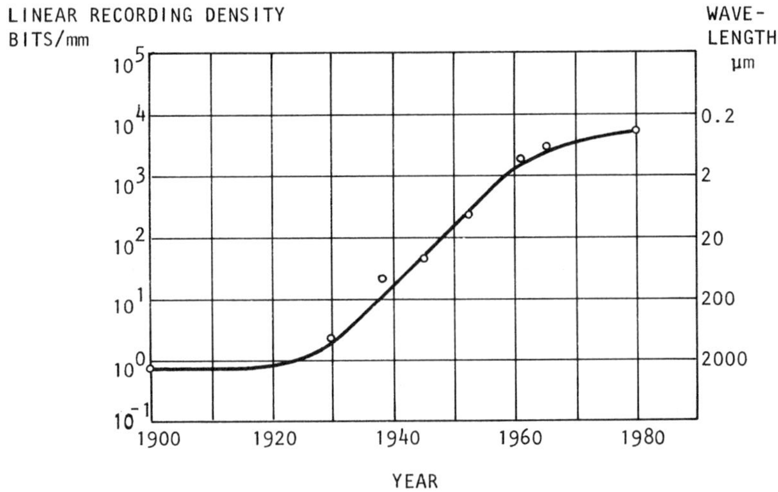

Figure 8-1. Growth of linear bit density in magnetic recording. Improvement in density has resembled a biological growth curve that levels off with maturity toward a saturation value (Camras 1965).

per in. and saying that 2000 per in. was probably the remotest limit, since we were approaching some vaguely undefined ultimate barrier. In 1980 they were saying the same thing, but this time they were talking about reported bit densities of 150000 per in. (75000 wavelengths). As a matter of interest, linear bit densities for magnetic recording are plotted in Figure 8-1, and it does look as if the growth curve is leveling off.

Area and Volume Density. Linear density is useful for bragging, but otherwise is a rather limited tool for comparing different systems. More useful is area density or bits per square millimeter. Figure 8-2 compares magnetic recording with other systems on an area basis. Going a step farther, we would expect that volume density is the ultimate measure, and Figure 8-3 therefore shows volume densities of different storage media. A theoretical upper limit is shown as 10^{20} bits per mm^3, which is approximately the number of atoms in a cubic millimeter of any material (Avagadro's number). It is unlikely that individual atoms could be set in stable *1* or *0* bit states; nor could they be fitted with wiring or other readout mechanism that did not use up some extra space. So this 10^{20} limit is on the high side. It is remarkable that genetic systems come pretty close at nearly 10^{17} bits per mm^3.

CDS Rating. Further thought shows that information storage has a fourth dimension, which is the time required to retrieve it, so our three-dimensional

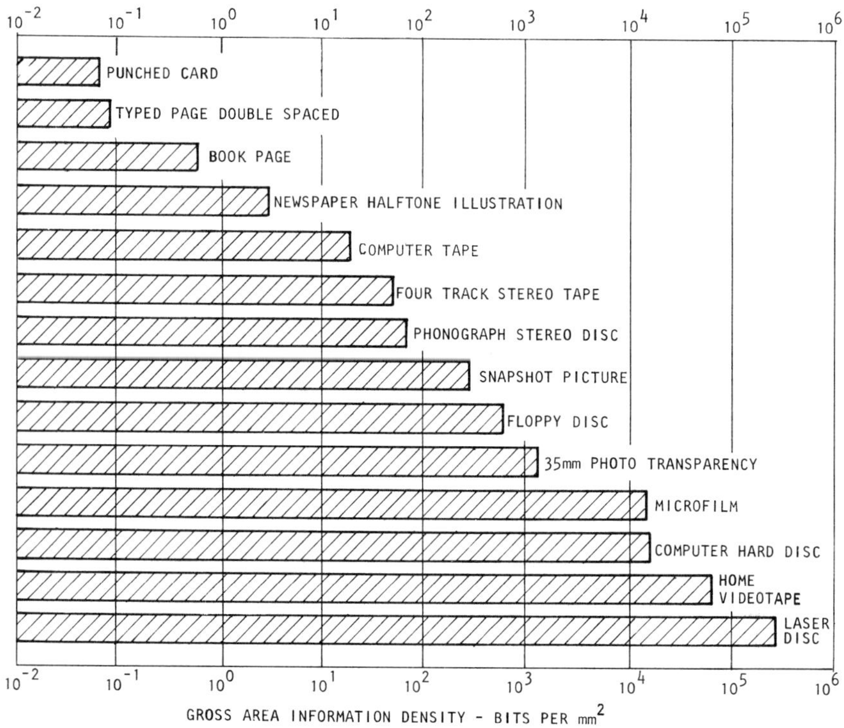

Figure 8-2. Area densities of common information storage media. The most compact systems devised by man are thin flat sheets, inscribed with language, picture, or coded bits. Gross values are given for representative units, including blank margins, etc. A snapshot has about 1000 times the information of language, verifying the old adage (Camras 1965).

measure of bits per mm^3 is not the ultimate one. A fifth dimension is the total storage capacity, which necessarily reduces the retrieval time, since with massive storage some bits must be farther away from the processor than others, and thus their retrieval time is greater even at the speed of light. A rating method that takes all of these dimensions into account (*capacity, density,* and *speed*) is the CDS rating used in Figure 8-4. The ratings are based on estimates as of about 1980, with assumptions as detailed in the original paper (Camras 1965). Advances in the state-of-the-art are continually changing the ratings. Computer tape storage has improved between one and two orders of magnitude in the years between 1960 and 1980. Figure 8-5 compares the capacity, access, and cost per bit of computer memories in 1980.

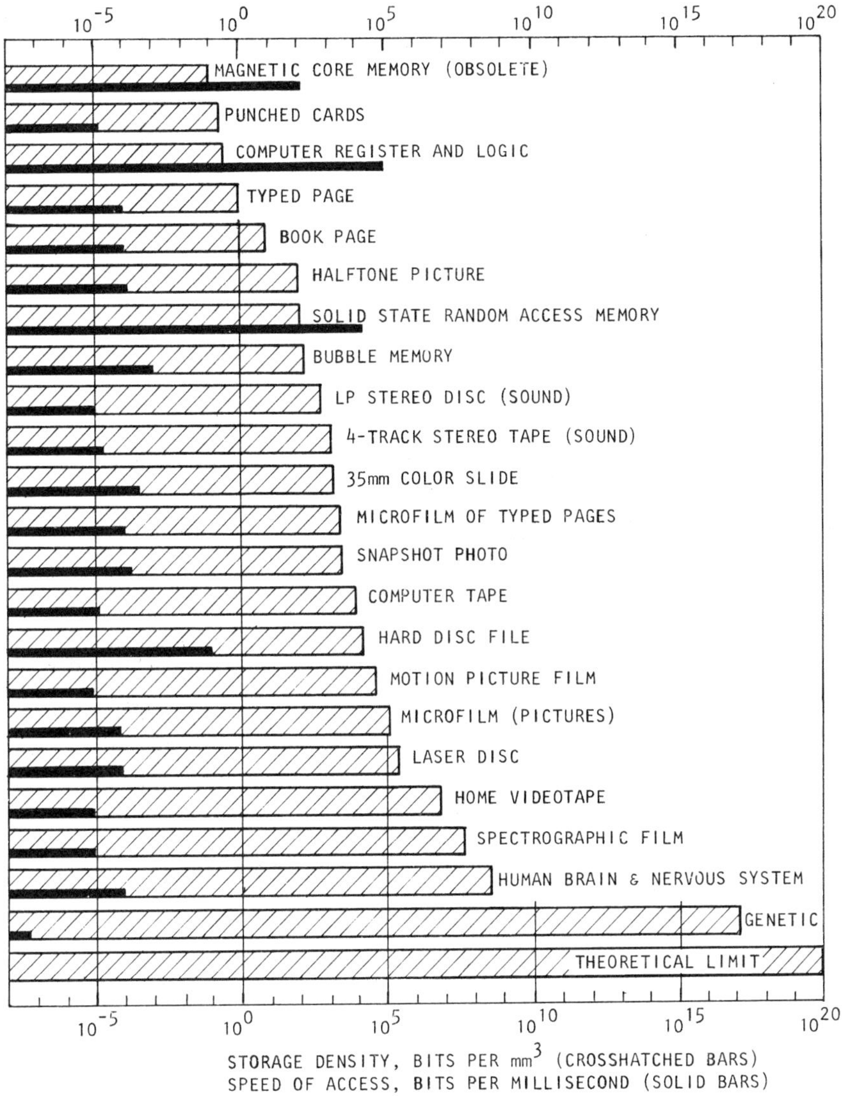

Figure 8-3. Volume storage densities and speed of access. The bits of information stored in a cubic millimeter (gross volume) of the record are compared with a theoretical limit of 10^{20} atoms per cubic millimeter (Avagadro's number). The DNA genetic system approaches the ultimate in compactness. For each kind of record the random access speed (not readout speed) is shown, except in the case of sound records and picture films (Camras 1965).

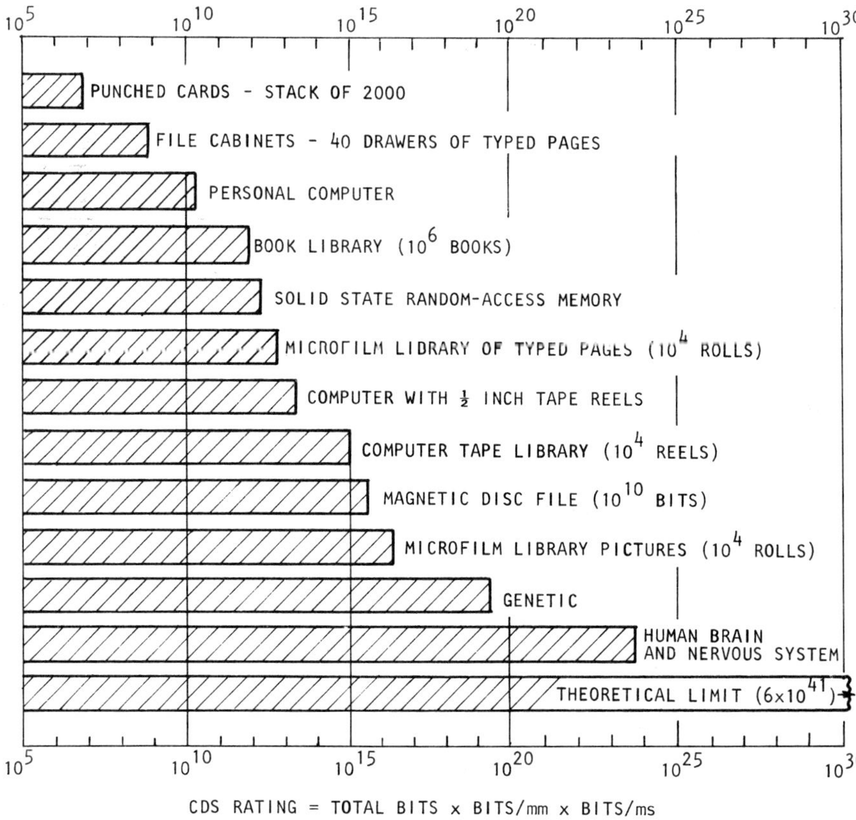

Figure 8-4. Overall figure-of-merit including capacity, density and speed (CDS rating). This is a measure of the overall merit of a system regardless of size, taking into account all three important factors. The human brain surpasses computers and book libraries by a ratio of about 10^8, mainly because of its compactness and high capacity coupled with reasonable access time (Camras 1965).

HIGH-DENSITY HEADS

Just as we can write smaller with a sharper pencil, we can record more wavelengths per millimeter with a smaller head gap. Figure 8-6 shows the concept of reducing the scale of all head and tape dimensions by factors of 2, 10, and 100. At $\frac{1}{10}$ scale, the linear recording density is $10\times$ and the surface density is $100\times$. But the magnetized volume, and hence the output energy, is reduced by $1000\times$. Evidently we cannot repeat this reduction process too many times without finally losing the signal. Somewhere we have to compromise.

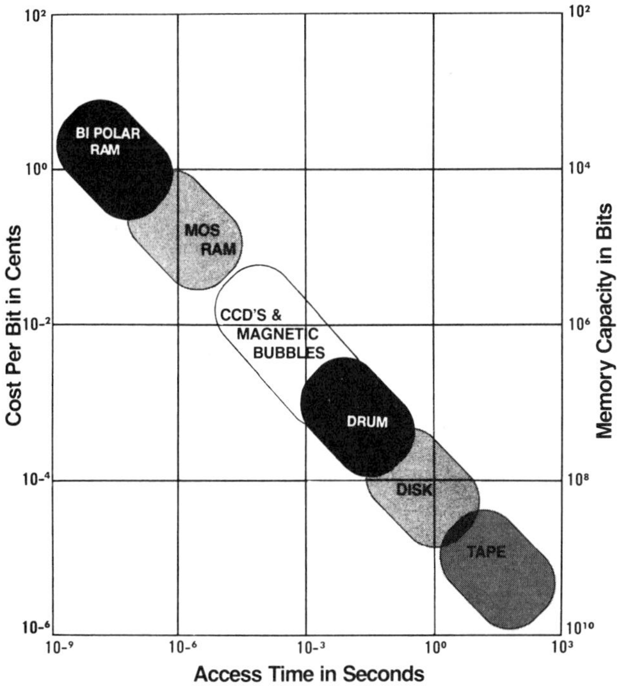

Figure 8-5. Computer memory technology hierarchy. Comparison of computer memories (3M Company 1980).

Effect of a Small Gap. Although everybody speaks with awe of super small head gaps, a small gap is actually one of the easiest of all factors to manufacture and quite worthless unless accompanied by subtle design considerations seldom mentioned. There is hardly a head manufacturer who has not tried to put a smaller gap into his best heads only to find that their performance became worse (Figure 8-7). With too small a gap, the output drops alarmingly while the recording density does not improve and in fact may become worse. A head that is used only for playback can have a much smaller gap than a recording head of the same design. In the small-gap playback head, the loss in output is compensated by more amplification. But a recording head with too small a gap will not bias the tape properly and will record weakly even when energized with a strong signal current. The behavior shown in Figure 8-7 is explainable as follows.

Magnetic Effects with a Smaller Gap Spacer. When a half-thickness gap spacer

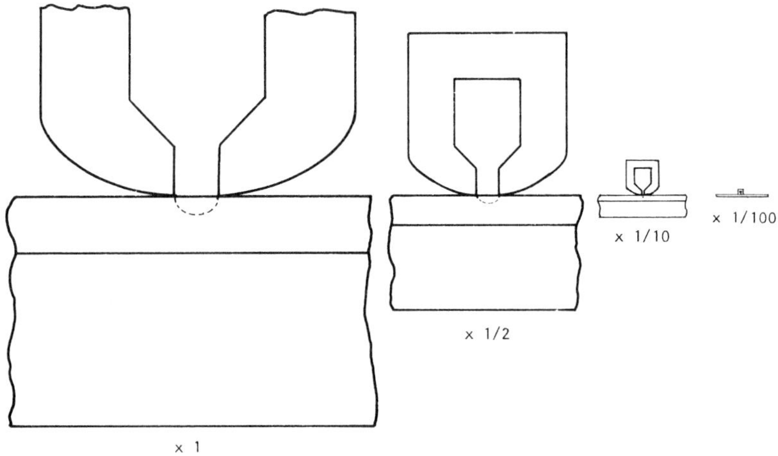

Figure 8-6. Reduction to scale of all head and tape dimensions. At $\frac{1}{10}$ scale the linear density is 10X, the surface density is 100X, but the magnetized tape volume is reduced by 1000X. Thus we are dealing with $\frac{1}{1000}$ of the original energy.

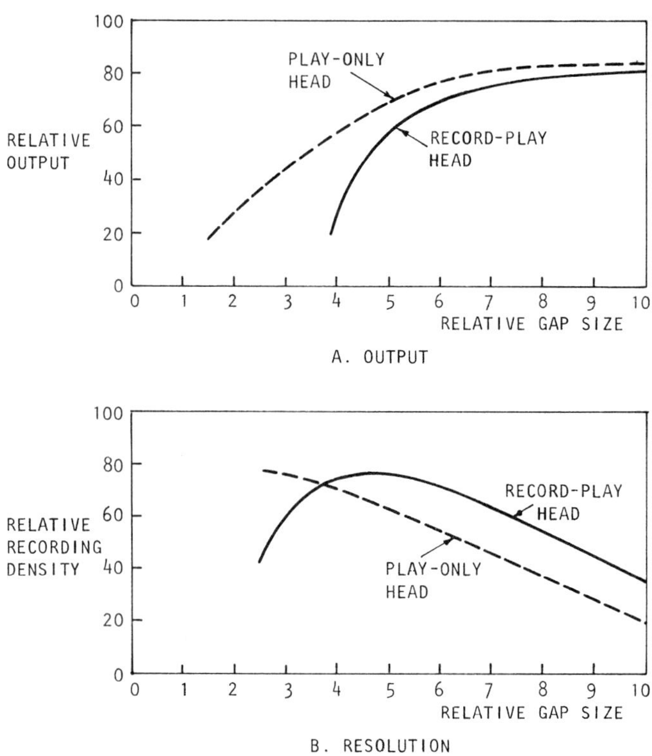

Figure 8-7. Effect of gap size on the output and resolution of record-play and play-only heads.

is placed in a given head, the immediate effect is to cut the front gap reluctance in half, greatly increasing the leakage across the gap (see Figure 4-9). In many well-designed heads, the playback output may be reduced about 6 dB. The head inductance will increase and thus impair the high-frequency response. If this head is used for recording, its core must carry much more flux to take care of the increased gap leakage and to penetrate the recording layer. The core may then saturate before it can magnetize fully a high-coercivity tape. Because of skin effects in the core, high frequencies are reduced more severely than lows when the head saturates. High-frequency bias suffers the most.

Dimensional Effects with a Smaller Gap Spacer. An economically successful commercial head seldom has accuracies and surface finishes better than necessary since these are expensive. The irregularities become more critical in proportion to a decrease in gap spacer size, so merely inserting a smaller spacer does not reduce the physical gap dimension proportionally if the core surface is rough. The physical gap also deteriorates by mushrooming at the gap of metal cores and by erosion at the gap corners, especially of ferrite cores. Such erosion is most detrimental with smaller gaps.

Surface Permeability of Head Cores. A so-called *dead layer* of low permeability has been reported to exist on surfaces of head cores. In permalloy cores, it has been attributed to surface stresses caused by shaping operations such as grinding, polishing, and even the rubbing of magnetic tape that passes over the head surface. In ferrites and stress-insensitive alloys, it may be due to subtle chemical changes at the head surface: oxidation or depletion of oxygen, absorption of gases, or reaction with lubricants or chemicals during fabrication, among other factors. It may even be that magnetic domains at the outer surfaces of permeable materials resist magnetic changes more than domains in the bulk material do. If such a dead layer exists on the gap faces, then the effective gap will obviously not be reduced as quickly as the gap spacer.

Head-to-Tape Interface. Whether or not a dead layer exists at the head surface, the observed effects can also be explained by imperfect contact between tape and head. Indeed the early magnetic tapes had surface protrusions sufficient to space the main magnetic layer on an average of 0.25 to 1.0 μm (10 to 40 micro inches) from the head gap. High-speed tapes, as in computers, carry a film of air with them, which produces a similar separation. Thc spacing (see Figure 2-47) reduces playback responses drastically at short wavelengths according to the spacing loss equation:

$$\text{Relative output dB} = 54.6 \frac{\text{spacing}}{\text{recorded wavelength}} \qquad (8\text{-}1)$$

but the effect can be even more drastic during recording because the spaced record calls for considerably more high-frequency bias, which erases short wavelengths. Biasless recording requires a higher recording current that gives peak shift and self-erasure at high-bit densities.

Overcoming Side-effects of Small Gaps. The cures for dimensional troubles mentioned above are more precise fabrication and a closer head-to-tape interface. In head design, the gap depth should be reduced in proportion to reduction of gap length to reduce wasted gap flux, but the head life may then be unsatisfactory, and a compromise must be made. A spacer of high electrical conductivity also helps to reduce wasted gap flux. In ferrite cores with a bonded glass gap-spacer, the glass fillet under the gap adds mechanical strength that is much needed with fragile polepieces of small depth. Similar reinforcement is desirable in other designs. Dead layers are avoidable by choosing core materials that are relatively free of this defect or by carefully removing surface layers, as for example by an etching process. Aerodynamic modification of head contours may be necessary to keep the tape close to the head at high tape velocities. But regardless of what is done to the head, it must be used with tape of sufficient smoothness and magnetic properties to receive and retain shorter wavelengths; otherwise, the recording density will not improve.

HIGH-DENSITY TAPES

The old 111A tape, standard for all-purpose recording in the 1950s, had a matte surface finish that burnished to a smoother surface during use, and it was observed that its recording density increased with such burnishing. Improvements in formulations, dispersions, and coating techniques now give a far smoother, finer-grained finish even without treatment, and there are a number of surface treatments that range from ironing out nodules to imparting a mirror finish to the tape surface. A mirror finish implies smoothness to within a wavelength of light, which is a necessary condition for high-density recorded wavelengths of the same order.

Interestingly enough, the magnetic oxide particles in tapes are also about the length of light waves (0.5 μm) and therefore their granularity should generate irregularities (noise) when magnetic recorded waves are of the same order. The overall granularity is usually much worse than of individual particles, since they are never perfectly dispersed, but are known to clump together in irregular patches of dozens of particles. As noted in Chapter 3, metal particles for tapes can be about 0.03 to 0.05 μm, an order of magnitude smaller than the oxides, thus allowing proportionally higher densities and lower noise levels. However oxide particles are also useful in sizes smaller than 0.5 μm if they are treated with cobalt; and very high recording densities have been reported for small unoriented oxide particle media (Lemke 1979). Highest densities have been

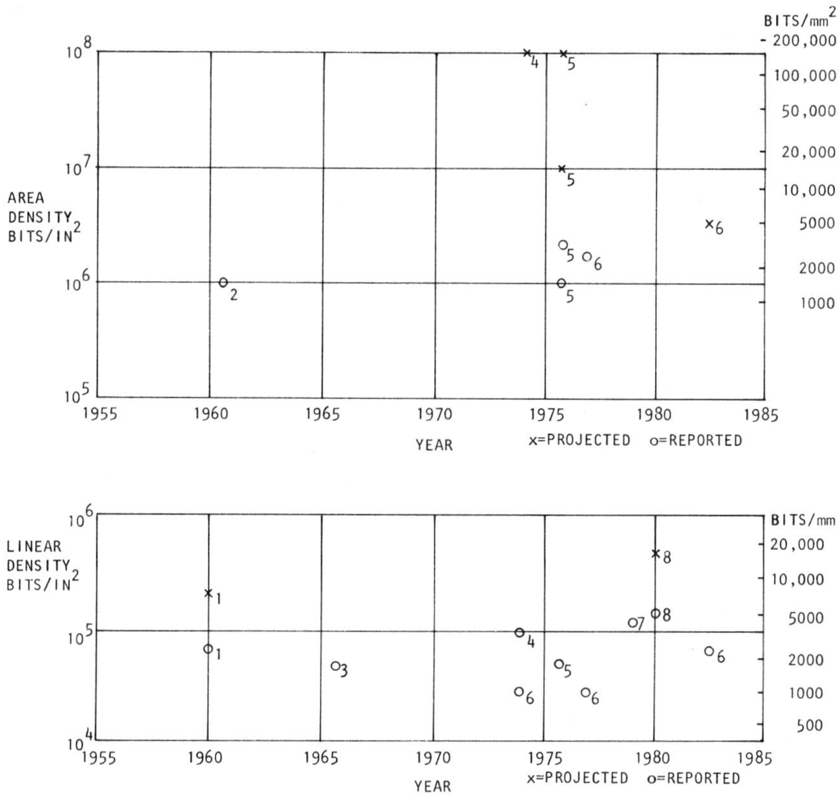

References
1. Brophy 1960
2. Eldridge 1961
3. Smaller 1967
4. Mallinson 1974
5. Mallinson 1976
6. Acosta 1977
7. Lemke 1979
8. Iwasaki 1980

Figure 8-8. Linear and area densities of magnetic recording.

achieved with sputtered layers of perpendicularly oriented cobalt-chromium by Iwasaki whose record media and heads are described in Chapters 3 and 5, respectively (Iwasaki 1980). Figure 8-8 summarizes some linear and area densities that have been reported or projected.

NOISE

Background noise sets a practical limit to information density of magnetic recording since higher density gives lower signal output without proportionally decreasing the noise. This degradation in signal-to-noise ratio cannot be remedied by amplifiers that amplify both signal and noise, and, in fact, the amplifiers may add some noise of their own. Therefore, as recording density is increased further and further, a point is reaches where the signal is lost in the noise and amplification becomes useless.

Noise differs from ordinary signals to which we are accustomed in that it is distributed across a broad spectrum instead of being concentrated at single frequencies and their harmonics. Random noise with equal power at all frequencies is called *white noise*. Noise level is expressed as RMS voltage or as noise power, not as peak voltage, since random peaks may rise briefly to very high values. Since noise power is distributed over a spectrum, the measured power is proportional to the spectrum bandwidth, which must therefore be specified. A convenient expression of noise is in *power per hertz*, which is obtained by dividing the total power by the bandwidth in Hz, giving a rather small number usually expressed in dB. Reversing the calculation gives the noise power level over whatever bandwidth is of interest. To convert a broad bandwidth measurement in dB to power per Hz, we subtract from it a dB amount:

$$\text{dB} = 10 \log_{10} \text{bandwidth, in Hz} \tag{8-2}$$

For example if the broadband noise reads −56 dB for a bandwidth of 10,000 Hz, we subtract 10 $\log_{10}$ 10,000 = 40 dB to give a −96 dB noise level on a per Hz basis.

The noise specification of a recording system is important and may be expressed as: (1) broadband noise for a flat equalized system; (2) weighted broadband noise for the flat equalized system, usually according to the A-*weighting curve* that accounts for human hearing sensitivity; (3) spectral distribution on a noisepower per Hz basis; and (4) spectral distribution on the basis of third octave bands. Weighted noise specs give a fair comparison among audio systems and are lower than unweighted noise figures because the A-weighted curve deemphasizes the highest and lowest frequencies. Of the spectral curves, the noise power per Hz is more basic, but the most popular instruments measure noise in third-octave bands, which superpose a 6 db per octave rising slope on the curve as shown in Figure 8-9 (McKnight 1960). The two are readily interconverted. Figure 8-10 shows the spectral characteristics of different kinds of noise on a per Hz basis and on a third-octave basis.

Response curves of frequency and of noise in a good audio recorder are shown in Figure 8-11. The system is equalized to give flat overall response for an operating level that gives about 3% third-harmonic distortion. The zero dB reference is chosen at this operating level. Notice that there is about 12 dB of headroom above operating level before the tape saturates at mid frequencies but no headroom at 15 or 20 kHz. Notice also that the relative high-frequency response improves when we record at a lower level, as for example 10 dB below

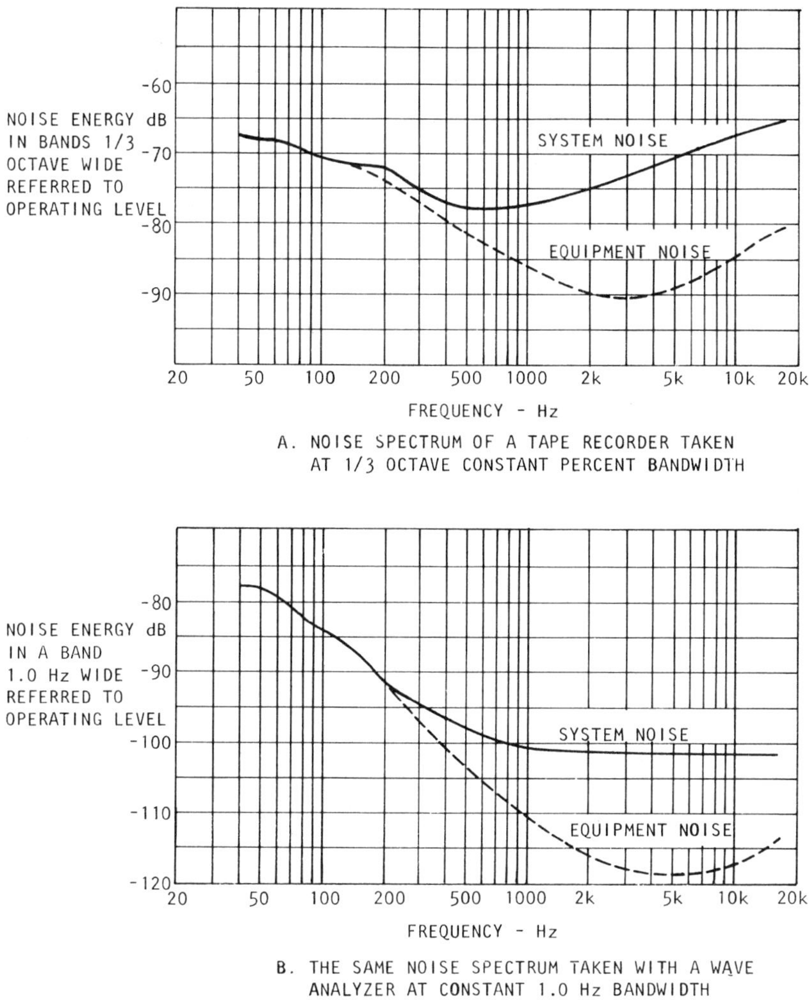

Figure 8-9. Noise spectrum of a recorder in one-third octave bands on a per Hz basis (McKnight 1960).

operating level (-10 dB response curve). Such high-frequency compression is a fundamental phenomenon of magnetic recording, always to be reckoned with, although it may be reduced in any given system by sacrificing the recording density.

The noise dB in Figure 8-11 is low because it is plotted for a 1 Hz bandwidth. For a 10000 Hz bandwidth, the noise power is 40 dB higher or 10000 times greater, and the RMS noise voltage is $\sqrt{10{,}000} = 100$ times greater. In well-

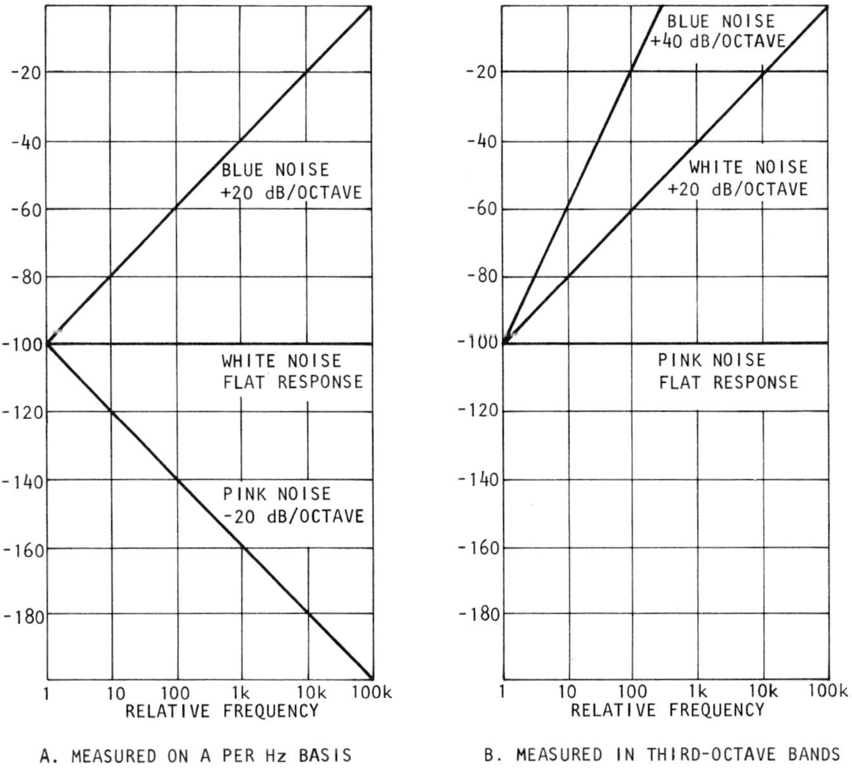

Figure 8-10. Spectral characteristics of white, pink, and blue noise (per Hz and in third-octave bands). A flat equalized tape recorder generates an approximately white noise spectrum. Without playback equalization its spectrum resembles blue noise (inductive type heads). The equalized playback amplifier has a pink noise spectrum at low and medium frequencies.

designed equipment, the signal-to-noise ratio is limited by tape noise (and possibly head noise) and not by amplifiers or other factors, since amplifiers basically have a higher dynamic range than magnetic recording components do. As is evident from the curves of Figure 8-11, the maximum output is limited by magnetic saturation of the tape, whereas the minimum is limited by noise due to magnetic irregularities in the tape layer. The lowest curve, obtained when the tape is stopped, is called the equipment noise.

Equipment Noise

Equipment noise, mostly originating in the amplifier input stage, is a minimum at midrange frequencies from 1000 to 3000 Hz. At lower frequencies the noise increases because of $1/f$ or flicker noise in transistors, resistors, and vacuum

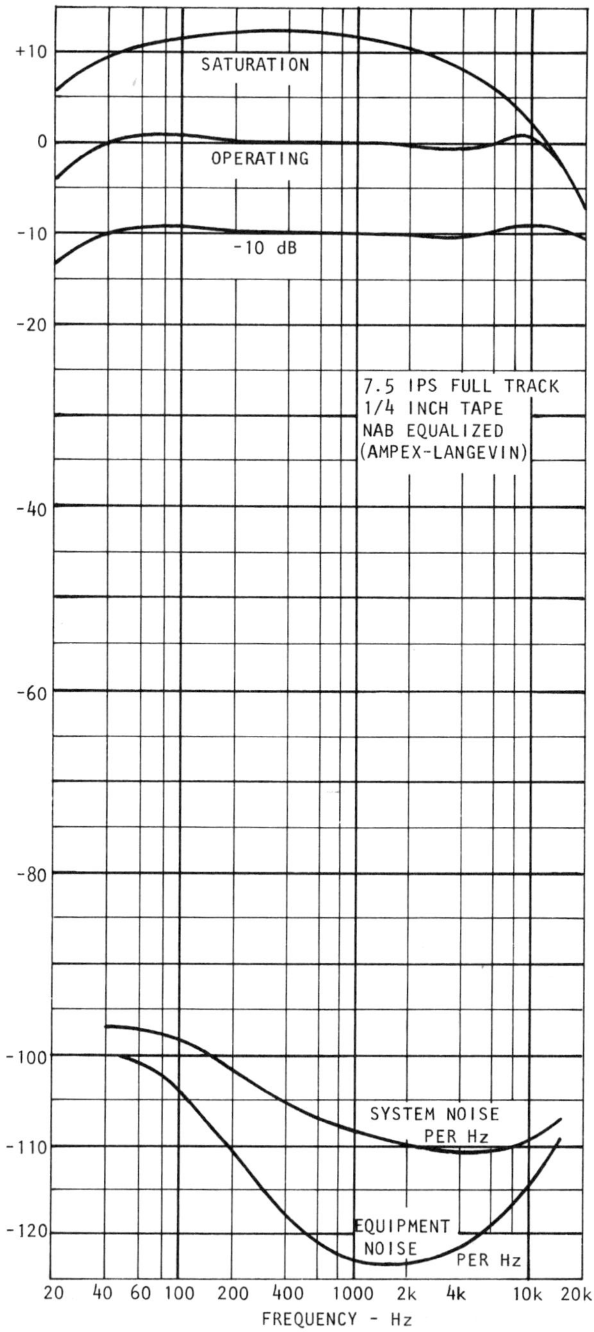

Figure 8-11. Response curves of frequency and noise in an audio tape recorder.

tubes. At higher frequencies the noise rises because of thermal noise generated by the effective head resistance, which increases rapidly due to core losses and skin effects. The thermal or Johnson noise generated by head resistance is

$$E_n = \sqrt{4kTR\Delta f} \tag{8-3}$$

where

E_n = thermal noise generated, RMS volts
k = Boltzman's constant = 1.37×10^{-23} joules per degree kelvin
T = absolute temperature, degrees kelvin
R = resistance, ohms
Δf = bandwidth in Hz

At room temperature, a head having 1000 ohms resistance generates a thermal noise voltage of 0.4 microvolts RMS in a 10 KHz bandwidth, which is about 68 dB below a signal level of 1 millivolt. For a 1 Hz bandwidth, the noise is 108 dB below 1 millivolt. Core losses may cause the same head to reach an effective resistance of 25,000 ohms in its high-frequency region, increasing its high-frequency thermal noise voltage per Hz by a factor of 5 ($\sqrt{25000/1000}$) to 94 dB below 1 millivolt.

Tape Noise

Tape noise is the playback output when a tape with no signal recorded on it passes over the head. It consists mainly of magnetic fluctuations in the tape itself, but can include noise of magnetostrictive or magnetized head cores when the tape rubs against them, and noise due to poor contact or vibration at the tape-head interface. Print-through and cross-talk also increase the noise level. In Figure 8-12 the curve labeled *bias noise* is a typical spectrum.

Erased Noise (AC Demagnetized Tape). Tape noise level depends greatly on the quality of the tape itself, but depends also on how it is erased or biased. As seen in Figure 8-12*A*, the lowest noise is produced by a tape that has been demagnetized with an AC bulk eraser. Erase heads can also approach this level. Virgin tapes made in a field-free environment might or might not be better, but are not available since even fresh unused tapes have been fluxed with strong orienting magnets during coating and are bulk-erased before leaving the factory.

A demagnetized record contains an array of tiny domains, each fully magnetized, but with the magnetizations in different directions so that statistically they cancel out. The cancellation is never quite perfect, thus giving theorists good exercise in predicting the background noise under various assumptions.

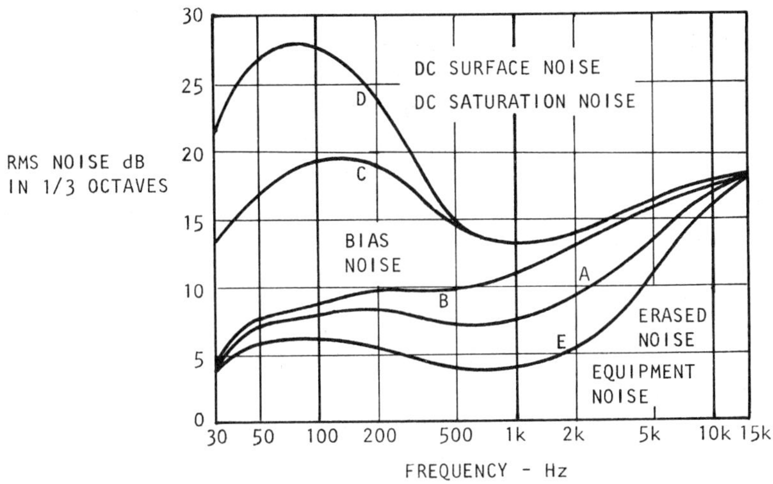

Figure 8-12. Noise levels in a tape recorder. Cassette recorder, 1.88 in/s (4.8cm/s) (Daniel 1972).

Some general observations are reasonable: (1) The smaller the volume of tape seen by the pickup head, the greater its relative statistical noise; (2) the smaller the particles, the more of them will be packed in a given volume and the less the noise; (3) agglomerations of particles act as larger units, increasing the noise; (4) nonuniform dispersions increase the noise.

In a fixed volume of *demagnetized* record medium, the noise due to magnetic randomness of single particle domains varies as $1/\sqrt{n}$, where n is the number of particles seen by the playback gap. The noise will therefore decrease 3 dB whenever we double the number of particles in a fixed volume by reducing the particle size.

Erased noise or *AC noise* is considered to be a measure of the size and distribution of magnetic domains. Theoretical calculations of noise and signal-to-noise ratio (SNR) have been based on Einstein's equation (Woolridge 1945), the Wiener-Kintchine theorem, and the Wiener auto-correlation theorem (Mallinson 1969). In general, they verify the experimentally measured figures and also verify the relation that tape noise voltage increases as the square root of track width while signal voltage increases directly with track width. Thus the SNR increases as the square root of increased track width, or 3 dB for every doubling of track width, and vice versa.

A typical record tape coated with gamma ferric oxide contains in the order of 10^{11} particles per mm^3. A full track head 6.3 mm ($\frac{1}{4}$ in) wide, scanned with a 2.5 μm gap, sees a volume containing about 10^7 particles and gives a signal-to-noise ratio of about 66 dB (2000:1). The noise is then equivalent to that of 0.5×10^4 unneutralized particles.

If we increase head width by a factor of 2, it covers twice as many particles and its SNR improves 3 dB, as verified by experience. The bandwidth is unchanged, but we use twice as much tape. We could keep the head width constant but double its gap length, again covering twice as many particles and improving its SNR by 3 dB. Now the useful bandwidth is cut in half but we do not increase tape usage. We can, of course, restore the bandwidth by doubling the tape speed and using twice as much tape. Either way we pay for a 3 dB improvement in SNR by doubling the amount of tape usage.

Bias Noise. The bias noise (Figure 8-12*B*) is usually 3 dB or more above the erased noise. Where bias is used, the bias noise is more realistic than erased noise as a measure of minimum noise because any tape passing through the recorder is subjected at least to the bias field. Bias noise is diminished by increasing the bias frequency so that its recording wavelength is far below the shortest wavelength that can possibly be recorded by the system. Otherwise the bias is modulated by irregularities that produce lower sidebands of noise within the useful recording band. For example, tape noise at 38 kHz is far above the cutoff of an audio recorder, but a bias frequency of 40 kHz can produce beats at $40 - 38 = 2$ kHz in the audible range. Similarly, any noise within the 30 kHz to 50 kHz region is heterodyned down to the 0–10 kHz audio band, but is recorded only if the head and tape can resolve frequencies in the bias range.

The 3 dB increase in bias noise above erased noise under best conditions is explained by a doubling of the noise power when the tape sees a reflected magnetic image of itself as it contacts the high-permeability recording head core. High-frequency bias serves to fix permanently this otherwise temporary increase.

Even a small DC component of magnetizing force in the presence of high-frequency bias will greatly increase the recorded noise. The DC component can result from, for example, a magnetized head core, magnetized elements near the head, the earth's field, or asymmetry of the high-frequency bias. Such noise is not true bias noise, of course, but is really a form of DC noise. Some recorders provide a tiny amount of adjustable DC in the head for counterbalancing any accidental DC component, bringing the recorded noise down to the minimum bias noise level.

DC Noise. A higher-level noise called *DC noise* is played back from a tape that is saturated with a DC field, as when the recording head has been energized with a sufficient DC current. DC noise, as shown in Figure 8-12C, is usually strongest at medium-long wavelengths. Unexpectedly, if the DC recording head field is reduced to below tape-saturation, the noise can be even higher than at saturation, as in Figure 8-12*D*. If high-frequency bias is present, the effects are similar but take place with much less DC head current.

The different character and spectrum of DC noise gives clues that its origin is not in the smaller domains themselves but in larger clumps of particles that are nonuniformly dispersed in the recording layer; or in the layer itself, varying in thickness irregularly as when coated on a rough substrate; or in the tape surface roughness, giving irregular magnetic contact with the head; or in little nodules protruding from the tape surface, temporarily lifting it away from the head gap. All of these large nonuniformities disappear magnetically when demagnetized, unlike the smaller magnetic domains, which remain magnetized perpetually. The nonuniformities emerge magnetically in the presence of a DC field just as a large bar magnet, originally demagnetized, becomes magnetized more and more fully in an increasing DC field.

An explanation for DC noises reaching its maximum at a DC field below saturation is that protruding nodules or *surface asperities* in the tape cause its surface to rise and fall irregularly away from and toward the head gap, thus modulating the strength of DC head field acting on the tape and adding this asperity-modulation to the inhomogeneity modulation. When the head field is increased until strong enough to saturate the tape regardless of its motion away from the gap, then the asperity-modulation ceases and only the inhomogeneity modulation remains.

Modulation Noise. Modulation noise is simply a dynamic manifestation of DC noise. Inhomogeneous spots on the tape are quiet as long as they remain demagnetized. Recorded waveforms bring out the nonuniformities. Modulation noise therefore produces a fuzz on a recorded waveform proportional to the instantaneous amplitude of the waveform.

OPTIMUM RECORDING DENSITIES

It is evident that increased area density of recording can be traded for signal-to-noise ratio (SNR) either by increasing the *track density* (number of tracks in a given tape width) or by increasing the *bit density* (number of magnetic poles in a given length of tape). And of course every improvement in the tape itself or in associated equipment that gives higher output and/or lower noise gives a net gain in SNR, which can theoretically be traded for higher area density.

Track Density

The simple relation that the SNR decreases 3 dB every time the track width is halved is described by the equation:

$$n = (n_o)\, 2^{1/3(SNR_o \; - \; SNR_n)} \qquad (8\text{-}4)$$

where n is the track density (tracks/in) that gives an SNR equal to SNR_n in dB, and n_o is the original track density that gave a measured SNR of SNR_o in dB.

With equation 8-4 we can investigate how narrow we can make each track $(\text{tracks/in})^{-1}$ if we are satisfied with an SNR of SNR_n. For example, we know that a professional recorder (Table 9-1) with $n_o = 8$ tracks/in. gives $SNR_o = 66$ dB per track for a given speed and bandwidth. Equation 8-4 then becomes:

$$n = (8)\ 2^{1/3(66\ -\ SNR_n)} \tag{8-5}$$

which tells us how many tracks/in (n) we can have for whatever SNR_n we specify under the same conditions of speed and bandwidth. If we are satisfied with $SNR_n = 60$ dB, the equation tells us we can have four times as many tracks/in., or $n = 32$, as expected. If we can tolerate an $SNR_n = 42$ dB, equation 8-5 gives $n = 2048$ tracks/in. (80.6/mm), a surprisingly high number. For digital recording a very respectable $SNR_n = 21$ dB gives $n = 262{,}144$, or $n^{-1} = 0.1\ \mu$m track width, which would tax the best mechanical systems for maintaining a head on its track. Figure 8-13 shows the enormous track densities that theoretically might be obtained for even lower SNR_n values.

Equation 8-4 can be rewritten to give the SNR when track density is specified.

$$SNR_n = SNR_o - 3 \log_2 \frac{n}{n_o}, \qquad \text{or} \tag{8-6}$$

$$SNR_n = SNR_o - 9.9658 \log_{10} \frac{n}{n_o} \tag{8-7}$$

where the symbols are the same as for equation 8-4

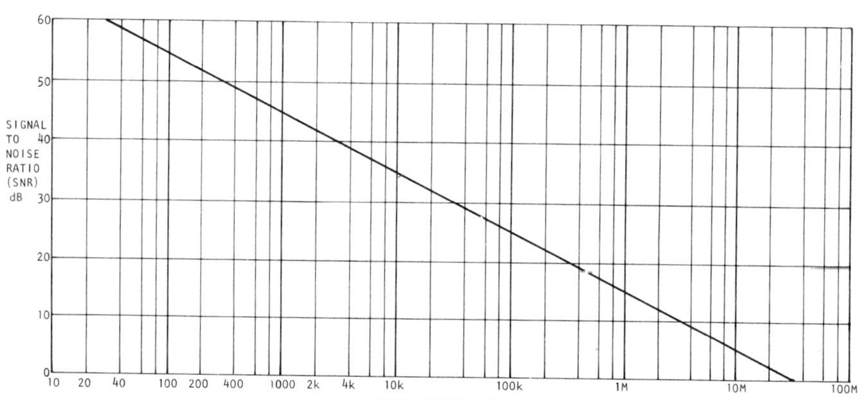

Figure 8-13. Signal to noise ratio as track density is increased.

Such densities are indeed too good to be true when one looks at other limitations. In addition to mechanical tracking difficulties, cross-talk between adjacent tracks becomes a greater problem with very narrow tracks. This is a wavelength effect that can be alleviated by shielding, by keepers (Chapter 4), and by restricting the recordings to very short wavelengths. Equation 8-4 gives only a theoretical SNR but tells nothing of absolute signal levels and noise levels. The signal flux, and therefore the induced head voltage, ordinarily decreases directly as the track width decreases. Doubling the track density of the above by reducing track width to half then reduces the signal by 6 dB, and soon the noise level becomes limited by equipment noise (standby noise is −68 dB in Table 9-1), which is mainly thermal or Johnson noise. With further increases in track density, the reduced tape noise gives no advantage in SNR because the thermal noise sets a fixed floor. The SNR then falls 6 dB instead of 3 dB every time track density is doubled and the equation describing it,

$$n = (n_o)\ 2^{1/6(SNR_o - SNR_n)}, \tag{8-8}$$

does not give the spectacular advantage assumed by equation 8-4.

Scaling Theory of Track Density. Scaling theory can be applied to good advantage in predicting how changes in head dimensions affect SNR. For a narrower track, the head core width shown in Figure 8-14*B* is reduced in the same proportion as track width but other dimensions are not altered from the original head (Figure 8-14*A*). Playback head inductance is then reduced in proportion to the narrowing factor ($1/n$) of the core, so it is possible to increase the number of turns on its coil by a factor ($\sqrt{n}$) to restore the inductance to its original value. The output voltage of the new head is therefore reduced by a factor of $1/n$ because of the narrower track, but is increased by a factor of $\sqrt{n}$ because it has more turns, giving a net loss of $\sqrt{n}/n$ (without changing the head inductance, which we assume is the maximum tolerable amount for circuit resonances, and so on). Unfortunately the increased turns are at the expense of multiplying the coil resistance n times for a given coil size, so the thermal noise voltage due to head resistance also rises by $\sqrt{n}$ and the SNR degenerates 6 dB for every doubling of track density when limited by thermal noise. Lower losses in the reduced core volume will alleviate this estimate as discussed below.

Another assumption of head dimensions is that all dimensions of core and coil, except the head gap, are reduced by the track width scale factor $1/n$, as shown in Figure 8-14*C*. With the same gap and a high permeability core the inductance remains the same as in Figure 8-14*B* and the coil turns can again be $\sqrt{n}$ times the original turns T. The smaller coil has to be wound with finer wire and its resistance is therefore n^2R. A head completely scaled down to track width (but retaining original gap size and optimum winding turns) therefore gives $\sqrt{n}/n$ times the voltage output but produces n times more thermal noise,

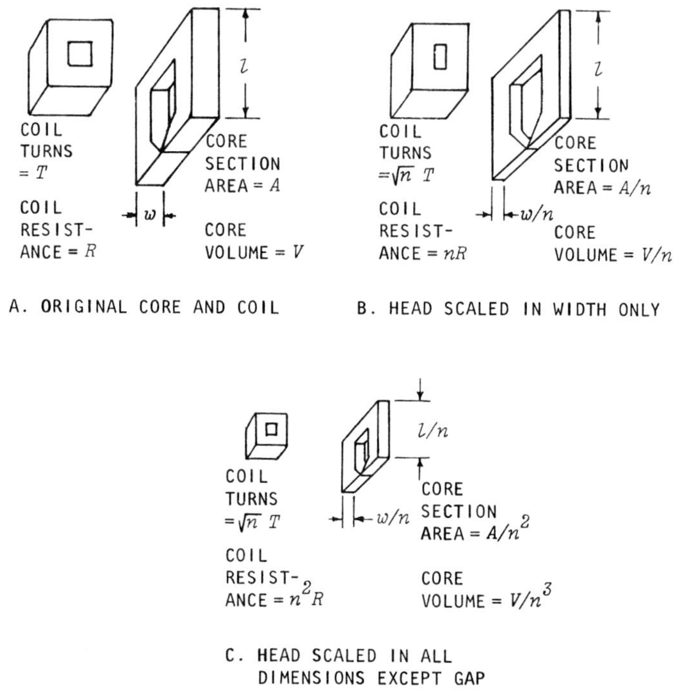

Figure 8-14. Scaling theory applied to track density and SNR.

and its SNR degenerates 9 dB for every doubling of track density. Again, core losses are neglected.

Complete scaling of all head dimensions including the gap gives the same result as Figure 8-14*C*, since the gap is $1/n$ as long but its area is reduced by $1/n^2$ so that the multiplying factor for gap reluctance is n as before. The inductance and resistance also remain the same as in Figure 8-14*C*. (Bit density and spectrum width are assumed constant in all examples for purposes of comparison.)

If core losses add to effective head resistance, the head cores of Figures 8-14*B* and *C*, which have less volumes of $1/n$ and $1/n^3$, respectively, will contribute less noise power in proportion to their reduced volumes. Core losses increase the effective head resistance in such a complex manner that it is best to measure the effective AC resistance (see Chapter 4) rather than to rely on calculations.

Bit Density

In the following discussion *bit density* also implies *wave density*, both being in the direction of scanning motion, as contrasted to *track density*, which is the number of separate tracks per mm, usually at right angles to the direction of scan. The bit density rating is ordinarily twice as high a number as the wave density rating because there are two bits per hertz.

Professional recordists know that they can increase their SNR by increasing tape speed, for example, from 7.5 in/s to 15 ins. By doubling the tape speed the useful wave density is cut in half, whereas the signal and noise retain the same ratio to one another (their waveforms and amplitudes unchanged after equalization), so any advantage in SNR comes only from cutting off the highest octave of reproduced noise. In other words, doubling the speed gives an extra octave of both signal and noise at the high-frequency end; and because this octave lies beyond the useful signal range of the recorder, it can be filtered out, cutting the noise power in half (3 db for white noise), since the highest octave accounts for half of the bandwidth.

Bit density can thus be traded for SNR in the same way as track density:

$$SNR_d = SNR_o - 3 \log_2 \frac{d}{d_o} \tag{8-9}$$

where d is the bit density (or wavelengths/mm) that gives SNR_d in dB and d_o is the original bit density that gave a measured SNR_o in dB.

By analogy the other equations given for track density can also be modified for bit density. As with track density, this analysis assumes that the quality of tapes, heads, and head-tape interface is fixed and that the tape noise is the limiting factor.

Scaling Theory of Bit Density. The same result is also obtained when scaling theory is applied to higher bit density along a track. For a head of constant width, the gap size can be scaled down and the bit density increased in the same proportion as in Figure 8-6. In the usual case where the recording layer is relatively thick, the area of the recorded magnets decreases in proportion to the wavelength. Hence, the magnetic flux and reproduced signal voltage fall in proportion to $1/d$ where d is the bit density, and the reproduced signal power falls as $1/d^2$. The noise power of the smaller wave-lengths falls as $1/d$ and therefore the SNR decreases as $1/d$ or 3 dB every time the bit density is doubled.

Track Density vs Bit Density

While track density can theoretically be increased to unbelievably high figures according to equation 8-4, similar increases in bit density encounter an impassable barrier very quickly, for when the wavelength or bit size is of the same order as tape roughness or head-to-tape spacing, the signal rapidly approaches zero. Head gaps much smaller than these magnitudes are of no avail; we assume here that equally small practical head gaps are used throughout these comparisons. The theoretical high track densities are also illusionary because when the head width is reduced to dimensions approximating gap size or tape particle size (0.5 μm), equations 8-4 to 8-7 no longer apply. Actually the larger clusters of particles and surface roughness set a much lower limit to track density.

The advantages of greater track density have not been utilized to fullest extent in tape recording and there is considerable room for progress. On the other hand, magnetic disc recorders have already achieved notable track densities and bit densities, since accurate tracking and multiple head problems are less severe than in tape recording.

All in all, it seems that the inconvenience of running tapes at high speeds has favored the setting of bit densities close to their upper limit as determined by state-of-the-art technology, and then selecting an appropriate track width (track density) that gives the desired SNR. Such track densities are usually far away from their ultimate density limit. For example, a typical audio recorder running at 15 in./s with a 70 mil track has the same theoretical SNR as a recorder running at 150 in/s with a 7 mil track, but the slower-speed machine has a longer uninterrupted playing time, is easier to drive, gives less tape and head wear, and runs more quietly, to mention a few advantages. Such considerations do not apply to magnetic discs or videotapes, and some of these use very high track densities of 100 tracks/mm or more.

INFORMATION THEORY

Living systems use information to harness their own benefit some of the energy and materials of inanimate nature, creating for themselves pockets of negative entropy or greater order within a vast sea of disorder. Information storage exists in a separate parallel world, apart from the world of action, energy, and movement. Information itself requires very little energy, yet can save enormous amounts, as anyone knows who has kept an appointment on the wrong day. World industries are based on knowledge printed on bits of paper that tell the best way to grow food, to extract metals, to combat disease, and even to devise better information storage for the future. Magnetic recording has become one of the most convenient and rapid methods of storing and recovering information.

Magnetic recording is not unique in its concern with signal fluctuations and noise that limit speed, accuracy, and density of information storage and transmission. Similar problems are universal in every other kind of recording as well as in all communication channels (telephone, radio, modulated light beams, etc.). Information theory deals with the general nature of communication systems; gives criteria for comparing different schemes, such as analog, digital, AM, FM, and PCM; and guides in the choice of the most efficient kind, or at least to a realization of why a less efficient system is better for the problem in hand.

Theoretical Limits. An important relation of far reaching consequences is that the maximum error-free rate at which information can be transmitted is proportional to the bandwidth and SNR of the transmission channel. Shannon's theorem (Shannon 1948) states that:

$$C = W \log_2 (1 + SNR) \tag{8-10}$$

where

C = the transmission rate in bits/sec (one Hz = 2 bits/sec)
W = the channel bandwidth in Hz
SNR = the signal-to-noise ratio (signal-power/noise-power or P_s/P_n).

According to this equation, it is possible to transmit a bit rate C equal to the bandwidth W even in the presence of such a high noise level that it equals the signal power, and at a lower rate if noise exceeds the signal power. At higher SNRs, more bits/sec can be transmitted because a larger number of signal amplitudes can be distinguished even when the bandwidth remains constant. Or expressed differently, more bits are needed to depict a "clean" signal of a given frequency and amplitude than to just barely indicate that a signal is present. As with other mathematical theories, equation 8-10 tells us what is possible, but does not tell how to do it. However it guides us in directions that are worthwhile, and steers us away from impossible efforts (but remember the Austin-Cohen formula).

Trading SNR for Bandwidth. One consequence of Shannon's theorem is that multiple channels of moderate SNRs are more efficient at transmitting information (more bits/sec) or at storing it than a single high-grade channel of better SNR. It is well known that frequency modulation (FM) allows "noise free" reception but uses up a wider frequency band than amplitude modulation (AM). How efficient is FM in trading SNR for bandwidth? Are there even more efficient modulation methods, and is there any that reaches the theoretical limit? We can rewrite equation 8-10 as follows:

$$W = \frac{C}{\log_2 (1 + SNR)} \tag{8-11}$$

Let C be fixed at any required bit/sec rate, which we can call 100%, and let an $SNR = 1$ be used as a basis for comparison. Figure 8-15 plots this relationship and shows that a very moderate improvement in bandwidth W has the same

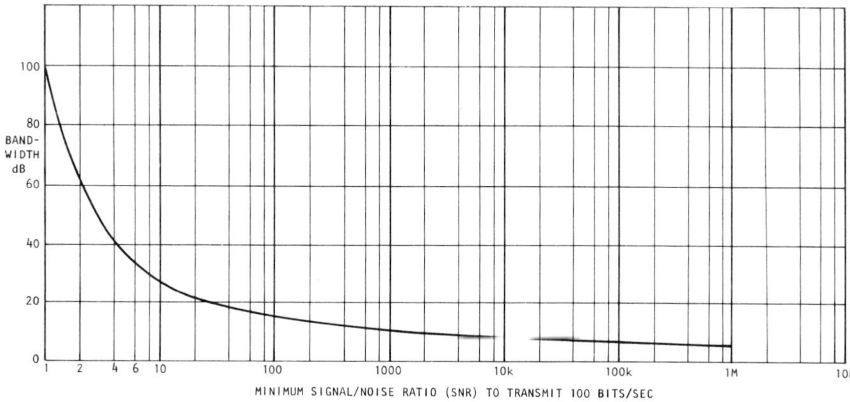

Figure 8-15. Tradeoff in bandwidth for signal/noise ratio.

effect as an enormous increase of *SNR*. This analysis tells us that, in a practical design situation, if improvements in bandwidth or in SNR are about equally difficult, it is better to work on the bandwidth.

Application to Magnetic Recording. In equation 8-10, the factor C giving *transmitted bits per second* can also represent *stored bits per unit area* on a tape. For when we move the tape past the playback heads at unit area per second, all the *stored bits in a unit area* are turned into *transmitted bits per second*.

On a tape of a given width, we have the option of recording a few channels of high SNR or many channels of lower SNR. Which carries the most information? Suppose we subdivide a recording track on a tape into m separate tracks. The gross number of bits handled by m tracks is m times as many as for a single track, but this gross number must be diminished by the logarithmic factor representing a smaller SNR for each narrower track. Theory and experience indicate that a track $1/m$ as wide has an SNR that is $1/m$ as much as the original track (under conditions of equation 8-8). Equation 8-10 for m tracks then becomes (Eldridge 1963):

$$C_m = mW \log_2 \left(1 + \frac{SNR}{m}\right). \tag{8-12}$$

Comparing equation 8-12 with 8-10 we see that the theoretical bit rate is increased by a factor m but decreased by $\log_2 (1 + SNR/m)$. Since the logarithm always decreases more slowly than m increases, the net result is that C_m for multiple channels is always greater than C for a single channel and that, in general, many low-grade channels handle more bits than a few high-grade chan-

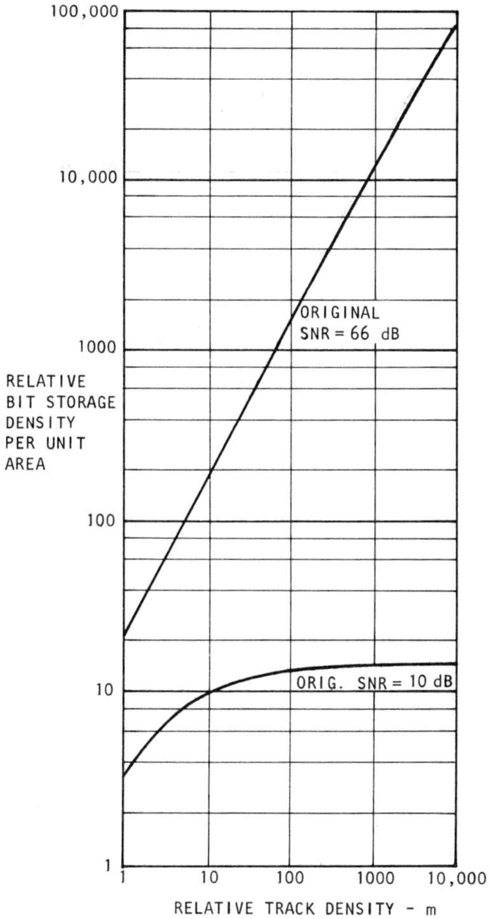

Figure 8-16. Bit density vs. track density according to information theory.

nels. Figure 8-16 is a comparison of bit density calculated from equation 8-12 for a 20 kHz bandwidth full-track recorder of 66 dB SNR and of 10 dB SNR.

Digital Recording Density of Binary and Higher-Base Systems. Although digital recording almost invariably uses the binary system, the question arises as to whether numbers might not be recorded more efficiently at several amplitude levels instead of as ones and zeros. A binary number base requires two signal levels; a trinary base, three levels; a decimal base, ten levels, and so on. The number of intervals between levels for any chosen base b is $(b - 1)$, so a decimal base requires $10 - 1 = 9$ times the dynamic range of a binary base where 2 − 1 = 1 interval. For perfect accuracy, the noise peaks must never be as high as half an interval, but the best we can do is to choose a SNR that has a very low statistical probability of such an excessive peak, say once every three months for a data rate of 10^5 per second. A gaussian distribution of white noise gives the above probability of error if the SNR is 23 dB. The probability changes rapidly with SNR; decreasing the SNR to 18.5 dB changes the probability to one error per second.

From these considerations Eldridge (1963) has derived a relationship for relative track width in a *b*-level recording system compared to a 2-level binary system:

$$\text{Relative track width for } b\text{-levels} = \frac{(b - 1)^2}{\log_2 b} \tag{8-13}$$

where b is the number of recorded levels that must be distinguished reliably by the system. This increases rapidly for a base greater than 2. A base of 3 requires 2.5 times the track width, while a decimal base 10 requires 24 times the track width of binary recording for the same reliability.

A somewhat different analysis (Oliver, Pierce, and Shannon 1948) gives the SNR needed for different number bases to be

$$\text{SNR} = k(b^2 - 1) \tag{8-14}$$

where k is a function of tolerable error rate. According to this equation, a base of 3 requires 2.67 times the SNR, and a decimal base requires 33 times the SNR of a binary recording.

Efficiency of Binary vs. Analog Recording. Eldridge (1963) has also derived a relationship for track width efficiency (area density) of binary-digital versus analog recording under specified conditions for purposes of comparison:

$$\frac{d_d}{d_a} = \frac{664 \log_{10} SNR_v}{(SNR_v - 1)^2} \tag{8-15}$$

where d_d and d_a are the respective track width for digital and for analog recording and SNR_v is the ratio of signal voltage to noise voltage.

Figure 8-17 is a plot of this relationship, showing that analog recording is better at lower SNRs, but as SNR requirements are increased there is a crossover point above which digital recording is superior. While the crossover point in Figure 8-17 may not seem realistic in current practice, the general concept agrees with the experience that it is not practical to record extremely high SNRs

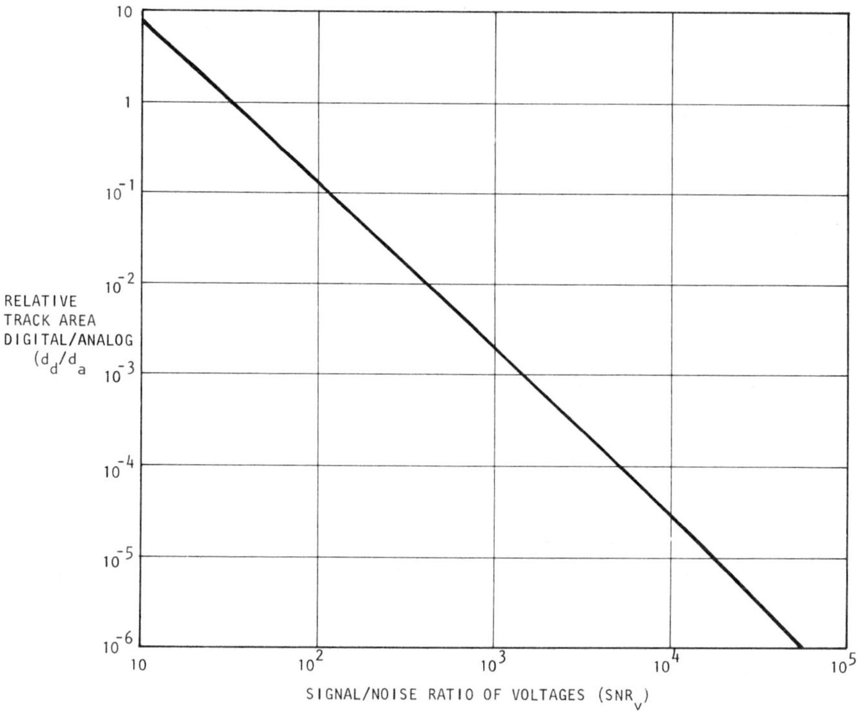

Figure 8-17. Comparison of track areas for analog vs. digital recording (After Eldridge 1963).

by analog means, whereas digital recording can give SNRs well above the analog limits.

Efficiency of Binary vs. FM Recording. Having shown that binary recording can exceed analog recording in SNR, it is of interest to compare it also with FM recording, which is widely used and which has the reputation of high SNR for wide frequency deviation systems. Such a comparison has been made by Oliver and colleagues (1948), who find that the SNR in dB for wide deviation FM using a bandwidth W is

$$\text{FM Transmission SNR(dB)} = 30 \log_{10} W + \text{constant} \qquad (8\text{-}16)$$

whereas for binary pulse-code modulation (PCM) under the same conditions of power and bandwidth the SNR is

$$\text{PCM Transmission SNR(dB)} = 6W + 10.8. \qquad (8\text{-}17)$$

Since the SNR of digital PCM transmission increases directly with the bandwidth while the SNR of FM transmission increases with the logarithm of the bandwidth, digital recording will always overtake and exceed FM recording as the bandwidth is increased.

APPLICATIONS OF INFORMATION THEORY AND NOISE REDUCTION

For maximum possible efficiency in using the available bandwidth, an ideal recording system would fill all frequencies in its spectrum with information, and on a time basis would eliminate all intervals when information was not being transmitted. Information efficiency experts therefore look for frequency or time domain gaps that they can fill up or cut out.

Preemphasis and Deemphasis. Long before information theory became a science, it was considered that music and speech did not have a flat frequency spectrum but was usually deficient at high and low extremes. Recorders could therefore increase the amplitude of these high and low frequencies without serious overloading during recording (preemphasis) and compensate for the frequency distortion during playback (deemphasis), thereby recording all parts of the spectrum as far as possible above the noise level. Frequency response corrections for inductive pickup, for the nature of frequency modulation, etc., are done as a matter of course. Details of correcting systems are described in Chapter 6.

Dynamic Noise Suppressors (Companders). In the time domain, analog signals that are far below the maximum amplitude level are wasteful of information density, resulting in a poorer SNR than is otherwise possible. The remedy is to compress the dynamic range toward high levels, thus recording low-level signals stronger than normal. The playback electronics expands the dynamic range back to the original, lowering the background noise with respect to the reproduced signal. The process is somewhat tricky because direct compression of waveform amplitudes gives severe distortion. We must therefore resort to automatic gain control (AGC), which is not instantaneous but takes time to react to changes in signal level. The time constant of AGC is audible and annoying unless carefully disguised as described in Chapters 9 and 10 for Dolby-type systems.

Dynamic Range Extenders. Dynamic range of recorders can be increased to any desired scope by recording the same signal at different levels on two or more tracks (Mullin 1965). Figure 8-18 shows a high-grade analog recorder with two tracks, each having a working range defined by a SNR of 65 dB,

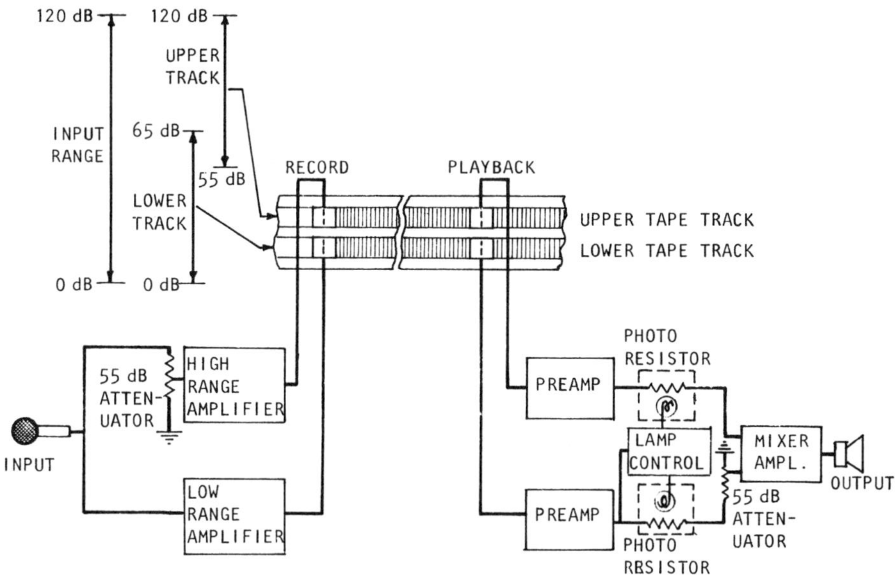

Figure 8-18. Dynamic range extender using multiple tracks (Mullin 1965).

arranged to cover an overall dynamic range of 120 dB. The lower track records properly for signals from 0 dB to 65 dB, above which it progressively distorts and becomes badly saturated. The higher track operates with an attenuated signal 55 dB weaker than the other and therefore its 65 dB range allows it to operate for inputs up to 120 dB. The playback system chooses one track or the other, depending on the level, fading between them in the overlap range by means of photosensitive resistors.

Dropouts. It is comforting to have a neat mathematical equation that fits known data and that lets us explore new territory without messy experiments that give confusing results. This chapter abounds in such equations. Alas, we sometimes have to bend the data to fit the mathematics; and sometimes the mathematical relevations about unexplored territory are invalidated by second-order effects (such as flux fringing), which become first-order effects under extreme conditions (such as very narrow tracks).

Dropouts are an example where nature does not follow the nice gaussian distribution we assign to tape noise level. The signal is lost during dropouts, which occur with a statistical distribution of their own, independent of the noise properties of the tape, being a function of tape wear, debris, and nodules, among other factors. Reliability therefore cannot be based on gaussian tape noise, and errors must be taken for granted. Corrective measures include use of parity-bits, use of sophisticated error correcting codes, verifying the readout immediately after recording, repeating questionable readouts, automatic skipping of defective record areas, and redundancy.

Redundancy. Age-old communication methods, spoken and written, are found by information theory to be redundant, containing more than the minimum information necessary to convey the ideas. We recognize the meaning of sentences not spoken clearly or with syllables or words missing. In printed sentences we can usually fill in missing letters and even words. Every language has letters of its alphabet that are used most often. Given certain letters, there is a high probability that they will be followed by some letters and not by others; in English *QU, TH, ED* are common. Many two-letter combinations are very likely to be followed by a third, for example *TH* and *E*. Information theory tries to utilize redundancies to abbreviate messages sent through reliable channels, or to reconstruct messages that have become imperfect due to errors or noise added during transmission.

Speech Compressors. An interesting discovery is that spoken language is so redundant that it is intelligible after half of it (or even more) is taken out at regular intervals. The speech waveform is divided into $\frac{1}{30}$ second segments, for example, and alternate segments are discarded. The remaining half of the segments are compressed together to fill the intervening spaces, as described in Chapter 10. The result sounds like someone talking twice as fast but is still quite understandable. An alternative is to fill the omitted segments with duplicates of the retained previous segment. Obviously, only half the information of the original speech is used. Music is also amenable to this method of compression, sounding quite interesting at the faster tempo. In aesthetic matters, however, one may question whether enjoyment is proportional to information density or to rapidity of perception in bits per second.

Video Dropout Correctors. A prime example of outrageous redundancy from the standpoint of information theory is the NTSC video signal used in television transmission and recording. Here the spectral content is crowded into narrow bands centered about the 15,734 Hz horizontal line frequency and its harmonics, while wide open spectral spaces exist in between. Much effort has been put into using information theory to decrease the bandwidth requirements for video, as for example to transmit only the *changes* between one picture field and the next instead of constructing a new picture field from scratch every $\frac{1}{60}$th of a second. While some techniques have been shown to be possible, however, nothing yet has been made simple or economical enough to displace existing methods.

Video dropout correctors are an important application, although somewhat trivial in information theory. They are based on the fact that the picture does not change much from one scanning line to the next one following it. When a tape dropout occurs, there is complete loss of picture information for a time interval corresponding to a horizontal dot or a dash along a scan line. This blank interval can be patched up (concealed) by substituting the corresponding information from the previous horizontal scan. The equipment includes a delay line that stores the previous horizontal scan, plus a fast switcher that connects the video reproduce circuits to the delay line temporarily whenever there is a dropout of video signal.

Digital Audio and Video. It is recognized that where SNRs of 75 dB or more are required, analog recording becomes impractical and one must go to a digital system using pulse code modulation (PCM), for example, and such a system is described in Chapter 9. The promises made by information theory for video have also stimulated a tremendous amount of research in digital video. Where recording is not involved, bandwidths of 100 MHz are feasible and studio accessories such as mixers and special effects equipment use digitized video to advantage. Economical recorders of 100 MHz have been a problem whose solution has been worked on for many years. Upon such recorders the future of digital video depends.

Digital Data Recording. Computer data is the classical example of binary digital recording where bit densities have increased at unbelievable rates toward the still out-of-reach limits predicted by information theory. Table 8-1 (page 372) lists some digital storage capabilities. Advances are due to better media, heads, and their interface; and also to error correcting codes, equipment, and procedures. Cobalt-chromium vertically oriented media promise another order-of-magnitude increase.

Instrumentation data recorders have not been digital except where the data to be recorded was already digital. The traditional data recorder has used FM recording of IRIG standards or direct recording for wideband data to about 25 MHz. The trend to digital data has changed these preferences and we can expect data recording of the future to be almost universally digital.

Part 2
APPLICATIONS OF MAGNETIC RECORDING

9. Professional Audio

Professional recording calls for the finest possible equipment to preserve valuable and irreplaceable performances and events. Cost is usually a secondary consideration. Professional audio includes master recording, mixing-editing, broadcast recording, motion picture sound, and copying of recordings, sometimes on a large scale. Analog recording is perfect enough for most applications, but digital recording now can eliminate residual distortions and allows perfect copies when editing or rerecording.

Recording brings to the performing arts the capability of editing a performance, omitting, adding, and changing it just as an artist with oil paints alters details while building a painting. Prior to the invention of recording, performances were live and fleeting, gone forever as soon as the last note was finished, the great moments as well as the mistakes. Today, a live performance is seldom presented: The vast majority of entertainment is recorded and edited to enhance the highlights and to eliminate the errors. Professional magnetic recording has therefore evolved along lines that give the producer of recorded programs great flexibility in improving the product.

MASTER RECORDING

At present when recordists complain that 32 tracks are not enough, it is hard to believe that a single track fulfilled nicely all mastering requirements in the 1948 era, with 6.3 mm ($\frac{1}{4}$ in.) tape recorders running at 381 mm/s (15 in./s).

Multiple Microphones and Tracks. Stereo sound requires at least two tracks. Multiple microphone techniques need far more. The philosophy behind multimicrophones is that each instrument is assigned its own separate track, allowing it to be blended during mixdown with other instruments in every possible combination until the golden ear of the music director is satisfied. An added bonus is a master record preserved unmixed, allowing future mixdowns to give different versions. During mixdown special effects such as reverberation or filtering may be added.

Track Formats. For professional recording it is best to have all tracks in the same direction and of related material, so that they are cut together when editing. A reverse direction recording on the same tape, as is common in home-use formats, would be ruined by cutting and splicing, and there is also a residual amount of pickup from the reverse tracks, incoherent with the main record.

Wider tracks give a better signal-to-noise ratio (SNR), so wide tapes are commonly used for multiple tracks. Standard tape widths are $\frac{1}{4}$, $\frac{1}{2}$, 1 and 2 in. (6.3, 12.7, 25.4, and 50.8 mm). Figure 9-1 shows some professional track formats; certain of these have been approved for international exchange of taped programs.

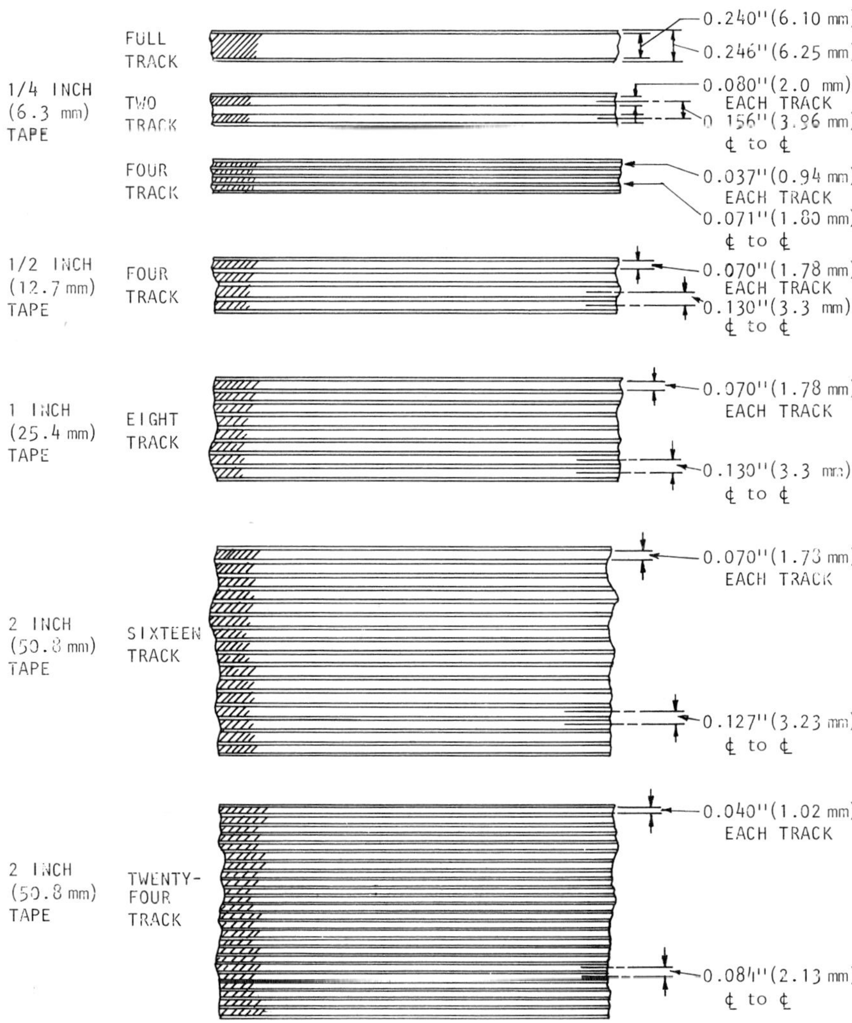

Figure 9-1. Track formats of professional audio recorders.

STUDIO TECHNIQUES

To this day studios still use concepts they borrowed long ago from the telephone art, such as 600 ohm balanced lines, attenuators, filters, equalizers, plug-in switchboards, and transformer coupled amplifiers. Through the years they have added their own unique components: faders, pan-pots, equalizers, and V.U. meters, evolving to the modern console. Along the way they have modified the matched impedance concept appropriate for long telephone lines but not critical for studio interconnections, finding that they could improve SNR by deliberate mismatching.

High performance in professional recording equipment is encouraged by competition for the best specifications, by ample size of the equipment, and by ample funds (although hardly anyone admits that they can not use more).

In most applications other than studio pickup, the signal is plugged directly into the recorder, but in studio work the audio sources generally go first through a control panel called a *console*, which enables the operator to select and mix the sources, and to modify the audio signals if he or she cares to do so.

Mixing Consoles

In the kingdom of audio, there is nothing more impressive than a large mixing console with hundreds of controls: knobs, dials, lights, and meters. The balance engineer who operates the console is like an airline pilot or a monarch, admired and respected by all, although there may be greater powers behind the prestigious throne.

The need for a control panel arises whenever two or more sound sources have to be selected, blended, or otherwise controlled. Crude mixing boards date back to early days of broadcasting and electronic recording. Present-day consoles (Figure 9-2) accept the feeble outputs from dozens of microphones or from recorded tracks, amplify them with almost no added noise or distortion; modify the frequency response as desired; mix them in any proportions; add reverberation and other effects; and direct the composite into other channels. Even available is a memory system that repeats a mix, changing all the controls automatically to repeat their settings for duplicating the original mix. Figure 9-3 shows some modules of a console and indicated their functions.

Outputs of any module are monitored either in headphones or in studio loudspeakers, whose volume is controlled or muted independently of line levels. Outputs (cue sends) are also fed to musicians so each one can hear the entire group while playing together. Some installations also provide for feeding a PA system for an audience.

Figure 9-2. A mixing console for recording studios. Advanced consoles for mixdown have servo controlled faders that repeat the settings of previous mixes, with override for updating. (Courtesy Rupert Neve, Incorporated)

Recording Engineers' Terms

Dry refers to a direct pickup without reverberation, while *wet* means a sound that has some reverberation either through room effects or added artificially. A *cue* is audio supplied to performers so they can hear the ensemble and keep in time. A *send* is an output to some processor or for a cue. A *return* is the audio received from a processor (such as a reverberation chamber). A *fader* is a gain control. A *pan-pot* is a control for stereo, which in center position gives a 50–50 mixture of right and left audio; off center it favors right or left in varying degrees up to 100%. *Busses* are lines into which several inputs or outputs can be

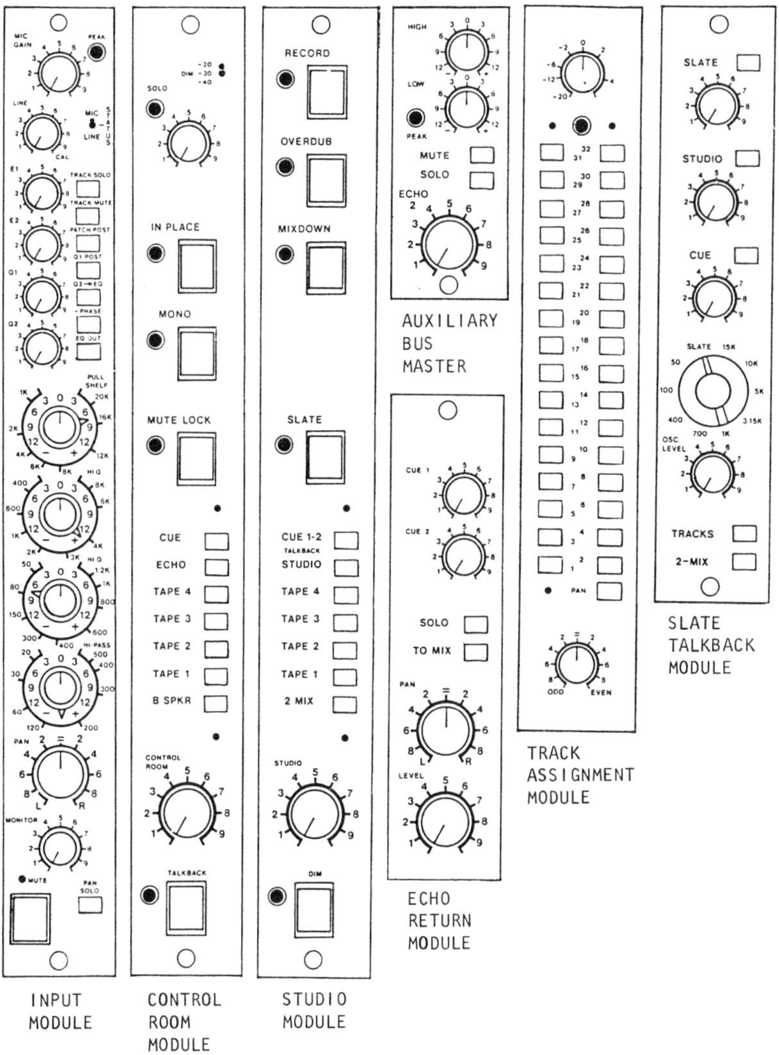

Figure 9-3. Modules of a mixing console and their functions. (Courtesy Neotek, Chicago, Illinois)

plugged. *Patching* means plugging in or switching to an audio line or buss. *Submasters* contain audio that is a component of the final mix. A *mute* switch silences the audio or sends it on a line. A *solo* switch isolates a single monaural channel for monitoring. *Phasing* refers to reversing the polarity of microphones or loudspeakers relative to one another. *Equalizers* boost or dip (cut) the frequency response in desired parts of the audio spectrum. *Talkback* enables the console operator to converse with artists in the studio. *Slating* involves recording a sound track sync mark that can be used in playback to start related parts together when they are on different tapes or films. A *clapboard* is a small blackboard with a hinged stick that is clapped against the edge of the board to generate a loud sound that serves as a sync mark, while the board itself is photographed with a code number written on it to designate the "take." A *take* is the serious recording or filming of a program, usually after preliminary setup and rehearsal.

Microphones

Most sounds that the professional deals with are ultimately picked up by microphones and through the years, various types have evolved. For professional studio work *condenser microphones* are a preferred type, widely used because their free diaphragm and small size gives high fidelity and because they are reasonably durable. Modern condenser microphones usually contain a built-in FET amplifier to give a relatively strong output signal into a balanced shielded cable. The nominal output impedance of such microphones is 150 ohms, so that the loading effects of microphone cables are negligible even when the cables are hundreds of feet long. Condenser microphones require a polarizing voltage, and their built-in amplifier requires DC power, which is conveniently supplied by maintaining the balanced audio lines at a DC potential with respect to the shield (simplexing or phantoming). To isolate the DC from the audio signals, center-tapped transformers or balanced RC circuits are used. The output of studio-type condenser microphones ranges from about −65 to −55 dB relative to 1 volt/microbar sound pressure, giving roughly a millivolt of output. A high-output microphone reduces noise problems in subsequent amplifier stages, but too high an output may overload a sensitive input stage, which ordinarily is not protected by a gain control.

Where a more rugged microphone is required, or where a microphone power supply is not provided, a *dynamic (moving coil) microphone* is chosen. The direct output of a dynamic microphone, unamplified, is about −75 dB, with a rated impedance of about 150 ohms.

Every kind of microphone imparts its own personality (or lack thereof) to the sound. Experts pride themselves on being able to recognize microphones by listening alone and, of course, each expert has his or her own preferences.

Microphone Pickup

If mixing the results of many channels to produce a pleasing result is an art, then the choice and placement of microphones is an art within an art.

Basic Microphone Acoustics. Although reproduced sound was monaural prior to the 1960s, the sound did not always seem to come from the loudspeaker itself. In good broadcasts or recordings, there was the illusion of breadth and depth to the sound sources many times larger than the loudspeaker and far behind it. The secret of "big sound" was known to expert engineers who could impart the pleasant illusion of sound perspective and room ambience by varying the proportion and kind of reverberation mixed with the direct sound perceived by the microphones. A master of the art could do this magic with great simplicity, using only one microphone but knowing where to place it and how to point it. Other engineers preferred multiple microphones, blending the close pickups and the far ones with gain controls or faders until they obtained the desired mixture. Multiple microphones are always more impressive to see, if not to hear.

The basis for the illusion of sound perspective is explainable by Figure 9-4. When a source on a concert stage generates a transient of sound, the listener first hears the direct wave (*1*) reaching him by the straight path, which is the shortest possible distance to the source. A few milliseconds later the listener hears a series of early reflections (*2a, 2b, 2c, 2d,* etc.) from the room walls,and somewhat later, the higher-order multiple reflections (*3a, 3b,* etc.). Human hearing does not perceive reflections as such unless the room is extraordinarily large; instead, the sound with added reflections is perceived as richer and fuller although less distinct. Also the reverberations make us conscious of room am-

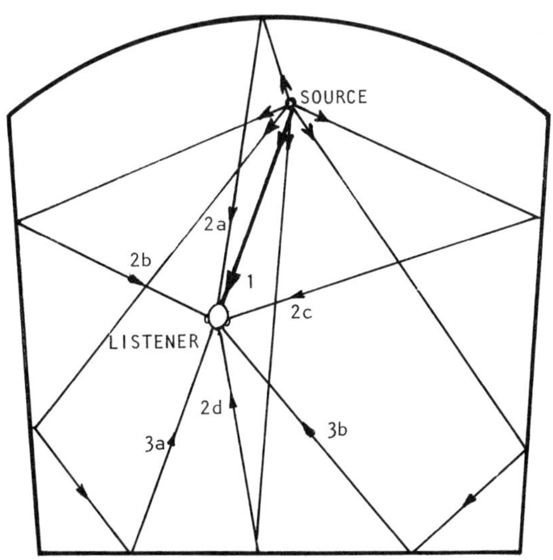

Figure 9-4. The basis for illusion of sound perspective.

bience telling us that we are listening in a small or large or lively or not-so-lively room.

A unidirectional mike close to the source picks up a high ratio of direct sound to reverberant sound. It gives a clear, concentrated, dry sound close to the listener, perhaps too close, too precise, and too intimate for comfortable listening. The other extreme is an omnidirectional mike at the rear of an auditorium, where the reverberant sound predominates and the direct sound is almost nil. Now the reproduced sound is far away, spread out, and somewhat smeared, giving a pleasant, unobtrusive effect for relaxing listening.

The most natural sound is between these extremes, depending also on the kind of music. A pipe organ sounds best in a large cathedral, whereas a pop singer is more at home in a small café. All these factors are instinctively weighed by the recording engineer and director, who after years of experience can glance at the situation, set the microphones very close to optimum, then zero in by shifting them a few inches after listening tests.

Stereo Pickup. Stereo pickup, which commonly uses two channels, spreads the sound image laterally from left to right, but depth perception fore and aft still depends on the reverberation ratio as described above. Stereo sound pickup should therefore be an extension of monaural techniques, with one microphone (or family of microphones) assigned to the left space and another to the right space. Indeed the early stereo recordings of the 1950s were made in this manner. In the 1960s, experimenters explored the almost endless combinations of microphone spread, directional patterns, and nearness to the performers and found that almost all gave unimpressive results. Common complaints were "hardly any stereo effect," "hole in the middle" with everything localized at the side loudspeakers, "indefinite localization" or even the jumping of some instruments from one side to the other, and "everything coming only from the nearest loudspeaker."

Studies showed that stereo can give only a crude approach to natural biaural hearing since stereo depends mainly on relative loudness of two spaced loudspeakers. Natural hearing takes into account not only loudness but also relative phase (arrival time at each ear), effects of moving one's head, differences in spectral content in the farther ear due to diffraction around the head, subtle resonances caused by the shape of ones external ears, and even the verification by seeing what one is listening to. There is also evidence that some listeners localize sound by different clues than others do. For example, some people who listen to reproduced binaural sound say that the sound always seems to come from in back of them, while others have no trouble visualizing it in front under identical conditions.

Phasing. When two spaced microphones pick up the same sounds they create phase problems when they are mixed together, as indicated in Figure 9-5*A*.

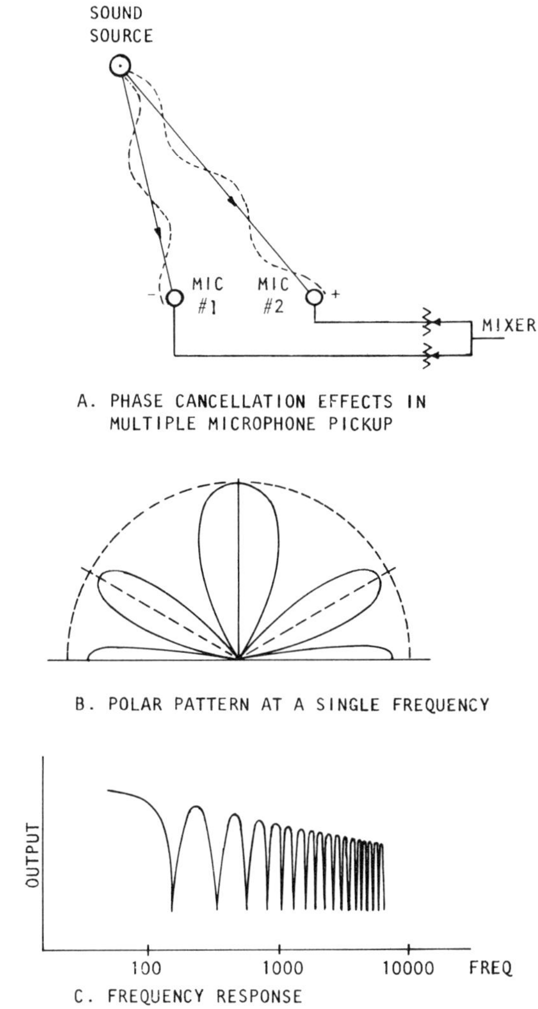

Figure 9-5. Interference effects in multiple microphone pickup.

Cancellation or reinforcement of the sound at different angles and different frequencies gives a polar pattern as in Figure 9-5*B* and an irregular response curve of Figure 9-5*C*. More that two microphones make the response even more irregular. The cancellations impart a hollow quality to the reproduction and this detracts from its naturalness, although the effect is masked by room reverberations. In Figure 9-5*C* the microphones are "in phase" at lowest frequencies. If the polarity (phasing) of one microphone is reversed, the low frequencies will cancel, giving almost no low-end response.

Instead of being mixed together the microphone outputs may go to separate stereo channels, which are later reproduced through spaced loudspeakers. The playback loudspeakers are usually in a different room than the microphones and are spaced apart at different distances. The listener is usually not on a center line between the loudspeakers. No wonder the acoustics are confused, and the "simple" stereo pickup gives a very complex product.

Stereo Techniques

Since the advent of stereo, thousands of practical experiments have been made in search of the best mode of pickup. Two distinct methods of recording have evolved: classical music recording and contemporary multitrack recording. Hybrids of these extreme methods are common.

Classical Music Recording. Recording of medium and large orchestras requires preserving the wide spread of the performers and eliminating phase interference that confuses the localization. Wide spread is enhanced by minimizing common pickup in the two channels, since common pickup tends to narrow everything toward the center. Common pickup is reduced by choosing directional microphones and pointing them to different parts of the orchestra without much overlap. The phase problem is solved by bringing both stereo microphones together, in the same housing if possible, but with their directional patterns differing. The close microphone concept shown in Figure 9-6 was used in

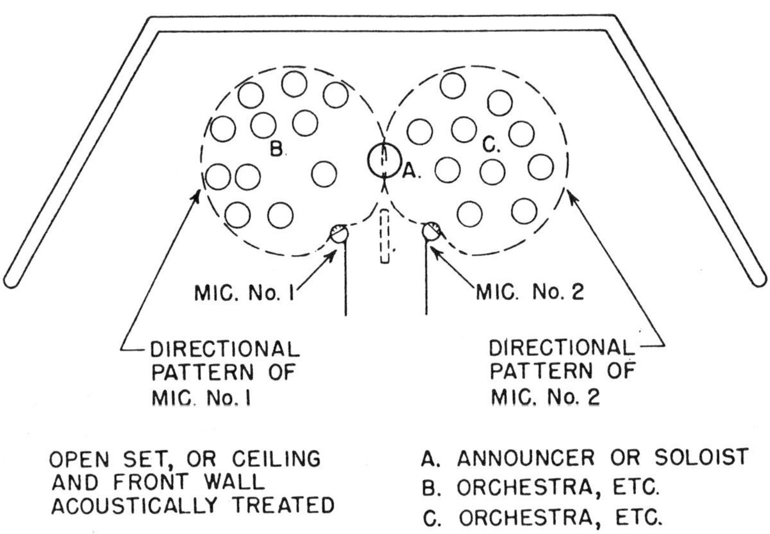

Figure 9-6. Closed spaced directional microphones for stereo pickup (Camras 1949).

some of the first magnetic stereo recordings by Camras (1949) and was suggested earlier by Blumlein, after whom such microphones are named. The idea has been expanded to quadraphonic (four-channel) pickup by B. Bauer, who called his version the Ghent microphone.

Many notable stereo recordings have been made with Blumlein-type microphones. This ''pure'' method is often supplemented by additional microphones for soloists and for enhancing weak areas or weak instruments such as the harp.

Contemporary Multitrack Recording. The artifices used during contemporary recording sessions are so foreign to the notion of a well-behaved orchestra or even of a rock group that the whole procedure is repugnant to the uninitiated observer. Musicians are isolated in soundproof cells and wear earphones to hear each other so they can play together. Microphones are literally stuffed inside some instruments, or vibration pickups sense the movement of bridges on string instruments instead of the sound itself. Soloists are absent. They will be recorded later by themselves. If a key musician is absent, that's okay too; the recording session goes right on and the absentee will be added in a few weeks at his or her convenience. The complete musical event does not exist at recording sessions—only a collection of parts, each on a separate magnetic track. The performance is created during mixdown, long after the performers have gone home.

A good precedent for such procedure is movie making, where scenes are filmed out of order. Special effects, trick photography, and music are added during assembly of the parts, so that the final version often surprises even the principal actors.

MIXDOWN

The mixdown in sound recording is like the cutting room in motion pictures. Here the straight, dry, separate parts are assembled and balanced, with reverberation and special effects added. Trial runs are made again and again while the musical chefs add a bit of sweetening here, a bit of spice there, until they lick their fingers and say, ''That's it.''

Mixdown Procedures. A console with special modules is used for mixdown. The sound from each separate playback tape is monitored and modified with equalizers, reverberation, and so on, then mixed with output from other tracks. Where there are many channels, the individual tracks are combined to give a lesser number; for example, a 32 × 8 mix condenses 32 tracks into 8. In the next mix the 8 tracks are combined into 2 tracks corresponding to a stereo master. Earphones or high-quality control-room loudspeakers help the balancer arrive at a desired result.

In classical music little or no mixing need be done if the microphone placement was optimum. Natural reverberation may be increased somewhat by adding from a far-away microphone, and soloists may be brought forward by mixing in a solo mike.

Mixdown Consoles. Advanced mixing consoles carry many refinements for convenient audio handling, claiming ergonomic (human engineering) design to place the operating controls where they have the right feel and where they give natural expected changes, with controls and indicators that tell what is happening at a glance. Arrays of linear faders show their position much better than rotary pots, and vertical luminescent bars for volume indicators are easier to read than meters.

Some consoles even have faders with programmed servos that repeat automatically the changes made in previous mixdown. The operator can override any automatic control simply by moving the fader where it is wanted, and the new position will then be recorded for the nest trial-balance. The controls for each program are digitally recorded on a floppy disc or a tape cassette, which is stored with the master audio tape for future reference.

Editing

Editing is the changing of a recorded program by deleting, adding, or altering parts of it. Ideal editing is imperceptible, usually being made during pauses in the program and without abrupt differences in volume of the sound or the background ambience or tempo of the music, and so on. The general concepts of editing described here apply also to television and motion picture editing.

Cutting and Splicing. The simplest and quickest editing method is to cut the tape physically and to splice together the desired parts. Good technique for analog recordings is a diagonal splice held together by special splicing tape of Mylar® with a very thin pressure-sensitive layer that does not ooze out. Multitrack digital recording may require a perpendicular splice. Splicing fixtures align and trim the tape in a professional manner.

Although cutting and splicing (razor blade editing) is considered primitive, it requires only one tape recorder and no mixing facilities. It is very fast, since the tape can be shuttled to the edit points at high speed. The sound quality of the original tape is retained. Disadvantages are that splices can fail in several ways; the character of the recorded sounds cannot be changed to give an ideal transition at the edit points; there cannot be much experimentation with different editing possibilities ("previews" are not practical); and one ends up with bits and pieces of tape that should not be patched and reused.

Rerecording. The most sophisticated editing method is to rerecord the desired parts on to a new tape, using a built-in computer to change over the programs at the edit points. A great advantage is that the edits can be rehearsed before rerecording and adjustments made in timing, sound quality, loudness, fade-in-fade-out, and so on. If the results are unsatisfactory, nothing is lost except time. The masters are unharmed and the copy tape can be erased for a retry.

Rerecording requires two or more recorders plus additional elaborate control equipment, a sizable investment at the mastering quality level. Analog copies always have more noise and distortion than the original does, but digital recording now has remedied this shortcoming. Editing by copying is time consuming, taking at least as long as the entire program plus the time for setup and rehearsal.

Computerized Editing. Advanced editing equipment uses an index code that is recorded along with the program, as for example the SMPTE time code on a parallel track (see Chapter 11). The editor shuttles the tape on each playback machine, chooses the edit point, notes the corresponding time code display, then keys the microprocessor computer to change over when, for example, tape No. 1 reads 1:21:32 time code, while tape No. 2 reads 11:02:05 time code. Both playback machines are started and must now be synchronized so the above numbers will occur at the same time, this being done by the computer, which regulates the tape speeds during this *preroll* period until their time codes have the correct spacing, then locks the machines in sync and changes over the programs at the edit point. The edit can then be rehearsed, with alterations made in timing and sound quality. Additional machines can be phased in and out. When the computer program is finalized, the complete edit is made under its control.

CONSOLE ELECTRONICS

Except for the sensitive microphone amplifiers, all signals and their processing in consoles are done at levels of about 1 to 2 volts (4 to 8 dBm), where most electronic circuits behave well, giving excellent linearity and low noise.

Preamplifiers. The best signal-to-noise ratio is obtained when the input impedance of preamplifiers is considerably higher than the 160 ohm nominal output impedance of the microphone. A good compromise is abut 1000 to 2000 ohms input impedance across the balanced line from the microphone. A transformer may be used at the preamplifier input to isolate the microphone lines, avoid ground loops, and give a high degree of common-mode rejection. Also, with the transformer's primary insulated from the rest of the system, the balanced micro-phone pair can carry DC relative to the grounded shield for a condenser microphone, which needs a polarizing voltage for its electrodes and power for its built-in amplifier. It is not uncommon for the microphone also to have a built-in transformer that matches its internal impedance to the microphone cable. These miniature low-level transformers add almost negligible distortion and noise, but their cost is not negligible and they must be shielded carefully to prevent hum pickup.

Alternatively, the microphone lines may go directly to transistor input circuits of push-pull or differential amplifier stages. The virtues of transformerless consoles are amply extolled by their manufacturers, but most of them offer optional modules with transformers for customers who prefer the tried and true.

The preamplifiers bring the audio signal to a working level where it can be switched, mixed, equalized, and so on. Zero-level (0 dBm) has been selected to be 1 milliwatt into a 600 ohm resistance, which corresponds to 0.775 volts across the 600 ohms. In practice the 600 ohm *standard impedance* is seldom observed in studio work, but voltage levels are monitored with a voltmeter having a dBm scale. A zero dBm reading, therefore, corresponds to 0.775 volts, and otherwise the dBm level is

$$\text{dBm} = 20 \log \frac{\text{voltage}}{0.775} .$$

More correctly the terms *dBv* or *dBu* are used for the above definition when the line impedance differs from 600 ohms, which is almost always true in modern equipment. The recordist is seldom disturbed by this distinction. The preferred working level (or line-level) in most recording studios is 4 dBm (1.29 volts). Some broadcasting studios work at a level of 8 dBm (1.95 volts). Disc cutters, broadcast transmitters and equipment that utilizes the audio (for example tape recorders) are designed with inputs that accept these working levels for normal operation.

Equalizers. Equalizers modify the frequency response of the audio, emphasizing or reducing the high, low, or mid-range tones. The boomy sound of a bad room may be tamed by an equalizer with a valley at the room resonance peak. A broad, easy lifting of audio response at 1500 to 2500 cycles improves the "presence" of voices or music. The sound can be brightened by boosting the highs or mellowed by reducing them. Weak bass instruments are brought up to par by bass boost.

The most versatile equalizers are so-called *parametric equalizers* with characteristics shown in Figure 9-7, where a peak or valley up to about 12 dB each way is inserted anywhere in the response spectrum, and where the peak can be moved frequency-wise without changing its proportional bandwidth. Alternatively, a shelving, rather than peaking, equalizer may be selected; this type of equalizer lifts or lowers one end of the response curve in a flat shelf above or below the rest, after a selected transition frequency region.

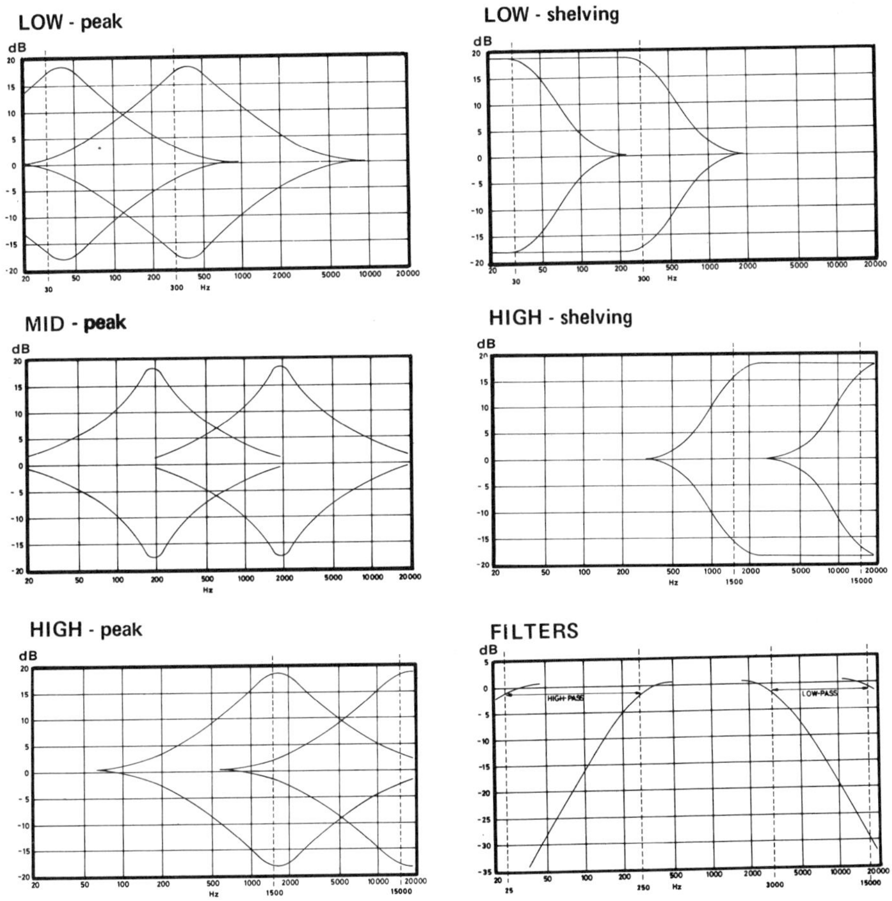

A. PEAKING EQUALIZERS. A peak or valley may be moved to any part of the spectrum, with the peak height and sharpness adjustable.

B. SHELVING EQUALIZERS. A desimated rise or fall of selected height modifies the response.

C. FILTERS. The response cuts off above or below a desired frequency.

Figure 9-7. Response of parametric equalizers. (Courtesy Rupert Neve, Inc)

Reverberation. Close microphone technique of multi-channel pickup is invariably too dry and will be improved by reverberation. Some console channels therefore "send" the signal to an "echo" reverberation unit and pick up a "return," which they can equalize, blend with the original, and mix into the right or left stereo channel as needed.

Noise Reduction. The dynamic range of the human ear is about 120 dB, and music dynamics in a quiet room can exceed 100 dB, far above direct capability of the best analog electronic reproducing systems, which are in the order of 65 dB. A compander (compressor-expander) is designed to increase the overall range by compressing the natural dynamic range before recording and expanding the range again after reproduction. In effect, when the enhanced range is heard at reasonable or even loud levels, the background noise is reduced greatly, becoming practically inaudible.

A well-known noise reduction system (compander) is the Dolby A system, which filters the audio into four spectral bands, as shown in Figure 9-8, compresses each band an appropriate amount and with an appropriate time constant, and then combines the bands again for recording. The playback is again divided into spectral bands, each band decompressed properly, and finally all are recombined to give a playback with noise reduction of about 15 dB. This procedure seems complicated, but the band-splitting technique is necessary: Otherwise the workings of the compander are accompanied by breathing and intermittent hissing, which are more annoying than the noise it eliminates.

Synchronization. Two or more recorders, or recorders and film projectors, can be locked to run together in synchronism by recording a control signal or time code on a track of each recorder. Modern recorders have so many tracks that one or two can be spared for sync signals, especially since the additional recorder tied into the system adds many more tracks. Figure 9-9 shows an arrangement for recorder synchronization using a digital time code. (See Chapter 11 for SMPTE Time Code). Other sync schemes use pilot signals, which may be superimposed on the main audio tracks if the pilot frequency is above or

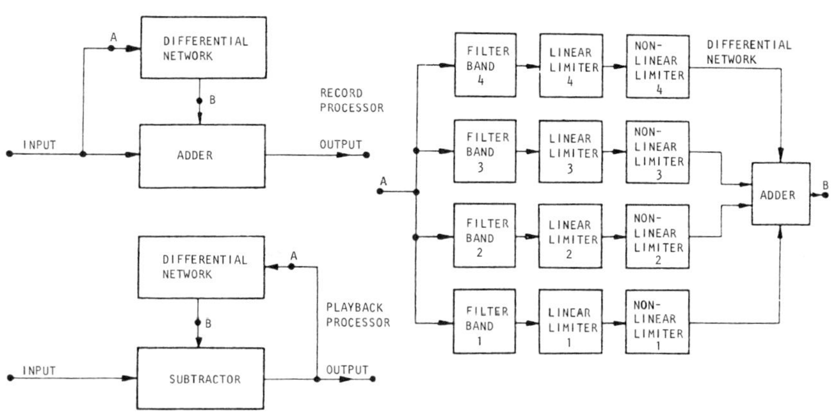

Figure 9-8. Elements of a four band noise reduction system. (Courtesy Dolby Corp.)

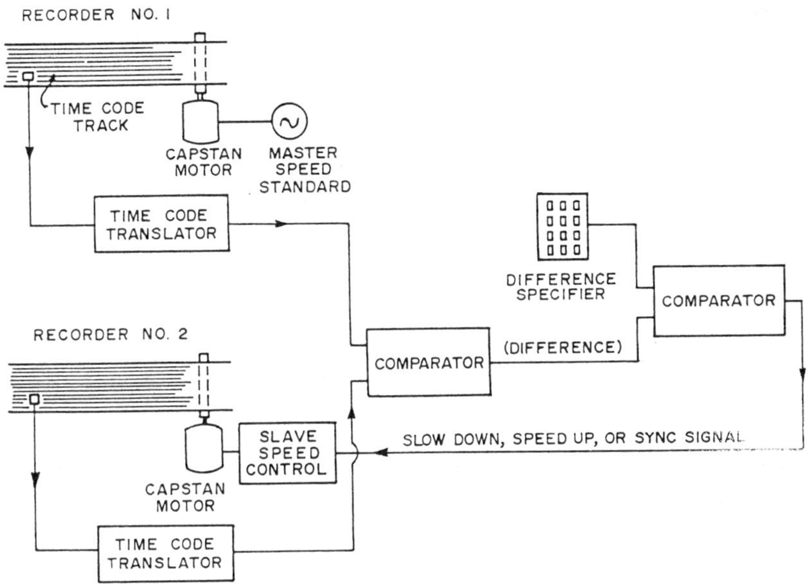

Figure 9-9. Synchronized recorders using time-code tracks.

below the useful audio range. One of the earliest systems recorded two low-level opposite-phased sync signals of 60 Hz line frequency superimposed on the audio track; the opposite phasing cancelled out in the audio head, but added in the sync pickup heads.

In motion picture work, the sound is usually recorded separately from the picture. Special arrangements are made to start the sound and picture together (slating) and to keep them in locked sync throughout the production.

PROFESSIONAL TAPE TRANSPORTS

Tape drives as well as the electronics of professional recorders have been refined to meet more and more stringent specifications, with keen competition among manufacturers to please buyers of new equipment.

Open Loop Drives. The most proven tape drive for professional recorders is the open loop system shown in Figure 9-10, with the capstan on one side of the heads and a tape-driven flywheel on the other side. There is hardly a studio that does not have such a deck, which fits 267 mm ($10\frac{1}{2}$ in) NAB reels with 6.3 mm ($\frac{1}{4}$ in.) tape. The multitrack version shown in Figure 9-11 is recognizable although it appears in many forms and under many brand names.

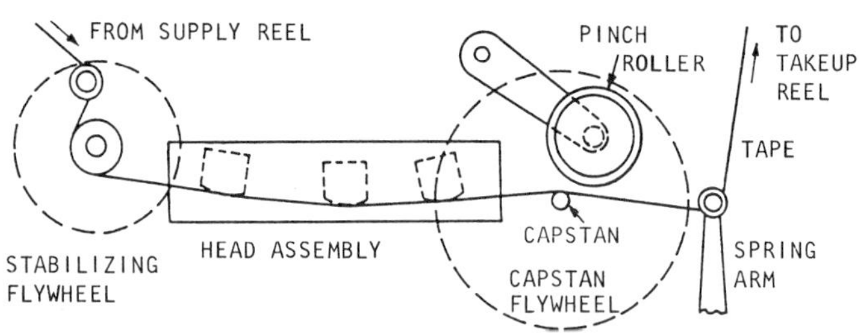

Figure 9-10. Open loop professional tape drive. This deck has been a familiar sight in recording studios for more than three decades. Quarter inch tape on ten inch reels is very common. At the left side of the deck is a tape-driven flywheel; on the right side are the capstan, pinch roller, and spring arm. These isolate the tape from irregularities in the supply and takeup reels. (Courtesy Ampex Corp.)

Figure 9-11. Multichannel two inch tape deck with open loop drive. (Courtesy MCI Corp.)

Closed Loop Drives. The closed loop tape drive (Figure 9-12) is found on some professional decks that use wide tapes for multitrack formats. It has a better flutter rating than open loop transports, but it is more expensive and somewhat more critical to adjust and to maintain.

Servo-Controlled Drives. Servo control logic elements and microprocessors are included in the latest generations of tape drives to regulate tape tension, minimize flutter, maintain long-term speed, synchronize the tape with other equipment such as cameras, and give convenient indexing and editing. As described in Chapter 7, servo-controlled capstans maintain precise speed without the need for heavy flywheels.

One application of servo control is in the tape transport shown in Figure 9-13, which guides the tape along a path where its magnetic coating touches only the heads, thus achieving a very long tape life free from wear, dropouts, and contamination. The drive has an unusually large-diameter capstan around

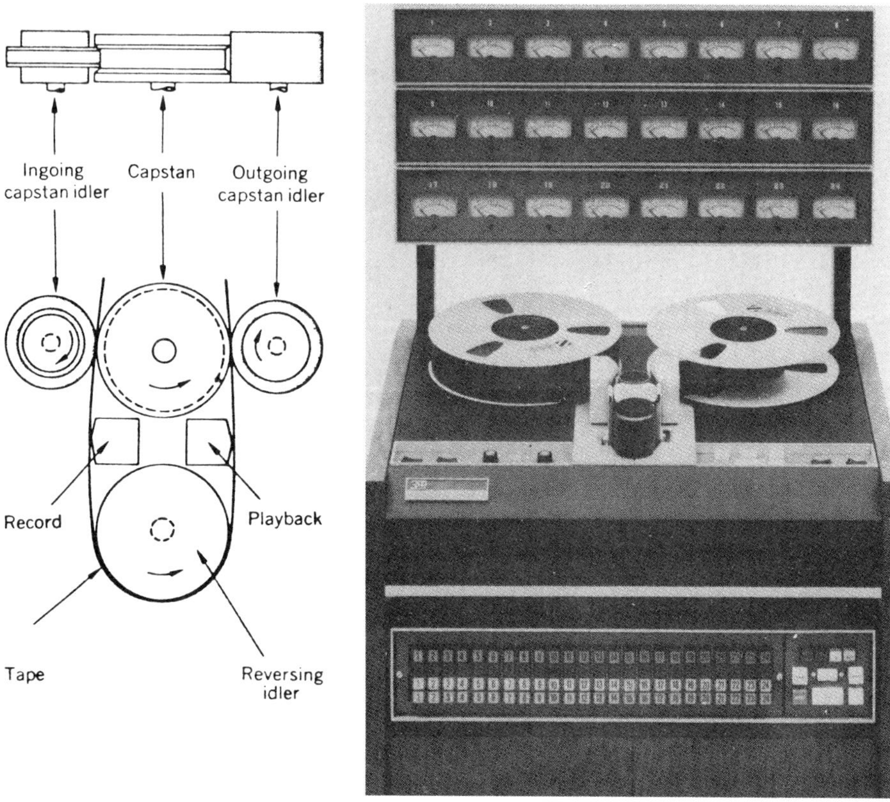

Figure 9-12. Closed loop tape transport. A. Pressure rollers against both sides of capstan isolate the heads from the reels. B. Twenty four channel recorder on two-inch tape using closed loop drive. (Courtesy 3M Corp.)

which the tape wraps. No pressure rollers are used, slippage at the capstan being avoided by keeping the tape tension practically equal at the input and output sides of the capstan. Thus the capstan supplies almost no driving action, in contrast with the usual situation where the capstan pulls the tape off the supply reel and across the heads. Correct tension is maintained by logic-controlled servos as the reels go from full to empty. Programmed control also ensures smooth acceleration during starting, stopping, high-speed winding, and shuttling for editing purposes.

Portable Transports. When it is possible to bring remote signals to the master recorder studio, they can be recorded under ideal conditions. Sometimes the master recorders are installed in vans, which are driven to remote locations.

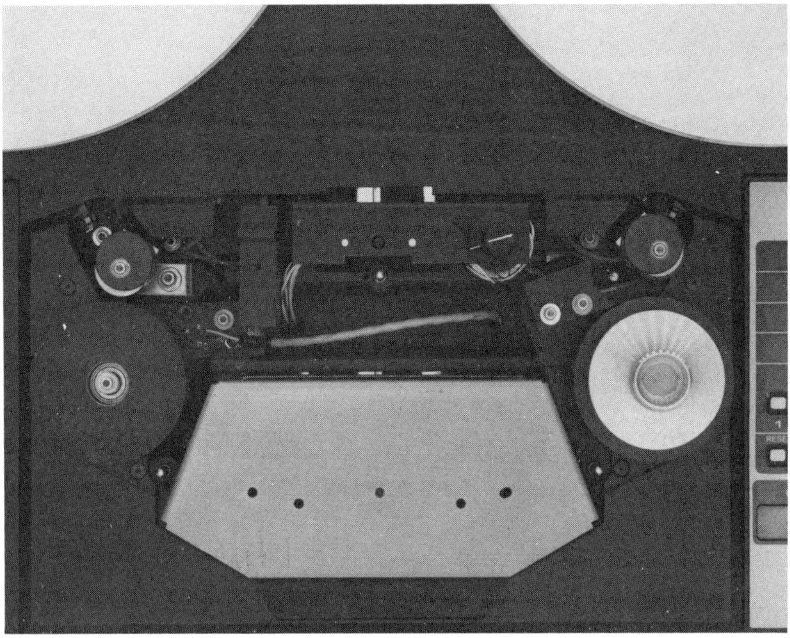

Figure 9-13. Servo controlled tape transport. ATR-100 Audio Recorder. (Courtesy Ampex Corp.)

Where AC power is available, lightweight semiprofessional equipment may be set up near the recording point.

In extreme cases, the recorder must be self-contained and self-powered. A popular tape recorder for such conditions is shown in Figure 9-14. For voice recording, as in news broadcasts, a good portable cassette deck may give acceptable quality.

RECORD-PLAY ELECTRONICS AND HEADS

In addition to the usual electronics associated with audio systems, tape recorders have special requirements for erasing and high-frequency biasing. Tape players need an integrating network or equivalent compensation for inductive heads.

Recording Electronics. Essential electronics for the recording mode are shown in the block diagram of Figure 9-15*A*. Operation and maintenance are intended for professionals; hence many controls are included for optimum adjustments that are omitted on ordinary machines because they are too expensive and too confusing for a casual user.

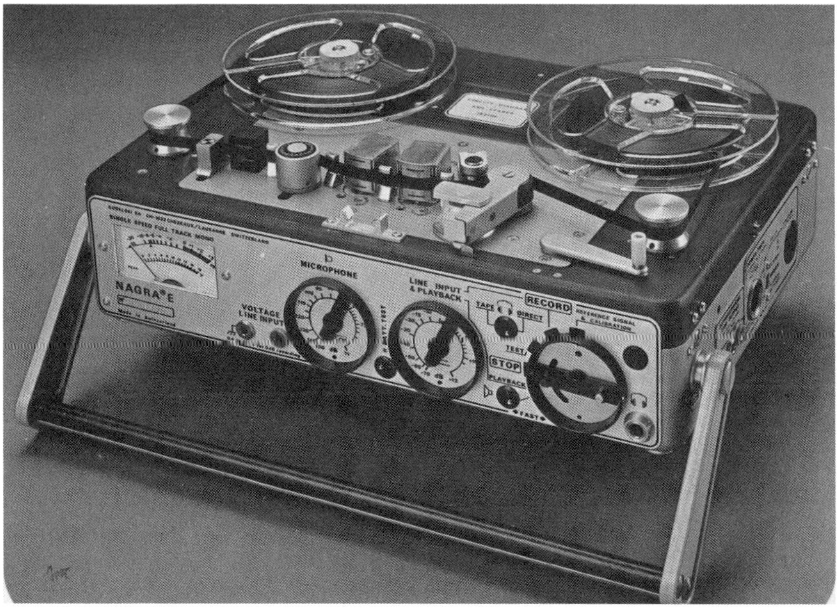

Figure 9-14. High quality battery operated portable recorder. (Courtesy Naagra Corp.)

Amplifiers. The preamplifier has sufficient gain to amplify normal studio microphones to operating levels of 1 to 2 volts. Since such microphones have an output of roughly 0.0001 to 0.001 volt at normal conversational sound levels, the preamplifier supplies a gain of about 80 dB, with a few extra dB for a margin of safety. The best noise figures and lowest hum levels require careful engineering of the first stages. The *power amplifier* for the head driving stage may have 10 or 20 dB additional gain, depending on the voltage required by the recording head and its equalizer network. Recording level is monitored by a *VU meter* calibrated in dB to indicate normal and overload levels. VU meters have fast responding movements and electronic damping networks that register peak levels quickly without excessive jitter of their pointer. Even faster are *peak indicator lamps* of the LED variety and luminescent bar indicators, which can be clustered in a small area for quick spotting of problem channels since close watching of 24 or more VU meters is a lively task.

Bias Supply. The bias supply of professional recorders is usually adjustable so bias can be optimized for a wide variety of tapes, or can be trimmed to compensate for head wear. Also, the bias can be compromised according to individual needs of lowest distortion, lowest noise, or best-high frequency response. The bias frequency is usually 100 kHz to 200 kHz or even higher (400

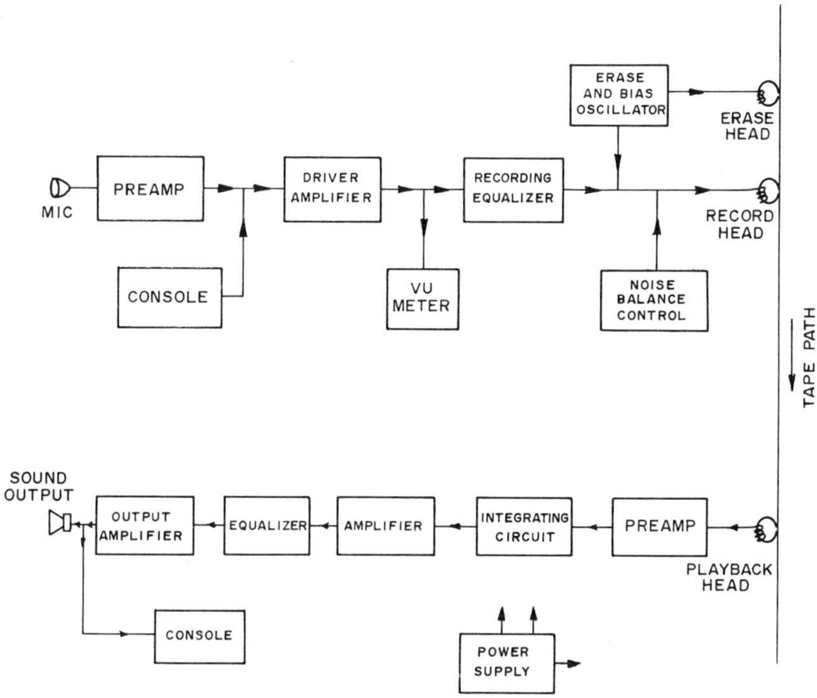

Figure 9-15. Recording and playback electronics for each channel of a professional tape machine. The power supply and bias oscillator are common to all channels.

kHz) if the tape speed is 15 or 30 in/sec (381 and 762 mm/sec). For minimum noise, the bias frequency should be high and the bias waveform should be symmetrical. A symmetrical waveform is obtained by choosing a push-pull oscillator or push-pull bias amplifier. Even under best conditions, there is apt to be a slight dissymmetry, which can be compensated by a *noise balance control.*

Noise Balance Control. The noise balance control introduces a very small DC current through the recording head to compensate for stray fields in or near it due to residual magnetization of the head core, magnetized objects near the head, the earth's magnetic field, or bias dissymmetry.

Playback Electronics. All but the simplest professional recorders have separate heads for recording and for playback. Playback heads give about the same output voltage as microphones (0.0001 to 0.001 volt), and therefore the playback amplifier can be the same or very similar to the recording amplifier. Essential playback electronics are shown in Figure 9-15*B*.

Playback Equalizers. The distinctive feature of playback amplifiers for the usual inductive heads is that they include an *integrating circuit* with a steeply falling frequency response that compensates for the 6 dB per octave $d\phi/dt$ rise of playback head voltage. Enough additional equalization is then added to give flat response with a standard reference tape.

VU Indicators. Professional recorders have done their duty when they deliver an audio voltage corresponding to +4 dBm or +8 dBm (1.29 or 1.95 volts) to equipment that uses these standard studio levels for manifold purposes such as mixing, monitoring, broadcasting, and rerecording. The same VU meters used for setting the recording level are generally switched over for watching playback level, their 0 dB reading now corresponding to standard studio level of 4 dBm or 8 dBm.

Power Supplies. Good engineering calls for a power supply located some distance from the sensitive input circuits and heads, usually at the bottom of the cabinet, with connecting cables distributing the power to each chassis.

Metering. In addition to the omnipresent VU indicator there are usually some panel-type meters that read key voltages and currents to warn of incipient trouble or to help pinpoint any that develops. Bias current is an important quantity that should be readable on a separate meter, or by a selector switch to a common meter.

Tape Heads. Figure 9-16 is a tape head from a professional recorder showing its rugged construction, triple shielding, and larger-than-average size. For

Figure 9-16. Tape heads of a professional recorder.

Table 9-1. Specifications of a Master Tape Recorder. AMPEX ATR-100

Frequency Response	at 95 mm/s (3.75 in./s) = ± **2 dB30–10000 Hz**
	at 190 mm/s (7.5 in./s) = ± **2 dB 30–15000 Hz**
	at 381 mm/s (15 in./s) = ± **2 dB 20–20000 Hz**
	at 762 mm/s (30 in./s) = ± **2 dB 35-28000 Hz**

Signal Noise Ratio*

Tape Speed and Equalization	Track Format	30Hz–18kHz	ANSI "A" Weighted	CCIR Rec 468 Weighted
30 in/s AES	full trk.	77dB	81dB	73dB
	2T & 4T	72dB	76dB	67dB
15 in/s	full trk.	74dB	78dB	70dB
IEC/CCIR	2T & 4T	70dB	74dB	65dB
15 in/s NAB	full trk.	73dB	77dB	69dB
	2T & 4T	69dB	73dB	63dB
7.5 in/s NAB	full trk.	75dB	78dB	70dB
	2T &4T	71dB	74dB	63dB
7.5 in/s	full trk.	71dB	76dB	67dB
IEC/CCIR	2T & 4T	68dB	71dB	62dB
3.75 in/s	full trk.	68dB	72dB	64dB
IEC/NAB	2T & 4T	64dB	66dB	57dB

Flutter and Wow

Speed	ANSI S4.3/ DIN 45507 Peak Weighted	ANSI/DIN Peak Unweighted	NAB RMS Unweighted
3.75 in/s	±0.1%	—	—
7.5 in/s	±0.05%	±0.12%	±0.06%
15 in/s	±0.03%	±0.08%	±0.04%
30 in/s	±0.03%	±0.08%	±0.03%

Distortion	at 190 to 762 mm/s, 370 nWb/m (0 VU) =**0.3%**
	at 190 to 762 mm/s, 1040 nWb/m (+9 VU) = **3%**
Erase depth	from 1040 nWb/m = more than **85dB**
Channel separation	at 15 in/s 100-15000 Hz = more than **45 dB**
Timing accuracy	= ± **0.05%** (1 sec in 2400 ft reel at 15 in/s
Start and stop time	= **0.5** s or less (approx)

Erase frequency = **144 kHz,** Bias frequency = **432 kHz**

Equalization: Choice of **AES, NAB, IEC, CCIR,** etc.

Size, weight, power: **483 × 400 × 305 mm; 56 kg; 600 VA** for basic machine

*(referred to 1040 nWb/m record level, 3% distortion except for 3.75 in/s where record level was 740 nWb/m)

long life, the gaps of professional heads are deeper than those of ordinary heads. On some machines the heads are mounted on a removable block with all parts precision aligned so that the entire assembly can be replaced without adjustment.

Specifications. The state-of-the-art possibilities for professional recorders are exemplified by their specifications. Table 9-1 gives the performance specifications for a typical high-quality machine.

MAGNETIC SOUND FOR MOTION PICTURES

There are enough differences in sound technique for the movie industry that it deserves a section of its own, although it has much in common with other studio methods. In motion pictures a scene seldom continues for long, but changes and merges and fades to other events, places, and times. The sounds must be juggled appropriately, with additions of music and effects. It is customary to record the sound on a recorder separate from the camera, with reference marks on both film and sound record so that they can later be combined in perfect synchronism.

Although sprocketed magnetic sound film has been widely used for such recordings, master sound recorders using ordinary tape are very attractive for motion picture studio work because they have superior sound quality and the tape costs less than sprocketed film. Synchronization with the picture is more difficult than with sprocketed film, and generally involves a recorded control signal that locks the sound tape drive together with the picture film drive. The control signal can be recorded on a separate track, or it can be superimposed on the sound track by transverse recording or by a supersonic pilot signal.

Sprocketed Film Transports. Sound recording systems for motion picture work have traditionally used sprocket hole films, first with photographic recording and later with magnetic sound. Sprocket holes are the most straightforward device for synchronizing the picture camera with a separate sound recorder; and during editing it is easy to keep several parallel sprocketed films together while moving them back and forth. With the great advances in multitrack tape recording and in sync tracks on ordinary tapes, it was inevitable that these alternative systems should make inroads in motion picture work. However, sprocketed film recorders have enjoyed wide use in sound on film production and are still an important factor.

Figure 9-17 shows two popular methods for driving and stabilizing sprocket hole sound film. The sprocket teeth introduce a film motion chatter that must be filtered out by a flywheel or servo system. The unusual feature of film transports is that the flywheel is driven by the film and is called a *stabilizer*. This

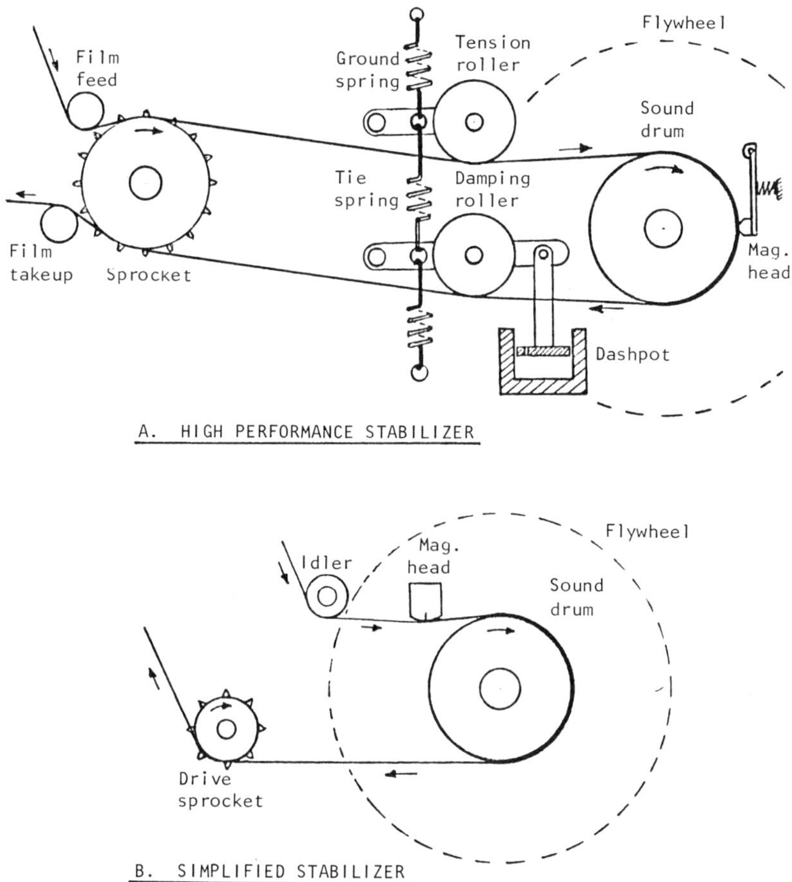

Figure 9-17. Drives and stabilizers for sprocket hole sound film.

contrasts with conventional transports where the flywheel capstan drives the tape.

Electronics. Electronic systems of sound film recorders are quite similar to those of tape recorders, with consoles and mixing facilities performing the same functions as described previously in this chapter.

Recording Techniques. Microphones for dialogue and other sounds in a film scene must not be in view of the cameras. In some cases the microphone is suspended above the actors by a long pole called a *boom*, which follows the action as it moves about the stage. Where possible, the microphone may be hidden in a nearby stage prop. Extremely directional microphones (gun microphones or parabolic reflectors) are necessary when the pickup must be from a distance, such microphones being used as a last resort since their fidelity is generally below par.

Instead of recording the sound while the picture is taken, it is often better to record the sound beforehand (*prescoring*) and to have the actor follow along, listening to a playback of the sound while he or she is concentrating on the acting. The final result is a singer or player who appears to do the most difficult musical passages with great ease and composure. In fact, the voice or music performance does not even have to be his own. Acting to a prescored sound is surprisingly easy. Amateurs will often put on a convincing pretense of singing while a record is playing, then flabbergasting their audience by stopping while the record keeps playing.

The opposite technique, called *postrecording*, is to shoot a silent movie and to record the sound afterwards, repeating the dialogue while watching the picture. Lip-sync passages of this nature are extremely difficult to do in a convincing manner and usually have to be repeated a dozen times or more. It is therefore helpful to splice the picture film into an endless loop and to run it through over and over again, with the speaker making a new try every time the scene comes around again, until the director is satisfied that the "take" is satisfactory.

All the tricks of multiple-track recording described at the beginning of this chapter for sound tapes are also available and useful in movie making. In fact, the sound recordists are Johnnies-come-lately who have borrowed their techniques from the movie industry instead of vice versa.

Single System. The simplest method for recording synchronized sound is to record a sound track in the camera on an edge of the film while taking the picture (Figure 9-18). This procedure, known as the *single system*, is used for newsreels and amateur movies. For serious movie making, however, the single system has several disadvantages: (1) It is difficult to edit because the sound pickup point is several frames away from the corresponding picture; (2) the original take is the final sound track, as there is no ready adaptability for adding music, special effects, and so on afterwards; and (3) flutter is a problem because camera film motion is intermittent, and it is impractical to filter out the flutter completely in a compact lightweight camera.

In the projector, the sound pickup corresponding to the frame being projected is ahead of that frame on the film, being 19 frames ahead on 35 mm film, 26 frames ahead on 16 mm film, and 18 ahead on super 8 mm film.

Double System. Professional recording of motion picture sound invariably uses the *double system*, where sound is originally recorded on a separate film or tape

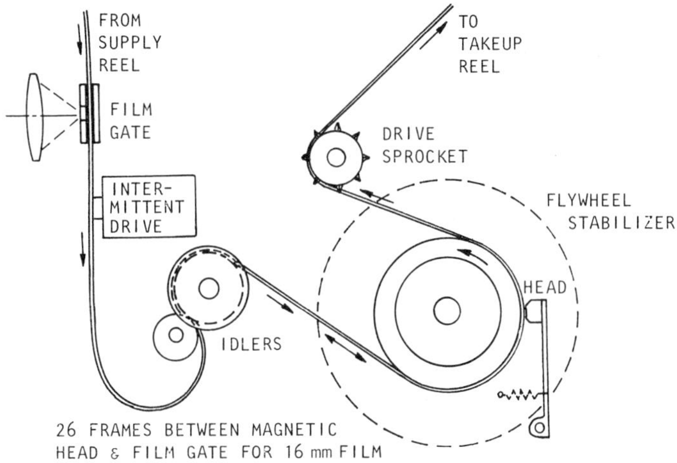

Figure 9-18. Single-system motion picture sound.

in equipment separate from the motion picture camera. The double system overcomes disadvantages of the single system, allowing great freedom of editing both sound and picture and of adding music and modifying the ultimate sound track. The main problem is in the coordination of the sound with the picture.

A sure way to achieve coordination is to use sprocketed, magnetic-coated film for the sound and to lock together the picture film drive sprockets and the sound film sprockets. Mechanical locking by gears or flexible shafts is easy to understand, but the same effect is obtainable electrically by sel-sync (synchro) motors. At the beginning of each take a clapboard ''plate'' creates a strong sound transient that enables the technician to later synchronize the sound and the picture. This familiar technique has been carried over from the days when film sound was recorded photographically. A magnetic sound track does not have visual cues traditionally used by the film editor. However, the magnetic film can be marked with a pen-and-ink recorder actuated by its magnetic track, and thereafter edited in the same familiar ways used for photo sound. Alternative methods such as chemical writing have also been used to make a visible sound envelope on a parallel track for editing purposes.

Editing. A basic editing table combines a working copy of the picture and of the magnetic sound film locked together on sprocket wheels for simultaneous review of picture and sound while reeling back and forth or stopping anywhere as desired. When the picture film is cut, the same amount is also cut from the sound film.

A handy accessory is an erase probe, which is a hand-held erase head that permits the skilled editor to blot out clicks and noises on the magnetic track. With an erase probe the editor can also fade out sound during transitions, bloop the noise at a film splice, and even remove an unwanted syllable from the dialogue. Blooping can also be done physically by adhering a blooping patch over the track to separate it from the head, or by cutting a diamond-shaped hole in the film to remove the track where it is to be silenced.

Many other parallel recordings can, of course, be linked together with the basic picture and sound films to give background music and effects mixed and faded in a console-type operation.

DIGITAL AUDIO

Digital recording brings to audio the possibilities of:

1. Immunity to mechanical flutter regardless of imperfect tape drive;
2. Freedom from recorded background noise to any degree desired;
3. Waveform accuracy as perfect as desired;
4. Perfect copies even after successive generations of copying;
5. A robust record immune from deterioration by aging, magnetic loss, or print-through.

The price paid for these qualities is the need for an enormous increase in recorded density. Digital audio requires about 100 or more times the bit density needed for the same signal recorded in analog fashion. In practice, this means that a video recorder or its equivalent must be used for digital audio. For a long time this video capability was a formidable obstacle, but in recent years efficient high-density recorders have made digital audio practical.

Digital Recording Principles

As shown in Figure 9-19, the audio wave to be recorded is sampled for amplitude at regular clock intervals. Each sample becomes a binary number representing the amplitude at that instant, and the numbers are recorded digitally on a tape. When played back, the binary numbers are reconverted to wave amplitudes. Irregularities in timing (tape flutter) are corrected by a buffer memory, which stores the readout and issues the sample numbers in perfect crystal-clock sequence, reconstituting a signal free of flutter and as accurate as the binary code itself.

It is apparent from Figure 9-19 that accuracy of reproduction requires a high sampling rate and high accuracy of each sample. Theoretically, the sampling rate must be at least twice the highest frequency contained in the audio (Nyquist rate); anything less produces interference called *aliasing*. In practice a rate of

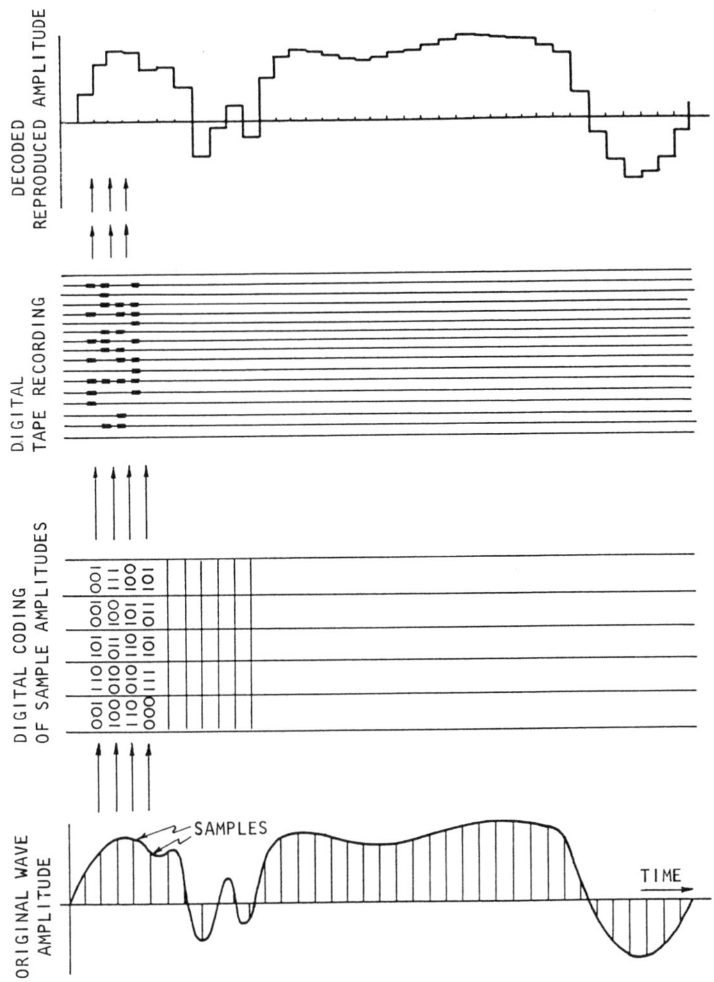

Figure 9-19. Digital recording of audio. Insufficient sampling gives imperfect reproduction, but accuracy can be made as perfect as desired by taking enough precise samples. A parallel format is shown for illustration. In practice the 16 bit words are in series (head to tail) on a single track, or interleaved on a few tracks.

about 48,000 samples per second is optimum for audio to 20 kHz. Sampling accuracy depends on signal-to-noise specifications. For example, if the accuracy is 1 part in 1,000 the sampling noise (uncertainty 1/1,000) is 60 dB below maximum signal. This is good but no better than ordinary analog recording, which we are trying to beat, so an accuracy of 1/2,000 (66 dB) to 1/64,000 (96 dB) should be chosen to take advantage of the superior capability of digital audio. A binary number of 11 bits (2^{11}) is needed for 66 dB signal-to-noise (SNR) ratio, while 16 bits (2^{16}) are required for 96 dB, usually with an extra bit or two for error-correcting codes. As a rough approximation,

$$\text{SNR, dB} = 6 \times \text{No. of bits per sample.}$$

Practical digital audio demands ultrarapid processing by its electronic circuits and extreme recording densities of its tape system. A 96 dB (16 bit) digital audio system at 48,000 samples per second generates 48,000 × 16 = 768,000 bits per second. More than these are required for "overhead," including parity and sync bits. Recording 768,000 bits/s on a single track at a tape speed of 762 mm/s (30 in/s) gives a recording density of 768,000/762 = 1008 bits/mm (25,600 bits/in). Reel-type recorders at lower speeds use comparable densities but divide the bits among two or more parallel tracks to get the same total of required bits per second. Video recorders have more than ample capability to record the above bit rate and are used by formatting the audio words to fit within the scanning mode of the rotating heads. Table 9-2 gives specifications of some reel-type digital audio recorders. Table 9-3 similarly lists video-cassette digital audio recorders.

Digital Audio Techniques

Digital audio requires analog-to-digital converters that operate at picosecond speeds since only about a microsecond is allocated for each sample. During this microsecond, the analog input is measured to an accuracy of about 0.001% by successive approximations and a 16 bit number is generated. The number is mixed with parity bits and sync words, then coded to a suitable format for energizing the recording head. Upon playback, the process is reversed, with digital-to-analog reconstruction of a very accurate replica of the original audio.

The techniques and concepts of digital recorders are described in more detail in Chapter 13 on digital data recording, including block diagrams, sampling theory, anti-aliasing filters, coding, error detection and correction, and so on. Digital recording for computers discussed in Chapter 14 is also of interest.

While the audio is in digital form, it can be modified by digital processing, which opens the door to effects that may be impossible or impractical by any other means. Some of these are time compression and expansion, pitch control, unique reverberation effects, delay without distortion (useful for preview and control of disc cutter pitch and depth: 1.8 s between 33.3 rpm grooves), digital filtering, noiseless distortionless attenuation, and fading, mixing and editing, including razor-blade editing. Digital consoles and accessories that otherwise resemble their analog counterparts are also available.

Table 9-2 Digital Audio Recorders - Reel to Reel

Number of Channels	Tape Width mm (in.)	Tape Speed mm/s (in./s)	Reel Size mm (in.)	Playing Time min	Sampling Rate* KHz	Bits per Sample	Freq. Resp. kHz	Distortion %	Flutter %	Dynamic Range dB	Manufacturer and Model
32	25.4 (1)	1143 (45)	318 (12.5)	32	50.4 48.0 44.1 40.1	16	20		~0	90+	3M DMS-32
4	12.7 (0.5)	1143 (45)	318 (12.5)	32	40.1 to 50.4	16	20		~0	90+	3M DMS-4
32	25.4 (1)	762 (30)	356 (14)	60	50.4	16	20	0.05	~0	90+	MITSUBISHI X-800
2	6.3 (0.25)	381 (15)	267 (10.5)	60	50.4	16	20	0.05	~0	90+	MITSUBISHI X-80
24	12.7 (0.5)	762 (30)	356 (14)	60	50.4	16	20	0.05	~0	90+	SONY PCM-3324
8	25.4 (1)	889 (35)	267 (10.5)	25	50	16	20	0.05	~0	92	DCR/SOUNDSTREAM DTR-$\frac{1}{8}$
4	25.4 (1)	889 (35)	267 (10.5)	50	50	16	20	0.05	~0	92	DCR/SOUNDSTREAM DTR-$\frac{1}{4}$
2	25.4 (1)	889 (35)	267 (10.5)	100	50	16	20	0.05	~0	92	DCR/SOUNDSTREAM DTR-$\frac{1}{2}$
4	6.3 (0.25)	381 (15)	267 (10.5)	60	50.4	16	20		~0	90	TECHNICS-Digital Audio Recorder

*A sampling rate of 48.0 KHz to replace the various others was agreed upon in 1982.

Table 9-3. Digital Audio Recorders - Helical Scan Cassettes.

Number of Channels	Tape Width mm (in.)	Tape Speed mm/s (in/s)	Cartridge	Playing Time min	Sampling Rate* KHz	Bits per Sample	Freq. Resp. KHz	Distortion %	Flutter %	Dynamic Range %	Manufacturer and Model
2	19 (0.75)	(3.75)	UMATIC $\frac{3}{4}''$	60	44.1	16	20		~0	90+	JVC BP-90**
2	19 (0.75)	(3.75)	UMATIC $\frac{3}{4}''$	60	44.1 or 44.056	16	20		~0	94	SONY PCM-1610**
2	12.7 (0.5)	VHS	VHS	120	44.056	14			~0		TECHNICS SV.P 100

*A sampling rate of 48.0 KHz to replace the various others was agreed on in 1982.
**Digital processors only, for use with the specified video taperecorder.

The high density of recording requires special smooth dropout-free tape designated as digital audio tape, comparable to the best tapes manufactured for instrumentation and computers. However, errors are inevitable at such high densities and are resolved by error detection and correction methods borrowed from digital computer and data recording described in Chapters 13 and 14. As an example, in an interleaved block of 6270 bits, crossword coding may allow complete correction of up to 2240 bits, while a greater burst of errors, up to 4480, are concealable by linear interpolation.

Tape transports for reel-to-reel digital recorders are practically identical to their analog counterparts except that they use different heads and electronic circuits.

Applications

The future of digital audio seems assured for the finest work and for archival preservation of recordings, unmarred by flutter or noise of the recording apparatus. In studios the working copies of second and higher generations are as good as the masters, allowing great flexibility in mixdown, editing, and special effects. For consumer digital audio on compact discs or on specially adapted home videotape recorders, nothing short of digital audio is adequate for the original master.

10. General Audio Recording

Even the visionary Mr. Fankhauser (Chapter 1) in his wildest predictions did not imagine in 1907 a recorder for anything except sound. To this day most magnetic recorders are still audio (sound) recorders. Low-cost recorders use cassettes or cartridges, while semiprofessional machines use reels. Recorded cassettes and cartridges now compete successfully with phonograph records.

CASSETTES AND CASSETTE RECORDERS

Compact, convenient, and cheap, the Philips compact cassette has replaced almost completely the 7- and 5-inch reels that were symbolic of home-use tape recorders prior to the middle 1960s. As shown in Figure 10-1, the cassette is nothing more than a small container that holds and protects a tape wound between a pair of hubs. Holes in the container allow the tape to be reeled and to be contacted by transducer heads.

In size, the compact cassette resembles a pack of playing cards. It is double faced (A side and B side); when turned over it exposes a second edge of the tape to the heads, doubling the playing time while rewinding the first recording. Tape is 3.81 mm (0.150 in.) wide and runs at 47.6 mm/s (1.875 in./s). Playing times are:

C45 cassette: 22.5 min/side
C60 cassette: 30 min/side
C90 cassette: 45 min/side
C120 cassette: 60 min/side

The C120 cassette uses an extra thin base and coating, giving less output, higher print-through, and possible handling problems in some tape devices. Official dimensions of the Philips compact cassette are given in Appendix F.

Portable Cassette Recorders

Portable cassette recorders still resemble the first (1964) Philips model (Figure A-25A, Appendix A) but are push-button operated and often have built-in microphones, making them completely self-contained.

Drive Mechanism. A typical cassette drive mechanism is shown in Figure 10-2. Most important is the capstan itself. Because it is only about 1.5 mm ($\frac{1}{16}$

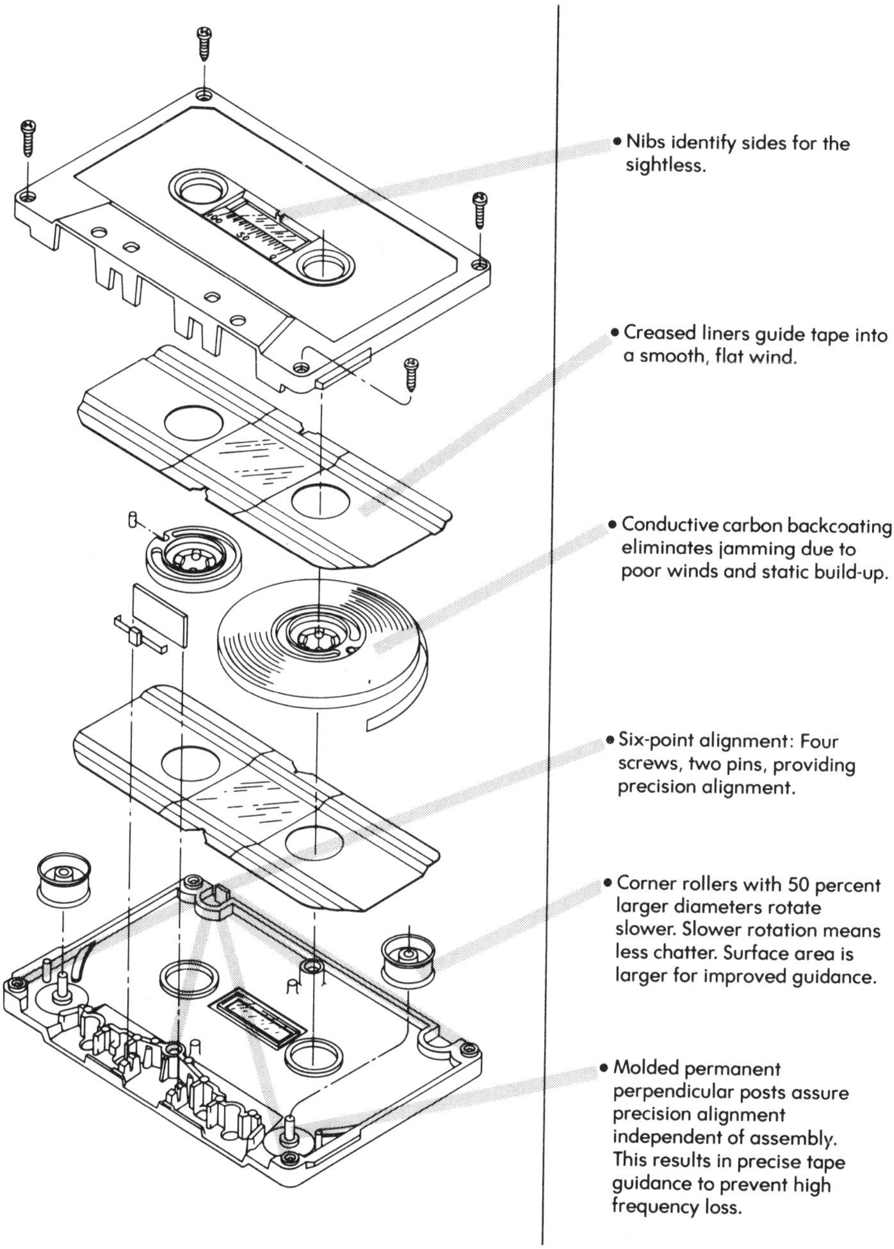

Figure 10-1. Interior of a cassette. Tape 0.15 inch wide runs between a pair of hubs. A pair of liners confine the tape laterally so that spool flanges are not needed. (Courtesy Capitol Records)

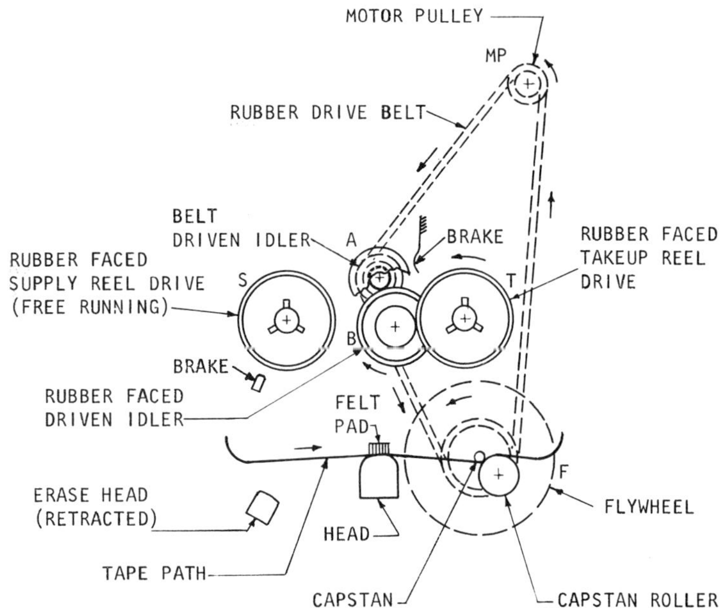

Figure 10-2. Typical cassette drive (in play position). The motor pulley MP belt drives idler A and capstan flywheel F all turning counter-clockwise. The capstan and its roller drive the tape from left to right past the head. Idler A drives rubber faced idler B which drives the takeup spindle T, with slippage compensating for differences as the tape winds up. Recording operation is the same except that the erase head is positioned against the tape, and the head is connected to recording circuits.

in.) in diameter, its accuracy (finish, eccentricity, wobble) must be within one or two wavelengths of light if flutter is to meet high-fidelity music standards, and its bearings and supports are designed accordingly. The capstan flywheel runs at relatively high speed, storing sufficient energy without excessive weight. A resilient capstan roller presses the tape against the capstan. Rubber belt and friction roll drives are the most popular coupling between motor, capstan, and takeup reels. The motor for portables is a dc type, with a governor to maintain constant average speed. Brushless DC motors are available, as are AC motors energized by inverter power supplies in more expensive recorders.

Operational Modes. Push buttons select the desired mode of operation by moving friction rollers into different positions as shown in Figures 10-2 and 10-3.

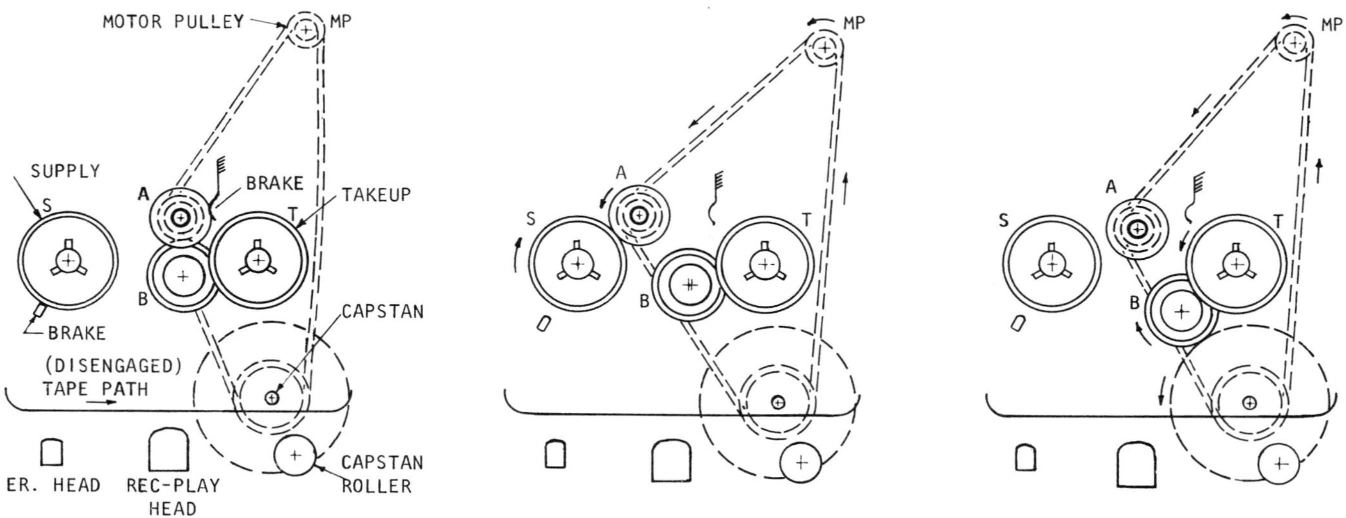

(a) (b) (c)

Figure 10-3. Cassette drive mechanisms in various modes of operation. (a) CASSETTE DRIVE IN STOP (OR LOADING) POSITION. The heads and capstan roller are retracted to give an unobstructed loading path. The motor is turned off. The idlers A and B are uncoupled. Brakes are applied to idler A and supply wheels. (b) CASSETTE DRIVE IN REWIND POSITION. Idler A is spring pressed against the rubber face of supply wheels, driving it for high speed rewind. Brakes, heads and capstan are all disengaged. (c) CASSETTE DRIVE IN FAST-FOWARD POSITION. Idler B rubber rim is driven from the flywheel, while its inner hub drives the takeup wheel at high speed. Brakes, heads, capstan are disengaged.

1. *Stop or load*: The reel drives are disengaged and the capstan roller and heads are retracted to allow the cassette to be set into place (Figure 10-3*A*).
2. *Play or forward.* The capstan drive and the heads are engaged with the tape, and the takeup reel is driven through slipping friction rollers. The supply reel is driven by the tape pulled off of it, the tape being tensioned by friction of the tape supply reel against its side plates (Figure 10-2). Driving is the same for recording as for playback—only the electronics are changed.
3. *Rewind or fast forward.* The supply reel or the takeup reel is driven by the motor while the other reel is not engaged (Figure 10-3*B* and *C*). The capstan drive and heads are retracted.
4. *Pause.* While the mechanism is otherwise in play position, the capstan pressure roller is disengaged, stopping the tape but having everything in readiness for quickly starting again.
5. *Eject.* Usually interlocked with the stop button, the eject button unlatches the cassette and opens its cover for easy removal.

Electronics. The circuit diagram for a portable cassette recorder is shown in Figure 10-4. In the recording mode the amplifiers, equalizers, and bias oscillator principles are as described in Chapter 6, the parameters being chosen to give a 120–3180μs recorded flux curve for the usual ferric oxide tapes. Playback amplifiers are correspondingly tailored to approximate the 120–3180μs playback curve shown in Figure 6-16. This response is often abbreviated as the "120μs cassette standard."

Newer tapes have better high-frequency response and use different playback standards, as in Table 10-1. IEC 2, 3, and 4 are interchangeable for playback only. But for recording each of them requires different bias and equalization, which may be selected by a knob, or automatically by notches in the cassette body.

Simple low-priced cassette recorders are monophonic, recording a single track for side A, and another for side B in the opposite direction. Monophonic tape track dimensions and location are as in Figure 10-5. Heads are necessarily very small to fit into the small openings of the cassettes. Their gap size is 2μm (80 microinches) or less for reasonable audio response at the low tape speed.

Overall Performance. Table 10-2 lists suggested specifications according to IEC Standards for domestic use at 47.6 mm/s (1.875 in./s). Though these cassette specifications would be scoffed at by high-fidelity buffs, the sound is quite acceptable and pleasing to unsophisticated listeners. The low frequencies would not, in any case, come through the small loudspeakers of small portables. Inexpensive cassette recorders seldom meet even these minimal standards, except in their advertising literature.

High-Fidelity Cassette Recorders

After a remarkable development effort, the best cassette recorders coupled with premium-grade tapes and noise reduction systems have achieved high-fidelity status. These recorders are fairly large and heavy, being part of a substantial stereo audio system. Stereo is recorded by dividing each track of Figure 10-5 into two narrower tracks (Figure 10-6). It was intended that stereo recordings should play monaurally on single-track players when the one head straddles both adjacent stereo tracks.

Mechanical Drive. A high-quality cassette recorder or *deck* is shown in Figure 10-7. The drive and general operation is as described for portable recorders except that it is heavier and more precise. The motor operates from AC and can be a silent-running hysteresis synchronous type. In the capstan and its roller, the highest possible precision is carefully maintained.

Electronics. Extended low-frequency response of the high-fidelity deck requires better head-shielding and hum reduction, while extended high-frequency response and dynamic range require low-noise preamplifier stages. Basic equalization for ferric oxide tapes is the 120–3180μs curve shown in Figure 6-16. Cassette recorders are handicapped by low tape speed, thin tape coatings, and narrow tracks. The state of the art is strained in every possible way to give best performance.

Dynamic noise suppressors. To increase dynamic range, high-quality cassette decks use dynamic noise suppressors. The most famous of these have been the Dolby-B and Dolby-C systems, which are simplified versions of the professional Dolby-A described in Chapter 9. As seen in Figure 10-8, Dolby-B is an automatic gain control for medium and high audio frequencies. Dolby-C is an improved version with somewhat greater noise reduction. Figure 10-8*A* shows flat response at high levels (0-VU) and 10 dB of preemphasis at low levels (−40 curve). A block diagram of the automatic gain network is at the top of Figure 10-8*B*, including a variable high-pass filter controlled by an integrator and rectifier according to the audio level. As shown at the bottom of Figure 10-8*B*, prior to recording this automatic gain network preemphasizes the high frequen-

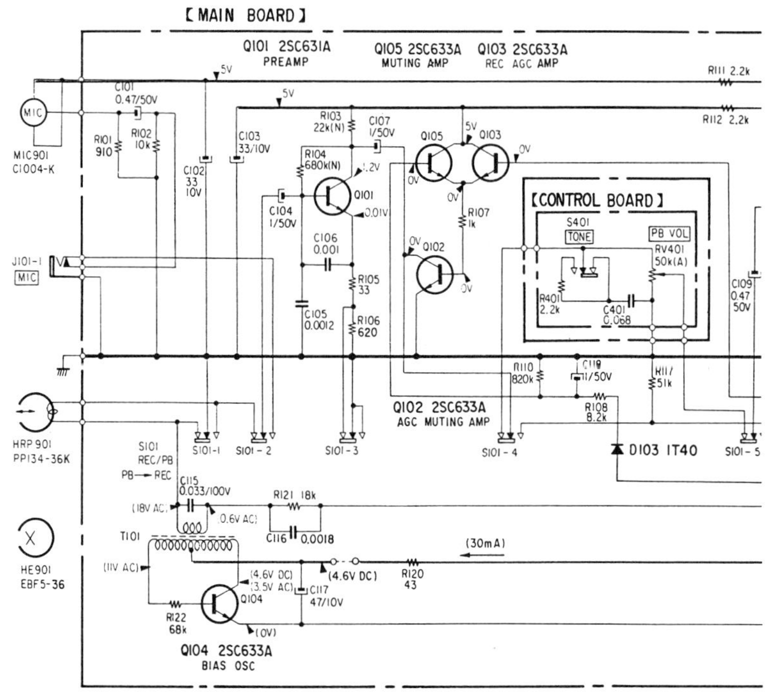

Note:

- All capacitors are in μF and ceramic unless otherwise noted. 50WV or less are not indicated except for electrolytics. pF = $\mu\mu$F, elect = electrolytic
- All resistors are in ohms, ¼W unless otherwise noted. kΩ = 1000Ω; MΩ = 1000kΩ
- (N) : low-noise resistor.
- ▭ : panel designation.
- Voltages are dc with respect to ground unless otherwise noted.
- Readings are taken under no-signal (detuned) conditions with a VOM (20kΩ/V).
  (V): REC
- Voltage variations may be noted due to normal production tolerances.
- Transistor base-emitter voltages are measured on the 2.5V range.
- Total current is measured with no cassette installed.

Figure 10-4. Circuit diagram of a portable cassette recorder. Features include a condenser microphone, automatic gain control (AGC) during recording, pause control, and battery or AC line operation (Sony Model TC-215).

cies in soft passages of music, bringing them well above the tape noise level. During playback of soft passages, the same network operates in a negative sense, reducing the preemphasized high frequencies back to normal, while also reducing high-frequency noise by the same amount. The net benefit is a 10 dB reduction in noise, enough to give an impressive demonstration in comparative

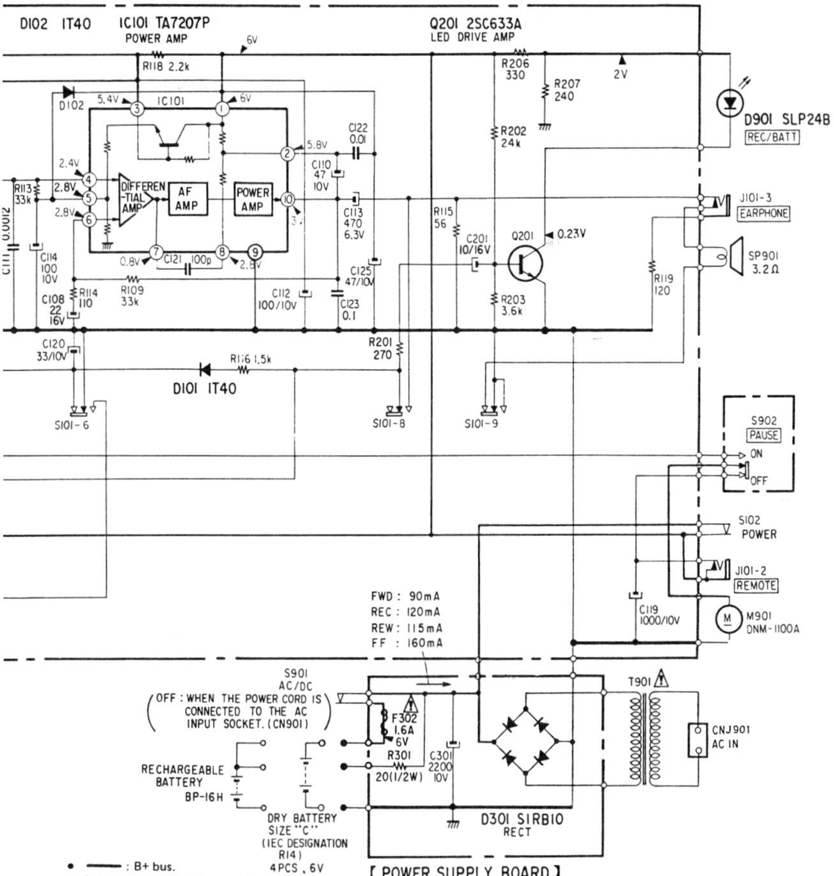

- ——— : B+ bus.
- ▭ : adjustment for repair.
- Switch

Ref. No.	Switch	Position
S101	REC/PB	PB
S102	POWER	OFF
S901	AC/DC	ON
S902	PAUSE	OFF

Figure 10-4. (*Continued*)

listening tests. When the Dolby suppressor is turned on, the background noise disappears, leaving only crisp, clear music.

Many recorded-music cassettes incorporate Dolby noise suppression during the duplicating process. For realistic reproduction these cassettes should be played on a deck that has the correct playback compensation. Actually the record producers have found that "Dolby-ized" recordings are quite acceptable

Table 10-1. Playback Equalization of Cassette Tapes*

IEC Type	Description	t_1 µs	t_2 µs	Identification Emblem
IEC 1	Fe_2O_3	120	3180	IEC I
IEC 2	CrO_2	70	3180	IEC II
IEC 3	Fe_2O_3/CrO_2 (Double Layer)	70	3180	IEC III
IEC 4	Metal Particle	70	3180	IEC IV

*See Chapter 6 for equalization curves having time constants t_1, and t_2.

when played on ordinary machines. They then have a brighter than normal sound and the dynamic range is more limited. Most listeners do not find this objectionable for, after all, they rarely hear a recording that has not been doctored in one way or another.

Tapes. To overcome the handicap of low speed, cassettes use the finest tapes that it has been possible to make. As explained in Chapter 2, the dynamic range is severely limited at short wavelengths by so-called *recording demagnetization*, which blurs these short waves before they move away from the gap field. Figure 10-9 shows the difference in range between a 315 Hz (long wave) re-

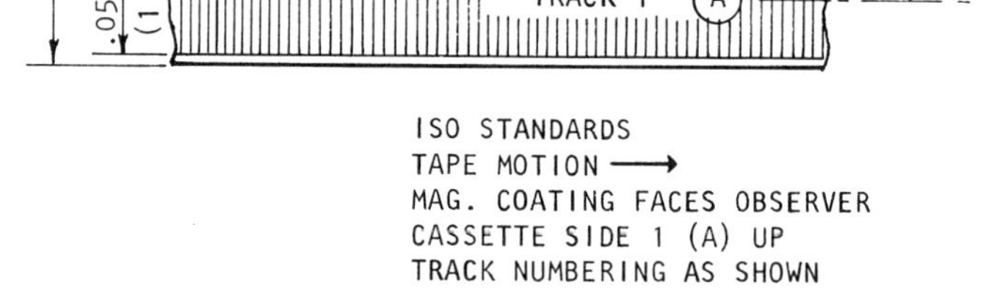

Figure 10-5. Track dimensions for monophonic cassette tape. A single head is used. It records or plays track A when face A of the cassette is upward. The B track cooperates with the head when the cassette is inverted.

Table 10-2. Frequency Response and Flutter of Low Speed Tape Systems Specifications Suggested by IEC.

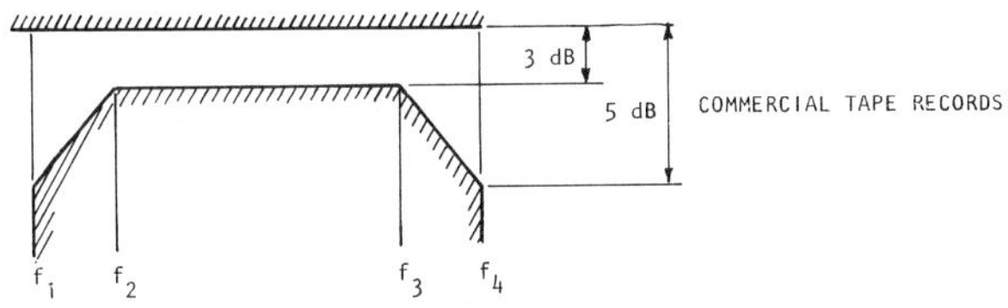

Tape speed	f_1 in Hz	f_2 in Hz	f_3 in Hz	f_4 in Hz
19.05 cm/s ($7\frac{1}{2}$ in/s)	40	80	8 000	14 000
9.53 cm/s ($3\frac{3}{4}$ in/s)	63	125	6 300	12 500
4.76 cm/s ($1\frac{7}{8}$ in/s)	80	160	4 000	8 000

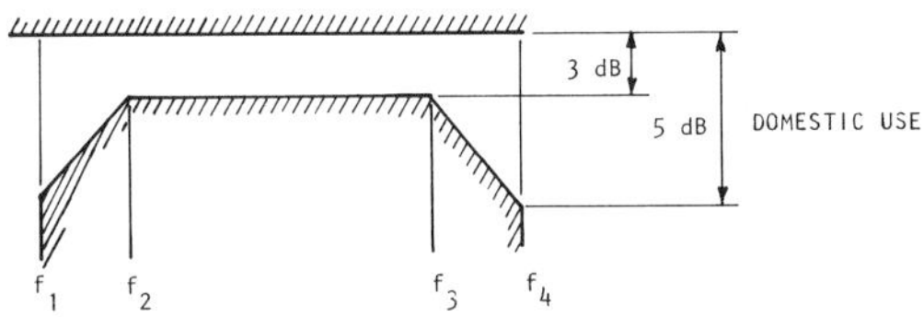

Tape speed	f_1 in Hz	f_2 in Hz	f_3 in Hz	f_4 in Hz
19.05 cm/s ($7\frac{1}{2}$ in/s)	40	250	4 000	12 500
9.53 cm/s ($3\frac{3}{4}$ in/s)	63	250	4 000	10 000
4.76 cm/s ($1\frac{7}{8}$ in/s)	80	250	4 000	6 300

Tape Carrier	Rated Speeds cm/s (in/s)	Tolerances on Reproduced Frequency	Wow and Flutter
	19.05 ($7\frac{1}{2}$)	±0.5%	±0.2%
reel-to-reel	9.53 ($3\frac{3}{4}$)	±0.5%	±0.3%
	4.76 ($1\frac{7}{8}$)	±0.5%	±0.5%
cartridge	9.53 ($3\frac{3}{4}$)	±0.5%	±0.3%
cassette	4.76 ($1\frac{7}{8}$)	±0.5%	±0.5%

cording and a 15 kHz (short wave) recording on a cassette tape. The 315 Hz signal has a dynamic range of about 55 dB between noise level and saturation. The 15 kHz signal has less than 15 dB range, and moreover the high-frequency output actually goes down instead of up when the input is above +5 dB. High-frequency inputs above 0 db cause intermodulation and bias beats (Figure 2-

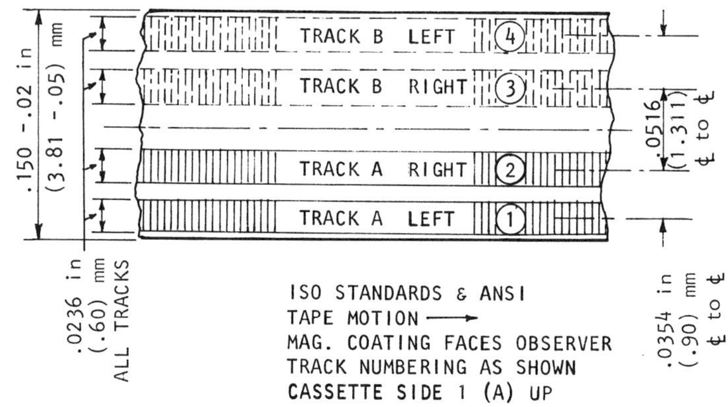

Figure 10-6. Track dimensions for stereo cassette tape. Both tracks of the stereo pair are close together. A monaural head straddles both tracks and mixes their output.

30), hashing-up the sound during transients and strong high passages. Recording engineers, cognizant of this fault, take extra precautions to avoid overloading the high end in cassette tapes.

Tapes of high coercivity have more range at the critical high frequencies. Chromium dioxide tapes with coercivity of 40 kA/m (500 Oe) were a boon to the cassette industry when first introduced. They were followed by cobalt-modified oxide tapes with similar coercivity and later by metal particle tapes with coercivity of 88 kA/m (1100 Oe). A hybrid double-coated tape called Ferri-

Figure 10-7. High fidelity cassette recorder. Sound quality is remarkable, overcoming handicaps of 1.875 IPS speed, narrow tape, and cassette confinement. Specifications indicate a weighted RMS flutter of 0.07%, frequency response from 30 to 16,000 Hz using CrO_2 tape, and signal to noise ratio of 60dB weighted with dolby-B noise reduction. (Courtesy TEAC Corp.)

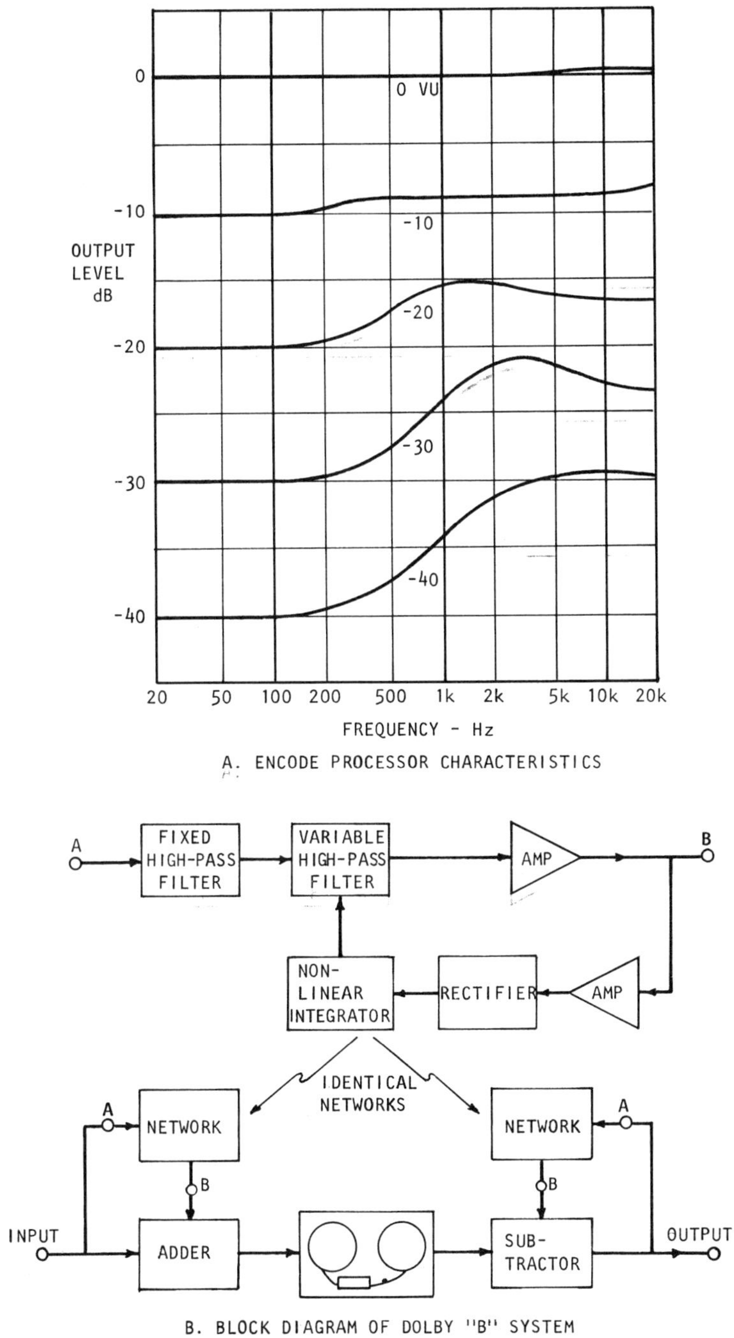

Figure 10-8. Dolby "B" noise suppressor. *A*. Encode processor characteristics. *B*. Block diagram of system (Dolby 1970).

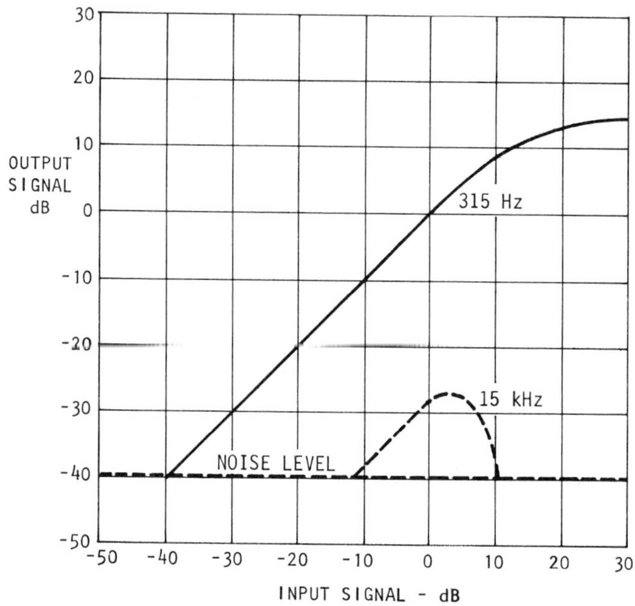

Figure 10-9. Dynamic range compression at short wavelengths.

chrome (FeCr) is also manufactured with an underlayer of ferric oxide and a thin top layer of chromium dioxide.

All of these high-energy tapes need special cassette decks that supply higher bias and erasing fields, and also different equalization that takes advantage of their better response. A switch on these decks selects ferric, CrO_2, metal, FeCr, and so on. The compromise equalization for high-energy tapes is the 70–3180μs curve, abbreviated as "70μs cassette equalization." Some decks have a sensor that registers with a notch in high-energy cassettes, automatically setting the electronics to bias and equalization appropriate for that cassette.

Heads. High-energy tapes need heads with high saturation densities and smallest possible gaps. Sendust is the preferred core material for heads in the most expensive decks, having higher saturation and a longer wear life than permalloy. Ferrite cores have been advocated for long wear, but their saturation density is too low for audio recording on high-energy tapes. Stereo is almost inseparably connected with high-fidelity audio. The stereo head pair conforms to track standards shown in Figure 10-6.

Accessories. Head-cleaning cassettes for periodic maintenance have a mildly abrasive tape designed to remove debris from heads and tape guides.

(Better still, clean them gently with alcohol and a wood stick swab while observing with a magnifying glass; also do this for the capstan and its roller.) Head-demagnetizing cassettes apply a diminishing AC field through the head, generated electronically or by a moving set of alternately opposite-pole magnets (a manual, general-purpose head-demagnetizer is also effective). Both functions, cleaning and demagnetizing, are often incorporated in the same cassette.

Special Cassette Recorders

For certain applications, the cassette principle needs to be modified — slightly, in some cases, or considerably — and some of these applications are illustrated in Figure 10-10. The microcassette, a *half-scale* version of the compact cassette, is used in tiny vest-pocket-sized memo and dictation recorders. The Elcaset is a *double-size* cassette with $\frac{1}{4}$ in. tape running at 95 mm/s (3.75 in/s) and is one of many attempts to achieve higher fidelity by overcoming the small size and low speed limitations of compact cassettes. Still another special cassette recorder, the learning machine, has been adapted for a special educational

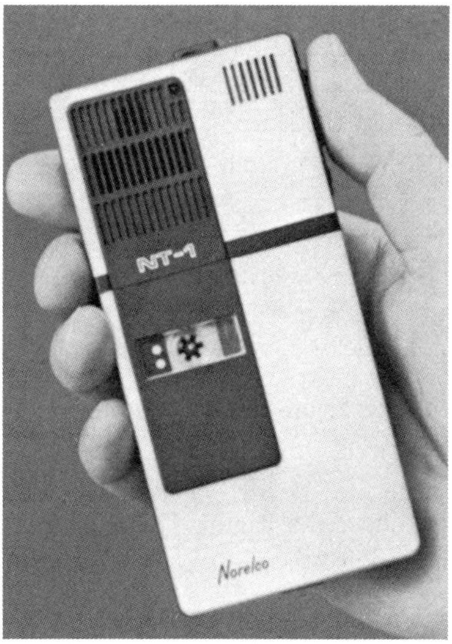

A. MICROCASSETTE RECORDER

Figure 10-10. Special cassettes. A. This eight ounce micro-recorder uses a "half-size" microcassette. B. A double sized cassette with $\frac{1}{4}$ inch tape opens for a complex tape path giving higher fidelity recording. C. In a learning machine format, teacher tracks are protected against accidental erasure.

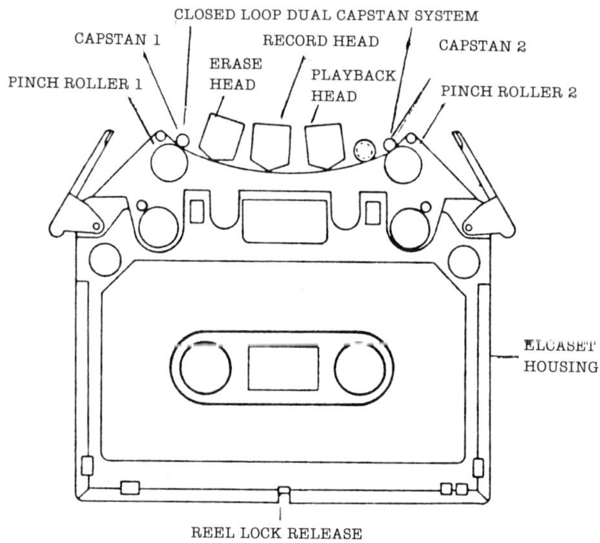

B. THE ELCASSETTE

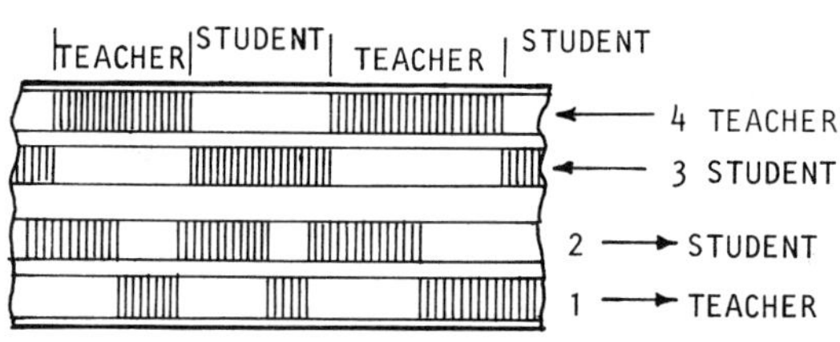

AAC System - Bi-directional monophonic

C. LEARNING MACHINE TRACK FORMAT

Figure 10-10. *(Continued)*

purpose. The variations shown in Figure 10-10 are examples of many modifications that have been made of the well known cassette principle.

CARTRIDGES AND CARTRIDGE RECORDERS

Eight-track stereo cartridges dominated the taped music market after they were introduced in about 1965 for automobile players. In size and shape, the eight-track cartridge resembles a paperback book. When inserted into a player, the cartridge plays stereo music for 15 to 40 minutes or more (equivalent to a

12-in. phonograph record) and then repeats without rewinding. The *eight-track* name comes from the four sets of stereo-pair tracks recorded across a tape 6.3 mm ($\frac{1}{4}$ in.) wide.

The endless loop cartridge operates on an unlikely principle that should not work, but does. As shown in Figure 10-11, tape is pulled off the inner hub of an open reel, is routed past heads and capstan, and then passes to the outer part of the tape winding. Inner and outer ends of the tape are spliced together to form an endless loop. Pulling the tape from the bottom rotates the entire winding, causing the "spent" tape to be pulled onto the top of the coil, gradually to work its way to the bottom again.

Cartridge Mechanism. Of all tape drives, the cartridge mechanism is the simplest, requiring only a capstan, a head, and a few guides. Eliminated are the supply, takeup, rewind, and fast-forward functions of cassette and open reel decks. For this simplicity, the endless cartridge principle trades off several features often regarded as the very essence of tape recording, namely convenient erasing, recording, and indexing.

The difficulties arise because the endless tape (Figure 10-11) can move in the forward direction only. It cannot rewind. If one makes a short recording ex-

Figure 10-11. Endless loop 8-track stereo cartridge. Inside an endless-loop cartridge a loosely packed winding of tape sits on a plastic turntable. During use, the inner end of the winding is peeled from the hub, drawn past the recording heads, and then deposited on the top of the coiled tape roll.

pecting immediate playback, one may have to wait 10 or 15 minutes before the tape comes around again to the recording. Short lengths of tape can be loaded into the cartridge, but now one must plan the recordings accurately; if they are a little too long, the tape runs out; if too short, there is a long blank space between the end and the beginning of the program. A faster forward speed is possible, but the drive becomes more complicated, the tape is apt to snarl, and if the beginning of the program is overshot, the operator cannot shuttle back but must go forward again through the whole loop. Because of this peculiarity, cartridges are most useful for continuous playback of recorded programs. Appropriately, their first large-scale use was for music entertainment in automobiles. Later, when cartridge players were incorporated into home audio systems, recording capability was added, which involved only an erase head and simple electronics. When used primarily for copying a series of broadcast programs or phonograph records, however, the one-direction capability of cartridges is not too serious a limitation.

Playing Time. The shorter the tape in an endless cartridge, the better it works. Endless loop windings have a tendency to jam which increases exponentially with the number of turns, for each turn must slip against its neighbor continuously and smoothly as the outside turn works its way down to the inner hub. To promote interlayer slippage, the back side of cartridge tape is coated with a low-friction layer such as colloidal graphite.

A good compromise between a short enough tape and a long playing time is a 10-minute tape loop at 95 mm/s (3.75 in./s), totalling 55 meters. The stereo-eight cartridge plays four times as long on this 10-minute loop by running it through four times using four different sets of stereo tracks played in succession (Figure 10-12). A 40-minute cartridge is equivalent to both sides of a 12-in. phonograph record. Tracks are selected by a cam that moves the heads sidewise to four different positions across the tape. The cam is activated by a short strip of electrically conductive foil on the tape each time the tape loop completes a cycle. Cartridges with more tape play up to 90 minutes, while larger cartridges hold as much as 366 meters of tape.

Drive Stability. Tape drive stability has always been a problem in endless cartridges. Peeling the tape sideways off the hub is irregular, the necessary loose winding gives uneven tension, and the limited free tape-path in the cartridge does not allow extra stabilization. Molded plastic cartridges contain inexpensive capstan rollers. But most important of all, there has not been a competitive effort toward superior fidelity in cartridges as was true with cassettes. Therefore, most cartridge systems, although falling short of high-fidelity standards, are considered "good enough." They give enjoyable music reproduction especially in automobiles where ambient noise masks any slight imperfections. As with pho-

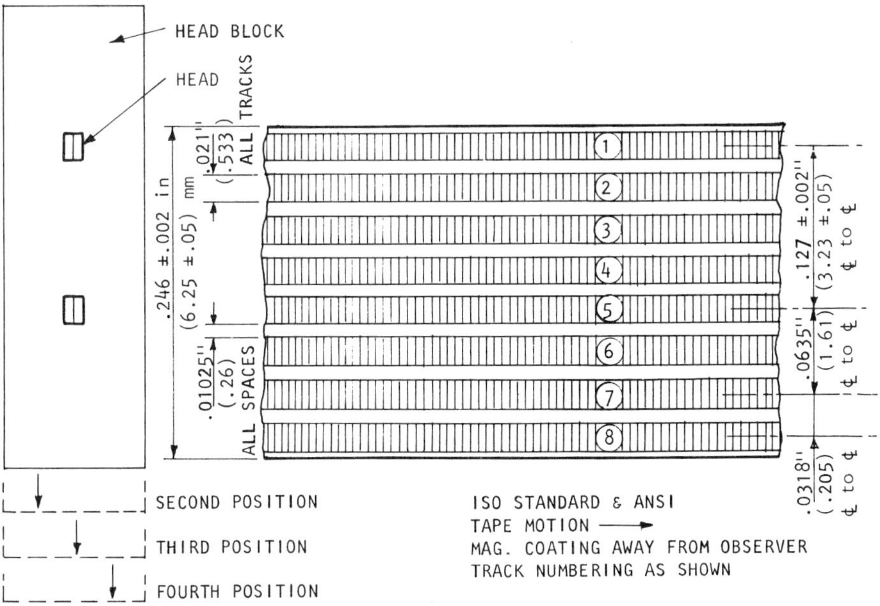

Figure 10-12. Track format of eight track stereo cartridge. Head at the left plays the first and fifth tracks as a stereo pair, then moves to tracks two and six for the second time around, etc. After all four positions have been played the sequence repeats. Track straddling allows spaced heads for economical construction.

nograph records, there is a wide range in recorded quality of purchased cartridges. In the 1980s cartridges have become less popular, while cassettes became the favorite format.

REEL-TO-REEL RECORDERS

Until the mid 1960s, reel-to-reel recorders reigned supreme. They are still the choice where performance must not be compromised for convenient handling. Reel-to-reel tape decks of recent years are considered to be semiprofessional or for the advanced high-fi enthusiast, with specifications that rival professional equipment and with prices to match. Figure 10-13 shows a deck of this kind.

Features of reel-to-reel decks include low flutter tape transports with isolation from the reels by auxiliary flywheel rollers or two capstans, and a choice of several tape speeds, usually 47.6, 95, and 190 mm/s (1.875, 3.75, and 7.5 in./s). Other features often included are extra speeds of 23.7 or 381 mm/s (0.9375 or 15 in./s), 267 mm (10.5 in.) reel capability, separate record and play heads, long-wearing heads with ferrite or sendust cores, playback monitoring while recording, sound on sound, automatic reversal with track changing, quadraphonic (four-channel) sound, and a case with handles for portability. Specifications of a reel-to-reel tape deck are included in Figure 10-13.

Drive Mechanisms. From 1945 to 1965, when there were nothing but reel-to-reel recorders that sold from $30 to $3000 or more, almost every conceivable drive mechanism could be found in various decks; examples are shown in Figures 7-7 to 7-12. Most of the poorer ones have been weeded out by a process of natural selection, and the better drives remain in the top-of-the-line market now served by reel-to-reel recorders. A favorite is the arrangement shown in Figure 7-7, where heads are isolated from the supply reel by a flywheel idler, and from the takeup reel by the capstan.

Tapes and Tracks. Tape for the reel-to-reel recorder is nominally $\frac{1}{4}$ in. wide (actually 0.246 in or 6.3 mm). Tracks conform to the *quarter-track* standard

Figure 10-13. High quality portable reel-to-reel recorder. Performance of good decks may equal or exceed professional standards. They include features such as sound-on-sound, echo effects, servo control drive, ferrite heads, automatic tape lifters, separate heads for record and playback, logic controlled transports. Specifications indicate a frequency response of 30–25000Hz ± 3 db, signal to noise ratio of 59 db, wow and flutter of 0.06% (0.03% at 15 IPS). The stereo electronics has 58 transistors, 46 diodes, 2 FETs, 1 integrated circuit. (Courtesy Sony Corp.)

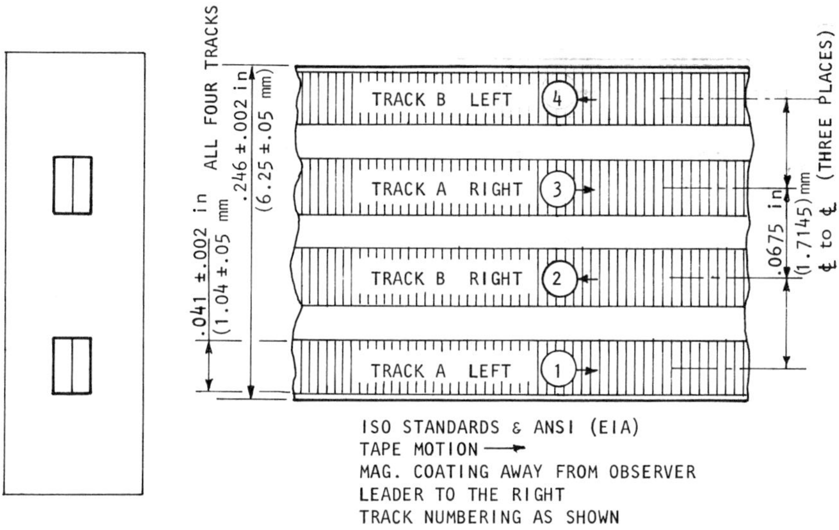

Figure 10-14. Four-track open-reel format. This so called quarter-track format is almost universal on $\frac{1}{4}$ inch reel to reel tape. The heads at the left scan odd numbered channels in the "forward" direction, and even numbered channels when the tape is turned over for "reverse" direction. Alternatively the four channels may be used separately for monophonic recording or simultaneously for quadraphonic recording.

shown in Figure 10-14 with interleaved track-pairs for stereo. The second pair of stereo tracks is utilized by "flopping over" and trading the supply and takeup reels at the end of the first recording, running the tape through in the backwards direction. Some deluxe machines record and play in either direction, with movable heads or with another set of heads for the reverse direction. Quadraphonic sound uses all four tracks simultaneously in one direction. Special-purpose track formats are available as options on some recorders.

SPECIAL RECORDERS

Ordinary tape recorders, or at least their mechanisms, are adaptable for a myriad of uses such as dictation, telephone answering, taking inventory, and even storing digital data. However there are unusual recorders and records that are better for rapid access or for easy correction by erasure, or for handling, mailing, storage, and so on. Not all of these are commercially available at present, but their concepts are valuable to designers.

Single-Spool Cartridges. Prior to the widespread acceptance of present-day cassettes and cartridges, there were attempts to introduce cartridges that did not compromise the recording quality of open-reel recorders and yet were convenient to load. An interesting solution is the ARF compatible cartridge shown in Figure 10-15. It loads without threading onto decks designed for it and also fits ordinary reel-to-reel recorders on which it is hand threaded like a reel of tape. Hence recorded music on the compatible cartridge would be universally playable, whereas at this time at least three versions are needed, all incompatible with each other.

Another form of cartridge (Figure 10-16) has a single reel inside a square casing with the tape drawn out of one corner, resembling a measuring tape. This form is also self threading; the tape path over heads and capstans can be as complex as necessary to allow multiple heads and to iron out any tape flutter.

In the single spool cartridges of Figures 10-15 and 10-16 the beginning of the tape usually hooks into a prethreaded leader that pulls the tape around its operating path. Alternatively a leader in the cartridge is guided through the tape path by an auxiliary threading system that retracts out of the way when the tape is threaded. A permanent takeup reel is built into the deck. The cartridge must be rewound before it can be removed from the machine, but this is a minor inconvenience since rewinding takes only a few seconds.

Figure 10-15. Compatible tape cartridge. It can be inserted in slot of a player, or stacked in an automatic changer; but it also fits on an ordinary reel-to-reel recorder with manual threading. The tape is protected by a wide leader held between beaded rims of the spool (Camras 1960).

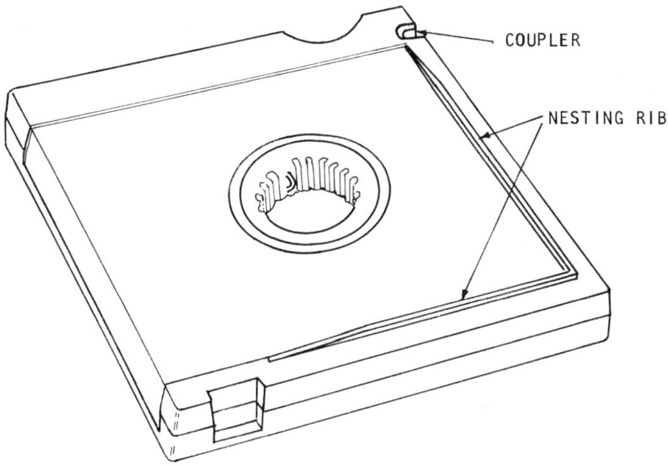

Figure 10-16. Single spool enclosed cartridge. Inside the plastic housing is a magnetic tape wound around a central hub, with its leader end terminating in a coupler at one corner. There is a resemblance to enclosures commonly used for measuring-tapes (Goldmark, Mee, Goodell, and Guckenberg 1960).

Dictation Machines. The ability to erase a spoken word or a sentence and to replace it with something different is unique to magnetic recording. An ideal dictation machine combines this feature with instant playback and the rapid access of a belt, disc, or drum recorder. A variety of such machines have been designed, including the flat sheet Dictorel recorder (see Figure A-20*B*, Appendix). But cassette machines have also been successful in this specialized market when modified with start-stop-backup-record-play buttons on the microphone, and with foot pedal controls for transcription. For conference recording and court reporting, the requirements are not as stringent and usually only playback controls are needed. Microcassette equipment (Figure 10-10*A*) is useful for memos and for light-duty dictation.

Voice recording does not require high standards of fidelity, so the tape can run very slowly, 23.87 mm/s ($\frac{15}{16}$ in/s) or less. In some cases the spools drive the tape, with the capstan omitted. The inclusion of speech compression or expansion is an interesting possibility (see Figure 10-20), allowing a slowdown of playback for convenient transcribing without changing the pitch of the voice. Such a feature is expensive, and there is some question of its value in view of the alternatives of intermittently stopping the recorder until the typist catches up, or of slowing the playback and accepting the change in voice pitch.

Pictures with Sound - 8 mm Soundstripe. Movie makers can add sound to their 8 mm movies by using film that is soundstriped. Soundstriped film has a narrow magnetic track on its edge, as shown in Figure 10-17 (see also Figure 9-18). Synchronized sound is recorded in a magnetic camera while the picture is taken, and the recording can be erased or modified after development to include music, sound effects, and so on. Soundstripes can also be added to film after it is developed. In this case the recording is made while the pictures are projected with a magnetic projector.

An alternative method for movies with sound is a cassette recorder equipped

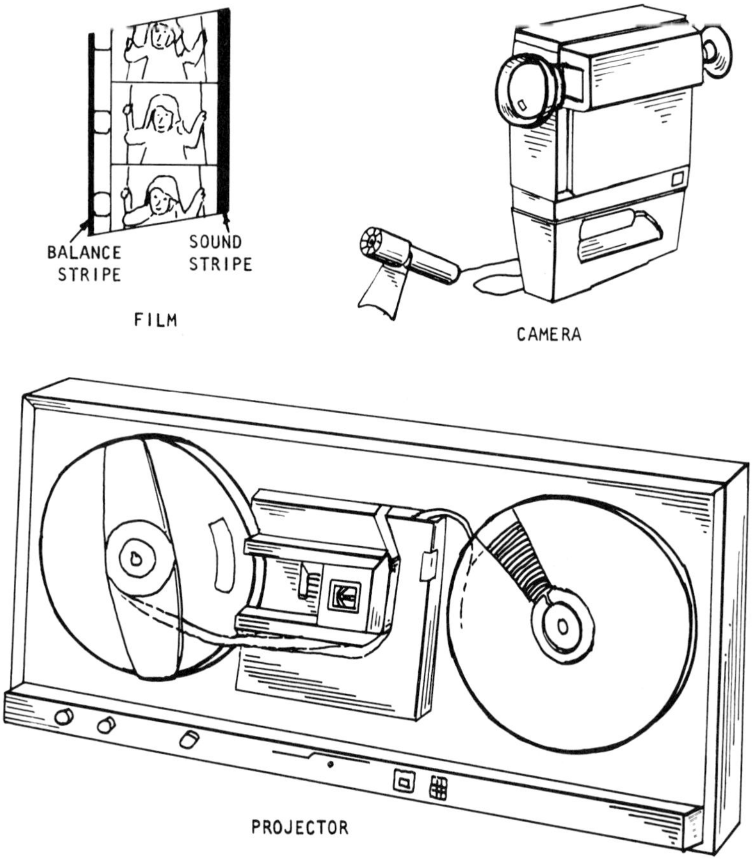

Figure 10-17. Magnetic soundstripe equipment. Super 8 mm film with a magnetic stripe may be purchased from photo dealers, recorded in a sound camera, and projected as with conventional amateur movies. The sound track may be edited in the projector, music added, etc. Sixteen mm equipment is available for professional use. (Courtesy Eastman Kodak)

with a synchronizer that plugs into the camera or projector. Audio-visual and teaching devices use a similar idea for synchronizing taped sound with a slide projector. An audible or inaudible signal on the tape changes the picture.

Audio Disc Recorders. Magnetic disc records have not been important in home-use or commercial recorders although introduced as novelties in the mail-a-voice recorder (Figure A-13*B*, Appendix) as a record-your-voice toy, and in dictation machines. Their sound quality and recording time are limited, machines are bulky, and tracking is a problem. Magnetic discs have been of importance, however, in professional fields (instant replay of video) and for computers (memory discs). Discs and flat sheet recordings can be duplicated readily by contact printing, a valuable feature for certain applications.

Rectangular Sheet or Card Recorders. The *Language-Master*, or pronouncing dictionary, was an early application of printed cards that carried a magnetic stripe to complement the printed word. IBM-size cards were placed on a reader that pronounced the words recorded on a 6.3 mm magnetic stripe while the printing was displayed. A second channel could also allow the student to record and play his or her own efforts at pronouncing words.

An attractive idea has been a larger printed page, book size or letter size, with a magnetic coated back that reproduces sound while the printing is readable from the front. This product has not been commercialized.

SPECIAL EFFECTS WITH MAGNETIC RECORDS

The capability of magnetically storing, retrieving, and erasing information in millisecond intervals has led to unusual and interesting effects that are produced with specially designed recorders: Time-delay, reverberation, and speech compression.

Time Delay. In sound reinforcement systems spanning large distances there is need for a delay of the audio to the farther loudspeakers, allowing them to blend with the airborne sounds arriving directly or from loudspeakers closer to the source. Delays range from tens of milliseconds up to a second or more. Magnetic recording is eminently suitable for such delays, as shown in Figure 10-18. A rigid disc record with spaced heads can run for years with little or no maintenance. This practically perfect solution has in recent years given way to an even more perfect solution: digital delay with no moving parts at all. The audio is converted to a sequence of digital numbers, stored in a memory, and read out of the memory after the desired delay, clearing the memory for continued sequencing.

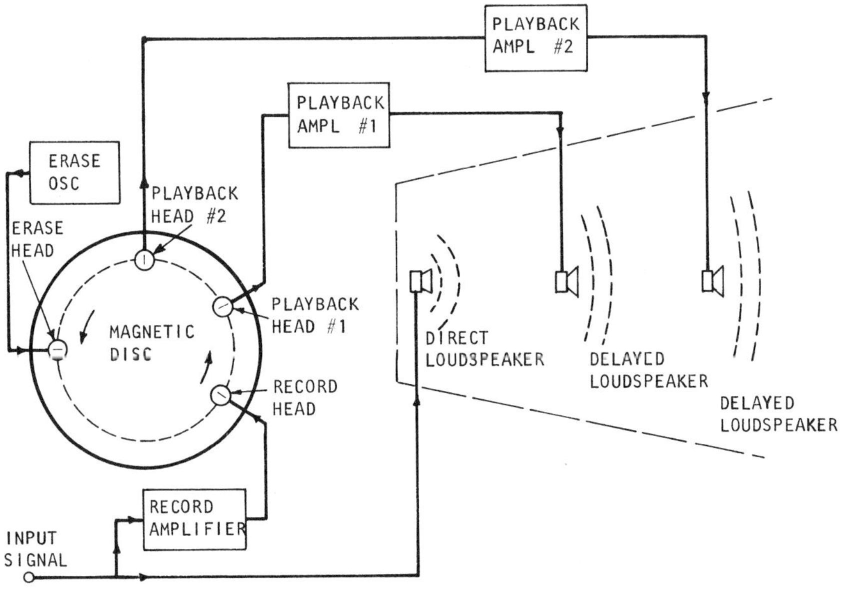

Figure 10-18. Time delay for sound reinforcement system.

An interesting experiment is to have a person who is recording his or her voice with a microphone listen to delayed playback of the voice through headphones, instead of hearing one's voice immediately as is usual while speaking. The resulting confusion often renders a person speechless. Some people are less sensitive than others, and there is a varying degree of tolerance developed through practice.

Reverberation Generator. A disc system such as the one shown in Figure 10-19 or a tape loop will add reverberation to the original sound by mixing it with some delayed playback. The mixture can be fed to the recording head again to give repeated reverberation. The reverberant component can be "colored" in frequency response to imitate reflections that occur naturally (as in a hard-walled room) and its intensity adjusted to give strong or weak reverberation. Multiple playback heads spaced irregularly give the most realistic results. Here again, magnetic recording has competition from digital delay systems as well as from acoustic echo chambers and metal-spring reverberators.

Speech Compression and Expansion. Speech compression is based on the discovery that portions of a spoken sound can be omitted at regular intervals,

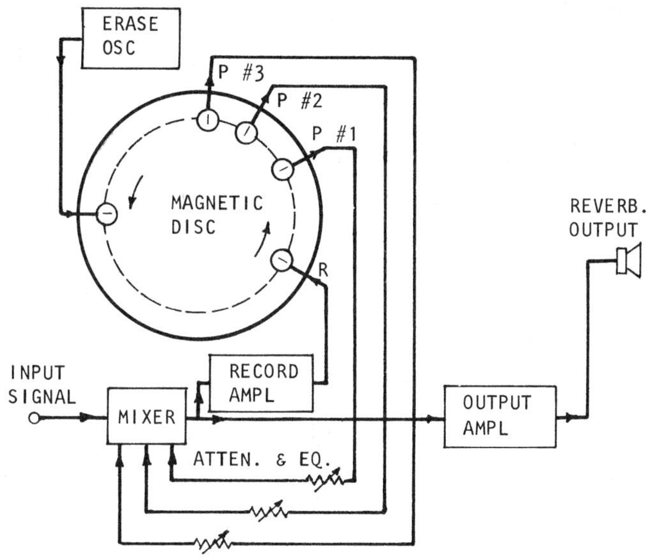

Figure 10-19. Reverberation generator with magnetic recorder.

yet the retained pieces will be almost perfectly intelligible if they are put together to close the spaces. The 2:1 compressor illustrated in Figure 10-20 uses four heads on a rotating drum; these heads omit recorded segments on the tape loop about 30 times per second while drawing out the remaining sound to half speed because the heads move in the same direction as the tape. The result is recorded, then played back at double speed. The played-back voice then talks twice as fast as originally recorded while the voice pitch remains normal. A 2:1 (100%) compression is somewhat extreme, but 50% is easily understandable and 10% to 20% hardly noticeable.

Speech can be expanded by rotating the head drum in the opposite direction, thereby adding sound segments that are repetitions of the neighboring ones. The playback recorder, slowed down, now speaks more slowly but at normal pitch. Compression or expansion is conveniently carried out with a recorded tape whose speed is controlled to fit the desired program time or talking rate while the rotating head acts as a pitch restorer. A digital memory device may be programmed to give the same results as the rotating heads.

Speech compressors-expanders are useful for adjusting programs to fit allocated time slots. They have been successful for music as well as speech. For dictation machines they can slow down the speaking rate while maintaining normal voice pitch.

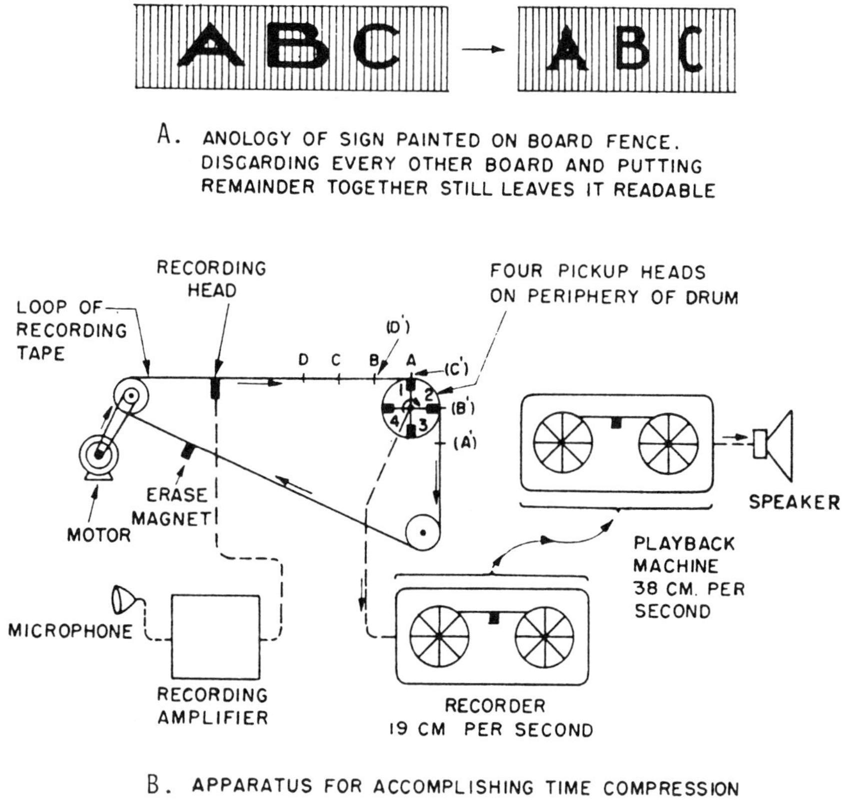

Figure 10-20. Speech compression and expansion.

DIGITAL AUDIO

Principles of digital audio were explained in Chapter 9. Digital audio is the wave of the future although its added perfection may not be necessary for most home listeners. It would probably not even be considered for home entertainment were it not for the popularity of home videotape cassettes that can store up to 8 hours of video in a medium-sized package. The same cassette will hold a comparable amount of digital audio, without excessive cost for the tape alone, although recording and duplicating are very expensive.

Table 9-3 specifies adapters that allow video cassettes to play digitally recorded audio. For home use there is competition from optical laser discs. Also other nonmagnetic discs have been proposed for marketing digital audio at a reasonable price. However, magnetic recording does offer recording capability as well as playback. This feature is most valuable when reproduced sound that comes up to digital standards is available, or when live recordings can be made.

DUPLICATION OF RECORDINGS

In 1949 a contact printing process for duplicating tape recordings (Figure 2-52) was announced but not commercialized because the market for recorded tapes was practically zero. In 1979, thirty years later, the market rose to hundreds of millions, but tapes were still not contact printed because high-speed re-recording gave somewhat higher quality and good enough economy. Equipment for high-speed re-recording is described next.

Tape-copying Machinery. A typical copying installation (Figure 10-21) consists of a *master* tape player fed into any number of *slave* recorders that copy the master program onto blank tape. In this installation the master player stores the tape as an endless loop in a glass-front bin, playing the same program over and over again at 8, 16, or 32 time normal speed. At these higher speeds all the heads and electronics operate at frequencies far above the usual 30 to 15000 Hz audio range. For example, at 16 times normal speed, the frequency band is from 480 to 240,000 Hz.

Multiple tracks, whether forward or backward, are all copied in the same pass by the appropriate number of heads, usually four for cassettes and reels, or eight for stereo cartridges. Stereo cartridges where all recordings are in the same direction are preferably copied backwards for better phase characteristics.

Figure 10-21 shows five slave recorders with large reels of tape on which the same program is recorded repeatedly. In a separate operation the recorded tapes are wound into cassettes, being cut apart and spliced to leaders at the start and finish of each recording. This installation, running at 100% capacity and 16× speed, can produce recordings for 2,560 cassettes per 8-hour day, where each cassette has 15 minutes per side of recording (5 copiers × 16 speed factor × 8 hours × 4 per hour = 2560). The capacity can be increased at any time simply by connecting more copiers to the master. One important reason for the success of these installations is that the tape transports are old reliable designs, in production for perhaps 30 years, in which time the bugs have been removed and all peculiarities are well known.

Copying, Copyrights, and Anti-Copying Inventions. Amateurs blightly copy anything and everything they can lay their hands on, and producers of recorded programs are increasingly concerned about exacting a toll from anyone and everyone who uses their material. In between are middlemen who have made huge profits from illegal or questionable copying. Aspects of these prob-

Figure 10-21. Tape duplicating system. This tape duplicator uses a "bin loop" for the master, and professional tape decks for the slave recorders to copy all four tracks simultaneously on a four track stereo tape where two tracks play in the forward direction and the other two in the reverse direction. (Courtesy Ampex Corp.)

lems are legal and technical, with weapons and counter-measures of almost military ingenuity.

One invention claims to make broadcast programs unrecordable on a tape recorder by adding to the broadcast a strong supersonic signal, which is inaudible to a listener but which beats with the bias oscillator to jam the recording.

No one has yet done it, but people talk about fame and fortune that awaits someone who invents a gadget that automatically cuts out commercials, advertisements, and messages from sponsors of broadcast programs. There is a device that distinguishes between speech and music by their difference in waveform, but all "talking" is not necessarily advertising and there are many "singing" commercials. If a silencer of commercials is ever invented, however, the program sponsors are not expected to stand by idly and allow it to be used.

Prior to 15 February 1972, recordings were not protectable by copyright, and anyone could copy and sell them without payment to the artists, producers, or original recording company; (a small fee was due the music publisher). Original recording made after that date can be copyrighted, but the recorded-tape industry has still been plagued by counterfeit tapes because they are so easily made in a concealed operation.

Still, many companies continue trying to prevent the use of recorded programs without payment. The Cartrivision company had a bright idea of leasing its recorded tapes *for one playing only* and therefore designed a cartridge that could be rewound only at the factory, not by the customer. This marketing idea never was tested for feasibility because the company discontinued operations.

EVR (Electronic Video Recording) records were supposed to be recordable only by high-resolution electron beams and special equipment owned by the parent company. Copies made by ordinary methods available to counterfeiters would be unacceptable in quality.

Videodiscs of MCA Philips will contain invisible coded digits, so that anyone who falsely claims to have made an independent recording will be identified by the secret code.

Some recorded videotape cassette programs are ''protected'' from copying by a ''copyguard'' process, which apparently produces annoying rolls and jumps in the picture. However, a copyguard stabilizer is offered for sale which removes ''copyguard'' from the tapes and eliminates the annoyances.

It is difficult to believe that any of the above schemes could not be counteracted by gadgets that someone could sell at a profit. In a free economy that legally sells ''fuzzbusters'' and bugging devices whose use is illegal, anticopying neutralizers will find a ready market.

Taxes, Subsidies, and Marketing. Copying by amateurs for their own use was of little concern until about 1970, when recorded tapes became big business. In Europe an attempted solution was to tax the sale of tape recorders, and to allocate part of the tax to musicians, producers, and so on. In the United States, the attempt has been to tax the sale of blank tape, since a reasonable amount of it is used for recording copyrighted material. While the broad issues are lobbied and litigated, program producers are fighting minor skirmishes by discouraging advertisements for blank tape and rewarding dealers who play down its use.

11. Professional Video Recording

In 1956 video tape recorders revolutionized the television industry, just as sound recorders had revolutionized the audio industry a decade before. Almost all television now is taped so it can be reviewed, edited, broadcast at a convenient hour, and repeated at any time or place in the future.

A video signal contains frequencies 300 times higher than a high-fidelity audio signal; therefore the recording of video requires a head-to-tape velocity 300 times greater if other conditions are comparable, which leads us to speeds in the order of 1000 in. per sec. The most successful method of achieving these velocities has been to mount the head on a high-speed rotating wheel, repeatedly scanning the tape from edge to edge while the tape moves forward at a relatively slow speed, enough to present a fresh surface to successive scans.

The original professional recorder, long the workhorse of broadcast studios, is the *quadruplex* video recorder shown in Figure 11-1, with four heads scanning transversely across a tape 2 in wide. More recently, less expensive recorders using slant scan and 1-in. tapes, have achieved professional status, the most popular of which are *Type C* using a single head and *Type B* with two heads.

Cassette recorders using 19.1 mm ($\frac{3}{4}$ in.) and 12.7 mm ($\frac{1}{2}$ in.) tape give a lower-quality picture with less resolution, less steadiness, less color fidelity, and more noise. However they are good enough for rehearsals, back-up, and off-line editing, and their picture can be upgraded for satisfactory broadcast of news and other location shots. The equipment is portable, inexpensive, and easy to use.

Table 11-1 gives essential information on the most common formats used in video recording.

Instant replay is a miracle that has become indispensable for sports events. Complex disc recorders for instant replay have given way to simpler helical scan types.

QUADRUPLEX VIDEO RECORDERS

Prior to the quadruplex recorder, no development in magnetic recording had ever used so many diverse and novel techniques: rotating heads, curved vacuum-shoes, servomechanisms, narrow-band FM signals, microsecond switching between multiple heads, megahertz electronic and magnetic components, tapes that withstand high head velocities, to name a few. The pioneers who made these all work together deserve exceptional credit (Ginsburg 1981).

A.

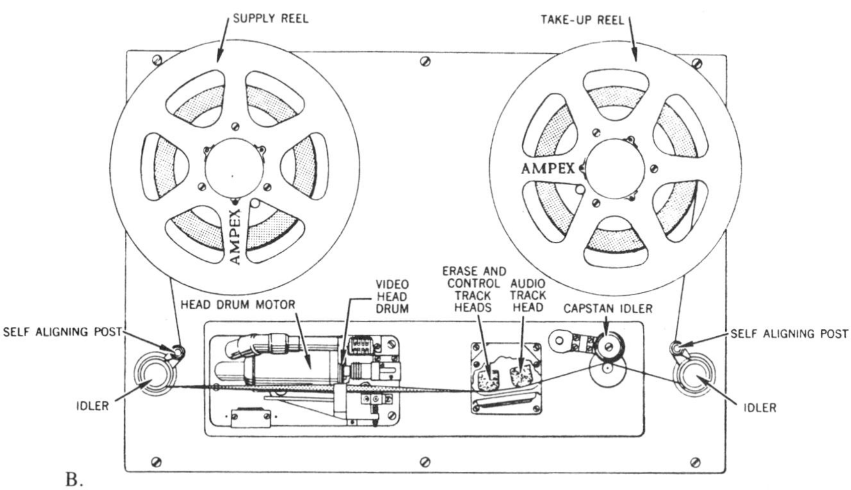

B.

Figure 11-1. The quadruplex video tape recorder. A. Type AVR-2 has push button editing facilities and a monitor bridge above, displaying the picture, waveform, and vectorscope. (Courtesy Ampex Corp.) B. Typical deck and threading path (Ginsburg 1957).

Table 11-1. Writing Speeds, Track Lengths, and Drum Diameters of Videotape Recorders

Format	Writing (Head to Tape Speed) in/s (m/s) NTSC (EBU)	Track Length Per Field NTSC (EBU)	Drum Diameter in (mm)	Drum Velocity RPM NTSC (EBU)	Tape Wrap Angle (Minimum Nominal)	No. of Heads (min. video)	Forward Tape Speed in/s (mm/s)	Track Width mils (mm)	Track Pitch mils (mm)
Quadruplex	1508 (38.3)	25.13 (638)	2.0 (5.08)	14,400	90° +	4	15 (381)	10 (0.25)	15.6 (0.40)
2 in tape	1570 (39.9)	31.4 (797)		(15,000)					
Quadruplex	1508 (38.3)	25.1 (638)	2.0 (5.08)	14,400	90° +	4	7.5 (190.5)	5 (0.12)	7.8 (0.20)
2 in tape	1570 (39.9)	31.4 (797)		(15,000)					
(half fwd-speed)									
Helical Type C	1007 (25.59)	16.8 (427)	5.30 (134.6)	3,600	346°	"1½"	9.61 (244)	5.2 (0.13)	7.1 (0.18)
1 in tape	842 (21.39)	16.2 (411)		(3,000)					
Helical Type B	934 (27.32)	15.6 (395)	1.98 (50.330)	9,000	190°	2	9.65 (245)	6.3 (0.16)	7.8 (0.20)
1 in tape		15.6 (395)		(9,000)					
Helical Open Reel	437 (11.1)	7.28 (185)	4.56 (115.82)	1,800	190°	2	7.5 (190.5)	3.9 (0.10)	6.8 (0.17)
½ in tape				(1,500)					
U-Matic Cartridge	404 (10.26)	6.73 (171)	4.33 (110.00)	1,800	190°	2	3.75 (95.3)	3.4 (0.085)	5.4 (0.14)
¾ in tape				(1,500)					
Betamax Cartridge II	272 (6.9)	4.53 (115)	2.93(74.5)	1,800	190°	2	0.787 (20)	1.5 (0.038)	1.1 (0.03)
½ in tape III				(1,500)			0.52 (13.3)		0.8 (0.02)
VHS Cartridge LP II	228 (5.8)	3.8 (97)	2.44 (62)	1,800	190°	2	0.657 (16.7)	1.0 (0.026)	1.1 (0.03)
½ in tape EP III				(1,500)			0.433 (11.0)	1.0 (0.026)	0.8 (0.02)

An understanding of quadruplex principles is basic to understanding the simpler slant-scan recorders and is especially useful because quad recorders have been the standard of perfection in professional recording. For simplicity the following explanation uses NTSC (American) standards. The description applies also to CCIR-EBU (European) standards if appropriate numbers are substituted, as given in Table 11-2. The functions described can be fulfilled with a wide variety of circuits, so only some examples have been chosen for illustration.

The Quadruplex Scanner

The heart of the quadruplex machines is its scanner (Figure 11-2), consisting of a 2-in. diameter head-wheel carrying four projecting heads spaced 90° apart,

Table 11-2. NTSC* (American) and CCIR-EBU (European) Video Standards and Corresponding Quadruplex Specifications**

	NTSC	CCIR-EBU
Horizontal picture lines/frame	525	625
Frames per second	30	25
Fields per second (2:1 interlace)	60	50
Bandwidth-Megahertz	4.2	5.5
Power line frequency-Hz	60	50
Capstan motor frequency-Hz	~60	~62.5
Head wheel (RPM)	14400	15000
Tach frequency	240	250
Scans (tracks) per frame	32	40
Horizontal picture lines/track (plus overlap)	16.4+	15.6+
Head to tape velocity	1508 in/s	39.9 m/s
Linear tape speed (full speed)	15 in/s	397 mm/s
Linear tape speed (half speed)	7.5 in/s	198 mm/s
Track width pitch (full speed)	10 mils	0.254 mm
Track width (half speed)	5 mils	0.127 mm
Track spacing (full speed)	15.6 mils	0.397 mm
Track spacing (half speed)	7.8 mils	0.198 mm
Track length (nominal total)	1.82 in	47.7 mm
Track length (non-redundant)	1.57 in	39.9 mm
Audio track width	75 mils	1.7 mm
Control track width	45 mils	1.0 mm
Cue track width	25 mils	0.6 mm
Tape width total	2.0 in	50.8 mm
Tape thickness	0.8 to 1.5 mils	20 to 38μm

*NTSC = National Television Standards Committee (USA).
**CCIR = Consultative Committee for International Radio.
**EBU = European Broadcasters Union.

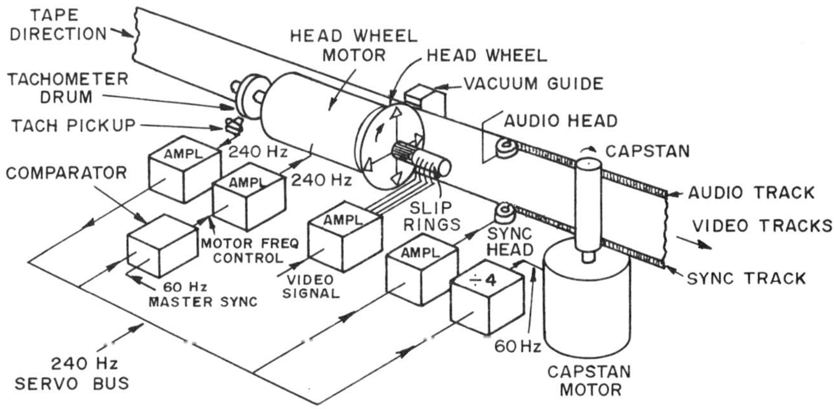

Figure 11-2. Quadruplex scanning system in recording mode (After Zahn 1979).

the wheel being mounted on the shaft of a synchronous motor rotating at 14,400 RPM (240 RPS) for NTSC (American) standards. A tape guide curls the 2-in. tape into the shape of a quarter of a cylinder in the region opposite the head wheel so that each head traces laterally across the tape from edge to edge. As seen in Figure 11-3 the tape guide is a curved shoe with a vacuum chamber that contours the tape to conform with the scanning head path.

Figure 11-4 shows the dimensions and track format of a quadruplex tape. Each quarter revolution of head wheel (90°) takes $\frac{1}{960}$ of a second, while each horizontal TV picture line (NTSC) takes $\frac{1}{15735}$ of a second. Thus a 90° travel of the head across the tape records about 16.4 horizontal picture lines. The information for a full NTSC video frame is contained in exactly 32 scans across the 2 in. tape. The 525 line NTSC picture divided by 32 tape scans verifies the result of approximately 16.4 picture lines per 90° head traverse.

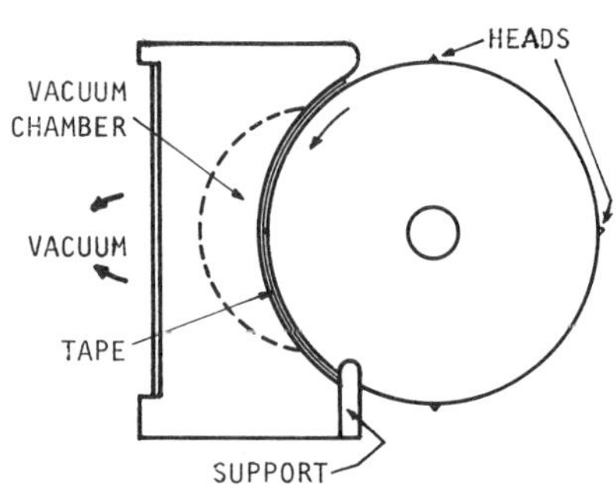

Figure 11-3. Vacuum guide for tape contour while scanning.

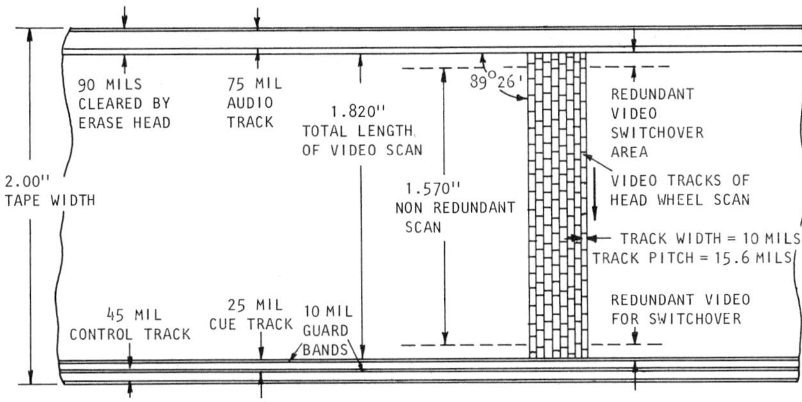

Figure 11-4. Tape and track dimensions of quadruplex tape (NTSC). See Table 11-2 for CCIR-EBU that apply to 625/50 PAL/SECAM.

Actually the tape wrap is more than 90°, giving some overlapped information for electronic splicing.

After every 16 or 17 picture lines, a different one of the 4 heads takes over the scanning of the tape, It is understandable that changes in picture quality can be noticed after every group of about 16 picture lines if the quadruplex machines are not properly adjusted. As seen in Figure 11-5, the misadjustments can be unmatched heads or head amplifiers, heads not displaced 90° from one another, or differences in tape guide shoe alignment between recording and playback.

Figure 11-6 is a detail of the head wheel and heads. The heads usually have a composite core, with pole tips of sendust and the main body or yoke of ferrite as shown in Figure 4-22. In earlier machines connections to the head windings were made through brushes which rode on slip rings. Later models eliminated the slip rings by coupling through a rotating transformer where the primary spins with the head wheel while the secondary is stationary.

Quadruplex Electronics

Recording the video spectrum is complicated by the fact that an ideal video signal has a flat frequency response from DC to 4.5 mHz with minimal phase shift. With DC restoration techniques, the low-frequency limit need not extend below 20 Hz but the phase shift at low frequencies must still be small. To cope with this broad band requirement of some 18 octaves, the recording is in the form of an FM carrier modulated by the video signal. Carrier recording necessarily requires more bandwidth than if the signal were recorded directly, but bandwidth is conserved as much as possible by using narrow band FM and

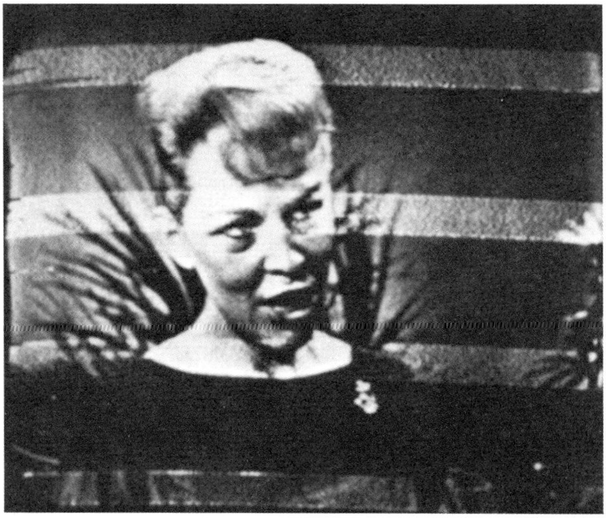

A. One of the four heads or amplifiers has low output

B. Venetian blind effect due to improper adjustment of vacuum guide (head penetration)

Figure 11-5. Effect of misadjustments in a quadruplex recorder. *A*, *B*, and *C* show various "banding" effects (Chinn 1957).

C. Heads not accurately 90° apart

D. Dropouts caused by insufficient head penetration, worn tape, etc.

Figure 11-5. (*Continued*)

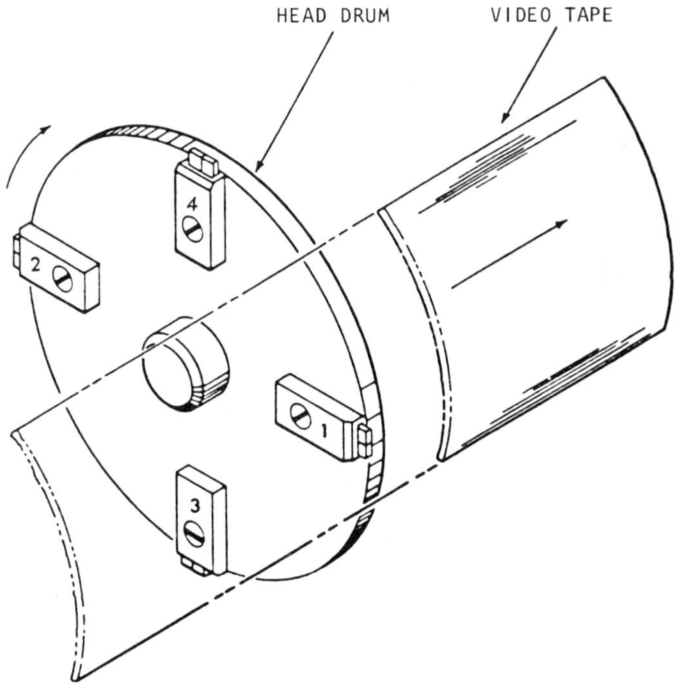

Figure 11-6. Head wheel of quadruplex recorder. (Courtesy Ampex Corp.)

vestigial sideband modulation. A great advantage of carrier recording is that long wavelengths are avoided, giving no perceptible print-through and minimum interference between adjacent recorded tracks. Heads and signal amplifiers do not handle low frequencies and hence are immune to hum and stray fields from nearby motors and transformers.

Recording Electronics. As shown in Figure 11-7, one way to generate the FM recording signal is with an oscillator whose frequency is voltage controlled, the control voltage being the video input signal. A multivibrator oscillator (Figure 11-7*A*) is widely used because it is simple and reliable. Its frequency varies with the control electrode bias and its somewhat rectangular waveform and coincident amplitude modulation can be filtered and shaped if necessary. An alternative FM generator is the oscillator shown Figure 11-7*B*, which has a tuned LC circuit where the tuning capacitance or inductance is modulated by an "amplifier" that simulates a variable reactance. Solid state variactors whose capacitance changes with bias voltage are also available as variable tuning capacitors for modulated FM generators.

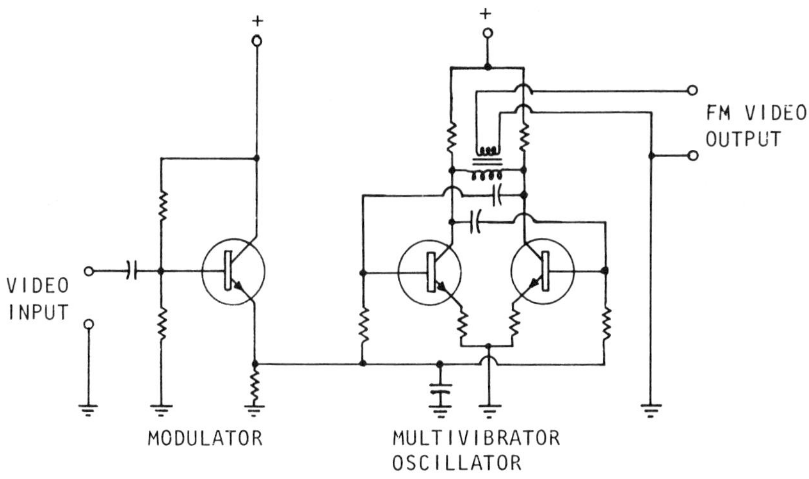

A. MULTIVIBRATOR FM GENERATOR

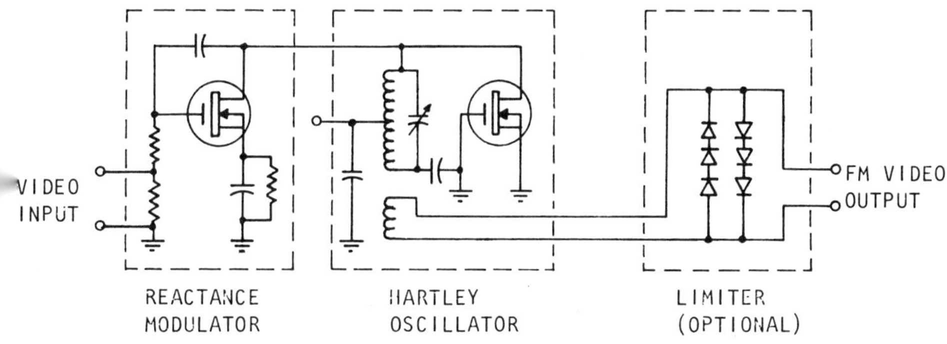

B. REACTANCE MODULATED FM GENERATOR

Figure 11-7. FM recording signal generated by voltage controlled oscillator (VCO).

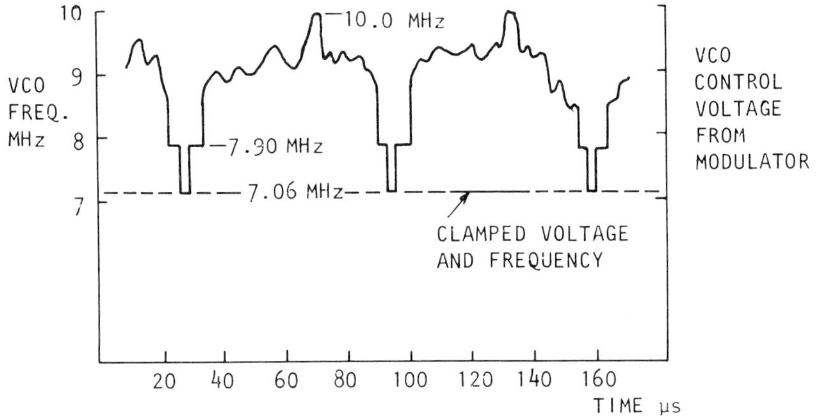

A. FREQUENCY OF VCO WHEN MODULATED BY A VIDEO SIGNAL

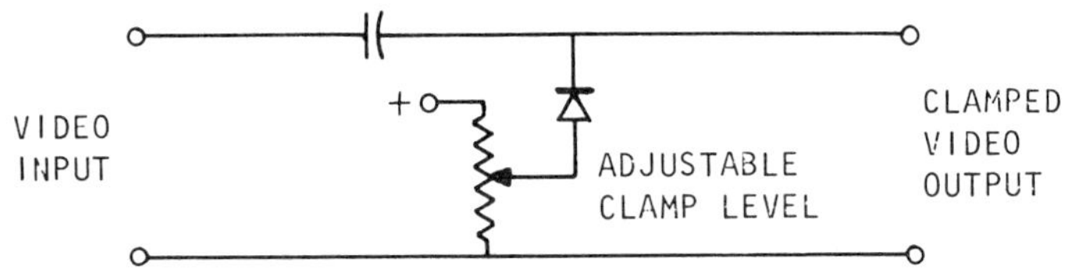

B. CLAMPING CIRCUIT FOR HOLDING SYNC TIPS AT DESIRED FREQ.

Figure 11-8. Frequency excursion of VCO modulated by a video signal.

The frequency of such oscillators may be maintained accurately by a phase-locked loop referred to a *crystal standard*. Because of the nature of television signals, the reference frequency is locked to the sync tip level or to the blanking level rather than to an average carrier, as seen in Figure 11-8*A*. The sync tips are represented by the lowest possible video frequency while the average carrier depends on the picture content. This locking is obtained by a clamper circuit (Figure 11-8*B*) that holds the modulator signal sync tips or blanking always to a fixed voltage that, when applied to the voltage controlled oscillator, gives the chosen reference frequency.

The modulated FM signal (Figure 11-9) is then divided among four similar

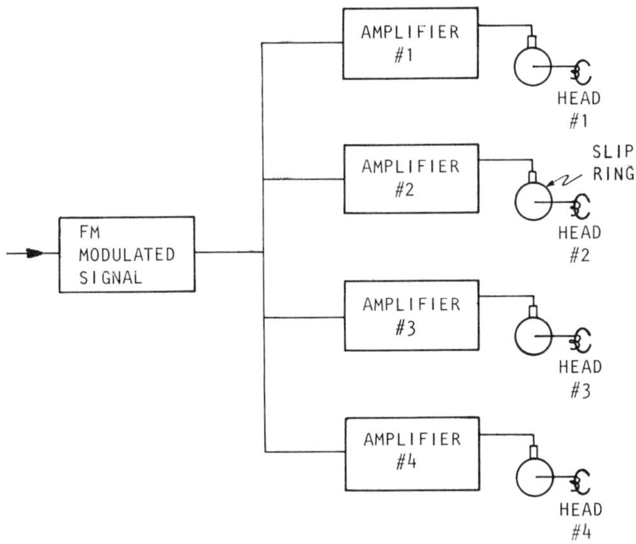

Figure 11-9. Distribution of recording signal to quad heads.

amplifiers, each connected to energize one of the four recording heads on the head wheel. By using four separate amplifiers each one can be adjusted to compensate for the efficiency of its own head, giving a uniform recording on the tape although made by four different heads. Not only amplitude and frequency response can be compensated, but also errors in head position on the wheel can be taken care of by adjustable delay lines in the driver circuit for each head.

Metering and Monitoring. A sample of the recording signal is demodulated and applied to monitor cathode ray tubes that display the waveforms, a vectorscope diagram, and a picture of the recording signal applied to the heads. A gating circuit samples the recording signal and indicates its frequency on a meter. Such monitoring, known as *E-to-E display*, tells an experienced operator where to make adjustments or where to look for troubles.

Playback Electronics. The important feature of played-back video is that it is discontinuous, having been broken by the recording heads into many parallel separate tracks, which must be patched together again during playback to give the same electronic signal that would be obtained from a single long continuous track. Fortunately the video signal itself is divided into horizontal and vertical scans separated by invisible blanked-out intervals, so that electronic splicing can be done in the blanking period as will be described.

Playback Switching. With heads 90° apart on the head wheel, any tape wrap beyond 90° records duplicate information at each edge of the tape since the second head begins its scan before the first head has left the tape, as seen in Figure 11-4. The effective tape wrap is about 113° to 120°, giving about 18.5 horizontal video lines per head-scan after the tape edges have been cleared for audio, sync, and cue tracks. There are exactly 32 head scans per NTSC video frame ($\frac{1}{30}$ second, eight revolutions of head wheel) and one scan averages $\frac{525}{32} = 16.4+$ horizontal video lines. The 18.5 line path therefore gives us about one redundant horizontal line at the beginning and one at the end of each track, the redundant lines being exact duplicates so that we can just as well use one or the other.

The fractional average number of needed lines per scan is not a problem because we switch only in the blanking period between whole lines to make the switching transients invisible. On some scans we use 17 of the 18.5 recorded video lines, on other scans only 16; but the sum total always adds up to the required 525 lines per frame.

A four-gate head switcher is illustrated in Figure 11-10, where heads *1, 2, 3, 4* on the head wheel are assigned respectively to gates *1, 2, 3, 4*. The gates block the head signals unless two conditions are met: (a) the head signal is raised on a positive pedestal, and (b) the switcher signal is positive for that head. To fulfill the first condition (a) a set of four 240 flat-top voltage waves are derived from the 240 Hz tach signal, as shown in Figure 11-11*B, C, E, G, I*. These waves, progressively 90° apart in phase, are added to the head signals such that whichever head is active, that head signal is elevated by the positive

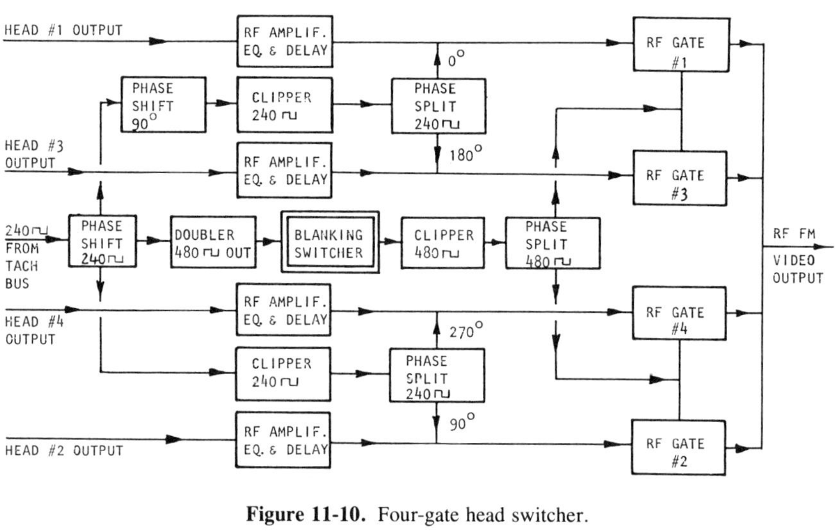

Figure 11-10. Four-gate head switcher.

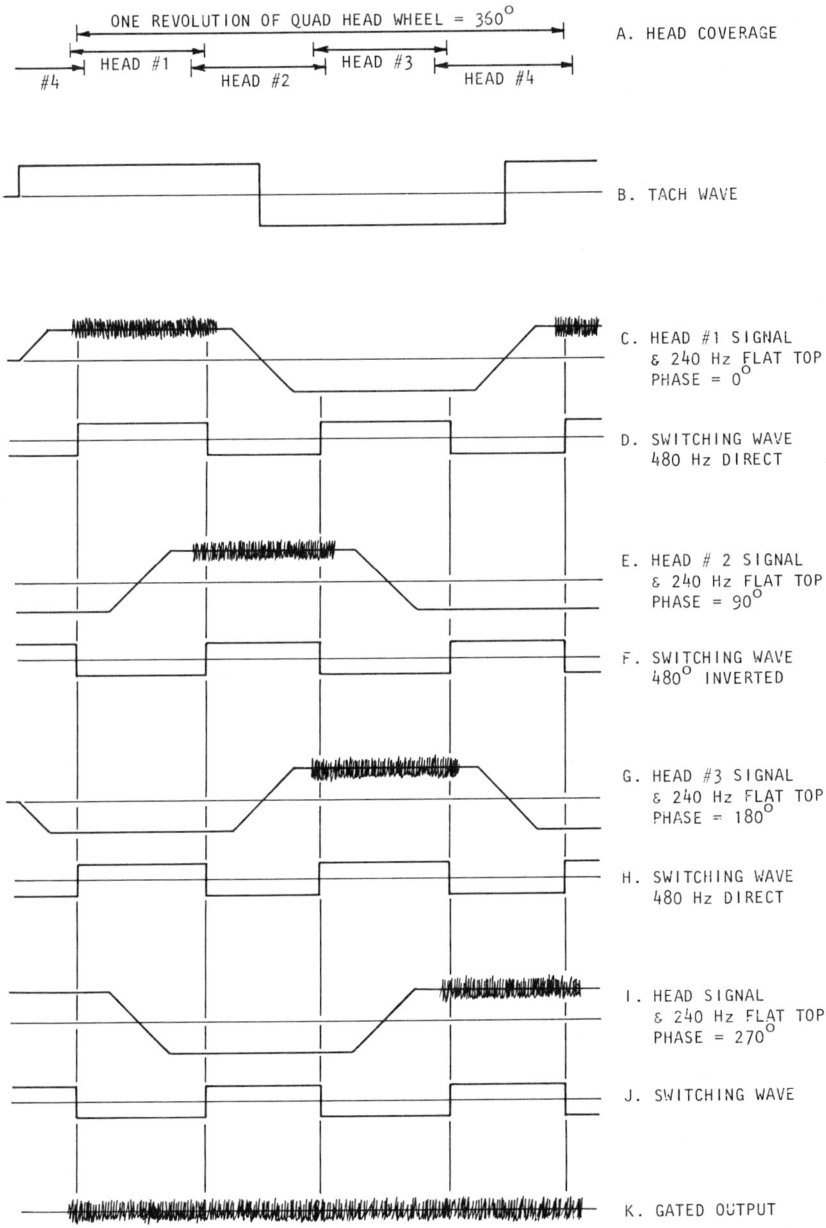

Figure 11-11. Waveforms of 4×1 head switcher.

pedestal of the added flat top wave. Because the head signals overlap, they cannot be combined by addition, but must await action of the switcher, condition (b), which determines the exact time during the overlap that a head is turned off while simultaneously the next head is turned on. The switcher wave (Figure 11-11*D, F, H,* and *J*) is a 480 Hz square wave derived by doubling the 240 Hz tach frequency and influencing the 480 Hz crossover points by played back sync pulses, such that crossover occurs only during the blanking period between horizontal video lines. The time between crossovers (polarity changes) of the switcher is therefore 16 or 17 horizontal intervals, depending on timing of the horizontal rate trigger pulses just when the 480 Hz generator waves are ready to reverse polarity. In Figure 11-11 the switching intervals are longer for heads *1* and *3*, so these give 17 picture lines each, while the shorter intervals for heads *2* and *4* give 16 picture lines apiece. The pattern of 16 and 17 changes gradually according to the incommensurate numbers 525 and 32.

During playback the tape runs through the same path as for recording. The same scanner is used, but heads are now connected to playback circuits and erase heads are turned off. Figure 11-10 shows that each head is connected to its own amplifier, equalizer, and delay corrector (if used) for balancing the separate outputs. These are fed in sequence into a single channel, the sequencing being accomplished by the switcher gate circuits that allow only one head signal at a time to come through. At this stage the recombined video is still a modulated FM carrier which next passes through amplifiers and limiters, then through a demodulator whose output is the video signal. Further stages clean up and reshape the sync pulses as well as remove transients and so on picked up during the complex record-play process.

Sync and Servo Methods

The ultimate video playback signal for a TV broadcast system must be in perfect synchronism line for line and frame for frame with the master sync generator of the broadcast system. Therefore the playback scanner must be locked into this external master sync signal, and the tape motion must be corrected constantly so that its recorded tracks will always register with the scanner heads. These requirements are met by using a sync track made on the tape simultaneously with the recorded video, and by controlling the playback tape speed so that its sync track locks with the external master. At the same time the heads on the head wheel are controlled to retrace the same paths they followed during recording. Note that slight errors in recording speed or rate are tolerable as long as the line and frame standards (as NTSC) are observed, since these deviations are corrected during play-back. Similarly, when playing back through monitors or individual TV sets not on a network system, the playback need not be at perfect standard speed as the monitors will lock to the tape player.

Recording Synchronization. During recording, a set of sync marks are imprinted adjacent to an edge of the tape by a 240 Hz square wave, giving one positive and one negative transition for each revolution of the head wheel. These magnetic marks act like invisible sprocket holes to enable exact synchronization of tape during playback. The sync or control track is shown on the lower edge of the tape in Figure 11-4. It is recorded with a head that superimposes 240 Hz sync signals right over the video recording, which otherwise extends completely across the tape. The sync signals are so strong that they clear the track without need for a prior erase head. Notice that one transition pulse is visible for every two video tracks imprinted by the four heads that move across the tape for every revolution of the head wheel.

Head wheel speed and tape forward movement in the recording mode are controlled by servomechanisms (servos) locked to the sync rate of the incoming video signal. Servomechanisms are seldom found in simple recorders, but are a most important component of professional video recorders. In video recorders servos are used as electromechanical gears and sprockets that lock together the tape, the capstan, the head wheel, and the vacuum shoe, keeping them all in step with each other and with a chosen master reference sync. To gain insight into the cooperation of various members of the servo chain, it is helpful to keep track of which is the controller, what is being controlled, and what is the control operating frequency.

Servo Control: Recording Mode. In the recording mode shown in Figure 11-12, switch *A* can choose the master reference to be the vertical frequency of NTSC color video (59.94005236 fields/s or 29.97002618 frames/s) or the black-and-white standard of 60 fields/s or the local power line frequency, less accurate at nominally 60 Hz. At *B* the signal is processed to give output pulses of vertical master frequency, designated by the round number 60 Hz. By means of comparator *C*, the signal from *B* controls the so-called tach (tachometer) frequency of 240 Hz, making it precisely four times the master frequency. This important 240 Hz tach signal is derived as a by-product of the head wheel rotation that is 240 revolutions/s. The head wheel *G* is driven by a three-phase 240 Hz motor (*F*) powered by an oscillator, amplifier, and phase splitter *D E*. The motor also carries a tachometer drum *H* with magnetic inserts that generate electric pulses in a pickup coil connected to the tachometer amplifier and shaper *I*. Tach output of *I* is a 240 Hz square wave, with its $\pm$ transitions governed accurately by the head drum pulses. A double insert on one side of the drum distinguishes the positive from the negative transitions so there is no ambiguity between head wheel position and polarity of tach signal. The head drum comparator *C* not only sets the frequency of oscillator *D* to 240 Hz but also controls its phase to a small fraction of a cycle such that the head wheel angular position locks to the 60 Hz sync pulses fed to *C*.

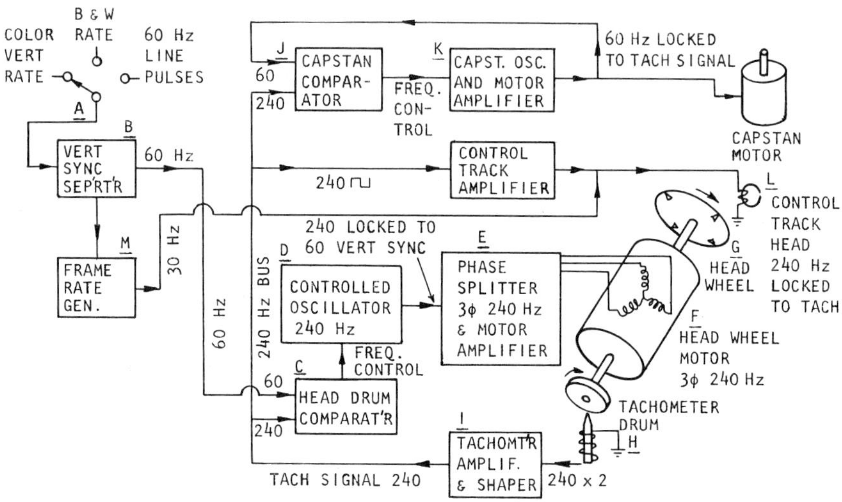

Figure 11-12. Recording mode. Quadruplex servo and sync.

While the head wheel rides across the tape the capstan must advance the tape 15.6 mils for each scan or 15.6 × 4 = 62.4 mils per revolution of the head wheel. The capstan motor is "geared" to the head wheel by means of comparator *J* which locks the motor oscillator-amplifier and therefore the capstan to the 240 Hz tach frequency. The capstan is driven by a 60 Hz single-phase hysteresis-synchronous motor. As the tape moves in the longitudinal direction, its control track is marked by head *L* with the 240 Hz tach reference signal. Superposed on the 240 Hz reference are some 30 Hz markers from the frame pulse generator *M*. These markers distinguish between odd and even TV fields and are useful for editing the video.

Servo Control: Playback Mode. If the playback is to be broadcast, its picture information must be read out in step with the master sync of the broadcast station. Figure 11-13 shows that the servo control for playback is similar to the recording servo except that the control track head *L* plays back through the amplifier *N* and phase (tracking) control *O*, regulating the capstan motor speed such that the control track gives a frequency of 240 Hz or exactly four times the external master sync. The control track frequency is locked by comparator *J* to the 240 Hz tach bus, which in turn is locked (in 4-to-1 ratio) to the external master sync from the vertical sync separator *B*. Thus the servos gear together all the essential elements: the head wheel, the forward tape movement, and the master sync.

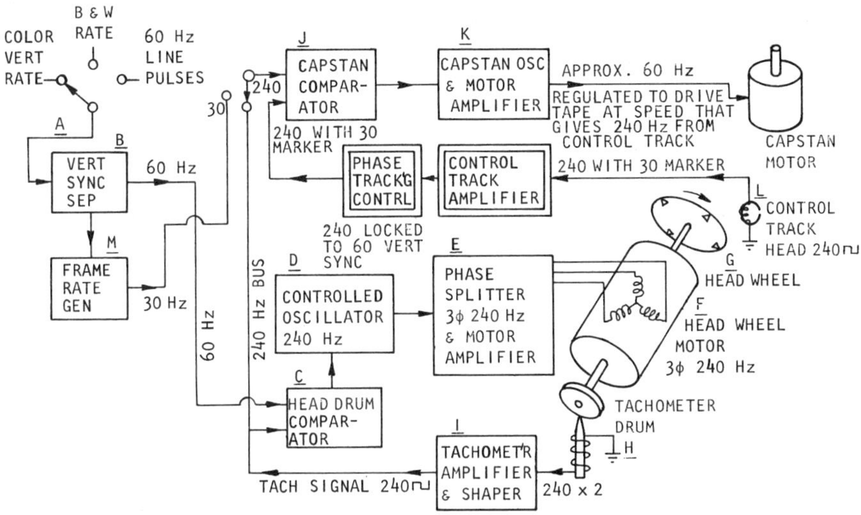

Figure 11-13. Playback mode. Quadruplex servo and sync.

Fine Tuning Phase Control. Although everything is moving at exactly the right speed, it is possible for the head wheel to be off-track if, for example, the control head on the playback machine is spaced a little differently from the head on the recording machine, or if the tape shrinks a little. We recognize this as a phase error readily corrected by adjusting the relative phase a few degrees forward or back with the phase controller *O*. The controller can be adjusted manually by turning the tracking knob until the best picture is tuned in, or in some situations the phase can be adjusted automatically by an additional servo for this purpose.

Small timing errors occur also in the other aspects of video playback. All of these are correctable by fine tuning the phase adjustment *O* in Figure 11-13. The details have purposely been omitted from Figure 11-12 and Figure 11-13 to prevent unnecessary complication in explaining basic servo action. In fact the early recorders did not have fine-tuning features, but as deficiencies were recognized, later models included more and more refinements either built-in or as optional accessories. Further details of these refinements are given in the next section.

Sync Regeneration

As the video goes through the various stages, it picks up noise and distortion. One way to improve the signal is to renew its blanking and sync pulses by discarding the old ones, literally stripping them off and replacing them with fresh, clean pulses. This can be done because the stream of sync pulses is regular and predictable, so we know in advance that noise and irregularities are spurious and were not in the original signal. Only the timing edges of the original sync are used as references in generating the fresh series. The module that regenerates the sync is called a *processing amplifier* (proc amp). In an analogous manner, the Colortec® system regenerates a color burst to replace a less perfect burst in the original video. Unwittingly perhaps we have applied information theory, using redundant information to remove chance imperfections that do not belong in the signal.

Frame Rate Sync. For perfect correspondence with studio master sync, the 30 Hz frame sequence must be observed since the sync sequence is different between the odd and even 60 Hz fields. Locking at the frame rate is accomplished by feeding the frame pulse to the control track by generator *M* in Figure 11-12, the frame pulse appearing in every eighth cycle on the 240 Hz tach rate control track. In playback of Figure 11-13, the capstan comparator *J* is automatically switched to the 30 Hz signal from *M* during start-up of the video player. As soon as the tape is locked to the 30 Hz marker, the comparator *J* changes to the more accurate 240 Hz mode while remaining correctly in sync with the frame rate.

Precision Lock-in. Stability of the basic head wheel servo as shown in Figure 11-12 is about ± 1 microsecond. This amounts to about eight horizontal picture elements, which would be quite noticeable except that the display unit follows and corrects such instabilities if they are reasonably slow, as is true of quadruplex servos (4 to 8 Hz rate). However, microsecond instabilities are intolerable in broadcast networks where the playback sync would be fighting the master sync of the network.

Closer control is obtained with additional servos available under trade names such as Intersync®, which is a phase-adjusting network inserted at *P* in Figure 11-13. Intersync is locked more finely to the tach frequency and improves the picture stability to ± 150 nanosecond, or within about one picture element.

Finer tolerance is impractical with mechanical servo correction, but is achieved electronically with a controlled delay line. The Amtec® corrector speeds up or slows down the passage of the signal through a delay line consisting of an inductance tapped at 88 points by *voltage controlled capacitors* (varicaps). The resulting timebase error is reduced to ± 30 nanoseconds, "perfect" for black-and-white video but not quite good enough for color video.

Colortec® refines this even further to ± 3 nanoseconds by a controlled delay line that adjusts the playback color bursts to be in phase with the 3.579545

MHz network standard. Such ultimate perfection is required when the playback colors must blend perfectly with live signals and studio originated effects such as superposed color titles.

Audio Recording

Audio that accompanies the video is recorded in a straightforward manner on the top edge of the tape after an erase head clears this space of superfluous video scans, as in Figure 11-4. The audio head is 9.0 inches upstream of the video wheel, so this separation of sound and picture must be considered when editing a tape by cutting and splicing. A monitor head is sometimes included for simultaneous listening while recording, but since the monitor point is displaced from the recording head, only the latter should be used for perfect sound sync during playback. High-fidelity recording is possible at the usual 15 in/s tape speed or at 7.5 in/s half-speed operation, even though the tape particles are oriented crosswise to favor the video recording. Stereo can be added by substituting an audio head having a pair of narrower tracks.

Cue Track. Just above the control track in Figure 11-4 is the cue track, reserved for voice remarks by the operator as to the nature of the recording, date, time, editing cues, and so on. It also receives, if desired, a time code for indexing to the exact video frame, either alone or together with other video or audio tapes or motion picture films. Automatic indexing and editing are possible by signals from the cue track.

Dropout Correction

Even the most perfect tape contains non-uniformities, mechanical and magnetic. Although they may hardly be noticeable in a fresh tape, they become worse with usage or successive rerecording, until their loss can no longer be corrected by the FM limiter action. Dropouts are then noticed on the playback picture, being white dots or streaks whenever the picture carrier is lost. Since the white blotches are particularly annoying, the earliest dropout correctors substituted a gray or a black streak in these intervals by switching quickly to an artificial carrier during the dropout or to a gray video level.

A superior way to improve dropout correction is to repeat the picture information of the previous scanning line or frame, substituting it for the lost information during the dropout. Figure 11-14 shows that this is done by storing the picture for 63.5μs in a delay line while constantly updating the delayed information. Tunnel diodes may be used for extremely fast switching to the delay line and back to current information. Here again we make use of information

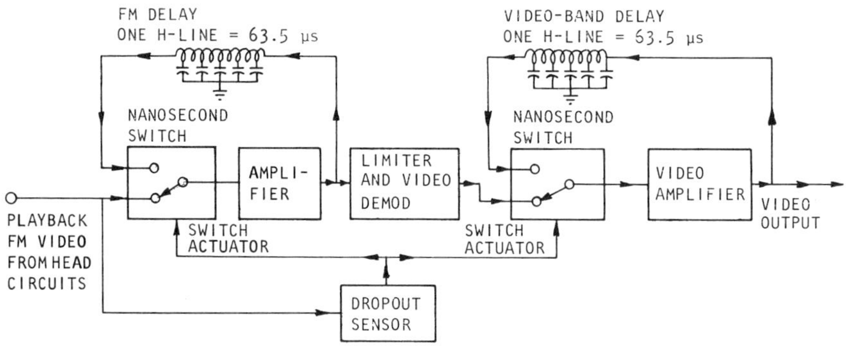

Figure 11-14. Dropout corrector for video.

theory to patch up a total loss of signal with something reasonable, and it is reasonable to assume that each successive scan line will not be much different from the previous line since pictures are somewhat redundant in content. Moreover the human mind is an excellent information processor that accepts, corrects, and supplies missing or offset details to give an overall satisfying impression.

Table 11-3. Quadruplex Characteristics, High Band and Low Band

System	Video Freq. Range MHz	RF Frequencies—MHz Sync Tip	Blanking (Carrier)	Peak White	Upper Sidebd	Head Wheel RPM	Head to Tape Speed	Track Angle from 90°
Low-band Monochr 525 line 30 frame	4.2	4.28	5.00	6.80	11.0	14,400	1508 in./s	0.57°
Low-band color 525 line 30 frame	4.2	5.50	5.79	6.50	10.7	14,400	1508 in./s	0.57°
High-band color 525 line 30 frame NTSC	4.5	7.06	7.90	10.00	14.5	14,400	1508 in./s	0.57°
Low-band 625 line 25 frame	5.0	5.00	5.50	6.80	11.8	15,000	39.9 m/s	0.55°
High-band 625 line 25 frame	5.5	7.60	7.80	9.30	14.8	15,000	39.9 m/s	0.55°

Editing. Quadruplex videotapes were at first edited by cutting and splicing, but this practice has been replaced by electronic editing that rerecords the desired parts on another tape. Editing procedure is described at the end of this chapter.

Quadruplex Specifications. As video recorders evolved from monochrome to color, low-band to high-band, vacuum tube to solid state, their specifications have improved accordingly. Most recorders in service are not the very latest models and therefore their operating specifications are still of interest. Table 11-3 lists some characteristics of low-band and high-band quad recorders. Table 11-4 summarizes other quad data of general interest.

Table 11-4. Quadruplex Video Recorder Data of General Interest

a) Angle of recorded track to forward tape direction, at 15 in./s tape speed = 89° 26′ (approx. 89.5°)
   Angle recorded track to forward tape direction at 7.5 in./s tape speed = 89° 43′ (approx. 89.75°)
b) Head wheel angle corresponding to a horizontal 63.5 μs video line ≅ 5.4°
c) Head wheel angle corresponding to a horizontal (525) 4.5 MHz picture element = 0.01° = 36″
d) Head wheel angle corresponding to maximum allowable (4°) phase shift of color burst = 0.0025° = 9.14″
e) Head gap size ranges from 0.08 mils (2 μm) for older equipment to 0.045 mils (1.18 μm) in later models
f) Heads on head wheel are adjusted 90° apart to a tolerance of ±30 sec of arc (0.0083°)
g) Inductance of heads, low band recorders ≅ 25μH; high band recorders ≅ <0.5μH
h) Self-resonant freq. (heads + wiring), low band recorders ≅ 6.5 MHz; high band recorders ≅ 15 MHz
i) The tachometer wheel on some older Ampex equipment is photoelectric rather than magnetic
j) Head wheel motors on some RCA equipment operate with 330 Hz, 3 phase instead of 240 Hz
k) A blower at the head wheel cools the motor and cleans the tape surface
l) A mechanical footage or "time" counter is built into some models for indexing
m) Air posts are used as tape guides to reduce tape friction on some RCA equipment
n) Approximately 9000 quad recorders were used worldwide in 1979; the principal manufacturers being Ampex, RCA and Fernseh (Bosch, Germany)
o) Standard tapes are available for "setting up" quad recorders, and are essential for adjusting the tape wrap vacuum guide.
p) Head tip penetration into the plane of the tape is approx. 0.002 in. (50 microns)
q) A complete video frame (two fields) occupies 32 quad tracks (NTSC 525 lines)
r) A complete video frame (two fields) occupies 40 quad tracks (CCIR-EBU 625 lines)
s) Standard recorded tapes are available and are essential for in-the-field adjustment of the vacuum guide, tape tension, etc.
t) Cue track response is modified by a notch filter tuned to sync track frequency (240 or 250 Hz) to remove interference from adjacent track. Early models use a 300 Hz low-frequency-cutoff filter.

HELICAL SCAN RECORDERS

Helical scan recorders are also known as *slant scan* and *oblique scan* machines. They cost less than quads, use less tape, are smaller and lighter, are easier to operate and to maintain, and are adaptable for slow motion and stop motion. These advantages were recognized early in the history of video tape recorders, and vigorous development programs led to commercial introduction of helical scan recorders soon after the quadruplex. The first helical scan machines used 2 in (50.8 mm) wide tapes like the quads, followed afterwards by 1 in., $\frac{1}{2}$ in., and even $\frac{1}{4}$ in. tapes (25.4, 12.7, 6.3 mm) as well as intermediate sizes like $\frac{3}{4}$ in. (19.1 mm), and 0.315 in (8 mm).

The original idea of a helical scan machine was to write a long enough track to record a complete video picture field in one uninterrupted headscan, as in Figure 11-15*A*. This track length is obtained by wrapping the tape helically around a cylindrical guide, (Figure 11-15*B*) while a head wheel inside the guide traces a circular path perpendicular to the cylinder axis. As the tape advances

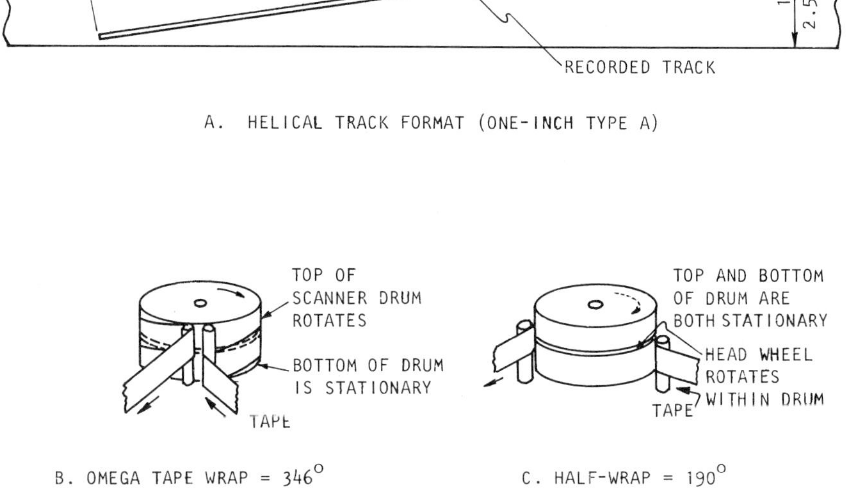

Figure 11-15. Track format and tape paths of a helical scan videotape.

in its own forward direction, the head wheel lays down a series of parallel slanted tracks.

The earlier years of helical recorders were chaotic: more than a dozen companies were producing recorders of different head speeds, track angles, and tape sizes according to their designers' whims, without interchangeability even among recorders of the same model and manufacturer. (Quadruplex recorders, on the other hand, were standardized almost immediately to give interchangeable tapes). With helical recordings, technical obstacles made interchange unattainable, and therefore the marketing was directed toward users who would record and play on the same machine. In this case, standardization was unimportant.

Experience proved that the numerous advantages of helical scan were accompanied by problems. In helical recording, there is a component of capstan-driven forward motion that gives serious amounts of flutter compared to perpendicular quad recording, where this component has no effect and where the flywheel action of the head wheel gives excellent stability. Also, on long scans, the slant track is not exactly a straight line because the tape is not tensioned uniformly from edge to edge and therefore narrow tracks do not register with the heads if the tape stretches or shrinks (skew error). The tracking errors, instability, and noninterchangeability were not too serious for amateur and closed-circuit work, but were unsatisfactory for studios and broadcasting.

For about 20 years helical scan recorders served a lower quality but very rewarding marketplace, with periodic announcements that certain models had made a breakthrough to quadruplex standards followed by equally regular disappointments with the new equipment. Eventually, helical recorders and appropriate corrective devices became good enough, and high-quality 1-in. recorders have been accepted. Two formats are most popular: Type C, with an omega wrap ($\approx 350°$) and single-head scanner; and Type B, with a half wrap ($\approx 190°$) and double-head scanner. The older 1-in. Type A format is considered obsolescent.

Design Considerations

For a complete NTSC field in one sweep, each slant track must contain 262.5 H-lines in the NTSC system, or 312.5 H-lines in the CCIR-EBU system. Two adjacent tracks constitute the odd and even fields, which together make an interlaced frame of 525 or 625 total lines. The length of the slant track is roughly equal to the circumference of tape wrapped around the drum. The required length has to accommodate 75,000 to 90,000 wavelengths (4,500,000 Hz video band $\times \frac{1}{60}$ sec $= 75{,}000$ NTSC/ while $4{,}500{,}000 \times \frac{1}{50} = 90{,}000$ CCIR-EBU). In fact, carrier recording requires at least twice as many wavelengths per track.

Track Length. Experience with state-of-the-art recording densities influences the track length decision. At one extreme is the quadruplex CCIR standard where each video field uses $20 \times 1.57 = 31.4$ in (797.6 mm), corresponding to about 6000 wavelengths/in (226/mm) with carrier recording under the most optimistic (2:1 basis) estimates. To record such a long track helically, we need a head wheel diameter of 10 to 20 in. (254 to 508 mm), giving a large clumsy machine. The small track angle (about 1°) requires fast tape movement, so there is little or no tape economy advantage over quadruplex.

At the other extreme, if we can achieve densities of 60,000 wave-lengths/in (2260/mm) then the total track length need be only about 3.14 in. (80 mm). Such high densities have been too critical for the normal variations in heads, tapes, and adjustments of manufacturable machines of consistent professional quality.

Practical machines have compromised on track lengths of about 7 in. (180 mm) to 18 in. (460 mm). The shorter tracks do not give the full video bandwidth and cannot record color video directly but must use a transposed color carrier.

Tape Economy vs. Performance. A narrow tape running at slow speed gives the greatest economy of tape use, but at the expense of performance. Narrower channels give noisier recordings and are difficult to track, while slow tape speed sacrifices picture resolution and adds noise and distortion. Table 11-5 compares the square feet (or square meters) of tape used per hour of re-

Table 11-5. Tape Economy of VTR Formats*

Format	Forward Tape Speed	Tape Consumption	
	in/s (mm/s)	ft/hr (m/hr)	ft^2/hr (m^2/hr)
Quadruplex 2-in. tape	15 (381)	4500 (1372)	750 (69.69)
Quadruplex 2-in. tape	7.5 (190.5)	2250 (686)	375 (34.84)
Helical Type C 1-in. tape	9.61 (244)	2880 (878)	240 (22.32)
Helical Type B 1-in. tape	9.65 (245) 2.41 (61.3)(LP)	2895 (882) 723 (221)	241 (22.41) 60.3 (5.61)
Helical Open Reel $\frac{1}{2}$-in. tape	7.5 (190.5)	2250 (686)	93.75 (8.71)
U-Matic Cartridge $\frac{3}{4}$-in. tape	3.75 (95.3)	1125 (343)	70.3 (6.53)
Betamax Cartridge $\frac{1}{2}$ in. tape	0.787 (20)	236 (72)	9.84 (0.914)
VHS Cartridge	0.657 (16.7)	197 (60)	8.21 (0.763)

*Calculated for 525 lines, 60 fields; but 625 line, 50 fields gives almost the same result.

cording for different formats. Quad recorders standardized in 1956 are somewhat wasteful of modern high-density tapes, while consumer recorders are 90 times more economical but have marginal picture quality. State-of-the-art helical recorders meeting broadcast specifications use about 22 m^2 of tape per hour, about a third as much as the full-speed quadruplex.

Omega Helical Scan. Only one head is necessary on the head wheel if the tape is wrapped nearly a full revolution around the drum guide to give the so called Omega (Ω) format of Figure 11-15*B*. A single head is less costly than two and eliminates problems of multiple heads: matching, balancing, aligning, connecting, switching, and so on. The drum diameter is minimal because the track length is nearly equal to its entire circumference. Still pictures and slow motion are readily obtained by stopping or slowing the forward tape movement.

The main problem with a simple Omega format is a vertical information gap. A complete 360° wrap is impractical for mechanical reasons and because some of the tape width must be reserved for audio and sync tracks. Fortunately, the video signal itself is blanked out at the end of each field for about 5% of the total vertical period; therefore about 18° (in practice 14°) of the 360° total can be dropped without losing any picture information. Unfortunately, the vertical blanking interval does contain synchronizing pulses needed for interlacing the fields and for correcting time base errors. Unless this lost information is recovered, these simple recorders are not satisfactory for high-quality studio use. Another problem with the high angle of Omega drum wrap is build up of friction, which stretches the tape unevenly and varies with humidity and temperature.

The vertical interval information gap of the Omega format can be filled in by a second head on the head wheel; this second head scans a narrow band on the tape edge during the time when the main video head is off the tape. This arrangement, called the *1½ head system* is the basis for the Type C Helical Scan format. The tape stretch problem is overcome by servo control of the tension to correct the track length during playback and by the selection of ample track dimensions.

Two-Head Helical Scan. In utilizing the single-head Omega scanner, so many engineering compromises have to be made that many designers have concluded that a two-head wheel is better after all, using a *half-wrap* as in Figure 11-15*C*. This is called a Type A format if each head records a full video field per scan. (More than two heads are not necessary for basic functioning of a helical scanner, though auxiliary heads can be added for erasing and so on). The critical dimensions and balancing that were so difficult at first have been brought under control by experience and precision fixtures so that head angles are routinely set within seconds of arc ($\frac{1}{3600}$ of a degree).

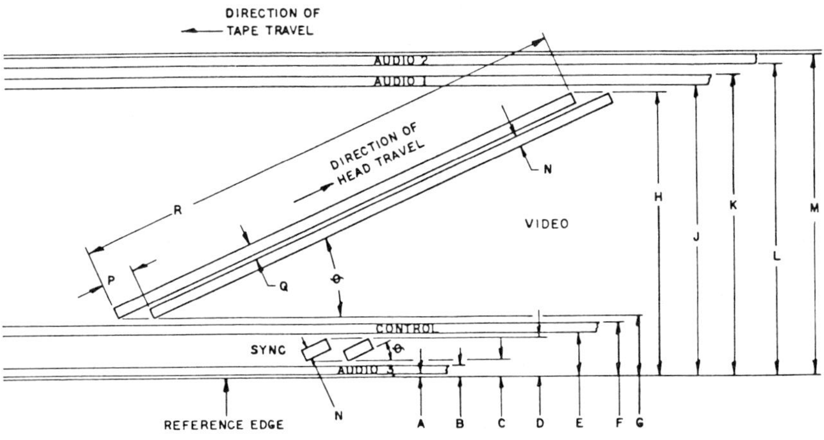

A. TRACK DIMENSIONS

(ANSI V98.28M)

Dimensions	Millimeters Minimum	Maximum
A Audio 3 lower edge	0.050	0.150
B Audio 3 upper edge	0.825	0.975
C Sync track lower edge	1.385	1.445
D Sync track upper edge	2.680	2.740
E Control track lower edge	2.870	3.130
F Control track upper edge	3.430	3.770
G Video track lower edge	3.870	3.910
H Video track upper edge	22.385	22.445
J Audio 1 lower edge	22.770	22.830
K Audio 1 upper edge	23.525	23.675
L Audio 2 lower edge	24.325	24.475
M Audio 2 upper edge	25.150	25.250
N Video and sync track width	0.128	0.132
P Video offset	4.067 ref (2.5H)	
Q Video track pitch	0.1823 ref	
R Video track length	410.764 ref (252.5H)	
S Control track head distance	101.8	102.2
T Vertical position, odd field	1.220 (0.75H)	2.030 (1.25H)
U Vertical position, even field	2.030 (1.25H)	2.850 (1.75H)
V Sync track length	25.620 (15.75H)	26.420 (16.25H)
W Vertical position, odd sync field	22.360 (13.75H)	23.170 (14.25H)
X Vertical position, even sync field	23.170 (14.25H)	23.980 (14.75H)
Y Vertical head offset	1.529 ref	
Z Horizontal head offset	35.350 ref	
θ Track angle	2°34′ ref	

Figure 11-16. Track patterns of a type-C helical tape recording. (Courtesy: Sony; BVH-1100 manual)

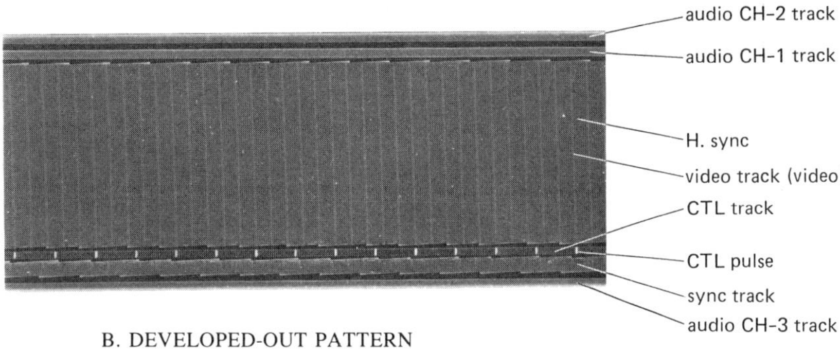

B. DEVELOPED-OUT PATTERN

(Courtesy SONY; BVH-1100 MANUAL)

Figure 11-16. *(Continued)*

To prevent gaps in video information the tape is wrapped about 190° around the drum. Two 180° opposed heads then give some duplication in video at the beginning and end of each oblique scan, the duplication being eliminated in playback by fast switchover as in quadruplex.

Still Framing and Slow Motion. In both the Omega scanner and the half-wrap Type A scanner, each oblique track is a complete field of video. A still-frame is then obtained by stopping the tape while the head wheel repeatedly plays back the same recorded track. Slow motion is obtained by slowing the forward tape movement, and reverse motion by backing up the tape. Tracking error caused by slowed or stopped tape is compensated as described later (Figure 11-21).

Segmented Scanning. To give the same quality of picture as an Omega wrap machine, the two-head simple scanner requires twice the head drum diameter, since only half the circumference is used. A broadcast-quality high-band recorder with a continuous track 16 in (406 mm) or longer needs too large a head wheel. One solution is to divide each picture field into several helical scans or segments instead of one long sweep. Upon playback the segments are electronically spliced together by the well-established techniques of quadruplex recording. The immediate benefits are a small diameter drum allowing a compact portable machine, and minimum tape friction and stretching. Time base and tracking errors are reduced with short large angle tracks, and compensation is easier with small head wheels. Still pictures and slow motion are no longer directly available, but for many applications these features are a minor consideration. Where required they are obtainable by the use of accessories described later for the Type B helical scan format.

TYPE C ONE-INCH HELICAL SCAN FORMAT

Helical scan machines using 1-in. tape are the successors to quad recorders for professional video recording. Such machines are less costly than quads and use only a third as much tape. A widely accepted standard is the Type C Helical Scan format, which is basically a single-head scanner with a full (Omega) tape wrap of about 346°. The sync information during the lost 14° is captured by

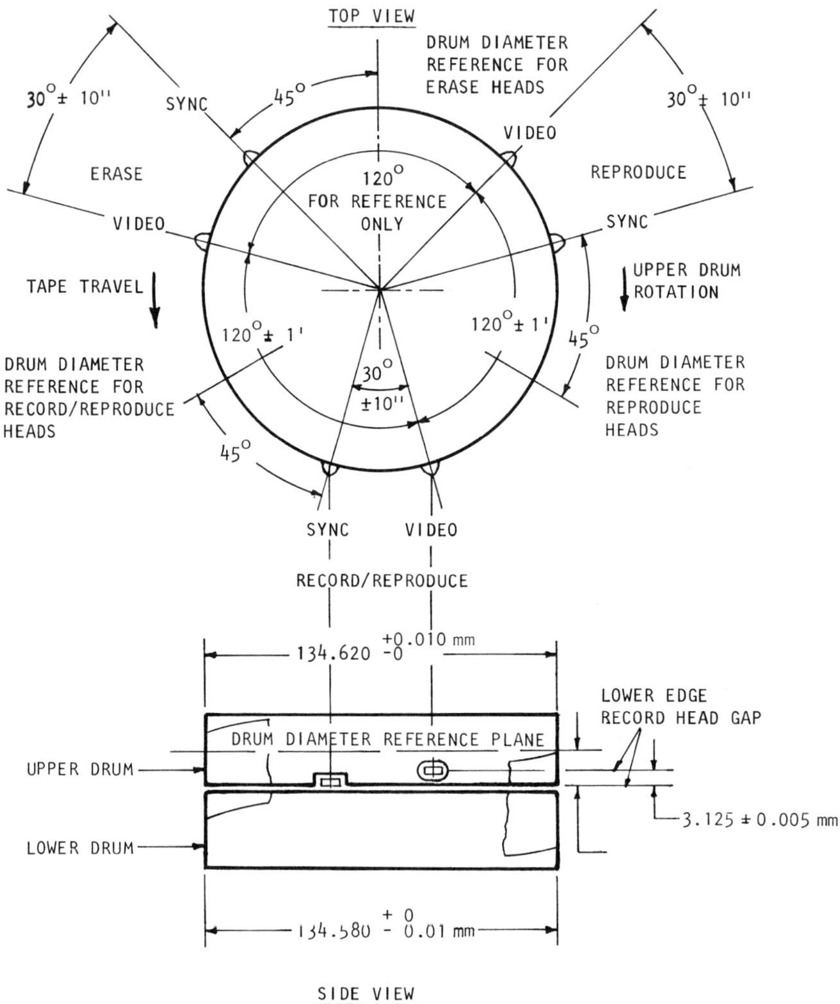

Figure 11-17. Head wheel of a type C helical recorder (from ANSI V98.27M).

an auxiliary displaced head on the scanner wheel and recorded as slanted segments on a separate narrow sync track, as shown in Figure 11-16. This arrangement, sometimes called a $1\frac{1}{2}$ head scanner, is shown in Figure 11-17. Commercial Type C recorders meet the standards of high-band color video and interface with accessories such as time base correctors, time code generators, editing devices, and slow motion controllers. A Type C recorder and its threading path is shown in Figure 11-18.

Type C Track Format. Figure 11-16*A* gives the track dimensions of a recorded Type C tape (the primary standard is in metric units), while Figure 11-16*B* is a pattern on a recorded tape developed out with fine magnetic particles. The bottom edge of the tape serves for guiding and reference. Two audio tracks at the top are for stereo recording and a third track at the bottom is for cues, time code, or a third audio channel. The control track records vertical sync rate, quite analogous to that of a quad recorder. A unique feature of the Type C format is the slanted segmented sync track just below the control track.

The main picture tracks are 16.2 in. (410.764 mm) long, at a 2° 34′ angle to the tape axis, the tape moving forward at 9.7 in/s (244 mm/s). The picture tracks contain 252.5 out of the 262.5 horizontal picture lines of an NTSC field.

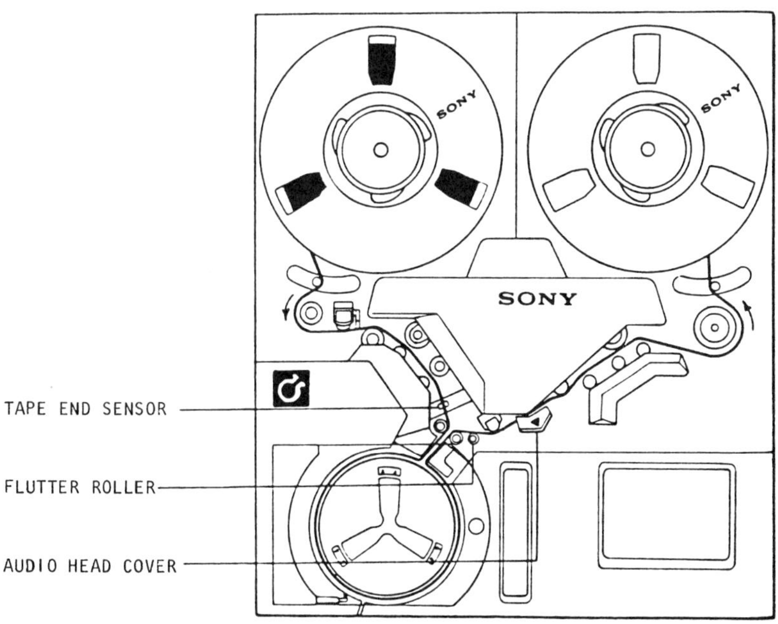

Figure 11-18. Type C helical recorder showing threading path. (Courtesy Sony Corp.)

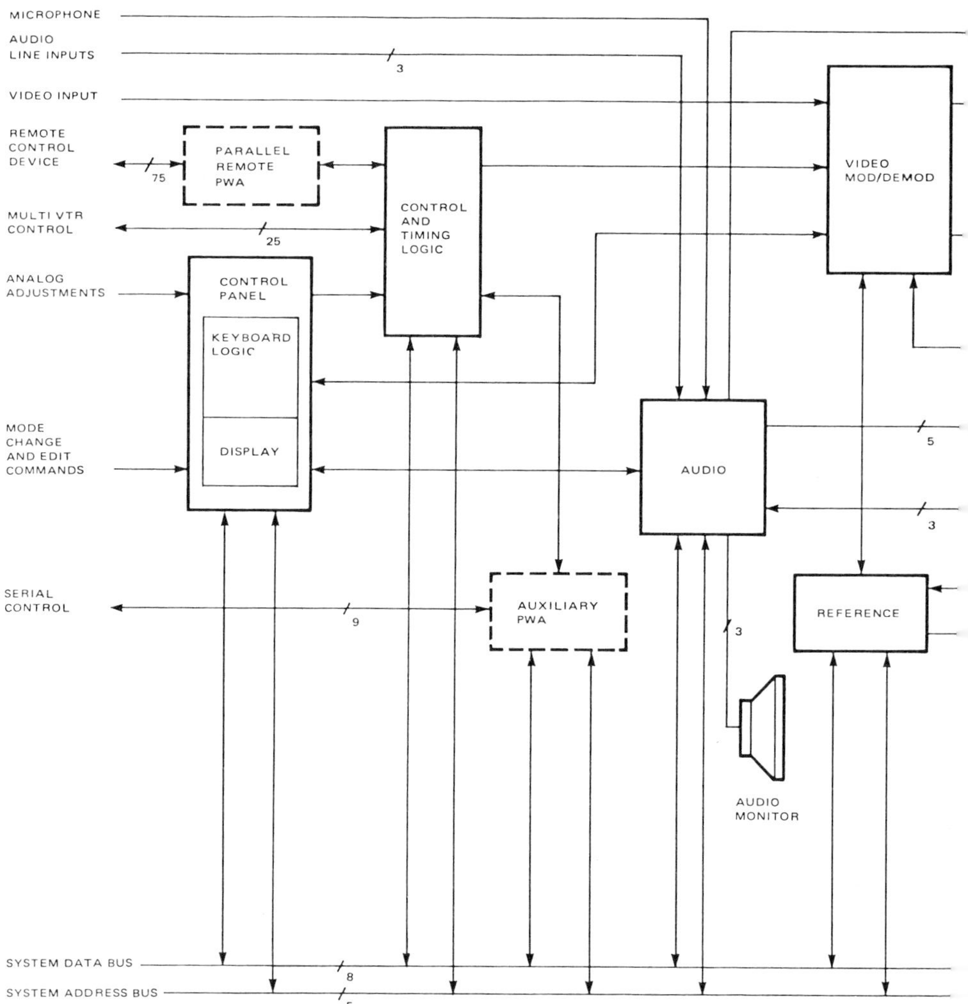

Figure 11-19. Block diagram of a type C video recorder. (Courtesy Ampex Corp.)

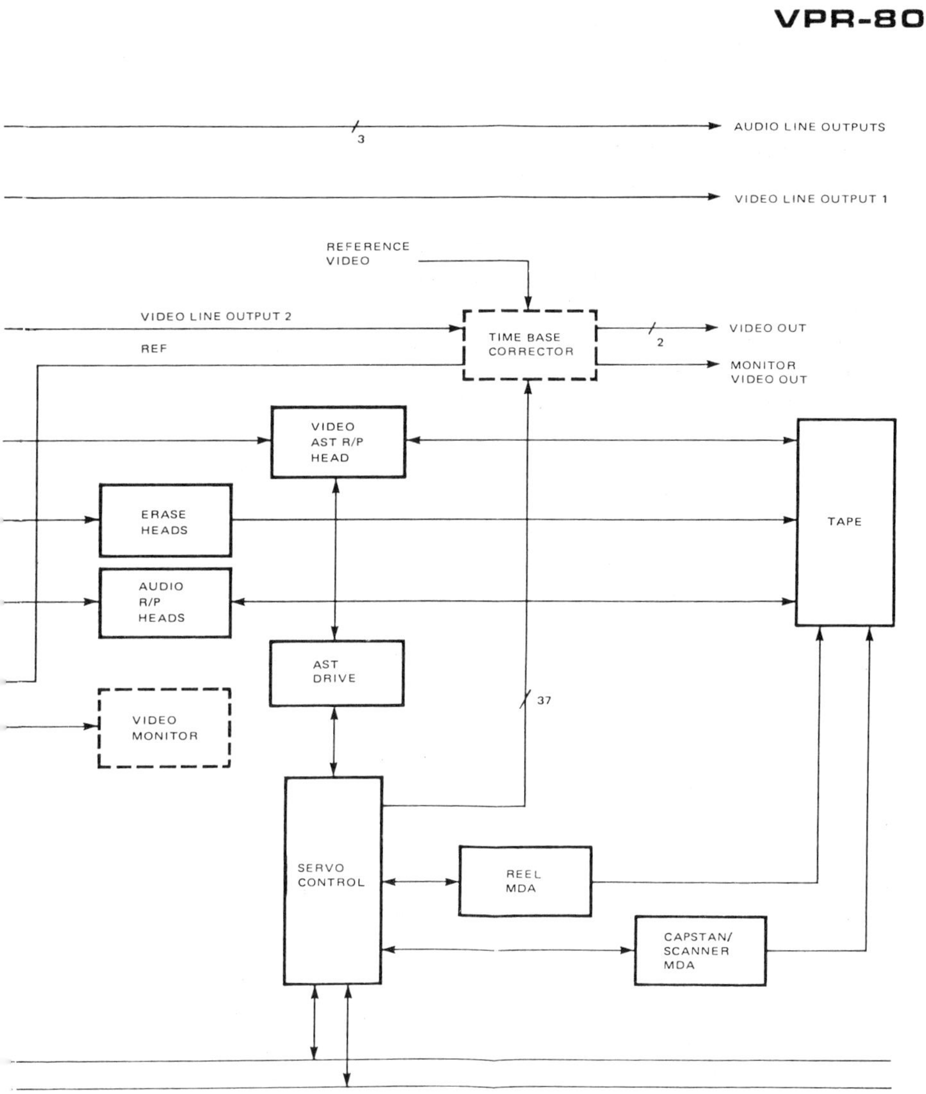

Figure 11-19. (*Continued*)

The segmented sync track contains 15.75 lines, centered around the 10 missing lines of the main picture tracks. Overlapped duplicate lines are switched out during playback.

Type C Scanning Wheel. Although only two heads are needed, the Type C standard allows for six heads around the head wheel as specified in Figure 11-17, two being for main record or play, two for monitoring, and two for erasing. In less complex machines, dummy pole tips can replace the extra heads. Dummy heads are essential because the protruding pole tips of heads set up mechanical waves as they strike the tape. The effect of these waves tends to cancel if played back on a wheel having the same head configuration as in recording. The head wheel spins at the field rate: 3600 RPM (60 RPS) for NTSC; 3000 RPM (50 RPS) for CCIR/EBU.

Electronics and Servo Control. A block diagram of Type C electronic and control circuits (Figure 11-19, see also Figure 11-25), shows that the basic functions and elements described for quad machines are present: VCO's for FM generation, rotary transformers for ''connecting'' to heads, head switchers, FM limiters and demodulators, drop-out compensators, time-base correctors, tach-wheel generators, and comparators. An essential feature is servo control of tape tension or tape skew to stretch the tape enough to compensate for dimensional changes due to humidity or temperature.

Type C recorders use integrated circuits for low power-level functions, each integrated circuit replacing dozens or hundreds of transistors of older generation VTRs, resulting in smaller machines, less power drain, and less heat dissipation.

Standard Tapes for Setup. The complexity and precision required for adjusting individual recorders to allow interchangeable records would be almost impossible for users in the field were it not for standard tapes that have been recorded to exacting specifications. These tapes are used on machines set to playback. An adjustment routine is then followed to optimize, among other things, the tape tension, synchronization, output, color, and audio. After the machine has been adjusted properly for playback of the standard tape, the adjustment will then follow the prescribed tracking mode when the machine is used for recording.

Operation and Accessories. Depending on usage of the machine, various accessories or options are useful or necessary. These include:

1. Monitoring units that display the picture, test patterns, and waveforms.
2. Processors for time base correction.

3. Time code generators and readers (SMPTE/EBU) for cue tracks, indexing, and editing.
4. Controllers for still frames, slow motion, jogging.
5. Editing consoles for synchronizing and controlling two or more tape decks to insert or assemble program material electronically by rerecording.
6. Special effects devices for fading, zooming, split displays, superposed titles, and so on.
7. Remote control units.

Time Base Correction. Slant scan machines record many more horizontal picture lines than quad machines do per revolution of head wheel and tachometer, and therefore the sync can stray farther from the master standard before correction takes place by electromechanical servo action. Further compensation is essential to bring the playback sync to broadcast standards and this is done electronically by analog or by digital processors. Analog methods use servo-controlled delay lines as have been described for quad recorders in connection with precision lock-in equipment. Digital processors convert the video signal to a digital code that can be similarly delayed by variable amounts to synchronize the final readout with the network sync. The digital coding must have enough bits so that negligible *digitizing noise* is added to the signal. A 9-digit code reduces such noise adequately to about 54 dB below the signal level. Figure 11-20 shows a digital time base corrector available as a module for the Type C recorder.

Still Framing and Slow Motion. Stop motion, slow motion, and jogging back and forth are obtained by stopping, slowing, or reversing the tape while

Figure 11-20. Digital time base corrector for helical scan recorders. (Courtesy, Sony Corp.)

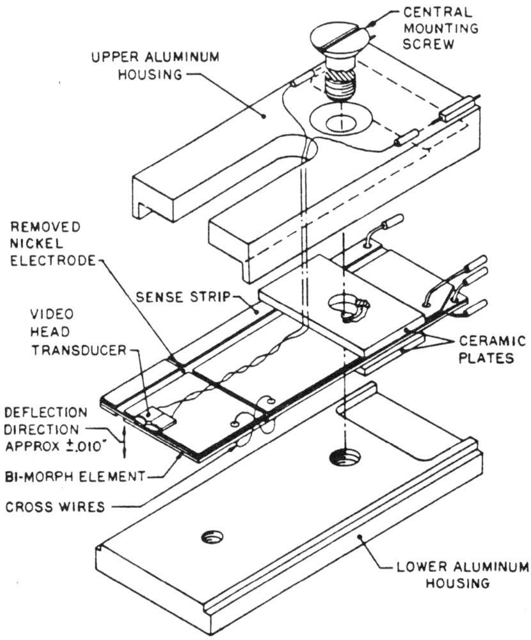

Figure 11-21. Tracking device for slow motion and stop motion with helical recorders. The piezoelectric bi-morph cantilever moves the head up or down to follow the tape track according to servo control. AST® automatic scan tracking (Hathaway and Ravizza 1980).

the head wheel spins at its normal rate. There is a complication in that when the tape is stopped on a track previously recorded when the tape had been moving, the track no longer fits the head path; in fact, the head actually moves between two adjacent tracks, giving a horizontal bar that divides the picture. Where picture quality is important, this bar can be eliminated by various means, as it is, for example, in Figure 11-21, where the head is shifted laterally by a piezoelectric element to make it follow the track.

TYPE B ONE INCH HELICAL SCAN FORMAT

The Type B format is a hybrid between helical and quad recording, giving the economy and convenience of 1-inch tape and the stability and multigeneration capability of quad. The small-diameter head wheel allows a relatively small transport. These benefits are obtained by doing away with the long complete-picture-field tracks that are considered the hallmark of helical recording and by breaking up each field into several parallel tracks (segments): five segments for

Figure 11-22. Type B helical recorder and threading path. (Courtesy Bosch Corp.)

NTSC video, or six segments for CCIR-EBU. No more heads are required than for the Type C ($1\frac{1}{2}$ head) format. Figure 11-22 shows a Type B recorder and its threading path.

An advantage of type B recorders is that up to seven generations (copies of copies) can be made, still retaining broadcast quality. This is a valuable feature for extensive editing and compilation. The small drum allows compact, portable and battery operated decks, and cassette loaded models. Also the head tip penetration is half as much as in type C giving long head life and reduced tape disturbances. The main disadvantage is that slow motion and stop motion effects are not obtainable directly, although they are available by special means.

Type B Track Format. Track dimensions of a Type B recorded tape are shown in Figure 11-23*A*, and a developed out pattern is pictured in Figure 11-23*B*. Going up from the bottom reference edge, there is a cue track (*Audio No. 3*), the segmented picture tracks at a relatively large angle of 14.288° and

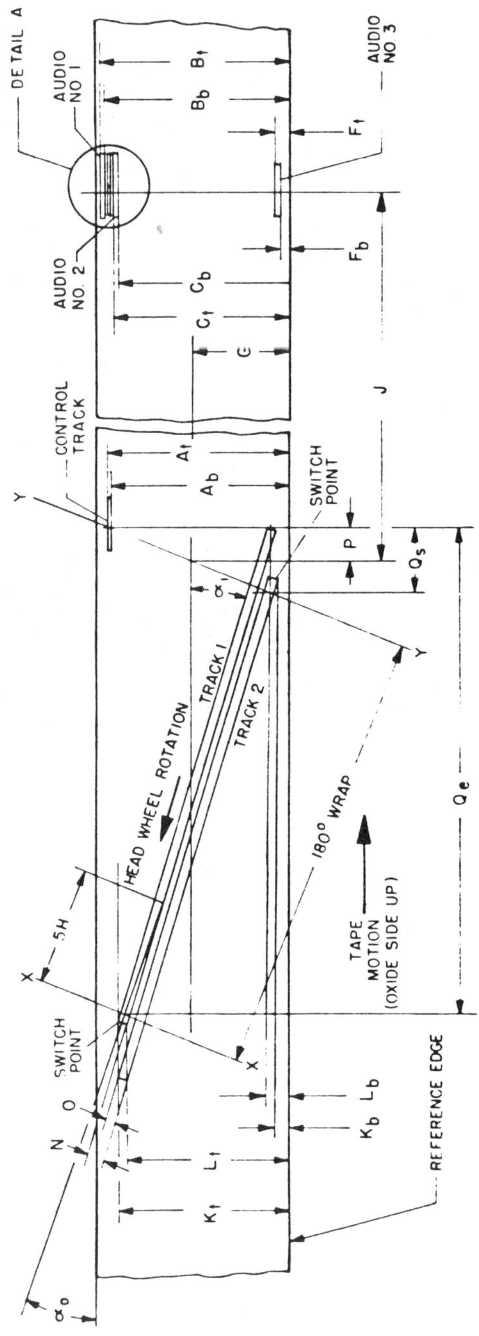

ANSI V98.30M

A. TRACK DIMENSIONS

	Dimensions	Millimeters Minimum		Maximum
A_b	Control track bottom edge	23.54		23.64
A_t	Control track top edge	23.94		24.04
B_b	Audio 1 track bottom edge	24.34		24.44
B_t	Audio 1 track top edge	25.14		25.24
C_b	Audio 2 track bottom edge	22.34		22.44
C_t	Audio 2 track top edge	23.14		23.24
F_b	Audio 3 track bottom edge	0.16		0.24
F_t	Audio 3 track top edge	0.96		1.04
G	Center of video tape	12.70	ref	
J	Position of audio heads	232.2		232.8
K_b	Full video width bottom edge	1.18		1.18
K_t	Full video width top edge	22.19		22.19
L_b	Video width (180°) bottom edge	1.82	ref	
L_t	Video width (180°) top edge	21.55	ref	
N	Video track pitch	0.200	ref	
O	Video track width	0.156		0.164
P	Position of control track head	2.845		2.875
Q_e	Switch point distance video 2 track	82.096		82.121
Q_s	Switch point distance video 1 track	5.523		5.533
S	Distance between control track head gap and center edit pulse at 180° switch point	0.040	ref	
α_0	Scanning angle (Helix angle)	14.434° ± 10″		
α_1	Video track angle (525/60)	14.288°		

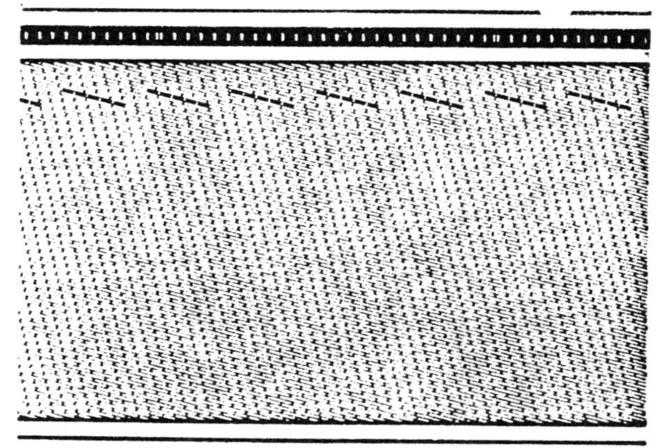

B. DEVELOPED-OUT PATTERN

Bosch Techn Berichte 6(1979)

Figure 11-23. Track patterns of a type-B helical recording.

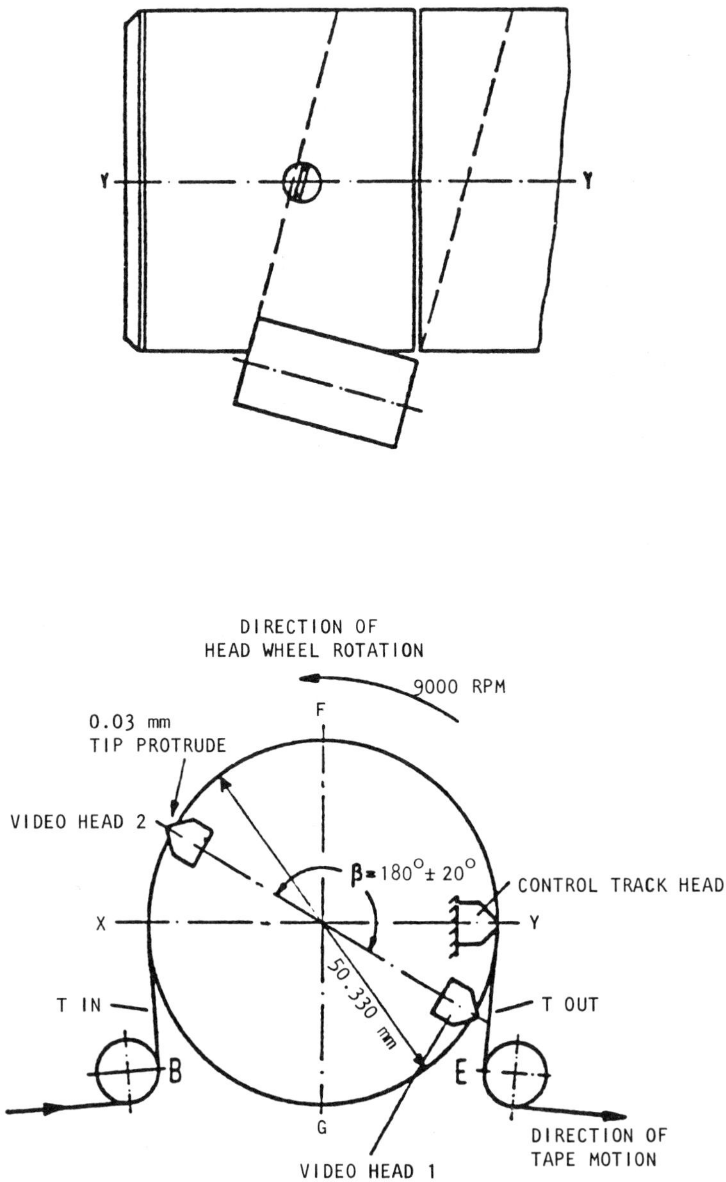

Figure 11-24. Head wheel of a type-B helical video recorder (After ANSI V98.29M).

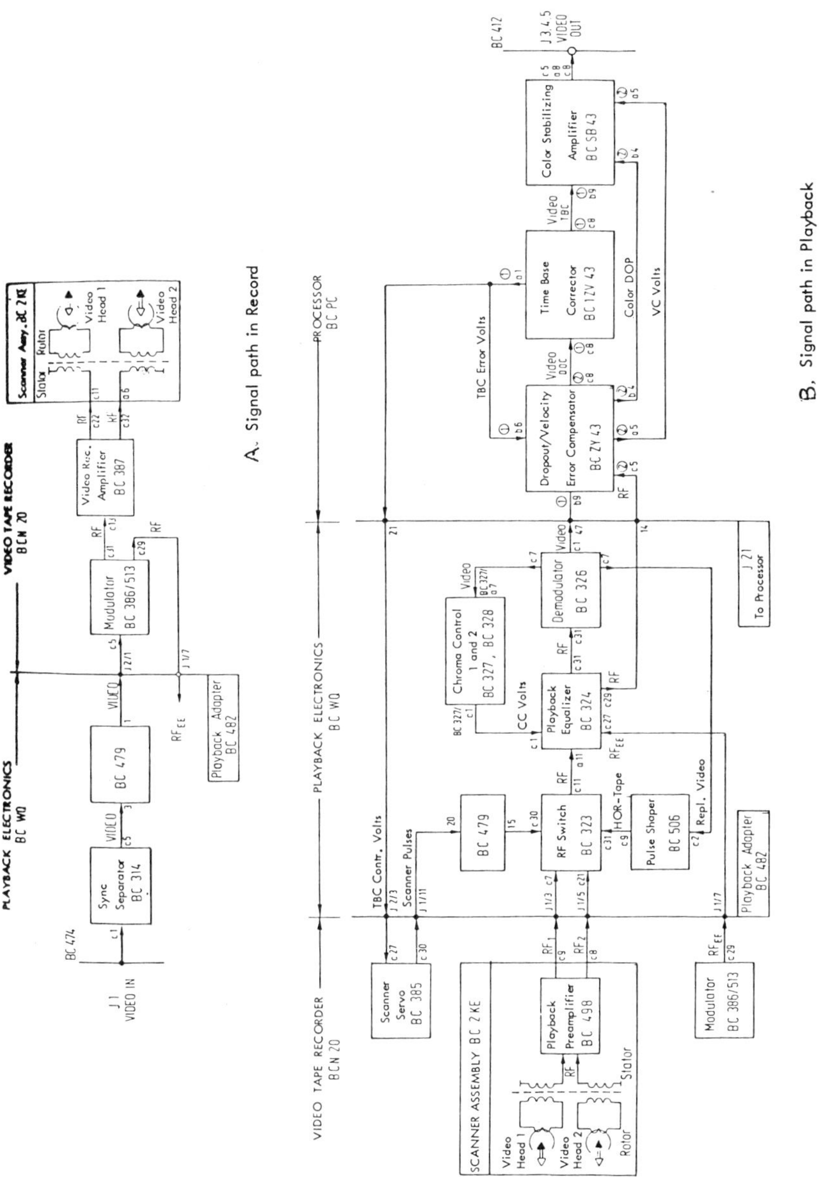

Figure 11-25. Block circuit diagram of type-B video recorder. (Courtesy Fernseh GMBH)

about 3.34 in. (84.7 mm) in length (NTSC), and finally at the top tape edge a pair of main audio tracks with a control track sandwiched between them. Note that the control track has a double pulse to distinguish a complete frame. On the video track the heavy dashes are the vertical blanking pulses.

Type B Scanning Wheel. As seen in Figure 11-24, the head wheel has two rotating video heads positioned 180° ± 20″ apart, spinning at 9000 RPM (150 RPS). A pair of erase heads may be added 90° from the video heads. The rotary heads are about 6.3 mils (0.16 mm) wide and leave a guard band of about 1.6 mils (0.04 mm) between tracks. A third head fixed to the tape guide drum serves for the control track, the control head being advantageously mounted close to the head wheel it controls. The tape wrap of 190° gives redundant video on adjacent tracks, eliminated by fast switching.

Electronics and Servo Control. The block diagram, Figure 11-25, of typical Type B circuits is quite analogous to that of quad and of Type C recorders, already described, and the same techniques are used appropriately modified for two-headed scan of segmented tracks.

Operation and Accessories. As with other video formats, it is essential to use a standard recorded tape for initial setup to insure that guides, tensions, head penetration, head alignment, tape path, and so on are proper. This procedure is especially important if the recorded programs will be played back on a different machine. Accessories are similar to those for other professional machines. For stop motion and slow motion, from the segmented format the tape is read into a storage device that holds two picture fields. These fields are read out repeatedly for a stopped picture, or updated during readout for slow motion.

INSTANT REPLAY

It is hard to imagine sportscasting without instant replay of important moments in stop motion and slow motion. This application is so important that special magnetic disc recorders were developed in the 1960s and 1970s when no other video recorder could do the job technically or economically. One version used both sides of a pair of concentric spaced 16-inch-diameter discs with four heads, one riding on each magnetic surface. The heads tracked in concentric circles and indexed by jumps to the next circle after every revolution. Disc speed was synchronized with picture rate to record one field per revolution, successive fields being switched to different heads in 1, 2, 3, 4, 1, 2, . . . fashion. There were enough tracks to recall 20 seconds of video, the heads moving back and forth constantly to update the recording. For instant replay of a still-frame, the recording was switched to playback and a head was held motionless while the desired track was played. For slow motion, the heads were stepped slowly in correct sequence.

The disc players became obsolescent with the upgrading of slant-scan recorders to color broadcast standards, and with accessories that allow them to fulfil all requirements of instant replay. Quad recorders can, of course, be used with memory type replay accessories but are too expensive and too clumsy to be tied up for this special application.

U-FORMAT CASSETTE RECORDERS

The U-Matic® was the first commercially successful narrow tape ($\frac{3}{4}$ in.) videocassette recorder and the forerunner of the Betamax® and VHS $\frac{1}{2}$ inch consumer-type recorders. None of these was intended for broadcast work: Their frequency response was substandard, their time base stability was poor, and their color information was unconventional since their bandwidth could not handle regular color signals. Nevertheless the U-Format recorders were so convenient and economical that they found a place in almost every video recording studio for rehearsals, portable news pickup, and other nonexacting duties. A professional U-Format machine is shown in Figure 11-26. There have been major improvements in tapes, heads, and electronic systems in the decade since the U-Format was introduced, and these improvements have upgraded the latest recorders to a point where they are challenging the more complex machines.

U-Type Track Format. Figure 11-27 shows the recorded track pattern of a U-format recorder, which is the simplest of all two-head wheel patterns (Type A) where every slanted track holds a complete video field without segmentation. Tracks are 6.97 in (177 mm) long at an angle of 4° 57′. Audio tracks, control tracks, and cue tracks are at the edges as in all other video tapes, although located differently.

A tape speed of 3.75 in./s (95.25 mm/s) and track writing speed of 404 in./s (10.26 m/s) are less than half the speeds of Type B or Type C 1 in. professional formats. The low speed gives economy and a compact tape cartridge but is the main reason for the lesser performance of the U-Format. The compressed information on short tracks is harder to stabilize, gives more noise and dropouts, and does not have reliable high-band response for direct color recording.

U-format Head Wheel. The head wheel shown in Figure 11-28 carries the familiar 180° head pair of a half-wrap machine, plus a set of erase heads, which are omitted in nonprofessional machines. The wheel turns at 1800 RPM to give

Figure 11-26. U-format cassette loading videotape recorder. (Courtesy Sony Corp.)

60 fields/s NTSC, and is driven by a DC motor unaffected by power frequency variations. Heads are 3.4 mils (0.085 mm) wide, writing at a track pitch of 5.4 mils (0.137 mm). The tape wrap is somewhat more than 180°, giving redundant H-lines.

To provide the tach frequency for head wheel servo control, there are eight magnetic inserts indicated on an inner circle, and a displaced single insert on a different inner circle that gives the frame rate once per revolution. Figure 11-28 also indicates the location and spacing of stationary heads for audio, control, and erasing.

Electronics and Servo Control. The U-Format has the usual servo and switching as described for other video formats (see, for example, Figure 11-25). A new feature is the mechanical threading mechanism (Figure 11-29), consisting of fingers that reach into the cartridge and wrap the tape around a rather devious path that includes the scanning drum, heads, and capstan. This operation takes a few seconds, giving a pause in operation after inserting a cartridge and before the cartridge can be removed.

The electronics for handling the color components are called the *color under* system, which dispenses with the usual 3.58 MHz color carrier and color bursts

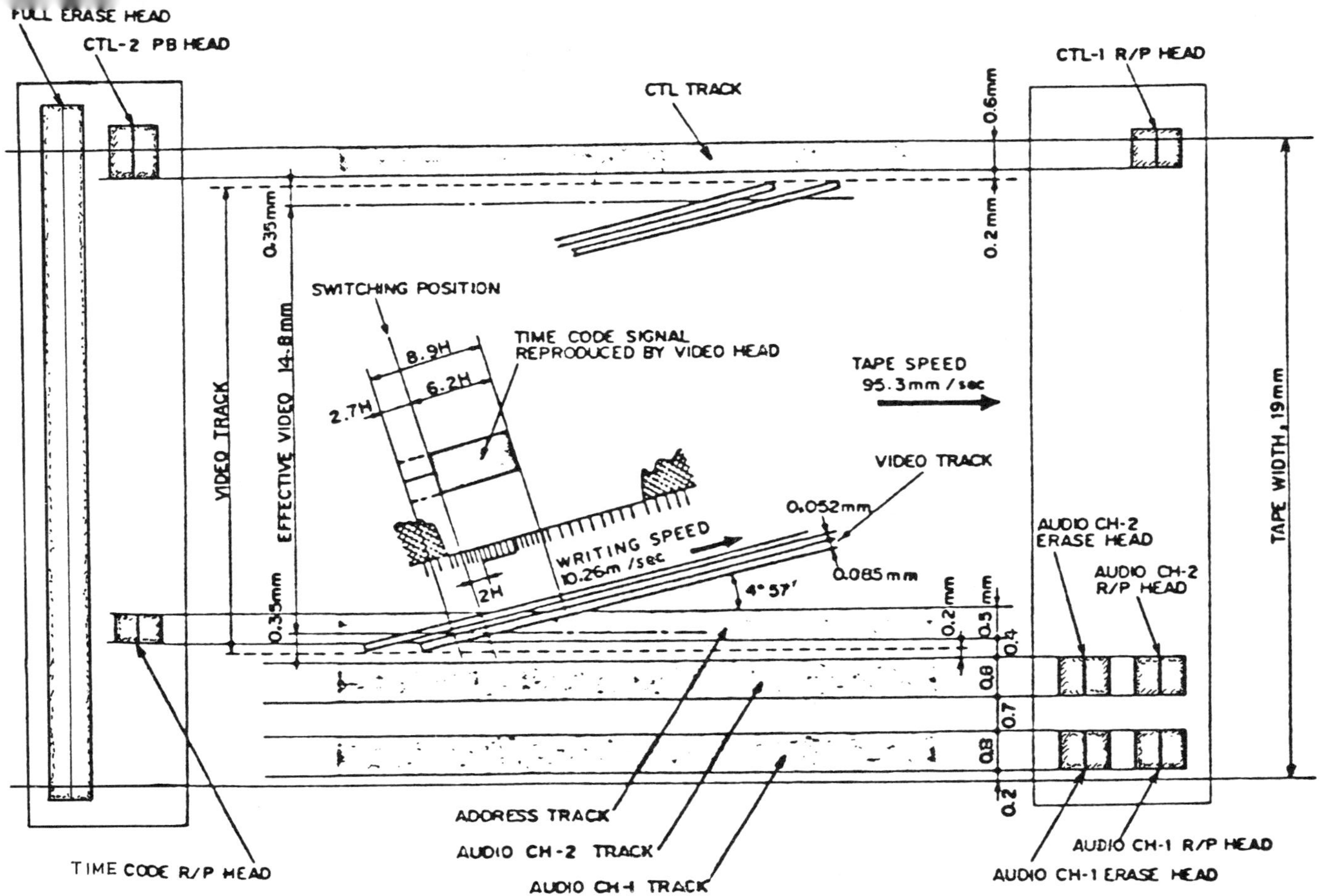

Figure 11-27. Track pattern of a U-format recorder (Sony BVU-200B manual).

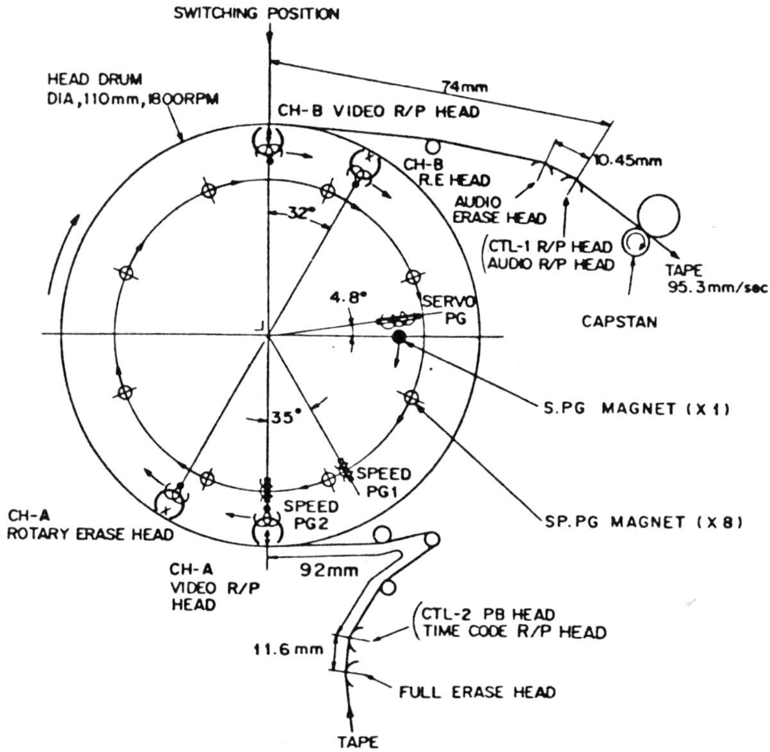

Figure 11-28. Head wheel of a U-format recorder (Sony BVU-200B manual).

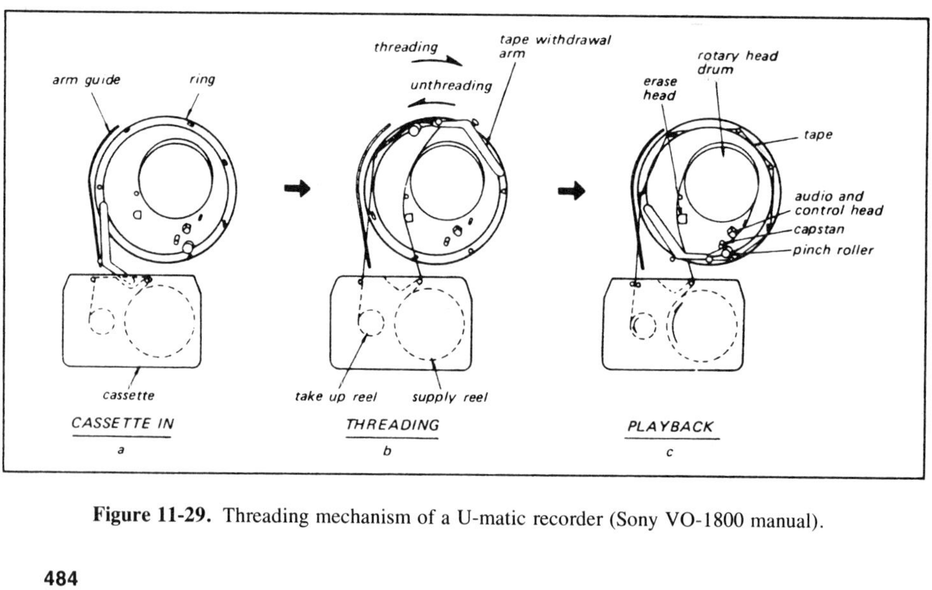

Figure 11-29. Threading mechanism of a U-matic recorder (Sony VO-1800 manual).

and substitutes a lower-frequency low-level subcarrier of 688 KHz for the color modulation. The *color under* system is described in Chapter 12.

Operation and Accessories. The professional versions of U-Format recorders are equipped with erase heads on the head wheel to allow for spliceless electronic editing. Various accessories for inserting and assembling programs are available, as are time-base correctors and color system converters that adapt these recorders for broadcasting, albeit at lesser quality than the more expensive machines.

VIDEO EDITING

Electronic editing, now an established discipline, enhances creativity in the television art by allowing quick, easy, and smooth assembly of pictures and sounds that would be impractical by older cut-and-splice methods. Editing is, of course, the life blood of television programming and had to be done at all costs, even in the earliest days of video recording.

With the first video recorders the picture quality was marginal, and rerecording added noise, distortion, and instability. Quality could be preserved by splicing together originals or early generation recordings. This technique was well-known in the motion picture and audio tape industry, from which tools and procedures were borrowed. Fortunately the design of quadruplex video recorders was so conservative that improvements after a few years allowed copies practically indistinguishable from the original recording, and hence made feasible electronic editing by rerecording. A secondary consideration is that electronic editing requires two or more video machines, a considerable investment.

Mechanical Splices. Although seldom done any more, quad tapes can be edited by splicing. To locate the cutting point, the magnetic pattern is developed out on the tape surface with a suspension of fine particles (Chapter 2, Figures 2-10 and 11). The 30 Hz marker on the cue track defines the cutting point between frames, where there will be no video disturbance. For audio continuity, the audio must be recorded after the splice has been made, either from a double system where the audio was originally made on a separate tape, or by copying the audio before splicing and afterwards rerecording the appropriate portion, since the audio track is spaced 9 in. from the video. Mechanical splices are apt to foul the heads with adhesive from the splicing tape and may break apart at an inappropriate time, so it is a good idea to copy them if possible and then use the copy. Mechanical editing is not practical for helical scan tapes with their long diagonal tracks.

Electronic Editing. In sheer volume of activity, video editing probably exceeds that of all other motion picture and audio media. The variety and versatility of editing equipment has kept pace with the needs of editors. A minimal setup includes a playback machine, a recorder, monitors, cueing indicators, and controls to start and end the recording when cue points are reached. Professional machines often build in the necessary controls for simple editing, since their equipment already includes rotary erase heads for editing helical scan tapes, cue tracks, and time code indexers. Two or more recorders with such built-in controls can then be connected for so-called *stand up editing*, where the operator manipulates controls on the actual machines.

Sit down editing brings a higher level of sophistication. Here there is a console, separate from the recorders themselves, with monitors, indexers, and controls. The operator searches both machines for their programs, synchronizes them to be together at the edit point, and records a cue where the slave machine will begin to record the picture and sound delivered by the master. Prior to the actual take, the operator can preview the proposed result and make major or minor changes in timing. The machines are then rewound and rolled forward in sync while the actual edit takes place. A similar cue ends the edit and restores conditions for the next editing point.

Computerized editing is perhaps the ultimate step. Up to 16 machines are controlled and hundreds of operations are stored in a microprocessor memory or on a flexible disc. Complex editing steps are prescheduled in the computer memory and viewed on a CRT display or printed out for permanent records. The schedule is revised and updated until satisfactory, then executed automatically when convenient. While the programs are being compiled, the recorders themselves are not needed but are free for other uses, so that maximum utilization of expensive equipment is possible. Computerized editing depends on accurate retrievable indexing of the recorded masters, made possible by coding such as the SMPTE time code that indexes each frame of the picture with an individual number.

Color Framing. Where editing from an old frame to a new frame of video pictures is involved, sometimes the edit will be perfect, but at other times there will be a visible disturbance. The cause is traceable to NTSC chroma frequency, which repeats its phase every *two* frames (four fields) of picture. If the splice is made at an odd frame, the color burst phase of the new sequence is 180° out of phase with the old, and the color oscillator will take some time to lock to the new phase. The remedy is to move over the splice by one frame or to use sync marks that distinguish between odd and even frames so an improper splice will not be made. The situation is even worse in the European EBU/PAL system, where the chroma frequency requires four frames (eight fields) before it repeats its phase.

The chroma frequency standards were purposely chosen to reduce visibility of the chroma carrier on high-resolution pictures by appearing in opposite phase on every odd frame. (Such frequency interleaving is explained in more detail in Figure 12-9; see also Tables 11-6 and 11-7 on worldwide television standards). The NTSC standard uses a color subcarrier of 3,579,545 Hz and a frame rate of 29.97002618 per second. This gives 3,579,545 ÷ 29.97002618 = 119,437.5 cycles of color carrier per frame, and requires two frames to give a whole number (238,875) of subcarrier cycles. Similarly the EBU/PAL standard uses a 4,433,618 Hz color subcarrier and 25 Hz frame rate, requiring four frames (practically) to give a whole number of color cycles.

Editing Accessories. For convenient and artistic editing a number of features and accessories are available:

- Slow motion, fast motion, jogging and freeze-frame generation
- Alarm and diagnostic malfunction indicators that give warnings and pinpoint troubles
- "Confidence" playback of a recording while it is being made, using an extra playback head on the head wheel
- Preroll that brings the machines up to speed prior to start of the program
- Character generator that displays time code, etc., superposed on the picture
- Search-to-cue that finds designated program and sets machine ready to play it
- Disc storage for instant recall of thousands of still pictures or short video subjects
- Split edits with separate in and out points for audio as compared to the video
- Frame-by-frame trimming of in and out points
- Combinations of SMPTE time code, Vertical Interval Time Code (VITC), and control track timing
- Memory that holds previous edit information in case edit is unsatisfactory
- Audio pitch control that maintains the correct sound pitch during slow motion, etc.
- Special effects, cuts, dissolves, keys, wipes, quad splits

SMPTE Time Code

An important help in editing has been the SMPTE time code, which has supplanted almost completely all other methods for indexing video programs. This code is also used in sound recording and in motion pictures, allowing an editor to find quickly a desired selection even in mixed media. Once indexed, the media can be locked to run together in synchronism, or can be switched over from one to another at designated intervals.

Table 11-6. Television Standards in Various Countries

CCIR Standard	Color Standard	Color Modulation Mode	Subcarrier Frequency in MHz	Bandwidth of the Video Channel in MHz	Sound/Vision Carrier Difference	Number of Lines/Fields per Second	Commonly Used in
M	NTSC	QAM*	3.58	4.2	4.5	525/60	USA, Canada Japan
M	PAL(M)	QAM	3.58	4.2	4.5	525/60	Brazil
B, G	PAL(B, G)	QAM	4.43	5	5.5	625/50	Europe
N	PAL(N)	QAM	3.58	4.2	4.5	625/50	Argentina
B, G	SECAM	NFM†	3.9 . . . 4.75	5	5.5	625/50	GDR, CSSR
K, L	SECAM	NFM	3.9 . . . 4.75	6	6.5	625/50	USSR, France

*Quadrature amplitude modulation.
†Narrow-band frequency modulation.
Courtesy: Bosch Technische Berichte.

Table 11-7. Color Subcarrier Frequencies of Different TV Systems and their Derivation

Color Standard	Vertical Frequency f_H in Hz	Line Frequency f_H in Hz	Subcarrier Coupling	Subcarrier Relationship	n in MHz	Subcarrier Frequency
NTSC	59.94	15734.26	Half-line offset	$(2n - 1)\frac{f_H}{2}$	228	3.579545
PAL (M)	59.94	15734.26	Quarter-line offset	$(4n - 1)\frac{f_H}{4}$	227.5	3.575611
PAL (B, G)	50	15625	Quarter-line offset + 25 Hz Precision offset	$(4n - 1)\frac{f_H}{4} + \frac{f_V}{2}$	284	4.433618
PAL (N)	50	15625	Quarter-line offset + 25 Hz Precision offset	$(4n - 1)\frac{f_H}{4} + \frac{f_V}{2}$	229.5	3.582056
SECAM (B, G)	50	15625	Zero offset, sequence	$n_1 \cdot f_H; n_2 \cdot f_H$	272; 282	4.250; 4.406

Courtesy: Bosch Techniche Berichte

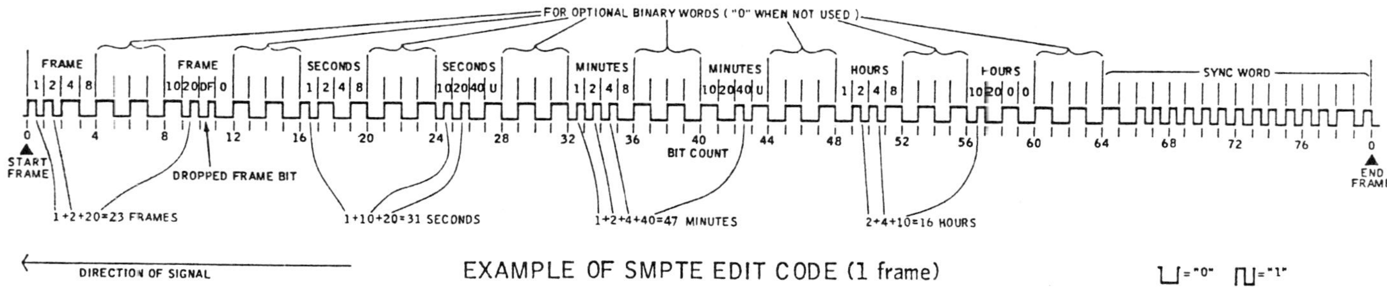

Figure 11-30. SMPTE/EBU time code. Eighty spaces are allocated to each marker. The presence of a "1" bit in designated spaces determines the readout count in hours, minutes, seconds, and frames ($\frac{1}{30}$ sec NTSC, $\frac{1}{25}$ sec EBU). The "sync word" in spaces 64 to 80 signifies end of a frame and ready to start the next, or it tells if the record is moving in reverse direction. In space 10 a dropped frame bit reconciles the NTSC color rate of 29.97002618 frames/sec with real time by omitting certain frame counts. (Courtesy EECO)

As shown in Figure 11-30, the SMPTE Time Code is a set of numbers representing hour, minute, second, and frame ($\frac{1}{30}$ sec NTSC or $\frac{1}{25}$ sec EBU). The numbers are in the form of a Miller type binary code as shown (with translation to decimal). The code is usually recorded magnetically on a cue track reserved for this purpose, or on any available audio track of a magnetic recorder. When the code is played back through a reader, an illuminated display shows the exact location on the tape, the numbering staying on count whether the tape moves fast or slow, backwards or forwards.

DIGITAL VIDEO RECORDING

Recording video digitally once was an impossible dream. However video in digital form (not recorded) has been used for several years in time base correctors and for signal processors where the digital format permits great versatility in special effects without degrading the picture.

Digital video recording removes practically all effects of recording and playback, giving the quality of direct, live video pickup without added distortion, noise, or flutter. It allows unlimited generations of copies without degradation, very important in electronic video editing and in copying for distribution of programs. It gives a picture free from dropouts because of its error-correcting capability. Archivally stored tapes will reproduce full picture quality regardless of expected dimensional changes or reasonable fading of the signal amplitude.

The obstacles to digital video recording are bit rate and recording density. For 5 MHz video, a sampling (Nyquist) rate of at least 10 million per second is required. For a signal-to-noise ratio of 48 dB, each sample must contain eight bits theoretically. Eight- or nine-bit samples have been very successful in electronic processors but, for recording, ten bits are more realistic to allow for sync words and error correction (or concealment) of dropouts that are common at high recording densities. We therefore need a bit rate of 100,000,000 per second (10^7 samples/sec $\times$ 10 bits/sample).

Feasibility of Digital VTR. Conventional single-track rotating-head recorders can record perhaps 1000 bits/mm (25000 bits/in) so we need a writing speed of 100 m/s (4000 in/s). This is not out of the question, being about three times the speed of quadruplex recorders. The heads have to respond to 50 or 100 MHz signals. Tape economy is another problem, but high track density (many tracks per mm) is a solution since digital recording requires only about 20 to 25 dB signal-to-noise ratio of the track itself.

An alternative to single-track recording is to divide the digitized video signal among two or more tracks, with a proportional reduction in bit rate and writing speed for each track and in the frequency response of the heads and associated

Table 11-8. International Listing of TV Broadcasting Systems

Country	TV Broadcasting Standards				
	Color	System	B/W	Hz	Power Voltage
Afghanistan	PAL	D	OIRT	50	220/380
Albania	SECAM(H)	B	OIRT		
Algeria	PAL	B	CCIR	50	127/220/380
Andorra	PAL	B	FRENCH	50	220
Angola		I	BRITISH	50	220/380
Antigua & Barbuda	NTSC	M	USA	60	230
Antilles	NTSC	M	USA	50+60	127/220
Argentina	PAL	N	USA	50+DC	220/380
Ascension Is		I			
Australia	PAL	B	CCIR	50	240/415
Austria	PAL	B	CCIR	50+DC	220/380
	PAL	G			
Azores	PAL	B	USA	50	110/220/380
(US Forces)		(M)			
Bahamas	NTSC	M	USA	60	110/220
Bahrain	PAL	B	CCIR	50	220
Bangladesh	PAL	B	CCIR	50	230/400
Barbados	NTSC	N	USA	50/60	110/220
Belgium	PAL	B	BELGIAN	50	220/380
	PAL	G(H)			
Bermuda	NTSC	M	USA	60	115/230
Bolivia	NTSC	N	USA	50+,60	110/220
		M	(CCIR)		
Botswana		I	BRITISH	50	220
Brazil	PAL-M	M	USA	50/60/DC	110/127/220
	PAL-M	M			
Brunei	PAL	B			
Bulgaria	SECAM(V)	D	OIRT	50	220/380

Burma	NTSC	N	USA	50	220/440
Burundi		K	OIRT	50	220/380
Cambodia		M	USA	50	120/208/220
Cameroun		K	OIRT	50	127/220
Canada	NTSC	M	USA	60	120/240
	NTSC	M			
Canary Is	PAL	B	CCIR	50	110/220
Cape Verde Is		I			
Central African					
Empire		B	OIRT	50	220
Chad		K	OIRT	50	220
Chile	NTSC	M	USA	50/DC	220/380
China	PAL	D	OIRT	50	220
Columbia	SECAM(V)	M	USA	60	110/220
Congo	SECAM(V)	D	OIRT	50	220
Costa Rica	NTSC	M	USA	60	120/220
Cuba	NTSC	M	USA	60	110/120/
(US Navy)		(M)			220/440
Curacao & Aruba	NTSC	M	USA	50+60	127/220
Cyprus		B	CCIR	50	240/415
		H			
Czechoslovakia	SECAM(V)	D	OIRT	50	240/380
	SECAM(V)	K			
Dahomey		K			
Denmark	PAL	B	CCIR	50/DC	220/380
Diego Garcia	NTSC	M	USA		
Dijibouti	SECAM	K1	OIRT		
Dominican Rep.	NTSC	M	USA	60	115/230
Ecuador	NTSC	M	USA	60	110/220
Egypt (United Arab					
Republic)	SECAM(V)	B	OIRT	50	110/220
El Salvador	NTSC	M	USA	60	115/230
Ethiopia		B	CCIR	50	127/220

TABLE 11-8. Continued

Country	TV Broadcasting Standards				
	Color	System	B/W	Hz	Power Voltage
Fernando Po		B			220/380
Fiji	PAL	B	CCIR	50	240/415
Finland	PAL	B	CCIR	50	220/230
	PAL	G			
France		E	FRENCH	50	110/127/ 220/230
	SECAM(V)	L			
Gabon	SECAM(V)	K	OIRT	50	220/380
Gambia		I	BRITISH	50	230
W. Germany	PAL	B	CCIR	50/DC	220/380
	PAL	G			
(US Forces)	(NTSC)	(G)			
E. Germany	SECAM(V)	B	OIRT	50	220
	SECAM(V)	G			
Ghana	PAL	B	CCIR	50	220/230/440
Gibraltar	PAL	B	CCIR	50	240
Greece	SECAM	B	CCIR	50+DC	127/220
		H			
Greenland	NTSC	M	USA	50	220/380
Guadeloupe	SECAM(V)	K1	OIRT	50	127/220/380
Guam	NTSC	M	USA	60	110/220
Guatemala	NTSC	M	USA	60	110/220
Guinea Bissau		I			
Guinea Dem. Rep.		B	OIRT		
Guinea Rep.		K	OIRT	50	220/380
Guyana, Fr.	SECAM(V)	K1	OIRT		
				50	127/220

Haiti	SECAM(V)	M	USA	60	115/230
Honduras		M	USA	60	110
Hong Kong	PAL	I	BRITISH	50	200/346
Hungary		D	OIRT	50	220/380
	SECAM(V)	K			
Iceland	PAL	B	CCIR	50	220/380
India		B	CCIR	DC+50	230/300/400
Indonesia	PAL	B	CCIR	50	127/220
Iran	SECAM(H)	B	CCIR	50	220/380
	PAL	B			
Iraq	SECAM(H)	B	CCIR	50	220/380
Ireland, Rep		A	BRITISH	50	220/380
	PAL	I			
Israel	PAL	B	CCIR	50	230/400
		G			
Italy	PAL	B	CCIR	50	127/220
	PAL	G			
Ivory Coast	SECAM(V)	K1	OIRT	50	220/380
Jamaica		M	USA	50	110/220
Japan	NTSC	M	USA	50+60	100/200
	NTSC	M			
Johnston Is	NTSC	M	USA		
Jordan	PAL	B	CCIR	50	220/380
Kenya	PAL	B	CCIR	50	240/415
Korea, N	SECAM	D	OIRT	60	100/200
Korea, S	NTSC	M	USA	60	100
(US Forces)	(NTSC)	(M)			
Kuwait	PAL	B	CCIR	50	240/415
Lebanon	SECAM(V)	B	CCIR	50	110/190
Leeward Is		M	USA	60	230
Lesotho		I		50	220
Liberia	PAL	B	CCIR	60	120
Libya	PAL	B	CCIR	50	127/220

TABLE 11-8. Continued

Country	TV Broadcasting Standards				
	Color	System	B/W	Hz	Power Voltage
Luxemburg	PAL	C,G	BELGIAN		
	SECAM(V)	L	50	110/220	
Macao		I	CCIR	50	110/220
Madeira		B	CCIR	DC+50	220/380
Madagascar		K		50	110/220
Malaysia	PAL	B	CCIR	50	230/400
Malawi		B	CCIR	50	230/400
Mali		K	OIRT	50	230/380
Malta		B	CCIR	50	240/415
		G			
Martinique	SECAM(V)	K1	OIRT	50	220
Maritania		K	OIRT	50	220/380
Mauritius	SECAM(V)	B	CCIR	50	230/400
Mexico	NTSC	M	USA	60	110/220
					120/208
					127/220
Midway Is	NTSC	M			
Monaco	SECAM(V)	C	FRENCH	50	127/220
	SECAM(V)	L			
	SECAM(V)	G			
Mongolia		D			
Morocco	SECAM(V)	B	CCIR	50	115/127/
					200/220
Mozambique		I		50	220/380
Netherlands	PAL	B	CCIR	50	127/220
	PAL	G			
New Caledonia	SECAM(V)	K1	OIRT	50	127/220/380
New Zealand	PAL	B	CCIR	50	230/400

Nicaragua	NTSC	M	USA	60	120
Niger		K	OIRT	50	220/380
Nigeria	PAL	B	CCIR	50	230/400
Norway	PAL	B	CCIR	50	230
	PAL	G			
Oman	PAL	B	USA	50	220
		G			
Pakistan	PAL	B	CCIR	50	230/400
Panama	NTSC	M	USA	60	115/230
Paraguay		N	USA	50+DC	220
Peru	NTSC	M	USA	60	220
Philippines	NTSC	M	USA	60	110/220
(US Forces)		(M)			
Poland	SECAM(V)	D	OIRT	50	220/380
		K			
Portugal	PAL	B	CCIR	50	220/380
	PAL	G			
Puerto Rico	NTSC	M	USA	60	120
(US Forces)		(M)			
Qatar	PAL	B	CCIR		
Reunion	SECAM(V)	K1	OIRT	50	
Romania	SECAM	D	OIRT	50	220/380
		K			
Rwanda		K	OIRT	50	220
Samoa	NTSC	M	USA	50	230
Saudi Arabia	SECAM(H)	B	CCIR (USA)	50+60	120/208, 230/415
Senegal	SECAM(V)	K1	OIRT	50	127/220
Seychelles		I			
Sierra Leone	PAL	B	CCIR	50	230/440
Singapore	PAL	B	CCIR	50	230/400
Somalia		B,G	CCIR	50	230

TABLE 11-8. Continued

Country	TV Broadcasting Standards				
	Color	System	B/W	Hz	Power Voltage
South Africa	PAL	I	BRITISH	DC+50	230/380
	PAL	I			
Spain	PAL	B	CCIR	50	127/220/380
	PAL	G			
Sri Lanka		B,G	CCIR	50	230/400
St. Helena		I			
St. Kitts	NTSC	M	USA	60	220
St. Pierre et Miquelon	SECAM	K1	OIRT	50	115
Sudan	PAL	B	CCIR	50	240
Surinam	NTSC	M	USA	50/60	110/115/127
Swaziland		I			
Sweden	PAL	B	CCIR	50 DC	127/220/380
	PAL	G			
Switzerland	PAL	B	CCIR	50	220/380
	PAL	G			
Syria		B	CCIR	50	115/200
Tahiti	SECAM(V)	K1	OIRT	60	120/240
Taiwan	NTSC	M	USA	60	110/200/220
Tanzania	PAL	I	CCIR	50	230
Thailand	PAL	B	CCIR	50	220/380
		M			
Tibet		D		DC/50	220
Togo	SECAM	K1	OIRT	50	127/220
Trinidad and Tobago	NTSC	M	USA	60	115/230
Trust Islands of the Pacific (Micronesia)	NTSC	M	USA		
Tunesia	SECAM(V)	B	CCIR	50	115/220/380
Turkey	PAL	B	CCIR	50	220/380
Uganda	PAL	B	CCIR	50	240/415
United Arab Emirates	PAL	B,G	CCIR		

United Kingdom	PAL	I	BRITISH	50	240/415
		A			
Upper Volta		K1	OIRT	50	220/380
Uruguay		N	USA	50	220
USA	NTSC	M	USA	60	120/240
	NTSC	M			
USSR	SECAM(V)	D	OIRT	50	127/220
	SECAM(V)	K			
Venezuela	NTSC	M	USA	50+60	120/240
Vietnam		D	USA	50	120/127/220/230
		M			
Virgin Is	NTSC	M	USA	60	110
Yemen Arab Rep		?B	CCIR	50	
Yemen Dem Rep		B	CCIR	50	220
Yugoslavia	PAL	B	CCIR	50	220/380
	PAL	H			
Zaire	SECAM(V)	K	CCIR	50	220
Zambia	PAL	B	CCIR	50	220/380
Zimbabwe		B	CCIR	50	230/240

Note: Indication of color system adopted does not necessarily imply regular or any service.

NOTES

NTSC: *National Television System Committee.* The color TV broadcasting system developed in the United States and used in North America, Japan, the Philippines, South Korea and Taiwan.

Modified NTSC or NTSC 4.43: A term used for NTSC recordings that can be replayed on European standard videocassette recorders. Although 525 lines/60 fields are replayed, the color subcarrier is shifted to 4.43 MHz, instead of the usual 3.58 MHz.

PAL: *Phase Alteration by Line.* The color TV broadcasting system developed by Telefunken and used in Western Europe and other parts of the world not using NTSC or SECAM.

PAL-M: A derivative of the normal PAL color system with a shifted subcarrier of 3.58 MHz, allowing it to be used on restricted vision bandwidth 525 line/60 field systems, such as M and occasionally on system N.

SECAM: *Séquential Couleur à Memoire.* The color TV broadcasting system developed in France and used in other parts of the world such as the USSR and Africa.

SECAM(V): *Vertical SECAM.* The old system used in France, the USSR, etc.

SECAM(H): *Horizontal SECAM.* The new system that is used especially in the Middle East.

*The color systems given in the table above are employed for broadcasting and not necessarily the originating system. For example: many countries use PAL for studio origination of the program (since SECAM can be difficult to work with), but change over to SECAM for broadcasting.

**A number of the listed countries are still in an experimental stage of TV broadcasting and in some cases the listed color system is only the proposed system.

(Courtesy JVC, Ltd.)

TV Broadcasting Standards-Color or B/W

Characteristics	A British VHF	B Gerber VFH	C Belgian VHF	D O.I.R.T. VHF	E French VHF	F Belgian VHF	G Gerber VHF/UHF
Number of lines per picture (frame)	405	625	625	625	819	819	625
Field frequency (field/second)	50	50	50	50	50	50	50
Picture (frame) frequency (picture/second)	25	25	25	25	25	25	25
Line frequency (lines/second)	10 125	15 625	15 625	15 625	20 475	20 475	15 625
Nominal video bandwidth (MHz)	3	5	5	6	10	5	5
Nominal radio-frequency bandwidth (MHz)	5	7	7	8	14	7	8
Sound carrier relative to vision carrier (MHz)	−3.5	+5.5	+5.5	+6.5	+11.15	+5.5	+5.5
Nominal width of vestigal sideband (MHz)	0.75	0.75	0.75	0.75	2	0.75	0.75
Type of polarity of vision modulation	POS.	NEG.	POS.	NEG.	POS.	POS.	NEG.
Type of sound modulation	AM	FM	AM	FM	AM	AM	FM
Pre emphasis	—	50/μs	50/μs	50/μs	—	50/μs	50/μs
Ratio of effective radiated power of vision and sound	4/1	10/1	4/1	2/1 ~ 5/1	4/1	4/1	10/1

Characteristics	H Belgium UHF	I British VHF/UFH	K Modified O.I.R.T. UHF	K1 Modified O.I.R.T. VHF/UHF	L French UHF	M R.T.M.A. VHF/UHF	N R.T.M.A. VHF/UHF
Number of lines per picture (frame)	625	625	625	625	625	525	625
Field frequency (field/second)	50	50	50	50	50	60	50
Picture (frame) frequency (picture/second)	25	25	25	25	25	30	25
Line frequency (lines/second)	15 625	15 625	15 625	15 625	15625	15 750	15 625
Nominal video bandwidth (MHz)	5	5.5	6	6	6	4.2	4.2
Nominal radio-frequency bandwidth (MHz)	8	8	8	8	8	6	6
Sound carrier relative to vision carrier (MHz)	+5.5	+6	+6.5	+6.5	+6.5	+4.5	+4.5
Nominal width of vestigal sideband (MHz)	1.25	1.25	0.75	1.25	1.25	0.75	0.75
Type of polarity of vision modulation	NEG.	NEG.	NEG.	NEG.	POS.	NEG.	NEG.
Type of sound modulation	FM	FM	FM	FM	AM	FM	FM
Pre emphasis	50/μs	50/μs	50/μs	50/μs	—	75/μs	75/μs
Ratio of effective radiated power of vision and sound	5/1 ~ 10/1	5/1	2/1 ~ 5/1	8/1	10/1 ~ 5/1	10/1 ~ 5/1 4/1*	10/1 – 5/1

Notes
- Systems A, C, E, F and H are becoming obsolete
- In general, PAL is used on B, D, G and I systems; SECAM on K,K1 and L systems; NTSC on M system; and PAL-M (3.58 MHz) on M and N systems.

B/W Broadcasting Standards

Item	System					
	USA	CCIR	OIRT	British	France	Belgium
Number of Scanning Lines	525	625	625	405	819	625
Channel & Frequency Width	6 MHz	7 MHz	8 MHz	5 MHz	14 MHz	7 MHz
Video Frequency Bandwidth	4	5	6	3	10.4	5
Position of Audio Carrier (as compared with Video Carrier) (MHz)	+4.5	+5.5	+6.5	−3.5	+11.15	+5.5
Position of Audio Carrier (as compared with Channel End)	−0.25	−0.25	−0.25	+0.25	+0.1	−0.25
Interlace	2/1	2/1	2/1	2/1	2/1	2/1
Line Frequency (Hz)	15,750	15,625	15,625	10,125	20,475	15,625
Field Frequency (Hz)	60	50	50	50	50	50
Picture (frame) Frequency	30	25	25	25	25	25
Video Modulation (Polarity)	Negative	Negative	Negative	Positive	Positive	Positive
Audio Modulation System	FM	FM	FM	AM	AM	AM
Synchronizm with Power Frequency	Synchronizable or not syncronizable	Not synchronizable	Not synchronizable	Not synchronizable	Not synchronizable	Not synchronizable

(Courtesy JVC, Ltd)

amplifiers. This method requires more complex and more expensive digital electronics to disassemble the recordings and to reassemble the playback.

All of the above techniques are feasible and, in fact, digital video has already been demonstrated on an experimental basis, using no more tape than with analog video recorders. It thus appears that digital video will become a commercial reality in the near future.

TELEVISION STANDARDS IN VARIOUS COUNTRIES

Different television standards have been adopted by various countries and therefore video recorders must be designed or modified for different scanning and color systems. Table 11-6 summarizes various standards in different countries. Table 11-7 lists color subcarriers and their derivation. Table 11-8 is an international listing of TV broadcasting systems.

12. Home Video Tape Recording

Three aspects of video recording by the general public have vast possibilities, which have already been fulfilled in recent years:

1. *Recording of broadcast television.* Programs are recorded automatically when someone is not at home or if watching something else. A person can collect entertainment or educational programs to be viewed whenever and as often as desired.
2. *Prerecorded cassettes.* Records can be purchased or rented with programs that are not available "off the air." The analogy to sound recordings is evident, but there is an important difference. Most people do not tire of hearing the same good music, occasionally or frequently, year in year out, but few would care to watch the same televised program dozens of times; hence the rental option. Educational and "how to do it" are analogous fields with interesting potential.
3. *Home movies.* With portable equipment and color camera, amateurs can make sound movies and watch them on any television receiver, immediately, without processing. They can erase what they do not like and re-record it. They can add or change narration and background music.

Electronics manufacturers have watched these fields with great interest because the dollar volume of consumer items far exceeds the combined totals for professional recorders. Over a 20-year period the low-cost helical recorders first gravitated toward open reel half-inch tape decks that eventually led to the EIAJ standard (variously called $\frac{1}{2}$ inch Type A and Type F). The next important step was the U-Matic® (U-Format) machine (Type E) with cartridge loading and color recording at half the tape speed of EIAJ recorders. (The U-Format is described in Chapter 11). Derived from the U-Format are the Betamax® (Type G) and VHS (Type H) machines, eagerly accepted by consumers, with millions of recorders and tapes sold world-wide. Figure 12-1 shows commercial versions of these popular recorders.

Operating principles of helical scan recorders are described in detail in Chapter 11; see particularly Figures 11 15, 11 27, and 11-28. Tablc 12-1 cxplains some abbreviations and nomenclature used in video recording. Table 12-2 describes video formats that have been in use.

(a)

(b)

Figure 12-1. Home videotape recorders of (a) Beta and (b) VHS types.

BETAMAX® AND VHS RECORDERS

Economy is the watchword in consumer electronics. A few pennies saved per machine makes a difference in profit and marketing of highly competitive products. The key to videotape economy is high-density recording made possible by super tapes and microgap heads.

Tapes, Heads, and Densities. Beta and VHS video cassettes use tapes having a coercivity of about 700, more than twice as high as in ordinary audio tapes (see Figure 3-12*B*). The video tape has an exceptionally smooth surface for close head contact and its fine-grain uniform dispersion insures freedom from dropouts. Heads are made of single crystal or hot-pressed ferrites with gaps 0.3 μm (12 microinches) long, and are only about 25 μm (one mil) wide.

Table 12-1. Terms Used in Home Video Recording.

Anhysterestic	Recording with the use of high frequency bias
Azimuth recording	Recording odd tracks with a different head gap angle than even tracks to eliminate guard bands
Cassette	A container with a pair of side by side reels of tape
Chevron tracks	Track pattern of azimuth recording
Chroma	Color signal
CS	Capstan servo
CTL	Control track used for synchronizing the tape speeds and heads
Dew point	Interlock and warning that cassette recorder is inoperative because of condensation on scanning drum
EP	Extra play, 6 hours, format for VHS cartridges
Guard band	Space that separates recorded tracks
HVR	Home video recorder
LP	Long play, 4 hours, format for VHS cartridges
PAL	Phase Alternation Line (European color video format)
Pitch	(Track pitch), center to center distance between the tracks
PG	Pulse generator
SECAM	Séquential Couleur Avec Mémoire (French and USSR color video format)
SLP	Super long play. Same as EP
SP	Standard play, 2 hours, format for VHS cartridge
VCR	Video cassette recorder
VHS	Video home system
VTR	Video tape recorder
Y	Luminance signal

Table 12-2. Video Tape Formats.

Format	Description
2 in. quad	Edge to edge scan using 2 in. tape and a drum with four heads
1 in. Type A	Helical scan on 1 in. tape with a drum and one head (obsolecent)
1 in. Type B	Helical segmented format on 1 in. tape with a drum and two heads
1 in. Type C	Helical scan on 1 in. tape with a drum and one main head (plus $\frac{1}{2}$ head)
$\frac{3}{4}$ in. Type E	Helical scan by 2 heads on $\frac{3}{4}''$ tape in a U-Matic format
$\frac{1}{2}$ in. Type F	Helical scan by 2 heads of $\frac{1}{2}$ in. tape in reel-to-reel recorders
$\frac{1}{2}$ in. Type G	Helical scan by 2 heads of $\frac{1}{2}$ in. tape in a Beta type cassette
$\frac{1}{2}$ in. Type H	Helical scan by 2 heads of $\frac{1}{2}$ in. tape in a VHS type cassette
$\frac{1}{2}$ in. Type J	Helical scan by 2 heads of $\frac{1}{2}$ in. tape in a camera type cassette
$\frac{1}{4}$ in. Type K	Helical format using $\frac{1}{4}$ in. tape
8 mm	Helical scan by 2 heads of 8 mm tape in a miniature cassette

The guard band is eliminated for long-play tapes, increasing the recording density still further.

With the advances described above, recording densities of about 800 wavelengths/mm (20,000/in) are achieved, and track densities are 50/mm (1200/in). These correspond to area density of 37000 wavelengths/mm^2 (24×10^6/in^2), unbelievable only a few years ago. Expressed as bit densities, the figures are twice as high since there are two bits per wavelength.

Tape Speeds, Playing Time, Cassette Sizes. When first introduced, the Beta and VHS machines were thought to be the ultimate in economy, long play and high density, giving $1\frac{1}{2}$ to 2 hours of recording per cassette with the tape moving at only 36 mm/s (1.4 in./s). These machines were later designated as operating in the Beta I and VHS-SP(I) modes.

Afterwards it was found that the playing time could be doubled by running the tape at half speed, keeping the head wheel speed unchanged, but modifying the heads and electronics to reduce interference between adjacent tracks. Newer machines then had a selection of two speeds and could record 3 to 4 hours per cassette: Beta II or VHS-LP(II). The success of the half-speed recorders encouraged further speed reduction to one-third speed, giving up to 6 hours playing time per cassette with a lower quality but still acceptable picture: Beta III or VHS-EP(III). Eight hours are possible with thinner tape.

Tables 12-3 and 12-4 list the writing speeds, tape speeds, track dimensions, head wheel dimensions, and so on of home VTR formats, contrasting them to quadruplex and to audio recorder quantities.

Table 12-5 gives the relative tape economy of different formats and also the cassette or reel dimensions and comparative volume. It is interesting that the newer video recorders actually use less area of tape per hour than a 7.5 in/s audio recorder does. VHS-EP uses less than an eight-track stereo cartridge, and almost as little as a compact cassette.

Video cassettes are somewhat more bulky (18% to 55%) than a 7-inch reel. The VHS cassette is about 30% larger than the Beta cassette and therefore can hold more tape than the latter, but at the expense of added bulk.

Track Format and Head Wheels. Figure 12-2 shows that the track format of Betamax I and VHS (SP) recorders are similar except that VHS tracks are somewhat shorter and at a greater angle. The shorter VHS track results from a smaller head wheel diameter (62 mm compared to 74.5 mm for the Betamax). Heads and tracks are 1.5 mils (38μm) wide with a track pitch of 2.3 mils (58.5 μm), giving a guard band of 0.8 mil (20.5 μm) that separates the tracks.

When the tape speed is reduced to half, the track pitch is halved to 29μm.

Table 12-3. Writing Speeds, Track and Drum Dimensions of Home and Other Video Recorders.

Format	Writing Speed in/s (m/s)	Forward Tape Speed in/s (mm/s)	Track Length Per Field in (mm)	Track Angle	Drum Diameter in (mm)	Head Wheel RPM
Beta I	272 (6.9)	1.57 (40)	4.53 (115)	5°	2.93 (74.5)	1800
Beta II	272 (6.9)	0.79 (20)	4.53 (115)	5°	2.93 (74.5)	1800
Beta III	272 (6.9)	0.52 (13.3)	4.53 (115)	5°	2.93 (74.5)	1800
Beta PAL/SECAM		0.73 (18.6)				
VHS-SP (I) 2 hr	228 (5.8)	1.31 (33.35)	3.8 (97)	6°	2.44 (62)	1800
VHS-LP (II) 4 hr	228 (5.8)	0.66 (16.67)	3.8 (97)	6°	2.44 (62)	1800
VHS-EP (III) 6hr[d]	228 (5.8)	0.44 (11.12)	3.8 (97)	6°	2.44 (62)	1800
VHS PAL/SECAM		0.92 (23.39)				1500
8mm (525 line-60f)	148 (3.76)	0.56 (14.345)	2.46 (62.6)	4.90°	1.57(40.0)	1800
8mm (525 line-50f)	123 (3.12)	0.79 (20.051)	2.46 (62.4)	4.92°	1.57 (40.0)	1500
U-Format[a]	404 (10.3)	3.75 (95)	6.73 (95)	4.96[c]	4.33 (110.0)	1800
Open reel (EIAJ)[b]	437 (11.1)	7.5 (190)	7.28 (185)		4.56 (115.8)	1800
Quad (NTSC)	1508 (38.3)	15 (381)	25.1 (638)	89.4°	2.0 (50.8)	14,400
Audio reel[c]	75 (0.190)	7.5 (90)	—	0°	—	—
Audio 8 tr Cart.	3.75 (0.095)	3.75 (95)	—	0°	—	—
Audio cassette	1.88 (0.048)	1.88 (48)	—	0°	—	—

[a] U-Format has been called "$\frac{3}{4}$ inch Type A" and "Type F".
[b] EIAJ has been called "$\frac{1}{2}$ inch Type A" and "Type E".
[c] Four track.
[d] Also called SLP.

Table 12-4. Track Dimensions of Home and Other Video Recorders.

Format	Tape Width in (mm)	Video Head Width mils (μm)[a]	Video Track Pitch[b] mils (μm)	Audio Track Width mils (μm)	Control Track Width mils (μm)
Beta I	0.5 (12.7)	1.5 (38)	2.3 (58.5)	41 (1050)	23.6 (600)
Beta II	0.5 (12.7	1.5 (38)[c]	1.1 (29.2)	41 (1050)	23.6 (600)
Beta III	0.5 (12.7)		0.8 (19.5)	41 (1050)	23.6 (600)
Beta PAL/SECAM	0.5 (12.7)				
VHS-SP(I) 2 hr	0.5 (12.7)	1.2 (30)	2.3 (58)	39 (1000)[e]	29.5 (750)
VHS-LP (II) 4 hr	0.5 (12.7)	1.0 (26)[c]	1.1 (29)	39 (1000)	29.5 (750)
VHS-EP (III) 6 hr[d]	0.5 (12.7)	1.0 (26)[c]	0.8 (19.3)	39 (1000)	29.5 (750)
VHS-PAL/SECAM	0.5 (12.7)				
8mm (525 line-60f)	0.32 (8.00)		0.81 (20.5)	23.6 (600)	23.6 (600)
8mm (625 line-50f)	0.32 (8.00)		1.35 (34.4)	23.6 (600)	23.6 (600)
U-Format	0.75 (19.0)	3.4 (85)	5.4 (140)	31.5 (800)	24 (600)
Open reel (EIAJ)	0.5 (12.7)	3.9 (100)	6.8 (170)	37 (940)	30 (760)
Quad (NTSC)	2.0 (50.8)	10.0 (0.25)	15.7 (400)	75 (1905)	25 (635)
Audio reel	0.25 (6.35)			41 (1040)	
Audio 8 track cartridge	0.25 (6.35)			21 (533)	
Audio cassette	0.15 (3.81)			24 (600)	

[a] One mil = 0.001 in. = 25.4 μm (micrometres)(microns).
[b] Pitch = center to center distance between tracks.
[c] Head overlaps tracks.
[d] Also called SLP.
[e] For stereo this width is allocated to two 350 μm tracks and a 300 μm guard band.

Table 12-5. Tape Economy and Cassette Volume of Home VTR Formats and Other Recorders.

Format	Forward Tape Speed in/s (mm/s)	Tape Consumption ft/hr (m/hr)	Tape Consumption sq ft/hr (sq m/hr)	Cassette (or reel) Size: in (mm) (L) (W) (H)	Relative Cassette Volume* (% of 7″ reel)	Playing Time (approx.) hours
Beta I	1.57 (40)	471 (144)	19.6 (1.829)	6.1×3.8×1.0 (156×96×25)	118%	1.4
Beta II	0.79 (20)	237 (72)	9.9 (0.914)	(Same)	118%	2.8
Beta III	0.52 (13.3)	156 (48)	6.5 (0.609)	(Same)	118%	5.5
Beta PAL/SECAM	0.73 (18.6)	219 (67)	9.1 (0.851)	(Same)	118%	3
VHS-SP (I) 2 hr	1.31 (33.35)	393 (120)	16.4 (1.524)	7.4×4.1×1.0 (188×104×25)	155%	2
VHS-LP (II) 4 hr	0.66 (16.67)	197 (60)	8.2 (0.763)	(Same)	155%	4
VHS-EP (III) 6 hr	0.44 (11.12)	132 (40)	5.5 (0.508)	(Same)	155%	6
VHS PAL/SECAM	0.92 (23.39)	276 (84.2)	11.5 (1.069)	(Same)	155%	3
8mm (525 line-60f)	0.56 (14.345)	168 (51.6)	4.45 (0.413)	3.7×2.5×0.6	28%	2
8mm (625 line-50f)	0.79 (20.051)	237 (72.2)	6.22 (0.578)	(95×62.5×15)	28%	2
U-Format	3.75 (95)	1125 (343)	70.3 (6.53)	8.7×5.5×1.3 (222×140×32)	315%	1
Open Reel(EIAJ)	7.5 (190)	2250 (686)	93.8 (8.68)	178D×19)	150%	1
Quad (NTSC)	15 (381)	4500 (1372)	750 (69.7)	(178D×60)	472%	0.5
Audio Reel	7.5 (190)	2250 (686)	23.5 (2.18)	178D×12.7)	100%	2
Audio 8tr. cart.	3.75 (95)	1125 (343)	5.8 (0.54)	(140×100×19)	84%	1.5
Audio Cassette	1.88 (48)	562 (172)	3.6 (0.33)	(100×64×12)	24%	1.5

*Volume relative to a standard 7″ reel of $\frac{1}{4}$ in. tape.

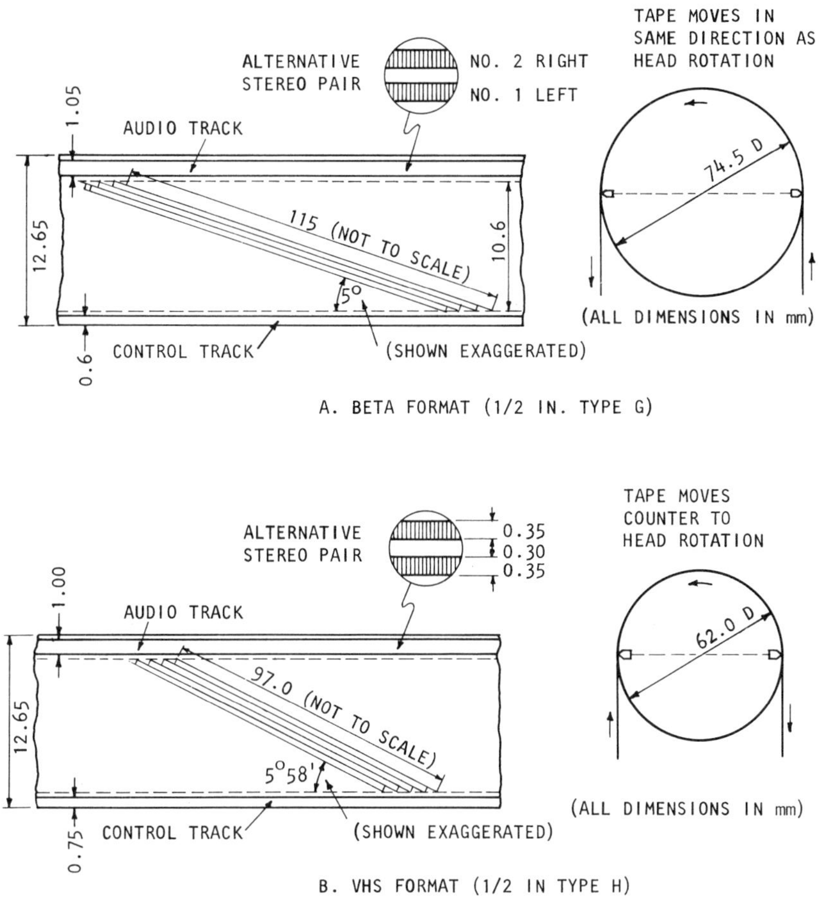

Figure 12-2A, B. Track format and head wheel dimensions of Beta & VHS recorders.

The simplest way to handle the closer track spacing would be to use a narrower head. Full speed and half speed recordings could both be played back with the narrow head, but the full speed tracks would be at a disadvantage and would play back noisier than with a full width head. It was decided that full speed recordings should be favored by retaining the wider head and that the longer play mode could be corrected as described below to give marginally satisfactory quality. The marginal slow speed quality turned out to be so good that full speed seemed like a waste of expensive tape, and slow speed (Beta II or VHS-LP) became a preferred mode for home recording. Even slower speeds of Beta III and VHS-EP are achieved by compromising still further, except that narrower

Track configuration and dimensions
(view on magnetic sensitive side)

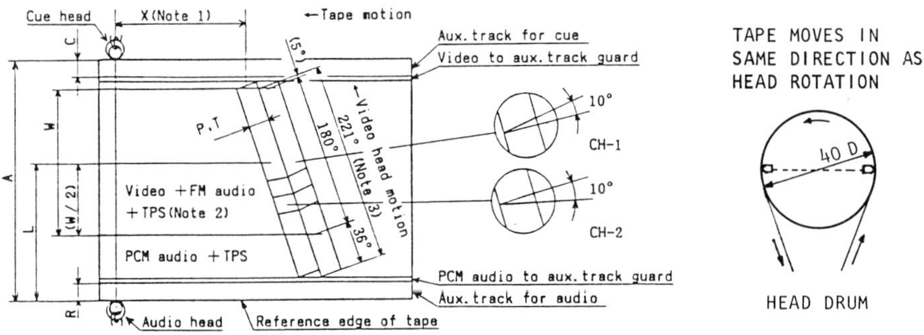

Note 1. X shall be measured from the end of the 180° scan of CH-2 to the audio and cue head on the tape.

Note 2. TPS: tracking pilot signal

units: mm

Item	525 line – 60 field system	625 line – 50 field system
1. (A) Tape width	8.000 ± 0.01	
2. (P) Video track pitch	0.020 5	0.034 4
3. (W) Video effective width (180°)	5.351	5.351
4. (L) Video track centre from reference edge of tape	4.461	4.461
5. (T) Video track width	0.020 5	0.034 4
6. (C) Auxiliary track for cue	0.6 ± 0.05	0.6 ± 0.05
7. (R) Auxiliary track for audio including optional edge guard (0.1)	0.6 ± 0.05	0.6 ± 0.05
8. (θ_0) Video track angle (tape stop)	4° 53' 06"	4° 53' 06"
9. (θ) Video track angle (tape runs)	4° 54' 13.2"	4° 54' 58.8"
10. (X) Position of audio and cue head	31 ± 0.24	31 ± 0.24

Figure 12-2C. Track format and head wheel dimensions of 8mm videocassette recorder. (Courtesy EIAJ)

heads are used. On three-speed recorders, the handicap of the extra narrow heads can be overcome by a head wheel with four heads: a narrow pair for slow speeds and a wide pair for high speeds.

Azimuth Recording and Chevron Tracks. With a 38μm head width and a 29 μm track pitch, the guard band becomes negative; in other words there is

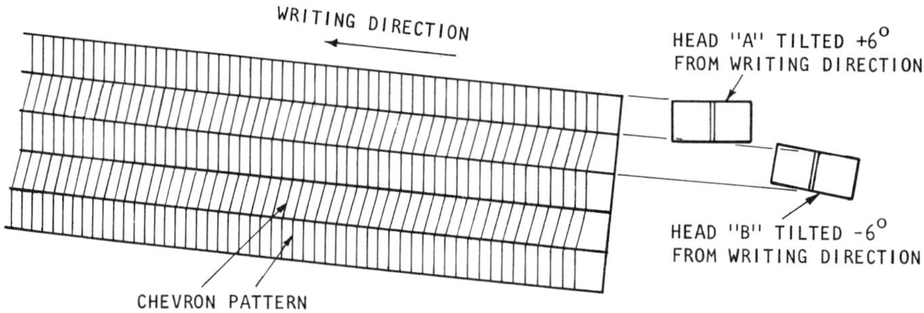

Figure 12-3. Tilted azimuth of gaps on opposite heads gives chevron track pattern.

a 9μm overlap or interference between adjacent tracks. To reduce cross-talk, the two heads on the VHS head wheel are tilted in opposite directions so the gap of one head makes an azimuth angle of $90 + 6 = 96°$ with respect to the writing direction, while the other gap makes an angle of $90 - 6 = 84°$ as shown in Figure 12-3. Each head gap reads correctly the track written at its own gap angle, but has a 12° azimuth error for the overlapped portion of the adjacent track, and hence discriminates against the adjacent track. On Beta recorders the gap angles are offset $\pm$ 7°.

The above principle is called *azimuth recording*, and the recorded tracks are in a chevron pattern (Figure 12-3). Azimuth recording does not give immunity against all cross-talk but acts as a comb filter where certain wavelengths cancel while other reinforce. In fact, longer wavelengths are picked up with almost no discrimination by gap tilt. Other methods must therefore be used to minimize interference at longer wavelengths, especially those at the color subcarrier, which is 688 KHz for Beta recorders and 629 KHz for VHS.

Electronics

The overall block diagram in Figure 12-4 indicates that the same FM recording method is used as described in Chapter 11 for professional recorders (Figures 11-7 and 8) but modified to take care of the *color under* principle, as described below. Head switching and servo control for tape drive are also as discussed in Chapter 11. In addition there is a rather elaborate sequence for threading and unthreading of the tape cassettes. Some machines also include a dew point sensor and possibly a heating element to alleviate conditions where moisture condensation prevents the tape from moving along the head wheel drum. For recording of broadcast television programs, a TV *front-end tuner* is included,

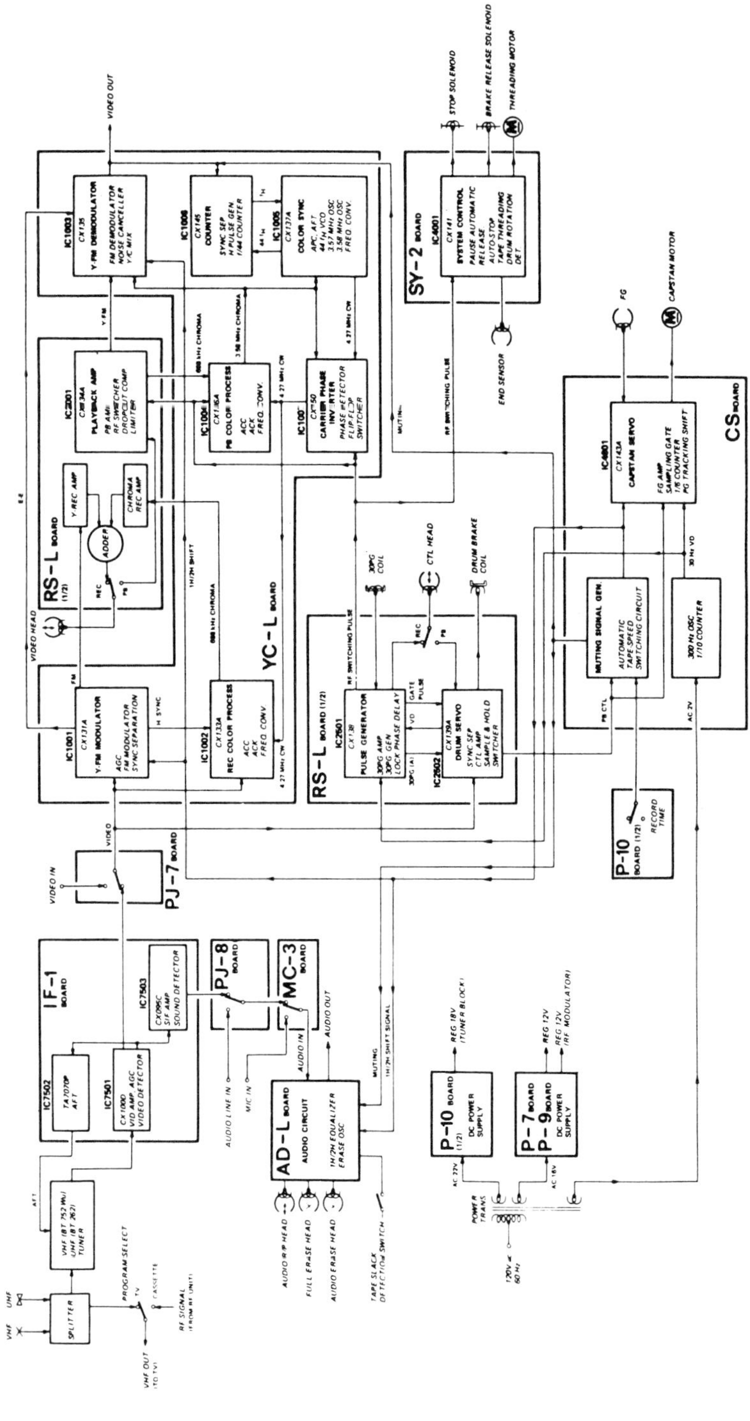

Figure 12-4. Overall block diagram of a Beta video recorder. (Courtesy Sony)

making the video recorder independent of any external television receiver. Programmable timers may also be built in to allow automatic recording of different TV programs according to a preselected schedule of station, time, and duration. Video player output is at television carrier frequencies, connected to the antenna terminals of any TV receiver for viewing on channel three or four. Altogether, the home video recorder is probably the most complex appliance that has ever been designed for home use.

Color Subcarrier. It has been impractical to record color information directly at the relatively slow writing speeds and high densities of home recorders because output and phase shift are unreliable at a color carrier frequency of 3.58 MHz. In these lower write speed recorders, which also include U-Format and EIAJ reel types, the color information is separated out before recording and heterodyned down to a lower subcarrier frequency, as shown in Figure 12-5. This method, sometimes called *color under* recording, uses a color subcarrier of about 0.7 MHz, easily recorded and played back by the recording system. During playback, the color under information is reconverted to a standard NTSC or other composite signal for television display. The color subcarrier is in a prominent and highly visible part of the spectrum and therefore is prone to causing interference effects.

Electronic Interference Reduction. Since azimuth recording does not reduce interference at subcarrier and lower frequencies, other methods must be used:

1. odd line inversion and cancellation for minimizing cross-talk between tracks (comb filtering)
2. frequency interleaving for overcoming effects of sync pulse intermodulation
3. use of FM for luminance while using modified AM for chrominance
4. recording the color information linearly at low amplitude compared to the luminance

It is interesting that the strong constant amplitude luminance carrier acts as high-frequency bias for the weaker color signal. Figure 12-6 shows the spectrum allocation for a typical recorder. Table 12-6 lists the carrier frequencies, modulation range and so forth of a family of recorders.

Cancellation of Spurious Pickup. An effective way to cancel unwanted cross-talk when a head overlaps an adjacent track is to combine in opposite phase the cross-talk from corresponding parts of two successive horizontal pic-

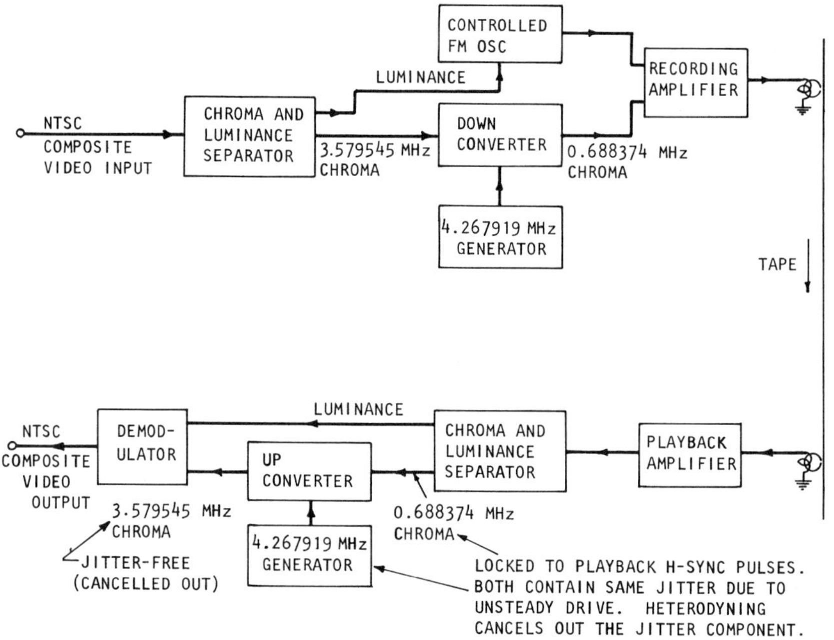

Figure 12-5. Simplified block diagram of "color under" sub-carrier method for low write-speed video recorders.

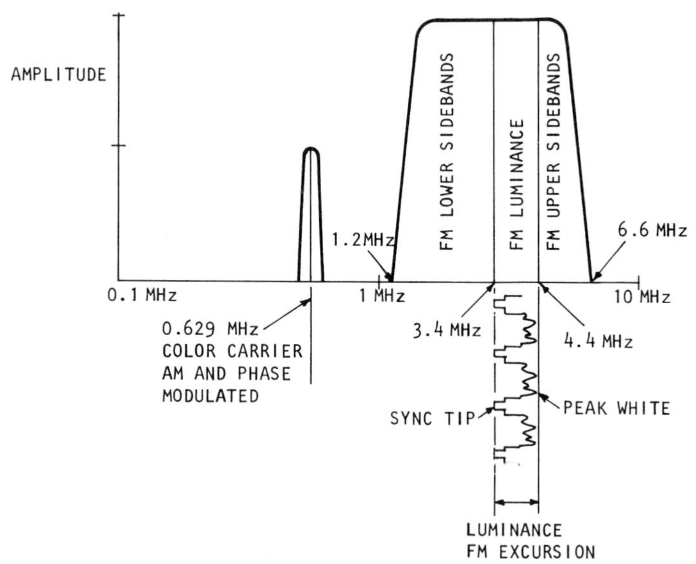

Figure 12-6. Spectrum allocation of a VHS recorder.

Table 12-6. Frequencies used in Home Video Recorders.

Format	Color Subcarrier MHz	Luminance Freq. MHz Sync. Tip	Luminance Freq. MHz Peak White	Picture Resolution Lines Luminance	Picture Resolution Lines Chrome	Audio Response Hz Range	S/N Ratio Video dB	S/N Ratio Audio dB
Beta I	0.688	3.5	4.8	270				
Beta II	0.688	3.5	4.8	250		50–10,000	45	40
Beta III	0.688	3.5	4.8			50– 7,000	45	40
Beta PAL/SECAM								
VHSI(SP)	0.629	3.4	4.4	270	230	50–10,000	45	40
VHS II(LP)	0.629	3.4	4.4	270	230			
VHS III(EP)	0.629	3.4	4.4	270	230			
VHS PAL/SECAM	0.627			300	250	70–10,000	40/43	40/50*
8mm (525 line-60f)	0.746	4.2	5.4					
8mm (625 line-50f)	0.732	4.2	5.4					
U-Format	0.688	3.8	5.4	330	260	50–15,000	46/50	48
Open reel (EIAJ)	0.767	3.1	4.5					
Quad (NTSC) (High band color)	3.58	7.06	10.00	to 5 MHz		50–16000	46	58

*With noise reduction.

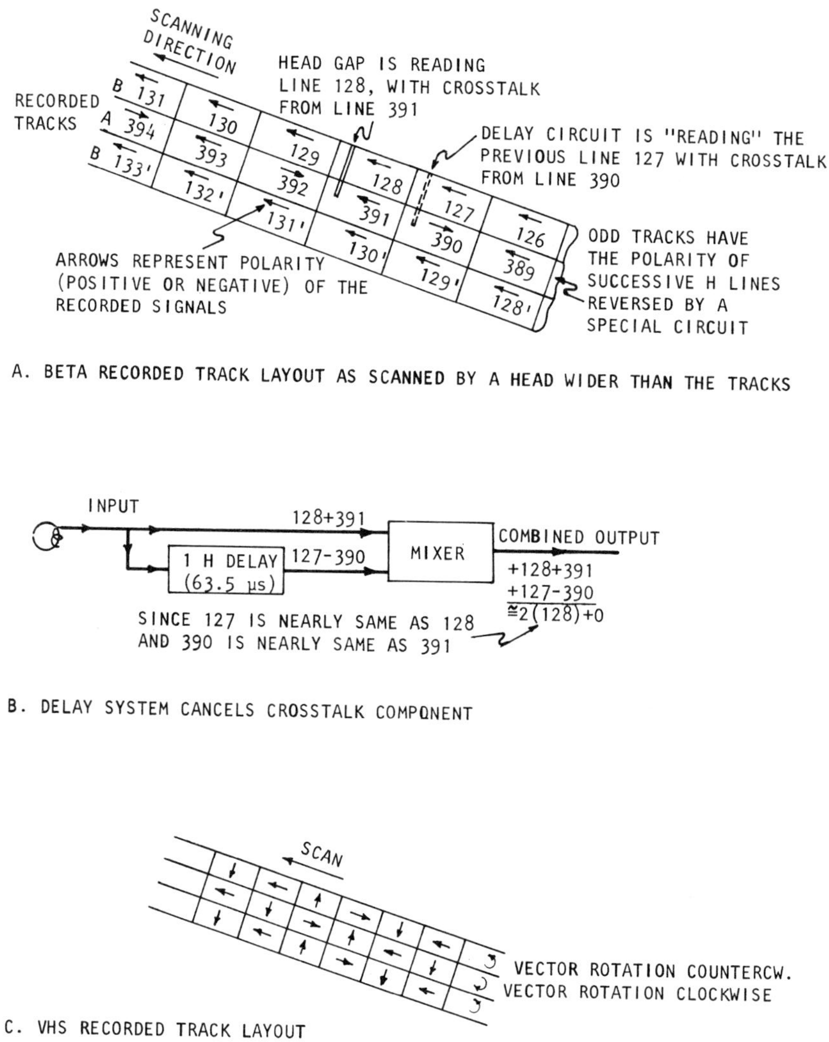

Figure 12-7. Electronic cancellation of crosstalk in "negative guard band" recording.

ture lines. Figure 12-7*A* shows a scan head picking up desired output from horizontal scan line 128 and undesired cross-talk from line 391 of the next track. The circuit of Figure 12-7*B* mixes this with delayed pickup from the previous scanned lines 127 and 390. In the combined output the adjacent lines 128 and 127, being nearly the same, are added. The adjacent cross-talk lines 391 and 390 are also nearly the same but since they are of opposite polarity they are subtracted from each other and cancel out. Success of this scheme is based on the fact that color information does not change much between horizontal picture lines. In effect we are combining two picture lines at a time for the information we want, while opposing two lines of a different track to get rid of cross-talk. Combining the lines cuts the vertical resolution in half, to 262 lines instead of 525, but this is more than enough since the color bandwidth is only about 1 MHz in any case. The luminance bandwidth, which is higher, is handled separately and is not affected by these color circuits.

Comb Filtering. The above discussion is phenomenological. A more elegant explanation is based on comb filter behavior. The circuit of Figure 12-7*B* has an unusual frequency response, plotted in Figure 12-8*A* with peaks at the 15734 Hz horizontal line frequency (H) and all of its harmonics (nH), and with zeros at odd harmonics of 7867 Hz or $mH/2$, where n is any integer and m is any odd integer.

In Figure 12-8*B* we examine the waveform of a repeated video picture horizontal line and observe that it contains only harmonics of the fundamental repetition rate according to Fourier's theorem. The NTSC fundamental repetition rate is 15734.264 Hz. Since all harmonics lie on peaks of the comb filter, the waveform of Figure 12-8*B* passes readily through this filter. Although the theory applies strictly when successive waves are alike, the broad upper peaks of the comb will tolerate slow changes, which are inevitable when painting a picture with horizontal scan lines.

In Figure 12-8*C* every other picture line signal has been inverted or negatived, which can be done with an electronic reversing switch. (Of course we would not display a negative picture until we reversed it again). The video pattern now repeats after every two lines, so its fundamental frequency has been cut in half to $H/2$ = 7867.132 Hz. Such a wave contains only odd harmonics of $H/2$, all of which occur at points where the comb filter output is zero, and therefore will be stopped by the filter. Note that the sharp rises of response near zero will let something get through if successive horizontal lines are not identical. Still a worthwhile cancellation takes place. Figure 12-8*D* shows comb filter response if an inverter is included.

As is evident from the line pattern in Figure 12-7*A*, the adjacent recorded tracks on a Beta video tape are indeed different, the odd tracks recorded by head *B* being unmodified, while the even tracks recorded by head *A* are negatived at every other horizontal line, as in Figure 12-8*C*. When head *B* is playing back, the video signal from its own track *B* passes through the comb filter, but the track *A* interference picked up by head overlap is stopped by the comb filter. When head *A* is playing back, an electronic switch reverses the alternate negative horizontal lines of track *A*, setting them right again so they pass through

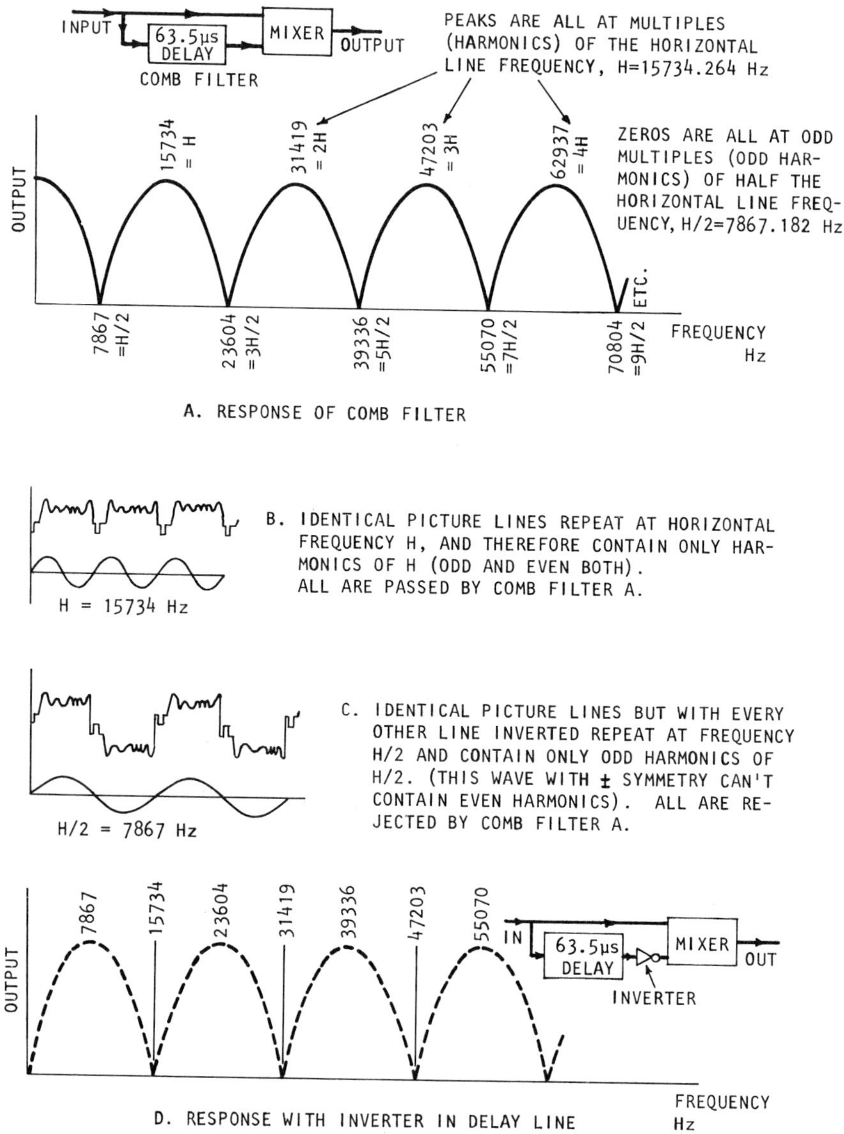

Figure 12-8. Comb filter action.

the comb filter. But the interference from overlapping track *B* has now become alternately reversed and will not go through the filter.

VHS recorders use a somewhat different scheme. As seen in Figure 12-7*C*, the horizontal line recordings on track *A* are caused to advance in phase, shifting 90° counter-clockwise on each successive horizontal line, to complete a cycle every four lines. On track *B* the recording phase is retarded in the same way, causing the phase vector to rotate clockwise 90° for each line. When playing back track *A*, the phase rotation of *A* is followed, making its output signal, in effect, stand still at zero phase for a positive picture display, which will of course, pass through a comb filter. Making the phase of track *A* stand still causes the relative phase of track *B* to rotate in the opposite direction (clockwise) twice as fast. Successive horizontal lines of track *B* are therefore 180° out of phase (negative to each other), as seen in the overlap pickup by the head *A* system, and are therefore stopped by the comb filter. When head *B* is active the overlap interferences from *A* tracks are similarly cancelled.

Frequency Interleaving. Despite the best efforts to cancel interference by azimuth tilt and electronic trickery, a certain amount still gets through. This residual is made as inconspicuous as possible by *frequency interleaving*, (Figure 12-9) which depends on visual cancellation on the picture screen. Interleaving is most effective for steady frequencies such as color carriers, sync pulses, and beat frequencies caused by carrier intermodulation.

As seen in Figure 12-9*A*, the most prominent patterns noticeable on a television picture are vertical bars generated by spurious frequencies that are multiples of the horizontal frequency *H*, especially even-multiples. Near-multiples give shifting rotating patterns that are very annoying. On the other hand, as shown in Figure 12-9*B*, spurious frequencies that are odd multiples of *H/2* reverse in phase on identical picture lines of successive frames, every $\frac{1}{30}$ sec in NTSC, because there is an odd number of lines (525) per frame. The alternate superposition of light and darkness at the same spots tends to average out to the same brightness as if the spurious frequency were absent. Persistance of the picture tube phosphor and persistance of vision help this averaging. The same choice of frequency also gives out-of-phase conditions between successive horizontal line scans on the screen, causing a checkerboard pattern that blends out any remaining irregularities.

Frequency interleaving is used in cassette recorders to clean up patterns generated by overlapped track intermodulation of strong and repetitive sync pulses. It is accomplished by raising the sync tip FM frequency (and all other FM frequencies) of the *A* tracks by 7867 Hz, leaving the *B* tracks unchanged.

VTR Refinements

Each new generation of videotape recorders has more and more complex refinements to give a higher definition picture with purer color and less noise. *AGC circuits* correct the signal on the basis of sync pulse height, which is known to be constant and therefore serves as a standard. *Head unbalance* of unmatched pairs or interchanged tapes is corrected by a circuit that automatically balances gain as playback shifts between *A* and *B* heads. *Nonlinear preem-*

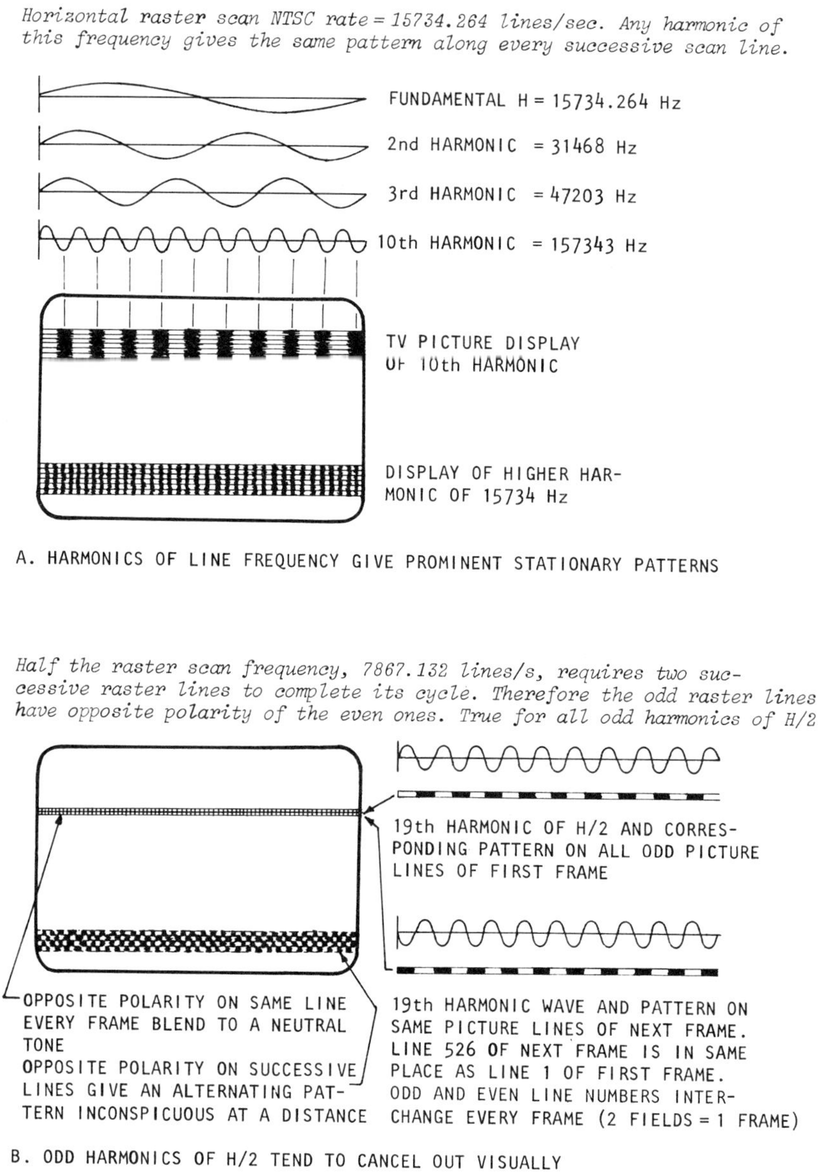

Figure 12-9. Frequency interleaving to reduce interference.

phasis and deemphasis overcomes noise by recording weak high frequencies at a higher level while preventing overload at strong high frequencies. *Noise cancellation* operates on high-frequency components of the signal; excessive spikes trigger an inverter circuit that cancels them out. *Dropout compensators* (Figure 11-14) replace missing signals with information from the preceding line.

Mechanical Threading. Figure 12-10 shows that the tape in Beta recorders is threaded by a simplified *U-load* scheme where one main finger pulls the tape out of the cassette and wraps it around the head drum, stationary heads, and capstan drive, as for U-Format recorders. VHS recorders use two fingers to perform a similar operation in a shorter *M-load* path. Interlocks prevent cassette removal unless the tape is unthreaded. Video cassettes operate in one direction only and take about three minutes to rewind.

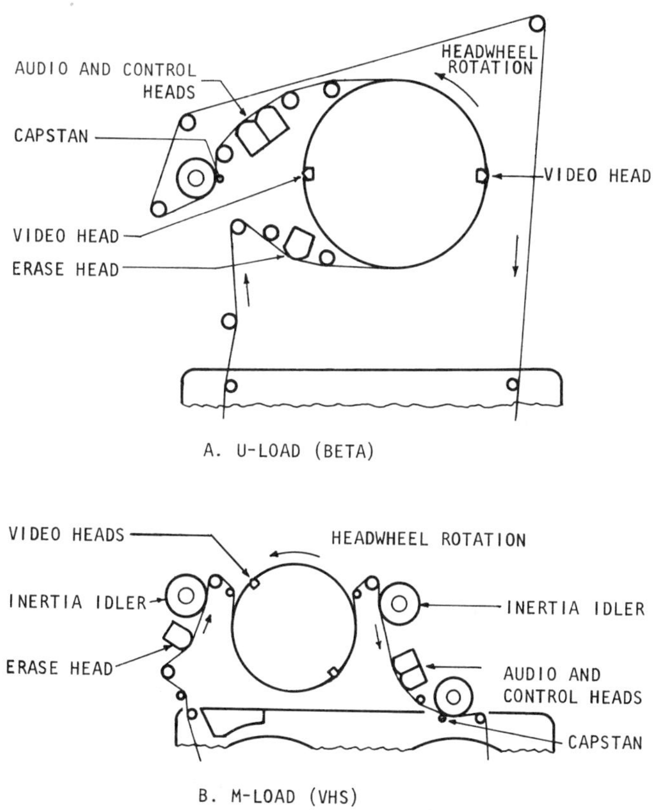

Figure 12-10. "U-Load" tape path in Beta recorders, and "M-Load" path in VHS.

Off-the-Air Recording. The widest use of home video cassette recorders has been to record television programs. Convenient features include automatic timers, pause control for skipping commercials while watching during recording, fast-forward for bypassing unwanted portions during playback, slow motion, and freeze-framing. Cost of blank tape is quite modest.

Prior to 1979 it was undecided whether the owners of video program material could prevent individuals from recording broadcasts for their own use. In 1979 a United States court decided that recording by individuals was permissible. An appeals court later reversed this decision. In 1984 the U.S. Supreme Court reversed the appeals court, again legalizing such home-recording. Program owners have asked Congress to pass laws whereby they can be compensated, as by taxing tapes or recorders. Regardless of the outcome, recording by individuals is almost impossible to regulate. It was always illegal, of course, to profit from copyrighted material without permission.

As with audio recording, the owners, performers, writers, composers, producers, and whoever else is involved in the original production worry about residual uses of programs, and quarrel among themselves on how to divide the income from the large new market and how to prevent any use without payment. Advertisers worry whenever people are able to watch something other than their messages.

Prerecorded Programs. The millions of new video cassette machines in use brought into bloom a satisfying market for recorded cassettes. As the novelty wore off, the market gravitated towards box office movie favorites, new shows, and X-rated entertainment. Rental libraries became popular because recorded cassettes were costly and people preferred variety.

For Beta and VHS recorded cassettes the two-hour formats have been chosen. These correspond to the Beta II and VHS-SP formats given in Tables 12-3 through 12-6. Two hours is sufficient capacity for most programs, and of course most cassettes will have shorter tapes tailored to the actual program length. Practically all video cassette machines of Beta or VHS type handle these cartridges.

Unauthorized copying of recorded cassettes is of concern. Ingenuity and invention have been directed toward preventing good copies from being made. One scheme called Copyguard garbled the synchronization to make the picture unstable. Another scheme was a rentable cartridge that could not be rewound by the consumer: It would be played only once and then returned for renewal at the factory.

Home Movies and Portable Recorders. The amateur movie market for photographic film is well-known and quite attractive to video recording. Video recording gives immediate playback with sound and does not require a projector or a darkened room. Tape gives long continuous recordings, costs pennies per minute instead of dollars, and can be erased after unsuccessful takes. The major drawbacks have been cost of equipment (a color camera is as expensive as the recorder itself); weight and bulkiness of a portable VCR and battery pack; and, to a lesser extent, a lower resolution picture than with super 8 film.

As with videocassette recording, the home movie VCR field has been in incubation for many years: Many experimental versions were not yet quite right, although demonstrated by bold pioneers who were handicapped by obvious flaws. Satisfactory results have been obtained with portable VCRs, battery operated and connected by a cable to a portable video camera, but such heavy, clumsy, costly equipment was really for the professional and serious amateur.

For the snapshot taker who uses a simple 8 mm film camera, the videocassette substitute has to be a self-contained camera recorder, small and lightweight, with a very small videocassette. To satisfy this market, manufacturers of Beta and VHS machines have offered video cassettes about a third the conventional size but still using 12.7 mm ($\frac{1}{2}$ in.) tape and standard format. For playback, an adapter fits these small cassettes to a standard-size home machine; or these cassettes fit a special machine with a detachable portable recorder. Camera recorders using full size cassettes are also available.

8 mm Videocassette. Because a small thin cassette was inevitable, an 8 mm Video Standardization Conference of 122 companies worldwide met in 1983 and approved a miniature video cassette using a special format on tape 8 mm wide. By fall of 1984 several companies announced camera recorders using this cassette (Figure A-28, Appendix A). Since the new cassette does not fit on ordinary VCRs for playback, special provisions must be made. In the version shown in Figure A-28 the camera fits into a cradle that adapts it for playback through any television receiver.

By changing to an entirely new format, the 8 mm system of Figure 12-2*C* allows radical improvements over previous videocassettes, such as compactness, digital audio, and higher density recording. Dimensions of the exterior and interior of the 8mm cassette are shown in Figure F-2 (Appendix F). Specifications of the 8mm cassette system are given in Tables 12-2 to 12-6.

One feature of the 8-mm cassette format is the possibility of high quality sound that accompanies the picture. This is accomplished by means of FM carrier recording with the high speed rotating heads, overcoming limitations of the very low forward speed when direct recording is used. The new generations of full size Beta and VHS videocassette recorders are also adopting this high quality sound method.

A different solution to portability is to separate the video camera physically from the recorder and to operate the recorder by radio control from the camera. The video camera with built-in transmitter is as light and versatile as a small movie camera. It controls a fixed recorder some distance away, which can be a conventional video cassette machine, line operated or with heavy-duty batteries in an automobile, for example. With a two-way radio link the monitor on the camera shows that the recorder is operating properly, even to the extent of displaying a played-back picture, insuring that the recording has been made (Camras 1978).

HIGH FIDELITY SOUND FOR SLANT SCAN RECORDER

The forward movement of tape in videocassette recorders is only about 1 in./sec (25 mm/s) or less, giving marginal sound quality from the audio edge tracks in Figure 12-2*A*, *B*, and *C*. However, excellent quality audio can be obtained by taking advantage of the video scanning speeds which are about 300 in./sec. The audio is effectively multiplexed by adding an FM sound carrier to the video heads.

FM Carrier Audio. In the 8mm format the FM carrier is 1.5 MHz, with a maximum deviation of ± 0.1 MHz, resulting in an FM bandwidth allocation of 1.5 ± 0.15 MHz that includes over 99% of the audio power spectrum. The FM audio occupies a narrow nitche in the spectrum which is just below the luminance band and just above the down converted chrominance band (See Figure 12-6). The specified recording current for FM audio is 13 dB less than the chrominance current. The rotating video scan heads do not record a continuous track but give a series of slanted segments that are spliced together electronically. Electronic splicing occurs in microseconds, producing no noticeable effect on the audio.

Beta and VHS videocassette systems originally used only conventional sound tracks on the edge of their videotapes. Now they may include the alternative high fidelity FM sound as described above, with stereo, which must of course be played or recorded on a later generation of equipment having the hi-fi capabilities.

Digital (PCM) Audio. The high scanning speeds of video recorders make them especially adaptable for digital audio (See Chapter 9). In the 8mm format of Figure 12-2*C* about 16% (36°) of each slant track has been allocated to digital audio. The scanning process results in bursts of digital audio information 16% of the time, separated by periods of silence 84% of the time while the picture is being picked up. The continuous audio signal must therefore be compressed into $\frac{1}{6}$ of its real time interval before being recorded. For recording, the analog audio is converted to digital form by an A/D converter. The digits are read into a buffer memory for $\frac{1}{60}$ of a second (NTSC) corresponding to one scan field; then the buffer is read out in $\frac{1}{360}$ of a second, emptying it at 6 times the read-in rate, and recording these speeded up digits on to the 36% of track (Figure 12-2*C*) allocated to digital (PCM) audio. During playback the process is reversed, with fast read-in to the buffer memory and slow readout, giving a continuous stream of real time digital audio that is finally D/A converted to analog output.

PCM standards for 8mm cassettes include a sampling frequency of 2 times the horizontal sync frequency, a ten bit code non-linearly compressed to an eight bit code for recording, and error correction using parity bits and CRC coding (See Chapter 13). To accommodate PCM audio, the recording system must operate at 5.8 MHz with an upper limit of 8.7 MHz.

Other Features of Videocassette Systems. The sophistication and complexity of videocassette systems are almost incredible. They are made possible by integrated circuits, by high recording densities of improved tapes and heads, and by miniaturization of mechanical components. A camera recorder may have automatic focussing using ultrasonic radar; tracking servo with pilot signals; automatic accomodation for different grades of tape, and for mono, stereo and bilingual modes; erase and dubbing protection; and compandor noise reduction. In addition there are complicated methods described in Chapters 11 and 12 for color video, FM and PCM audio, still-frame and slow motion, and other features such as unattended recording of selected TV broadcasts.

OTHER VIDEO FORMATS

Scientific merit alone does not assure commercial success. Many ingenious designs have been offered that did not achieve the popularity of Beta or VHS recorders. Table 12-7 describes other helical videotape recorders.

SPECIAL EFFECTS

Special effects include freeze framing, slow motion and fast motion backwards and forwards, and viewable pictures while searching the tape in either direction at very high speeds. To accomplish these with reasonable picture quality and without bars of interference across the picture (except at extremely high speeds) requires special methods including heads wider than the tracks and head pairs that are unmatched in width. Noiseless stills, slow motion, and double speed play are an unexpected bonus of the electronic schemes that allow overlapped tracks.

Noiseless Still Frame

When the tape is stopped during playback while the head wheel spins at normal speed, the head scan angle no longer matches the recorded track angle since the

Table 12-7. Other Helical Videotape Recorders.

Format	Tape Width (Inches)	Playing Time (Hours)	Tape Container	Remarks
Cartrivision	0.5	1	Cartr.	Large head wheel with 3 heads. Only 1 video field out of every 3 was recorded. Every recorded field was played 3 times, cutting tape consumption to $\frac{1}{3}$.
Great time machine, VR1000	0.5	2	Cartr.	Cartridge fits over head drum eliminating threading. Alpha wrap. Concentric tape hubs.
V-Cord II	0.5	2	Cassette	
Akai, $\frac{1}{4}$ in. helical	0.25		Reel	One-fourth inch tape advocated as economical format.
Technicolor	0.25		Cassette	Smaller cassette than Beta. "Miniaturized" VCR for portable camera work.
Philips, Video 2000		8	Cassette	Turnover cassette, 4 hour per side for 8 hour total.

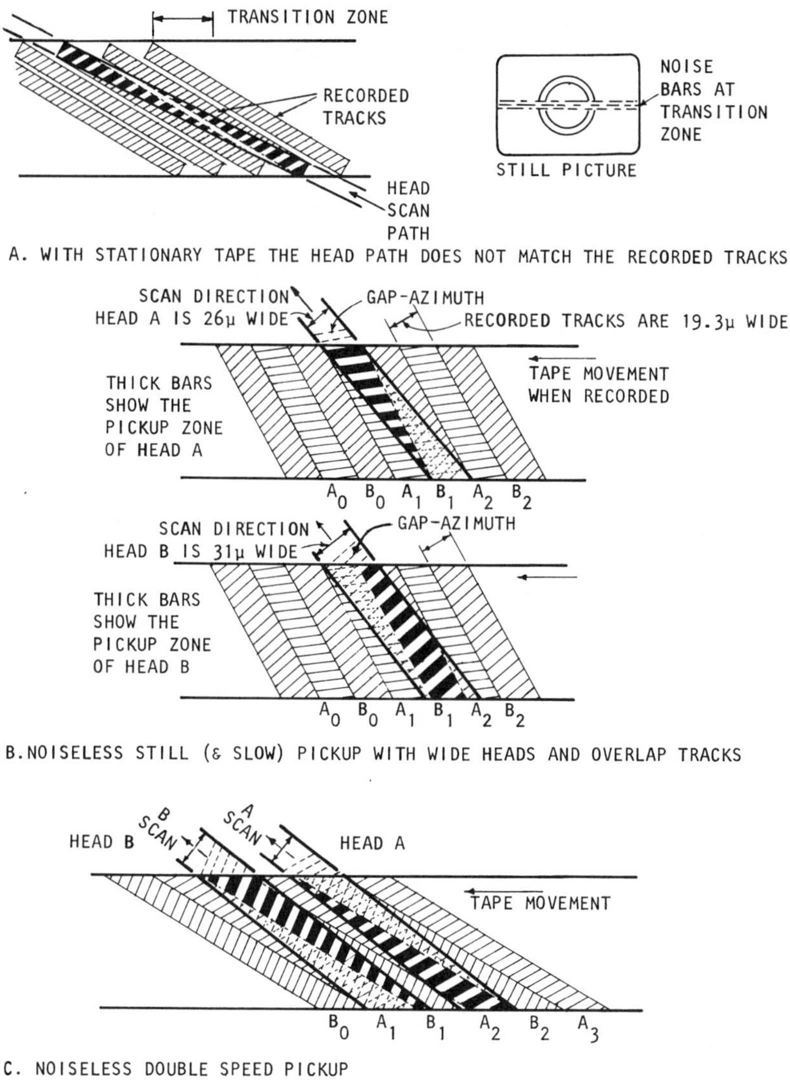

Figure 12-11. Noisless still-frame, slow motion, and double speed motion.

forward component of tape motion is lost. Figure 12-11*A* shows this situation in a recorder with normal heads and guard bands. An extended region of interference between two adjacent tracks gives a band of confusion and noise in the still picture thus obtained. Professional recorders with an AST head (see Figure 11-21) restore the head to the proper track path to give a noiseless still.

Overlapped head recorders can also give a noiseless still because each head senses only one of the two overlapping tracks and rejects the other. In Figure 12-11*B*, the first head *A* of the head wheel pair follows a path that picks up track *A*, whose azimuth angle fits head *A* but ignores track *B*, whose azimuth angle is wrong. Head *B* follows the same path as *A* (the tape is not moving) but senses only track *B* and ignores track *A*. Thus our picture repeats field *A* alternately with field *B*, and although not perfect in vertical registration or signal-to-noise ratio, the result is a great improvement over the interference-marred condition of Figure 12-11*A*. As each track is scanned in Figure 12-11*B*, the output level varies considerably because of changing track width but this is taken care of by the FM limiters; the picture brightness depends only on the FM frequency and is unaffected by reduction in level, although there is a slight increase in noise content. Head *B* is extra wide to span the tracks properly, while the narrower head *A* gives a cleaner signal because it does not cope with as much adjacent track clutter. Together they give an optimum still picture.

Noiseless Slow Motion. When the scanning of Figure 12-11*B* is analyzed for other positions of the tape tracks, a few disadvantageous positions where noise is highest are noticeable. These must be traversed when the tape moves slowly for slow motion pictures, but the noise effects can be avoided by moving the tape intermittently, jogging it rapidly past the noisy portions while parking it for a few still frames at the quiet portions. This tricky business is handled by the capstan and reel servos with settings as low as $\frac{1}{30}$ real time motion backwards and forwards.

Double Speed Playback. Figure 12-11*C* shows the effective scanning path when the tape at playback moves at twice the recording speed. Heads *A* and *B* now follow spaced paths picking up in succession track A_2, track B_1, track A_4, track B_3, and so on, but skipping entirely tracks A_1, B_2, A_3, B_4. . . . Skipping half the fields while showing completely the other half gives noiseless double-speed pictures. As with stills, there is an optimum positioning of the tape which can be set by the framing control. Similarly noiseless playback is obtained at double speed reverse.

Fast-Forward (Cue) and Fast-Reverse (Review). Pictures can still be discerned at tape speeds of up to 10 times normal, but color is lost and noise bars appear. Neither of these factors harms the searching, which is usually slowed down for a better picture after roughly locating the required place. In Figure 12-12*A*, which depicts fast-forward, the head-scan paths follow a higher angle than the recorded tracks on the tape. Each scan then crosses several tracks giving output in bursts separated by interference regions (noise bars) where two tracks are picked up simultaneously.

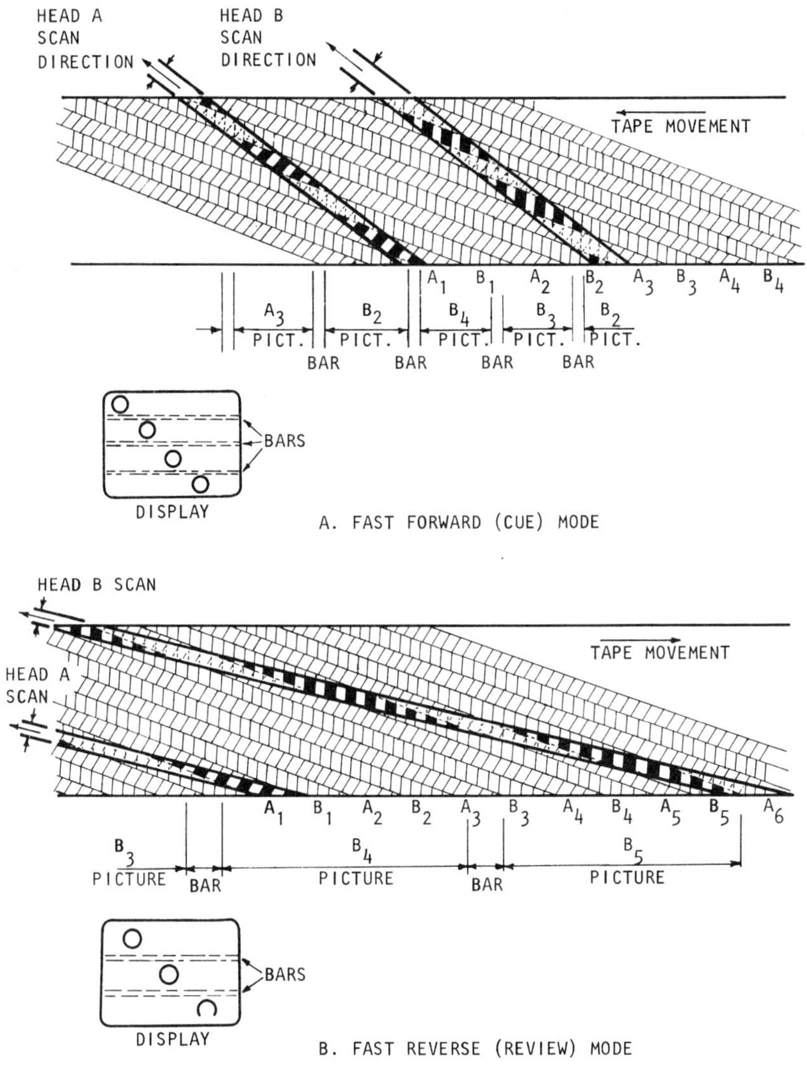

Figure 12-12. Fast forward (cue) and fast reverse (review) displays.

In fast-reverse (Figure 12-12*B*), the head scan paths have a lower angle than the recorded tape tracks, giving bursts of signal separated by blank spots or interference, which appears as noise bars. The above analyses are based on same-direction scan and tape movement but can be easily modified if tape and head movement is opposite.

FIXED-HEAD VIDEO RECORDERS

Fixed-head video recorders resemble audio recorders except that the tape moves at much higher speeds to give response at video frequencies. A succession of parallel tracks gives a long playing time with a single fixed head that changes to the next track whenever a track has been used up.

Simplicity is the main advantage of fixed-head video recorders, but they also have the advantage of long, continuous recordings uninterrupted by switching. The recordings are readily interchangeable among different machines because tracks are parallel to tape motion. Home recorders do not require servo controls, switching, rotating heads, slip rings, control tracks, and so on, which are essential to rotating head recorders. It has been estimated that a simple fixed-head video deck could be manufactured for a third of the cost of a basic rotating head deck.

In the search for ultimate economy in mass-produced home video recorders, the fixed-head principle seemed very attractive, especially when competitive helical scan machines were beset with expensive mechanical and electronic problems. Fixed-head recorders, however, have also had problems, mostly in performance rather than cost: (1) Reel-to-reel recorders at high speeds run through the tape in a few minutes and when they reverse and change to the next track, there is a gap in recorded information; (2) recording densities are pushed to the limit and electronic handling of the signal is compromised in order to reduce tape speed for longer running time; and (3) dropout compensators, capstan servos, and other sophistications have not been generally used to improve picture quality.

Fixed-Head Formats. Figure 12-13 shows an early reel-to-reel recorder for applications where long, continuous runs are not required. Playing time was about five minutes per track at a tape speed of 120 in./s (3.05 m/s), 10 minutes at half speed, and 20 minutes at quarter speed. At the end of the reel the tape direction was reversed and the head was ratcheted sidewise to the next track position for another run. Ten successive tracks were used at first on a $\frac{1}{4}$ in. tape, which seemed like a large number in the early 1960s, although, afterwards, 50 and even 100 tracks were fitted successfully across the same tapes by substituting narrower heads. Though not commercialized this format attracted attention at a time when rotary head recorders were far too expensive for home use.

Later, BASF designed an improved version that overcame the information gap when reels were reversed. A modified Newell tape transport (see Figure 7-17) was operated by a printed circuit motor of such high torque that it reversed the tape in a few milliseconds, too small an interval to be noticed by the viewer. High tape speed of 180 in./s (4.57 m/s) with high-density tape and a microgap

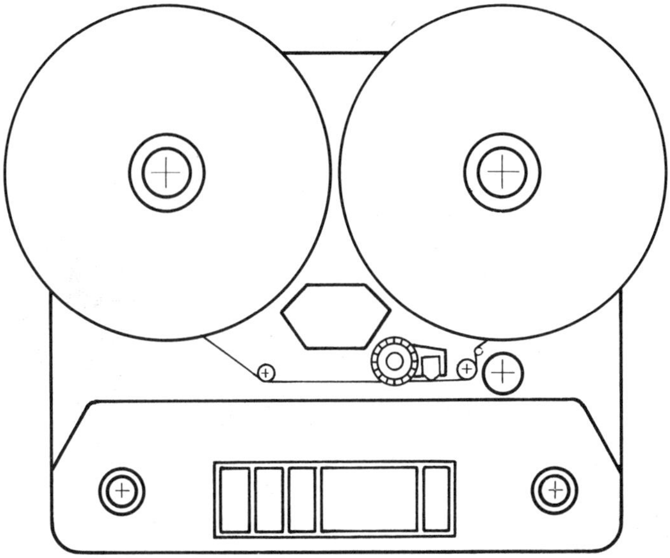

Figure 12-13. Reel type fixed head video recorders (IIT Research Institute).

head gave frequency response as in helical home recorders, so that similar circuits could be used for the video signals.

Figure 12-14*A* shows a different concept for a long-playing fixed head video recorder where no time is lost in reel reversal. The tape, in an endless loop cartridge, travels only in the forward direction. When the first track is finished, a reflective foil on the tape triggers a motor drive that moves the head to the next track over a slanted path a few feet long so there is no discontinuity in the recorded information, which can last for an hour or more. The format has the advantage that a channel can be repeated over and over again by not indexing to the next one. In effect, there can be 100 short messages each 30 seconds long, for example, and each available immediately by moving the head to the selected track. This option is useful for educational purposes, slide shows with moving pictures and so on.

A production prototype of the endless loop concept is shown in Figure 12-14*B*. The endless tape is contained in a simple cartridge with an opening in the bottom to receive the capstan drive, while the tape moves around the periphery of its container.

Fixed-Head Electronics: Direct Recording. Electronics of utmost sim-

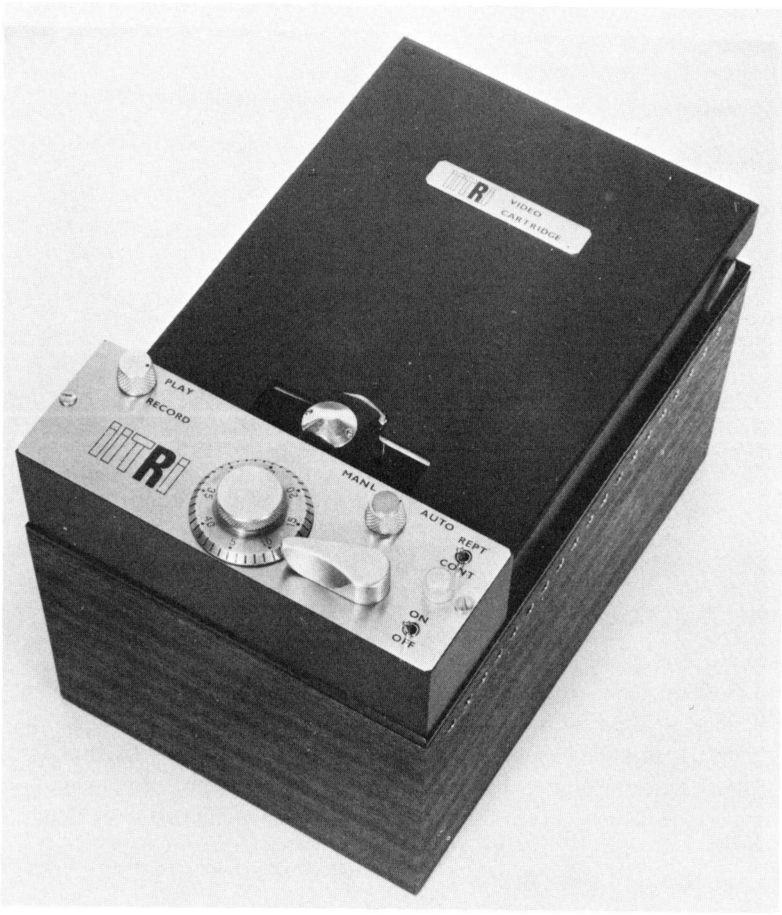

(a)

Figure 12-14. A. Endless loop cartridge fixed head video recorder (IIT Research Institute). B. Endless loop cartridge fixed head video recorder. (Courtesy Toshiba Corp.)

plicity is obtained in a fixed-head recorder by connecting it to the circuits of a television receiver or monitor, which must, in any case, be used for displaying the playback picture. As shown in Figure 12-15, no amplifiers are required in the recorder, but only a matching network between the recording head and the picture tube driver, which usually has ample power output to drive the added head circuit. For playback, a 3-transistor voltage amplifier suffices to bring the head output to the level needed by the video circuits of the receiver or monitor.

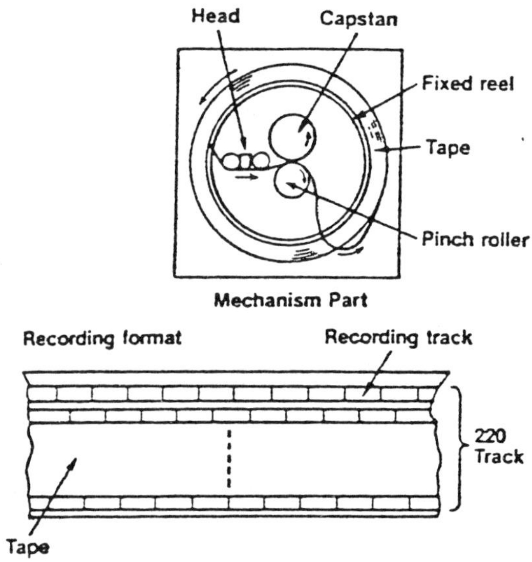

(b)

Figure 12-14. (*Continued*)

These economies require a television receiver equipped with an outlet that allows plugging into its video circuits, or a TV console with a built-in recorder. Indeed, both these alternatives were examined in home video tape development and are still possibilities for special applications. The more complex FM recording and playback electronics are also feasible in fixed-head recorders.

DUPLICATION OF CASSETTES

Supplying the demand for recorded cassettes has become an important industry, quite analogous to the recorded audio tape industry described in Chapter 10. Video recordings are made in real time, all components and procedures are critical, and every product must be spot-checked individually.

Multiple Copying. Production of recorded video cassettes is far less mechanized than for audio cassettes, which can be loaded with tape that has previously been bulk-duplicated at high speed. Video cassettes are usually recorded on ordinary or slightly modified Beta or VHS decks fed in multiple from a single master recorder and a suitable distribution amplifier. Hundreds of decks are mounted on racks or shelves for simultaneous duplication of a popular program. The leading manufacturers of recorders provide ruggedized decks with hardened

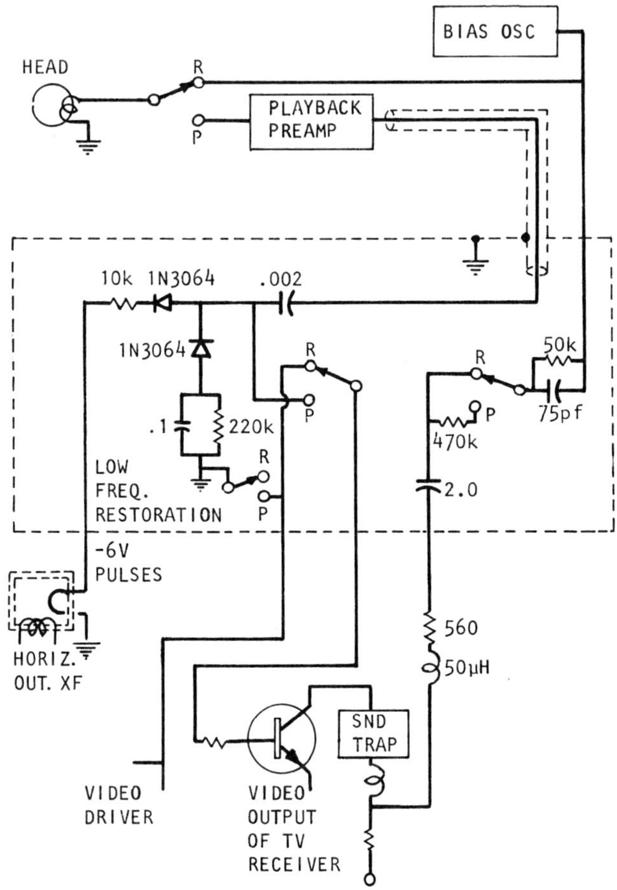

Figure 12-15. Electronic circuits for direct recording.

guides for long life, omitting timers, tuners, and other superfluous accessories. They also provide carefully made alignment cassettes to ensure that the recorded product will play on home recorders.

Recorded cassettes are made as SP(I) speed for VHS machines, and at Beta II (medium) speed for Beta machines. Playback is checked near the end, middle, and beginning at an inspection and rewinding station before a cassette is released. The master video source is usually a professional recorder, as for example a high-band Type C, quadruplex, or Type B. Audio that accompanies the picture is recorded simultaneously as usual.

Contact Printing. At least one manufacturer has developed a high-speed video duplicator that uses contact printing. The printing master tape is of extra high coercivity, recorded on an opposite format machine so its format is the mirror image of that in Figure 12-2*B*. The recorded master and the copy tapes are wound together at high speed on a hub with their active surfaces in contact. The wound roll is next immersed in a magnetic transfer field as described in Figure 2-53. The roll is then unwound rapidly, after which the master is ready to be used again. The copy tape is loaded into a cassette for playing in the usual manner. Quality of reproduction from a contact print is not yet equal to that of the best direct recordings, the most noticeable difference being a slightly increased noise level. The overall result, however, is satisfactory for home entertainment.

MAGNETIC RECORDING OF STILL PICTURES

Video recording is not only an alternative to motion pictures but its techniques can also be applied to still-picture photography. Magnetic photography does not require processing. It can be viewed immediately on a color TV screen and its color balance and contrast adjusted electronically. The record does not have to be kept in the dark. It can be erased and used again. A printer is available for making color print snapshots.

The magnetic camera shown in Figure A-27 (Appendix A) resembles an ordinary 35 mm reflex camera but has in its image plane a charge-coupled matrix (CCD), which is scanned to generate an electrical video signal. This signal is stored on a circular track of a magnetic disc, approximately 50 mm (2 in) in diameter, which rotates within a protective jacket like a miniature floppy disc. Fifty pictures can be recorded on concentric tracks.

The scanning rate is 525 lines, 60 fields/sec for playback on NTSC color TV receivers. The charge-coupled image device has 570 horizontal and 490 vertical elements to give resolution capabilities comparable to an excellent TV picture.

13. Data Recording

Data or instrumentation recording is a major specialty of tape recording. Its professionals have their own societies and their favorite equipment, and they speak their own lingo so they can puzzle the uninitiated. Table 13-1 defines some terminology commonly used. Data recorders and computers have much in common, especially in digital recording, and therefore both Chapters 13 and 14 should be considered together.

Data recorders preserve readings of sensors on rocket engines during space flights and collect data of weather satellites and of deep space probes. They speed up for analysis the slow geophysical data of tides, and they slow down the fast events of an explosion. They monitor electrocardiograms, radar signals, oil well data logs, and chemical process plants. They simulate complex motions of flight tests or road tests for training or research. They interact with people to revise writings or drawings. They operate typewriters, machine tools, robots, and even control the recorder itself. Table 13-2 lists other applications of data recorders.

Many applications that require only a small amount of data storage formerly used magnetic recording, but now can be handled expeditiously by solid state memories storing perhaps 4096 to 256K bytes (or characters) at low cost, rapid access, and without moving parts. Such replacements have already been mentioned and also include word processing, robotics, numerical machine tool control, and time delay. However magnetic recording is still needed along with these solid state devices for permanent storage of programs and data that can be loaded and unloaded into the working memory. It is also used for overflow beyond the capacity of the built-in memory, and for backup to ensure against loss of valuable data in volatile memories. Flexible magnetic disks are most useful for this auxiliary storage, although cassettes or cartridges are sometimes chosen.

INSTRUMENTATION RECORDERS

The earliest data recorders were audio machines with little or no alteration, and even now an elementary cassette recorder suffices for rough data. Modern instrumentation recorders evolved when users specified response to megahertz frequencies, dozens of parallel tracks, a gamut of drive speeds, and electronic circuits for FM multiplex and predetection recording. Several manufacturers supply high-performance instrumentation recorders such as the laboratory unit shown in Figure 13-1*A* and portable units of the same format in Figure 13-1*B*,

Table 13-1. Terminology in Instrumentation Recording.*

Term	Definition
Analog recording	Recording proportional to signal amplitude or frequency.
ANSI	American National Standards Institute.
BPI	Bits per inch.
Coding	Translating the signal by special rules before or after recording.
Crossplay	Variation in frequency response when a tape is recorded or played with different size head gaps.
Crosstalk	Interference from adjacent tracks, heads, etc.
CTG	Control track generator, as segmented disc for speed control. See Tach.
Dropout	Temporary loss of playback signal.
DR	Direct Recording.
FFT	Fast Fourier Transform - Converts time base to frequency base data and vice versa in real time.
File protect	A tab, or ring etc., on a spool or cassette that prevents erase-record operation.
FM/FM Telemetry	Data link using a main FM carrier modulated by several FM subcarriers.
Foldover	Spurious signal generated by insufficient bandwidth at low frequencies during modulation.
Gap scatter	Departure of multi-track head gaps from a straight line.
HBR	High Bit Rate Digital Recording. See HDR, HDDR, PCM.
HDR	High Density Digital Recording. See HDDR, HBR, PCM
HDDR	High Density Digital Recording. See HDR, HBR, PCM.
IRIG	Inter-Range Instrumentation Group.
ISO	International Standards Organization.
MTBF	Mean time before failure.
NAB	National Association of Broadcasters. Formerly NARTB. (Reel Standards).
PAM	Pulse Amplitude Modulation.
Peak detector	Circuit for detecting maximum slope or zero crossings of binary recordings.
Pre-detection recording	Recording of an entire wideband RF or IF spectrum that may contain several modulated carriers.
PCM	Pulse Code Modulation. General term including HDR, HDDR, or HBR.
PDM	Pulse Duration Modulation. Same as PWM.
PWM	Pulse Width Modulation. Same as PDM.
RFI	Radio Frequency Interference.
Scramble Bin	Planar bin that stores tape in random back and forth loops.
Servo	Negative feedback arrangement for regulating speed, tape tension, etc.
Skew	Tilt of tape with respect to head gaps due to errors in guidance, tensioning, etc.
SNR	Signal-to-noise ratio. Same as S/N ratio.
S/N ratio	Signal-to-noise ratio. Same as SNR.
SSB	Single sideband.
Squelch	Amplifier cut-out to prevent noise when signal is not coming through.
Tach	A tone wheel generated reference frequency.

Table 13-1. (*Continued*)

Telemetry	A radio or wire link that transmits (instrumentation) data.
TBE	Time Base Error.
TDE	Time Displacement Error.
THD	Third Harmonic Distortion.
THIC	Tape Head Interface Committee.
Track Density	Number of tracks per unit tape width.
VCO	Voltage Controlled Oscillator.
Wide Band	Extended high frequency response.

*See also Tables 13-11 and 13-13 for more specialized terms.

C, and *D*. Instrumentation recorders of lesser capabilities, smaller size, and much lower price are available for less stringent applications. Although such recorders appear complicated, anyone who is familiar with audio recording soon realizes that the principles and procedures are almost identical, at least for analog recording.

The development of uniform standards for data recording was encouraged by U.S. defense and space programs, which set telemetry standards in 1948 and

Table 13-2. Applications of Instrumentation Recording.

AIDS (Aircraft Integrated Data System) recording	Process monitoring
Analog modelling	Recording for later analysis
Aircraft flight testing	Satellite recording
Amplitude analysis	Search (of tape) and control
Autocorrelation	Simulation
Continuous time monitoring	Slowing down or speeding up real time data
Control signals for processes	Spectral analysis
Data logging	Statistical plots, as of peak amplitudes
Engine testing (piston, jet, rocket)	Structural analysis
Fault recorders	Surveillance
FFT (Fast Fourier Transform) analysis	Tape loops for repeating data
FM/FM subcarrier recording	Test data processing
Geophysical shock wave correlation	Telemetry of remote data
Medical and biophysical data	Time and frequency domain (Fourier analysis)
Noise analysis	Time base compression or expansion (time scale changes)
On board recording (spacecraft, aircraft, vehicles)	Time code recording (time signals)
Particles count analysis	Transient capture
Physical quantities transduced to electrical (pressure, temperature, acceleration, velocity, position, barometric, humidity, radiation)	Vibration analysis
Pre-detection recording	Voice cue recording
	Wideband FM (and AM) recording

(a)

Figure 13-1. High-performance instrumentation tape recorders. (Courtesy Ampex Corp.)

which in 1951 appointed IRIG (Inter-Range Instrumentation Group) for establishing telemetry and recording standards. IRIG standards have become almost universally accepted and allow interchange of recordings on a wide range of equipment.

Three modes of data recording are important: direct recording, FM recording, and digital recording. All of these can use the same tapes, heads, and drives. These common features will therefore be described, followed by details of the three special modes. More general descriptions of tapes, heads, drives, electronics, and basic theory found in previous chapters also apply to data recorders.

(b)

Figure 13.1 *(Continued)*

Track Format and Tape Width. Fourteen tracks on a 1-in. (25.4 mm) tape (Figure 13-2*B*) have been accepted for many years, although this format is now unduly conservative. Two separate head stacks are common: one for the odd-numbered tracks, and one for the interleaved, even-numbered tracks. The gap lines of the stacks are precisely 1.5 in. (38.1 mm) apart. The heads within each stack are thus doubly spaced, allowing more room for windings and for internal shielding, without wasting valuable tape area with excessive guard bands between tracks.

When critical correlation is required between data on different tracks, such data are recorded with heads of the same stack, since flutter, tape stretch, and skew can cause greater errors between stacks than within a stack, especially if the tape is read on different machines.

Tapes $\frac{1}{2}$-in. (12.7 mm) wide are also standard (Figure 13-2*A*), giving worthwhile savings in tape and head cost where fewer channels are needed. For more

(c)

Figure 13.1 *(Continued)*

channels the 1 in. (25.4 mm) format can also give 28 or 42 tracks, as shown in Figure 13-3*A* and *B*. Wider tapes are feasible for more tracks. It has even been proposed to make machines that handle tapes 4-inches (101.6 mm) wide. However, it is better to use several machines when many channels are required, since multiple machines are more flexible than a single complex recorder, and spare units can guard against failure.

Improvements in heads and tapes have led to denser track formats: 28, 32, or 42 tracks on 1-in. (25.4 mm) tape, and proportionally more or less for wider and narrower tapes, with obvious advantages in tape economy and data capacity and little sacrifice of performance. The main disadvantage is that so many tapes exist in the old format. The newer standards are shown in Figure 13-3.

(d)

Figure 13.1 (*Continued*)

Many instrumentation recorders allow conversion to narrower or wider tapes. Adapter kits are available with different guide widths, reel mounts, and head blocks. For special applications, many off-standard data recorders are custom built for use with unusual tape widths and head arrays.

Heads. Tape heads are described in Chapters 4 and 5. Head stacks as shown in Figure 13-4 are critical and expensive items that wear out and must be replaced after 1,000 to 3,000 hours of use, depending on tape speed, tension, and abrasiveness. Some machines allow rewinding with the tape separated from the heads to increase head life. Head stacks are usually mounted on precision blocks, replaceable as a unit, factory preset for azimuth and track alignment. However, the azimuth is so critical at highest densities that IRIG standards have allowed for adjustment. Adjustment may also be beneficial when playing tapes recorded on a different machine.

The high price of instrumentation heads has attracted many manufacturers,

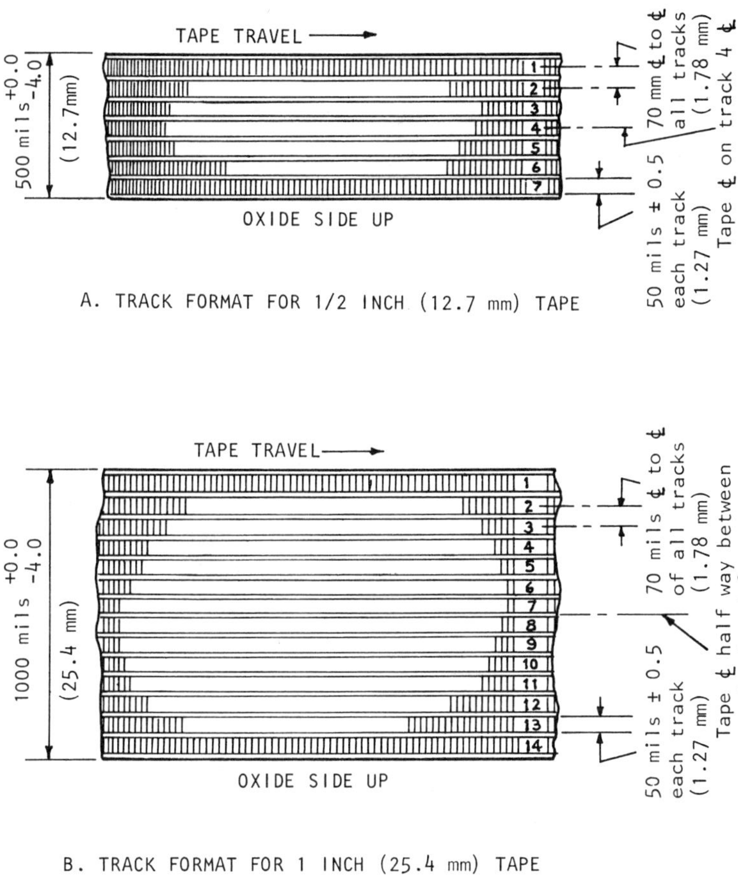

Figure 13-2. Seven and fourteen-track IRIG formats on half-inch and one-inch tapes.

each claiming unique advantages for their product, which is usually fabricated by secret methods. A favored construction for highest-density recording is the metal-faced head shown in Figures 4-22 and 4-30, using ferrite for the main body to give low losses at high frequencies, and a long-wearing erosion resistant metal alloy, sendust, to define the head gap, which is as small as 12 microinches (0.3 μm). Another favored construction uses 100% ferrite of hot-pressed or single-crystal material with a glass-bonded gap. Ferrite is even harder than sendust and wears longer, but is brittle and therefore its gap edges can break, causing failure before actual wearout. For lower densities, slower speeds, and less critical applications, the head construction can be less exotic and less expensive.

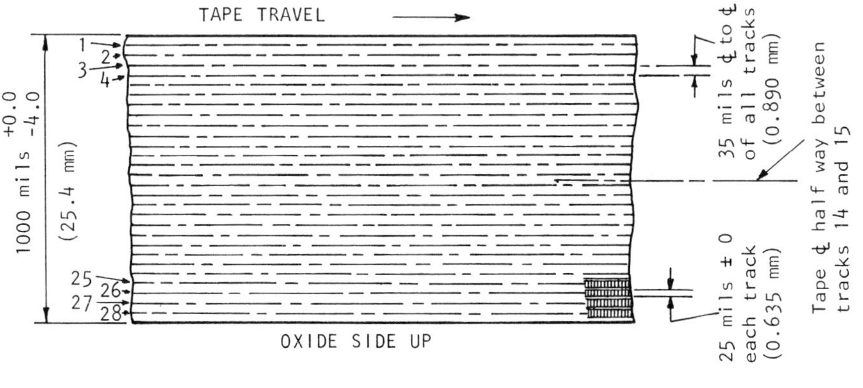

A. TRACK FORMAT FOR 28 TRACK 1 INCH TAPE

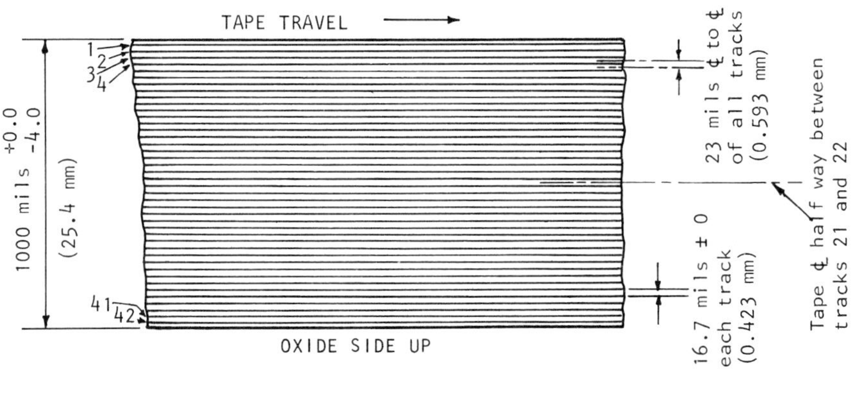

B. TRACK FORMAT FOR 42 TRACK 1 INCH TAPE

Figure 13-3. 28 and 42 track IRIG formats on one-inch tapes.

Recording Density. A maximum recording density of 656 wavelengths/mm (16,667/in) implied by IRIG standards has for many years been a comfortable although challenging specification, corresponding to 2.0 MHz response at 120 in./sec and to a bit density of 1312 bits/mm (33,333/in) at two bits per wavelength. Earlier specifications were less critical (1.5 MHz), whereas modern recorders are capable of exceeding the 2.0 MHz IRIG benchmark by 25% to 100%. In actual use, for sake of reliability, the recorders are seldom pressed to utmost capability.

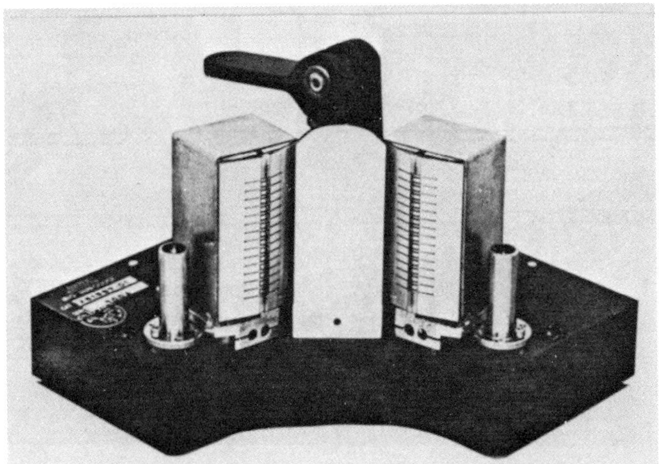

A. HEAD-STACK ASSEMBLY: 2 x 14 = 28 HEADS

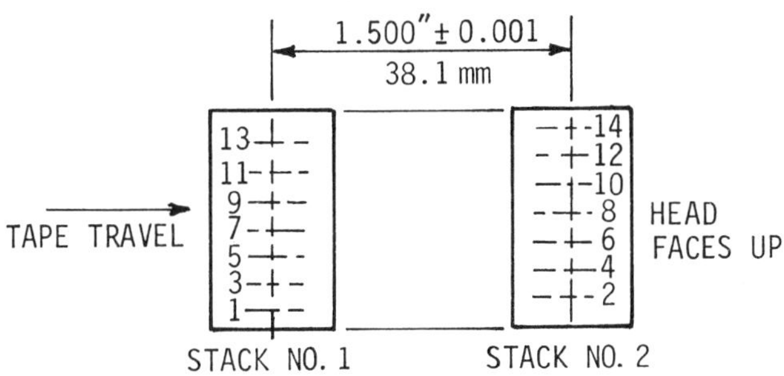

B. IRIG 14-TRACK HEAD-STACK CONFIGURATION

Figure 13-4. Head stacks for instrumentation recorders. (Courtesy Ampex Corp.).

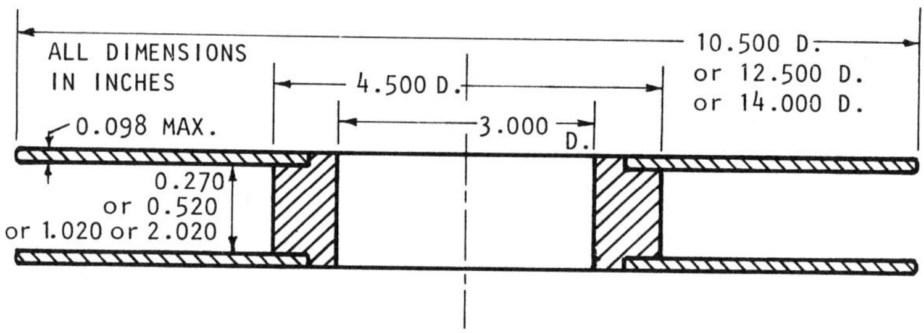

PRECISION REELS*—3 INCH CENTER HOLE

REEL SIZE IN (mm)	CAPACITY 1.5 MIL TAPE FT (m)	RUNNING TIME-MIN @ 120 IPS @ 15 IPS	CAPACITY 1.0 MIL TAPE	RUNNING TIME-MIN @ 120 IPS @ 15 IPS
7 (178)	1200 (366)	2 16	1800 (549)	3 24
10.5 (267)	2400 (732)	4 32	3600 (1097)	6 48
12.5 (318)	3600 (1097)	6 48	5400 (1646)	9 72
14.0 (356)	4800 (1463)	8 64	7200 (2195)	12 96

*Precision reels, Federal Specification W-R-175/4B
For Glass flange reels, Federal Specification W-R-175/6T
For NAB reels, Federal Specification W-R-175/3B

Figure 13-5. Reel dimensions, capacity, and running time.

Reels, Tapes, Handling, Storage. The most common instrumentation reels have an outer diameter of 267 mm (10.5 in) and 356 mm (14 in), with other dimensions and tape capacity as in Figure 13-5. These reels will also fit professional audio recorders and are available in various degrees of precision and ruggedness.

Special grades of tapes are designated as *instrumentation grade*, with wide band grades being of premium quality for highest recording density. Each manufacturer has several types to satisfy different applications, and the type numbers multiply indefinitely as new kinds are offered while old kinds are not withdrawn because they are still in demand. Specifications of typical gamma oxide wideband tapes are given in Tables 13-3 and 13-4.

Despite the confidence displayed by bold tape salespeople, tape manufacture is still an art, with variable quality found even among batches of same type. There are all kinds of faults peculiar to different types, including dropouts, excessive head wear, adherence to heads, shedding of coating or backing, poor short wave response, high noise level, to name a few. This is understandable since we are pressing outer limits of state of the art where recorded waves and

Table 13-3. Specifications of Wideband Instrumentation Tape (H_{ci} = 310).
799 High Bit Density PCM Tape
(Measured per W-T-001553)

	TEST NOTES	
PHYSICAL PROPERTIES		
Base Film:	1	1.0 mil polyester (nominal)
Oxide Coating Thickness:	1	0.20 mil nominal
Backcoat Thickness:		0.04 mil nominal
Tape Width:		Nominal; +0.000, −0.004"
Curvature:		0.125" per 48" of tape
MAGNETIC PROPERTIES		
Retentivity, (Br, Gauss):	2	1000 ±10%
Coercivity (Hci, Oersteds):	2	310 ± 5%
Erasure Field (Oersteds):	3	1000 for 60 dB minimum erasure
PERFORMANCE		
Sensitivity (200 KHz at 120 ips):	4	0 ± 1 dB
Wavelength Response		
1 mil:	5	0 ± 1 dB
0.25 mil:	5	0 ± 1 dB
0.125 mil:	5	0 ± 1 dB
0.10 mil:	5	0 ± 1 dB
0.08 mil:	5	0 ± 1.5 dB
0.06 mil:	5	0 ± 2.0 dB
1.5 MHz Slot Signal-to-Noise Ratio:		>46 dB
Uniformity, Long & Short Term — (1 MHz signal):	9	2.0 dB maximum
PCM Dropouts*:		Less than an average of 2 per 100' per track measured across the entire tape width and its entire working lengths, randomly distributed.
Oxide Surface Resistivity:	8	1 x 10^8 ohms/square maximum
TEMPERATURE		
Operating Environment:		4 C (40 F) to 49 C (120 F)
Storage Environment:		4 C (40 F) to 32 C (90 F)
HUMIDITY		
Operating Environment:		50% RH ± 15%
Storage Environment:		Less than 40% RH

*A PCM dropout is a 75% (12 dB) or greater loss in signal output for a period of 1 microsecond when reproducing a 1 MHz test signal at 120 ips, on 25-mil wide tracks.
(Courtesy Ampex Corp.)

tape roughness have dimensions the size of light waves. Tape manufacturers often work vigorously within their factory to remedy complaints, although outwardly they seem uncooperative because they want to keep their processes secret. For these reasons an activity called *tape certification* has arisen, where critical users are compelled to test incoming tapes 100% to ensure suitability for their purpose. Certification is discussed in the section on digital recording.

Some precautions for handling and storing tapes are summarized in Table 13-5; more details are discussed in Chapter 3. Proper tensioning of tape by driving system is of utmost importance especially when starting and stopping large, heavy reels. Tapes are damaged if stored with portions of the winding under excessive tension. At the time the tensioning occurs, the tape may acquire

Table 13-4. Specifications of Wideband Instrumentation Tape (H_{ci} = 650).
721 High-Energy PCM Tape

PHYSICAL PROPERTIES	
Base Film:	1.0 mil polyester (nominal)
Oxide Coating Thickness:	0.20 mil nominal
Backcoat Thickness:	0.04 mil nominal
Tape Width:	Nominal; +0.000, −0.004"
Curvature:	0.125" per 48" of tape
Magnetic Layer Material:	Cobalt doped Fe_2O_3
MAGNETIC PROPERTIES	
Retentivity (Br, Gauss):	1000 ±10%
Coercivity (Hci, Oersteds):	650 ±10%
Erasure Field (Oersteds):	2000 for 60 dB minimum erasure
PERFORMANCE SPECIFICATIONS	
Sensitivity (200 KHz at 60 ips):	0 ±2 dB
Wavelength Response	
1 mil:	0 ±2 dB
0.06 mil:	0 ±2 dB
0.04 mil:	0 ±3 dB
0.03 mil:	0 ±3.5 dB
1.5 MHz Slot Signal-to-Noise Ratio:	>46 dB
Uniformity, Long & Short Term — (1 MHz signal):	2.5 dB maximum
PCM Dropouts*:	Less than an average of 1 per 100' per track measured across the entire tape width and its entire working lengths, randomly distributed.
Oxide Surface Resistivity:	1 x 10^8 ohms/square maximum
TEMPERATURE	
Operating Environment:	4°C (40°F) to 49°C (120°F)
Storage Environment:	4°C (40°F) to 32°C (90°F)
HUMIDITY	
Operating Environment:	50% RH ± 15%
Storage Environment:	Less than 40% RH

*A PCM dropout is a 75% (12 dB) or greater loss in signal output for a period of 1 microsecond when reproducing a 1 MHz test signal at 120 ips, on 25-mil wide tracks.
(Courtesy Ampex Corp.)

a permanent stretch and may cinch the winding by pulling together a number of layers. Newer machine designs incorporate extraordinary means to give even tension: vacuum buffers, tension sensors, and electronic controls. During running, the tape tension is kept at about 4.45 newtons (1 pound) per inch of tape width for the usual tapes with 25 μm (1 mil) backing. Some machines use about half this tension. Thinner tapes, seldom used, require even less tension. There is evidence that a programmed tape tension is beneficial, with higher tension near the beginning and end of the winding (bathtub tension curve).

Drive Systems. Standardization leaves little room for creativity in most aspects of data recorders, but tape drives are a designer's paradise since cost is (almost) no consideration and there are few constraints of space or weight in a

Table 13-5. Suggestions for Handling and Storage of Tapes.

- Handle the reel only at the hub, be careful not to bend the reel flanges
- Don't touch the tape surface with your fingers, handle the leader only
- Be sure that winding tension is correct
- Check that edges of the tape winding are smooth and not touching the flanges
- Avoid gummy tapes for splicing and fastening at ends of the reel
- On the tape deck clean guides, heads, capstans, and everything that touches the tape
- Avoid magnets and magnetic field sources
- Smoking and food are prohibited in tape operating areas
- Operating areas should be vacuum cleaned, not swept, dusted, or waxed
- Store reels on edge, not flat, in protective containers
- Store on approved reels, avoiding plastic hubs that have different expansion coefficient than the tape
- Store at low humidities and low temperatures, about 40% RH, 20°C or lower
- Allow enough time to reach operating room equilibrium (20°C, 40% RH) if stored at low temperatures, etc.
- A forward wound condition is preferred to a high speed rewound condition for storage
- Occasional rewinding is recommended during long storage to equalize the winding "set"
- Clean tape guides and rollers preferably with Freon TF solvent, or with isopropyl alcohol (see Chapter 7)

laboratory unit. The objectives are low flutter, synchronous speed, and least possible stressing of tape during starting, stopping, driving, and reeling. Commercial data recorders have used dual capstan drives (Figure 7-8), closed loop differential capstans (Figure 7-9), modified zero-loop drives (Figure 7-11*B*), vacuum capstans (Figure 7-12), servo-controlled capstans (Figure 7-20), vacuum bin isolators (Figure 7-13), and capstans without pressure rollers (Figure 9-13).

Servo Control. The trend has been to sophisticated servo controls to counteract flutter, and to low inertia DC and brushless DC motors for direct capstan drive. Servos are attractive because their once-complex electronics are now compressed into inexpensive integrated circuits. Brushless DC motors and printed circuit motors respond quickly to servo control. When they drive the capstan directly, they free the designer from speed reducers (belts or rollers), which were always a nuisance and source of unreliability, and they give high torque even at lowest speeds.

Capstan Servo (Tach Servo). As shown in Figure 7-20, a tone wheel (tachometer or tach) is mounted on the capstan shaft and its generated frequency is compared with a reference such as a frequency-divided crystal oscillator. Any difference between the generator and reference produces a positive or negative error signal that corrects the capstan speed. Dual servos are useful: a rate servo giving a proper error signal during start-up when frequencies differ widely, and a position or phase servo sensitive to small angular errors after the capstan is locked in synchronism with the reference.

Tape Servo. Slippage between tape and capstan can never be avoided; therefore, the capstan alone is not capable of assuring perfectly synchronous tape speed. However, if a time code or pilot signal is recorded on the tape, it can be used instead of the tone wheel to regulate the capstan in order to counteract slippage and all other timing errors, running the tape in exact synchronism according to its recorded control track (see Figure 11-13).

Tape servo should be closer to ideal than capstan servo, and indeed it reduces time displacement error to ± 0.5 microseconds between any two points on the tape; it is therefore useful for perfect, long-term timing for spectrum analysis, and for predetection recording. The main flaw is that the control track pickup point is separated from the capstan drive point, giving a time delay before the capstan correction is propagated along the tape to the pickup point where it can be acknowledged. Fast correction therefore gives instability and the servo must be stabilized by making it respond slowly, which prevents it from correcting fast flutter components. An improvement would be to pick up the control track signal from the tape while it is on the capstan. Modern instrumentation drives allow either the capstan servo or the tape servo to be selected, whichever is most advantageous for a particular task.

Tape Speeds. Providing a wide range of easily selected tape speeds is an important feature of instrumentation recorders because it allows slow data such as daily variations to be speeded up to bring them within the operating range of oscilloscopes and spectrum analyzers. Similarly, nanosecond data is slowed down for convenient analysis. Table 13-6 gives IRIG speeds and frequency specifications.

Many schemes have been used for shifting speeds, including stepped pulleys, multiple motors, pole-changing motors, electronic frequency control of AC motors, servo control of DC motors, and combinations of the above. Traditional speeds have been in binary (2:1) submultiples of 120 in/sec (3.048 m/s) as in Table 13-6, but decimal submultiples are also convenient for high ratio conversion of data. Binary speed ratios permit frequency multiplication of tachometer or control track frequencies; the multiplied frequency will then lock to the fixed reference standard at the new tape speed. Continuous speed variation is also possible by variable frequency control or with DC motor drive.

Changing the tape speed also changes tape output and response so that amplifier gain and equalization should be adjusted correspondingly. At one time this involved a change of circuit boards, but later units gang together necessary

Table 13-6. IRIG Speeds and Response of Instrumentation Recorders.

mm/s	(ips)	±3 dB Pass Band kHz[a]	Record Bias Set Frequency (UBE) kHz	Record Level Set Frequency (0.1 UBE) kHz
Intermediate Band			(Overbias 3 dB)[b]	
3048.0	(120)	0.3–600	600	60.0
1524.0	(60)	0.3–300	300	30.0
762.0	(30)	0.2–150	150	15.0
381.0	(15)	0.1– 75	75	7.5
190.5	($7\frac{1}{2}$)	0.1– 37.50	37.5	3.75
95.2	($3\frac{3}{4}$)	0.1– 18.75	1.875	1.875
47.6	($1\frac{7}{8}$)	0.1– 9.38	9.38	0.938
Wideband			(Overbias 2 dB)[b]	
6096.0	(240)	0.8–4,000	4,000	400
3048.0	(120)	0.4–2,000	2,000	200
1524.0	(60)	0.4–1,000	1,000	100
762.0	(30)	0.4– 500	500	50
381.0	(15)	0.4– 250	250	25
190.5	($7\frac{1}{2}$)	0.4– 125	125	12.5
95.2	($3\frac{3}{4}$)	0.4– 62.5	62.5	6.25
47.6	($1\frac{7}{8}$)	0.4– 31.25	31.25	3.12
Double Density			(Overbias 2 dB)	
3048.0	(120)	2–4,000	4,000	400
1524.0	(60)	2–2,000	2,000	200
762.0	(30)	2–1,000	1,000	100
381.0	(15)	2– 500	500	50
190.5	($7\frac{1}{2}$)	1– 250	250	25
95.2	($3\frac{3}{4}$)	.5– 125	125	12.5

[a]Passband response reference is the output amplitude of a sinusoidal signal at the Record Level Set Frequency recorded at Standard Record Level. The Record Level Set Frequency is 10 percent of the Upper Band Edge Frequency (0.1 UBE).
[b]When setting record bias level, a UBE frequency input signal is employed. The signal input level is set 5 to 6 dB below Standard Record Level to avoid saturation effects which could result in erroneous bias level settings. The record bias current is adjusted for maximum reproduce output level and then increased until the output level decreases by the number of dB indicated in the table (see subparagraph 4.1.3.3 of Volume III, RCC Document 118-79-IRIG).

mechanical and electronic changes for operation by a convenient speed selector control.

Recording Methods. Three methods of data recording in common use are described in the following sections:

1. *Direct recording* is simple and efficient. Here the tape magnetization fol-

lows the signal amplitude and duplicates it on playback, just as it does in an ordinary sound recorder.

2. *FM recording* is much less economical but more precise, giving accuracies to a fraction of a percent and better signal-to-noise ratio, at the cost of using perhaps five times the bandwidth of direct recording. Low frequencies including DC are readily recorded and played back by FM.
3. *Pulse recording*, which has become almost exclusively binary digital recording, is even less economical but, in principle, *pulse code modulation* (PCM) can give perfect reproduction (to any specified accuracy at the expense of more bandwidth). It has high immunity to noise, can be copied and recopied without degeneration, and can be processed to remove flutter of drive systems. Pulse recording also includes *pulse duration* or *pulse width modulation* (PAM), which are of minor importance in data recording at this time.

DIRECT RECORDING

Direct recorders translate electrical waveforms into magnetic waveforms of the same shape and store them on tape for later reproduction. Data recorders try to reproduce the waveforms as accurately as possible. From previous discussion we know that the accuracy will be marred by noise, by nonlinearity of electromagnetic processes, by unsteadiness of tape motion, and by nonuniformity of tapes. However, the percentage of these imperfections is small, and the overall results are highly satisfactory for many applications, considering that more accurate methods consume 5 to 20 times as much tape and have poorer high-frequency limits.

Direct data recorders have much in common with professional audio recorders or even home recorders, which were described in Chapters 9 and 10. The resemblance is noticeable in the block diagram given in Figure 13-6. Important differences are:

1. High-frequency bias of 5 to 10 MHz is used, keeping it $3\frac{1}{2}$ or more times the highest recorded frequency to avoid beat notes (Chapter 2).
2. Preequalization is limited to compensating for recording head losses at high frequencies. Preemphasis cannot be used as in audio work (Chapter 6), where the program material lacks highest frequencies; general-purpose instrumentation must handle full intensities anywhere in its spectrum.
3. Phase equalization is included. Otherwise phase shift causes the playback wave shape to depart considerably from the recording wave shape, especially at high recorded densities.

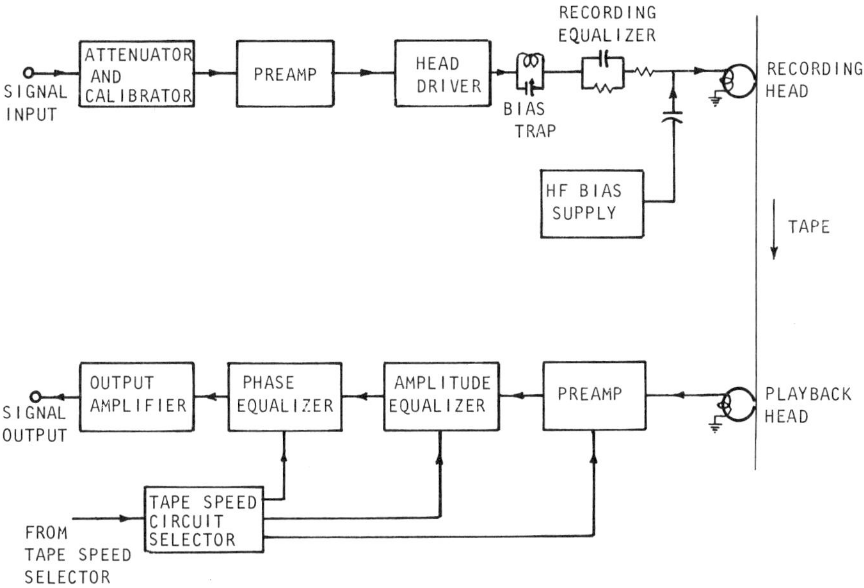

Figure 13-6. Block diagram of a direct instrumentation recorder.

Frequency Response. Flat frequency response to 2.0 MHz or more is obtained by running the tape at a fast linear speed of 3.05 m/s (120 in./s) with high recording densities that utilize special tapes and microgap heads. Figure 13-7 shows the response of a typical recorder, before and after equalization. Lower tape speeds give proportionally less high-frequency response, less output, and less noise, in keeping with theory explained by Figure 6-12. Table 13-6 gives IRIG specifications of direct recorder response at tape speeds that are provided in most instrumentation recorders. These calculate to 656 wavelengths/mm (16,667/in.) and are comparable to the 800/mm (20,000/in) for home video recorders. The IRIG specifications are conservative; wide-band recorders with the best heads and tapes equal or exceed the VTR figures, giving flat response to above 2.5 MHz and even to 4.0 MHz at 3.05 m/s (120 in/s). Low-frequency response is purposely cut off below 400 Hz to reduce hum and to avoid head contour effects (Figure 4-26).

Dynamic Range, Distortion, Noise. The signal-to-noise ratio (SNR) of direct instrumentation recorders is not impressive, being in the order of 25 or 30 dB with the signal at 1% THD (third harmonic distortion) level compared

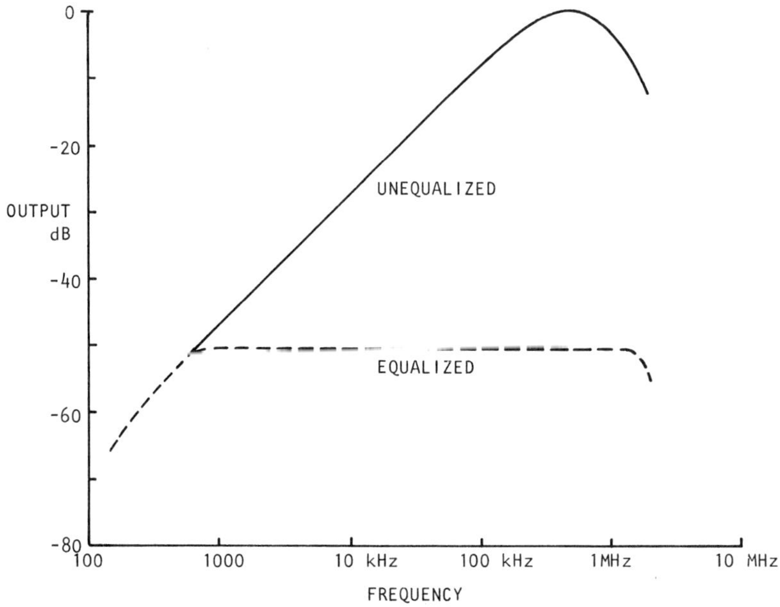

Figure 13-7. Frequency response of direct recorder at 120 in/sec (3.05 m/s)

to wideband noise. This contrasts with the SNR of 65 dB for professional audio recorders. Yet the performance of all components in the instrumentation recorder is better, if anything, than in audio recorders. The main reason is that the 2.0 MHz bandwidth is 100 times as wide as the 20 KHz band of audio machines, and for this reason alone the noise voltage is increased by $\sqrt{100}$ = 20dB. Data recorders cannot take advantage of preemphasis and post deemphasis (Chapter 6) and so are penalized an additional 10 or 15 dB in SNR. The remaining 5 to 10 dB are accounted for by higher core losses in heads at megahertz frequencies, by fewer turns on head windings to prevent self-resonance, by miscellaneous inefficiencies of wideband amplifiers, and by the 1% THD operating level of data recorders instead of 3% THD of audio decks. Nevertheless the SNR figure is adequate for most applications of analog recording and can be improved if necessary by limiting the bandwidth.

In other respects the direct instrumentation recorder behaves as described in previous chapters. The bias amplitude is set by increasing it until a low-level signal of about 1 kHz shows maximum output, and then increasing the bias further until the signal output falls 1 dB below its maximum. IRIG specifications give detailed procedure for bias adjustment. The saturation level is about 14 dB above the 1% THD normal maximum. The recorder overloads gracefully at low and medium signal frequencies, but gives bias beats at high frequencies, which can be very disturbing (Figure 2-30).

Phase Equalization. Although the record-playback response is flat over a wide band, recorded waveforms are not reproduced faithfully unless the phase is also correct. In terms of delay, the fundamental and all harmonics should be delayed an equal amount of time so they will add together in the same relationship as in the original wave, which is the same as saying that the phase angle of all harmonics should be delayed in proportion to their frequency.

Ordinary circuits involving amplifiers, attenuators and R, L, or C components form networks that have an interesting property called *minimum-phase*. Minimum-phase networks always have the same phase shift associated with a rising or falling frequency response. Therefore when we equalize or correct minimum phase networks with one another, we find that by correcting to give a flat frequency response, we automatically correct all the phase shifts and restore the original wave shape.

Magnetic recording does not behave like a minimum phase network because its basic response is not governed by RLC behavior but by head gap *aperture* effects, wavelength effects in the magnetic layer, Hilbert transforms, and so on. These attenuate the output at short wavelengths (high frequencies) without introducing the phase lag of RLC circuits. When we correct for the magnetic recording losses with the usual minimum phase equalizer networks, we find that the resulting phase curve is not flat but has excessive phase advance at high frequencies. Such phase advance gives an overshoot at leading edges of a reproduced square wave, as shown in Figure 13-8*B*. Non-minimum-phase distortion also takes place at very low frequencies where output falls because long wavelengths are not picked up efficiently by small heads. The above discussion does not apply to the 6 dB per octave rise inherent in coil type heads nor to the usual core losses in heads, both of which are correctable by minimum-phase (RLC) networks.

Special arrangements are needed to correct the phase of magnetic recordings. These include transmission lines and bridge circuits, as in Figure 13-9. Transmission lines can introduce a time lag or phase lag independently of frequency discrimination, overcoming the troublesome phase advance when head gap and wavelength effects are corrected. Bridge circuits also depart from minimum-phase behavior. The all-pass-network of Figure 13-9*B* is a special form of bridge circuit that introduces up to 180° phase lag at high frequencies, although it has a flat frequency response.

A critical way to test an analog system is to record and play back square waves. Figure 13-8 shows the appearance of waveforms associated with differ-

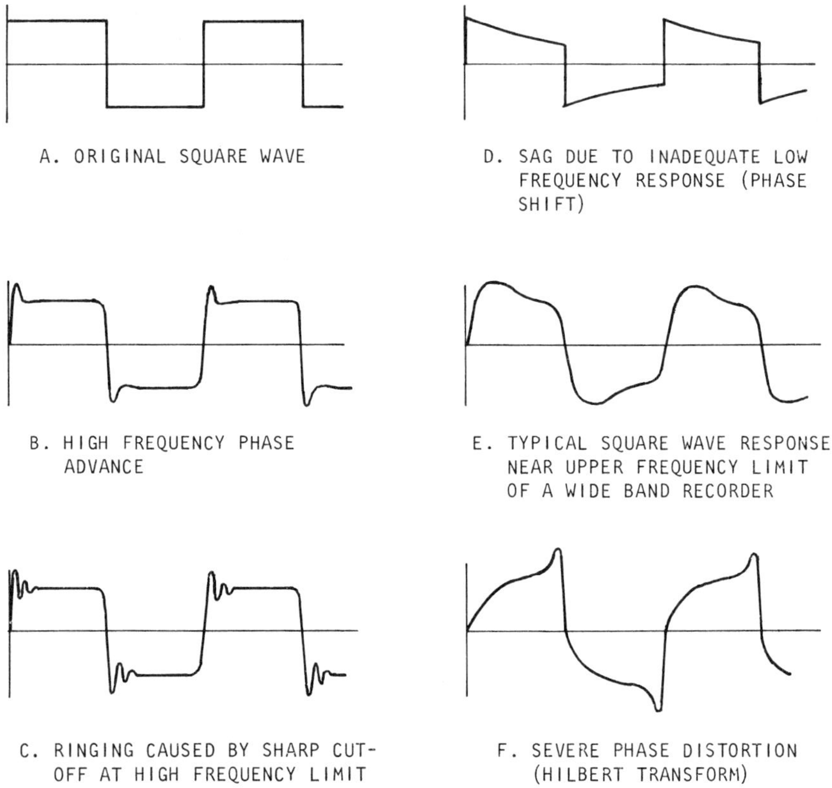

Figure 13-8. Square wave response of direct recorders.

ent characteristics of the system. For good reproduction of square waves, the frequency response must extend from about 10 times below the frequency of the square wave to 10 times above, so a wide-band recorder of 400 Hz to 2.0 MHz will give good square waves only from 4000 Hz to 200 kHz, unless non-minimum-phase correction is incorporated.

Table 13-7 lists the main advantages and disadvantages of direct instrumentation recording.

Applications of Direct Recording. Major applications of direct instrumentation recording include predetection recording, spectral analysis, correlation analysis, and transient capture.

Predetection Recording. In predetection recording, an entire wide-band radio frequency (RF) or intermediate frequency (IF) spectrum is recorded. The wide

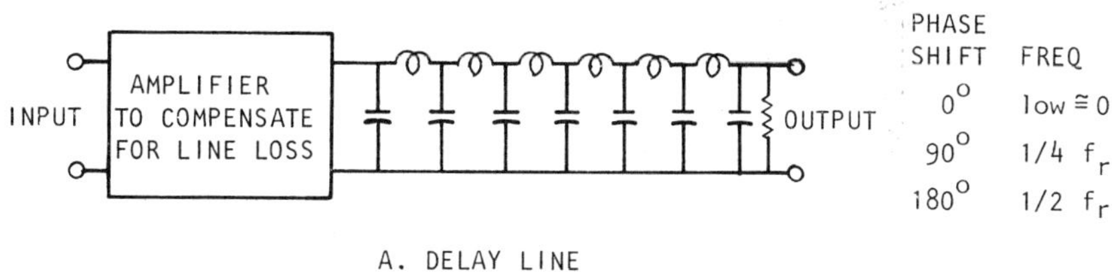

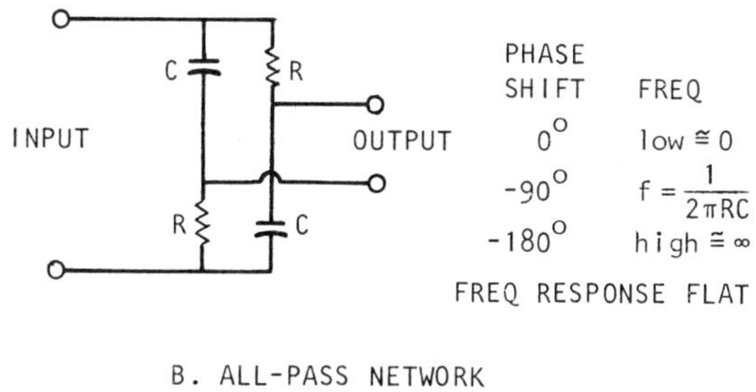

Figure 13-9. Non-minimum-phase networks.

Table 13-7. Instrumentation Direct Recording.

Advantages of Direct Recording

Highest frequency recordings for a given tape speed
Maximum tape economy
Simplest electronics
Gradual limiting as saturation is approached
Operates with reduced response under conditions where deteriorated high frequency response would cause failure of carrier recordings
Versatility for multiplexing, pre-detection recording, etc.

Disadvantages of Direct Recording

Limited signal to noise ratio
Limited linearity to 1% or 2%
Limited low-frequency response
No response to dc
Uneven low frequency response due to head contour effects
Amplitude variations due to non-uniform tape (magnetic & mechanical), head-tape interface, skew, and tracking
Sensitive to print through (inter layer cross talk)
Sensitive to interference from adjacent tracks at long wavelengths
Sensitive to stray magnetic fields from nearby equipment

band preserves all of the AM and FM carriers that were originally present in a broadcast band, for example. When the tape is played back, any of the many carriers can be demodulated separately or simultaneously to recover the information as if tuning in on the original broadcast. By running the tape repeatedly and tuning to different carriers, one can gather data from many different channels and have many chances to optimize the reception. Without such a recorder, a large number of receivers and operators would be required, and they might still miss important signals.

In telemetry the predetected signals are on FM carriers for high accuracy and freedom from noise. All elements of the communication link are adjusted for strong reception and for matching the bandwidth and signal level of the recorder. In surveillance work the conditions are often adverse, with strong and weak signals separated irregularly. Useful recording levels depend on skill of the operator and on automatic gain control (AGC). When monitoring AM signals the recording band is cut off below 100 KHz, as frequencies below 100 KHz give spectrum foldover and other undesirable interferences. Any part of the RF spectrum is brought within the scope of the recorder by heterodyne translation. During playback, the RF signals can be translated back to their original frequencies for processing, or they can be demodulated directly to recover the signal information.

Spectral Analysis. For detailed analysis of a spectrum, the signal is recorded on a loop of tape and played over and over again through a wave analyzer that responds to a narrow frequency band. The narrow band response is changed incrementally for each cycle of the loop until the entire spectrum has been covered. Such analysis is time consuming but can be speeded up where appropriate by recording at the lowest practical speed and then playing back the loop at high speed, correcting the resulting spectrum for change of frequency scale. Various schemes are used for centering the data on the loop and for muting the effects of the tape splice. Small-sample spectral analyzers have been devised with rotary heads, but these have been replaced with analyzers that store the sample in a digital memory.

Correlation Analysis. Spaced playback heads along a tape allow a delayed signal to be picked up from the same track or from a different track. The delayed signal can be compared to the undelayed with a double trace oscilloscope. Two or more pickups can be combined in opposition or aiding, with appropriate filters to bring out time correlations that would not otherwise be evident. The delay is adjusted to synchronize the correlated components.

Transient Capture. In power systems a breakdown is a rare event, but when it occurs engineers would like to have a record of transients, and other distur-

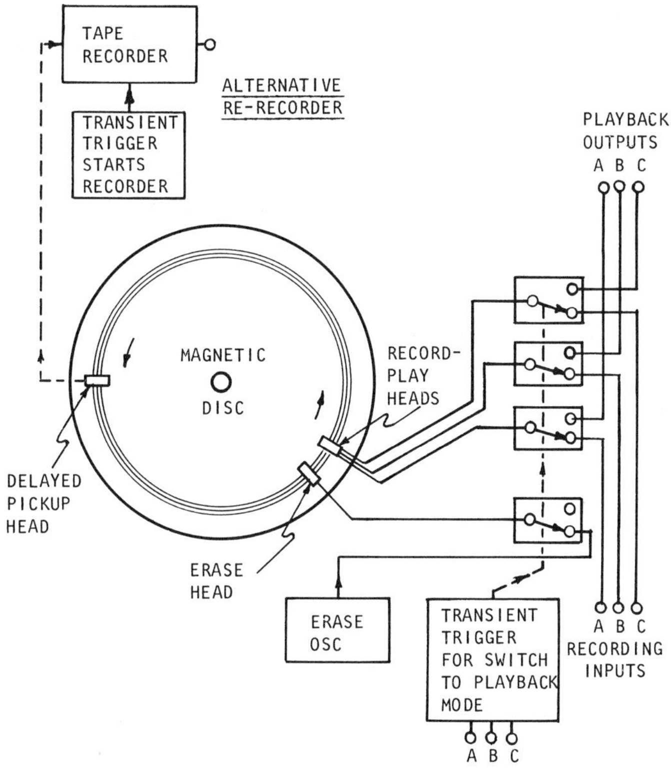

Figure 13-10. Transient capture recorder.

bances that took place immediately before the breakdown. An endless magnetic record is ideal for such an application, preferably in the form of a disc or drum with an out-of-contact head to prevent wear. In Figure 13-10 the disc holds a required amount of information, which is erased and updated every revolution. When an event occurs it turns off the erase head and switches the heads to playback so the transient can be analyzed. Alternatively the heads are not turned off; instead the actuator quickly turns on a second recorder attached to a delayed pickup head. The initial transient is thus rerecorded, as is everything that follows it until the second recorder is stopped, which can be after the fault is remedied.

FM AND CARRIER RECORDING

FM (frequency modulation) recording overcomes many limitations of direct recording. It records signals down to DC, gives accurate reproductions of low frequency waves, and is insensitive to amplitude fluctuations. *FM recording* and *direct recording*, although dissimilar, are sometimes classified together as *analog recording* to distinguish them from *digital recording*.

Two different kinds of FM recording are useful in instrumentation: *Multiplexed telemetry FM* records more than 20 subcarrier channels of limited response on each tape track; *single carrier or wideband FM* records a single carrier per track with high accuracy, low noise, and extended bandwidth. Figure 13-11*A* is a block diagram of an FM multiplex recorder, while Figure 13-11*B* shows a single carrier or wideband FM system.

Engineering Considerations. Important operating parameters that must be chosen for any FM system are *deviation ratio* and *percentage deviation*, defined as follows:

$$\text{Deviation Ratio (modulation index)} = \frac{\text{carrier deviation}}{\text{highest signal frequency}} = \frac{\Delta f_c}{f_s(\text{max})}$$

$$\text{Percentage Deviation} = \frac{\text{carrier deviation} \times 100\%}{\text{carrier center frequency}} = \frac{\Delta f_c}{f_c} \times 100\%$$

Ordinary FM broadcasting has chosen a deviation ratio of 5, which confers an excellent immunity to noise but which generates broad sidebands, taking up more bandwidth than tape recording can usually afford. Telemetry FM operates at a deviation ratio of 5 because its signals are of low frequency, but wideband FM must use a deviation ratio as low as 0.675.

On the other hand, the percentage deviation in telemetry FM is only 7.5% to 30%, while wideband FM uses 40% deviation. A high percentage deviation reduces the effect of tape flutter at the expense of bandwidth.

Multiplexed Telemetry FM. Telemetry was originally designed to transmit, by wire or by radio, the readings of sensors on a test device such as a rocket. A convenient set of 25 FM channels are standardized for FM/FM Telemetry, as in Table 13-8, based on a 7.5% deviation series. Broader channels based on a 15% and 30% deviation series are also listed in Table 13-8. Channel 1 is good only to 30 Hz, but each higher channel has a better response, up to 168,000 Hz for channel LL. The lettered channels overlap and therefore are not used simultaneously but according to a schedule where the closest adjacent channel or channels are omitted.

All the frequencies listed in Table 13-8 are well within the capabilities of a direct recorder. The entire spectrum is recorded on one tape track during a test,

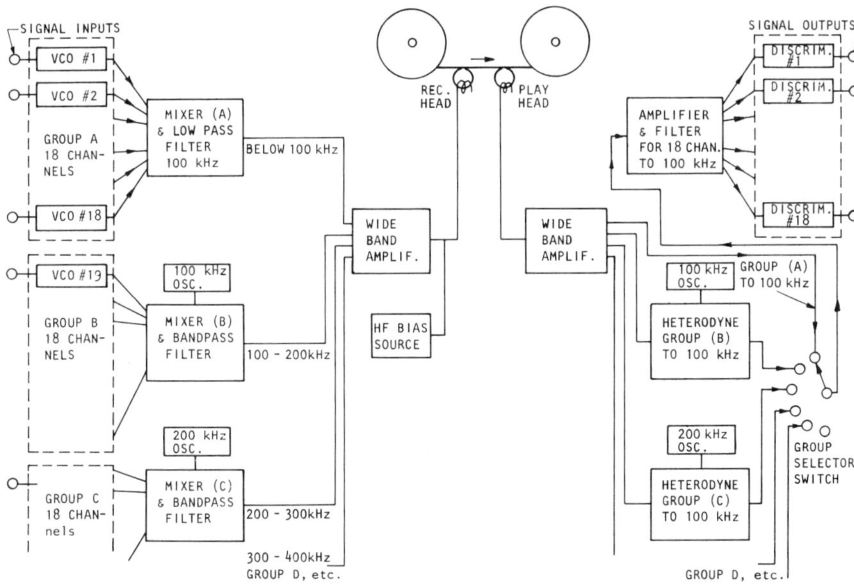

A. FM MULTIPLEX (FM/FM) SYSTEM HANDLES 18 SIGNALS PER GROUP AND UP TO 20 GROUPS PER WIDEBAND RECORDER TRACK. A 42 TRACK TAPE HOLDS 15,120 SIGNALS

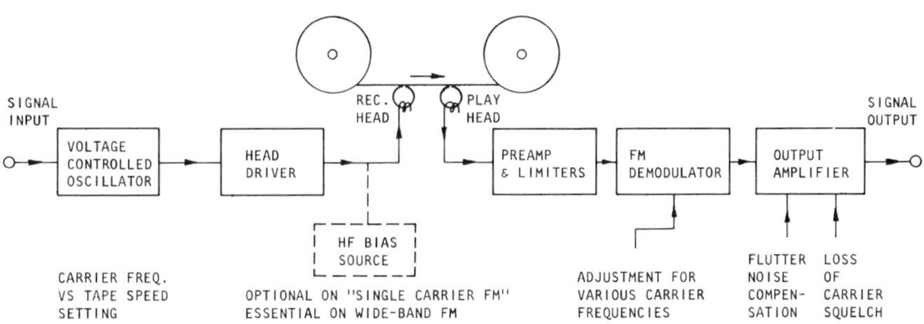

B. SINGLE CARRIER OR WIDEBAND FM SYSTEM (SIMPLIFIED BLOCK DIAGRAM)

Figure 13-11. FM instrumentation recording systems.

including all the subcarriers, and the results are played back later into the FM demodulators to recover the data. With 25 channels per track and 42 tracks, a recorder can log 1050 channels. The recorder must be linear to about 1% to prevent intermodulation among the simultaneous FM channels, and this linearity is attained by using high-frequency bias as described for direct recording. When low percentage deviation is chosen for FM/FM telemetry, the recordings

Table 13-8. FM/FM Telemetry Subcarrier Channels (IRIG 106-86) Proportional-Bandwidth FM ± 7.5% Channels Subcarrier Channels.

Channel	Center Frequencies (Hz)	Lower Deviation Limit (Hz)	Upper Deviation Limit (Hz)	Nominal Frequency Response (Hz)	Nominal Rise Time (ms)	Maximum Frequency Response (Hz)	Minimum Rise Time (ms)
1	400	370	430	6	58	30	11.7
2	560	518	602	8	44	42	8.33
3	730	675	785	11	32	55	6.40
4	960	888	1,032	14	25	72	4.86
5	1,300	1,202	1,398	20	18	98	3.60
6	1,700	1,572	1,828	25	14	128	2.74
7	2,300	2,127	2,473	35	10	173	2.03
8	3,000	2,775	3,225	45	7.8	225	1.56
9	3,900	3,607	4,193	59	6.0	293	1.20
10	5,400	4,995	5,805	81	4.3	405	.864
11	7,350	6,799	7,901	110	3.2	551	.635
12	10,500	9,712	11,288	160	2.2	788	.444
13	14,500	13,412	15,588	220	1.6	1,088	.322
14	22,000	20,350	23,650	330	1.1	1,650	.212
15	30,000	27,750	32,250	450	.78	2,250	.156
16	40,000	37,000	43,000	600	.58	3,000	.117
17	52,500	48,562	56,438	790	.44	3,938	.089
18	70,000	64,750	75,250	1,050	.33	5,250	.067
19	93,000	86,025	99,975	1,395	.25	6,975	.050
20	124,000	114,700	133,300	1,860	.19	9,300	.038
21	165,000	152,624	177,375	2,475	.14	12,375	.029
22	225,000	208,125	241,875	3,375	.10	16,875	.021
23	300,000	277,500	322,500	4,500	.08	22,500	.016
24	400,000	370,000	430,000	6,000	.06	30,000	.012
25	560,000	518,000	602,000	8,400	.04	42,000	.008
			±15% Channels				
A	22,000	18,700	25,300	660	.53	3,300	.106
B	30,000	25,500	34,500	900	.39	4,500	.078
C	40,000	34,000	46,000	1,200	.29	6,000	.058
D	52,500	44,625	60,375	1,575	.22	7,875	.044
E	70,000	59,500	80,500	2,100	.17	10,500	.033
F	93,000	79,050	106,950	2,790	.13	13,950	.025
G	124,000	105,400	142,600	3,720	.09	18,600	.018
H	165,000	140,240	189,750	4,950	.07	24,750	.014
I	225,000	191,250	258,750	6,750	.05	33,750	.010
J	300,000	255,000	345,000	9,000	.04	45,000	.008
K	400,000	340,000	460,000	12,000	.03	60,000	.006
L	560,000	476,000	644,000	16,800	.02	84,000	.004

Table 13-8. ***(Continued)***

Channel	Center Frequencies (Hz)	Lower Deviation Limit (Hz)	Upper Deviation Limit (Hz)	Nominal Frequency Response (Hz)	Nominal Rise Time (ms)	Maximum Frequency Response (Hz)	Minimum Rise Time (ms)
			±30% Channels				
AA	22,000	15,400	28,600	1,320	.265	6,600	.053
BB	30,000	21,000	39,000	1,800	.194	9,000	.038
CC	40,000	28,000	52,000	2,400	.146	12,000	.029
DD	52,500	36,750	68,250	3,150	.111	15,750	.022
EE	70,000	49,000	91,000	4,200	.083	21,000	.016
FF	93,000	65,100	120,900	5,580	.063	27,900	.012
GG	124,000	86,800	161,200	7,440	.047	37,200	.009
HH	165,000	115,500	214,500	9,900	.035	49,500	.007
II	225,000	157,500	292,500	13,500	.026	67,500	.005
JJ	300,000	210,000	390,000	18,000	.019	90,000	.004
KK	400,000	280,000	520,000	24,000	.015	120,000	.003
LL	560,000	392,000	728,000	33,600	.010	168,000	.002

are very sensitive to flutter in the tape drive; mechanical and electronic flutter compensation are necessary. A 17 KHz pilot signal is recorded in the frequency gap between channels 13 and 14 and used as a control signal for the capstan servo. In addition, a higher-frequency reference signal is recorded for electronic flutter compensation in the FM demodulator. IRIG reference frequencies are 2:1 submultiples of 200 KHz for servo and electronic compensation; or 960, 480, and 240 KHz for flutter compensation only.

Some applications require more channels of constant bandwidth instead of the assortment in Table 13-8. IRIG has standardized the series given in Table 13-9 for constant bandwidth requirements.

Pulse Duration Modulation (PDM). The term *pulse duration modulation* (PDM), also called *pulse width modulation* (PWM) is somewhat misdescriptive of its use in instrumentation because here PDM is really a time-division multiplex system rather than a modulation method. It evolved from the common laboratory practice of using a single meter plus a multicontact switch to select readings of different thermocouples located throughout a test engine, for example. In standardized PDM, a rotating selector switch reads 900 different pieces of information per second, which may be 90 readings repeated 10 times per second, or other combinations such as 30 readings 30 times per second. The samples are all in succession and can therefore be recorded on a single

Table 13-9. Constant-Bandwidth FM Subcarrier Channels (IRIG 106-86).[1]

A CHANNELS Deviation limits = ±2 kHz Nominal frequency response = 0.4 kHz Maximum frquency response = 2 kHz[2]	*B CHANNELS* Deviation limits = ±4 kHz Nominal frequency response = 0.8 kHz Maximum frequency response = 4 kHz[2]	*C CHANNELS* Deviation limits = ±8 kHz Nominal frequency response = 1.6 kHz Maximum frequency response = 8 kHz[2]	*D CHANNELS* Deviation limits = ±16kHz Nominal frequency response = 3.2 kHz Maximum frequency response = 16 kHz[2]	*E CHANNELS* Deviation limits = ±32 kHz Nominal frequency response = 6.4 kHz Maximum frequency response = 32 khz[2]	*F CHANNELS* Deviation limits = ±64 kHz Nominal frequency response = 12.8 kHz Maximum frequency response = 64 kHz[2]	*G CHANNELS* Deviation limits = ±128 kHz Nominal frequency response = 25.6 kHz Maximum frequency response ± 128 kHz[2]
Center Frequency (kHz)	Center Frequency (kHz)	Center Frequency (kHz)	Center Frequency (kHz)	Center Frequency (kHz)	Center Frequency (kHz)	Center Frequency (kHz)
		32	64	128	256	512
16	32	64	128	256	512	1024
24	48	96	192	384	768	
32	64	128	256	512	1024	
40	80	160	320	640		
48	96	192 [3]	384	768		
56	112	224	448	896		
64	128	256	512	1024		
72	144	288	576			
80	160	320	640			
88	176	352	704			
96	192	384	768			
104						
112						
120						
128						
136						
144						
152						
160						
168						
176						

[1]The constant bandwidth channel designation shall be the channel center frequency in kilohertz and the channel letter indicating deviation limit, for example, 16A, indicating f_c = 16 kHz, deviation limit of ±2 kHz.

[2]The indicated maximum frequency is based upon the maximum theoretical response that can be obtained in a bandwidth between deviation limits specified for the channel. (See discussion in Appendix B of IRIG 106-86 for determining practical accuracy versus response tradeoffs.)

[3]Prior to using a channel outside the enclosed area, the user should verify the availability of range assets to support the demodulation of the channel selected.

track of tape. Since amplitudes do not record accurately on magnetic tape, the data sample voltages are first converted to pulses whose durations (pulse width) represent the sample voltages, and these variable-width pulses can be recorded at saturation levels without bias. Naturally the pulse duration must not exceed its allocated time slot of $\frac{1}{900}$ second and therefore ranges from 60 to 750 microseconds. Some of the pulses are used for synchronization and calibration. Greater accuracy is obtained by recording the derivitives of the pulses rather than the pulses themselves and restoring the pulses after playback if necessary. Eventually the data are demultiplexed with a controlled synchronous rotating switch and reconverted to analog readings for analysis.

PDM standards limit it to slowly varying data (upper limit 1 to 3 Hz) from, for example, thermocouples, barometers, position sensors. It is about seven times more wasteful of bandwidth than FM. For these reasons and others its popularity has waned in favor of FM multiplex recording, although, where applicable, PDM is simpler than alternative methods. A modern version would use electronic gating instead of the rotating selector.

Single-Carrier and Wideband FM Recording. Single-carrier FM recording (Figure 13-11*B*) is an alternative to direct recording, chosen when better accuracy, better SNR, or response to DC, is required. FM recording can also be played at an extremely wide range of speeds (1000 to 1 or more), unlike direct recording where the playback becomes lost in noise if speed is reduced greatly. But FM uses more tape and cannot record as high a signal frequency as direct recording.

High-frequency bias is not required for single-carrier FM, which ignores magnetic nonlinearities of the recording process and the intermodulation that would interfere with multiplexed FM. Only the zero crossings of the FM signal are important. Moreover the carrier frequencies are high enough so that objectionable distortion products fall outside the high-frequency capability of the recording system while high pass filters remove troublesome products at the low-frequency end. Nonbiased FM is usually recorded at or near tape saturation. However, high-frequency bias is not precluded and becomes necessary again in wideband (single-carrier) FM recording where densities of 397 or more wavelengths/mm (10000/in) bring recording capability within the range of distortion products. It is also optional and beneficial in the other forms of single-carrier FM recording that have been discussed in this section.

IRIG standards given in Table 13-9 reflect compromises determined by practical experience in recording densities, accuracy, noise, and insensitivity to flutter. What are now called "low band" and "intermediate band" recordings were once considered "wide band" recordings, and as densities improved still farther the "wideband group 2" was added. Even in wideband group 2 the data bandwidth (modulation frequency) has been extended 25% above the listed figures.

A deviation ratio of about 2.1 and a percentage deviation of ±40% have been chosen for single-carrier FM recording, as can be verified from Table 13-10. Recorders following these IRIG standards are capable of linearity within 1% and SNRs as high as 56 dB when flutter compensation is used.

Wideband FM recording systems, Group 2 of Table 13-10, have a deviation ratio of only 0.675 and a percentage deviation of ± 30%. These parameters allow their bandwidth to be extended by a factor of five to six times as high as for ordinary single-carrier recording, reaching 400 or 500 Khz at a 3048 mm/s (120 in.s) tape speed instead of 80 KHz. The tradeoff is a reduction in SNR, which is only about 25 dB instead of 56 dB. High-frequency bias is used to reduce distortion, and the electronic systems: VCO, filters, and demodulators must be designed more carefully than for narrower-band single-carrier FM recording. The VCO is considered to be the most critical FM component in the electronic chain. Other elements, including cycle-counting FM demodulators, have been perfected to a degree where they contribute negligible distortion.

Electronic Flutter Compensation. After the flutter in tape motion has been reduced as much as possible by servos and flywheels, there is still enough residual flutter to increase the noise of FM recordings. When the signal has 30% deviation, a flutter content of 1% gives a SNR of 30, or only about 30 dB, where otherwise the SNR might be 50 dB. Fortunately there is a simple way to cancel out the noise due to tape flutter. As seen in Figure 13-12, the desired signal on Track 1 is demodulated by discriminator No. 1 as usual. Track 2 has a recording of an unmodulated carrier, which will pick up the same tape flutter and hence the same noise as Track 1 when demodulated. The noise component of Track 2 is inverted and added to Track 1, thus cancelling (subtracting) the noise of Track 1 while the signal remains. Second-order effects prevent complete cancellation by this subtractive method, and in fact better cancellation is obtained by using the reference discriminator to control the sensitivity of the data discriminator. Still, a worthwhile reduction in noise of about 10 or 15 dB is obtained by using the system of Figure 13-12, and compensation methods of this kind are used extensively.

Discriminator compensation reduces the noise but does not reduce the signal flutter itself. The actual flutter can be reduced electronically by more complex methods, such as by variable delay lines described in Chapter 11, but these are seldom used in instrumentation recording.

Rotating Head Recorders. Higher frequencies and higher data rates can be obtained by speeding up ordinary tapes but this has limitations, including relatively short running time even with large reels. Rotating-head video recorders (Chapter 11) are capable of response to 6 MHz or higher for long recordings. They use narrow-band FM, yet have high SNRs of 40 to 50 dB as

Table 13-10. Intermediate, Wideband and Double Density FM Record Parameters (IRIG 106-86)

Tape Speed mm/s (ips)		Carrier	Carrier Deviation Limits[a]			
Intermediate Band FM	Wideband FM	Center Frequency kHz	Plus Deviation kHz	Minus Deviation kHz	Modulation Frequency kHz	Response At Band Limits dB[b]
47.6 ($1\frac{7}{8}$)		3.375	4.725	2.025	d.c. to 0.625	±1
95.2 ($3\frac{3}{4}$)	47.6 ($1\frac{7}{8}$)	6.750	9.450	4.050	d.c. to 1.250	±1
190.5 ($7\frac{1}{2}$)	95.2 ($3\frac{3}{4}$)	13.500	18.900	8.100	d.c. to 2.500	±1
381.0 (15)	190.5 ($7\frac{1}{2}$)	27.000	37.800	16.200	d.c. to 5.000	±1
762.0 (30)	381.0 (15)	54.000	75.600	32.400	d.c. to 10.000	±1
1524.0 (60)	762.0 (30)	108.000	151.200	64.800	d.c. to 20.000	±1
3048.0 (120)	1524.0 (60)	216.000	302.400	129.600	d.c. to 40.000	±1
	3048.0 (120)	432.000	604.800	259.200	d.c. to 80.000	±1
Double Density	Wideband FM Group II					
	47.6 ($1\frac{7}{8}$)	14.062	18.281	9.844	d.c. to 7.810	+1, −3
	95.2 ($3\frac{3}{4}$)	28.125	36.562	19.688	d.c. to 15.620	+1, −3
95.2 ($3\frac{3}{4}$)	190.5 ($7\frac{1}{2}$)	56.250	73.125	39.375	d.c. to 31.250	+1, −3
190.5 ($7\frac{1}{2}$)	381.0 (15)	112.500	146.250	78.750	d.c. to 62.500	+1, −3
381.0 (15)	762.0 (30)	225.000	292.500	157.500	d.c. to 125.000	+1, −3
762.0 (30)	1524.0 (60)	450.000	585.000	315.000	d.c. to 250.000	+1, −3
1524.0 (60)	3048.0 (120)	900.000	1170.000	630.000	d.c. to 500.000	+1, −3
3048.0 (120)	6096.0 (240)	1800.000	2340.000	1260.000	d.c. to 1000.000	+1, −3

[a]Input voltage levels per subparagraph 6.4.2. (IRIG 106-80)
[b]Frequency response referred to 1-kHz output for FM channels 13.5 kHz and above, and 100 Hz for channels below 13.5 kHz.

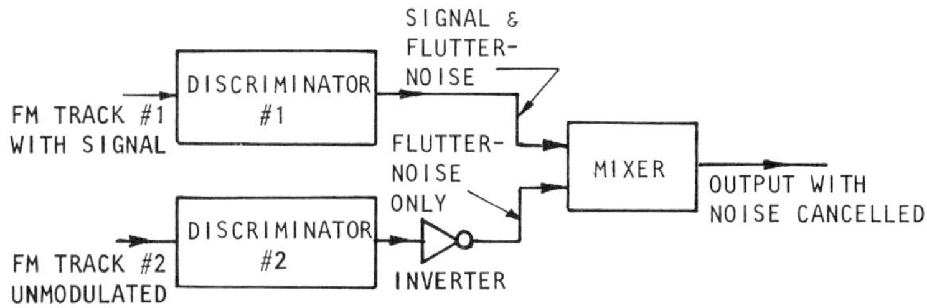

Figure 13-12. Compensation for flutter noise in FM recorders.

their direct bandwidth extends to 15 MHz or higher. Their chief problem for instrumentation use has been to splice together electronically the overlapping segments of data picked up by multiple heads, which has been solved by special fast-switching circuits. Rotating-head instrumentation recorders are relatively expensive and require more skill to use and to maintain than linear recorders. Their potential is still unexplored, and it has been suggested that they are the most promising approach to gigabit/second data output rates.

Carrier Erase Recorders. In contrast to the above recorders, which are the most complex that have ever been conceived, there is the carrier erase recorder, which is probably the simplest. A carrier erase recorder holds a length of tape that has previously been recorded with a steady tone. During the event to be recorded, the tape runs past an erase head, usually energized by DC. A surprisingly small head current partially erases the recorded tone; direct connection of the head to a thermocouple or strain gauge is enough without amplification. A movable magnet or a current carrying wire can be substituted for the head. Playback output response to the sensor is easily calibrated. Carrier erase recorders have been used in nose cones of missiles and in other extreme situations where compactness, ruggedness, and low power is required.

DIGITAL DATA RECORDING

All the instrumentation recordings discussed so far are analogs or scale models of the data. In direct recording, we regulate the tape magnetization in proportion to the amplitude of signal (or its waveform) that we want to represent. In FM we regulate the frequency of tape magnetization in proportion to signal amplitude. This proportional representation has been true of audio, video, and everything else, and so might be considered a universal requirement.

A completely different way was used accidentally, in the early days of audio recording when people found it convenient to take inventory with a portable tape recorder. They would count the number of pencils, then turn on the recorder and say, "12,683 Eagle brand yellow pencils" and turn it off while they counted "8,304 small writing pads," which they recorded next, and so forth. Scientists could record data in the same way by dictating into an audio recorder: "At time 20 seconds the Geiger counter reads 6,047,383; at time 30 seconds 7,924,215; 40 seconds 8,621,902; 50 8,812,777; 9,320,456; 9,718,212. Notice that data readings finally boiled down to a series of pure numbers, since the explanations do not have to be repeated because we understand that these are Geiger readings 10 seconds apart.

The inventory method seems like a crude, indirect way to record scientific data and is very wasteful of time and tape, using up perhaps 100 mm of tape for each voice-recorded reading, while a direct analog impression takes a small fraction of one millimeter. Moreover the analog recording is continuous, while the inventory gives only intermittent readings and for data recording poses the possible danger of missing data between readings. But a bright person will perceive two advantages that are impossible with analog recording (1) The readings are of great accuracy and, in fact, there is no limit to the accuracy, because we can specify a quantity to any number of decimal places; and (2) the full accuracy is retained as long as the quality or fidelity of the recording is enough to remain understandable, so we can tolerate a very low-quality recorder having a high noise level, high distortion, poor frequency response, high wow and flutter and yet *none of these affects the ultimate data* as long as there is enough fidelity to make out the recorded number. It is almost unbelievable but true that we can have perfect data recording with a very poor recorder and that all of the recorder's imperfections will disappear from the final result. The data can be rerecorded time after time and will not deteriorate, since the numbers emerge fresh after each recording.

These possibilities, when refined for efficient recording on magnetic tape, have led to digital data recording. The concept of digital data recording is simple. We measure the height of the waveform that represents the data; we express our measurement as a number; then we record the number. For a complex waveform, we need many close-spaced measurements, with a recorded number for each one. The recording is then a long list of disconnected numbers that bear no physical resemblance to the original waveform. It is really a code from which we can reconstruct the waveform if we know the rules. An analogy is printed language: the letters and words on this page do not look like the pictures and ideas they convey—they must be translated in our minds.

PCM and HDR

Digital data recording is also called *Pulse Code Modulation* (PCM), but recent workers in instrumentation recording prefer the term *High-Density Digital Re-*

cording (HDR or HDDR) as more descriptive, since high densities are necessary to make digital recording commercially economical.

The merits of HDR are many. It allows (in principle at least) unlimited accuracy and dynamic range: freedom from noise, distortion, flutter, cross talk, stray pickup, to name just a few features. It rerecords without degradation. Its time base is readily expanded and contracted. Its digits are accepted by computers and processors.

HDR also has difficulties, enough so that it was the last of all methods to become feasible for instrumentation recording. It requires more bits, typically two or three times as many as FM and eight times as many as direct recording, to record the same data at a SNR of about 30 dB. It requires complex and expensive circuitry, especially the A/D (analog to digital) and D/A (digital to analog) converters. HDR is sensitive to dropouts and therefore uses expensive certified tapes and requires extra care and maintenance. For these reasons, the simpler FM or DR recordings have been considered more than adequate for run-of-the-mill data.

Although digital recording requires more bits or waves to record the usual kind of data, it is satisfied with such a low grade of signal that the tracks can be compressed greatly and the tape drive is less exacting. When optimized with high track and bit densities, HDR becomes more attractive and competitive. Digital instrumentation recording is being used increasingly and is becoming the most important of all methods of data recording. Table 13-11 gives some terminology used in digital instrumentation recording.

Elements of an HDR System. The basic elements of a digital data recorder are shown in Figure 13-13*A*. The waveform to be ''recorded'' is sampled at closely spaced time intervals. In the A/D converter, each sample generates a binary number that represents its amplitude, and the streams of binary numbers are recorded on tape in a parallel format, or series format, or some combination of the two. (See Figure 9-19 for an example of digital audio recording). In playback (Fig. 13-13*B*), the process is reversed; the numbers are converted to analog voltages, which are integrated to reconstruct a replica of the original waveform. The digital processors use techniques borrowed from computers for assembling the raw data and feeding this to the tape according to chosen format, for deskewing the readout, detecting errors, correcting the clocking, and deciphering the format to restore data in proper sequence (see Chapter 14).

Digital Coding. Binary numbers are universally chosen for digital recording, although it is feasible but less efficient to use number systems of other bases, as discussed in Chapter 8. A direct approach is to record a positive pulse for a *one* and a negative pulse for a *zero*, as in Table 13-12 under the RZ heading.

Table 13-11. Terminology of Digital (PCM, HDR, HDDR, HBR) Instrumentation Recording.

Term	Definition
A/D Converter	Converts analog samples of a waveform to digital numbers.
Aliasing	False signal generated by too low a sampling rate.
Base line wander	Changes in average signal level due to dc content of coded data.
BCD	Binary coded decimal.
BER	Bit error rate.
BLW	Base line wander.
Binary	Number system to the base 2, using only *ones* and *zeros*.
Bit parallel	Data word is distributed across a set of parallel tracks.
Bit serial	Data word is a sequence of bits on a single track.
Bit slippage	Erroneous bits or bit streams due to dropouts, transients, etc.
Buffer	A device for temporarily storing and usually upgrading a digital data stream.
Clocking	A uniform series of pulses that time the digital data sequence.
Code	A system for representing a number, as by a set of impulses.
CRC	Cyclic Redundancy Check (Hamming Code).
D/A Converter	Converts numbers into analog voltages.
Data word	A number made up of a designated multiplicity of bits.
Dropout	Temporary loss of data.
Deskew	To correct the distortion due to tape skew.
DSV	Digital Sum Variation. A measure of dc content of a code.
Dynamic distortion	Irregularities due to nonuniform tape motion.
EDAC	Error detection and correction.
Error detection code	A redundant code, as with parity bits that signal the presence of an error.
Error rate	Allowable or expected errors in a specified tape length or number of bits.
Eye patterns	Oscilloscope display showing degree of clutter in a data stream.
HBR	High bit rate digital recording. Same as HDR.
HDDR	High-density digital recording. Same as HDR.
HDR	High-density digital recording. Same as HDDR.
ITDE	Interchannel time displacement error.
LSB	Least significant bit.
MSB	Most significant bit.
Nyquist rate	Minimum sampling rate = twice the highest frequency in the data.
Overhead	Extra bits added to data words for parity, clocking, etc.
Parity	Extra bit added to a code word to make sum of bits odd, for example.
PSD	Power spectral density of a coded group.
RLL	Run Length Limited. Codes where a string of ones or zeros is interrupted by opposite polarity digits.
Reclocking	Regenerating effective clock pulses for timing irregular playback data.
SSH	Simultaneous Sample and Hold.
Stepper	Drive system that moves tape in stepped increments.
Stuffing	Extra bits inserted in a data stream for sync (clocking), error detection, dc correction, etc.
TBE	Time base error.
TDE	Time displacement error.

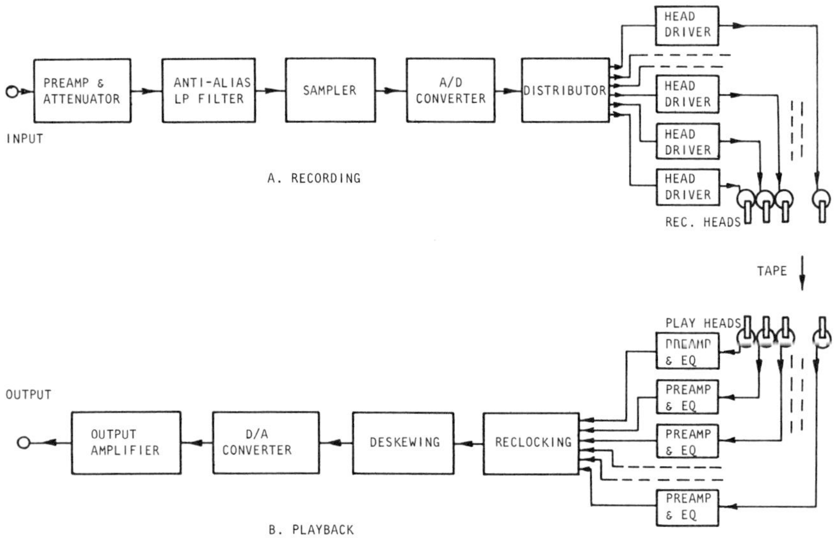

Figure 13-13. Elements of a digital data recorder.

The RZ (return to zero) code is more complicated than necessary because when data are clocked (as is customary in computers), we can omit all the zero pulses and if nothing happens in a clock interval we call it a zero. Instead of recording the pulse itself we can record a transition from + to − or vice versa to signify the presence of a pulse. This doubles our economy since a pulse requires two transitions and now we need only one, as shown in Table 13-12 in the NRZ-M (non return to zero) column.

The above reasoning follows that of pioneers of digital recording, and arrives at a useful variety of NRZ coding called NRZ-M. (the *M* stands for mark), widely used in computers. If this were the only efficient code it would be universally adopted, avoiding one hurdle to standardization. Unfortunately, there are many codes, each having a special advantage or avoiding some disadvantage. Table 13-13 lists some of the codes that have been adopted by manufacturers of PCM/HDR recorders.

No code has yet been discovered that does not have disadvantages. Furthermore, although any one disadvantage can be avoided, this is done at the expense of generating a different one. Some disadvantages are (1) not self-clocking; (2) containing a DC component or unbalance; (3) requiring more bits to represent a given number; (4) requiring more ''overhead'' bits for synchronization; (5) throwing all subsequent bits out of kilter after a single dropout. The choice of a code is difficult and depends on the sensitivity of specific circuits and logic

Table 15-12. Elementary Digital Codes.

Abbreviation	RZ	NRZ-M NRZ-I	NRZ-L NRZ-C	BiφL Manchester PE-L FM	MFM Miller Code DM DM-M
NAME	Return to Zero	Non-Return to Zero (M = Mark)	Non-Return to Zero (L = Low or Level)	Bi-Phase Level Freq. Shift Keying Phase Encoding Level Freq. Modulation	Modified Freq. Mod. Jacobi Modulation Delay Modulation Phase Modulation
CODING SCHEME	Positive pulses for ones, negative (or no) pulses for zeros	Transitions only when ones are present	Transitions mark a change from zero to one or vice versa	Positive transition for a one, negative transition for a zero	Pos. polar for a one, neg. tr. for a zero, clock tr. between zeros only
BINARY NUMBER RECORDING CURRENT AND APPROX. TAPE FLUX PLAYBACK VOLTAGE DATA RECOVERY x = Clock pulse	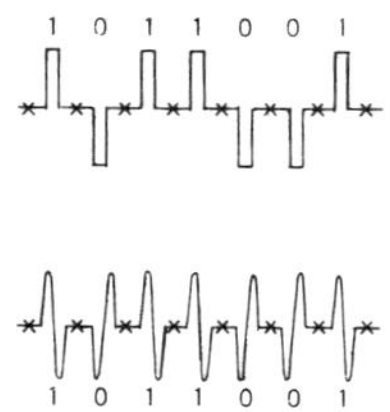	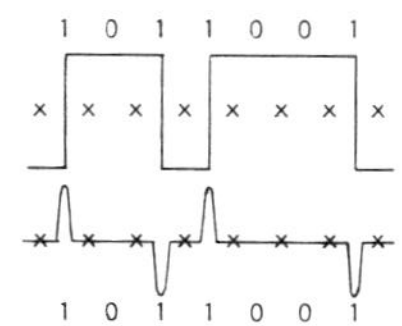	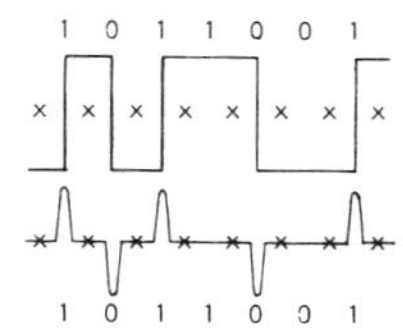	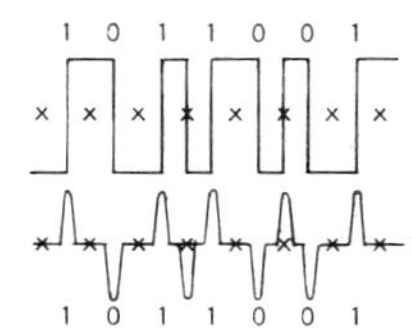	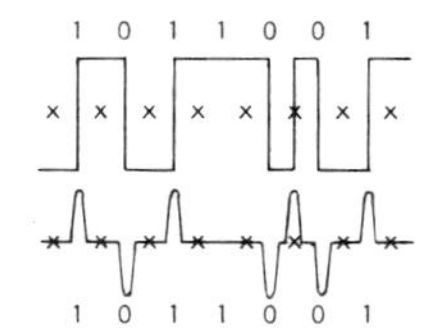
BIT DENSITY	High	Very low	Low	Medium-High	Medium
CLOCKING	Self-Clocking	Requires Clocking	Requires Clocking	Self Clocking	Self Clocking*
DC CONTENT	High dc up to 100%	High dc up to 100%	High dc up to 100%	Negligible dc	Low dc-33% max.
OVERHEAD	Low	High	High	Low	Low
SPECTRAL CONTENT	Dc to clock freq.	Dc to $\frac{1}{2}$ clock freq.	Dc to $\frac{1}{2}$ clock freq.	$\frac{1}{2}$ clock freq. to clock f.	$\frac{1}{2}$ clock freq. to clock f.
ERRORS	Next bit is corrected	Continue until next "one"	Next bit is corrected	Next bit is corrected	Next bit is corrected
REMARKS	Most elementary code Not economical Seldom used	Has been the most widely used code especially in digital computers	Generated by A/D converters. Popular code for HDR	Used in high density systems. Self-clocking allows higher data packing density than NRZ codes	Higher data packing density than BiφL (fewer transitions)

*depends on 101 transitions.

Table 13-13. Digital Codes.

BiφL	Bi Phase, Level (or Low). Manchester. See Table 13-12.
BiφM	Bi Phase, Mark. Level changes at every clock pulse. Ones cause a midbit level change. Zeros, no change.
BiφS	Bi Phase, Space. Same rules as BφM except designations for ones and zeros are interchanged.
Block Code NRZ	Words are translated into blocks of bits, each block having an average value of zero. (Ampex).
DM-M	Modified Delay Modulation. See Table 13-12.
DM-S	Delay Modulation, Space. Same rules as DM-M except designations for ones and zeros are interchanged.
ENRZ	Enhanced NRZ. Seven bit code plus eighth bit to give odd parity. (Bell & Howell).
Manchester	Same as BiφL. See Table 13-12.
MFM	Modified Frequency Modulation. Miller Code. See Table 13-12.
Miller	See Table 13-12.
M^2, $Miller^2$	Miller code modified to remove the dc component by deleting the final transient of an even number of ''ones''.
NRZ	Non-return to Zero. See Table 13-12.
NRZ-I	Same as NRZ-M. See Table 13-12.
NRZ-L	Non return to Zero, Level. See Table 13-12. (Honeywell, EMI).
NRZ-M	Non-return to Zero, Mark (or Low). See Table 13-12.
NRZ-S	Non-return to Zero, Space. Same rules as NRZ-M except designations for ones and zeros are interchanged.
PE-L	Phase Encoding, Level. Manchester. See Table 13-12.
PE-M	Phase Encoding, Mark. Same as BiφM. See Table 13-12.
R-NRZ	Randomized NRZ. Data is modulated and demodulated by pseudo-noise. (Sangamo, Weston).
RB	Return to Base (or Bias). Positive pulses for ones, no pulses for zeros.
RZ	Return to Zero. See Table 13-12.
ZM	Zero Modulation Coding.

to some faults and not to others. Different codes have been selected by major suppliers of digital recorders, making proposed standardization difficult, since everybody prefers his own code in which he has a vested interest in hardware, software, and experience.

Sampling Theory

Information theory reassures us that we will not lose anything if we take readings often enough to give at least two samples at the highest frequency that is present in our data. Intuitively this is reasonable because by definition nothing in the data can change at a rate faster than its highest frequency content, so there is no possibility of missing anything between samples. Thus if the original data is an audio waveform with an upper frequency limit of 20,000 Hz, we must take 20,000 × 2 = 40,000 or more samples per second (Nyquist rate). To be conservative, a ratio of 2.5, 3, or even 5 is sometimes chosen, so we would probably sample at a 50,000 rate. How can these brief samples tell us that at 20 kHz we have a sine wave and not some other shape? Again, any variation from a sine wave implies that the wave contains higher harmonics, which would require a response to at least 40 kHz for the second harmonic, well beyond the range of our data.

Each sample is merely a number, and we are recording these numbers one after the other in a continuous stream. We have chosen to use binary numbers rather than decimals for efficiency reasons given in Chapter 8. How accurate should our samples be? Greater accuracy requires a longer number with more places. In the binary system, each additional place or extra binary digit (bit) doubles the accuracy. For an accuracy of 0.1%, ten binary digits are needed, since $2^{10} = 1024$ and there is an uncertainty of about ± 1 part in 1024, theoretically ± 0.5 part in the A/D conversion and ± 0.5 part in the D/A conversion. This uncertainty is really a rounding off to the nearest digit of the sampled number and manifests itself as a *sampling noise* or *digitizing noise*, which can be reduced by using longer, more accurate numbers. The size of the numbers is a compromise between long numbers that use up more tape and require more expensive and slower D/A converters, and shorter numbers that are more economical but less perfect in accuracy and SNR. The compromised number length (word) must be chosen carefully. Once built into the system, it cannot be changed.

Aliasing. What if we sampled at a 40,000 Hz rate and our data contained some components above 20,000 Hz, in other words more than half the sampling rate? As shown in Figure 13-14, the samples of a 23,000 Hz wave are exactly the same as the samples of a 17,000 Hz wave. In reconstructing the data from the samples, we would obtain a 17,000 Hz component when none was present in the original data, if the data had something at 23,000 Hz. This effect is called *aliasing* and arises from the fact that sampling cannot distinguish between two frequencies that are separated equal amounts from half the sampling rate, one above and one below.

$$\tfrac{1}{2}\text{ sampling rate} - f_1 = f_2 - \tfrac{1}{2}\text{ sampling rate},$$

where f_1 is the aliased or phantom output frequency that was not in the original data and f_2 is the too-high frequency that caused it. To prevent aliasing (1) we must remove all frequencies above half the sampling rate with a low-pass filter, or (2) we can increase the sampling rate until its half-rate is above the highest frequency that could possibly be present in the input, including noise. The first

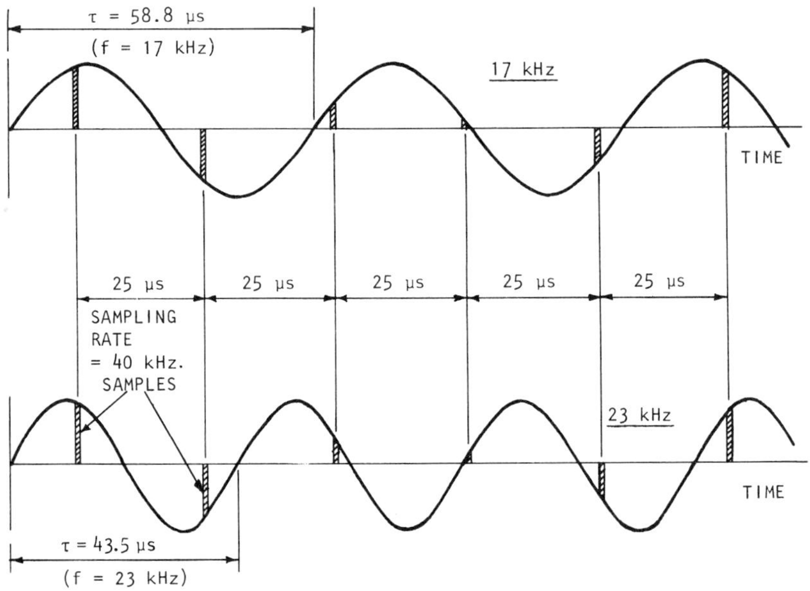

Figure 13-14. Aliasing of frequencies above half the sampling rate. Both frequencies yield identical samples.

remedy is less wasteful of bit rate, but filters do not have a perfectly sharp cutoff and so we must increase the sampling frequency to a reasonable compromise.

Digital Formats and Bit Rate

Some extra bits, in addition to the data bits, are usually needed for service functions such as marking the beginning or end of a word, parity bits for error detection, or reducing the DC content. These extra bits are appropriately called *overhead*. The total word length is then the sum of data bits and overhead bits. A typical code word is 14 bits long, made up of 12 data bits and 2 service bits.

This word length might seem just right for parallel recording of all its bits on the old 14-track IRIG format, but it is spoiled somewhat by the tracks' being divided between two spaced head stacks whose accuracies are much less than the micron-packed bits, to say nothing of variable tape stretch and skew between heads. In any case, digital recording does not need the SNR and freedom from cross-talk of head stacks designed for analog systems but would benefit from close-spaced narrow heads in a single stack. The later IRIG standards recognize and recommend special heads and formats for digital recording, including 14, 16, and 31-head arrays.

Data Compression. Digital recording needs several times as many bits as any other system to record the same data. For economy the bits are squeezed together. When bits are only 1 or 2 micrometers apart, mechanical inaccuracies of this order become important. Gap scatter, tape skew, and drive flutter are the main offenders, as shown in Figure 13-15, causing bits of a parallel word to stray from their cell during playback, or words from different tracks to be mistimed from one another. Computer techniques can overcome these faults by storing the displaced bits in a buffer, then reading them out with delays that counteract the timing errors. If the errors are considerable, some overhead bits must be added to the signal to set the buffer. There is a limit to the correctable bit-displacement depending on the logic circuits, coding, and tape-head geometry.

Noise causes erratic amplitude variations of the bits, which becomes more serious as the bits become smaller when the data is compressed. Limiters and sensors of zero crossings confer some immunity, but the tolerance and correction of amplitude errors has limits, just as there are limits to correctable time displacement errors.

Eye Patterns. On an oscilloscope display of a playback waveform, the tim-

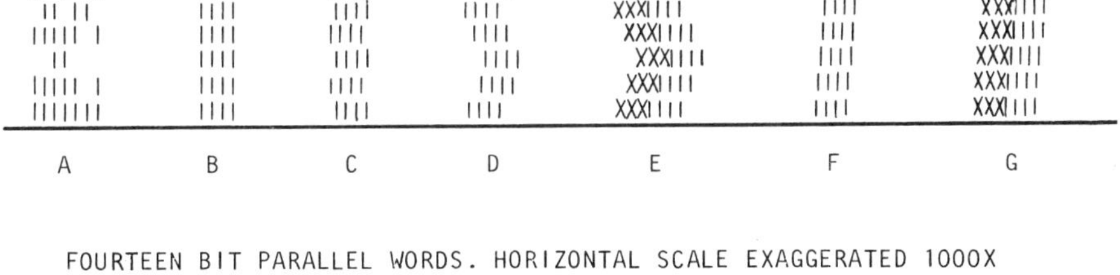

Figure 13-15. Effects of gap spacer and tape skew on high density recording. A. A set of seven 14-bit words, ones and zeros, in perfect alignment. B. A set of four 14-bit words, all ones, in perfect alignment. C. Gap scatter has displaced the bits, but they are still within their own word-cell. D. Greater gap scantter moves bits to wrong cells. Words can no longer be recognized. E. A code word represented by *XXX* precedes each set of words to synchronize a buffer that descatters the words. The greater the scatter, the more often a code or sync word will have to be inserted. F. Skew or tilt of the tape displaces the bits. Angle shown exaggerated 1000×. Dynamic skew changes continually as tape moves with slight irregularities. G. Code word *XXX* gives clue to deskewing buffer circuits that resynchronize the words.

ing errors show up horizontally while the amplitude errors give vertical variations. This is the basis of an interesting test for evaluating a recording system. The oscilloscope triggering rate is set at $2f$, $2f/3$, $2f/5$, or in general $2f/n$, where n is an odd integer and f is the frequency of a sine wave reproduced by the system under test. Each successive sine wave display is inverted as shown in Figure 13-16*A*, giving familiar patterns that resemble an open eye if the wave is noise-free and flutter-free. Noise on the reproduced sine wave causes its amplitude to fluctuate, partially closing the eye vertically, as in Figure 13-16*B*. Time-base error or flutter causes the eye to contract and expand horizontally, as in Figure 13-16*C*. Both these defects of the tape system close the eye, reducing the clear space in its center. The amount of amplitude and time base errors can be estimated from the calibration of the oscilloscope. The higher the recorded frequency, the more sensitive the test. A high enough frequency closes the eye completely, indicating that error-free data are difficult or impossible to extract at that density.

Eye patterns are also useful for analyzing digital waveforms independently of the recording process. An eye pattern of an NRZ wave is shown in Figure 13-16*D*. Patterns will vary with the coding. An experienced observer can quickly estimate the effects of pulse crowding, low-frequency sag, overshoot, dissymetry, unfavorable equalization, and malfunctions in data processors. For quantitative data, one can measure the *eye ratio* (clear center space/peak-to-peak excursion) and the nodal intersections.

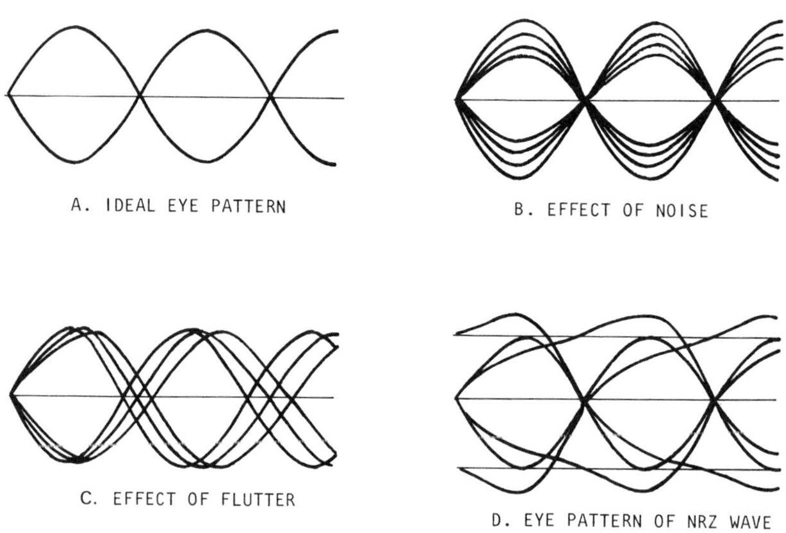

Figure 13-16. Eye patterns for evaluating digital systems.

Dropouts. Digital recording has an all-or-nothing aspect: perfection if we can perceive the numbers at all; catastrophe if we lose any. Lost bits are caused by dropouts due to imperfections in the tape or, more often, to particles that momentarily separate the tape from the head. These particles, called *debris*, can be airborne dust, loosened pieces of tape coating or tape backing, or splinters from slit tape edges. They can be wear products of the heads, tape guides, rollers, or any other parts of the drive mechanism, or of the reels, tape packages, and so on. A momentary tape-to-head separation of only 0.75 μm (30 microinches) causes a 27 dB dropout at state-of-the-art packing densities of 1300 bits/mm (33,000/in.). The same separation would give a dropout of only 13 dB at 650 bits/mm (16,000/in). Microscopic particles can therefore cause great damage but are less serious if the packing density is reduced.

In instrumentation recording, a loss of one bit out of every million is considered tolerable. It can be caught and discarded or corrected by the logic in the bit processors. Even if the mistake got past them into the output, it would likely give an absurd data point that would be recognized as a dropout error. In computer recording, the standards are about a million times higher, one bit error in every 10^{11} or 10^{12}. To achieve this accuracy, the bit densities are much less than for instrumentation, redundant systems are used, and very sophisticated error correcting codes have been devised (further described in Chapter 14 on computer recording).

Tape Certification. Regardless of other causes of dropouts, we should start with a faultless tape if possible. High-density digital tapes are therefore tested 100% on special machines that count the dropouts at different bit packing densities and make a printout of any defects. They will even mark the location of the dropouts. The marked tapes are afterwards indexed and inspected microscopically, and the dropout is corrected by appropriate means that range from brushing off a loose speck, to burnishing down a nodule, to filling a cavity with a tiny brush full of magnetic dispersion. After inspection, or reinspection after correction, the tapes are certified to meet the appropriate standards, as for example less than one 12 dB dropout per million bits at 33,333 BPI density.

SPACECRAFT RECORDERS

Satellite recorders gather data during all parts of their orbit but are in position to deliver it for only that small fraction of the orbit when they are within range of a receiving station. They carry a multispeed tape recorder that records in real time but plays at high speed to unload the data. Deep-space probes transmit data very slowly to give a narrow frequency band that improves SNR of extremely weak signals. Again the capability of changing time scale and high storage capacity make tape systems useful.

Figure 13-17. Lightweight spacecraft recorder (USAF) (DMSP) DDS-5000. (Courtesy NASA)

Under sponsorship of GSFC/NASA (Goddard Space Flight Center/National Aeronautics and Space Administration), the earlier designs of spacecraft recorders have evolved to lightweight units, such as the one shown in Figure 13-17, that store 4.5×10^8 data bits, operate while hermatically sealed for two years, consume only 10 to 20 watts and are small in size and weight.

Spacecraft Recorder Design

The lightweight recorder of Figure 13-17 uses 6.3 mm ($\frac{1}{4}$ in.) tape on coaxial reels 190 mm (7.5 in.) in diameter. Weight is saved by web construction and magnesium parts. The sealed enclosure contains 90% nitrogen/10% helium at slightly above atmospheric pressure. Energy is conserved by using a negator spring to tension the tape.

The negator is a spiral clockspring through which the concentric reels are connected to each other. As the capstan pulls the tape from the supply reel, this reel's rotation is transmitted through the negator spring to turn the takeup reel, while the negator spring winds or unwinds enough to take care of the differential speeds of the reels as one empties while the other fills. Thus the negator spring maintains tape tension, saves energy, and eliminates the need for motor drive of the takeup reel.

A different unit (SPOT) has been designed for much higher capabilities at the expense of more weight, larger size, and higher-power consumption. The SPOT unit transmits 50 megabits per second and stores 6.6×10^{10} bits.

Table 13-14 gives further specifications for the lightweight spacecraft re-

Table 13-14. Specifications of a Lightweight Spacecraft Recorder.

Tape - $\frac{1}{4}$ in. width × 1 mil thick (nominal) × 500 usable ft. long
Required tape passes - 25,000
Standard number of tracks - Eight Data
Record reproduce time:

Operating Mode	Record	Reproduce
8 tracks in sequence, minimum rate	40.7 hr.	119.5 min
8 tracks in sequence, maximum rate	23.7 min	23.7 min
4 tracks in parallel (2-passes), minimum rate	10.1 hr.	29.8 min
4 tracks in parallel (2-passes), maximum rate	6.0 min	6.0 min
8 tracks in parallel, minimum rate	5.0 hr.	14.9 min.

Input data rates - 3.125 to 2560 kbps
Output data rates - 64 to 2560 kbps
Tape packing density - 9560 BPI/Track
Total data storage - 4.58×10^8 bits
Input data format - Serial, NRZ or Bi-Phase-Level PCM
Output data format - Serial bi-phase level PCM (NRZ, optional)
Error rate (max.) - 5 in 10^7 bits (Beginning of Life)
Error rate (max.) - 1 in 10^6 bits (End of Life)
Jitter - (dejittred) .01% peak-to-peak or equal to external reference
Input voltage - 28 V ±35%
Input power (record) - 10 to 14 watts (depending upon input data rate)
Input power (playback) - 17 to 21 watts (depending upon output data rate)
Transport size (Including container) - 6 × 7.8 × 3 in. (excluding mounting lugs)
Electronics size - 6 × 7.8 × 5 in. (excluding mounting lugs)
Transport weight - 7.4 pounds
Electronics weight - 7.9 pounds (excluding interconnecting cables)
Tape speed range - .327 ips to 33 ips in 21 increments.

(Courtesy NASA)

Table 13-15. Spot General Parameters.

Tape size	1 inch × 9200 ft
Storage capacity	6.6 × 10^{10} bits
Record/reproduce rate	50 Mb/s (2-25 Mb channels)
Maximum record time	22 min
Number of data tracks	36 plus 2 spare
Tape speed	78.9 in./s
Input voltage	25-37 VDC
Power consumption	280 watts maximum
Input data	Dual 25 Mb serial streams
Input code	NRZ-L
Output data	Dual 25 Mb serial streams
Output code	NRZ-L
Bit error rate	≤ 10^{-6} with error correction system
Size - TU:	508 × 508 × 350 mm
EU (Less cables):	508 × 508 × 439 mm
Mass:	70 kg maximum

(Courtesy NASA)

corder. Table 13-15 gives parameters of the SPOT high-performance spacecraft recorder.

SPECIAL MAGNETIC DATA RECORDING

Magnetic Cards

A card with a magnetic surface is a convenient format to use for storing a limited amount of information that can be erased and updated. For identification and credit cards the invisibility of the information is an advantage.

Credit Cards. The credit card shown in Figure 13-18 has a magnetic stripe 6.35 mm (0.250 in.) wide or 10.20 mm (0.405 in.) wide, recorded with two or three tracks as shown. Such cards are used in retail merchandising, travel and entertainment, automatic teller banking, commuter ticketing, and security and identification systems.

Information Content. Track 1 of Figure 13-18 contains the name and title of the cardholder, the account number, and miscellaneous additional data up to 79 characters. Track 2 contains numerical data only, including the account number and up to 40 additional numbers. Track 3, if present, contains banking data

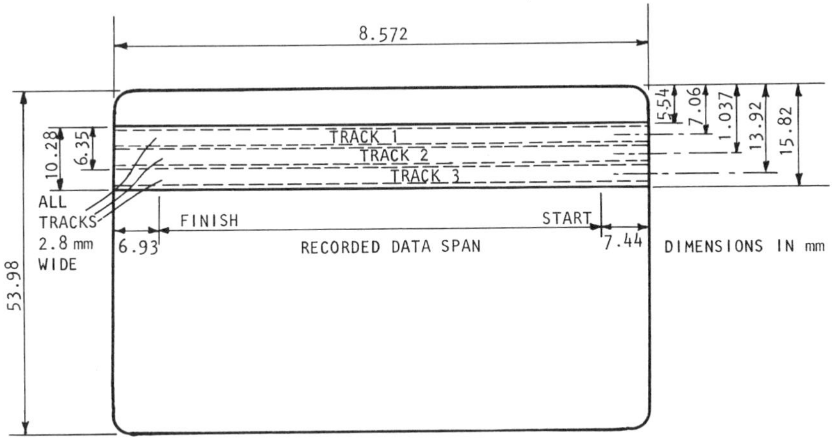

Figure 13-18. Magnetic-striped credit card. The three-track card has a stripe 10.28 mm (0.405 in) wide. The two-track card stripe is 6.35 mm (0.250 in) wide.

including Primary Account Number (PAN), country, type of currency, amount authorized and amount remaining per cycle period, date next cycle begins, length of cycle, retry count, Personal Identification Number Control Parameter (PIN-PARM), interchange, type of account (TA), Service Restriction (SR), card sequence, card security, subsidiary account numbers, Crypto Check Digits (CCD), and discretionary data.

Recording Characteristics. The magnetic coating is similar to that of computer tapes with approximately H_c = 20 kA/m (250 Oe). The encoding is *two-frequency coherent phase recording*, which is a self-clocking binary code. On Track 1 and Track 3, the maximum bit density is 16.5 reversals/mm (420/in.). Their alphanumeric code is a 6-bit adaptation of ASCII (see Table 14-2). Track 2 has a lower bit density maximum of 5.91 flux reversals/mm (150 in.) and carries only numerals in a 4-bit code.

Calculator Programs and Audio Cards. A software library can be stored on compact magnetic cards of programmable hand calculators. (TI-59 cards measure 74 × 16 mm). The calculator program in digital form is scanned by a magnetic head that transfers it to or from the calculator.

The Language-Master and other audio applications of magnetic cards are described in Chapter 10.

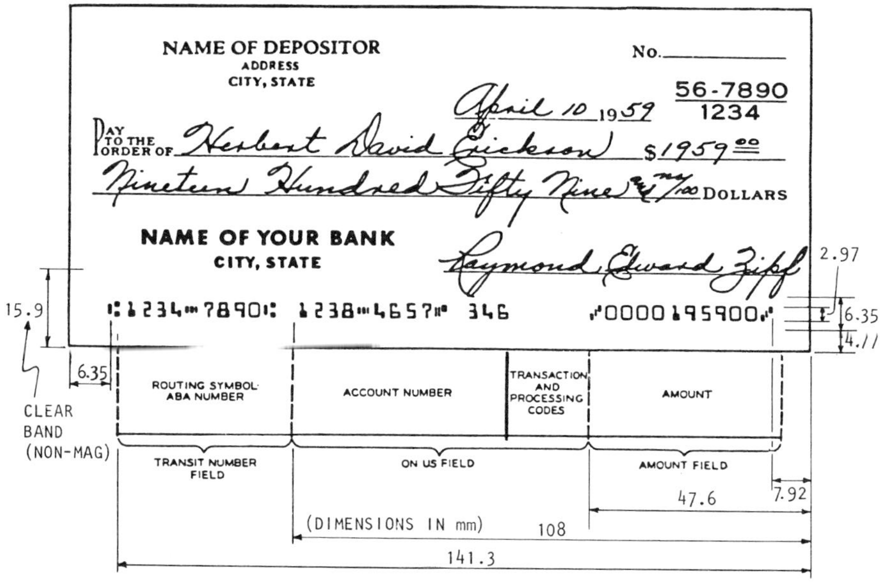

A. MICR Symbols and Field Locations

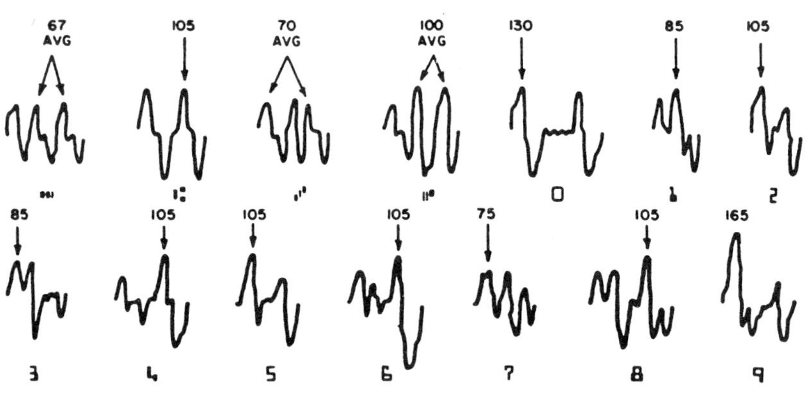

B. Readout Wave Patterns

Figure 13-19. Magnetic ink character recognition for bank checks.

Magnetic Ink Character Recognition (MICR) for Bank Checks

The numerals at the bottom of bank checks are printed with magnetizable ink and have the distinctive shape known as *E-13B Font*, as shown in Figure 13-19*A*. Such imprinting is also useful for coupons, tickets, receipts, bills, and credit cards.

The symbols are read from right to left at a speed of 3810 mm/s (150 in/s) with a reading head having a 75 μm (0.003 in.) gap and a width of 15.88 mm (0.625 in.), the printing being presaturated magnetically in the direction of movement by a DC field. On readout each numeral generates a different wave pattern, as in Figure 13-19*B* that is recognized by the computer circuits.

The information format is indicated in Figure 13-19*A*. At the lower left is the preprinted *transit number field* with a routing symbol and ABA (American Bankers Association) number to identify the drawee bank. The middle area *on us field* is left to the banker's choice; he can preprint or postprint anything here, usually the account number and/or processing codes. The right hand area is reserved for the *amount field*, where the amount of the check is post-printed. Extra long checks have room for an *auxiliary on us field* printed to the left of the transit number field. Notice the special symbols that identify the beginning or end of each kind of field.

14. COMPUTER RECORDING

Cartoonists like to picture a computer as a kind of tape recorder with large reels on its face. The advent of personal computers has changed this popular image and most people now think of a computer as a display screen, a keyboard, and a module for software bits and bytes recorded on magnetic discs. Actually the recorder portion does not compute anything but is merely a convenient memory that stores millions of numbers. Blocks of numbers are picked up from the recorder, then processed by the computer, and the result is again recorded. Access to a chosen block is slow on a reel-to-reel transport if the information happens to be at the other end of the reel. Where greater speed is needed, the information is stored on discs, reducing access time to milliseconds. Even this is not fast enough for the actual computer, which therefore utilizes solid state memories. Fast-access memories do not hold enough information, so the entire hierarchy of solid state, disc, and tape storage is needed for large general-purpose computers.

The main difference between a computer recorder and a digital instrumentation recorder is that the computer input is already numerical and does not require an A/D (analog to digital) converter; and similarly its output does not require D/A (digital to analog) conversion. Another difference is that computers generally use 12.7 mm ($\frac{1}{2}$ in.) tapes with seven or nine tracks, written by parallel heads all in the same block; while instrumentation uses a variety of IRIG (Inter-Range Instrumentation Group) and special formats. A third difference is that a computer tape transport must start, stop, and reverse the tape motion in a matter of milliseconds for rapid updating and error correction; while instrumentation recorders normally operate with a steady flow of tape.

The technique of coding, writing, and reading binary numbers on tapes were described for digital instrumentation recording in Chapter 13. Historically digital computer recording came first, and digital instrumentation borrowed its techniques from computers. Table 14-1 is a vocabulary of computer terms. Table 14-2 gives the ASCII code for translating between plain language and recorded digits. Table 14-3 gives the Hollerith code for translating IBM cards into digital representation.

REEL-TO-REEL STORAGE

A standard 267 mm (10.5 in.) reel of computer tape holds up to 180 megabytes of digital data (8 bits/byte) in a package that is readily removed and stored. Tape handlers with vacuum bins utilize tape in an ideal manner for many kinds

Table 14-1. Terminology of Computer Recording. (See also Table 14-8, Disc Storage Terminology)

Algorithm Procedure for solving a mathematical problem
Alphanumeric Using or convertible to alphabetical letters and numbers
ALU Arithmetic Logic Unit. Data processing portion of a computer
ASCII Code A code for translating binary numbers to alphanumeric characters (American Standard Code for Information Interchange)
Baud A measure of transmission rate of alphanumeric data (bits/sec ≅ words/min)
BER Bit error rate
Bit density Bits per inch or per mm recorded along a track (see BPI, CPI, FCI, FRPI)
Block A separate unit of recorded data, usually containing an identifying index code
BOT Beginning of tape; a marker at end of a leader signalling beginning of data tape
BPI Bits per inch
Buffer A device which temporarily stores a limited amount of computer data
Busy signal Indication that certain computer circuits are temporarily not available
Byte A group of bits that represent a character; eight bits per byte in most computers.
Code Representation of data bits by a sequence of changes (electrical, magnetic, etc.).
Computer peripheral Equipment auxiliary to the central processing unit, such as tape or disc memories
CPI (Magnetic) changes per inch
CPU Central processing unit
CRC Cyclic redundancy check. Error detection by parity bits added to a data-block.
Daisy chain Multiple tape or disc drives controlled together to increase on-line computer memory. See string.
Detection window Short time interval in which a recorded bit can be read out reliably
Disk file One or more disks on which data is recorded
ECC Error Correcting Code
Encoding Converting information, as alphanumeric, to a binary code
EOF End of file; a tape marker that signals the end of stream of data
EOT End of tape (data portion), usually followed by a marker and trailing leader
FCI Flux changes per inch
File Stored data on tape, disc, etc.
File protect ring A ring, removable from a tape reel flange to prevent the recorder from altering data
Formatting Arranging the data in patterns, usually with bits added to identify and check each block of data
FRPI Flux reversals per inch (see BPI, CPI, FCI)
GCR Group coded recording, used at 6250 bits per inch
Hard Copy Printed or written data
Hard Error An error that cannot be corrected by rewriting, due to a tape flaw, for example
Hollerith Code A code for translating "IBM" punched card data into digital characters
IBG Inter-block gap - Separation between recorded blocks of data
Main frame Data processing and control section of a computer
Microcomputer, microprocessor Solid state chip that performs computer functions as ALU, RAM, ROM, etc.
Minicomputer A small scale computer using separate wired components
MTBF Mean time before failure. Expected trouble-free life
MTTR Mean time to repair. An estimate of expected down-time if failure occurs
NRZI Non return to zero code widely used in computers. (See Table 13-12)
On line Immediately available for processing

Table 14-1. (*Continued*)

Overwrite Rewrite over previous recording
Parity Adding an extra bit to give an odd (or even) bit count to every word.
Peripheral Equipment auxiliary to the main frame of a computer
PE Phase Encoding. A digital code especially favored for high bit densities.
Postamble A series of bits following a block of data, identifying its end
Preamble A series of bits prior to a block of data, for identification, synchronization, etc.
RAM Random Access Memory
Recovery window Time interval during which data can be read reliably
Reel Servo A servomechanism that operates tape reels in response to signals from the tape handler.
Reread A repeated reading of data, usually if error is present in previous reading
Rewrite Overwriting of data, usually in attempting a more perfect recording if previously in error
ROM Read Only Memory
Run length limited A code which limits the number of successive ones or zeros in any word.
Skew Timing error between bits on different tracks due to tilting errors in tape motion
Soft error Error caused by a temporary imperfection, which disappears when tape is re-read or re-written.
String A group of tape or disc units tied together so they are all reachable by the computer
Strobe A brief clock pulse that triggers data reading at the most significant instant
Vacuum buffer Vacuum bins that isolate reels from tape drive, allowing rapid tape start and stop
Word A multi-bit unit of computer language. See byte
Write-enable ring Same as file protect ring

of data processing. But for random access to data, reel storage is inherently at a disadvantage compared to discs. Prior to the 1970s, both drum and disc storage were somewhat primitive and expensive. At that time computers relied mainly on tapes, and there was feverish development activity in tape handlers. Since then the action has shifted to magnetic disc storage, which features rapid random access with megabyte capacity rivaling tape reels. However, reel-to-reel storage still plays an important role in computers and is utilized in every large installation. For an introduction to computer tape systems, a general purpose medium-density type is described followed by a more sophisticated system to introduce later trends.

General-Purpose Reel-to-Reel System

Figure 14-1 shows a computer recorder that accepts 267 mm (10.5 in) reels holding 732 m (2400 ft) of 12.7 mm ($\frac{1}{2}$ in.) tape. To illustrate its mode of operation, a moderate tape speed and recording density is appropriate. Since the same transport offers many other options, these are summarized in Table

Table 14-2. ASCII Code (Digital to Alphanumeric).
7-BIT ASCII CODE

Octal Code	Char	Octal Code	Char	Octal Code	Char	Octal Code	Char
000	NUL	040	SP	100	@	140	
001	SOH	041	!	101	A	141	a
002	STX	042	"	102	B	142	b
003	ETX	043	#	103	C	143	c
004	EOT	044	$	104	D	144	d
005	ENQ	045	%	105	E	145	e
006	ACK	046	&	106	F	146	f
007	BEL	047	'	107	G	147	g
010	BS	050	(	110	H	150	h
011	HT	051	)	111	I	151	i
012	LF	052	*	112	J	152	j
013	VT	053	+	113	K	153	k
014	FF	054	,	114	L	154	l
015	CR	055	–	115	M	155	m
016	SO	056	.	116	N	156	n
017	SI	057	/	117	O	157	o
020	DLE	060	0	120	P	160	p
021	DC1	061	1	121	Q	161	q
022	DC2	062	2	122	R	162	r
023	DC3	063	3	123	S	163	s
024	DC4	064	4	124	T	164	t
025	NAK	065	5	125	U	165	u
026	SYN	066	6	126	V	166	v
027	ETB	067	7	127	W	167	w
030	CAN	070	8	130	X	170	x
031	EM	071	9	131	Y	171	y
032	SUB	072	:	132	Z	172	z
033	ESC	073	;	133	[	173	{
034	FS	074	<	134	\	174	|
035	GS	075	=	135	]	175	}
036	RS	076	>	136	∧	176	~
037	US	077	?	137	—	177	DEL

14-4, which shows higher and lower possibilities. With a modest bit rate requirement, the tape speed is chosen to be 635 mm/s (25 in./s) and the bit density 800 CPI (changes per inch or bits per inch) (31.5/mm). This gives a data transfer rate of 200,000 Hz. Nine tracks on the tape allow simultaneous parallel recording of eight bit words plus an extra bit for parity check. At this density, the 10.5 in. reel of tape has a gross capacity of 23×10^6 words or bytes. Figure 14-2 shows the nine-track format commonly used. The older seven-track format is shown in Figure 13-2*A*.

Table 14-3. Hollerith Code ("IBM" Punched Card to Digital) Standard and Compressed Hollerith Code.

Character	Rows Punched	Standard Code (Octal)	Compressed Code (Octal)
Space	None	0000	000
0	10	1000	040
1	1	0400	001
2	2	0200	002
3	3	0100	003
4	4	0040	004
5	5	0020	005
6	6	0010	006
7	7	0004	007
8	8	0002	010
9	9	0001	020
&	12	4000	200
—	11	2000	100
A	12,1	4400	201
B	12,2	4200	202
C	12,3	4100	203
D	12,4	4040	204
E	12,5	4020	205
F	12,6	4010	206
G	12,7	4004	207
H	12,8	4002	210
I	12,9	4001	220
|	12,11	6000	300
{	12,10	5000	240
}	11,10	3000	140
J	11,1	2400	101
K	11,2	2200	102
L	11,3	2100	103
M	11,4	2040	104
N	11,5	2020	105
O	11,6	2010	106
P	11,7	2004	107
Q	11,8	2002	110
R	11,9	2001	120
/	10,1	1400	051
S	10,2	1200	042
T	10,3	1100	043
U	10,4	1040	044
V	10,5	1020	045
W	10,6	1010	046
X	10,7	1004	047
Y	10,8	1002	050
Z	10,9	1001	060
`	8,1	0402	011
:	8,2	0202	012
#	8,3	0102	013
@	8,4	0042	014
'	8,5	0022	015

Table 14-3. *(Continued)*

Character	Rows Punched	Standard Code (Octal)	Compressed Code (Octal)
Space	None	0000	000
=	8,6	0012	016
″	8,7	0006	017
a	12,10,1	5400	241
b	12,10,2	5200	242
c	12,10,3	5100	243
d	12,10,4	5040	244
e	12,10,5	5020	245
f	12,10,6	5010	246
g	12,10,7	5004	247
h	12,10,8	5002	250
i	12,10,9	5001	260
j	12,11,1	6400	301
k	12,11,2	6200	302
l	12,11,3	6100	303
m	12,11,4	6040	304
n	12,11,5	6020	305
o	12,11,6	6010	306
p	12,11,7	6004	307
q	12,11,8	6002	310
r	12,11,9	6001	320
[	12,8,2	4202	212
·	12,8,3	4102	213
<	12,8,4	4042	214
(	12,8,5	4022	215
+	12,8,6	4012	216
! or ↑	12,8,7	4006	217
~	11,10,1	3400	141
s	11,10,2	3200	142
t	11,10,3	3100	143
u	11,10,4	3040	144
v	11,10,5	3020	145
w	11,10,6	3010	146
x	11,10,7	3004	147
y	11,10,8	3002	150
z	11,10,9	3001	160
]	11,8,2	2202	112
$	11,8,3	2102	113
*	11,8,4	2042	114
)	11,8,5	2022	115
;	11,8,6	2012	116
ˆ or	11,8,7	2006	117
\	10,8,2	1202	051
,	10,8,3	1102	052
%	10,8,4	1042	054
—	10,8,5	1022	055
>	10,8,6	1012	056
?	10,8,7	1006	057

Figure 14-1. Tape handler for computer applications. Rectangular vacuum bins isolate the reels from tape at the capstans and head assembly (left side) allowing the tape to start, stop or reverse in milliseconds. A knife edge scraper and vacuum cleaner remove debris from tape surface. (Courtesy Kennedy Corp.)

The drive has features familiar to digital computer specialists but not ordinarily found in recorders for other purposes. The reel hubs are equipped with a write-enable ring (file-protect ring), Figure 14-3. Removing this ring prevents the machine from accidentally altering the recorded data. A playback (read) head-stack picks up the recorded data shortly after recording and compares them with the original. If there is any error in a block of data, the tape is automatically backspaced, the data block erased, and the data rewritten, possibly with an increased recording current. If the second try is unsuccessful, the tape is backspaced and another trial is made with still a different threshold. After four trials it is assumed that a permanent defect exists on that segment of tape, which is then erased and passed over for a fresh segment.

Table 14-4. Specifications of a Reel Type Computer Recorder.

Tape - Half inch (12.7 mm) wide, 1.5 mil (0.0381 mm) thick, Mylar base
Tape tension - 8 oz for half in. tape
Tape buffers - Vacuum chamber
Tape speed - 25, 37,5, 45, 75, 125 in/s (tolerance ± 4% or less)
Rewind speed - 200, 300 in./s
Start-stop time - 3 to 5 millisec, 16 ms @ 25 in./s, 5.3 ms 75 in./s
Start distance - 0.17 ± 0.02 in.
Stop distance - 0.19 ± 0.02 in.
Interchannel displacement error - 150 microinches at 800 CPI
Reel size + tape capacity - 10.5 in. diam., 2400 ft, (1.5 mil tape); 8.5 in. diam., 1200 ft.; 7 in. diam., 600 ft
Track format - 7 track IBM compatible or 9 track ANSI
Heads - Dual stack, read after write
Bit densities 200, 556, 800, 1600 CPI (Bits per inch)
Character rate - 20,000 to 250,000 characters/sec
Block lengths - 512, 1024, 2048 characters
Coding - NRZI for bit densities to 800 CPI, PE for higher densities
Logic - true = logic 1 = 0.4 volt (25 ma sink)
false = logic 0 = 3 volts (or more)
Electronics - Parity generator; cyclic redundancy check character generator; gap control; read after write with automatic backspace, erase and rewrite, re-read cycled through different thresholds; electronic skew compensation for NRZI.

(Courtesy Kennedy Corp.)

One might expect dramatic differences in computer tape formats, coding, and so on by various manufacturers, who would, of course, take advantage of improved tapes and heads as they developed. In instrumentation recording such differences have indeed taken place and created a chaotic situation where a half-dozen different digital codes are in use and where special-purpose recorders have unconventional track widths and spacing. Nonstandardization in the instrumentation field is tolerated because installations usually are self-sufficient, and, in fact, the manufacturers would like customers to be tied to their own systems. The situation differs for computers, where there is much more interchange of computer data and, therefore, a great incentive for it to be "IBM compatible." Some data are old and so even the newest drives make provision for 200 CPI (7.87/mm), which has been obsolete for decades.

Advanced Reel-to-Reel System

A high-performance unit runs its tape at 200 in/s (5080 mm/s) to give a data rate of 1,250,000 bytes/sec, making it effective for transferring data to and from

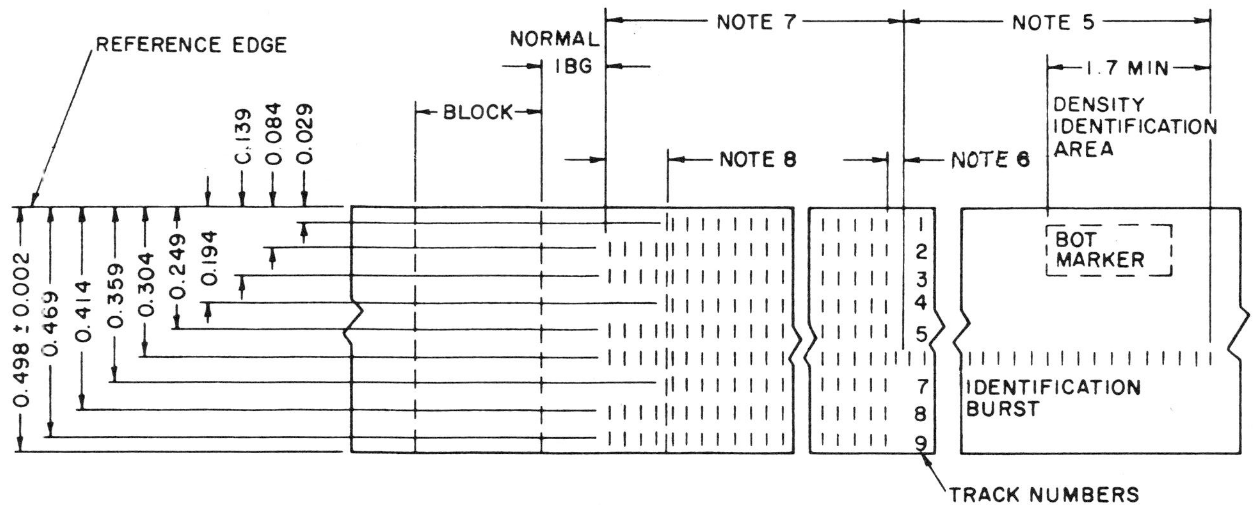

NOTES:
(1) Tape is shown in 6250 mode, oxide side up.
(2) All dimensions are given in inches.
(3) Track placement tolerance is ± 0.003 for each track.
(4) Tape to be fully saturated in the erase direction in the interblock gap and the ID area.
(5) ID burst (see 5.8).
(6) Undefined gap (see 5.9).
(7) ARA burst (see 5.9).
(8) ARA ID characters (see 5.9).

Figure 14-2. Computer format using nine-track half-inch tape. (Seven track format is shown in Figure 13-2*A*) (ANSI St'd X3.54 – 1976).

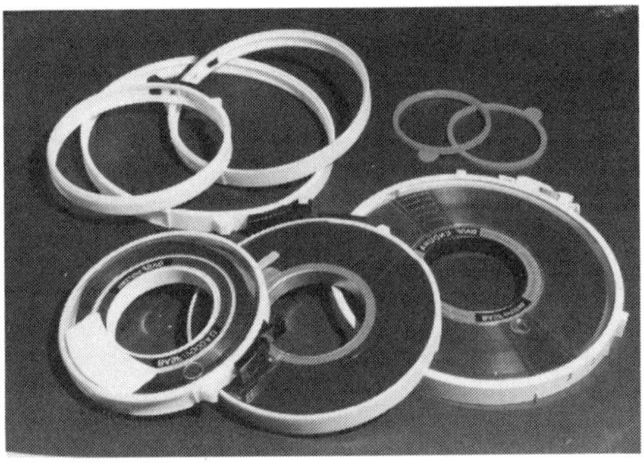

Figure 14-3. Half inch tape reels and accessories. All reels fit on a standard 3 inch (76.2 mm) hub and have solid flanges. Tape seals for outer periphery keep out dust. Write-enable file protect rings (with tab) protect against erasure when removed from groove in side of reel. Reflective tabs are adhered to tape to mark beginning and end of active region. $10\frac{1}{2}$ inch, $8\frac{1}{2}$ inch and $7\frac{1}{2}$ inch reels hold 2400, 1200 and 600 feet of tape respectively (267 mm reel, 732 m; 216 mm reel, 366m; 190mm reel, 183 m).

large-capacity disc drives. It has automatic load and unload operation, vacuum capstan, air bearings, tape cleaner, and an error counter that marks the tape reel if its tape deteriorates enough to give excessive errors. Table 14-5 gives specifications of an advanced reel-to-reel system. A complete system usually includes a tape handler, read-write heads, and a controller with electronics that coordinate the drive and data transfer with the rest of the computer.

Tape Handlers. As described in Chapter 7, computer tape handlers must be nimble in shuttling tape back and forth; even a millisecond is thousands of times longer than mainframe computations. The best drives use vacuum bins (buffers) to isolate the reels from a free length of tape at the heads. The free length is light enough to start or stop in about a millisecond, using fast-acting pinch rollers against capstans that run continuously at full speed. In some designs a vacuum capstan is used. Where quick start action is not as important, spring arm tape isolators are sufficient and less expensive. In either case, the supply and takeup reels are servo operated to replenish or take up tape from the isolators. Other features of tape handlers include reverse operation for faster data retrieval during shuttling (with electronics that re-reverse the data blocks) and automatic loading/unloading (self-threading) for operator convenience.

Computer Tapes. Improved tapes with smoother surfaces have made possible the increased bit densities of modern computers. Specifications of a typical

Table 14-5. Specifications of an Advanced Reel Type System.*

Bit densities (CPI)	6250	1600
Coding	GCR	PE
Data transfer rate (bytes / sec)	1250 k	320 k
Tape speed (in./sec)	200	200
Inter block gap (in.)	0.3	0.6
Track format ($\frac{1}{2}$ in. tape)	9	9
Rewind time (2400 ft)	45 sec	45 sec
Unload time	7 sec	7 sec
Auto-loading time	7 sec	7 sec
Start time	0.8 msec	0.8 msec
Stop time	1.1 msec	1.1 msec
Power required	230v 3 phase 2.8 kva	
Weight	992 lb (450 kg)	
Size	30.7W 30.7D 66.9H inches 780W 780D 1700H (mm)	

(*Memorex 3228)

computer tape are given in Table 14-6. The tape is certified to be essentially error-free, which is achieved by careful manufacturing and testing as well as microscopic inspection and repairing of accidental defects (see Chapter 13).

Computer tape is usually supplied on precision-molded reels of high-strength, dimensionally stable plastic. Reel flanges are solid so that they can be sealed against dust by a removable edge-seal ring, as shown in Figure 14-3, which also shows the write-enable rings that fit into a slot in the reel hub to protect against accidental erasure. Half-inch (12.7 mm) width is standard for tapes, and their reels conform to the dimensions listed in Figure 13-5 such that they all fit on the standard 3 in. (76.2 mm) computer hub. Outer diameter of reels vary for different tape capacities. Usual reel sizes are 10.5 in. for 2400 ft of tape, 8.5 in. for 1,200 ft, and 7.5 in. for 600 ft (Figure 14-3). Tape-winding direction, reflective marker placement, and usable area are specified in Figure 14-4.

When a tape is used and handled repeatedly it accumulates debris that cause loss of head contact and dropout of data bits. Eventually the data should be transferred to another tape. In most cases, a tape can be rejuvenated to some extent by cleaning. The simplest cleaner is a miniature vacuum cleaner that removes loose particles. Removal of stubborn embedded matter is facilitated by a scraper blade, but it must be used with great care not to injure the good tape surface. Many computer drives have built-in vacuum cleaners, some with additional gentle-action scrapers. Professional tape cleaning services are available; these may use liquid solvents in addition to the above remedies and are equipped to recertify the tapes. Cleaning machines for in-house rejuvenation of tapes are also available.

Table 14-6. Specifications of a Computer-Grade Tape.*

Dimensions
Width: 0.498″ ±0.002″ (1.26cm ± 0.005cm)
Length: 2400′ + 50′ − 0′ (731.5m + 15.3m − 0)
600′ − 1200′ + 50′ − 0′
(182.9m − 365.8m + 15.3m − 0)
Marker Placement
BOT: 16′ ± 2′ (4.88m ± 0.61m)
EOT: 25′ + 5′ − 0′ (7.62m + 1.62m − 0)

Thickness
Base Film: 1.6 mils nominal
Total Thickness: 1.9 ± 0.3 mils
E-Value: min. 0.25″ (0.64cm)

Physical Properties
Splices: None
Creases: None
Base Film: Oriented Polyethylene-terephthalate
Toxicity: Non-toxic
Flammability: Self-extinguishing in a still CO_2 atmosphere
Compatibility: Not an inhibitor
Longitudinal Curvature: 0.125″ (0.32cm) maximum per 36″ (91.4cm)
Yield Strength: Min.10lbs. at 3%
Moment of Inertia: max. 0.091 in. lb. sec^2
Layer-to-Layer Adhesion: None
Surface Roughness: 1.5μ″ CLA nominal
Cupping: Less than 0.010″ (0.025cm) of 0.25″ (0.64cm) length/0.5″ (1.27 cm) width
Flexibility: Nominally stiffer than standard tapes

Magnetic and Electrical Properties
Signal Level: 100% + 25% − 10% at 3200 FRPI relative to NBS SRM-3200; 100% + 50% − 35% at 9042 FRPI
Surface Resistance:
(Magnetic Coating); 0.5-500 megohms per square
Uniformity—
End-to-End: Max ± 3.0% (2400′)
Short Term: Max ± 2.0%
Ease of Erasure: Less than 4% signal at 9042 FRPI
Note added: $H_c \cong 250/300$Oe
$B_r \cong 1000$
for gamma oxide tape materials

Performance Characteristics
Prior to shipment, each reel of tape has been 100% tested to assure minimum system performance as follows:
Dropout Performance: 9042 FRPI, 25% signal remaining
- No single tape to exceed 2 two-track errors, where a two-track error is defined by two defects occurring within 0.177″ (0.45cm) on different tracks.
- Lot average maximum of 1.5 two-track errors per reel, based upon a 100 reel sample.

Exclusive Noise Errors: 0
Dynamic Skew: Max. 1.1 μsec at 200 ips (5.08 mps) or 1.5 μsec at 112.5 ips (2.86 mps)
Start/Stop Times: Meets all IBM 729, 2400, 3400 requirements

Environmental Characteristics
Operating
Temperature: 60°-90°F (15.6°C-32.2°C)
Humidity: 20-80% RH
Storage—Recorded/Unrecorded Temperature: 40°-120°F (4.4°C-48.9°C)
Humidity: 20-80% RH
Maximum Wet Bulb: 80°F (26.7°C)

(*Memorex Cubic® HD)
(Courtesy Memorex Corp.)

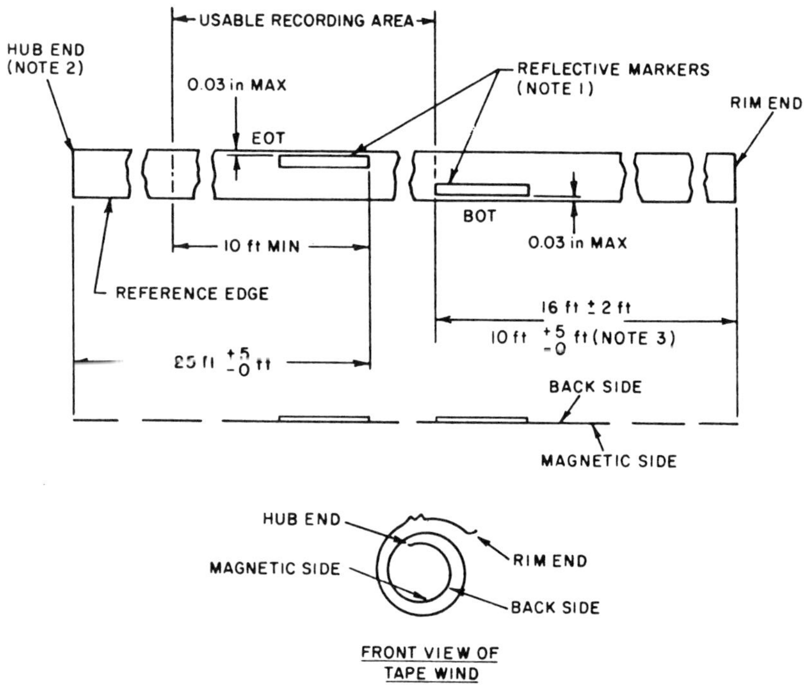

Legend

BOT: Beginning-of-tape marker
EOT: End-of-tape marker

NOTES:

(1) Photoreflective markers shall not protrude beyond the edge of the tape and shall be free of wrinkles and excessive adhesive. Marker dimensions: length, 1.1 inch ± 0.2 inch; width, 0.19 inch ± 0.02 inch; thickness, 0.0008 inch maximum.

(2) Tape shall not be attached to the hub.

(3) Two values for placement of the BOT marker are given, both of which can be handled by most tape units. The indicated value of 16 feet ± 2 feet is the current specified dimension.

Figure 14-4. Tape winding direction, markers and useable area (ANSI St'd X3.54-1976).

Heads. Computer heads (Figure 14-5) are built in a nine (or seven) track stack by techniques described for multitrack heads in Chapter 4. Cores are made of permalloy or ferrite, with possibilities for sendust inlays or for film-deposited cores in advanced designs. Windings are of relatively few turns to give the low inductance needed for megahertz readout. Commonly a second set of heads is included in the same mounting to read the data block immediately after it is written and thus ensure an error-free recording. The read heads must be shielded carefully to prevent direct pickup from the write heads.

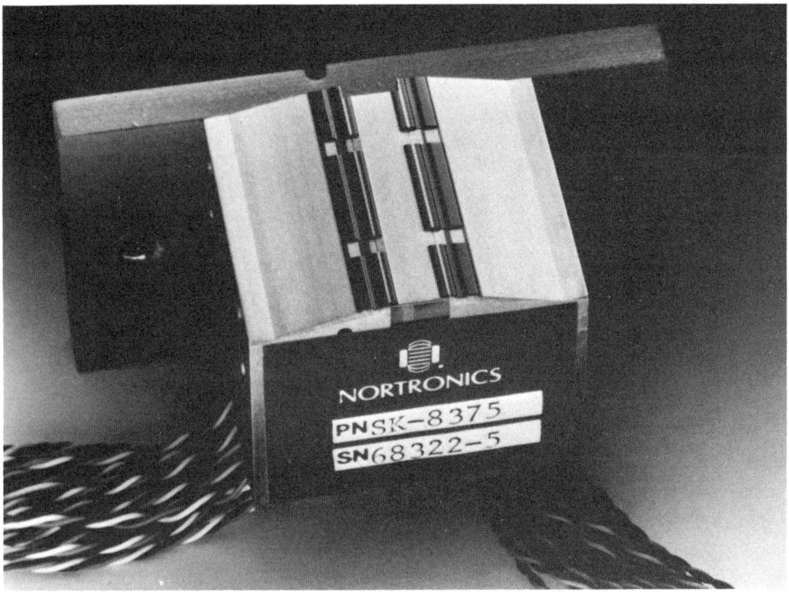

Figure 14-5. Nine-track computer head assembly with read-after-write capability (Courtesy Nortronics Corp.)

Electronics

Write Electronics. A favorite circuit for write heads, shown in Figure 6-18A, is a flip-flop or multivibrator that energizes one side or the other of a center-tapped head winding according to the ones and zeros of the binary coding. Other logic circuits produce an equivalent result. In modern high-density recorders, the effects of peak shift and pulse crowding can be alleviated by appropriate opposite changes in timing of the recording pulses by means of sophisticated logic arrays.

Read Electronics. The sharp transitions in tape magnetization produce positive and negative voltage pulses from the playback head. Pulse peaks are reliable indicators of the zero crossings in the recorded code, but all playback peaks do not occur at the same absolute voltage levels since the levels are perturbed greatly by adjacent bits in the recorded data. Hence the playback circuits (Figure 6-18*B*) are designed to single out the relative peaks rather than the

Table 14-7. Track Assignments of Nine-Track Computer Tapes.

Track (Starting from reference edge)	1	2	3	4	5	6	7	8	9
Binary weight	2^2	2^0	2^4	P	2^5	2^6	2^7	2^1	2^3
ASCII Bits	b_3	b_1	b_5	P	b_6	b_7	Z*	b_2	b_4
Environment	E3	E1	E5	P	E6	E7	E8	E2	E4
CRC CHARACTER (800 CPI)	C_7	C_9	C_5	C_1	C_4	C_3	C_2	C_8	C_6
CRC Polynomial (6250 CPI)	X^6	X^8	X^4	X^0	X^3	X^2	X^1	X^7	X^9

*Zero, or higher order than ASCII

absolute peaks by action of the diodes and capacitor that shift the bias level after each bit is detected. More elaborate circuits have been devised that respond to bits crowded in the recording process, giving them the separation that restores the perfect clock timing of the original binary code.

Computer Formats, Recording Densities and Codes

Nine tracks on $\frac{1}{2}$ in (12.7 mm) tape (Figure 14-2) are most common, with each digital number occupying all nine tracks in parallel. The digital bits are assigned as shown in Table 14-7, such that the most important bits are on the more reliable center tracks, while the least critical bits use edge Tracks 1 and 9. The recorded nine-bit number is called a *character* or *word* or *byte*, which is primarily the number itself or which can represent an alphabetical letter or symbol according to the ASCII code given in Table 14-2.

For bit densities of 200, 556, and 800 CPI the NRZI code (change-on-ones) is standard. At 1600 CPI phase encoding (PE) has been adopted, while 6250 CPI systems use group-coded recording (GCR). The latter codes are more complex than NRZI but less sensitive to skew errors. Table 13-12 describes the NRZI and PE codes, while GCR is described in the following discussion on error detection and correction.

Error Detection and Correction

Information theory and experience teach us that it is more efficient to make an occasional error and to correct it than to overdesign to a point where errors never occur. The high densities achieved in digital recorders would not be possible unless an occasional recording error were allowed. But automatic correction is so effective that literally years may go by without an undetected error's slipping through. To allow error detection the digital data are formatted into words and blocks according to certain rules before being recorded. The playback is continuously checked by digital logic circuits, and whenever a rule is violated the circuits know where the error has occurred and how to possibly correct it.

The concept of error correction is analogous to making an array as in the "crossword puzzle" shown in Figure 14-6*A* where the words must make sense both horizontally and vertically. If the playback is as in Figure 14-6*B*, then Word 3 and Column 2 are indicated to be in error, so we can pinpoint their intersection as having the letter that is wrong. In correcting an alphabetical intersection, there are 26 (English) letters to choose from, but in the digital language of Figure 14-6*C*, there are only ones and zeros, so we know that if a *zero* is wrong the only alternative is a *one*. The digital logic makes this correction and then tests the corrected group and allows it to go through if it passes scrutiny (Figure 14-6*D*). If something else is wrong and cannot be corrected automatically, then the circuits indicate the location of an uncorrectable error. In this elementary example it may be argued that other mistakes could, by coincidence, give intelligible words, which is a finite possibility in all error detection schemes, but which approaches zero with good coding.

In ordinary language only certain combinations of letters form valid words, but in numerical data any combination is permissible. Therefore in numerical data each group is assigned a password-number that allows testing whether that group is valid. The simplest password-number in digital notation is a *parity bit* included with the number and chosen so that the sum of the ones in each recorded number (including the parity bit) will always be odd. Every played-back number is tested and if its parity sum is even, we know there is a mistake. Of course there could be more than one mistake, which would again give odd parity, but at least this simple device will reject all numbers with even parity as being surely in error.

Of the nine tracks on a computer tape, eight are used in parallel to define a character (word, byte, or number) and one is used for the parity bit. Eight bits give a vocabulary of 256 different characters ($2^8 = 256$). Adding the ninth bit would expand the vocabulary to $2^9 = 512$ characters, but half of these do not fulfil parity and hence are invalid. One bit per character (about 11%) is sacrificed for security. We could add two parity bits per character, in which case only 25% of the possible vocabulary would be valid, to give even greater protection. However after adding a single parity bit to a parallel word, it is more advantageous to add bits to the columns of small blocks of words, since this vertical (serial) parity check allows error correction as explained in Figures 14-6*C* and *D*. Instead of a simple parity check in the vertical direction, a more sophisticated mathematical scheme is chosen for generating a *cyclic redundancy*

A. Coded block before recording. Columns and rows spell recognizable words.

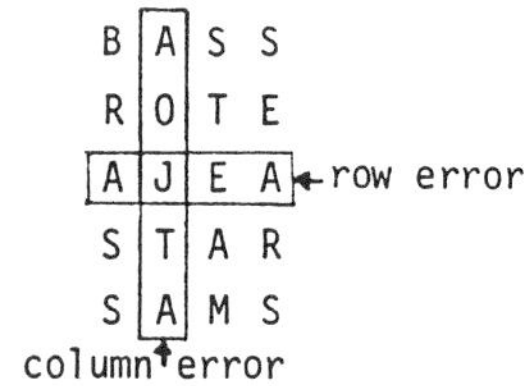

B. Playback block. Row 3 and column 2 don't spell words. Hence their intersection (J) must be in error.

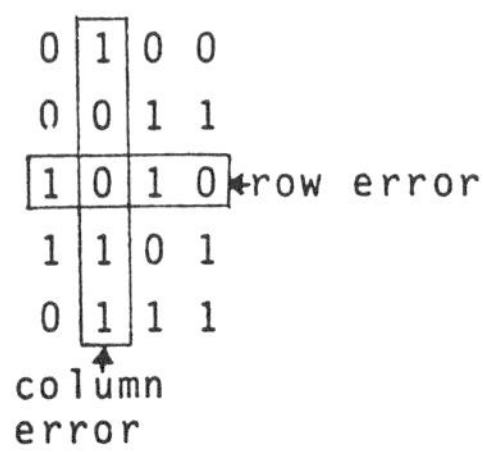

C. A played back digital block has an error detected in row 3 and column 2. Therefore the zero at their intersection is in error.

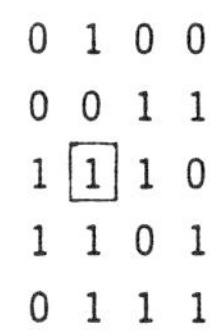

D. Corrected block of digital numbers, where the zero has been replaced by a one, the only alternative.

Data Values (Group Positions: 1234/5678)	Record Values (Group Positions: 12345/678910)
0000	11001
0001	11011
0010	10010
0011	10011
0100	11101
0101	10101
0110	10110
0111	10111
1000	11010
1001	01001
1010	01010
1011	01011
1100	11110
1101	01101
1110	01110
1111	01111

D. A 4-5 code translates four-bit numbers in the left column to corresponding five-bit numbers in the right column. The latter are run-length-limited and allow for error detection in 6250 group-coded recording. (From ANSI × 3.54-1976)

Figure 14-6. Error detection and correction.

check (CRC) word, which serves the same purpose and which is convenient and economical of computer circuitry.

At higher densities, the recorded data becomes more error-prone, requiring the insertion of error-checking bits and synchronizing bits at closer intervals. In group-coded recording (6250 CPI), the information stream prior to recording is subdivided into four bit blocks and each four-bit sequence is coded or translated into a five-bit sequence according to the coding scheme in Figure 14-6*E*, called a *4-5 code*. Expanding the bit content by a $\frac{5}{4}$ ratio is analogous to adding parity bits but has additional advantages. The coding in Figure 14-6*E* is run-length-limited, since no more than two zeros in sequence are allowed in the right-hand column. Besides its error-correcting qualities, such coding reduces the DC and low-frequency content of the playback signal, overcoming problems in response of magnetic heads and of amplifiers and equalizers.

In addition to the 4-5 coding and CRC characters, the 6250 CPI recording standard also includes *error-correcting code* (ECC) characters, auxiliary CRC characters, residual characters, resync bursts, and automatic read amplification (ARA) bursts for synchronizing the read-back amplifiers and adjusting their gain. All these measures to ensure reliability add redundancy, which appears to be wasteful (increases ''overhead''), but they make it possible to compress the recording bits enough to give an impressive increase in overall recording density.

RAPID ACCESS STORAGE

The longest-felt need in computers has been for an economical on-line memory where vast amounts of information are almost instantly accessible. Tape reels are not suitable for quick random access, although they are economical, hold hundreds of megabytes of information, and can be stored on the shelf. Solid state memories and previously popular core memories are just the opposite, with fast access but of high cost and limited information content. Solid state memories are also volatile, their contents being lost in case of a power failure, while magnetic memories are quite permanent.

For intermediate memories, storage systems using magnetic drums and discs have evolved. Early systems used drums, but drums have become obsolete in favor of discs because discs are more compact and less expensive. Also, in some designs the discs themselves can be removed for permanent storage. Disc file storage applications include airline reservations, savings accounts, credit ratings, data banks for literature searching, and ready-access (on-line) computer memories. All of these can be loaded from desired portions of reel-to-reel transports for computation, the results being unloaded to reels again for permanent storage.

Access Time and Storage Capacity. Access time, typically 10 to 70 milliseconds, is increased or reduced by choice of record speed and indexing time of movable heads. Storage capacity may be as low as 0.1 megabytes for a single-face 5.25 in. (133 mm) flexible disc to 500 megabytes for a 14 in. (356 mm) disc pack having 16 recording surfaces and to 10,000 megabytes or more for "strings" of many disc machines. Readout rate may be several megabits per second for each head, multiplied by the number of heads used simultaneously.

Flying Heads. Heads that float on an air film above the recording surface are called *flying heads* (Figure 14-7*A*). Originally developed for drums, flying heads are now used for practically all hard-disc records. Earlier fears that heads would rub against the record have been dispelled; in fact, it is so difficult to break through the protective air stream that the heads must be designed with minimal exposed surfaces, as with the narrow skids in the head shown in Figure 14-7*B*. A skid pressure profile is shown in Figure 14-7*C*. Figure 14-8 shows details of heads and mountings.

Flying heads are tolerant of the slight warpage inevitable in disc records, which soon became the preferred form when the precision of drums was no longer necessary. For a given amount of storage surface, a stack of discs (disc-pack) is more compact, more economical, and more versatile than an equivalent drum.

Magnetic Drums and Evolution of Flying Heads. For early computers, magnetic drums were chosen because tracks on the outer drum surface all ran at the same speed, the heads could be mounted solidly with great exactness, and the drum itself was a rigid structure that maintained very close tolerances when run at high speed in precision bearings. To prevent wear, the heads were held in out-of-contact relation with the drum but as close as the tolerances would allow, since the recording density fell off rapidly as the head spacing was increased.

With the closest tolerances humanly possible to maintain, the heads could not reliably be held closer than about 1 μm from the drum. Even this short distance was far too much to take full advantage of available heads and tapes. The situation was improved greatly by mounting the heads resiliently and pressing them toward the magnetic surface so they could follow slight imperfections in the coating and the mechanism. Under these conditions, the head and surface wear were much less than expected because the heads were separated by a thin layer of air carried along with the high-speed surface of the drum. The only contact occurred during startup or stopping of the drum, and this contact (head crashing) could be avoided by lifting the heads wherever the drum was not up

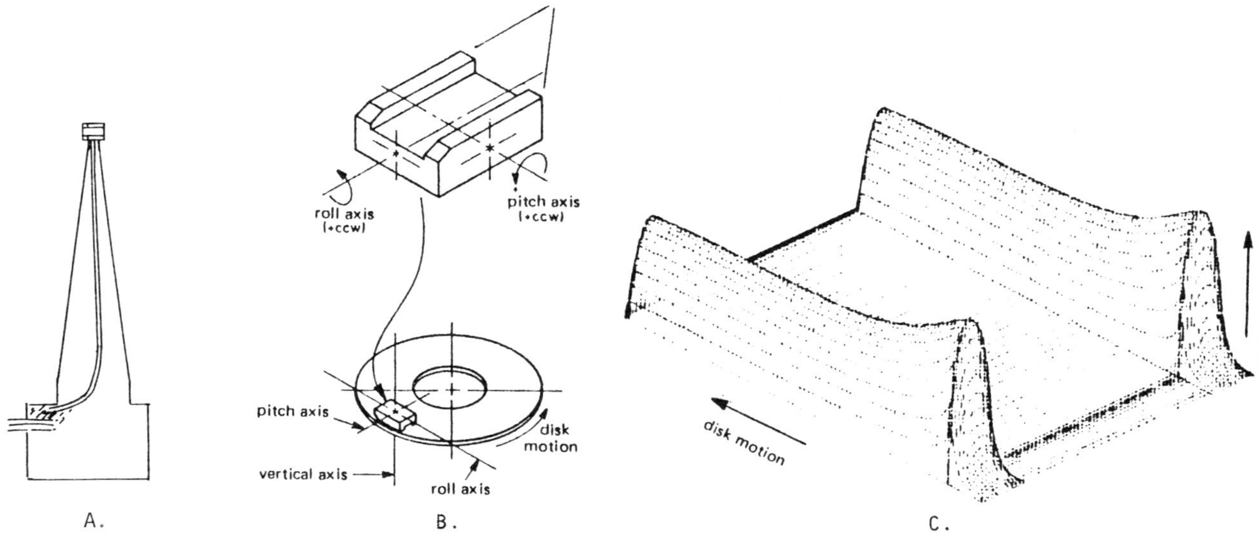

Figure 14-7. Flying heads. A. Head mount with wiring for miniature head, showing skids that ride on air film, separating head from disc. B. Skid (or slider) design and its relation to high-speed disc surface. C. Air pressure profile under skids. (Courtesy IBM Corp.)

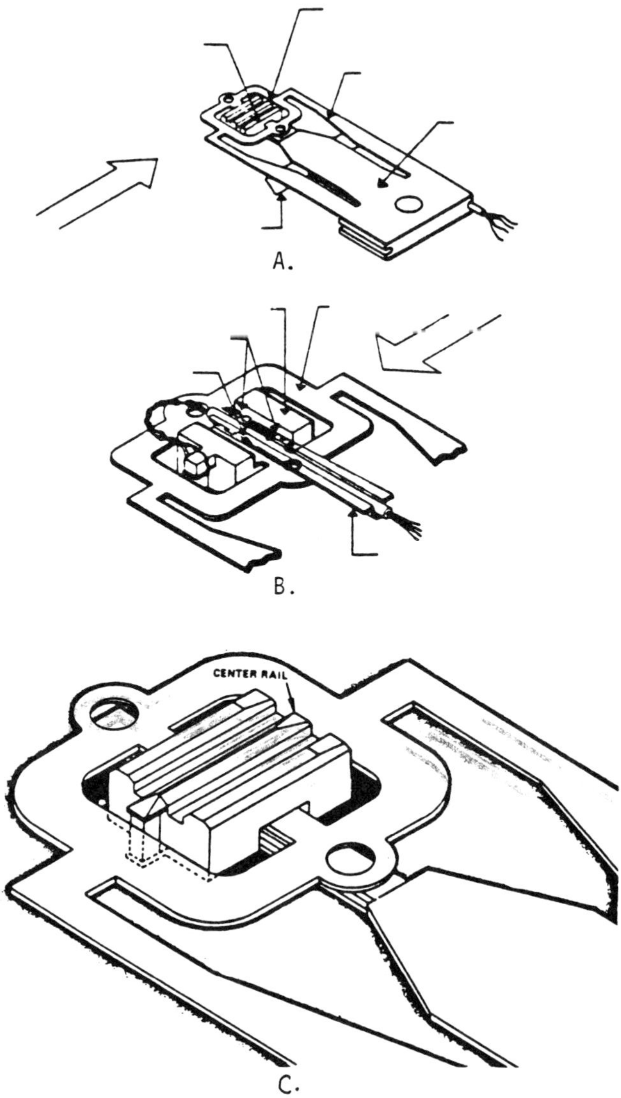

Figure 14-8. Flying heads for Winchester discs. A. Bottom view showing head and three sliders. B. Top view detail of suspension. This "gimbal" mount allows the head to follow slight warpage and irregularities of the disc. Aerodynamic sliders float the head a fixed distance from the disc surface (*IBM J. Res. Develop.* Sept. 1981). C. Magnified bottom view (*Ampex Databits* 1981).

to speed. This evolution to flying heads was transferred with great benefits to magnetic disc technology.

MAGNETIC DISC SYSTEMS

Magnetic discs are the most important computer memory for on-line accessibility to vast banks of data. Cost per data bit on discs, however, is higher than for efficient storage on tape reels, and access time although small, is thousands of times greater than with solid state memories. Table 14-8 gives some terminology used in magnetic disc storage.

In comparing magnetic discs with reel-to-reel tapes, it is apparent that bit capacity is proportional to the area of usable recording surface (square inches or square meters) and to the recording density in bits per unit area. As shown in Table 14-9, the surface area of a 2,400-ft (732 m) roll of computer tape is 14,400 in^2 (9.29m^2), and its gross bit capacity for nine-track 6200 BPI format is therefore 1,600 megabits (about 180 megabytes). For comparison, a 14-in. (356 mm) two-sided magnetic disc has about 150 in^2 (0.097 m^2) of effective area so that 96 such discs are required to give the same area as a 2,400-ft roll of tape. However, the track density of disc tracks has increased to over 900/in, so that two high-density discs can theoretically hold as much data as a 2,400 ft (10.5 in.) roll of tape. Figure 14-9 shows track dimensions of a high-density disc.

Originally intended as a compromise between fast processor memories and slow megabit storage on tape reels, disc storage has developed so effectively that it is the computer's most valuable memory, without which today's computers would be impossible. Large disc memories have a tremendous capacity to 10,000 megabytes or more, any of which can be reached in milliseconds, making possible data banks with "instant" access to information that would fill thousands of books.

Disc systems evolved to capabilities and price ranges that fit large mainframes, mini and microprocessors. Full-size rigid discs first developed for large computers are of aluminum, 14 in (356 mm) in diameter and about $\frac{1}{16}$ in. thick (1.27 or 1.90 mm). These are somewhat clumsy, so a smaller 8 in. (200 or 210 mm) disc became popular for smaller computers. Later a 5.25 in. (133 mm) disc was introduced, still more compact and with bit capacity rivaling the 8 in size, and more recently 3.5 in (89 mm) discs have come into use.

Because rigid discs were at first impossibly expensive for small computers, the cheap and versatile 8 in. floppy disc was invented. The 8-in. floppy disc and its smaller versions have persisted in spite of shorter life and lower reliability than their rigid cousins, and in spite of threatened competition from bubble memories, which have taken a long time to materialize.

Table 14-8. Terminology Used in Magnetic Disc Storage.

Access time Time to recover a desired piece of information from storage
Address Designation of data by its location on the record
Areal density Bits per unit of recorded area
Backup recording A duplicate recording insuring against failure of primary system
Bit density Bits per unit of track length
Block A sequence of related data bits set apart from adjacent blocks
Buried servo Deeply recorded ''non-erasible'' signals for positioning a head on its track (See embedded servo)
Byte A character formed by a designated number of bits, usually eight bits for computer tapes or discs.
Cache A computer memory based on solid state bipolar junction devices
Controller Electronic circuitry that actuates drives and heads for data retrieval and interfacing
Cylinder Tracks on several disc surfaces accessible without repositioning the heads
Diskette A thin disc that rotates within its protective jacket
Disk file A data storage device or memory using (magnetic) disks
Embedded servo Deeply recorded ''non erasible'' servo signals (See buried servo)
Flexible disc Floppy disc (See diskette)
Floppy disc Flexible disc (See diskette)
Flying head A head which utilizes a thin air film to separate it from the record
Flying height Separation of a flying head from the record surface
Front load Removable disc pack that loads into front of drive unit
Hard sectoring Using punched holes to index data on a flexible disc
Index cylinder Cylinder 00 (on a flexible disc) used for storing information about the disc
Initialization Pre-recording a set of signatures for indexing on a data disc
Megabyte A million bytes, or eight million bits. A measure of data storage capacity
MFM Modified frequency modulation. A code used for high density recording See Table 13-12
OEM A subunit that is incorporated into finished units by other manufacturers
Rotary Positioner A tone arm arrangement for head positioning
Rotational delay Average time for data to reach a head after it is placed on a disc track
Sectoring Dividing a disc revolution into sub-units for indexing purposes
Segments Fractional divisions of disc revolution. Sectors
Servo disc (or surface) A disc surface reserved for head positioning information
Settling time Recovery time for stable tracking after a head is moved
Soft sectoring Using recorded information rather than punched holes to locate information on a disc
Top Load Removable disc pack that loads into top of drive unit
TPI Tracks per inch (a measure of track density)
Track density Number of tracks per unit of length
Stepping motor A motor actuated by pulses, each pulse advancing it through a small angle
Streamer A simplified backup recorder, usually on reels of tape
String An interconnected set of data storage units, all accessible through a controller
Voice coil A head position acutator using a moving coil as in a loudspeaker
Winchester A hard disc and head assembly that operates in a sealed chamber

Table 14-9. Effective Surface Area, Bit Density, and Storage Capacity of Various Media.*

Record Medium	Effective Surface Area in^2 (m^2)	Linear Density bits/in (b/mm)	Track Density tr/in (tr/mm)	Areal Density bits/in^2 (b/mm^2)	Gross Capacity megabits	Rated Capacity megabytes
$\frac{1}{2}$ in. tape 2400 ft. (732 m)	14,400 (9.3)	6250 (246)	18 (0.71)	112,500 (174)	1620	180‡
14″ Disc cartridge 2 Surfaces (DEC RL-02)	150.7 (0.097)	3725 (147)	512 (20.2)	1.91×10^6 (2956)	288	10.4
14″ Disc pack 19 Surfaces MEMOREX XV	1431 (0.923)	6038 (238)	384 (15.1)	2.32×10^6 (3596)	3320	300
14″ Winchester 16 Surfaces (DEC RP-07)	1381 (0.89)	11139 (439)	537 (21.1)	5.98×10^6 (9263)	8256	516
8″ Winchester 10 Surfaces (IBM 62-PC)	246e (0.159)e	8534 (336)	457 (18)	3.9×10^6 (6045)	959	70e
5.25″ Winchester 5 Surfaces	21.2e (0.014)e	8290e (326)e	700 (27.6)	5.8×10^6e (9042)e	240	30
3.5″ Winchester 4 Surfaces	9.4e (0.006)e	13902 (547)	804 (31.7)	11.2×10^6 (17360)	205	25.6

(continued on next page)

Table 14-9. *(Continued)*

Record Medium	Effective Surface Area in^2 (m^2)	Linear Density bits/in (b/mm)	Track Density tr/in (tr/mm)	Areal Density bits/in^2 (b/mm^2)	Gross Capacity megabits	Rated Capacity megabytes
8″ flexible disc 2 Surfaces	49.2e (0.032)e	3600/7100 (145/280)	48/150 (1.9/5.9)	$0.17/1.1 \times 10^6$ (1705)	12.8	1.6
5.25″ flexible disc	21.2e (0.014)e	2900/6000 (117/233)	48/96 (1.9/3.8)	$0.14/0.57 \times 10^6$ (884)	4/8	0.5/1.0
$\frac{1}{4}$″ Streamer (3M DC300A)	1350 (0.87)	1600 (63)	16 (0.63)	25,600 (39.7)	34.6	4.3
$\frac{1}{4}$″ Streamer High Capacity (3M DC600HC)	1800 (1.16)	10,000 (393)	64 (2.51)	640,000 (992)	1152	75
Compact Cassette (3M 835)	506 (0.338)	1600 (63)	26.7 (1.05)	42,720 (66.1)	21.6	2.7

*Representative data ‘82
‡9 parallel bits/byte
e = estimated

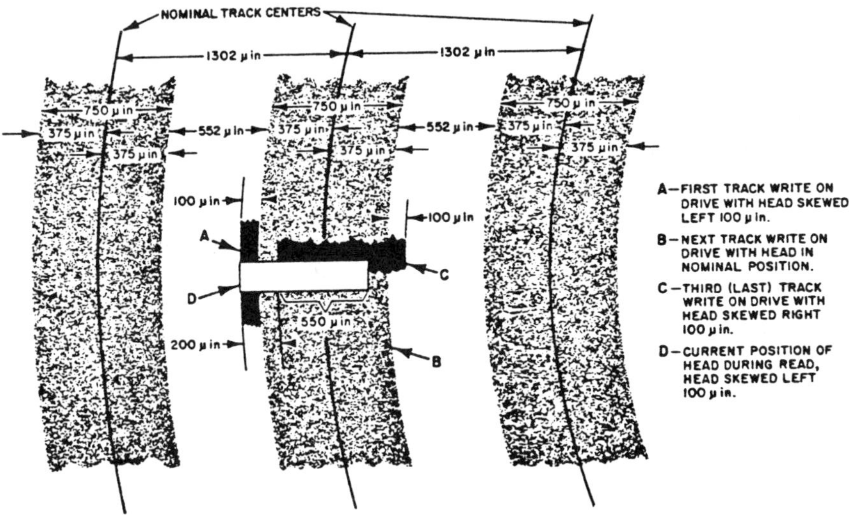

Figure 14-9. Track dimensions and tolerances of a high density disc. At a density of 30.2 tracks/mm (768 tracks/in) a misalignment of 2.5 μm (100μin) as shown can give serious crosstalk from previous non-erased signals (*Ampex Databits* 1981).

Disc storage is so efficient that reel-to-reel systems are used less and less for actual data processing but have become relegated to archival storage and to a new role of disc backup.

Types of Discs

In the evolution of modern disc memories (disc files), there has been an enormous variation in sizes and formats. These eventually boiled down to three major classes, each of which will be discussed more fully later in this chapter:

1. The *Winchester* disc system is a permanent, nonremovable array of rigid discs, heads, and drives sealed in a chamber that protects them from dust and handling.
2. The *disc-pack* is a similar set of discs, which are removable from their heads and drives for storage while another disc-pack is in use.
3. The *floppy disc* (flexible disc, diskette) is a small, thin removable plastic-jacketed disc of limited capacity and durability, but low in cost for both the disc drive and for the disc itself.

Heads

Flying heads for Winchester discs and for disc-packs are shown in Figures 14-7 and 14-8. The head core is very thin, two mm or less in its largest dimension, to reduce its mass. The lower the head mass, the lower the spring pressure needed to keep it against the record surface. Ferrite core heads have been the most popular, with glass or SiO gaps constructed as described in Chapter 4. A head gap of 0.5 μm is theoretically useful for bit densities up to 1000/mm (25,000/in.). Larger gaps may be used for lesser densities with a worthwhile gain of efficiency and reliability. Smaller gaps are required for super-high densities of 1300 bits/mm (33,000 BPI) or more.

After many years of development, film heads were introduced commercially in about 1980 on several disc drives. Figure 14-10 shows the construction of a film head made by techniques of photolithography, sputtering, and etching commonly used in manufacture of integrated circuits. Such heads are producible simultaneously in large numbers by multiple masking and are extremely small

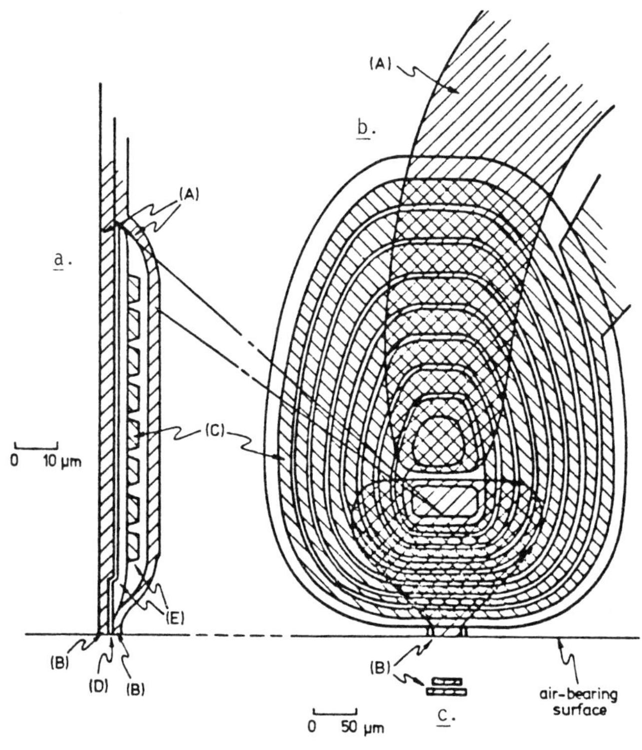

a. A schematic cross section of the film head showing the magnetic layers (A), pole tips (B), conductor turns (C), gap layer (D), and insulation layers (E)

b. A planar view of the film head

c. Pole-tip structure at the air-bearing surface

Figure 14-10. Film head made by integrated circuit techniques. The core and windings are formed by sputtering, plating, and etching through miniature photo masks. (Courtesy IBM Corp.)

in size and light in weight. Their thin magnetic cores retain permeabilities of above 1000 at 100MHz, while ferrites become deficient above 10 MHz. With the head of Figure 14-10, an areal density of 7,800,000 bits/in^2 (12,090/mm^2) is achieved in a commercial disc file.

Head Mounting. In order to "fly" stably a fraction of a micron from the disc surface, the head is mounted, gap face down, in a slider having pair of metal or ceramic skis that glide above the disc surface, riding on a film of air and separating the head gap a nominal distance of 19 microinches (0.5 μm) or less from the record surface. The head-and-skid assembly is held in a gimbal mount stamped from flat springy metal, which allows it to turn freely along two axes for self-alignment with the record while pressing the head toward the record with specified tension. Relative velocity between the head and a 356 mm (14-in.) 3,600 RPM disc is about 140 miles per hour (224 km/hr), as seen in the magnified scale model shown in Figure 14-11.

Head Positioning. One head (or sometimes two) must index to all the tracks on a disc surface, typically 315 to 2,240 different tracks. For this purpose the head is mounted on an arm that moves radially to different positions, or on a pivoted arm that resembles the tone arm of a phonograph. A *radial* movement has the advantage of keeping the head gap always perpendicular to the track and of spacing the tracks apart by equal increments. The *pivoted* arm has a less perfect action but has the advantage of much lower effective inertia because only its outer extremity moves at full speed. The lower inertia allows quickness of indexing which is of prime importance in determining the speed of a computer.

A simple way to drive the head positioner is with a stepping motor (Figure 14-12), the number of tracks moved being proportional to the number of pulses fed to the motor by the electronic circuits. A faster but more complex head positioner is a *voice coil* movement similar to that of a loudspeaker but with greater linear motion and more power. The voice coil does not of itself remain in a fixed position but must be retained by a current controlled by a servomechanism. Figure 14-13 shows a voice coil actuator for pivoted head arms. Great pains are taken to speed up the track seeking operation since indexing wastes valuable computing time. Therefore one finds powerful head-positioning motors energized with "Bang-Bang" algorithms that give maximum acceleration followed by maximum deceleration to the head arm such that it arrives at a new track in a minimum time. Although the mass of moving parts is made as small as possible, it is still substantial for a multihead actuator and the shock and vibration during normal operation cause wonder at how it can hold together at all. The shock does require a settling time before the heads are reliably on

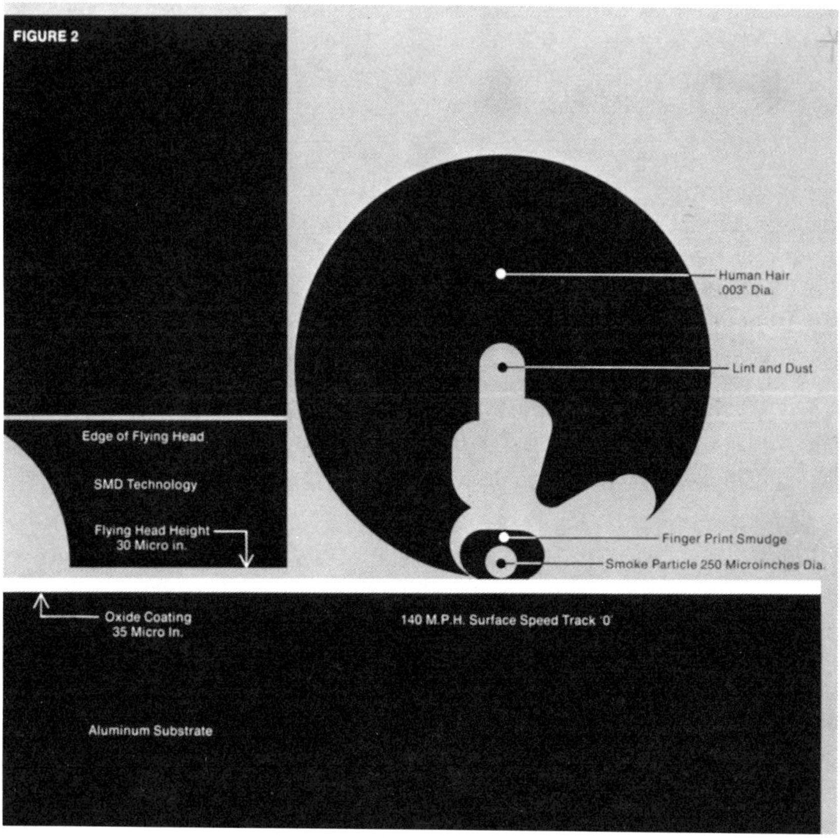

Figure 14-11. Head-to-disc magnified scale model. Dust, smoke, and smudges are enormous compared to the flying-head-height separation. The flying height of Winchester systems is even smaller (19 microinches or 0.5 micron, or less). Note the high surface speed. (Courtesy CDC Corp.)

track, and this also adds to indexing time. Typical index times are listed in Table 14-10, where capabilities of various disc media are compared.

Cylinders. In multiple disc systems, one disc surface is devoted to a set of servo tracks that operate the indexing for all of the discs. All of the heads are mounted on the same (branched) arm and therefore move together, so that all 19 data heads on both sides of an 11-disc-pack, for example, will be indexed to track number 357 at the same time. This action gives rise to the term *cylinder* since all 19 of the simultaneously scanned tracks lie on an imaginary cylinder whose surface passes through track 357 of every disc. *Cylinder* and *track* are

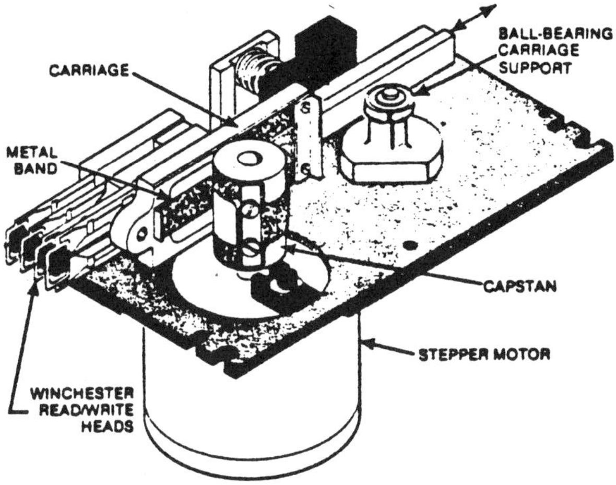

Figure 14-12. Stepper motor drive for straaight line head movement. A split metal band converts rotary to linear motion. The gimbal mounted heads are ganged to index to the same cylnder of the disc pack. (Courtesy Shugart Corp.)

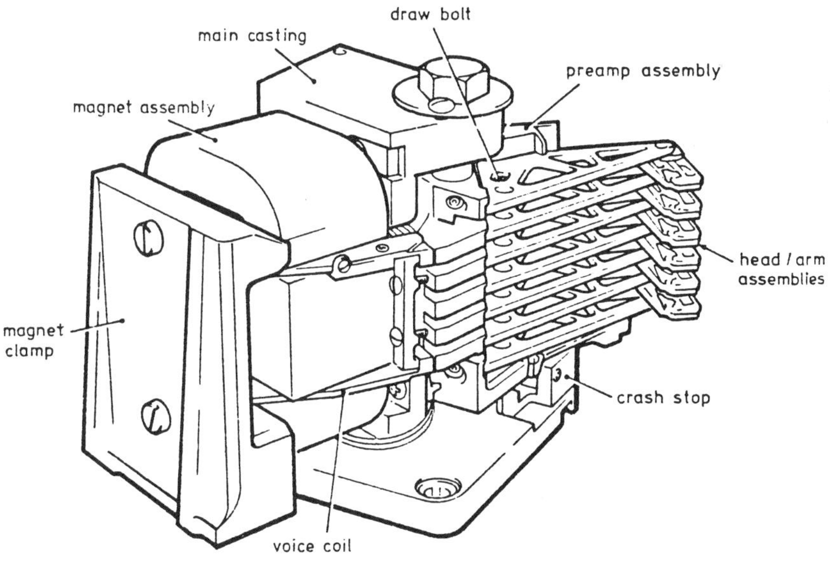

Figure 14-13. Voice coil actuator for pivoted head arms. (Courtesy IBM Corp.)

Table 14-10. Winchester Disc Specifications.

Capacity megabytes	Disc Size	Surfaces	Access Time ms	Data Transfer Rate megabytes/s	Track Density tr/in (tr/mm)	Bit Density bits/in (b/mm)	Tracks or Cyl.	Bytes/Track	Tracks/Cyl	By/Cyl	Remarks
25	14″	2	32	0.885	300	5636					Memorex 601
					(11.8)	(222)					Single disc
677.8	14″	15	22	1.207	935	6350	2240				Memorex 659 OEM
					(36.8)	(250)	(2 heads)				Area density
											= 6 Megabits/in^2
1260	14″		16	3.00			885	47476	15	712140	IBM 3380 with
							(2 heads)				1.4 megabyte fixed-
											head option
14e	3″	2									Shugart single disc
70e	3″	11			457	8534	630				IBM 62-PC
					(18)	(336)					
30	5.25″	5	40	0.62	700		3665	10416			Tandon TM703AT
					(27.6)						
25.6	3.5″	4	80	0.62	804	13902	615	10416			Tandon TM 362
					(31.7)						

often used interchangably, the term cylinder implying that more than one disc surface is indexed.

Embedded or Buried Servo. Some disc systems cannot afford a separate surface devoted only to servo information, and rely on a servo recording that is superimposed on the recorded data itself. The servo recording is not erasable by the ordinary data recording heads and therefore persists regardless of changes and updating of the usual recorded data.

Embedded servo is recorded with a special head having an extra large gap, using longer wavelengths and higher recording flux than is present in the superimposed digital data recording. The longer wavelength, stronger servo recording penetrates deep into the magnetic layer and cannot be erased by the data-recording head, whose smaller gap, weaker field, and shorter wavelengths affect only a very thin layer at the recording surface. The long-wave servo information is at a frequency below that of the digital data recording and is separated from it by filter circuits that direct each kind of information to its proper channel.

Embedded servo in often used for fine adjustment of the head position after the dedicated servo on a separate disc brings the head within a half track of the desired position. This arrangement can reduce the access time from 50 ms to 30 ms for example. Embedded servo can also follow eccentricities of the track during a revolution of the disc. Such tracking becomes important for the closely spaced tracks of high-density recordings. or for discs of poor dimensional stability, or for interchangeability among drives where mechanical tolerances and fits can easily accumulate to the distance between tracks.

Drive Mechanisms

Disc spindles are driven directly on the shaft of a DC or AC motor turning at 3,600 RPM, for example, or can be driven by a pulley and belt from a motor that is offset from the disc spindle. The direct drive is simpler, with less parts that can fail; it can even be sealed into the closed chamber of a Winchester drive. On the other hand, an external separated motor creates less electrical interference pickup in the heads, does not generate heat near the critical magnetic discs, and can be serviced or replaced without disturbing the discs or heads. Also, a standard motor can be chosen for a belt system, while a direct drive usually requires a special motor and/or bearing design.

Fixed-Head Direct-Access Storage. For fastest possible access to data stored on a disc, the movable head arms are dispensed with and a large number of fixed heads are used, each head dedicated to its own single track. Average access time on 3,600 RPM discs is then cut to 8.33 millisecond, compared to about 30 milliseconds for movable head recorders, the shorter time giving proportionally faster computations. The tradeoff is the higher cost of a large number of separate heads; also the number of tracks per disc is reduced and therefore its data storage capacity is less. These disadvantages may be minimized with the introduction of integrated circuit heads, since such heads do not have to be individually manufactured.

Drives are available that combine the usual movable head system with some additional discs on the same spindle that have fixed heads. An alternative arrangement assigns a few tracks on each disc to dedicated fixed heads, while the rest of the tracks are indexed by movable heads. These systems can be programmed so that the most used data is transferred to the fixed head surfaces where it is available quickly, while less used data is on the higher density but slower indexable surfaces.

WINCHESTER SYSTEMS

Winchester discs, the first of the three major classes of discs, are sealed in a chamber together with their own heads. The protected atmosphere eliminates contamination from dust particles, handling of the records, and such, allowing thin recording layers and a characteristic head-flying height of only 19 microinches (0.5 micron) or less and achieving recording densities of about 11,000 bits per inch (433/mm) or more. Since the heads and tracks retain a close relation with one another, a higher track density is possible than would be if the discs had to be removable and interchangeable among different machines.

Full-Sized Winchesters. Figure 14-14 shows a full-sized Winchester disc-stack (14 in or 356 mm in diameter) with heads and motor in cutaway representation of the sealed unit. Figure 14-15 gives dimensions of each of the 11 discs in this stack. Together with the external electronics and drive control, the disc unit makes a free-standing memory unit; however, for more capacity it is customary to combine two independent Winchester disc systems in one cabinet.

The number of discs on a spindle varies from 1 to 12, but not all surfaces are used for data. The topmost and bottommost surfaces of a stack are generally left blank, because they are more exposed to possible injury and because they would need additional arms carrying only a single head instead of two. Another surface is dedicated to servo information that enables the servo to position the heads accurately on the closely spaced tracks, because the disc tracks themselves are used as reference. Alternative methods are not as accurate since they rely on external sources which could easily accumulate positioning errors greater than the track pitch. An 11-disc-stack therefore uses 19 surfaces for data in a typical recorder. Table 14-10 includes performance specifications of various full-sized (14-in.) Winchester systems.

Figure 14-14. Winchester disc system. The sealed chamber gives high reliability of 25000 hours MTBF and highest recording densities. (Courtesy Memorex Corp)

Multidisc memories are used with large mainframe computers. A 14-in. (356 mm) Winchester stack may store 317.5 megabytes. Two such stacks (two spindles) are usually mounted in each cabinet, giving a 635 megabyte capacity. A larger installation includes strings of these units assigned a controller that can access any of them to make available 10,080 megabytes of memory. These

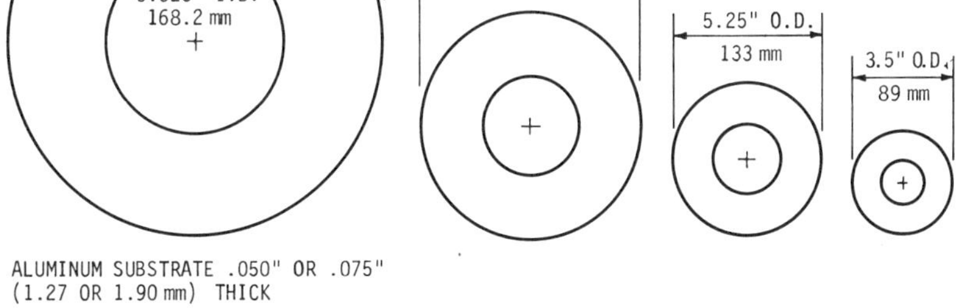

Figure 14-15. Dimensions of rigid discs.

figure as are merely representative. Each year the prominent manufacturers of computer peripherals announce higher densities and higher megabyte capacities for their new models.

Small Winchester Systems. Smaller-diameter Winchester disc systems have been designed to meet the needs of mini and micro computers. Economy and compactness were the original objectives of 8-in-diameter (200 and 210 mm) Winchesters. Competition among suppliers has increased the capacity of 8-in systems so greatly (to about 40 megabytes) that even smaller 5.25-in (133 mm) and $3\frac{1}{2}$-in (89 mm) Winchesters have been introduced with capacities of up to about 10 megabytes.

Advantages. The Winchester system has many advantages over the removable disc pack option:

- Greater reliability (mean time before failure (MTBF) is 25,000 hours or more)
- Lower flying heights of heads, allowing greater bit densities

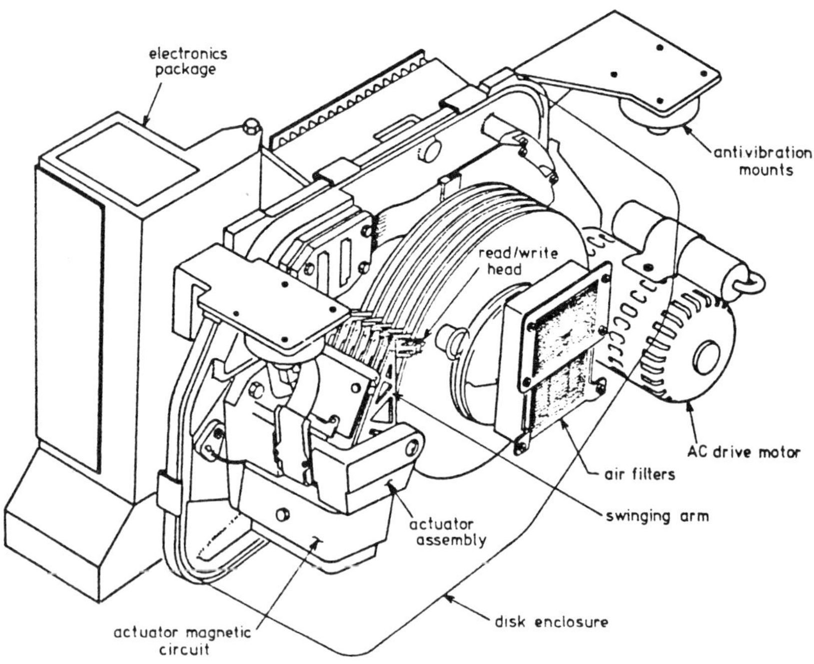

Figure 14-16. Eight-inch Winchester assembly. (Courtesy IBM Corp.)

- Faster access time, averaging 32 ms compared to 53 ms for removable discs
- Less power consumption
- Lower mass heads of less than 0.25 gm
- Simpler and fewer mechanical parts

The Winchester disc systems specified in Table 14-10 include the smaller 8 in (203 mm) and 5.25 in (133 mm) diameter Winchester discs. Figure 14-16 shows an 8-in Winchester disc assembly.

REMOVABLE DISC-PACK SYSTEMS

In many applications such as data bases it is convenient and economical to be able to remove the magnetic discs and store them, while replacing them with a different set of discs full of other data. This is possible with disc-packs, the second type of disc storage available. Loading a disc-pack into the machine makes a large bank of data immediately accessible, in contrast to a Winchester drive, where data would first have to be unloaded electrically into a backup device, then reloaded electrically with new data, a process that takes many minutes or even hours (dump time) and requires extensive storage equipment and supplies.

Figure 14-17. Multiple disc-pack for a top loader. (Courtesy Memorex Corp.)

Table 14-11. Removable Disc Packs and Cartridges.*

Type	No of Discs	Capacity Mybtes	Data Surfaces	Access Time (ms)	Data Transfer Rate KBytes/s	Track Density TPI	Bit Density	Cylinders or Tracks	Remarks
Top-load	1	5	2			200	2200		Memorex III T‡
Top-load	1	10	2			200	6060		Memorex V T
Top-load	5	80	5		1209	370	6060	815	Memorex VIII
Top-load	3	160	5	30	1200	768	6060		Ball BDA 160 V
Top-load	3	80	5		1209	384	6060	815	Memorex XIV
Top-load	6	58	7			400	4040	815	CDC 884-80
Top-load	11	58	20			200	2200		Memorex VI
Top-load	12	100	19			192	4040		Memorex X
Top-load	11	120	20			200	4400		CDC 874-80 32
Top-load	12	300	19			384	6038	815	3M 949/300
Front-load	1	2.5	2			200	1100		Memorex III F‡
Front-load	1	10	2			200	6060		Memorex V F
Front-load	4	69.9	6	25	885	300	5636	696	Memorex 70F

*14-inch diameter discs.

‡ Available with sector hub slots 12, 14, 16, 20, 24, 32, 48,50.

Removable disc-pack units use the same sizes and kinds of discs, and they operate on the same principle as Winchester drives except that the heads retract far enough to allow loading and unloading of the discs. The disc chamber cannot, of course, be permanently sealed; it must be openable for receiving and removing the disc-pack. Since there is less protection from such things as dust, and since there is a greater chance for head misalignment, the removable disc-packs operate at lower bit densities, lower track densities, and higher flying heights than do the heads of Winchester systems.

Two noncompatible systems are currently in use: front loading and top loading. Figures 14-17 and 14-19*B* show a multiple disc-pack for a *top loader*. The disc-pack is stored in a plastic container that resembles the protector for a layer cake. The bottom of this container is unlatched and removed, after which the disc-pack together with the top portion of its container fits into the top of the machine. As indicated in Table 14-11, top loaders are designed for packs with various numbers of discs. The simplest pack is for a single disc, shown in Figure 14-18*A*.

Front loaders take a different pack design (single disks, Figure 14-18*B*; multiple discs, 14-19*A*) with an articulated side door that opens for the heads to enter. The spindle for driving the disc-pack attaches to the bottom, while the pack's container remains in place at all times. Front-loading packages containing single discs are very popular; in addition to the side door they have an air

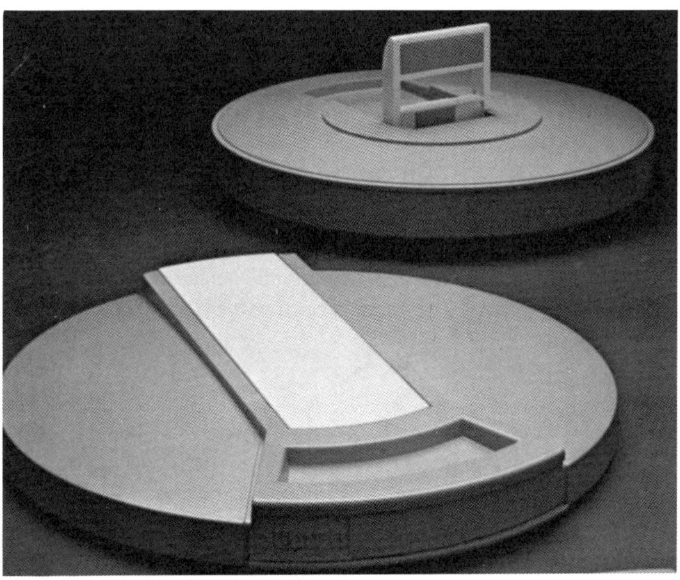

A.

B.

Figure 14-18. Single disc packs. A. Top loading. B. Front loading. (Courtesy 3M Co.)

A.

B.

Figure 14-19. A. Front loading and B. Top loading multiple disc packs. (Courtesy 3M Co.)

door in the bottom to allow ventilation for proper operation. Table 14-11 gives operating specifications of front-loading disc devices.

With the increasing popularity of smaller disc storage devices using 8-in. hard discs, there has been a need for removable discs analogous to 14-in. removables. One 8-in. version combines a removable disc with a Winchester disc on the same spindle.

FLEXIBLE DISCS

Flexible discs, the third class, also known by the more dignified name *diskettes* and by the lesser names *floppy discs* or *floppies*, are the poor man's disc file and have met with tremendous success in the mini and microcomputer fields. The traditional 8-in. size shown in Figure 14-20 is a thin plastic disc with magnetic-coated surfaces, permanently enclosed in a square vinyl (PVC) jacket with

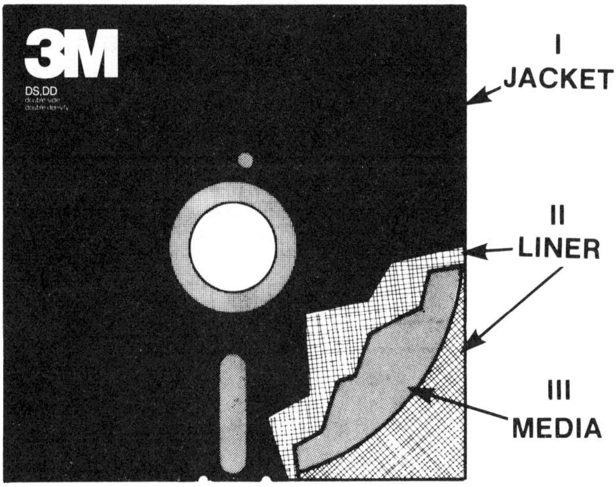

Figure 14-20. Diskette or floppy disc. The thin flexible magnetic coated disc turns within its jacket. Read-write heads contact the recording surface through slots. (Courtesy 3M Co.)

a soft lining of polyester fibers. In operation, the disc turns within its jacket liner, which cleans it by wiping action. The disc jacket has a center hole somewhat larger than the disc opening, which allows the drive mechanism to grip the disc for turning. A radial slit receives the heads, which can move to all tracks on the disc. Additional holes and notches provide for indexing and stress relief. Figure 14-21 shows a head that is used for flexible discs.

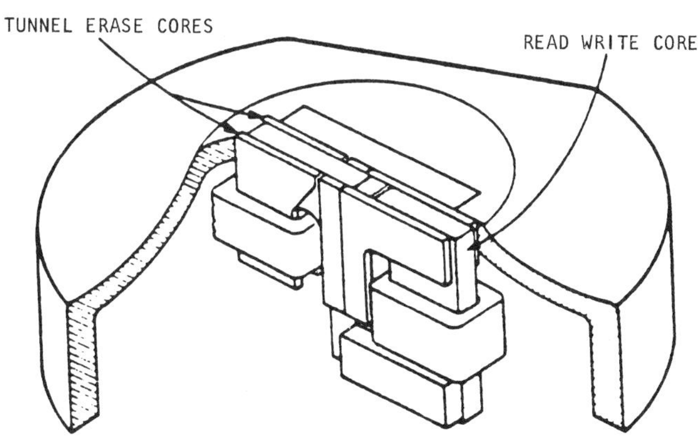

Figure 14-21. Diskette head with tunnel-erase. Auxiliary erase heads on both sides of the main head trim off any side-fringe recordings, old or new (IBM J. Res. Develop., Engh 1981).

Applications. Flexible discs are the most versatile of all disc memories:

- They are a substitute for punched cards: One disc stores more information than several large, heavy trays of cards; disc information is accessible almost instantly; the information can be erased or corrected, while to change a card file a new card must be punched.
- They serve as a rapid access memory of reasonable capacity in small computers where tape or hard disc is not justified.
- They provide permanent storage for computer programs and for data.
- They are handy for temporary off-line storage.
- They store Microcode and diagnostic data.
- They are capable of data processing for small systems.

Offsetting these manifold advantages of flexible discs are their fragility, limited life, and limited recording density. The disc itself is usually made of polyester 0.003 in. (76 μm) thick, about twice as heavy as used for magnetic tapes but still very thin to give it flexibility to make good contact with the head gap. Its surfaces are coated with gamma iron tape-oxide, circumferentially oriented or unoriented and containing a surface lubricant. The head-tape interface is similar to that of a tape recorder, and, therefore, the head and tape are both subject to wear, in contrast to the no-wear flying-head condition of rigid discs.

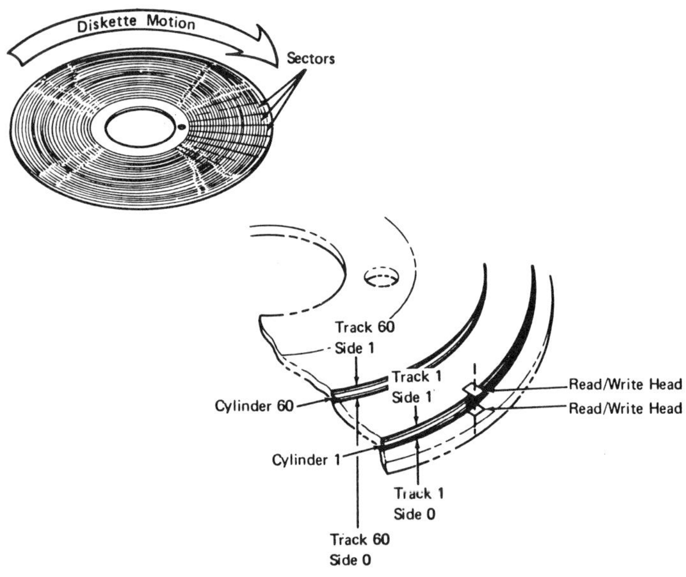

Figure 14-22. Formatting of flexible discs by sectors. (Courtesy IBM Corp.)

Recording density of flexible discs is kept low enough to give reliability considering that the discs are handled frequently and interchanged among machines without protection from such elements as dust. Higher track densities are also limited by dimensional changes in the plastic disc with temperature, humidity, and time. The polyester used for the base is made by a stretching process that renders it anisotropic, so that it expands unevenly and the circular tracks be-

Table 14-12. Diskette Formats.

Sides	Type Density	Sector*	No. of Tracks (Cyls)	No. of Sectors	Bytes/Sector	Total Bytes	Remarks
1	1D	S	77	26	128	246,272	8 in. disc size. In all of these formats the index cylinder (00) has 26 sectors with 128 bytes/sector. Cyl 75/76 reserved for corrected data.
1	1D	S	77	15	256	284,160	
1	1D	S	77	8	512	303,104	
2	1D	S	77	26 × 2	128	492,544	Index cyl on each side has 26 sectors with 128 bytes per sector
2	1D	S	77	15 × 2	256	568,320	
2	2D	S	77	26 × 2	256	985,088	Index cyl on side 0 has 26 sectors @ 128 bytes/sector; on side 1, 26 sectors @ 256 bytes/sector.
2	2D	S	77	15 × 2	512	1,136,640	
2	2D	S	77	8 × 2	1024	1,212,416	Cyl 0 has 256 sectors each with two 128 byte data set labels.
1	1D	S	reversible				Hard sectored 8-in. discs have 32 sector holes
1	2D	S					
1	1D	H					
1	1D	H	reversible				
1	1D	H	reinforced hub				
1	2D	H					
1	2D	H	reinforced hub				
2	2D	H					
1	1D	S	-				5.25 in. mini
1	1D	H	-				5.25 in. mini
1	2D	S	40				5.25 in. mini
1	2D	H	40	10			5.25 in. mini 10 holes
1	2D	H	40	16			5.25 in. mini 16 holes
2	2D	S	40/77				5.25 in. mini
2	2D	H	40/77	10			5.25 in. mini 10 holes
2	2D	H	40/77	16			5.25 in. mini 10 holes

*S = soft
H = hard

come ellipses. Tracking becomes unreliable at track densities above 100 tracks/in. (4/mm) unless servos are used, which adds complexity and cost.

Flexible Disc Formats. The format of data arrangement on flexible discs is well standardized, being divided into wedge-shaped sectors spaced around the circumference of the disc, as shown in Figure 14-22. A hard-sectored disc contains 10, 16, or 32 holes punched through it on a circle that registers with the index hole so that their presence is detected photoelectrically for locating the recorded data. A soft-sectored disc contains a single punched index hole and the sectoring is recorded magnetically on the data tracks. Discs may carry in-

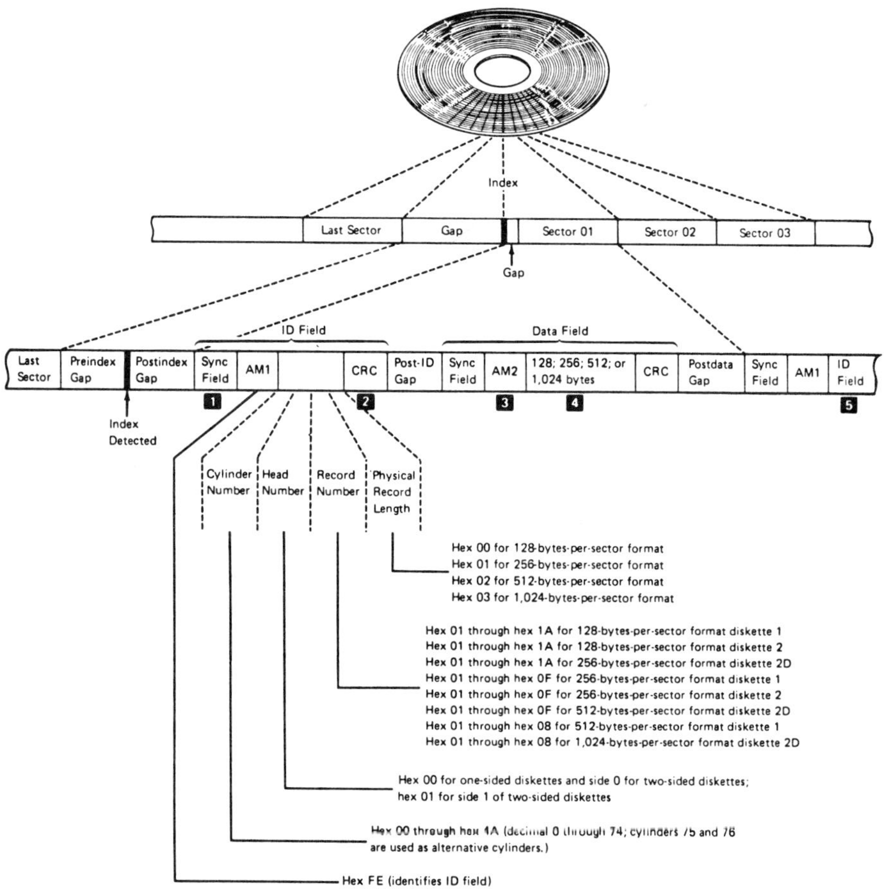

Figure 14-23. Distribution of digital data along tracks and sectors. (Courtesy IBM Corp.)

formation on only one side or on both sides, and at single, double, or quadruple bit-density. Various combinations are listed in Table 14-12.

Figure 14-23 shows how the digital data is distributed along the tracks and sectors. On double sided discs successive information is distributed on a single cylinder (same number top and bottom tracks) to save seeking-time when the head must be changed from track to track.

Initialization. Diskettes are usually recorded with data addresses and label identification on the 00 index cylinder prior to use. This step called *initialization* may be carried out by the manufacturer of the disc. Such data may be altered by *reinitialization* if it is desired to change the sector size or to instruct the head to bypass defective cylinders.

Capacity. A rule of thumb has been that an 8-in (203 mm) diskette holds about one megabyte of information, with possible upgrading by an order of magnitude as technology advances. There are 35 to 96 concentric tracks per inch and, as seen in Table 14-12, a typical format has 77 cylinders numbered from 0 to 76, plus a 00 cylinder for index data.

Flexible discs are also available in a smaller-size 5.25 in (133 mm) diameter *minidisk* and even a 3.5 in (89 mm) diameter *microdisk*. Specifications for these sizes are listed in Table 14-13. Interestingly enough, the byte capacity of smaller flexible discs does not go down in proportion to their area because their lesser dimensional changes (due to expansion, etc.) allow the use of narrower tracks.

Table 14-13. Specifications of Flexible Disc Systems.

Disc Size in. (mm)	Capacity Mbytes	Data Surfaces	Track Density tr/in (tr/mm)	Bit Dens. b/in (b/mm)	Data Transfer Rate Mbits/s	Access Time ms	Latency Time ms	Remarks
8 (203)	1.6	2	48 (1.89)	6536/ 3672 (257/145)	0.5	174	83	Decitek 8302/T
8 (203)	6.3	4 (2 discs)	150 (5.91)	7100 (280)	0.125	100	57/25	Burroughs MD-122
5.25 (133)	0.5	2	48 (1.89)	2768 (109)	0.25	240	100	BASF 6108 OEM Drive
5.25 (133)	0.5/ 1.0	2	48 96 (1.89/3.78)	2961/ 5922 (117/233)	0.125 0.250	92 158	100	Ex-Cell-O Remex RFD 480/960
3.5 (89)	0.438	1	135 (5.31)	7610 (300)	0.5	35	50	SONY Series 35WP

BACKUP STORAGE

Stored data on disc files is so valuable and so vulnerable that special precautions are taken to prevent its loss or unavailability at a critical time. Obviously, all the data could be recorded on duplicate equipment, which would be on-line even if some of the units broke down. Unfortunately, most users can hardly afford one file, and if they had a second they would put it to better use than merely as a duplicator.

Good enough insurance in most cases is to transfer (dump) the information on to standard $\frac{1}{2}$ in. rolls of computer tape. These are put into semipermanent storage if the file data never changes, as for example it if is a data base of published articles or court decisions. On the other hand, if the data are continually updated, as in airline reservations or financial ledgers, the backup storage must be renewed at appropriate intervals and would also be duplicated somewhere as hard copy to be available as a final resort in extreme emergencies. Although magnetic data on discs are quite permanent, they are out of reach while equipment is being repaired. And in spite of all precautions, valuable data can be erased accidentally, or physically damaged or sabotaged.

Backup on Half-Inch Reels. Installations that already have standard computer drives with idle capacity can easily transfer disc data to 10.5-in. reels. Otherwise vacuum capstan drives are more formidable than necessary for backup recording and are wasteful of tape at their low track density of only 14 or 18 tpi. Inexpensive drives have been designed especially for steady running of the tape while absorbing data from a disc file. Such drives, called *streamers*, omit the niceties of rapid starting, stopping, and blocking functions of a regular drive, serving only as efficient backup recorders.

Tape Cartridges. A much more convenient backup system uses a special cartridge designed for the purpose (Figure 14-24). This cartridge holds 23 megabits on 300 ft (91 m) of $\frac{1}{4}$-in. (6.3 mm) tape, or 34.5 megabits on 450 ft (137 m) of thinner tape. Streamer drives for these cartridges provide adequate backup for medium sized disc memories.

Tape Cassettes. The ubiquitous compact cassette of audio fame was adapted for various computer uses early on and serves admirably for backup recording to the extent of its more limited capacity. It holds about 0.1 megabyte of formatted data. Heavy-duty *precision* cassettes containing computer-grade

Figure 14-24. Tape cartridge for streamer backup. (Courtesy 3M Co.)

tapes are recommended for digital recording, these cassettes otherwise being interchangeable with audio cassettes.

Floppy Discs. Where one megabyte of capacity is adequate, a floppy disc makes an excellent backup device, but must have logic and buffer interfaces to store data while the head shifts between tracks and to conform the data to the disk sectoring. Of course, on large systems a great many discs are needed to equal the 180 megabyte capacity of a tape reel, and the dump time and cost of diskettes become prohibitive.

TRENDS IN COMPUTER RECORDING

Cost per megabyte of stored information has decreased exponentally as recording density has increased each year. As always, there is a premium for rapid

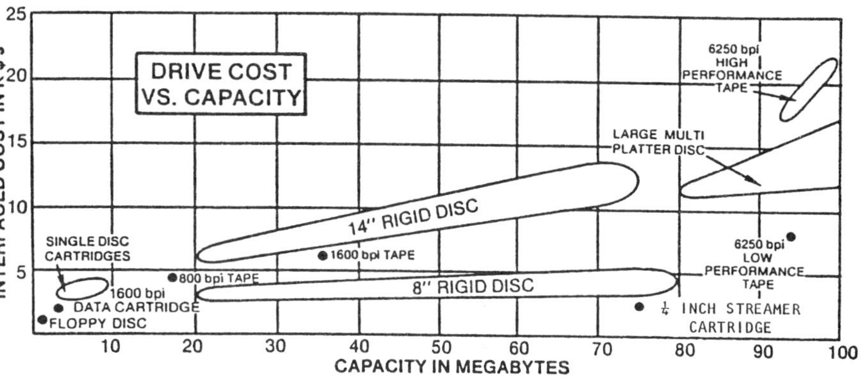

Figure 14-25. Cost comparison of disc and tape storage systems. (Courtesy 3M Co.)

access, so the most economical bit cost is not necessarily the best choice. Figure 14-25 compares the relative costs of available disc and tape systems.

Table 14-14 tabulates diverse improvements in computer recording in the years 1957 to 1981. In disc files using rigid discs, the recording density and overall capacity in megabytes have increased so steadily that for a long time they doubled every three years. Obviously this rate cannot continue forever, but some very promising developments have not yet been explored fully: thin film and metal particle coatings for discs, printed circuit heads, and embedded servo to enable tracking at increased track densities. Long-term stability of new recording materials should be confirmed before replacing the extremely permanent gamma iron oxide. The trend to smaller discs and compact machines continues.

For flexible discs, all the improvements apply that are proposed for rigid discs, especially servo controls that enable the head to follow narrow tracks that become dimensionally distorted by instabilities of the plastic record medium.

Reel-to-reel drives have not taken advantage of increased track densities, which have tripled to 42 tracks per inch in instrumentation recorders and could increase an order of magnitude before they become comparable to disc track densities. A new format would have to be adopted that would conveniently be contained in a large-size cassette package.

The ultimate computer memory would be smaller, would have no moving parts, be more rapidly accessible, would store more megabytes, would be selectively alterable yet permanent, and would not necessarily be magnetic. Such a memory pack was proposed in 1962 without disclosing what should be in it (Camras 1962). Already in use are magnetic bubble memories, solid state memories, laser disc memories, and electrochemical memories (human brain). None of these is as yet ideal, so the frontiers of progress are more challenging than ever before.

Table 14-14. Evolution, Specifications and Capabilities of IBM Computers, 1957 to 1981.

Year of first ship	*1957*	*1961*	*1962*	*1963*	*1966*	*1971*	*1973*	*1976*	*1979*	*1979*	*1981*
Product	350	1405	1301	1311	2314	3330	3340	3350	3310	3370	3380
Recording density											
Areal density(Mb/in.^{2})	0.002	0.009	0.026	0.051	0.22	0.78	1.69	3.07	3.8	7.8	>12
Linear bit density (bpi)	100	220	520	1025	2200	4040	5636	6425	8530	12134	15200
Track density (tpi)	20	40	50	**	100	192	300	478	450	635	>800
Key geometric parameters (microin.)											
Head-to-disk spacing	800	650	250	125	85	50	18	**	13	**	<13
Head gap length	1000	700	500	250	105	100	60	50	40	25	**
Medium thickness	1200	900	543	250	85	50	41	**	25	41	<25
Air bearing & magnetic element											
Bearing type	hydrostatic		hydrodynamic		**	**	**	**	**	**	**
Surface contour	flat	**	cylindrical		**	**	taper flat		**	**	**
Slider material	Al	**	stainless steel		ceramic	**	ferrite	**	**	ceramic	**
Core material	laminated mu-metal		**	**	ferrite	**	**	**	**	film	**
Slider/core bond	epoxy	**	**	**	**	glass	integral	**	**	deposited	**

(*continued on next page*)

Table 14-14. (*Continued*)

Disk											
Diameter (in.)	24	**	**	14	**	**	**	**	8.3	14	**
Substrate thickness (in.)	0.100	**	**	0.050	**	0.075	**	**	**	**	>0.075
Rpm	1200	**	1800	1500	2400	3600	2964	3600	3125	2964	3620
Fixed/removable	fixed	**	**	removable pack		**	module fixed		**	**	**
Data surfaces / spindle	100	**	**	10	20	19	6	15	11	12	15
Actuator											
Access geometry	x-y	**	linear radial		**	**	**	**	rotary	linear	**
Heads	2 heads/actuator		1 head/surface		**	**	2 heads/surface		1 h/s	2 h/s	**
Positioning	motor-clutch		hydraulic	**	**	voice coil motor			**	**	**
Final position	detent	**	**	**	**	servo surface			(+ sector)	servo surface	
Actuators/spindle (max. no.)	3	**	2	1	**	**	**	**	1	2	**
Avg. seek time (ms)	600	**	165	150	60	30	25	**	27	20	16
Read/write electronics											
Data rate (Kbytes/s)	8.8	17.5	68	69	312	806	885	1193	1031	1859	3000
Encoding	NRZI	**	**	**	2f	mfm	**	**	mfm	2.7	**
Detection	ampl	**	**	**	peak	delta	**	**	**	delta clip	
Clocking	2 osc	**	clk trk	osc	vfo	**	**	**	**	**	**

**Same as in preceding column. (Harker, IBM Jour. Res. Dev. 9-81)

Bibliography and References

BOOKS

Athey, Skipwith W. *Magnetic Tape Recording.* Washington: NASA SP-5038, U.S. Gov. Printing Office, 1966.

Begun, S. J. *Magnetic Recording.* New York: Rinehart Books, Inc., 1949.

Bernstein, Julian. *Video Tape Recording.* New York: John F. Rider Publications, 1960.

Bycer, Bernard. *Digital Magnetic Tape Recording.* New York: Hayden Book Co., 1965.

Camras, Marvin, (ed). *Magnetic Tape Recording.* Stroudsburg, Pennsylvania: Van Nostrand Reinhold Co., 1985.

Davies, Gomer. *Magnetic Tape Instrumentation.* New York: McGraw Hill, 1961.

EMI Technology Inc. *Modern Instrumentation Tape Recording.* EMI Technology Inc. 1978.

Eargle, John. *Sound Recording.* New York: Van Nostrand Reinhold Co., 1980.

Hoagland, Albert. *Digital Magnetic Recording.* New York: John Wiley, 1963.

Howard, James A. and L. N. Ferguson. *Magnetic Recording Handbook.* Mountain View, CA: Hewlett Packard Co. AN 89, 1966.

Jorgensen, Finn. *The Complete Handbook of Magnetic Recording.* Blue Ridge Summitt, PA: Tab Books Inc., 1980.

Kalil, Ford (editor). *Magnetic Tape Recording for the Eighties.* NASA Reference Publ. 1075. Washington: U. S. Gov. Printing Office, 1982.

Lowman, Charles E. *Magnetic Recording.* New York: McGraw Hill, 1972.

Mee, C. D. *The Physics of Magnetic Recording.* New York: John Wiley and Sons, 1964.

McKnight, J. G., and Morrison. *Level and Frequency Response Standardization in Magnetic Sound Recording.* Redwood City CA: Ampex Corp., 1968.

McGinty, Gerald P. *Videocassette Recorders, Theory and Servicing.* New York: McGraw-Hill Inc., 1979.

Pear, C. B. *Magnetic Recording in Science and Industry.* New York: Reinhold Publishing Corp., 1967.

Sebastian, L. G. *Digital Magnetic Recording for Computer Applications.* London: Chapman and Hall, 1973.

Snel, D. A. *Magnetic Sound Recording.* Eindhoven: Philips Technical Library, 1958.

Speliotis, Dennis E. and Clark E. Johnson Jr., editors. *Advances in Magnetic Recording.* Annals of the New York Academy of Sciences, vol. 189, 1971.

Spratt, H. G. M. *Magnetic Tape Recording.* London: Temple Press Books Ltd., 1964.

Stewart, W. Earl, *Magnetic Recording Techniques.* McGraw Hill Book Co., NY 1958.

Tall, Joel. *Techniques of Magnetic Recording.* New York: Macmillan & Co., 1958.

Univac Field Eng. Dept. *Fundamentals of Digital Magnetic-Tape Units.* Howard Sams Publ. 20580, 1966.

Winckel, F. *Technik der Magnetspeicher.* Berlin: Springer, 1959.

Chapter 1

Angus, R. 75 years of magnetic recording. *High Fidelity* **23:** 42-50 (1973).

Angus, R. History of magnetic recording. *Audio* **68:** 27-33, 96, 97 (August 1984), and *Audio* **68** 33-39 (September 1984).

Barrett, A. E. and C. J. F. Tweed. Some aspects of magnetic recording and its application to broadcasting. *J. Institution of Elec. Engrs.* **82:** 265-88 (1938).

Begun, S. J. *Magnetic Recording*, New York: Rinehart Books, Inc., 1949.

DeForest, L. The audion detector and amplifier. *Proc. of the IRE* **2:** 15-36 (1914).

Fankhauser, C. K. The telegraphone. *J. Franklin Inst.* **167:** 37-47 (1909).

Ginsburg, C. P. The first VTR: a historical perspective. *Broadcast Engineering* 22-42 (May 1981).

Hickman, C. N. Sound recording on magnetic tape. *Bell Lab. Record* **16:** 165–77 (1937).

Howell, H. A. Magnetic sound recording on coated paper tape. *J. Soc. Mot. Pict. Engrs.* **48:** 36-46, 46-9 (1947).

Larsen, Absalon. *Telegrafonen og den traadlose.* Ingeniorvidenskabelige Skrifter, 1950 Nr. 2, Akademiet for de Tekniske Videnskaber og Dansk Ingeniorforening, I Kommission Hos, Teknisk Forlag, Kopenhavn (1950).

Lübeck, H. Magnetische shallaufzeichnung mit filmen und ringköpfen. *Akustische Zeit.* **2:** 273-95 (1937).

Mallina, R. F. A mirror for the voice. *Bell Lab. Record* **13:** 200-2 (1935).

V. Poulsen, U. S. Patent 661,619 (Nov. 13, 1900).

Poulsen, V. The telegraphone: a magnetic speech recorder. *Electrician* **46:** 208-10 (1900).

Rust, N. M. Marconi-Stille recording and reproducing equipment. *Marconi Review* **46:**1-11 (1934).

Smith, O. Some possible forms of phonograph. *Electrical World* **12:** 116-7 (1888).

Stille, C. Die electromagnetische shallaufzeichnung. *Elec. Zeit.* **51:** 449-51 (1930).

Tiffany, G. S. U. S. Patent 1,142,384 (June 8, 1915).

Chapter 2

Bauer, B. B. and C. D. Mee. A new model for magnetic recording. *IRE Trans. Audio* **AU-9:** 139-45 (1961).

Bertram, N. Long wavelength ac bias recording theory. *IEEE Trans. on Magnetics* **MAG-10:** 1039-49 (1974).

Bertram H. N., M. K. Stafford, and D. R. Mills. The print-through phenomenon. *J. Audio Eng. Soc.* **28:** 690-705 (1980).

G. W. Carpenter and W. L. Carlson, U. S. Patent 1,640,881 (August 30, 1927).

M. Camras, U. S. Patent 2,351,004 (June 13, 1944).

Camras, M. Theoretical response from magnetic wire record. *Proc. IRE* **34:** 597-602 (1945).

Camras, M. and Herr R. Duplicating magnetic tape by contact printing. *Electronics* **22:** 78-83 (1949).

Camras, M. Graphical analysis of linear magnetic recording using high frequency excitation. *Proc. IRE* **37:** 569-73 (1949).

Camras, M. Current Problems in magnetic recording. *Proc. IRE* **50:** 751-761 (1962).

Camras, M. Experiments with electron scanning for magnetic recording and playback of video. *IEEE Transactions on Audio* **AU-11:** 93-96 (1963).

Camras, M. An X-field microgap head for high density magnetic recording. *IEEE Trans. on Audio* **AU-12:** 41-52 (1964).

Daniel, E. D. and P. E. Axon. Accidental printing in magnetic recording. *B.B.C. Quarterly* **5:** 241-56 (1951).

Daniel, E. D. and I. Levine. Experimental and theorctical investigation of the magnetic properties of iron oxide recording tape. *J. Acoust. Soc. Amer.* **32:** 1-15 (1960).

Daniel, E. D and I. Levine. Determination of the recording performance of a tape from its magnetic properties. *J. Acoust. Soc. Amer.* **32:** 258-67 (1960).

Eldridge, D. F. Magnetic recording and reproduction of pulses. *IEEE Trans. Audio* **AU-8:**42-(1960).

Eltgen, Jean-Jacques P. and J. G. Magnenet. Magnetic printer using perpendicular recording. *IEEE Intermag Conference Proc.* (1980).

Fayling, R. E. Edge profile studies of recorded flux-transitions. *IEEE Trans. Mag.* **16:** 1249- (1980).

Geller, S. B. Erasing myths about magnetic media. *Datamation:* 65-70 (March 1976).

Greiner, Joachim. Der untershied swischen vor- und nachecho beim kopiereffekt. *Nachrichtentechnik* 1-7 (1953).

Herr, R., and R. A. Von Behren. Selective erasure of magnetic tape cross talk. *Electronics* **25:** 114-5 (1952).

Holmes, L. C. and D. L. Clark. Supersonic bias for magnetic recording. *Electronics* **18:** 126-36 (1945).

Iwasaki, S. and Y. Nakamura. An analysis for the magnetization mode for high density magnetic recording. *IEEE Trans. Mag.* **MAG-13:** 1272-7 (1977).

Karlqvist, O. Calculation of the magnetic field in the ferromagnetic layer of a magnetic drum. *Trans. Royal Inst. of Tech.* Stockholm Sweden **86:** 3-27 (1954).

Lindholm, D. A. Spacing losses in finite track width reproducing systems. *IEEE Trans. Mag.* **MAG-14:** 55-9 (1978).

Magna-see. Reeves Soundcraft Co., New York, N.Y.

Mayer, L. Curie point writing on magnetic films. *J. Appl. Phys.* **28:** 1003- (1958).

Recorded pattern viewer. Minnesota Mining and Mfg. Co., St. Paul, Minnesota.

McKnight, J. G. The fringing response of magnetic reproducers at long wavelengths. *J. Audio Eng. Soc.* **20:** 100-105 (1972).

Mallinson, J. C. Tutorial review of magnetic recording. *Proc. of IEEE* **64:** 196-208 (1976).

Nagai, K., S. Sasaki and J. Endo. Experimental consideration upon the a-c erasing on the magnetic recording and proposition of the new recording method. *Denki Gakkwai* (J. Institute of Electrical Engineers of Japan) **180:** 445-7 (1938).

Schwantke, G. The magnetic recording process in terms of Preisach representation. *J. Audio Eng. Soc.* **9:** 37-47 (1961).

Stoner, Edmund Clifton. Magnetism, *Encylcopedia Britainnica,* Chicago, 1944.

Sugaya, H. Newly developed hot-pressed ferrite head. *IEEE Trans. on Magnetics* **MAG-4:** 295-301 (1968).

Sugaya, H. and F. Kobayashi. Magnetic tape duplication by contact printing. *Annals of the New York Academy of Sciences.* **189:** 214-238 (1972).

Suits, J. C. A review of some recent work in magnetooptical recording. *IEEE Trans. Mag.* **MAG-8:** 421-5 (1972).

Teer, K. Investigation of the magnetic recording process with step functions. *Philips Res. Rep.* **16:** 469-91 (1961).

Thornley, F. R. M. and J. A. Williams. Switching speeds in magnetic tapes. *IBM J. Res. Develop.*: 576-8 (1974).

Tjaden, D. L. A. and J. Leyten. A 5000:1 scale model of the magnetic recording process. *Philips Tech. Rev.* **25:** 319-29 (1963-1964).

Togami, Yuji. Magnetooptic disk storage. *IEEE Trans. Mag.* **MAG-18:** 1233-7 (1982).

van Herk, A. Side-fringing response of magnetic reproducing heads. *J. Audio Eng. Soc.* **26:** 209-11 (1978).

Wallace, R. L. Reproduction of magnetically recorded signals. *Bell System Tech. Jour.* **30:** 1145-73 (1951).

Westmijze, W. K. Studies on magnetic recording. *Philips Res. Rep.* **8:** 148-157, 161-183, 245-269, 343-366 (1953).

Wetzel, W. W. Review of present status of magnetic recording theory. *Audio Engineering* **31:** 14-17 (Nov. 1947); 12-16, 37 (Dec. 1947); **32:** 26-30, 46-47 (Jan. 1948).

Wooldridge, D. E. Signal and noise levels in magnetic tape recording. *Trans. Amer. Inst. of Elec. Engrs.* **65:** 343-52 (1946).

Woodward, J. G. and E. Della Torre. Magnetic characteristics of recording tapes and the mechanism of the recording process. *J. Audio Eng. Soc.* **7:** 189-95, 222 (1959).

Yeh, N. H. Ferrofluid Bitter patterns on tape. *IEEE Trans Mag.* **MAG-16:** 979-81 (1980).

Zenner, R. E. Magnetic recording with AC bias. *Proc. IRE* **39:** 141-6 (1951).

Chapter 3

Allen, H. R., M. G. Peterson, J. M. Carr and J. S. Sallo, *Proc. 1964 Intermag Conf.* p. 641 (1964).

Ampex Bulletin. Care and handling of magnetic tape. *Ampex bulletin:* (1976).

J. W. Ayers, U. S. Patent 2,133,267 (Oct. 18, 1938).

Bate, G. Thin metallic films for high-density digital recording. *IEEE Trans. Mag.* **MAG-1:** 193-205 (1965).

Bate, G. and J. K. Alstad. A critical review of magnetic recording materials. *IEEE Trans. Magnetics* **MAG-5:** 821-39 (1969).

Bate, G. and L. P. Dunn. On the design of magnets for the orientation of particles in tapes. *IEEE Trans. Mag.* **MAG-16:** (1980).

Beardsley, I. A. Effect of particle orientation on high density recording. *IBM Research Rept.* RJ3552 (**41389**): 1-31 (1982).

Camras, M. Magnetic recording tapes. *Amer. Inst. of Elec. Engrs. Winter Genl. Meeting.* **Paper 48-75:** Jan. 26-30 (1948).

M. Camras. U. S. Patent 2,694,656 (Nov. 16, 1954).

M. Camras. U. S. Patent 3,185,775 (May 25, 1965).

M. Camras. U. S. Patent 3,256,112 (June 14, 1966).

Chen, T. and P. L. Cavallotti. The microstructure and magnetic properties of electroplated cobalt film for perpendicular magnetic recording. *IEEE Trans. Mag.* **MAG-18:** 1125-7 (1982).

Corradi, A. R. Progress in recording materials: a critical review. *IEEE Trans. Mag.* **MAG-14:** 655-60 (1978).

Cuddihy, E. F. Aging of magnetic recording tape. *IEEE Transactions on Magnetics* **MAG-16:** 558-568 (1980).

Dismukes, J. P., D. F. Martin, L. Ekstrom, C. C. Wang and M. D. Coutts. Ferromagnetic chromium dioxide for magnetic tapes. *Ind. Chem. Prod. Res. Develop.* **10:** 319-29 (1971).

Foner, S. Versatile and sensitive vibrating sample magnetometer. *Rev. Sci. Instr.* **30:** 548-57 (1959).

Fukuda, K., Y. Kitahara, F. Maruta and J. Ezaki. Co-V films for perpendicular recording. *IEEE Trans. Mag.* **MAG-18:** 1116-18 (1982).

Fujiwara, T., M. Isshiki, Y. Koike and T. Oguchi. Recording performances of Ba-ferrite coated perpendicular magnetic tapes. *IEEE Trans. Mag.* **MAG-18:** 1200-02 (1982).

Geller, S. B. Archival data storage. *Datamation:* 72, 75, 76, 80 (October 1974).

Geller, S. B. Care and handling of computer magnetic storage media. *NBS Special Publication 500-101:* U.S. Dep't of Commerce, National Bureau of Standards, Washington D.C. 20234 (June 1983).

Grove, E. Chemical analysis of iron oxide tape coatings. *IIT Res. Inst.* Private Communication (1980).

Howell, H. A. Magnetic sound recording on coated paper tape. *J. Soc. Mot. Pict. Engrs.* **48:** 36-46, 46-9 (1947).

Iwasaki, S. and K. Nagai. Some considerations on the design of high output magnetic tape for short wave-length recording. *Research Inst. for Elec. Communication*, Tohoku University, Sendai, Japan (1962).

Iwasaki, S. and Y. Nakamura. An analysis for the magnetization mode for high density magnetic recording. *IEEE Trans. Mag.* **MAG-13:** 1272-77 (1977).

Iwasaki, S., K. Ouchi and N. Honda. Studies of the perpendicular magnetization mode in CoCr sputtered films. *IEEE Trans. Mag.* **MAG-16:** 1111-13 (1980).

Iwasaki, S. Perpendicular magnetic recording. *IEEE Trans. Mag.* **MAG-16:** 71-6 (1980).

J. Y. Johnson and G. W. Johnson, Improvements in the manufacure and production of sound record carriers. British Patent No. 466023 (Nov. 18, 1936).

Judge, J. S., J. R. Morrison, D. E. Speliotis, and G. Bate, Magnetic properties and corrosion behavior of thin electroless Co-P deposits, *J. Electrochem Soc.* **112:** 681-684 (1965).

Kubo, O., T. Ido and H. Yokoyama. Properties of Ba ferrite particles for perpendicular magnetic recording media. *IEEE Trans. Mag.* **MAG-18:** 1122-4 (1982).

Kubota, Y., H. Morita, T. Tokuoka, and Y. Imaoka, A new approach to the synthesis of cobalt modified iron oxide. *IEEE Trans. Mag.* **MAG-15:** 1558-60 (1979).

Minn. Mining. The handling and storage of magnetic recording tape. *Sound Talk* **3:** (1970a).

Minn. Mining. A discussion of the physical characteristics of Scotch brand #900 wideband instrumentation tape. *Minnesota Mining* (1970b).

Ohta, Y., K. Nakamura, A. Itoh and C. Hayashi. Recording and reproducing properties of magnetic tape by high incidence angle nucleation method. *IEEE Trans. Mag.* **MAG-18:** 1209-11 (1982).

Ouchi, K. and S. Iwasaki. Perpendicular magnetization structure of Co-Cr films. *IEEE Trans. Mag.* **MAG-18:** 1110-12 (1982).

Ransom, L. D. and V. Zentner. Magnetic films of electroless cobalt-phosphorous with uniaxial anisotropy, *J. Electrochem. Soc.* **111:** 1423-1525 (1964).

Sischka, F. J. Development of an advanced γFe_2O_3 cassette tape coating. *Audio Eng. Soc.* (1973).

Speliotis, D. E. and J. R. Morrison. Recent advances in magnetic recording materials. *Tr. New York Acad. Sci.* **28:** 1005-19 (1966).

Suganuma, Y., H. Tanaka, M. Yanagisawa, F. Goto and S. Hatano. Production process and high density recording characteristics of plated discs. *IEEE Trans. Mag.* **MAG-18:** 1215-20 (1982).

Swoboda, T. J., P. Arthur Jr., N. L. Cox, J. N. Ingraham, A. L. Oppegard and M. L. Sadler. Synthesis and properties of ferromagnetic chromium oxide. *Central Res. Dept. E. I. Dupont de Nemours, Wilmington, Del:* Contribution No. 593 (1961).

Tobin, H. G. and C. Powell. Surface lubrication of magnetic tape. *Audio Eng. Soc.* 55th Convention: **Preprint No. 1189 (N-1)** Oct. 29-Nov. 1 (1976).

Uehori, T., A. Hosaka, Y. Tohuoka, T. Izumi and Y. Imaoka. Magnetic properties of Iron-cobalt alloy particles for magnetic recording media. *IEEE Trans. Mag.* **MAG 14:** 852-4 (1978).

Watanabe, A., T. Uehori, S. Saitoh and Y. Imaoka. Fine metal particles having super high coercivity. *IEEE Trans. Mag.* **MAG-17:** 1455-8 (1981).

Wielinga, T., J. C. Lodder and J. Worst. Characteristics of RF-sputtered CoCr Films. *IEEE Trans. Mag.* **MAG-18:** 1107-9 (1982).

Wohlfarth, E. P. A review of the problem of fine particle interactions with special reference to magnetic recording. *J. Appl Phys.* **35:** 783-90 (1964).

Wright, C. D. and H. G. Tobin. Surface lubrication of magnetic tape. *Int. Telemetering Conf. Proc.* Instrum. Soc. Amer., Pittsburg, Pa. 336-44 (1976).

Chapter 4

Adams, E. Recent developments in soft magnetic alloys. *Jour. Applied Phys.* **33** Supp: 1214-20 (1962)

Anderson, M. E., J. A Granath and D. C. Reukauf. A wideband magnetic recording system. *IRE Wescon Convention Record* **5:** 40-5 (1960).

Arnold, R. R., L. J. Ananka, and S. V. Marsov. An investigation of the magnetic head-magnetic tape contact. *Jour. Audio Eng. Soc.* **20:** 470-4 (1972).

Bendson, S. A. New wear-resistant permalloy material for magnetic recording-head application. *Jour. Audio Eng. Soc.* **24:** 562-7 (1976).

M. Camras, U.S. Patent 2,351,003 (June 13, 1944).

M. Camras, U.S. Patent 3,582,572 (June 1, 1971).

Eckstein, H., and T. L. Gilbert. Field of a magnetic head. *Armour Research Foundation* **Internal Report:** (1947).

Fritzsch, K. Long-wavelength response of magnetic reproducing heads. *IEEE Tr. on Audio & Electroac.* **AU-16:** 486-94 (1968).

Ketori, T., N. Suzuki, I. Ohshima, and M. Yamada. A high performance combination head for cassette metal tapes. *Hitachi*, Japan (1979).

Kornei, O. A magnetic head for the megacycle range. *IRE Convention Record* **Part 5:** 141-149 (1956).

Lindholm, D. A. Long-wavelength response of magnetic heads with beveled outer edges. *Jour. Audio Eng. Soc.* **27:** 542-7 (1979).

McKnight, J. G. Erasure of Magnetic Tape. *Jour. Audio Eng. Soc.* **11:** 223-8 (1963).

McKnight, J. G. Tape flux measurement theory and verification. *Jour. Audio Eng. Soc.* **18:** 250-62 (1970).

McKnight, J. G. The permeability of laminations for magnetic recording and reproducing heads. *57th Conv. Audio Eng. Soc.* **Preprint 1265** (1977).

McKnight J. G. How the magnetic characteristics of a magnetic tape head are affected by gap length and a conducting spacer, *Jour. Audio Eng. Soc.* **26:** 930-4 (1978).

McKnight, J. G. Magnetic design theory for tape recorder heads. *Jour. Audio Eng. Soc.* **27:** 106-15 (1979).

Meguro, T., Y. Sawada, Y. Ogata, T. Miyazaki. High magnetic induction, high permeability Co base amorphous alloys for magnetic recording heads. *Proc. 4th Int. Conf. on Rapidly Quenched Metals*, Sendai (1981).

Mersing, R. Wear characteristics of ultrasil® vs alfesil instrumentation heads. *THIC Steering Committee Meeting*, Laurel, MD (Oct. 4, 1982).

Nortronics. Design digest, digital recording. *Nortronics Co.*, Minneapolis MN. 55427 (1976).

Otto, R. The effect of conducting gap spacers on the impedance of magnetic heads. *Jour. Audio Eng. Soc.* **20:** 258-64 (1972).

Peterson, E. and L. R. Wrathall. Eddy currents in composite laminations. *Jour. Audio Eng. Soc.* **25:** 1027-32 (1977).

Sugaya, H. Newly developed hot-pressed ferrite head. *IEEE Trans. on Magnetics* **MAG-4:** 295-301 (1968).

Tsukagoshi, T., Ito, H., Ogasawara, K., Namiki, T., Yasuda, S., Masumoto, Y., Ota, H., and Yamamoto, T. Ribbon sendust magnetic tape heads. *Jour. Audio Eng. Soc.* **29:** 867-72 (1981).

Unger, E. and K. Fritzch. Calculation of the stray reluctance of gaps in magnetic circuits. *Jour. Audio Eng. Soc.*, **18:** 641-3 (1970).

Yasuda, I., Yoshisato, Y., Kawai, Y., Koyama, K., and Yazaki, T. Newly developed sendust video head for high coercive tape. *IEEE Tr. on Magnetics* **MAG-16:** 870-872 (1980).

Chapter 5

Bozarth, Richard M. *Ferromagnetism.* New York: D. Van Nostrand Co., 1951.

M. Camras, U.S. Patent 2,351,003 (June 13, 1944).

M. Camras, U.S. Patent 2,479,308 (August 16, 1949).

Camras, M. New magnetic recording head. *Soc. of Mot. Pict. and Telev. Eng. Jour.* **58:** 61-6 (1952).

Camras, M., Some experiments with magnetic playback using Hall-effect sensitive elements, *IRE Transactions on Audio,* **AU-10:** 84-8 (1962).

Camras, M. An electron cloud head for reproduction of tape recordings. *J. Audio Eng. Soc.* **11:** 354-8 (1963a).

Camras, M. Experiments with electron scanning for magnetic recording and playback of video. *IEEE Transactions on Audio* **AU-11:** 93-96 (1963b).

Camras, M. An X-field micro gap head for high density magnetic recording. *IEEE Trans. Audio,* **AU-12:** 41-51 (1964).

Camras, M. Magnetic scan head for high frequency recording. *IEEE Trans. Mag.* **MAG-3:** 96-100 (1967).

Camras, M. and R. Sears. Outside-coil magnetic head improves high frequency recording. *Electronics* 89-91 (1961).

Daniels, H. L. Boundary displacement magnetic recording. *Electronics* **25:** 116-120, (1952).

deNiet, E. and R. Vreeken. A magnetoresistive head with magnetic feedback. *IEEE Intermag Conference* (July 1979).

Hunt, Robert P. A magnetoresistive readout transducer. *IEEE Transactions on Magnetics.* **MAG-7:** 150-154 (1971).

Iwasaki, S. and Y. Nakamura. The magnetic field distribution of a perpendicular recording head. *IEEE Trans. Mag.* **MAG-14:** 436-8 (1978).

Iwasaki, S., Y. Nakamura and K. Ouchi. Perpendicular magnetic recording with a composite anisotropy film *IEEE Trans. Mag.* **MAG-15:** 1456-8 (1979).

Kotera, N. J. Shigeta, T. Oi, M. Nakashima and K. Sato. Transverse l/f noise in InSb thin films and the signal-to-noise ratio of related Hall elements. *J. Appl. Phys.* **49:** 5990-96 (1978).

Kuijk, K. E., W. J. van Gestel and F. W. Gorter. The barber pole, a linear magnetoresistive head. *IEEE Trans. Mag.* **MAG-11:** 1215-17 (1975).

Lopez, J. L. and J. C. Licini. Characterization of magnetic transistors. *IEEE and MMM Joint Conf.;* Montreal (1982).

Schurch, E. C. and Schleif, F. R. A magnetic tape oscillograph for power system analysis. *Electrical Engineering* **70:** 999-1007 (1951).

Schwarz, T. A. and S. K. Decker. Comparison of calculated and actual density responses of a magnetoresistive head. *IEEE Trans. Mag.* **MAG-15:**1622-4 (1979).

Skellett, A. M., L. E. Leveridge and J. W. Gratian. Electron-beam head for magnetic tape playback. *Electronics* **26:** 168-171 (1953).

Thompson, W. Effects of magnetization on electrical conductivity of Ni and Fe. *Proc. Roy. Soc.* (London) **8:** 546-550 (1857).

Wiegand, D. E. and R. E. Zenner. A turn-in-gap erase head for magnetic recorders providing intense high-frequency fields. *AIEE Trans.* **67:** 507-10 (1948).

Wiegand, D. E., Magnetic modulator playback head. *Armour Research Foundation* Bulletin No. 74, Aug. 1952; Bulletin No. 77, Dec. 1952.

Zenner, R. E. Type 407 bump-buck playback head. *Armour Research Foundation* Bulletin No. 37 (1947).

Chapter 6

AES (Audio Engineering Society). Proposed recommended practice for 760 mm/s (30 in/s). *J. Aud. Eng. Soc.* **19:** 68 (1971).

DIN (Deutsche Industrie Normen) 45 513 (1965).

Doyle, N. P. Sources of noise in magnetic tape reproduction and their minimization. *Fairchild Semiconductor Co.* Consumer Applications (1967).

Graveraux, D. W., A. J. Gust and B. B. Bauer. The dynamic range of disc and tape records. *J. Audio Eng. Soc.* **18:** 530-5 (1970).

IEC (International Electrotechnical Commission) Publication 94, 1968. (Usual European Standard) (1968).

C. D. Mochtenbacher and F. C. Fitchen. *Low-Noise Electronic Design.* New York: Wiley, 1973.

MRL Calibration Tapes. Catalog 1982-01. *Magnetic Reference Laboratory.* Mountain View, CA (1982).

NAB (National Association of Broadcasters). Standard for reel- to-reel recording and reproducing, 1965). (Usual USA standard) (1965).

Sivian, L. J., H. K. Dunn and S. D. White. Absolute amplitudes and spectra of certain musical instruments and orchestra. *IRE Transactions on Audio* **AU-7:** 47 75 (1959).

Chapter 7

Balducci, D. The measurement of instantaneous speed stability of instrumentation-type, magnetic-tape transport systems. *Audio Eng. Soc. 15th Ann. Conv.* New York (Oct. 18, 1963).

Comerci, F. Perceptibility of flutter in speech and music. *J. Soc. Mot. Pict. Engrs.* **64:** 117-22 (1955).

IEEE Standard: Method for measurement of weighted peak flutter. *IEEE Trans. Audio Electroac.* **AU-20:** 81-88 (1971).

McKnight, J. G. Mechanical damping in tape transports. *J. Audio Eng. Soc.* **11:** 140-6 (1964).

McKnight, J. G. Speed, pitch and timing errors in tape recording and reproducing. *J. Audio Eng. Soc.* **16:** 266-74 (1968).

McKnight, J. G. Development of a standard measurement to predict subjective flutter. *IEEE Tr. on Audio and Electroac.* **AU-20:** 75-78 (1972).

Sear, A. W. Wire recorder wow. *J. Acoust. Soc. Amer.* **19:** 172-8 (1947).

Sink, R. L. and J. G. Frayne. Design tape recorders for minimum size, weight and power. *Electronic Design* (Nov. 23, 1960).

Stott, A. and P. E. Axon. Subjective discrimination of pitch and amplitude fluctuations in recording systems. *Proc Institution of Elec. Engrs.* **102:** 643-56 (1955).

Strickland, J. C. A new tape transport design. *Audio Eng. Soc. 40th Conv.* Preprint No. 804 (Apr. 27-30, 1971).

Werner, P. H. The mechanical properties of various magnetic tapes and their influence on recording quality. *Technische Mitteilungen PTT* **30:** 173-80 (1952).

Wolf, W. Electromechanical analogs of the filter systems used in sound recording transports. *IEEE Trans. Audio Electroac.* **AU-14:** 66-85 (1966).

Chapter 8

Acosta, W. L. and J. P. Ramult. High density digital recorders. *Countermeasures*: 34-37 (1977).

Brophy, J. J. High density magnetic recording. *IRE Trans. Audio* **8:** 58-61 (1960).

Camras. M. Information storage density. *IEEE Spectrum:* 98-105 (July 1965).

Chapman, D. W. Theoretical limit on digital magnetic recording density. *Proc IEEE* **51:** 394 (1963).

Daniel, E. D. Tape noise in audio recording. *J. Audio Eng. Soc.* **20:** 92-9 (1972).

Eldridge, D. F. and A. Baaba. The effects of track width in magnetic recording. *IRE Trans. on Audio* **AU-9:** 10-15 (1961).

Eldridge, D. R. A special application of information theory to recording systems. *IEEE Trans. Audio* **AU-11:** 3-6 (1963).

Iwasaki, S., K. Ouchi, and N. Honda. Studies on the perpendicular magnetization mode in CoCr sputtered films. *IEEE Trans. Mag.* **MAG-16:** 1111–13 (1980).

Lemke, J. U. Ultra-high density recording with new heads and tapes. *IEEE Trans. Mag.* **MAG-15:** 1561-63 (1979).

Mallinson, J. C. Maximum signal-to-noise ratio of a tape recorder. *IEEE Trans. Mag.* **MAG-5:** 182-6 (1969).

Mallinson, J. C. On extremely high density digital recording. *IEEE Trans. Mag.* **MAG-10:** 368-73 (1974).

Mallinson, J. C. A unified view of high density digital recording theory. *IEEE Trans. Mag.* MAG-11: 1166-9 (1975).

Mallinson J. C. Tutorial review of magnetic recording. *Proc. of IEEE* **64:** 196–208 (1976).

McKnight, J. G. A comparison of several methods of measuring noise in magnetic recorders for audio applications. *IRE Trans. on Audio* **AU-8:** 39-42 (1960).

Mullin, J. T. Advanced tape mastering system: electronic features. *IEEE Trans. Audio* **AU-13:** 31-5 (1965).

Oliver, B. M., J. R. Pierce, and C. E. Shannon. The philosophy of PCM. *IRE Proceedings* **36:** 1324-31 (1948).

Ragle, H. V., and P. Smaller. An investigation of high-frequency bias-induced tape noise. *IEEE Trans. Mag.* **MAG- 1:** 105-110 (1963).

Shannon, C. E. A mathematical theory of communication. *Bell System Tech J.* **27:** 623-56 (1948).

Smaller, P. Reproduce system noise in wide-band magnetic recording systems. *IEEE Trans. Mag.* **MAG-1:** 357-63 (1965).

Smaller, P. An experimental study of short wavelength recording phenomena. *IEEE Trans. Mag.* **MAG-2:** 242-6 (1966).

Su, J. L. and M. L. Williams. Noise in disc data recording media. *IBM J. Res. Develop.* **18:** 570-5 (1974).

Wooldridge, D. E. Signal and noise levels in magnetic tape recording. *Trans. Amer. Inst. of Elec. Engrs.* **65:** 343-52 (1946).

Chapter 9

Busby, S. B. Jr. Digital audio recording on videotape. *J. Soc. Mot. Pict. Telev. Engrs.* **89:** 508-12 (1980).

Camras, M. Magnetic sound for motion pictures. *Soc. of Mot. Pict. Engrs. Jour.* **48:** 14-28 (1947).

Camras, M. A stereophonic magnetic recorder. *Proc. IRE* **37:** 442-7 (1949).

Camras, M. Approach to recreating a sound field. *J. Acoust. Soc. Amer.* **43:** 1425-31 (1968).

Camras, M. State of the audiotape art. *IEEE Spectrum* **14:** 28-35 (Oct. 1977).

Dolby, R. M. An audio noise reduction system. *Audio Eng. Soc. Convention* Preprint No. 543: (Oct. 1967).

Eargle, John. *Sound Recording*, Second Ed. New York: Van Nostrand Reinhold Co., 1980.

Hare, D. G. C. and W. D. Fling. Picture synchronous magnetic tape recording. *J. Soc. Mot. Pict. Telev. Engrs.* **54:** 554-66 (1950).

Harshberger, R. P. A closed loop transport servo system for pressure roller-less drive. *J. Audio Eng. Soc.* (1977).

Heaslett, A. M. Proposal of 48 kHz sampling rate for digital audio. *Ampex* (Sept. 17, 1982).

Lindsay, H. and M. Stolaroff. Magnetic tape recorder of broadcast quality. *Audio Engineering.* **32:** 13-16 (1948).

Narma, R. and W. M. Fujii. Requirements in performance and reliability with design solutions for a master tape recorder. *Audio Eng. Soc.* 15th Annual Convention, New York, (Oct. 14-18, 1963).

Selsted, W. T. Synchronous recording on $\frac{1}{4}$ inch magnetic tape. *J. Soc. Mot. Pict. Telev. Engrs.* **55:** 279-84 (1950).

Warnock, R. B. Longitudinal digital recording of audio. *Audio Eng. Soc.* Paper No. 1169 (L-3) 55th Convention (Oct. 29- Nov. 1, 1976).

Weisser, A. The digital recording of sound in broadcasting. *J. Soc. Mot. Pict. Telev. Engrs.* **89:** 520-4 (1980).

Chapter 10

Camras, M. Magnetic sound for 8 mm projection. *Soc. of Mot. Pict. Eng. Jour.* **48:** 348-56 (1947).

Camras, M. A compatible tape cartridge. *IRE Trans. Audio* **AU-8:** 62-7 (1960).

Dolby, R. M. A noise reduction system for consumer applications. *Audio Eng. Soc. Convention,* **Paper M-6,** (Oct. 1970).

Goldmark, P. C., C. D. Mee, J. D. Goodell, and W. P. Guckenburg. A $1\frac{7}{8}$ - IPS magnetic recording system for stereophonic music. *IRE Trans. Audio* **AU-8:** 161-7 (1960).

IEC (International Electrotechnical Commission) **Publication 94:** (1980, 1982).

Lee, B. S. Effect of delayed speech feedback. *Acoust. Soc. Amer. J.* **22:** 824-6 (1950).

Lubinski, L. D. A high fidelity $1\frac{7}{8}$ ips recording system. *Audio Eng. Soc. 15th Annual Meeting* **Preprint No. 287:** (Oct. 14 to Oct. 18, 1963).

Robinson, D. P. Production of Dolby B-type cassettes. *J. Audio Eng. Soc.* **20:** 835-41 (1972).

Schiesser, H. Devices for time extension of sound reproduction. *Funk und Ton* **3:** 256-60 (1949).

van der Lely and G. Missriegler. Audio tape cassettes. *Philips Tech. Review* **3:** 77 (1970).

Wolf, S. K. Synthetic production and control of acoustic phenomena by a magnetic recording system. *Proc. IRE* **29:** 365-71 (1941).

Chapter 11

Anderson, C. E. The modulation system of the Ampex video tape recorder. *J. Soc. Mot. Pict. Telev. Engrs.* **66:** 182-4 (1957).

Bernstein, Julian. *Video Tape Recording.* New York: John F. Rider Publications, 1960.

Chinn, H. A. Status of video tape in broadcasting. *J. Soc. Mot. Pict. Telev. Engrs.* **66:** 453-8 (1957).

Dolby, R. M. Rotary-head switching in the Ampex video tape recorder. *J. Soc. Mot. Pict. Telev. Engrs.* **66:** 184-8 (1957).

Ginsburg, C. P. Comprehensive description of the Ampex video tape recorder. *J. Soc. Mot. Pict. Telev. Engrs.* **66:** 177-82 (1957).

Ginsburg, C. P. The first VTR: a historical perspective. *Broadcast Engineering:* 22-42 (May 1981).

Hathaway, R. A. and R. Ravizza. Development and design of the Ampex auto scan tracking (AST) system. *J. Soc. Mot. Pict. Telev. Engrs.* **89:** 931-4 (1980).

Hausdörfer, M. Transcoding techniques. *Bosch Techn. Berichte* **6:** 99-107 (1979).

Heitman J. K. R. An analytical approach to the standardization of digital videotape recorders. *J. Soc. Mot. Telev. Engrs.* **91:** 229-32 (1982).

Kerr, R. J. *Video the Better Way. Vol. 5.* Yokohama: Victor Company of Japan, Ltd., 1980.

Lowman, Charles E. *Magnetic Recording.* New York: McGraw Hill, 1972.

SMPTE (ed.) *One-Inch Helical Video Recording.* New York: Society of Motion Picture and Television Engineers, 1979.

Stratton, L. Reviewing slow motion disc principles. *Broadcast Eng.* **11:** 14-18 (1969).

Yokoyama, K., S. Nakagawa and H. Katayama. An experimental digital video recorder. *J. Soc. Mot. Telev. Engrs.* **89:** 173-80 (1980).

Zahn, H. L. The BCN system for magnetic recording of television programs. *Bosch Techn. Berichte* **6:** 176-88 (1979).

Chapter 12

Buchsbaum, W. H. Video cassette recorders. *Popular Electronics.* **14:** 39-46 (August 1978).

M. Camras. U.S. Patent 4,097,893 (June 27, 1978).

Kihara, N. The Mavica-magnetic video still camera system. *IEEE Magnetics Conference,* Montreal (July 20-23, 1982).

McGinty, Gerald P. *Videocassette Recorders, Theory and Servicing.* New York: McGraw Hill Inc., 1979.

Morio, M., Y. Matsumoto, Y. Machida, Y. Kubota and N. Kihara. Development of an extremely small video tape recorder. *IEEE Spring Conference on Consumer Electronics,* Chicago (June 3, 1981).

Sadashige, K. An overview of longitudinal video recording technology. *J. Soc. Mot. Pict. Telev. Engrs.* **89:** 501-4 (1980).

Sato, S., K. Takeuchi and M. Yoshida. Recording video camera in the beta format. *IEEE Int'l Conf. on Consumer Electronics,* Chicago (June 9, 1983).

Sugaya, H. Home video tape recording and its future prospects. *Proc. Institution of Electronic and Radio Engineers* **54:** 75-83 (1982).

Toshiba. Toshiba LVR. Japan: Toshiba Co. (1980).

Universal City Studios v. Sony Corp. of America. 480 F. Supp. 429 (1979).

Chapter 13

Athey, Skipwith W. *Magnetic Tape Recording.* Washington: NASA SP-5038, U.S. Gov. Printing Office, 1966.

Davies, Gomer. *Magnetic Tape Instrumentation.* New York: McGraw Hill, 1961.

EMI Technology Inc. *Modern Instrumentation Tape Recording.* EMI Technology Inc. 1978.

Favin, D. L. and T. C. Anderson. Data system performance as evaluated by eye patterns. *AIEE Conference Paper CP 62- 363* (1962).

Howard, J. A. and L. N Ferguson. *Magnetic Recording Handbook.* Mountain View, CA: Hewlett Packard Co. AN 89, 1966.

IRIG. *Telemetry Standards, Document 106-77 (Revised January 1977).* New Mexico: Inter-Range Instrumentation Group, Secretariat Range Commanders Council, White Sands Missile Range, 1977.

Jorgensen, F. Phase equalization is important. *Electronic Industries* 98-101 (Oct., 1961).

Jorgensen, Finn. *The Complete Handbook of Magnetic Recording.* Blue Ridge Summitt, PA: Tab Books Inc., 1980.

Kalil, Ford (editor). *Magnetic Tape Recording for the Eighties.* NASA Reference Publ. 1075. Washington: U.S. Gov. Printing Office, 1982.

Langland, B. J. Phase equalization for perpendicular recording. *IBM Corp., San Jose, Ca.* (1982).

Lowman, Charles E. *Magnetic Recording.* New York: McGraw Hill, 1972.

Mullin, J. T. Flutter compensation for FM/FM telemetering recorder. *IRE Natl. Conv. Record* **Pt 1:** 57 (1953).

Pear, C. B. *Magnetic Recording in Science and Industry.* New York: Reinhold Publishing Corp., 1967.

W. W. Peterson and E. J. Welden. *Error Correcting Codes, 2nd ed.* Cambridge, MA: MIT Press, 1972.

Ratner, V. A. Predetection recording - solution to the telemetry dilemma? *Electronics* **36:** 30-3 (1963).

Stiltz, H. L. *Aerospace Telemetry.* Englewood Cliffs, N. J.: Prentice Hall, Inc., 1961.

Chapter 14

Bernett, W. A. Current state-of-the-art and future trends in flexible and rigid disk media for minicomputer systems. *Data Recording Products Division, 3M Center,* St. Paul, Minnesota, 55101 (1980).

Camras, M. Magnetic recording: 2012 A.D. *Proc. IRE* **50:** (1962).

Eldridge, D. E. magnetic recording and reproduction of pulses. *IRE Trans. Audio* **8:** 42-57 (1960).

Engh, J. T. The IBM diskette and diskette drive. *IBM J. Res.Develop.* **25:** 701-10 (1981).

Harker, J. M., D. W. Brede, R. E. Pattison, G. R. Santana and L. G. Taft. A quarter century of disc file innovation. *IBM J. Res. Develop.* **25:** 677-89 (1981).

Harris, J. P., W. B. Phillips, J. F. Wells and W. D. Winger. Innovations in the design of magnetic tape subsystems. *IBM J. Res. Develop.* **25:** 691-9 (1981).

IBM. *IBM Disk Storage Technology.* San Jose: IBM Corp. San Jose, California (1980).

IBM. The IBM diskette general information manual. *IBM Corp., Rochester, Minnesota,* Publ. GA21-9182-5 (1982).

Lieberman, D. Shirt-pocket floppies diversify. *Electronic Products* **26:** 43-6 (1983).

Mulvany, R. B. and L. H. Thompson. Innovations in disc file manufacturing. *IBM J. Res. Develop.* **25:** 711- 723 (1981).

White, R. M. Magnetic Disks: storage densities on the rise. *IEEE Spectrum* **20:** 32-8 (Aug. 1983).

Appendixes

Barrett, A. E. and Tweed C. J. F. Some aspects of magnetic recording and its application to broadcasting. *J. Institution of Elec. Engrs.* **82:** 265-85, 285-8 (1938).

Begun, S. J. Recent developments in magnetic recording. *J. Soc. Mot. Pict. Engrs.* **28:** 464-72 (1937).

Begun, Semi J. *Magnetic Recording.* New York: Rinehart Books, Inc., 1949.

Camras, M. Magnetic recording on a steel wire. Master of Science Degree, *Illinois Inst. of Tech.* (1942).

Camras, M. A stereophonic magnetic recorder. *Proc. of the IRE* **37:** 442-7 (1949).

Coombs, J. M. Storage of numbers on magnetic tape. *Proc. of Nat'l Electronics Conference:* 201-9 (1946)

DeForest, L. The audion detector and amplifier. *Proc. of the IRE* **2:** 15-36 (1914).

Fankhauser, C. E. The telegraphone. *J. Franklin Inst.* **167:** 37-47 (1909).

Ginsburg, C. P. The first VTR: a historical perspective. *Broadcast Engineering:* 22-42 (May 1981).

Hickman, C. N. Sound recording on magnetic tape. *Bell Lab. Record* **16:** 165-77 (1937).

Howell, H. A. Magnetic sound recording on coated paper tape. *J. Soc. Mot. Pict. Engrs.* **48:** 36-46, 46-9 (1947).

Larsen, Absalon. *Telegrafonen og den traadlose.* Ingeniorvidenskabelige Skrifter, 1950 Nr. 2, Akademiet for de Tekniske Videnskaber og Dansk Ingeniorforening, I Kommission Hos, Teknisk Forlag, Kopenhavn (1950).

Lübeck, H. Magnetische schallaufzeichnung mit filmen und rinköpfen. *Akustische Zeit.* **2:** 273-95 (1937).

Mayer, L. Curie point writing on magnetic films. *J. Appl Phys.* **28:** 1003 (1958).

Mallina, R. F. A mirror for the voice. *Bell Lab. Record* **13:** 200-2 (1935).

Meyer, E. and E. Schuller. Magnetische schallaufzeichnung auf stahlbänder. *Zeit fur Tech. Phys.* **13:** 593- 99 (1932).

Mullin, J. T. A video magnetic tape recorder. *IRE Nat'l Convention Record* **Pt. 7:** 120- (1954).

Olson, H. A magnetic tape system for recording and reproducing standard FCC color TV signals. *RCA Rev.* **15:** 3- (1954).

V. Poulsen, U.S. Patent 661,619 (Nov. 13, 1900).

Poulsen, V. The telegraphone: a magnetic speech recorder. *Electrician* **46:** 208-10 (1900).

Pugsley, C. W. Wire recording. *Electrical Engineering* **65:** 316-21 (1946).

Rust, N. M. Marconi-Stille recording and reproducing equipment. *Marconi Review* **46:** 1-11 (1934).

Smith, O. Some possible forms of phonograph. *Electrical World* **12:** 116-7 (1888).

Stille, C. Die electromagnetische schallaufzeichnung. *Elec. Zeit.* **51:** 449-51 (1930).

Note: Later references may be found in the appropriate chapters.

APPENDIX A
HIGHLIGHTS OF MAGNETIC RECORDING DEVELOPMENT

A comprehensive history of magnetic recording would fill a volume in itself. The following outline, although incomplete and imperfect, attempts to document the events, companies, and individuals associated with advances in magnetic recording. Many of the first attempts were crude, of course, and not commercially acceptable. Figures A-1 through A-28 show some of the early recorders. Table 14-14 highlights some important computer developments.

Geological Time Earth's field was recorded in different regions of the world by magnetized minerals, whose direction of permanent magnetization indicate the earth's field direction at the time those minerals were formed.

1888 "Some Possible Forms of the Phonograph" appeared in the 8 September 1888 issue of *The Electrical World*. In this article Oberlin Smith, an American engineer, suggested that a thread or ribbon of magnetizable material, solid or impregnated with magnetic dust, could record and play back sound electromagnetically. Figure A-1 is a reproduction from Oberlin Smith's article (Smith 1888).

1898 Magnetic recording was successfully demonstrated by a Danish engineer, Valdemar Poulsen. On 1 December 1898, Poulsen filed for a Danish patent, which was issued as No. 2653 on 31 October 1899.

1899 U.S. Patent was applied for by Poulsen on 8 July 1899; issued as Patent No. 661,619 on 13 November 1900.

1900 Telegraphone was exhibited by Poulsen at the Paris Exposition, where it won the Grand Prix. Numerous articles were published in scientific journals such as *Annalen der Physik, Electrician, Comptes Rendus,* and *Scientific American* (Poulsen 1900).

1900–1903 Peter Jensen demonstrated Telegraphones to European royalty and financiers for backing to manufacture the machines, without success.

1903 American Telegraphone Co., incorporated November 1903, was capitalized at $5,000,000 and located in Springfield, Mass. It manufactured wire recorders, which were acquired mainly by the curious and by experimenters who tried to improve them. C. K. Fankhauser, Charles D. Rood.

1906 Record media of thin layers were electroplated on a substrate. U.S. Patent 836,339 (20 November 1906) P. O. Pederson.

1907 DC bias for the telegraphone was patented by Pederson and Poulsen in U.S. Patent 873,083, issued on 10 December 1907.

1913 Experiments by Dr. Lee deForest with synchronized sound for motion pictures used a telegraphone. Dr. deForest also combined the telegraphone with his audion vacuum tube amplifier, without commercially successful results (deForest 1914).

1921 A form of AC bias was described in a patent application dated 26 March 1921 by Wendell L. Carlson and Glenn W. Carpenter of the U.S. Naval Research Laboratory, which issued as U.S. Patent 1,640,881 on 30 August 1927. They used an AC bias to increase sensitivity and decrease

THE ELECTRICAL WORLD. SEPTEMBER 8, 1888.

Some Possible Forms of Phonograph.

BY OBERLIN SMITH.

There being nowadays throughout the scientific world great activity of thought regarding listening and talking machines, the readers of THE ELECTRICAL WORLD may be interested in a description of two or three possible methods of making a phonograph which the writer contrived some years ago, but which were laid aside and never brought to completion on account of a press of other work.

One of these methods is rudely shown in Figs. 1, 2 and 3, the construction and operation being as follows: *A* is a mouth piece and diaphragm, with spring and indenting needle, as in the Edison machine. *B* is a reel, carrying a thin ribbon *E* of iron, steel or other substance capable of being temporarily softened by heat. This ribbon is unwound from *B* and wound on to another reel *C*, which is revolved slowly by clock work or other means. *D* is a supporting roller (or stationary bar) with a flat groove the width of the ribbon *E*, and having a V-groove in the bottom of it for the needle to descend into, as seen in Fig. 2. *F* is a heating lamp, which, of course, must be protected from draughts, etc. All this is the recording apparatus or transmitter. The ribbon *E* being short at the point where, for the time being, it is hot, receives the indentations as easily as the tin-foil, or more so. It cools by the time it gets to reel *C*, and is then much harder and more durable than tin foil. The same apparatus can be used for the "talker," as in Edison's machine, but advantage may be taken of having the indented ribbon made of a hard substance by using a special talking diaphragm *G*, Fig. 3, which will augment the vibrations in amplitude by means of a lever *H*, the ribbon *E* being hard enough not to lose its form by the increased pressure due to the leverage, as tin-foil would do.

The probable advantages of this form of apparatus are: 1. The loudness of voice produced by the increased amplitude of vibration. 2. The simplicity and cheapness of the whole machine—requiring no accuracy in "registering" devices beyond having the groove in roller *D* to about fit the width of ribbon *E*. 3. The cheap material of which the ribbon may be made. 4. Durability of ribbon, even with oft-repeated use. 5. Convenience and freedom from injury in handling and transporting the ribbon-record when wound upon spools like thread. This ribbon would, if of iron or steel, probably be about $\frac{1}{10}$ inch wide and $\frac{1}{200}$ inch thick.

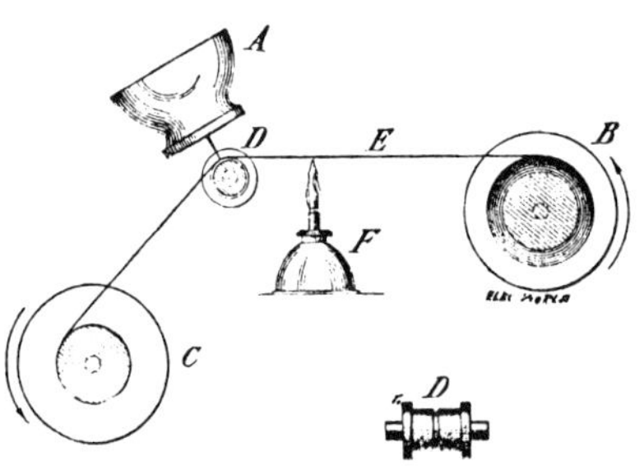

FIGS. 1 AND 2.—SOME POSSIBLE FORMS OF PHONOGRAPH.

Its disadvantages, possibly fatal ones, would be the difficulty of evenly heating the metal ribbon and the probable rasping noise which would occur in the diaphragm when the sound was reproduced. A modified and somewhat simpler form of the above process might be employed by using an ordinary wire instead of the ribbon *E*, and allowing a chisel-shaped needle to indent it into a flattened and somewhat widened form, wherever it was struck.

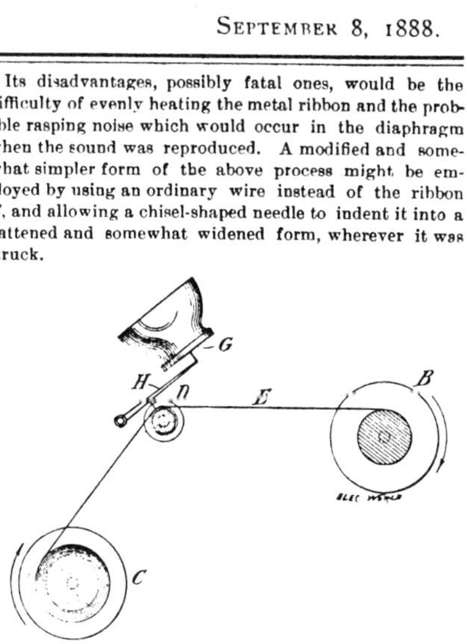

FIG. 3.

The above two methods are, of course, wholly mechanical, as in the ordinary phonograph. The following proposed apparatus is, however, *purely electrical*, and is, as far as known to the writer, the only one fulfilling such conditions that has been suggested. Fig. 4 is the recording part of an electrical phonograph. Fig. 5 is the talking part of the same. Many of the pieces, as

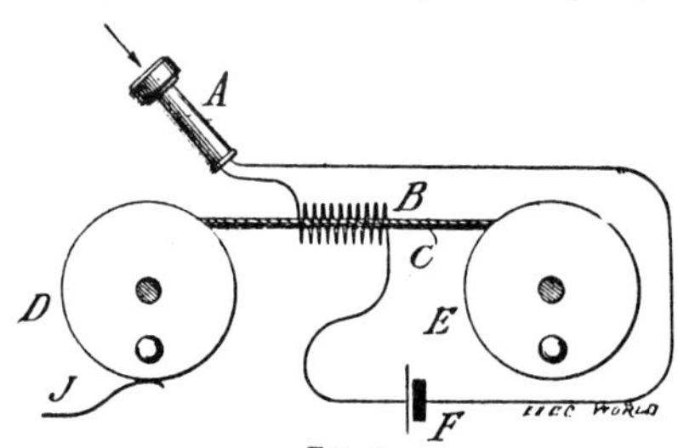

FIG. 4.

D, *E*, *B*, *C*, etc., can be the same ones as are used in Fig. 4. Fig. 6 shows the same ideas applied to a telephone line wire, so as to speak at a distance and at the same time record what is said, thus making a *recording telephone*. The sketches show only the essential parts, without the supporting framework, etc.

In Fig. 4 the voice or other sound is delivered into an ordinary telephone *A*. Preferably, this should be a carbon transmitter so as to have a battery *F* in the circuit, and thus use as strong a current as practicable. Possibly, however, a Bell telephone without a battery would answer the purpose. In either case the current, broken into waves of varying lengths and intensities corresponding with the vibrations of the diaphragm in the telephone, passes in its

Figure A-1. Oberlin Smith's concept. Ten years before Poulsen proved it possible, magnetic recording was suggested in 1888 by Oberlin Smith, an American engineer.

circuit through the helix B, converting into a permanent magnet any piece of hardened steel which may be at the time within the helix. Through this helix B passes a cord, string, thread, ribbon, chain or wire C, made wholly or partly of hardened steel, and kept in motion by being wound on to the reel E from off the reel D, E being revolved by hand, clock-work or other means. J is a tension spring or brake pressing against D to keep the cord C taut.

When in operation with the undulatory current from the telephone A passing through the helix, the cord C becomes, so to speak, a series of short magnets grouped into alternate swellings and attenuations of magnetism

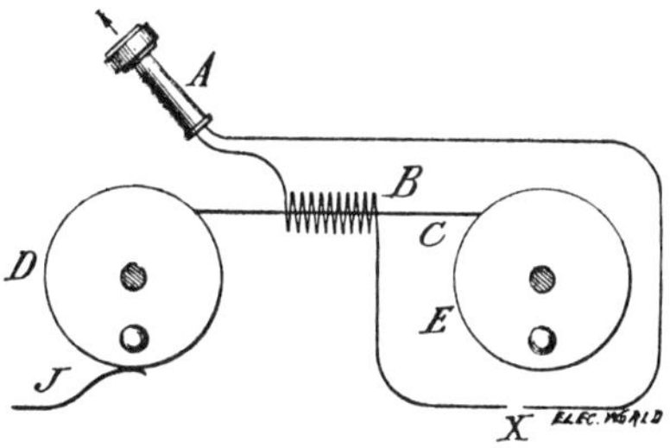

FIG. 5.

The actual lengths of these groups depends upon the speed of their motion, but their *relative lengths* depend upon the *relative lengths* of the sound wave; and their *relative intensities* depend upon the *relative amplitudes* of these waves. The cord C therefore contains a perfect record of the sound, far more delicate than the indentations in the tin-foil of the mechanical phonograph. The probable construction of C would be a cotton, silk or other thread, among whose fibres would be spun (or otherwise mixed) *hard steel* dust, or short clippings of very fine steel wire, hardened. Each piece would, of course, become a complete magnet. Other forms of C might be a brass, lead or other wire or ribbon through which the steel dust was mixed in melting—being hardened afterwards in the case of brass or any metal with a high

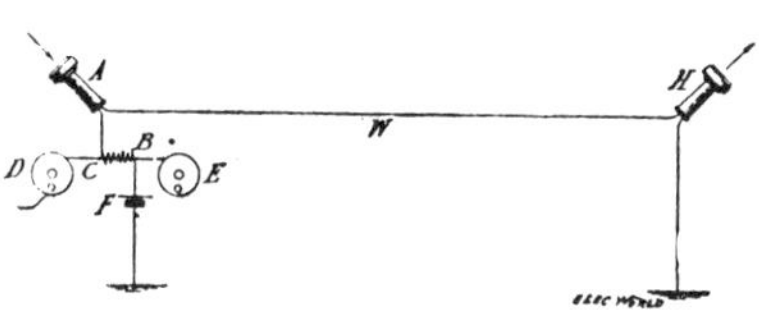

FIG. 6.

melting point. Another (but too expensive) form of C would be a chain with each link a magnet; or, if the magnets affected each other too much when in contact, each alternate link could be of non-magnetic material. This chain would not be as delicate as the dust magnets, because the effects of a given vibration might extend but part way along a link. Another imaginable form of C would be simply a hard steel wire, but it is scarcely possible that it would divide itself up properly into a number of short magnets. The magnetic influence would probably be distributed along the wire in a most totally depraved way, with nodal points just where they were not wanted. If it could be made to work it would obviously be the simplest thing yet suggested.

The cotton thread above mentioned would seem to be preferable to anything else on account of its cheapness, lightness and flexibility. The Lord's Prayer could be written upon a few feet of thread or string, while a young lady receiving a small spool of cotton from her lover would think herself abominably neglected if it was not "warranted 200 yards long."

In Fig. 6 the arrangement is precisely the same as in Fig. 4, except that the circuit is made through the telegraph wire W and the receiving telephone H in Boston or some other distant place. Of course the record might be made at the receiving instead of the transmitting end of the line, and thus our hypothetical young lady might, while listening to the impassioned pleadings of her chosen young man, be preparing the evidence for a future breach-of-promise suit.

To make the thread or cord C "talk back" it is, after having been rewound on to reel D again drawn through a helix B, Fig. 5, in whose circuit is the "talking" telephone A, probably a Bell receiver. Of course it is drawn through at approximately the same speed as before. In passing, the small permanent magnets in the cord C induce currents of electricity in their enveloping helix analogous to the currents in the field of a magneto-electric machine, or a dynamo with permanent magnets in its armature. A more exact analogy would, however, be the currents in the helix of a solenoid if its ordinary action were reversed, and its core made a permanent magnet. These waves of current will correspond in length and relative intensity with the original wave currents, and will therefore reproduce the vibrations of the original sound in the diaphragm of the telephone at any time in the future. If such induced currents are not strong enough to produce sufficiently loud sounds it may be possible to insert at X, Fig. 5, some intensifying apparatus, such as a battery, but which has not yet been thought out.

Like the two mechanical methods first mentioned, this electrical method has never been worked out to completion. The writer went far enough with it to build a temporary apparatus and to develop a successful machine for spinning metallic dust into a cotton cord, but was obliged to lay aside the whole thing before arriving at any *acoustic* results. His experiments showed that it was difficult, with ordinary tools, to harden steel filings on account of excessive oxidation. Experiments with hardened steel wire, broken in a special machine into very short pieces, showed that they must not be too short—say not less than three or four times their diameter—or they could not be saturated with magnetism to any appreciable degree. Possibly this is because the poles (or points of maximum polarity) of a magnet lie at some distance from the ends of the bar, and consequently neutralize each other when the bar is too short. If this theory is correct it would prevent making magnets of steel dust, the grains of which are supposed to be about as broad as they are long.

To digress a little, it may be remarked that such a theory does not seem to agree with the fact of a magnetic polarity in approximately spherical or cubical bodies, like the earth, or a chunk of loadstone. Possibly, however, they would be much stronger magnets if elongated; and the tiny pieces of wire above referred to, may possess as much strength in proportion, though it be scarcely perceptible on account of their smallness.

The writer confesses to a good deal of ignorance upon the subject, but he was somewhat surprised to find an equal amount in several well known electricians whom he consulted; and also to find that none of the books he had at hand gave any definite data regarding the best proportions for permanent magnets or their actual strength (when saturated) in pulling power. Surely, there is in this department of electrical science a good sized (magnetic) field for a number of lines of force—mental ones—to work in, in the way of careful experimenting.

To return to our magnetized cord as a "phonogram," it is possible that an insuperable objection to it would be found

Figure A-1. *(Continued)*

A

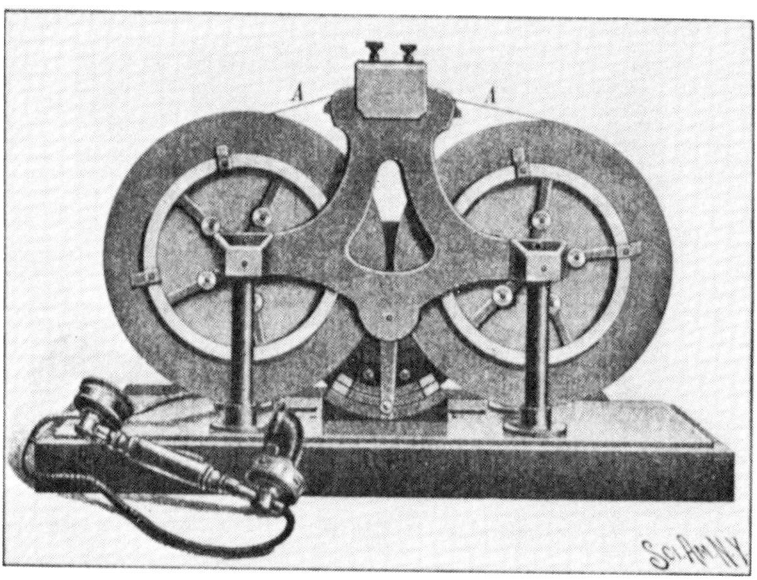

B

Figure A-2. Telegraphones of the 1900-1910 era. A. Cylinder or drum recorder of about 1900. A telephone receiver for each ear helped to make audible the faint sounds (Courtesy BASF). B. Steel tape recorder of 1900 (Scientific American). C. Wire recorder (*The Electrician*, July 1, 1903). D. Steel-disc recorder (*The Electrician*, July 1, 1903).

C

D

Figure A-2. (*Continued*)

A

B

Figure A-3. Wire telegraphones. A. American telegraphone, Springfield, Mass. (ca 1900-1910). B. Record-O-Phone, Washington, D.C. (1925). C. Remote control speech unit and copying unit of VOX wire recorder (*Electrical Review*, 1925). D. The Textophone (1933) (from *The Electrician*, 1939).

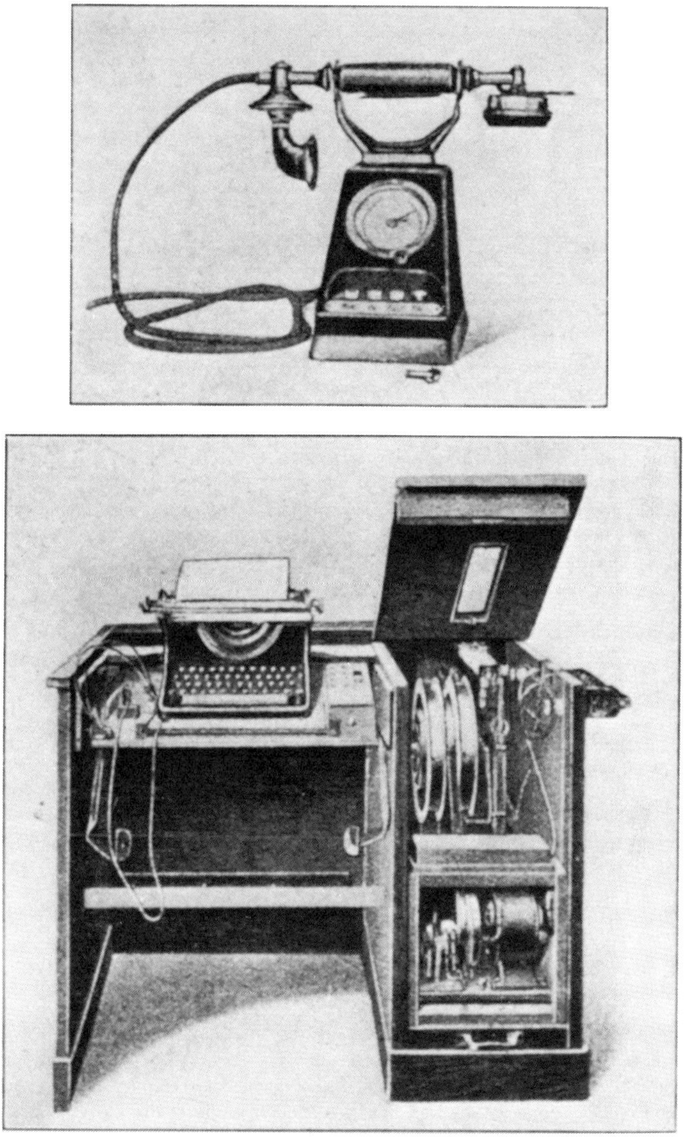

C

Figure A-3. (*Continued*)

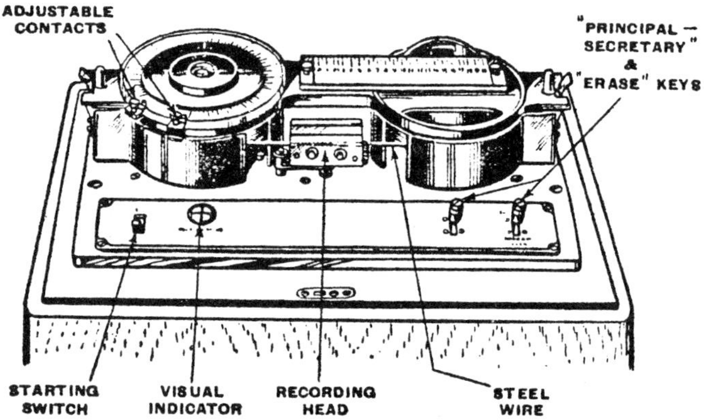

D

Figure A-3. (*Continued*)

noise in a "Radio Telegraph System." Since their system was tuned to a single note for Morse code reception, it is doubtful that anyone realized that low-distortion sound recording could result from AC bias. For this reason, no one paid attention to their work until more than a decade later, when the linearization aspects of AC bias were discovered independently.

1925 Remote-controlled dictation used wire recorders. Vox Co., Berlin, Germany. Kurt Stille.

1925 to 1930 Telegraphie-Patent-Syndikat was organized in Germany by Kurt Stille and others to license magnetic recording. Prominent licensees were Karl Bauer and Ludwig Blattner (Stille 1930).

1927 Recording Tape of Powdered Magnetic Material was suggested in U.S. Patent 1,653,467 issued on 20 December 1927 to J. A. O'Neill.

1928 Recording Tape of Powdered Magnetic Material was described in German Patent 500,900 issued on 31 January 1928 to Fritz Pfleumer, who in 1931 interested I. G. Farben (BASF) and AEG to develop a powder-coated plastic tape.

1930 The Blattnerphone was a steel tape recorder introduced in England by Ludwig Blattner for experiments with sound for motion pictures. The British Marconi Co. acquired Blattner's company and improved the recorder with the help of Dr. Heising of the Stille laboratories to produce a sophisticated Marconi-Stille machine used by the BBC (Rust 1934).

1932 A speech by King George V was broadcast on Christmas day from a recording on BBC's steel tape machine.

The Dailygraph was a wire recorder for dictation and telephone recording made in Europe by the Echophone Company organized by Karl Bauer. It featured cartridge loading.

1933 The Textophone, an improved version of the Dailygraph for dictation and telephone recording, was made in Europe by the C. Lorenz Co., a division of IT&T who in 1932 took over the Echophone and their Dailygraph. Karl Bauer.

1935 The Magnetophone was announced by A.E.G. (German General Electric Co.). It was exhibited at the 1935 Radio Exposition in Berlin and described in German publications. The Magnetophone's main feature was the use of plastic tape 6.5 mm wide, coated at first with a carbonyl iron dispersion, later with magnetite, and running at approximately 1 meter per second. Cost of coated tape was much lower than of solid metal tape. The early Magnetophone was intended for dictation

A

B

Figure A-4. Steel-tape recorders, Great Britain. A. The BBC Marconi-Stille steel tape machine (1934). B. Improved BBC recorder, Barrett and Tweed (*I.E.E. Journal:* v. 82, 1938).

A

B

Figure A-5. Steel-tape recorders, Germany. A. The Lorenz Stahltonmachine (1935). B. Lorenz recorder mounted in a van for on-the-spot recordings.

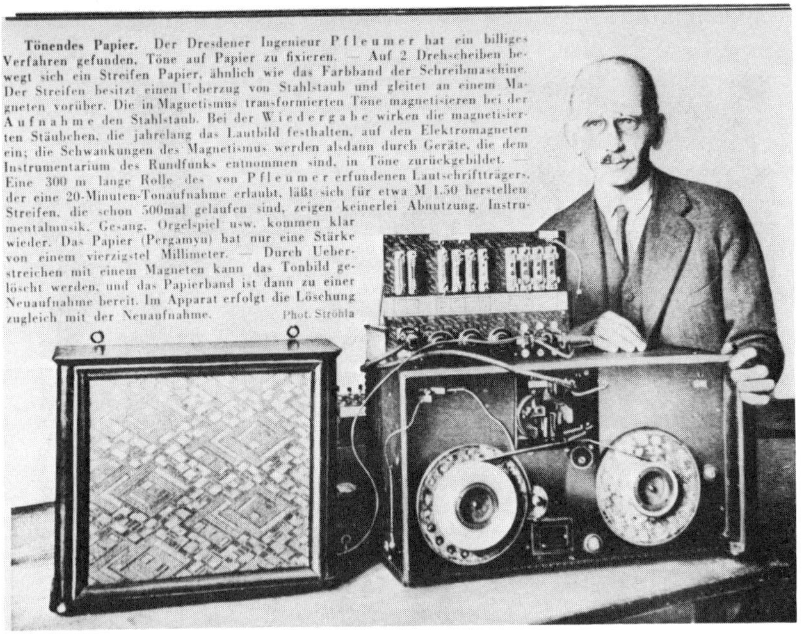

Tönendes Papier. Der Dresdener Ingenieur Pfleumer hat ein billiges Verfahren gefunden, Töne auf Papier zu fixieren. — Auf 2 Drehscheiben bewegt sich ein Streifen Papier, ähnlich wie das Farbband der Schreibmaschine. Der Streifen besitzt einen Ueberzug von Stahlstaub und gleitet an einem Magneten vorüber. Die in Magnetismus transformierten Töne magnetisieren bei der Aufnahme den Stahlstaub. Bei der Wiedergabe wirken die magnetisierten Stäubchen, die jahrelang das Lautbild festhalten, auf den Elektromagneten ein; die Schwankungen des Magnetismus werden alsdann durch Geräte, die dem Instrumentarium des Rundfunks entnommen sind, in Töne zurückgebildet. — Eine 300 m lange Rolle des von Pfleumer erfundenen Lautschriftträgers, der eine 20-Minuten-Tonaufnahme erlaubt, läßt sich für etwa M 1.50 herstellen. Streifen, die schon 500mal gelaufen sind, zeigen keinerlei Abnutzung. Instrumentalmusik, Gesang, Orgelspiel usw. kommen klar wieder. Das Papier (Pergamyn) hat nur eine Stärke von einem vierzigstel Millimeter. — Durch Ueberstreichen mit einem Magneten kann das Tonbild gelöscht werden, und das Papierband ist dann zu einer Neuaufnahme bereit. Im Apparat erfolgt die Löschung zugleich mit der Neuaufnahme. Phot. Ströhla

A

B

Figure A-6. Evolution of the magnetophone. A. Fritz Pfleumer with early paper tape recorder. (Courtesy H. Thiele). B. Earliest Magnetophone (1935) (Courtesy BASF). C. Improved Magnetophone (ca 1945) (Courtesy J. Mullin).

C

Figure A-6. (*Continued*)

and had mediocre sound quality. In subsequent years, it was improved greatly due mainly to improved tape invented by Dr. Rudolf Brill of I.G. Farben Company. Th. Volk., W. H. Hansen, Heinz Lübeck, R. Brill (Lübeck 1937).

The Lorenz Stahltonmachine (steel tape sound recorder) was adopted by the German Broadcasting Company. One version was installed in vans for on-the-spot recording. It was a product of C. Lorenz Co. and judged superior to the Magnetophone.

"Mirror for Voice," a tape recorder using a short endless loop of metal tape, was built by Bell Laboratories and described in *Bell Labs Record*. It was demonstrated at fairs as a "Hear-Your-Telephone-Voice" recorder (Mallina 1935).

1936 The London Philharmonic Orchestra, conducted by Sir Thomas Beecham, was recorded at Ludwigshafen Germany, I. G. Farben (BASF) headquarters.

1937 The soundmirror was an endless-loop tape recorder that could handle a one- to two-minute message on $\frac{1}{4}$-inch-wide tungsten-steel tape for announcements and voice training. It was manufactured by Acoustic Consultants and by Brush Development Co., Cleveland, Ohio, after development by Magnetone, Inc. Semi J. Begun, S. K. Wolf (Begun 1937).

Vicalloy, a new alloy for tapes, and improved recording heads made high-quality sound recording possible at a tape speed of only 16 inches per second. Bell Laboratories (Hickman 1937).

1938 A high-quality steel tape recorder used for BBC broadcasts (improved Marconi-Stille) was described. A. E. Barrett and C. J. F. Tweed (Barrett and Tweed 1938).

A new home-use wire recorder had innovations that led to the ARF (Armour Research Foundation) Model 50 wire recorder. Marvin Camras, William Korzon.

Synthetic reverberation using delayed pickup from a tape recorder was described by S. J. Begun and S. K. Wolf.

AC erasing and biasing experiments in Japan were published in an article by Kenzo Nagai, S. Sasaki, and J. Endo. (Nagai, Sasaki, and Endo 1938)

A

Figure A-7. A. London Philharmonic at Ludwigshafen, and B. program of 19 November 1936. (Courtesy BASF).

1939 The Brush Development Co., Cleveland, Ohio, marketed the Soundmirror, an endless steel tape machine, and under the direction of Dr. Semi J. Begun carried on research in subsequent years that led to many advances in magnetic recording. Otto Kornei.

Tape recorders were used in German radio stations.

Speech compression or expansion with rotating heads was described by E. Schuller in U.S. patent application 291,620.

1940 Wire recorders at Armour Research Foundation and Armour Institute of Technology were in use for research in submarine detection, language classes, and music recording. Model 50 wire recorders were developed. Marvin Camras, Donald Richardson, Thomas C. Poulter, George Ziegler, Harold Vagtborg, Haldon A. Leedy, Jesse Hobson.

The Mirrophone, an endless-loop tape recorder running for one minute, was manufactured by the Western Electric Co. for short repetitive announcements and for voice training. C. N. Hickman.

1941 High-frequency bias was used in Model 50 wire recorders. Application for patent was granted as 2,351,004 on 13 June 1944. Armour Research Foundation. Marvin Camras.

H. J. von Braunmühl and W. Weber of the German Rundfunk also applied for a patent, published in 1943 as a "Method of Magnetic Sound Recording," Serial 413,380, U.S. Alien Property Custodian.

1942 Model 50 wire recorder, manufactured by H. G. Fischer Co., Chicago. Donald Richardson, Howard Osbourne, H. G. Fischer, Peter Musket.

F E I E R A B E N D H A U S
Ludwigshafen a. Rh.

Donnerstag, den 19. November 1936, abends 8 Uhr

Einziges Konzert

des

Londoner Philharmonischen Orchesters

unter Leitung von

Sir Thomas Beecham

I.

Die Wespen des Aristophanes, Ouvertüre . . V. Williams

II.

Sinfonie Es-Dur Nr. 39 . . . . W. A. Mozart

Adagio-Allegro
Andante con moto
Menuett
Finale

III.

a) Sommernacht am Fluß
b) Beim ersten Kuckucksruf im Frühling . . . . Fr. Delius

IV.

Einleitung und festlicher Zug aus
„Der goldene Hahn" . . . . N. Rimsky-Korssakow

—— **P a u s e** ——

V.

Sinfonie Nr. 4 G-Dur op. 88 . . . . A. Dvorak

Allegro con brio
Adagio
Allegretto grazioso
Finale (Allegro ma non troppo)

K o n z e r t l e i t u n g: Bildungsausschuß der I. G. Farben Ludwigshafen/Rh. in Gemeinschaft mit der Mannheimer Konzertdirektion Heinz Hoffmeister, Mannheim O 7. 16

Preis 30 Pfennig

B

Figure A-7. (*Continued*)

A

B

Figure A-8. Endless loop steel tape machines. A. The Soundmirror (*Electronics,* Sept. 1938). B. The Mirrophone (*Wireless World,* February 1942).

A

B

Figure A-9. Improved wire recorders. A. Armour Research Foundation unit (1939-1940). B. Camras with Model 50 recorder (1942).

A

B

Figure A-10. Military wire recorders. A. Model 20 airborne recorder (1942). B. Model 50A ground unit.

Model 50 wire recorder manufactured by Armour Research Foundation for U.S. military forces. George Ziegler, Raymond Zenner.

Model 20 airborne recorder manufactured by Peirce Wire Recorder Co., Evanston, Illinois. Charles P. Peirce, Emil C. Steinbach.

General Electric as a licensee of Armour Research Foundation began research that led to Model 50A recorder and advanced models (Pugsley 1946). D. W. Pugsley, Arthur W. Sear, I. J. Kaar, Fred A. Ray, H. B. Marvin, W. R. G. Baker.

Brush Development Co., Cleveland, Ohio, designed and built quantities of wire recorders for the U.S. military forces. The KS-12016 had a magazine that totally enclosed the spools of wire, level winders, heads, indicators, etc.; in fact, it contained almost the entire recorder except for the motor drive and amplifiers. The wire was of bronze with a magnetic plating of NiCo as in Fig.

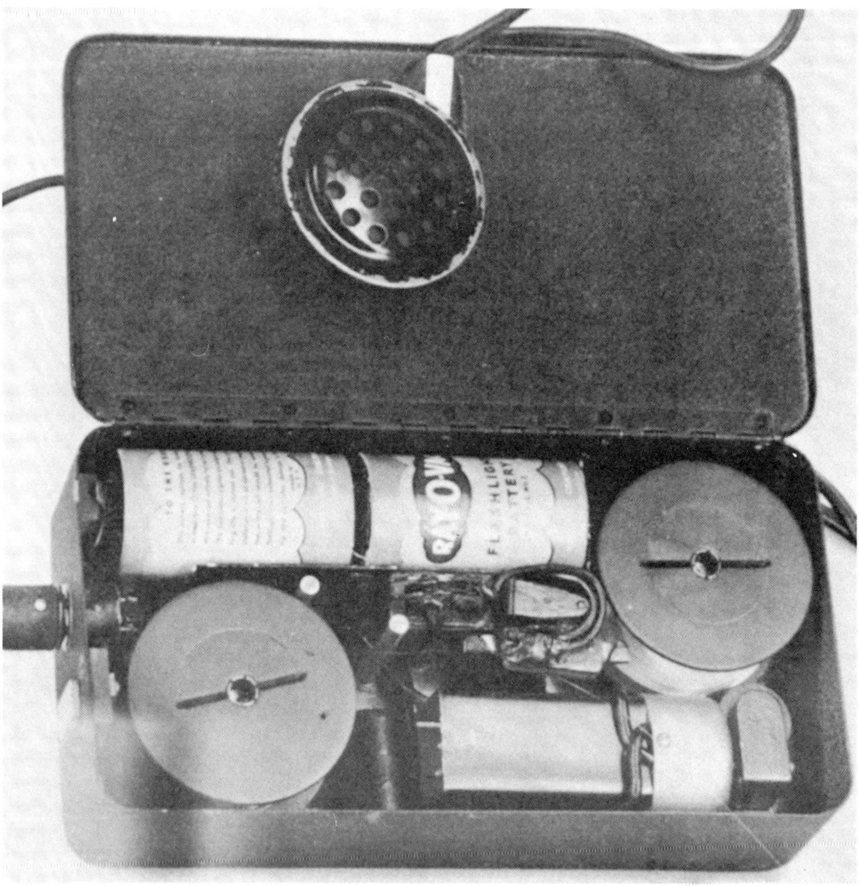

A

Figure A-11. A. Pocket model wire recorder (1944). B. Dictation wire recorder, Peirce Model 55B (1945). C. Automobile wire recorder, Model PA, by Wirecorder Corp. of Detroit.

2-13*D*. The plating, patented by Zapponi, was later used on memory drums and discs. O. Kornei, S. J. Begun.

Armour Research Foundation continued research and development program in magnetic recording on heads, high-frequency bias, drives, wires, tapes, cassettes, pocket-size recorders, motion picture sound, stereo sound, and so on. Staff included: R. E. Zenner, D. E. Wiegand, R. Vaile, Jr., F. Tyzzer, R. J. Tinkham, F. Rest, E. S. Perrine, C. Pederson, J. Motz, J. Markin, L. A. Lembach, F. S. Kirn, J. S. Kemp, W. W. Hansen, T. L. Gilbert, J. J. Fischer, H. Ekstein, W. J. Carr, C. Claras, M. Camras, J. S. Boyers, H. Barnett, A. V. Appel.

1943 Stainless Steel (18-8 type) for recording wire and tape was a major advance in metallic media. It was magnetically superior to previously used carbon steel and even to chromium and tungsten alloy magnet steels. It did not rust or corrode. Developed at the Armour Research Foundation (Marvin Camras, Hyrum E. Flanders, Ardell Glaze, George Ziegler) in cooperation with the National Standard Company of Niles Michigan (Richard E. Koontz).

Model 50 wire recorders were manufactured by Utah Radio Co., Chicago. George E. Ziegler, W. Austin Ellmore, Stan Danisch, Fred R. Tuerk.

Magnetic recorder licensing program by the Armour Research Foundation included George E.

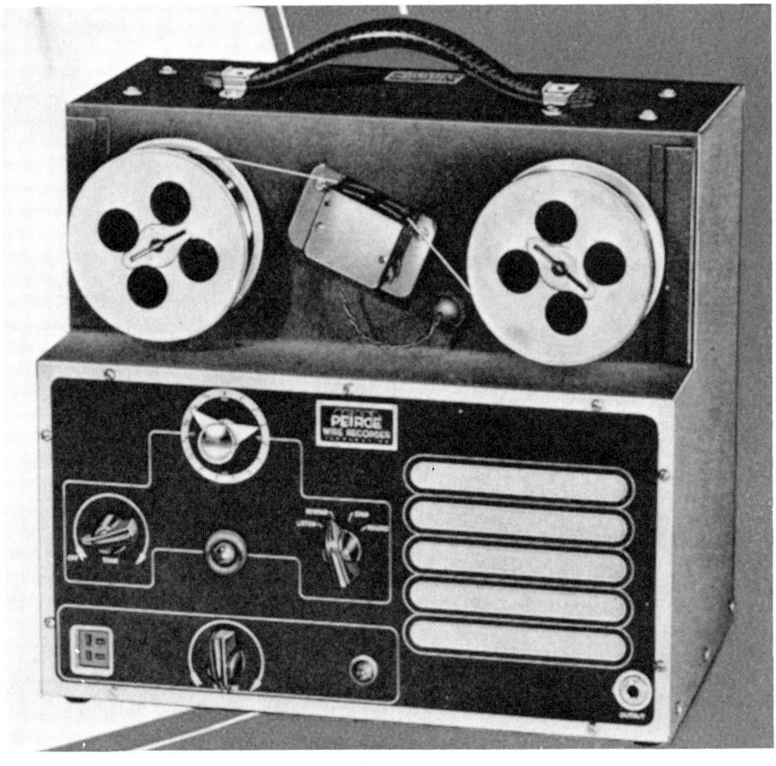

B

Figure A-11. (*Continued*)

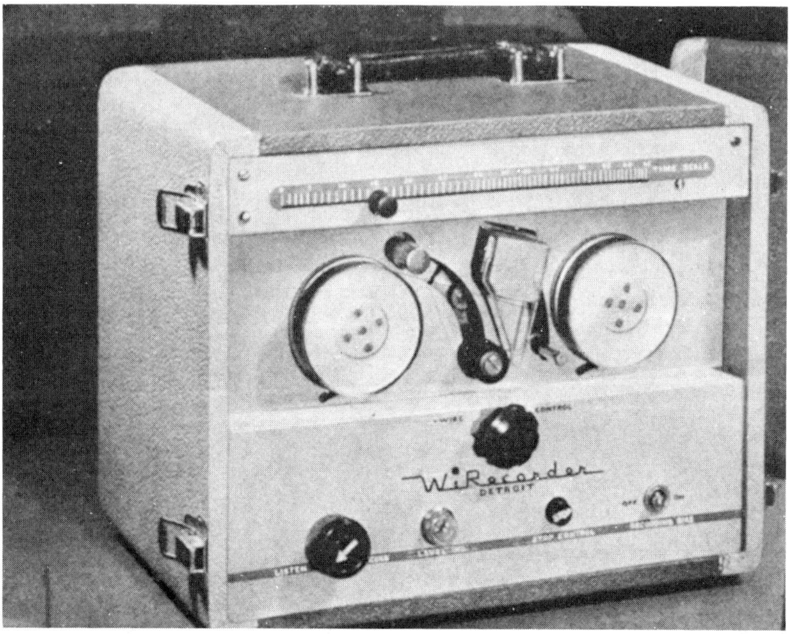

C

Figure A-11. *(Continued)*

Ziegler, Donald J. Simpson, Lucius Crowell, Carl L. Titus, Jack S. Kemp, Eugene Clears, John P. Skinner.

1944 Pocket-size wire recorder was battery operated, self-contained. Armour Research Foundation. M. Camras, P. D. Bennett.

1945 Home-use wire recorders on sale. Peirce wire recorder (Model 55B) was based on the Model 50 design and was adapted for dictation. Later models featured magazine loading.

1946 Magnetic Tape Oxide of high coercivity acicular particles was developed and manufactured by Armour Research Foundation, including Fe_3O_4 and gamma Fe_2O_3. Marvin Camras. This oxide is used almost universally in tapes up to the present day. Annual dollar value of tape production since 1968 has exceeded that of any other material used for magnetic purposes, and amounted to $2.5 billion in 1979.

Model BK401 Soundmirror, introduced by Brush Development Co., Cleveland, Ohio, used $\frac{1}{4}$-inch paper tape coated with low coercive oxide. It had many features of modern reel-type home-use recorders.Semi J. Begun, Otto Kornei, F. J. Reed. Brush also manufactured a long-play recorder using large reels of brass wire plated with a magnetic alloy.

Ampex Corporation was founded. Their first recorders were made for transcribing the Bing Crosby Show. *AMP* are initials of the founder Alexander M. Poniatoff, and *EX*, excellence. A. M. Poniatoff, Harold Lindsay, Myron J. Stolaroff, Walter T. Selsted.

Magnecord Corporation was formed for manufacture of professional recorders by staff members

A

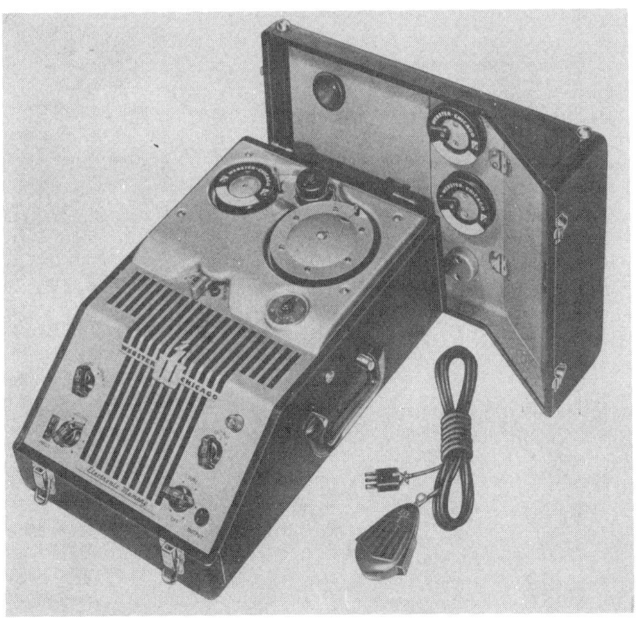

B

Figure A-12. A. Silvertone (Sears-Roebuck) combination phono-radio-recorder (1947). B. Webster-Chicago Model 80 wire recorder (1947).

A

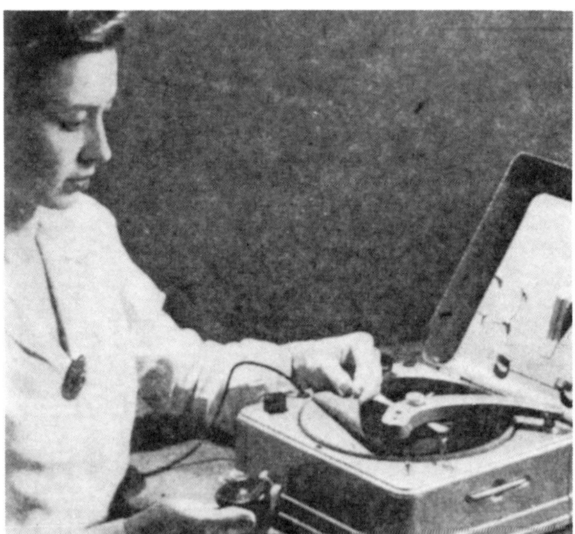

B

Figure A-13. A. Brush sound-mirror tape recorder BK-401 (1946). B. Brush mail-a-voice disc recorder BK-505 (*Communications*, April 1947).

A

B

Figure A-14. A. Eicore tape recorder (1947). B. Webcor two direction tape recorder, Model 210 (1948).

from Armour Research Foundation. Russell J. Tinkham, John S. Boyers, Clayton G. Barker, and Robert L. Landon.

The Magnecord Model SD-1, a master wire recorder with response from 35 to 15000 Hz and flutter below 0.1%, was manufactured by the Magnecord Co., Chicago, for broadcast audio recording. SD stood for Super Duper as suggested by John Boyers. It was based on the design of Armour Research Foundation's master recorder.

Magnetic Sound for Motion Pictures, using sprocketed magnetic film and soundstriped film, was introduced in October 1946 at the Society of Motion Picture Engineers convention in Hollywood, California, by Armour Research Foundation. Marvin Camras.

Magnetophone recorders, modified and improved by Jack Mullin, were demonstrated to major recording studios in Hollywood, attracting attention of Crosby Enterprises as an ideal way to record the Bing Crosby Show for transmittal at a later hour. (John T. Mullin, William A. Palmer.)

Hyflux tape was developed (Howell 1947). It was made with high coercivity metal particles (iron). Indiana Steel Products Co. Hugh A. Howell. W. E. McKibben.

Digital information recording on a magnetic surfaced drum was reported by J. M. Coombs, and also by S. M. Rubens of Engineering Research Associates (Coombs 1946).

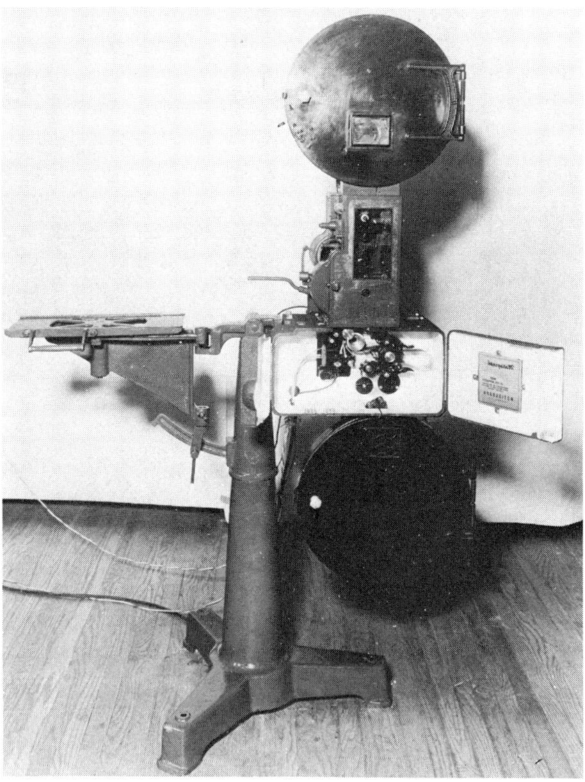

A

Figure A-15. A. Magnetic sound projector using sprocketed 35mm film (1946). B. Eight mm sound projector with soundstriped film (M. Camras, J. S. Kemp, ca 1946).

B

Figure A-15. (*Continued*)

1947 High coercivity magnetic oxide tape, the first commercially available, was made and sold by Minnesota Mining and Manufacturing Co. (3M). It included Type 100 (acicular gamma oxide coated on paper backing), Type 111 (same, coated on cellulose acetate backing), and magnetic-coated sprocketed film for the motion picture industry. Robert Herr, William W. Wetzel, Robert von Behren.

The Brush Mail-A-Voice, Model BK-503, recorded spiral tracks on a 9-inch-diameter magnetic-coated disk. The record could be folded and mailed in an ordinary envelope.

Home-use wire recorders were in mass production. Sears Roebuck sold the Silvertone, which combined a wire recorder, phonograph and radio, manufactured by Colonial Radio Corp.

Webster-Chicago Model 80 became the most popular of all wire recorders because of its low cost, portability, and relative reliability.

Crescent Industries, Chicago, manufactured a similar recorder. W. Austin Ellmore.

WiRecorder Corp. of Detroit designed a compact recorder for automobiles, but it attained only limited production. McLough Steel Co. Mort Neff, William Habousch.

The Lear Dynaport was the "Cadillac" of wire recorders, with many deluxe features for those who could afford the best. Lear Corp. William Lear.

1947 Stereo tape recording, three channel, two channel, and headphone binaural recording were

A

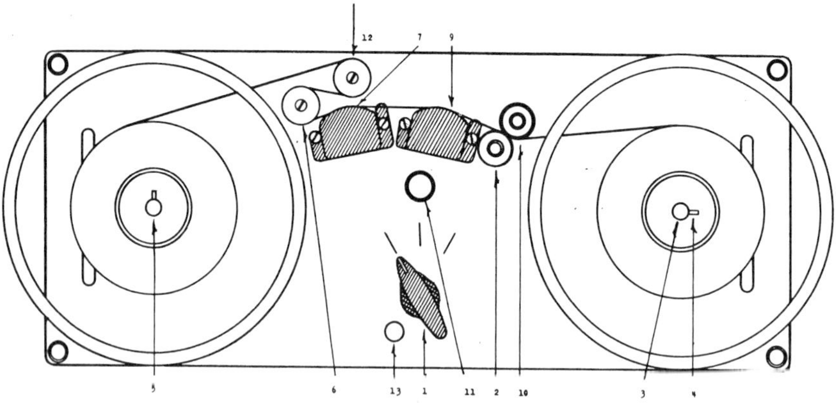

B

Figure A-16. A. Earliest Ampex broadcast audio recorder, Model 200, with H. W. Lindsay, Project Engineer. B. Magnecord professional recorder PT6-A (1948).

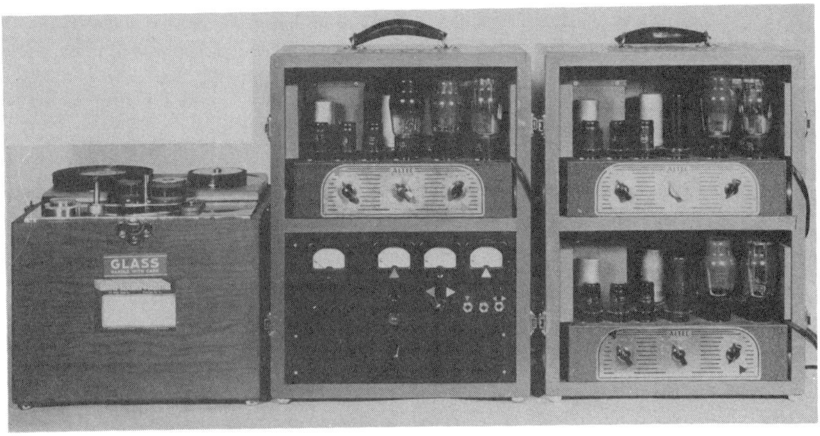

A

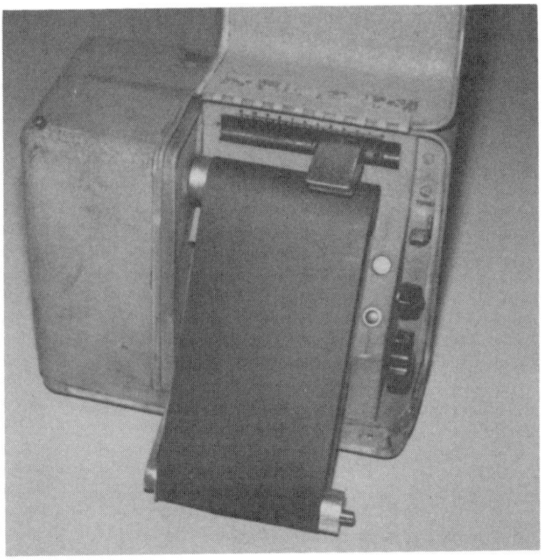

B

Figure A-17. A. Stereo tape recorder, ARF three channel (1947). B. Dictation machine, belt type (1948) (Courtesy Comptometer Corp. U.S.A.).

demonstrated by Armour Research Foundation. Early recordings were made at Arthur Murray Studios, Chicago. M. Camras, E. Clears, J. Lekas. "Debut" to the press, with Chicago Musical College, May 1947. Marvin Camras, Raymond Zenner (Camras 1949).

1948 The Eicore, an inexpensive two-direction tape recorder licensed by Armour Foundation, was marketed. It was designed by Raymond Zenner.

The Crestwood was another early tape recorder manufactured in Canada, through the efforts of Stan Danisch.

Ampro, a Chicago movie projector company, also pioneered the manufacture of early two-way home-use tape recorders. A. Shapiro.

The Audi-Ad, a compact one-minute tape loop scramble-bin recorder, was marketed for point-of-sale advertising by Magnecord Co.

Digital recording on tape was used for computer memories.

The Magnecord Model PT6-A was introduced. It became the most popular professional recorder because of its convenience, moderate price, and good sound quality.

FM carrier data recording was developed by Armour Research Foundation and others. H. Ekstein, A. A. Gerlach, T. L. Gilbert, E. H. Schulz, F. G. Rest, C. E. Barthel.

1948-49 Numerous home-use tape recorders were introduced by companies including Webcor (Webster-Chicago), Revere-Wollensak, Pentron, V. M.

A

Figure A-18. A. Mohawk message repeater (ca 1949). B. Telephone answering machine (ca 1949). C. Magnetic drum storage for computer memory (Engineering Research Associates, St. Paul, Minn., ca 1950).

B

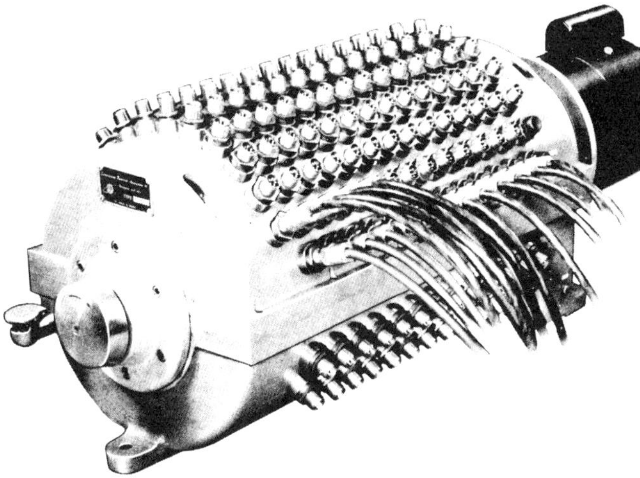

C

Figure A-18. *(Continued)*

1949 Contact printing of magnetic tapes was demonstrated. Printing from master to copy tapes was effected by an AC field. Armour Research Foundation. 3M Company. M. Camras, Robert Herr.

The PT-63 BAN, a two-channel stereo tape recorder, was manufactured by the Magnecord Company. John S. Boyers.

Figure A-19. Rotating head video recorder. Armour Research Foundation prototype (ca 1949).

A 16 mm movie projecter with magnetic soundstriped film was marketed by Bell and Howell. Malcolm Townsley. RCA later introduced a similar model.

1949–50 Numerous special-purpose recorders were introduced: The Ipsophone was a very sophisticated telephone answering machine (Oerlikon, Switzerland); Mohawk's *Telemagnet* telephone answering machine was moderately priced; the *Dictorel* flat sheet recorder (ACEC, Belgium) used

A

Figure A-20. Recorders of about 1949-50. A. Sony's first tape recorder (G-type) weighed approximately 60 kg (132 lb). B. Dictorel flat sheet recorder (Ateliers de Constructions Electriques de Charleroi). C. Peirce's belt-on-drum dictation machine.

B

C

Figure A-20. (*Continued*)

a letter-size magnetic-coated sheet wrapped around a drum during recording or playback for scanning by a head; *Peirce Dictation Systems* sold an endless-belt machine; the Mohawk *Message-Repeater* had a compact endless-loop tape cartridge.

1950 Sony Corporation marketed their first (G-type) tape recorder and magnetic tape in Japan.

Armour Research Foundation demonstrated a prototype rotating-head video recorder.

1951 The Cross-Field (X-Field) head was described at the October SMPTE (Society of Motion Picture and Television Engineers) Convention. ARF. M. Camras.

Flux responsive playback using magnetic-modulator heads was developed. ARF. David E. Wiegand, E. C. Schurch, F. R. Schlief, L. W. Ferber.

Drums and discs were in use for computer memories. Recording layer was a plated film of nickel-cobalt or magnetic oxide dispersion.

Stereo tape recorders and prerecorded stereo tapes on reels were available. High-quality stereo recorders were manufactured by Ampex and Magnecord for professional and domestic use.

1952 Video recording on magnetic tape was demonstrated by John T. Mullin and Wayne R. Johnson, Crosby Enterprises.

Magnetic Recording Rubber (impregnated with oxide particles) was made for drums and belts. Brush Development Co.

1953 Vacuum-column digital tape drive was used for computers. IBM 726. W. S. Buslik.

RCA demonstrated color video recording using large, high-speed reels. Harry F. Olson, Joseph Zenel.

Speech compression or expansion recorder used rotating heads. University of Illinois. G. Fairbanks, W. L. Everitt, R. Jaeger.

1954 An endless-loop tape system was introduced by George Eash. Magnetic drums were used commercially for data storage. IBM 650. F. E. Hamilton, E. C. Kubie.

1955 First home-use sterophonic tape recorder (TC-551) was marketed in Japan by Sony Corp.

A composite head with ferrite body and metal pole tips was described by Otto Kornei in U. S.

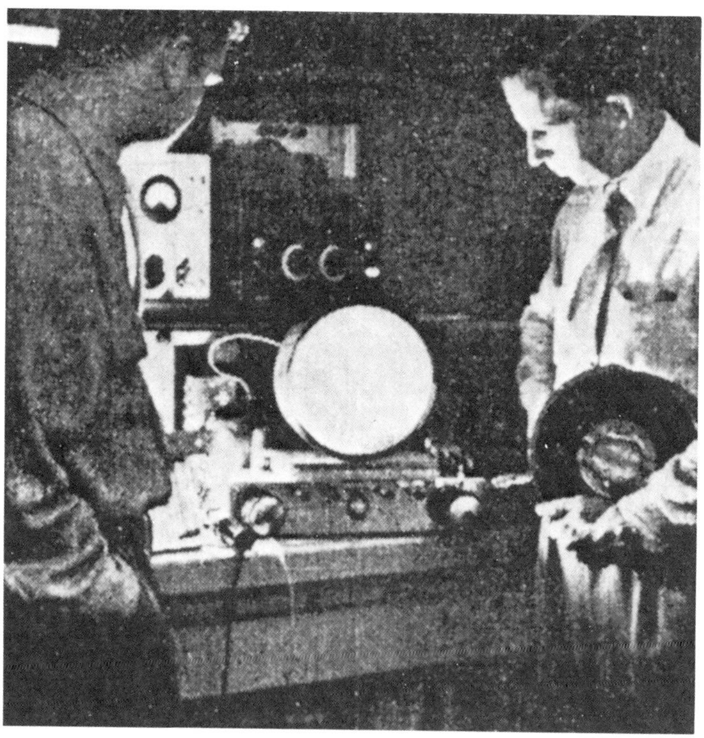

Figure A-21. Videotape recorder. Demonstrated in 1952 by Jack Mullin and Wayne R. Johnson, this equipment showed the possibility of videotape recording. (Courtesy J. Mullin)

Patent 2,711,945 (1955) and in an article (Kornei 1956). Such heads have been widely used for video and instrumentation. Brush Development Co.

1956 Quadruplex videotape recorder was introduced by Ampex Corp. for broadcast VTR. It used tape 2-inches wide scanned by a high-speed rotating drum carrying four heads. It became the standard professional video recorder. Charles P. Ginsburg, Charles E. Anderson, Ray M. Dolby, Walter Selsted, Alex Maxey, Shelby Henderson, Fred Pfost.

A helical-scan recorder for video recording (more economical than quadruplex recorders) was demonstrated. Ampex. A. Maxey.

1957 A disc-file was marketed that had movable heads for rapid access of computer data on 50 discs, 24-inches in diameter, with heads spaced by hydrostatic air pressure. IBM 350. W. A. Goddard (1953), N. A. Vogel, T. C. Cowan.

1958 Magnetic Curie-point writing and readout with electron beams was described by Mayer Ludwig of General Mills Corp (Mayer 1958)

1959 The NAB (National Association of Broadcasters) endless-loop cartridge was adopted for broadcast commercials, from which evolved the four-track automobile cartridge and, in 1965, the Lear Stereo-Eight cartridge.

A two-head VTR was developed by Victor Company of Japan (JVC).

Figure A-22. The quadruplex videotape recorder. Introduced in 1956 by Ampex, this "Mark IV" video recorder revolutionized the television broadcasting industry. The Ampex development team (left to right) included Fred Pfost, Shelby Henderson, Ray Dolby, Alex Maxey, group leader Charles Ginsburg, and Charles E. Anderson (Courtesy Ampex Corp.)

Figure A-23. Early helical-scan video recorder. Alex Maxey of Ampex made video recordings on this prototype helical-scan recorder in1956.

1960 Wideband data recorders to 2 MHz at 120 in/sec became available from Ampex, Mincom, Honeywell, and others.

Single-spool cartridges announced by ARF and by CBS, and a large cassette by RCA stimulated development of convenient home-use cassettes and cartridges. (P. Goldmark).

Satellite recorders (in satellites Tyros, Nimbus) were in operation. Data recorded on tape during entire orbits were played back at high speed when the satellite was in a position favorable for radio reception.

1962 Flying heads (hydrodynamic) used the air film carried by the surface of rapidly-moving discs to separate heads from the record surface. IBM 1301. J. J. Hagopian, W. A. Gross, R. K. Brunner, J. M. Harker, K. E. Hauton, A. G. Osterlund.

Figure A-24. Wide band data recorder. The Ampex FR-600 was a pioneer machine in the data recorder category, achieving a 250 KHz frequency limit in 1959. (Courtesy Ampex Corp.)

Fixed-head low-cost video was demonstrated by Telcan. Nottingham Valve Company, England.

The 1-$\frac{1}{2}$ head videotape principle (see Chapter 11) was introduced by Sony on a 2 inch VTR.

1963 Removable disc packs were used for off-line computer storage (six 14-inch discs, 2.68 megabytes). IBM 1311. W. E. Tibbetts, E. R. Solyst, J. D. Carothers, R. K. Brunner. J. L. Dawson, M. O. Halfhill, R. E. Kubek. The disc-packs included track-following servo developed in 1961 by A. S. Hoagland.

Skip-field VTR increased tape economy by omitting every second field of a TV recording. JVC (Japan Victor Company).

1964 The Philips compact cassette was introduced.-N. V. Philips Co., The Netherlands. G. Missriegler.

A recorder using flexible discs for broadcast commercials was demonstrated at the NAB show by Ampex. Skov.

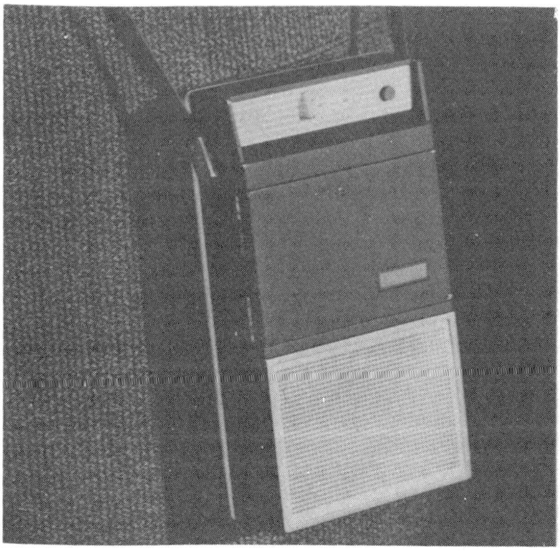

A

B

Figure A-25. A. Audio cassette recorder (1964). The Philips Carrycorder and its compact cassette became the most widely used audio format. B. Audio cartridge interior. The Lear stereo-eight cartridge has been a favorite for recorded music in automobiles. C. Evolution of the cassette. The small cassette of 1964 is in universal usage, while its larger predecessor of 1960 is almost forgotten.

C

Figure A-25. *(Continued)*

1966 Fixed-head low-cost color TV (LVR) demonstrated by IIT Research Institute, Chicago, Illinois.

Home-use economical black-and-white video tape recorder was marketed by Sony Corp. The CV-2000, a $\frac{1}{2}$-inch helical scan recorder sold for $800.

Magnetic Ink Character Recognition for check handling was introduced. (See Chapter 13).

1967 Chromium dioxide tape was introduced by DuPont Co.

1969 Magnetic tape production exceeded in dollar value any other magnetic material, hard or soft, produced in that year (also true after 1969).

1970 Magnetoresistive readout head was described. Ampex Corporation. Robert F. Hunt.

Cartrivision home-use cartridge-videotape recorder was announced by Cartrivision/Avco Corporation. In following years it was sold by Sears Roebuck and Company.

''Color-under'' VTR chroma system was adopted as an EIAJ standard. U-format VTR with $\frac{3}{4}$-inch tape cassette was developed by JVC-Panasonic-Sony.

1971 Super-Eight amateur motion picture cameras, projectors, and film with magnetic soundstriping were marketed by Eastman Kodak Company.

A

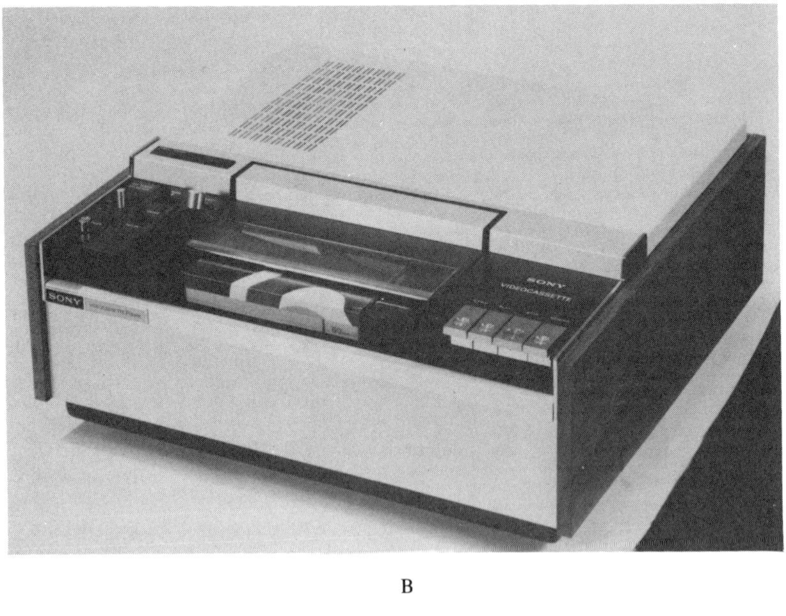

B

Figure A-26. Video cassette recorders (VCR). A. The Sony U-Matic color video cassette recorder of 1971 used a $\frac{3}{4}$ inch tape. B. The Sony Betamax SL-7300 was more economical with a $\frac{1}{2}$ inch tape (1975). (Courtesy Sony Corp.)

1972 U-Matic video cassette system was introduced by Sony Corporation. It was first announced in 1969.

1973 Winchester system using a sealed chamber for data disc files was marketed. IBM 3340. Data disc Corp. J. T. Ma, W. S. Buslik, M. W. Warner, R. B. Mulvany.

Flexible (floppy) discs with read-write capability while in a protective jacket were introduced commercially. IBM 33FD. E. V. W. Zschau. A read-only version had been used as early as 1970 and was under development since 1966.

Cobalt-adsorbed gamma ferric oxide was introduced by TDK in Avilyn Video tape. Y. Imaoka.

1975 Betamax® home-use videotape recorder was marketed by Sony Corporation.

A mass data storage system automatically selected numerous 50 megabyte slant-scanned data cartridges. IBM 3850. C. T. Johnson.J. P. Harris, R. S. Rhode, N. K. Arter.

1976 V-cord Mark-H VTR sold by Toshiba and Sanyo.

Digital audio recorders for professional use demonstrated. Soundstream. T. Stockham. Digital audio recording was in experimental use at BBC.

Figure A-27. Magnetic camera. A still-picture camera stores fifty color pictures on a small magnetic disc contained in a 60×54×3 package. The pictures can be displayed on a television receiver or can be printed as color snapshots. (Courtesy Sony Corp.)

1977 Home-use video recorders were in mass production, with more than two-hour recording capability of television programs.

Vertically oriented CoCr tape with super-high-density recording was described by S.I. Iwasaki and Y. Nakamura. CoCr perpendicular isotropy was described in 1975. Tohuko University, Sendai, Japan.

1978 VHS home-use VTR marketed. Development finished in 1976 by JVC (Japanese Victor Corp.) was adopted by many companies.

Metal particle tape with coercivity of about 1100 was introduced by 3M.

Figure A-28. 8mm videocassette camera recorder. This camcorder is used like a photographic motion picture camera. It is a self-contained unit that records up to 120 minutes of color pictures and sound on an 8-mm magnetic tape in a cassette approximately the size of an audio cassette.

Specifications of the CCD-V8

Video recording system	Rotary two-head helical scanning/FM color recording
Audio recording system	Rotary two-head helical scanning FM recording
Tape speed	Approx. 1.43 cm/sec.
Record/playback time	120 minutes (with P6-120)
FF/REW time	Approx. 4 minutes (with P6-120)
Imager	2/3″ interline CCD
Viewfinder	Electronic viewfinder (1″ black & white)
Lens	Motorized 6 : 1 zoom lens (11.5–70 mm) with macro
Minimum illumination	19 lux
Power requirements	6.0V (with battery)/8.5V (with AC power)
Power consumption	6.6W (camera recording)
Dimensions (W × H × D)	117 × 193 × 325 mm (with hood retrieved)
Weight	1.97 kg (Approx. 2.3 kg with battery and tape)

(Courtesy SONY Corp.)

1979 Thin-film (printed circuit) recording heads were used in a commercial machine. Earlier designs were tried since 1970. IBM 3370. L. T. Romankiw, I. M. Croll, M. Hatzakis, D. A. Thompson, R. E. Hones.

1981 MAVICA camera took 50 different snapshots, recorded on a disc about 50 mm in diameter, for display on a TV receiver or for color prints (see Chapter 12). Sony Corp. N. Kihara.

Portable $7\frac{1}{2}$-pound VTR used $\frac{1}{4}$-inch tape in mini-cassette of audio size. Technicolor 212.

1984 8mm videocassette camera-recorder using a small videocassette containing tape 8mm wide was introduced to the American market. Manufactured by JVC, Hitachi, and others (Figure A-28).

APPENDIX B

MAGNETIC RECORDING ALBUM

Individuals who contributed to magnetic recording over many years are so numerous that this brief album can hardly begin to include everybody. We apologize for inaccuracies and omissions.

VALDEMAR POULSEN

Copenhagen Telephone Co.
1890s Denmark

SEMI J. BEGUN
Brush Development Co. 1930s U.S.A. Germany

MARVIN CAMRAS
Armour Research Fdn. 1930s U.S.A.

EDUARD SCHULLER
AEG-Telefunken 1930s Germany

KENZO NAGAI
Tohoku University
1930s Sendai, Japan

WALTER WEBER
Deutches Rund-
funk 1930s
Germany

OTTO KORNEI
Brush Development
Co. 1930s U.S.A.
Germany

LYNN C. HOLMES
Stromberg Carlson
1940s

W. W. WETZEL
Minnesota Mining
Co. 1940s

AUSTIN ELLMORE
Utah Radio, Cres-
cent Industries
1940s

GEORGE E. ZIEGLER
Armour Research
Fdn. 1940s

CHARLES P. PEIRCE
Peirce Recorders
1940s

A.M. PONIATOFF
Ampex
1940s

HAROLD LINDSAY
Ampex
1940s

JOHN MULLIN
Crosby Enter-
prises 1940s

DAVID E. WIEGAND
Armour Research
1940s

RAYMOND E. ZENNER
Armour Research
1940s

HUGH A. HOWELL
Indiana Steel
Products 1940s

WALTER SELSTED
Ampex
1940s

RUSSELL J. TINKHAM
Armour Research Fdn
Magnecord 1940s

JOHN S. BOYERS
Armour Research Fdn
Magnecord 1940s

C.G.(SPEC) BARKER
Armour Research Fdn
Magnecord 1940s

ROBERT F. LANDON
Armour Research Fdn
Magnecord 1940s

BENJAMIN B. BAUER
Shure Bros., CBS
1940s

C. DENIS MEE
CBS, IBM, 1950s

ERIC DANIEL
Memorex 1950s

DONALD F. ELDRIDGE
Ampex, Memorex
1950s

CHARLES GINSBURG
Ampex 1950s

RAY DOLBY
Ampex 1950s

MASARU IBUKA
SONY Corp. (Founder)
1950s Japan

AKIO MORITA
SONY Corp. (Founder)
1950s Japan

SHIGEO SHIMA
SONY, NHK
1950s Japan

NOBUTOSHI KIHARA
SONY Corporation
1950s Japan

A.S. HOAGLAND
IBM 1950s

J.G. McKNIGHT
Ampex 1950s

W.K. WESTMIJZE
Philips 1950s
Netherlands

SHUN-ICHI IWASAKI
Tohoku University
Japan 1960s

HIROSHI SUGAYA
Matshushita
Japan 1960s

MYRON STOLAROFF
Ampex
1940s

APPENDIX C

ADVANCES IN BASIC MAGNETISM

Basic magnetism relates to magnetic theories, laws, effects, and their measurement.

BC

600 Thales, a Greek, wrote of the attractive power of lodestones.

AD

1269 A treatise on magnetism, attraction, repulsion, and compasses was written by Peter Peregrinus.

1600 The classic book *De Magnete*, was published by William Gilbert.

William Gilbert (1540-1603)

Figure C-1. William Gilbert. His famous treatise on magnetism was published in the year 1600.

1785 Inverse square law established by C. A. Coulomb.

1819 Magnetic effect of electric current discovered by Hans Christian Oersted.

1825 Ampere published his ''beautiful'' theory of electromagnetism

1831 Electromagnetic induction was explained quantitatively by Michael Faraday.

1845 Faraday discovered that all substances have magnetic susceptibility and classified them as paramagnetic or diamagnetic.

1873 Accurate quantitative measurements of ferromagnetism were made by Henry A. Rowland.

1879 Hall effect was discovered by Edwin Hall, a student of Rowland.
1890 Interaction effects of molecular magnet models were demonstrated by J. A. Ewing.
1895 Curie Law $\chi = C/T$ (suceptibility inversely related to absolute temperature) was announced by Pierre Curie, who measured and reported the susceptibility of many substances.
1898 Heusler alloys (Al, Cu, Mn) discovered; ferromagnets made of nonmagnetic constituents.
1905 Curie's Law explained theoretically by P. Langevin.
1906 Domains and spontaneous magnetization concepts introduced by P. Weiss.
1907 Curie-Weiss Law $\chi = C/(T - a)$ and molecular field theory formulated by P. Weiss.
1908 Gyromagnetic effect proposed by O. W. Richardson.
1915 Anhysteretic magnetization demonstrated by W. Steinhaus and E. Gumlich.
1919 Barkhausen effect discovered.
1925 Zeeman effect explained by electron spin theory.
1930 Single-domain particles postulated by Frenkel and Dorfman.
1932 Theory of antiferromagnetism developed by Louis Néel.
1935 Distribution of magnetic particle interactions shown by F. Preisach.
1948 Ferrimagnetism of ferrites explained by L. Néel.
1949 Theory of domains in silicon iron single crystals published by Williams, Bozarth, and Shockley.

APPENDIX D

ADVANCES IN APPLIED MAGNETISM

Applied magnetism deals with magnetic materials and with practical devices utilizing such materials.

BC

2637 A chariot equipped with a "compass" was said to have been constructed by Hoang-ti in China.

AD

121 A Chinese dictionary referred to the south-pointing property of magnets.
1100 The compass was known and used extensively by navigators (AD 1100–1200).
1600 William Gilbert (Fig. C-1) noted that iron lost its magnetic powers at red-hot temperatures, and that it was sensitive to being magnetized while cooled from red heat.
1821 Simple DC electric motor made by Michael Faraday.
1825 First electromagnet was made.
1831 Electromagnetic induction laws were discovered by Michael Faraday.
1835 Magnetic telegraph invented by Samuel F. B. Morse.
1842 Magnetostricion discovered by J. P. Joule.
1876 Telephone invented by Alexander Graham Bell.
1879 Hall effect discovered by Edwin Hall.
1886 Transformer introduced by William Stanley
1886 Induction motor invented independently by Nikola Tesla and Galileo Farraris.
1898 First magnetic recorder made by Valdemar Poulsen.
1900 Silicon-iron was invented by Robert Hadfield in England.
1913 Magnetic annealing discovered by Pender and Jones. Cooling in a magnetic field from a high temperature raised permeability.
1923 Permalloy was described by H. D. Arnold and G. W. Elman.
1931 Alnico-type permanent magnet made by Mishima in Japan.
1933 Grain-oriented silicon iron was developed by Norman Goss.

1933 Ferrites of high magnetic permeability were developed by J. L. Snoek of Philips Co. (1933 to 1945).

1935 Magnetic tapes coated with iron particles were introduced by German AEG.

1936 Sendust was prepared by Masumoto in Japan (magnetically soft, mechanically hard alloy of Fe, Al, Si).

1940 Alnico V was developed in England and Holland by heat-treating high cobalt alloys in a magnetic field.

1946 Magnetic tapes with acicular particles of high coercivity were introduced by M. Camras, Armour Research Foundation.

1946 Metal particle high-coercivity recording tapes were demonstrated by H. A. Howell, Indiana Steel Products.

1952 Barium ferrite ceramic magnets were developed by Philips Company, Eindhoven.

1967 Bubble domain memories were described by A. H. Bobeck.

APPENDIX E
TRUE POSITION DIMENSIONING

(Notation for dimensioning tape-tracks, cassettes, etc.)

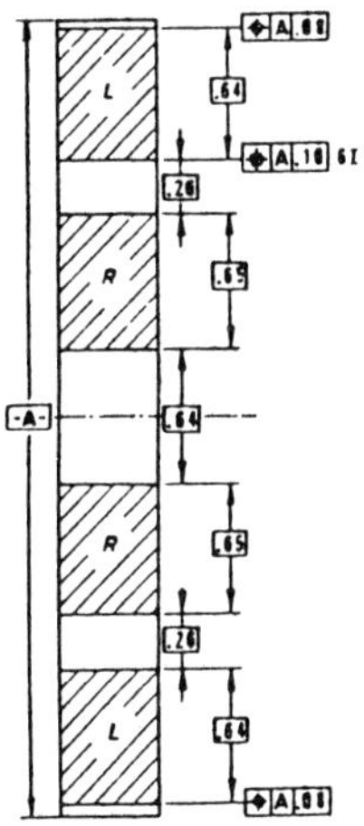

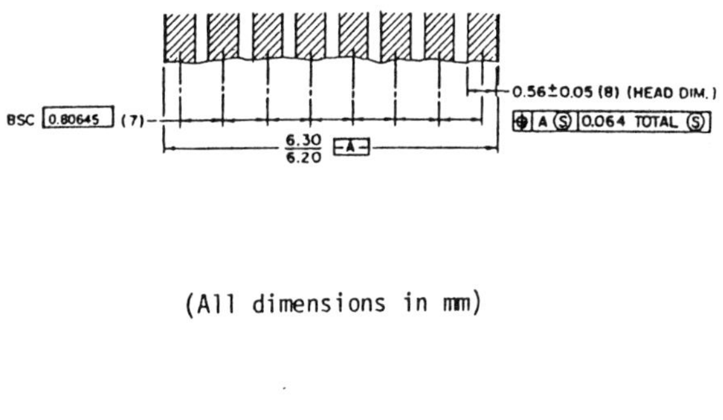

(All dimensions in mm)

⌖ A .10 6I means that the basic reference is the center-line dividing the dimension **A.** The allowable deviation (tolerance) of any dimensioned location is 0.10 mm (±0.05 mm) from the marked dimensions (true position) referred to the center line. **6I** means that there are 6 locations having the stated tolerance.

(continued on next page)

⊕ | A Ⓢ | 0.064 TOTAL Ⓢ	specifies tolerance of additional center-lines (of tape tracks, for example), which are symmetrically placed with respect to the center-line of basic dimension **A.**
BSC [0.80645] (7)	means that there are 7 of such additional center-lines each spaced 0.80645 mm apart, center to center.

APPENDIX F

COMPACT CASSETTE DIMENSIONS

AMERICAN NATIONAL STANDARD X3.48-1977

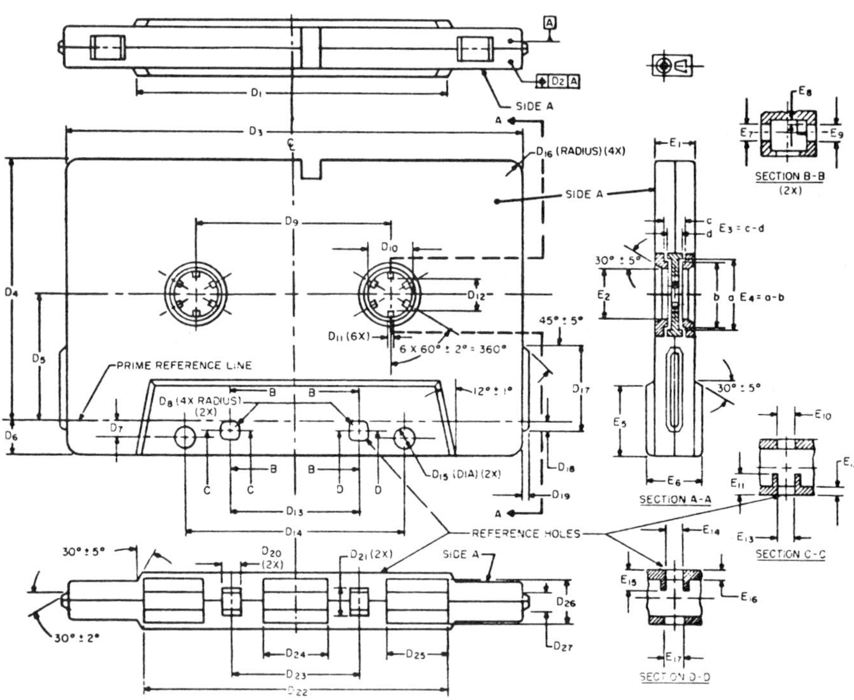

Dimension	Inches Min	Inches Max	Millimeters Min	Millimeters Max
D_1	2.658	2.696	67.5	68.5
D_2	0	0.0078	0	0.2
D_3	3.941	3.964	100.1	100.7
D_4	2.2284	2.2519	56.6	57.2
D_5	1.0926	1.1043	27.75	28.05
D_6	0.2638	0.2795	6.7	7.1
D_7	0.1260	0.1417	3.2	3.6
D_8	0.0512	0.0590	1.3	1.5
D_9	1.662	1.685	42.2	42.8
D_{10}	0.3937	0.4094	10.0	10.4
D_{11}	0.0512	0.0669	1.3	1.7
D_{12}	0.3071	0.3149	7.8	8.0
D_{13}	1.0965	1.1082	27.85	28.15
D_{14}	1.8819	1.8976	47.8	48.2
D_{15}	0.1733	0.1889	4.4	4.8
D_{16}	0.099	0.137	2.5	3.5
D_{17}	0.693	0.724	17.6	18.4
D_{18}	0.067	0.098	1.7	2.5
D_{19}	0.0355	0.0433	0.9	1.1
D_{20}	0.1497	0.1653	3.8	4.2
D_{21}	0.237	0.267	6.0	6.8
D_{22}	2.626	2.649	66.7	67.3
D_{23}	1.093	1.112	27.75	28.25

Dimension	Inches Min	Inches Max	Millimeters Min	Millimeters Max
D_{24}	0.548	0.570	13.9	14.5
D_{25}	0.508	0.535	12.9	13.6
D_{26}	0.3859	0.3937	9.8	10.0
D_{27}	0.1182	0.1338	3.0	3.4
E_1	0.3347	0.3503	8.5	8.9
E_2	0.4292	0.4448	10.9	11.3
E_3	–	0.0511	–	1.3
E_4	0.0394	0.0748	1.0	1.9
E_5	0.571	0.610	14.5	15.5
E_6	0.4685	0.4842	11.9	12.3
E_7	0.1457	0.1496	3.7	3.8
E_8	0.016	–	0.4	–
E_9	0.1457	0.1496	3.7	3.8
E_{10}	0.154	0.177	3.9	4.5
E_{11}	0.166	0.188	4.2	4.8
E_{12}	0.0748	0.0826	1.9	2.1
E_{13}	0.1418	0.1456	3.6	3.7
E_{14}	0.1418	0.1456	3.6	3.7
E_{15}	0.166	0.188	4.2	4.8
E_{16}	0.0748	0.0826	1.9	2.1
E_{17}	0.154	0.177	3.9	4.5

8-mm Cassette Dimensions

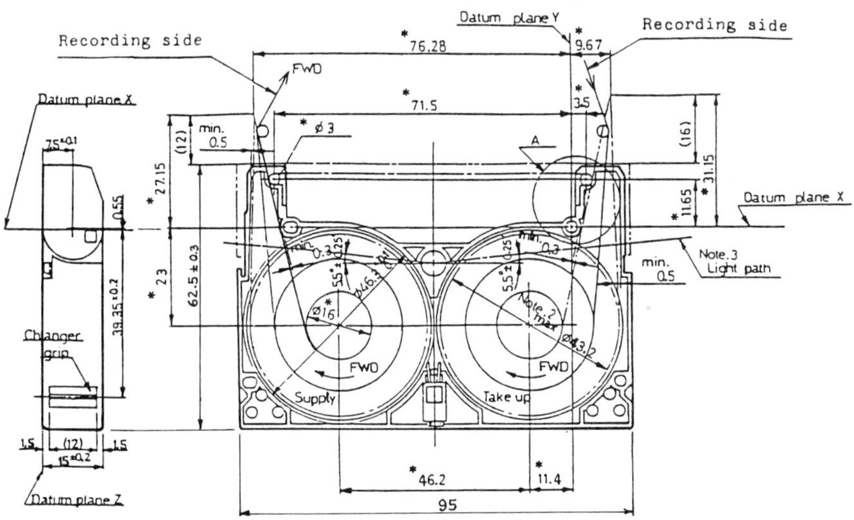

Note 1. Dimensions marked with * are nominal values specifying the tape path.

Note 2. Max. diameter of wound magnetic tape including leader and trailer tape.

APPENDIX G

STANDARDS AGENCIES*

Catalogs of standards may be obtained from these organizations:

ANSI American National Standards Institute, 1430 Broadway, New York, NY 10018 (Broad fields, audio, video, business, computers.) (Also supplies standards of other organizations listed below.)

ARD Arbeitsgemeinschaft der Rundfunkanstalen der Bundesrepublik Deutschland, Institute fur Rundfunktechnik, 200 Hamburg, Mittelweg 113, West Germany.

BS British Standards Institution British Standards House, 2 Park Street, London W. 1, England. (Available through ANSI.) (SL 10 Acoustics; SL-1 Cinematography.)

CCIR International Radio Consultative Committee, International Telecommunication Union, Place des Nations, Geneva, Switzerland. (Broadcasting and television.)

DIN Deutsche Industrie Normen, Beuth-Vertrieb GmbH, 1 Berlin 30, Burggrafenstrasse 4-7, West Germany. (Available in U.S.A. from ANSI; also in E/DIN English translations.) (High-fidelity, magnetic recording, amplifiers.)

IBTO, OIRT International Broadcasting and Television Organization (OIRT in French), Liebknechtova 15, Prague, 5 Czechoslovakia.

EIA Electronic Industries Association, 2001 Eye Street, N. W., Washington, DC 20006 (Some EIA standards have also been approved as ANSI standards.)

IEC International Electrotechnical Commission, 1, rue de Varembé, Geneva, Switzerland. (Standards listed and available from ANSI.)

IEEE Institute of Electrical and Electronics Engineers, Inc., 345 East 47th Street, New York, NY 10017 (Some approved also as ANSI standards.)

IRIG Inter Range Instrumentation Group, Sandia National Labs., P.O. Box 5800 Albuquerque, NM 87185

ISO International Organization for Standardization, 1, rue de Varembé, Geneva, Switzerland. (Standards listed and available from ANSI.)

JIS Japanese Standards Association (JSA), 1-24 Askaska 4, Minato-Ku, Tokyo, Japan. (Available from ANSI.)

MRIA Magnetic Recording Industries Association (merged with EIA).

NAB National Association of Broadcasters (formerly NARTB), 1771 N. Street, N.W., Washington, DC 20036 (Broadcast industry standards.)

PPI Philips Phonographic Industries, Baarn, The Netherlands. (Standards on cassette recording.)

RIAA Record Industry Association of America, Inc., One East 57th Street, New York, NY 10022. (Tape and disc record standards.)

SMPTE Society of Motion Picture and Television Engineers, 9 East 41st Street, New York, NY 10017. (Video, sound, motion pictures) (Approved standards available from ANSI only.)

USA FED United States of America, Federal Specifications. Procurement standards for Federal Agencies. (a) Bureau of Ships, Department of the Navy, Naval Ship Engineering Center, Code 6665.2M, Washington, DC 20360. (b) General Services Administration, Washington, DC

UTE Union Technique de l'Electricité, 20, rue Hamelin, Paris (16e), France. (Available from ANSI, in French only.)

*From *Journal of the Audio Engineering Society* 18:317-8 (1970).

APPENDIX H

ANSI STANDARDS RELATING TO MAGNETIC RECORDING (from ANSI catalog, see Appendix G)

AUDIO

Magnetic Tape Records: Four Track Open-Reel Stereophonic Records at 3.75 and 7.5 in/s (9.5-19 cm/s), ANSI/EIA RS-434-1976, [S4.14]

Magnetic Tape Records, Four Channel Sound, ANSI/EIA RS-387-1971, [C83.65]

Weighted Peak Flutter of Sound Recording and Reproducing Equipment, ANSI/IEEE 193-1971, [S4.3]

†Audio-Visual and Educational use of Coplanar Magnetic Cartridge, Type CP II (Compact Cassette), ANSI PH7.4-1975.

†Cards for Audio-Visual And Educational Application, Audio Recorded Magnetically Striped Information, ANSI PH7.15-1979

†Counters Used in Instructional Audio-Visual Cassette Recorders and Players, Requirements for, ANSI PH7.8-1979

†Audio Operating Level and Multifrequency Test Tape for Quadruplex Video Magnetic Tape Recorders Operating at 15 in/s (381 mm/s), Specifications for, ANSI C98.8-1977

CASSETTES AND CARTRIDGES

Magnetic Tape Records: Compact Cassette (EIA RS-399-A) with Four-Track Mono/Stereo Compatible Records at 1.875 in/sec (4.76 cm/sec), ANSI/EIA RS-433-1976 [S4.15]

Magnetic Tape Records: Endless-Loop Cartridges for Eight-Track Stereophonic Records at 3.75 in/s (9.53 cm/s), ANSI/EIA RS-432-1976 [S4.13]

†Magnetic Tape Cassette for Information Interchange, 3.81 mm (0.150 in) Tape at 32 bpmm (800 bpi), PE, ANSI X3.48-1977

♦†Magnetic Tape Cassettes for Information Interchange, Dual Track Complementary Return-to-Bias (CRB) Four-States Recording on 3.81-mm (0.150 in) Tape, ANSI X3.59-1981

♦†Recorded Magnetic Tape Cartridge for Information Interchange (4 Track, 0.250 Inch) 1600 bpi Phase Encoded), Parallel, ANSI X3.72-1981

†Recorded Magnetic Tape Cartridge for Information Interchange (4-Track, 0.250 Inch, 1600 BPI, Phase Encoded), ANSI X3.56-1977

Dimensional Standard Coplanar Magnetic Tape Cartridge, Type CP II (Compact Cassette), ANSI/EIA RS-399-A-1975 [S4.8]

Dimensional Standard Eight Track Endless Loop Cartridge, ETA Type III, ANSI/EIA RS-332-A-1978 [S4.16]

DIGITAL

One Half Inch (12.7 mm) Magnetic Tape Reel for Computer Use (Requirements for Interchange), ANSI/EIA RS-352-1968(R1978) [C83.43]

♦†One Half Inch Magnetic Tape Interchange Using a Self-Loading Cartridge, ANSI X3.85-1981

†Recorded Magnetic Tape for Information Interchange (1600 CPI, PE) ANSI X3.39-1973

†Recorded Magnetic Tape for Information Interchange (200 CPI, NRZI), ANSI X3.14-1973

†Recorded Magnetic Tape for Information Interchange (6250 CPI,Group Coded Recording), ANSI X3.54-1976

† Recorded Magnetic Tape for Information Interchange (800 CPI, NRZI), ANSI X3.22-1973

†Take-Up Reels for One-Inch Perforated Tape for Information Interchange, ANSI X3.20-1967(R1974)

†Unrecorded Magnetic Tape Cartridge for Information Interchange (0.250 Inch, 1600 BPI, Phase Encoded), ANSI X3.55-1977

†Unrecorded Magnetic Tape for Information Interchange (9-Track 200 and 800 CPI, NRZI, and 1600 CPI, PE), ANSI X3.40-1976

†Code Extension Techniques for Use with the 7-Bit Coded Character Set of American National Standard Code for Information Interchange (ASCII), X3.4-1968, ANSI X3.41-1974

DIGITAL DISC

†Unrecorded Single Disk Cartridge (Front Loading, 22000 BPI), General, Physical, and Magnetic Requirements, ANSI X3.52-1976

♦†Unrecorded Single-Disk Double-Density Cartridge (Front Loading, 2200 BPI, 200 TPI), General, Physical, and Magnetic Requirements for, ANSI X3.89-1981

♦†Unrecorded Twelve-Disk Pack (100 Megabytes)(General, Physical, and Magnetic Requirements, ANSI X.3.63-1981 ▲

♦†Unformatted Single Disk Cartridge (Top Loading, 200 TPI, 4400 BPI)-General, Physical, and Magnetic Requirements, ANSI X3.76-1981

♦†Unformatted Twelve-Disk Pack (200 Megabytes) (General, Physical, and Magnetic Requirements), ANSI X3.84-1981. ▲

†Unrecorded Eleven-Disk Pack-General, Physical, and Magnetic Requirements, ANSI X3.58-1977

†Unrecorded Magnetic Six-Disk Pack (General, Physical, and Magnetic Characteristics), ANSI X3.46-1974

†Cartridge, One-sided Single-Density Unformatted 5.25 Inch Flexible Disk, ANSI X3.82-1980

†Cartridge, Single-Sided Unformatted Flexible Disk (for 6631 BPR Use), ANSI X3.73-1980

♦†Interfaces between Flexible Disk Cartridge Drives and Their Host Controllers, ANSI X3.80-1981

INSTRUMENTATION

Recorded Tape Formats for 7, 14, and 21 Tracks on $\frac{1}{2}$-Inch Magnetic Tape and 14, 28, and 42 Tracks on 1-Inch Magnetic Tape for Instrumentation Recording, ANSI/EIA RS-394-1971(R1979) [C83.71]

Flutter Measurement of Instrumentation Magnetic Tape Recorder/Reproducers, Test Method for, ANSI/EIA RS-405 1972 (R1979)[C83.99]

Recommended Test Method - Timing Error Measurements of Instrumentation Magnetic Tape Recorder/Reproducers, ANSI/EIA RS-413-1973(R1979) [C83.94]

MISCELLANEOUS

Magnetic Shield Efficiency in Attenuating Alternating Magnetic Fields, Test for, ANSI/ASTM A698-74 (1980)

†Magnetic Tape Labels and File Structure for Information Interchange, ANSI X3.27-1978,

†Bank Check Specifications for Magnetic Ink Character Recognition, ANSI X3.3-1970 (R1976)

REELS

Precision Reel for Magnetic Tape, ANSI/EIA RS-254-A-1967 [C83.31]

Type A Hubs and Reels for Magnetic Tape (Requirements for Interchange), ANSI/EIA RS-346-1968[C83.38]

Type B Plastic Reel for Magnetic Tape (Requirements for Interchange), ANSI/EIA RS-351-1968 [C83.39]

†Plastic Reels for $\frac{1}{2}$-Inch (12.7 mm) Video Magnetic Tape, Dimensions of, ANSI C98.14-1975(R1981)

TAPES

Measuring Recorded Flux of Magnetic Sound Records at Medium Wavelengths, Method of, ANSI/IEEE 347-1972

Dimensions for Unrecorded Magnetic Sound Recording Tape, ANSI/EIA RS-355-1968 [C83.45]

Test Method - Layer-to-Layer Adhesion of Magnetic Tape, ANSI/EIA RS-339-1967 [C83.35]

Test Method - Magnetic Tape Electrical Resistance Coating, ANSI/EIA RS-342-1967 [C83.36]

Test Method - Tensile Properties of Magnetic Tape, ANSI/EIA RS-362-1969 [C83.56]

Unrecorded Magnetic Tape for Reel-to-Reel Instrumentation Applications, ANSI/EIA RS 338-1967 [C83.34]

VIDEO RECORDING

†Basic System Parameters for 1-Inch Type B Helical-Scan Video Tape Recording. ANSI C98.15M-1980

†Color Video Magnetic Tape Leader, Specifications for, ANSI C98.9-1967(R1975)

♦†Dimensions of Cartridge Spools for 2-Inch Quadruplex Video Magnetic Tape, ANSI V98.13-1981

†Dimensions of 2-in Video Magnetic Tape, ANSI C98.1-1978

†Dimensions of 2-in Video Magnetic Tape Reels, ANSI C98.5-1970 (R1976)

♦†Frequency Response and Operating Level of Recorders and Reproducers for Audio 1 Record for 2-inch Quadruplex Video Magnetic

Tape Operating at 15 and 7.5 in/s, ANSI V98.3-1980

†Frequency Response and Operating Level of Recorders and Reproducers for Audio Records for 1-inch Type B Helical-Scan Video Tape Recording, ANSI C98.17M-1980

†Frequency Response and Reference Level of Recorders and Reproducers for Audio Records for 1-in Type C Helical-Scan Video Tape Recording, ANSI C98.20M-1979

†Geometry Parameters for 1-in Type C Helical-Scan Video Tape Recording, Basic System and Transport, ANSI C98.18M-1979

†Monochrome Video Magnetic Tape Leader, Specifications of, ANSI C98.2-1963(R1975)

†Records and Basic Electrical Parameters for $\frac{1}{2}$-inch Type F Helical-Scan Video Tape Recording, Dimensions and Location of, ANSI C98.23M 1980

†Records for 1-in Type C Helical-Scan Video Tape Recording, Dimensions and Location of, ANSI C98.19M-1979

†Records for 1-inch Type B Helical-Scan Video Tape Recording, Dimensions and Location of, ANSI C98.16M-1980

†Records for $\frac{3}{4}$-inch Type E Helical-Scan Video Tape Cassette Recording, Dimensions and Location of, ANSI C98.21M-1980

†Speed of 2-in Tape for Quadruplex Video Magnetic Tape Recording, ANSI C98.4-1970(R1976)

†Test Tape for Quadruplex Video Magnetic Tape Recorders Operating at 7.5 in/s (190.5 mm/s), Specifications for an Audio Operating Level and Multifrequency, ANSI C98.11-1977

♦†Time and Control Code for Video and Audio Tape for 525 Line/60 Field Television Systems, ANSI V98.12M-1981

†Video Cassette for $\frac{3}{4}$-inch Type E Helical-Scan Video Tape Recording, Dimensions of, ANSI C98.22M-1980

♦†Video, Audio, and Tracking Control Records on 2-inch Video Magnetic Tape Quadruplex Recorded at 15 and 7.5 in/s, Dimensions of, ANSI V98.6-1981

APPENDIX I SMPTE MAGNETIC RECORDING STANDARDS (from SMPTE Catalog, see Appendix G)

2-in. video magnetic tape
Monochrome mag tape leader
Audio record for 2-in. quad
(15 and 7.5 in/s).
Tape speed, 2-in. quad
2-in. video mag tape reels
Track record dimension, 2-in.
Audio multifrequency, 15 in./s
Color leader
Audio multifrequency, 7.5 in./s
Time and control code
Cartridge spools, 2-in. quad
$\frac{1}{2}$-in. plastic reels
Basic parameters, 1-in. Type B
Track record dimension, 1-in., B
Freq response of audio, 1-in., B
Basic parameters, 1-in., Type C
Records, 1-in., Type C
Audio records, 1-in., Type C
Records, $\frac{3}{4}$-in., Type E
Cassette, $\frac{3}{4}$-in., Type E
Records, $\frac{1}{2}$-in., Type F
1-in. Helical Reels
1-in. Tapes
1-in. Ref Tape, raw stock
1-in. Type C ref recorders
1-in. Type C ref records
1 in. Type B ref recorders
1-in. Type B ref records
$\frac{3}{4}$-in. Type E small cassette
35-mm photo sound record
16-mm photo sound record
Intermodulation tests
16-mm edge numbering
35/17$\frac{1}{2}$-mm magnetic record
16-mm (100-mil) mag striping
8-mm (R) mag striping
16-mm magnetic (200-mil)
16-mm (2-30 mil) striping
35-mm, 150-mil magnetic sound record
16-mm (100 mil) magnetic sound record
Sound track density
16-mm mag striping with picture and photo record

APPENDIX I (*Continued*)

8-mm (R) mag sound record
16-mm perf S8 mag striping
35-mm 4-track mag records and picture
S8 Model 1 silent and sound 50-ft camera run length
Mag striping super 8 Film
Mag striping 16-mm, perf S8
Mag striping 35-mm, perf S8
Super 8 mag sound records
Mag striping 35-mm 4-track with picture
70-mm mag sound records
35-mm six-track mag records
S8 Mod 1 sound 50-ft cartridge
S8 Model 1 sound 50-ft cartridge film position
S8 Model 1 sound 50-ft cartridge aperture
S8 Model 1 sound cartridge-camera interface
S8 Model 1 sound cartridge aperture profile
Splices in 2-in. mag tape
Modulation levels, 2-in. quad
Vacuum guide radius, 2-in. quad
Tracking control, 2-in. quad
Sound sync
Quad and helical-scan labels
Video test tape, 15 in./s
8-mm (S) test film
Video test tape, HB, 15 in./s
Video test tape, HB, 7.5 in./s
Dropout detection
Edge numbering on 16-mm
2-in. cartridge labels
Super 8 mag azimuth test film
Super 8 mag flutter test film
35-mm mag flutter test film
16-mm mag flutter test film
35-mm mag azimuth test film
16-mm mag azimuth test film
35-mm mag 4-track flutter TF
35-mm mag 4-track azimuth TF
Tracking control record, 1″, B
Ref carrier and pre-emphasis, 1″, Type B
Tracking control, 1″, Type C
Video record parameter, 1″, C
Frequencies, $\frac{3}{4}$″, Type E
Frequencies, $\frac{1}{2}$″, Type F
Dual-program audio, 2-in.
16-mm mag, multifrequency TF
S8 mag, multifrequency TF
Time code in 1-in. tape
Video reference tapes
Video/audio ref tape, 1″, C
Interchange ref tape, 1″, C
Time code requirements (quad)
Audio response (quad)
1-in, Type B reference tape
Vertical interval time code
Spectral response of 8-mm reproducers
Cartridge/cassette spool use

THE FOLLOWING STANDARDS WERE IN STUDY BY SUBCOMMITTEES (1982)
Precision TV reels
Two-track 16-mm magnetic time and control code
Sound splices
35-mm mag reproducing charact.
Sound sync for super 8
35-mm mag 3-track MF test film
35-mm mag 4-track MF test film
16-mm mag recorded charact.
S8 mag recorded characteristic

70-mm mag recording charact.
Sound record indentification
16-mm mag pink noise test film
35-mm mag pink noise test film
Digital audio
Comtrak
Guideline for Super 8 audio reproducing
TV Audio characteristics
Polyester base study
Digital television
Digital video standards
Digital tape recording
Audio/Video performance
Nomenclature

$\frac{1}{2}$-in. Type G track record, Beta
$\frac{1}{2}$-in Type G tape and cassette, Beta

$\frac{1}{2}$-in Type G audio pre-emphasis and reference signal
$\frac{1}{2}$-in Type H dimension track records, VHS

$\frac{1}{2}$-in Type H tape and cassette
$\frac{1}{2}$-in Type H audio pre-emphasis and reference signal

APPENDIX I (*Continued*)

Test tapes	Reference record, 1-in, Type B
Helical recording	$\frac{1}{4}$-in helical format, Type K
Tapes and reels	
Editing procedures	Archival storage (study)
Dropout on 1-in. tape	Duplicating piracy (study)
$\frac{1}{2}$-in camera/cassette, type J	Multi-channel TV sound

APPENDIX J PHYSICAL CONSTANTS

Permeability of a vacuum, magnetic constant	$\Gamma_m = 4\pi \times 10^{-7}$ H/m
Permittivity of a vacuum, electric constant	$\Gamma_e = 8.854 \times 10^{-12}$ F/m
Speed of light in a vacuum	$c_o = 2.9979 \times 10^8$ m/s
Electronic charge	$e = 1.602 \times 10^{-19}$ C
Standard gravitational acceleration	$g_n = 9.807$ m/s^2
Boltzmann constant	$k = 1.38 \times 10^{-23}$ J/°K
Planck constant	$h = 6.626 \times 10^{-34}$ J·s
Faraday constant	$F = 9.649 \times 10^4$ C/mol
Normal atmospheric pressure	atm = 101.3 kPa

APPENDIX K CONVERSION FACTORS

LENGTH
1 in. = 25.4 mm*
1 ft = 0.30480 m
1 m = 3.280840 ft
1 yd = 0.9144 m*
1 mile = 1.6093 km
1 km = 0.62138 mile
1 mile (nautical) = 1.852 km*
1 Å = 0.1 nm (nanometer)*

AREA
1 in.^{2} = 645.16 mm^2*
1 ft^2 = 0.0929 m^2
1 yard^2 = 0.8361 m^2
1 circ. mil = 0.0005067 mm^2

VOLUME
1 in.^{3} = 16387 mm^3
1 in.^{3} = 0.0164 L
1 ft^3 = 0.02832 m^3
1 yd^3 = 0.7646 m^3
1 barrel (US petro) = 0.1590 m^3
1 fl oz = 29.57 ml
1 qt = 0.9464 L
1 L = 0.264171 gal

VOLUME (*Continued*)
1 L = 0.219976 Imp. gal
1 Imp. gal = 4.545960 L

FUEL RATE
1 mile/gal = 0.4251 km/L
1 gal/mi = 2.3527 L/km
1 km/L = 2.3527 miles/gal

VELOCITY
1 ft/min = 5.080 mm/s*
1 mi/hr = 0.4470 m/s
1 mi/hr = 1.6093 km/hr
1 km/hr = 0.2778 m/s
1 knot = 0.5144 m/s

ACCELERATION
1 in/s^2 = 0.0254 m/s^2
1 ft/s^2 = 0.3048 m/s^2

FORCE
1 dyne = 10^{-5} N (Newton)*
1 oz = 0.278 N
1 lb = 4.448 N
1 kg = 9.807 N

APPENDIX K (*Continued*)

MASS
1 oz (avoir) = 28.349527 gm
1 oz. (troy) = 31.103481 gm
1 grain = 0.064798918 gm
1 lb = 0.45359 kg
1 kg = 2.2046 lb
1 ton = 907.18 kg
1 ton = 0.907 tonne (metric ton)

DENSITY
1 lb/ft^3 = 16.02 kg/m^3

PRESSURE
1 in of H_2O = 249.1 Pa (pascal)
1 mm of Hg = 133.3 Pa
1 lb/in.^{2} = 6.894757 kPa
1 kPa = 0.145037 lb/in^2 (PSI)
1 kg/m^2 = 9.807 Pa
1 mb (millibar) = 100 Pa*

ENERGY, TORQUE
1 ft-lb = 1.355818 N-m
1 N-m = 0.737562 ft-lb
1 N-m = 1 J (joule) (watt-second)
1 BTU = 1055 J
1 calorie = 4.184 J*
1 kW-hr = 3.6 × 10^6 J

POWER
1 N-m/s = 1 W (watt)
1BTU/hr = 0.2931 W
1 hp = 746 W*

ILLUMINATION
1 ft-candle = 10.764 lumens/m^2 (lux)
1 ft-lambert = 3.426 candelas/m^2

* Indicates exact value.

APPENDIX L REFERENCE INFORMATION OF SPECIAL INTEREST

Magnetic units and conversion factors	Table 2-1; p. 19
Magnetic properties of hard and soft magnetic materials	Figure 2-5; p. 22
B-H curves of commercial tapes	Figure 3-11, 12; pp. 116–117
Tests of magnetic record media	Table 3-8 to 3-12; pp. 138 to 142
Tape winding and storage	Table 13-5; Chapter 3; p 551, p. 143
Capacity, playing time and dimensions of tapes, reels, cassettes	Table 3-2; p. 98
Wire size	Tables 4-4, 4-5; p. 200, 201
Multichannel track formats	Figures 4-28; 10-5; 10-6; 10-12; 10-14; 13-2; 13-3; 13-18; 14-2; 14-9; Table 10-2
Head wear estimates	Chapter 4; p. 218, 219
Recommended response standards	Table 6-3; 10-2; pp. 281 to 286; p 419
Flutter weighting curve (IEEE)	Figure 7-30; p 339
Video recorder writing speeds, tracks, drums, etc.	Tables 12-2; 12-3; 12-4; p 506 to 510
Video tape formats	Table 12-1; p. 506
IRIG standards	Chapter 13; p. 564–566, 569
Instrumentation reel dimensions	Figure 13-5; p. 548
Elementary digital codes	Table 13-12; p 575
Magnetic credit cards	Figure 13-18; p. 585

APPENDIX L (*Continued*)

Magnetic ink character recognition for bank checks	Figure 13-19; p. 586
Dimensioning (code of symbols)	Appendix E
Compact-cassette dimensions	Appendix F
Standards agencies	Appendix G
ANSI standards	Appendix H
SMPTE standards	Appendix I
Physical constants	Appendix J
Conversion factors	Appendix K

INDEX